# Lecture Notes in Computer Science 7951

Commenced Publication in 1973
Founding and Former Series Editors:
Gerhard Goos, Juris Hartmanis, and Jan van Leeuwen

T0223406

Chengan Guo  Zeng-Guang Hou
Zhigang Zeng (Eds.)

# Advances in Neural Networks – ISNN 2013

10th International Symposium on Neural Networks
Dalian, China, July 4-6, 2013
Proceedings, Part I

 Springer

Volume Editors

Chengan Guo
Dalian University of Technology
School of Information and Communication Engineering
A 530, Chuangxinyuan Building
Dalian 116023, China
E-mail: cguo@dlut.edu.cn

Zeng-Guang Hou
Chinese Academy of Sciences
Institute of Automation
Beijing 100864, China
E-mail: zengguang.hou@mail.ia.ac.cn

Zhigang Zeng
Huazhong University of Science and Technology
School of Automation
Wuhan 430074, China
E-mail: zgzeng@mail.hust.edu.cn

ISSN 0302-9743                          e-ISSN 1611-3349
ISBN 978-3-642-39064-7                   e-ISBN 978-3-642-39065-4
DOI 10.1007/978-3-642-39065-4
Springer Heidelberg Dordrecht London New York

Library of Congress Control Number: 2013940936

CR Subject Classification (1998): F.1.1, I.5.1, I.2.6, I.2.8, I.2.10, I.2, I.4, I.5,
F.1, E.1, F.2

LNCS Sublibrary: SL 1 – Theoretical Computer Science and General Issues

*Typesetting:* Camera-ready by author, data conversion by Scientific Publishing Services, Chennai, India

Printed on acid-free paper

Springer is part of Springer Science+Business Media (www.springer.com)

# Preface

This book and its sister volume collect the refereed papers presented at the 10th International Symposium on Neural Networks (ISNN 2013), held in Dalian, China, during July 4–6, 2013. Building on the success of the previous events, ISNN has become a well-established series of popular and high-quality conferences on neural network and its applications. The field of neural networks has evolved rapidly in recent years. It has become a fusion of a number of research areas in engineering, computer science, mathematics, artificial intelligence, operations research, systems theory, biology, and neuroscience. Neural networks have been widely applied for control, optimization, pattern recognition, signal/image processing, etc. ISNN aims at providing a high-level international forum for scientists, engineers, educators, as well as students to gather so as to present and discuss the latest progresses in neural network research and applications in diverse areas.

ISNN 2013 received a few hundred submissions from more than 22 countries and regions. Based on the rigorous peer reviews by the Program Committee members and the reviewers, 157 papers were selected for publications in the LNCS proceedings. These papers cover major topics of theoretical research, empirical study, and applications of neural networks.

In addition to the contributed papers, three distinguished scholars (Cesare Alippi, Polytechnic University of Milan, Italy; Derong Liu, Institute of Automation, Chinese Academy of Sciences, China; James Lo, University of Maryland - Baltimore County, USA) were invited to give plenary speeches, providing us with the recent hot topics, latest developments, and novel applications of neural networks. Furthermore, ISNN 2013 also featured two special sessions focusing on emerging topics in neural network research.

ISNN 2013 was sponsored by Dalian University of Technology and The Chinese University of Hong Kong, financially co-sponsored by the National Natural Science Foundation of China, and technically co-sponsored by the IEEE Computational Intelligence Society, IEEE Harbin Section, Asia Pacific Neural Network Assembly, European Neural Network Society, and International Neural Network Society.

We would like to express our sincere gratitude to all the Program Committee members and the reviewers of ISNN 2013 for their professional review of the papers and their expertise that guaranteed the high qualify of the technical program! We would also like to thank the publisher, Springer, for their cooperation in publishing the proceedings in the prestigious series of *Lecture Notes in Computer Science*. Moreover, we would like to express our heartfelt appreciation to the plenary and panel speakers for their vision and discussion of the latest research developments in the field as well as critical future research directions, opportunities, and challenges. Finally, we would like to thank all the speakers, authors, and participants for their great contribution and support that made ISNN 2013 a huge success.

July 2013                                                    Chengan Guo
                                                         Zeng-Guang Hou
                                                           Zhigang Zeng

# ISNN 2013 Organization

ISNN 2013 was organized and sponsored by Dalian University of Technology and The Chinese University of Hong Kong, financially co-sponsored by the National Natural Science Foundation of China, and technically co-sponsored by the IEEE Computational Intelligence Society, IEEE Harbin Section, Asia Pacific Neural Network Assembly, European Neural Network Society, and International Neural Network Society.

## General Chair

Jun Wang      The Chinese University of Hong Kong, Hong Kong, China
Dalian University of Technology, Dalian, China

## Advisory Chairs

Marios M. Polycarpou      University of Cyprus, Nicosia, Cyprus
Gary G. Yen      Oklahoma State University, Stillwater, USA

## Steering Chairs

Derong Liu      Institute of Automation, Chinese Academy of Sciences, Beijing, China
Wei Wang      Dalian University of Technology, Dalian, China

## Organizing Chairs

Min Han      Dalian University of Technology, Dalian, China
Dan Wang      Dalian Maritime University, Dalian, China

## Program Chairs

Chengan Guo      Dalian University of Technology, Dalian, China
Zeng-Guang Hou      Institute of Automation, Chinese Academy of Sciences, Beijing, China
Zhigang Zeng      Huazhong University of Science and Technology, Wuhan, China

## Special Sessions Chairs

Tieshan Li                          Dalian Maritime University, Dalian, China
Zhanshan Wang                       Northeast University, Shenyang, China

## Publications Chairs

Jie Lian                            Dalian University of Technology, Dalian, China
Danchi Jiang                        University of Tasmania, Hobart, Australia

## Publicity Chairs

Jinde Cao                           Southeast University, Nanjing, China
Yi Shen                             Huazhong University of Science and
                                        Technology, Wuhan, China

## Registration Chairs

Jie Dong                            Dalian University of Technology, Dalian, China
Shenshen Gu                         Shanghai University, Shanghai, China
Qingshan Liu                        Southeast University, Nanjing, China

## Local Arrangements Chair

Jianchao Fan                        National Ocean Environment Protection
                                        Research Institute, Dalian, China

## Program Committe

| | | |
|---|---|---|
| Alma Y. Alanis | Jixiang Du | Junhao Hu |
| Tao Ban | Haibin Duan | He Huang |
| Gang Bao | Jianchao Fan | Tingwen Huang |
| Chee Seng Chan | Jian Feng | Amir Hussain |
| Jonathan Chan | Jian Fu | Danchi Jiang |
| Rosa Chan | Siyao Fu | Feng Jiang |
| Guici Chen | Wai-Keung Fung | Haijun Jiang |
| Mou Chen | Chengan Guo | Min Jiang |
| Shengyong Chen | Ping Guo | Shunshoku Kanae |
| Long Cheng | Shengbo Guo | Rhee Man Kil |
| Ruxandra Liana Costea | Qing-Long Han | Sungshin Kim |
| Chuangyin Dang | Hanlin He | Bo Li |
| Liang Deng | Zeng-Guang Hou | Chuandong Li |
| Mingcong Deng | Jinglu Hu | Kang Li |

Tieshan Li
Yangmin Li
Yanling Li
Jinling Liang
Hualou Liang
Wudai Liao
Jing Liu
Ju Liu
Meiqin Liu
Shubao Liu
Xiaoming Liu
Wenlian Lu
Yanhong Luo
Jiancheng Lv
Jinwen Ma
Xiaobing Nie

Seiichi Ozawa
Qiankun Song
Norikazu Takahashi
Feng Wan
Cong Wang
Dianhui Wang
Jun Wang
Ning Wang
Xin Wang
Xiuqing Wang
Zhanshan Wang
Zhongsheng Wang
Tao Xiang
Bjingji Xu
Yingjie Yang
Wei Yao

Mao Ye
Yongqing Yang
Wen Yu
Wenwu Yu
Zhigang Zeng
Jie Zhang
Lei Zhang
Dongbin Zhao
Xingming Zhao
Zeng-Shun Zhao
Chunhou Zheng
Song Zhu
An-Min Zou

## Additional Reviewers

Aarya, Isshaa
Abdurahman, Abdujelil
Bai, Yiming
Bao, Gang
Bao, Haibo
Bobrowski, Leon
Bonnin, Michele
Cao, Feilong
Chang, Xiaoheng
Chen, Fei
Chen, Juan
Chen, Yao
Chen, Yin
Cheng, Zunshui
Cui, Rongxin
Dai, Qun
Deng, Liang
Dogaru, Radu
Dong, Yongsheng
Duan, Haibin
Er, Meng Joo
Esmaiel, Hamada
Feng, Pengbo
Feng, Rongquan
Feng, Xiang

Fu, Siyao
Gan, Haitao
Garmsiri, Naghmeh
Gollas, Frank
Guo, Zhishan
He, Xing
Hu, Aihua
Hu, Bo
Hu, Jianqiang
Huang, He
Huang, Lei
Huang, Xia
Jeff, Fj
Jiang, Haijun
Jiang, Minghui
Jiang, Yunsheng
Jin, Junqi
Jinling, Wang
Lan, Jian
Leung, Carson K.
Li, Benchi
Li, Bing
Li, Fuhai
Li, Hongbo
Li, Hu

Li, Kelin
Li, Lulu
Li, Shihua
Li, Will
Li, Xiaolin
Li, Yang
Li, Yongming
Li, Zifu
Lian, Cheng
Liang, Hongjing
Liang, Jianyi
Liang, Jinling
Lin, Yanyan
Liu, Cheng
Liu, Huiyang
Liu, Miao
Liu, Qingshan
Liu, Xiao-Ming
Liu, Xiaoyang
Liu, Yan-Jun
Liu, Yang
Liu, Zhigang
Lu, Jianquan
Lu, Yan
Luo, Weilin

Ma, Hongbin
Maddahi, Yaser
Meng, Xiaoxuan
Miao, Baobin
Pan, Lijun
Pang, Shaoning
Peng, Zhouhua
Qi, Yongqiang
Qin, Chunbing
Qu, Kai
Sanguineti, Marcello
Shao, Zhifei
Shen, Jun
Song, Andy
Song, Qiang
Song, Qiankun
Sun, Liying
Sun, Ning
Sun, Yonghui
Sun, Yongzheng
Tang, Yang
Wang, Huanqing
Wang, Huiwei
Wang, Jinling
Wang, Junyi
Wang, Min
Wang, Ning
Wang, Shenghua

Wang, Shi-Ku
Wang, Xin
Wang, Xinzhe
Wang, Xiuqing
Wang, Yin-Xue
Wang, Yingchun
Wang, Yinxue
Wang, Zhanshan
Wang, Zhengxin
Wen, Guanghui
Wen, Shiping
Wong, Chi Man
Wong, Savio
Wu, Peng
Wu, Zhengtian
Xiao, Jian
Xiao, Tonglu
Xiong, Ping
Xu, Maozhi
Xu, Yong
Yang, Bo
Yang, Chenguang
Yang, Feisheng
Yang, Hua
Yang, Lu
Yang, Shaofu
Yang, Wengui
Yin, Jianchuan

Yu, Jian
Yu, Wenwu
Yuen, Kadoo
Zhang, Jianhai
Zhang, Jilie
Zhang, Long
Zhang, Ping
Zhang, Shuyi
Zhang, Weiwei
Zhang, Xianxia
Zhang, Xin
Zhang, Xuebo
Zhang, Yunong
Zhao, Hongyong
Zhao, Mingbo
Zhao, Ping
Zhao, Yu
Zhao, Yue
Zheng, Cheng-De
Zheng, Cong
Zheng, Qinling
Zhou, Bo
Zhou, Yingjiang
Zhu, Lei
Zhu, Quanxin
Zhu, Yanqiao
Zhu, Yuanheng

# Table of Contents – Part I

## Computational Neuroscience and Cognitive Science

# Neural Network Models, Learning Algorithms, Stability and Convergence Analysis

## Kernel Methods, Large Margin Methods and SVM

## Optimization Algorithms / Variational Methods

# Feature Analysis, Clustering, Pattern Recognition and Classification

# Vision Modeling and Image Processing

# Table of Contents – Part II

## Control, Robotics and Hardware

## Bioinformatics and Biomedical Engineering, Brain-Like Systems and Brain-Computer Interfaces

## Evolutionary Neural Networks, Hybrid Intelligent Systems

## Data Mining and Knowledge Discovery

## Other Applications of Neural Networks

# Information Transfer Characteristic in Memristic Neuromorphic Network

Quansheng Ren, Qiufeng Long, Zhiqiang Zhang, and Jianye Zhao

School of Electronics Engineering and Computer Science, Peking University,
Beijing 100871, China

**Abstract.** Memristive nanodevices can support exactly the same learning function as spike-timing-dependent plasticity in neuroscience, and thus the exploration for the evolution and self-organized computing of memristor-based neuromorphic networks becomes reality. We mainly study the STDP-driven refinement effect on memristor-based crossbar structure and its information transfer characteristic. The results show that self-organized refinement could enhance the information transfer of memristor crossbar, and the dependence of memristive device on current direction and the balance between potentiation and depression are of crucial importance. This gives an inspiration for resolving the power consumption issue and the so called sneak path problem.

**Keywords:** Memristor, Neuromorphic Computing, Mutual Information, STDP.

## 1 Introduction

Recently, the limitation of CMOS scaling and the re-discovery of the memristor [1,2] have brought neuromorphic computing architectures new attention. An important discovery is that many adaptive nanoelectronic devices (such as memristor) can support exactly the same learning functions in neuroscience, such as the spike-timing-dependent plasticity (STDP) [3,4] as well as short-term plasticity and long-term potentiation [5]. Several schemes for neuromorphic computing and learning have been proposed and fabricated by using the crossbar architecture [6,7,8]. However, much yet remains to be done, such as the power consumption issue and the so called sneak path problem [9,10]. Below this setting, it is necessary to study the role of resistance distribution from the perspectives of STDP adaptation and information transfer for further exploration of the relationship between the demands of hardware integration and information processing.

In order to address this issue, we should revisit some related researches in neuroscience. The STDP learning rule says that a synapse is strengthened if the presynaptic neuron fires shortly before the postsynaptic neuron, and weakened when the temporal order is reversed. The brain has a very dense population of synapses after birth and most of these synapses are irreversibly pruned during development when their strength falls below a certain threshold [11]. A sparse network structure will be achieved at last. STDP is likely to be a crucial factor in the synaptic pruning process, for it can refine a fully connected neural network to be a simplified one with

C. Guo, Z.-G. Hou, and Z. Zeng (Eds.): ISNN 2013, Part I, LNCS 7951, pp. 1–8, 2013.

a bimodal distribution of synaptic weights [12], which could bring several non-trivial topological characteristics [13-16], and can be used to systematically improve and optimize synchronization properties [17]. In addition, by inputting temporal-correlated input signals, the information transfer efficiency of neural network could be improved obviously [18].

In this paper, we establish an equivalent model of memristor-based neuromorphic networks according to the characteristics of the memristor and some latest relevant experimental results [3,4,5,12]. We mainly focus on the mutual information between the input and output layers of the crossbar network. To compute the mutual information between the input and output layers, we adopt a classical algorithm [17,19] to achieve this purpose. Besides, Bayes estimation [20] based on the context tree method was put to use to calculate the entropy rate. Our results suggest that self-organized refinement effect can increase the information transfer efficiency of the crossbar network.

## 2      Analysis and Modeling of Memristic Neural Networks

### 2.1      Neuromorphic Analysis of Memristor

The neuromorphic network studied in this paper is the memristor-based crossbar structure firstly promoted by HP Labs [1]. This kind of crossbar network consists of some horizontal and vertical metal nanowires, the intersections of which are connected by memristor. And it has been proven that the change of the memristive resistance is related with the direction, amplitude and duration of the voltage across it.

There is a one-to-one relationship between memristive characteristics and STDP rules in neuroscience. The aforementioned voltage direction, duration and amplitude are respectively equivalent to the order of the pre- and postsynaptic action potentials, the time interval between pre- and postsynaptic spikes and the STDP-driven learning efficiency in neuroscience. The only difference between the memristive and the neural network is the form of processed signals. For this reason, to establish an equivalent model of memristor-based crossbar network, we should transform the time difference between a pair of neural spikes into a PWM electrical signal with the equal time interval and corresponding direction of the two spikes. These works are not difficult to deal with by using analog digital mixing circuit, such as the integral-and-fire (I&F) neuron circuit [3].

### 2.2      Equivalent Model of Memristor-Based Neuromorphic Network

On the basis of neuromorphic analysis of memristic circuit, we adopt the aforementioned crossbar structure network as a simulation model. In addition, multi-parameter expression is utilized for more accurate research (see Fig. 1).

CMOS I&F neuron circuit could read and record the time parameter of input and output spikes. According to the time difference of the two time parameters, the circuit can generate a PWM signal with corresponding polarity and duration of the pre- and postsynaptic timing. The PWM signal can be used to regulate and control the memristive synaptic conductance. The internal potential parameter $V_j$ of the $j$th neuron and the conductance $g_i$ satisfy

$$C_m \frac{dV_j}{dt} = g_L(V_{rest} - V_j) + g_i(t)(E_{ex} - V_j) \tag{1}$$

This formula has the same parameter setting as a classical document [21], the capacitance $C_m$=200pF, leakage conductance $g_L$=10nS, resting potential $V_{rest}$=−70mV and excitatory reversal potential $E_{ex}$=0 mV. When potential parameter reach the threshold $V_{th}$=−54mV, the neuron model will generate an action potential. After a refractory period $\tau_{ref}$=1ms, potential will decrease to $V_{reset}$=−60mV. Furthermore, synaptic conductance parameter $g_j(t)$ in LIF model is determined by

$$g_j(t) = g_m \sum_{i=1}^{n} w_{ij}(t) \sum_k f(t - t_i^k) \tag{2}$$

where $n$ is the number of excitation sources, $g_m$ is the maximal synaptic conductance, $w_{ij}$ is the synaptic weight between the $i$th and $j$th neuron circuit and $t_i^k$ is the $k$th effective output time of the $i$th neuron circuit. $f(x)$ usually adopts $\alpha$ function [22].

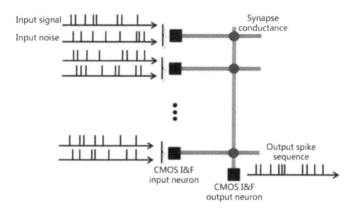

**Fig. 1.** Model of memristor-based neuromorphic network and input excitation

According to the above analysis, the memristor synapse conductance are controlled by PWM signals in a STDP-like way. The expressions describing the STDP-driven learning mechanism depend on the time difference between pre- and post-synaptic timing $\Delta t = t_j - (t_i + \tau_d)$, where $t_i$ and $t_j$ respectively represent the output time of the pre-synaptic neuron circuit $i$ and the post-synaptic circuit $j$, $\tau_d$ is the time for processing and transmitting the effective signals from the input circuit $i$ to output circuit $j$.

Multiparameter expressions are utilized to describe various of STDP learning curves by setting the parameters with different values. The variation of connecting weights $\Delta w_{ij}$ satisfys

$$\Delta w = \begin{cases} \lambda(1-w)^{\mu_+} \exp(-|\Delta t|/\tau_+), & \text{if } \Delta t > \tau_d \\ -\lambda a w^{\mu_-} \exp(-|\Delta t|/\tau_-), & \text{if } \Delta t \leq \tau_d \end{cases} \tag{3}$$

Here, $\lambda$ is the learning rate corresponding to the amplitude of the output PWM signal, $w$ determines the memristic conductance $g_m w$, $(1-w)\mu_+$ and $w\mu_-$ reflects the dependencies of potentiation and depression on the current synaptic weights. Then we can restrict $w$ in the range [0,1] so that the synaptic conductance $g_m w$ is positive and can't exceed $g_m$=2.0nS. When $\mu_+=\mu_-=0$,$w$ should remain unchanged once approaching the maximum 1(or minimum 0), which is equivalent to the restricted condition promoted by R. S. Williams [1].

In Eq. (3), $\alpha$ is a asymmetric coefficient of the ratio between potentiation and depression while $\tau_+$ and $\tau_-$ controls the width of effective time-window of the potentiation and depression. To balance $\alpha$, we utilize asymmetric STDP time window ($\tau_+$=16.8ms and $\tau_-$=33.7ms), which is fairly close to the STDP parameters observed in neuroscientific investigation [23]. As what has been suggested in [21], to keep system steady, the integrated result of $\Delta w_{ij}$ must be negative, i.e. $\alpha(\tau_-/\tau_+)>1.0$.

## 3    Input Scheme and Mutual Information Computation

The input scheme is as follows: As shown in Fig. 1, 100 uncorrelated periodic spike trains with mean firing rate $f_s$ and 100 stochastic Poisson spike trains with mean firing rate $f_n$ are respectively regarded as input signals and noise of the 100 input neuron circuits (see Fig. 1).

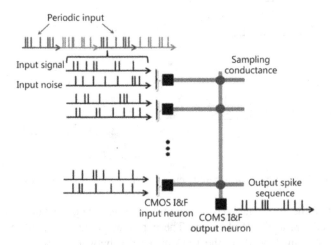

**Fig. 2.** Input setting for computing mutual information

For the investigation of the information transfer characteristics in the simplified network, we calculate the mutual information between the output and input signals with a classcal method [17,19]. One hundred 1.5-s-long spike sequences with mean firing rate $f_s$ are generated by independently truncating from a single Poisson spike trains. These spatiotemporal spike patterns are then respectively repeated 1000 times at regular intervals, and finally interleaved among one hundred 1500-s-long stochastic Poisson stimuli. As shown in Fig. 2, the input signals of each synapse is composed of alternating repeated spike segments and stochastic spike segments, each lasting 1.5s. In addition, noise is still stochastic Poisson spike process with mean firing rate $f_n$. This indeterminacy of both the input signals and input noise can be quantified by total entropy $H_{total}$. In our input scheme, the input repeated spike segments are identical, so the indeterminacy of output signals exactly equal to the input noise entropy $H_{noise}$. Finally we could know that the mutual information between the output and input signals is just the difference of the two entropies, i.e. $I_{info}=H_{total}-H_{noise}$.

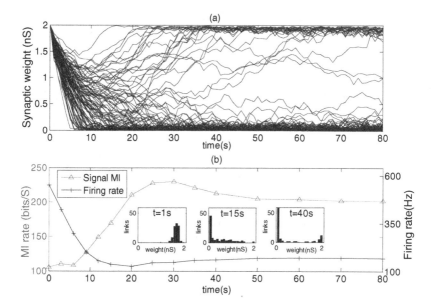

**Fig. 3.** Analysis **of information transfer in a refinement process. Here** $f_s$=30Hz, $f_n$=10Hz, $a$=0.51. (a) The changing curves with time of the 100 memristor conductances. (b) Different mutual information rate (red triangle line) and mean firing rate (black cross line) of output signals in the refinement evolution. Inset figures: The distribution of memristor conductances at 1s,15s and 40s.

To directly achieve the entropy rate estimators of $H_{total}$ and $H_{noise}$, we adopt a well-known algorithm proposed in Ref. [20]. In addition, with regard to the first-order Markov Chain $\sigma$, if a new character $r \in A$ can be produced with a stationary transition probability $P(r|\sigma)$, $R=\{r_1,r_2,r_3,...\}$ could be regarded as a time series. According to the entropy calculation formula $h=\lim_{D\to\infty}H(r_{i+1}|r_i,...r_{i-D})$ ($D$ is the length of the sequential characters obtained from time series), the entropy rate of the time series $R$ can be obtained by

$$h = \sum_{\sigma} H[p(r|\sigma)]\mu(\sigma) \tag{4}$$

Here, $H[...]$ is the Shannon Entropy and $\mu(\sigma)$ is the probability of state $\sigma$. According to the output series, we can establish a context tree weighting model for binary character streams. Every node record the occurrence probability of the next character and the number of its current character in the time series. The entropy rate of this context tree method can be calculated with Eq. (4). Then by using Minimum Description Length (MDL) method, we can obtain a topological structure of the context tree, each terminal node of which corresponds to a sample value $P(H|\bar{c})$, i.e.

$H^*(n)$ calculated with Markov Chain Monte Carlo, where $\bar{c}$ is the number of the next generated character in the time series. Then, we can calculate the entropy estimator $h_i^* = \sum_n H^*(n)\mu(n)$, where $\mu(n) = N(n)/N = \sum_{i=1}^{A} c_i / N$ in which $c_i$ and $N$ respectively represent the occurrence number of the corresponding character of $n$th node and that of all characters in the time series. Finally by repeating this process for $N_{mc}$ times, we can obtain the entropy estimator with Bayesian confidence interval, $h_{Bayes} = \sum_{i=1}^{N_{mc}} h_i^* N_{mc}$.

## 4      Influence of Refinement Phenomena on Information Transfer

According to the method proposed in 3rd paragraph, we can calculate and contrast the information transmission rate of the network at different times in a simplification process. The simulation results are displayed in Fig. 3. After a long-time simulation, all synaptic conductances form a stable bimodal distribution. Simulation results show that the mutual information rate increase along with the refinement process while the output firing rate undergo a decrease process.

We conduct many simulation experiments with different α. We can see from Fig. 4(a): When 0.49<α<0.5, the scale of potentiation and depression ($τ_+/ατ_-$) tends to be a

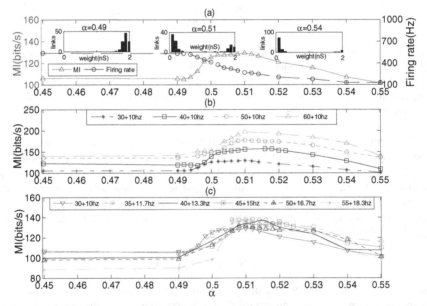

**Fig. 4.** Influence **of different parameter setting on information transfer.** (a) In the condition of $f_s$= 30Hz, $f_n$=10Hz, the change curves with α of mutual information rate (red solid line with triangle marker) and the mean output firing rate (black dash line with circle marker). (b) The mutual information rate with different $f_s$, $f_n$=10Hz. (c) The mutual information with different $f_s$ and $f_n$ ,SNR = 3/1.

balance state, during which the information transfer rate grows. If $0.5 < \alpha < 0.51$, the information transfer rate keeps in a higher level. However, if depression is stronger than potentiation (such as $\alpha > 0.51$), information transfer rate will markedly decline mostly because of the lack of output response. As what have been stated in 2nd paragraph, memristor STDP learning characteristics are directly related with: (1) The dependence on current direction of memristor conductance, (2) The balance between potentiation and depression. Consequently, the two factors play an important role in optimizing information transfer in memristor-based neural networks.

To get a general conclusion, we conduct simulations with constant firing rate of input noise and different firing rate of input signals for the implementation of different SNR configuration. Simulation results with different $\alpha$ and SNR are demonstrated in Fig. 4(b), from which we can see that the greater the input SNR is, the higher the maximal mutual information transfer rate becomes. In addition, we also experiment with constant SNR (3/1) and different firing rate $f_s$ and $f_n$, which, however, have been proved to be not of critical to change the information transfer characteristics (see Fig. 4(c)).

## 5      Conclusions

In this paper, we mainly research the effect of STDP-driven pruning on the structure and information transfer characteristics of the neuromorphic crossbar network. Simulating results show that, by inputting periodic signals, the conductance of most memristor synapses will falls to 0nS (HRS) and remain unchanged due to the STDP learning mechanism, which is the so-called synaptic pruning phenomena. We adopt a classical method to compute the mutual information, and calculate the entropy rate with a method based on the context tree weighting (CTW). It is shown that the simplification process driven by STDP learning mechanism can improve information transmission rate. This means that the reduction of LRSs is naturally achieved by STDP-based learning, which can improve information transfer rate and give an inspiration for resolving the power consumption issue and the so called sneak path problem. Investigations with many kinds of parameter configurations prove that the aforementioned results are by no means accidental events.

Though the memristor was for the first time discovered just in 2008, some researchers have made a great progress on hybrid nano-devices/semiconductor circuits. Based on the CMOS I&F neuron circuit and nano-memristor synapse, the days of "Electronic brain" are numbered [6]. In this background, theoretical exploration from the aspect of information theory and system architecture should be conducted urgently. This paper for the first time study the memristive neuromorphic networks from the angle of the information transfer characteristic. Above-mentioned research conclusions may be instructive for crossbar architecture designing and neuromorphic computing.

## References

1. Strukov, D.B., Snider, G.S., Stewart, D.R., Williams, R.S.: The Missing Memristor Found. Nature 453, 80–83 (2008)
2. Chua, L.O.: Memristor-The Missing Circuit Element. IEEE Trans. Circuits Syst. 18, 507–519 (1971)

3. Jo, S.H., Chang, T., Ebong, I., Bhadviya, B.B., Mazumder, P., Lu, W.: Nanoscale Memristor Device as Synapse in Neuromorphic Systems. Nano Lett. 10, 1297–1301 (2010)
4. Kuzum, D., Jeyasingh, R.G.D., Lee, B., Wong, H.-S.P.: Nanoelectronic Programmable Synapses Based on Phase Change Materials for Brain-Inspired Computing. Nano Lett. 12, 2179–2186 (2012)
5. Ohno, T., et al.: Short-Term Plasticity and Long-Term Potentiation Mimicked in Single Inorganic Synapses. Nat. Mater. 10, 591–595 (2011)
6. Likharev, K.K.: CrossNets: Neuromorphic Hybrid CMOS/Nanoelectronic Networks. Sci. Adv. Mater. 3, 322–331 (2011)
7. Snider, G.S.: Self-Organized Computation with Unreliable, Memristive Nanodevices. Nanotechnology 18, 365202 (2007)
8. Zamarreno-Ramos, C., et al.: On Spike-Timing-Dependent-Plasticity, Memristive Devices, and Building a Self-Learning Visual Cortex. Front Neurosci. 5, 26–47 (2011)
9. Flocke, A., Noll, T.G.: Fundamental Analysis of Resistive Nano-Crossbars for the Use in Hybrid Nano/CMOS-Memory. In: Proc. 33rd Eur. Solid-State Circuits Conf., pp. 328–331 (2007)
10. Kügeler, C., Meier, M., Rosezin, R., Gilles, S., Waser, R.: High Density 3D Memory Architecture Based on the Resistive Switching Effect. Solid-State Electronics 53, 1287–1292 (2009)
11. SyNAPSE: Systems of Neuromorphic Adaptive Plastic Scalable Electronics, http://www.darpa.mil
12. Strukov, D.B.: Nanotechnology: Smart connections. Nature 476, 403–405 (2011)
13. Shin, C.-W., Kim, S.: Self-Organized Criticality and Scale-Free Properties in Emergent Functional Neural Networks. Phys. Rev. E 74, 045101 (2006)
14. Jost, J., Kolwankar, K.M.: Evolution of Network Structure by Temporal Learning. Phys. A 388, 1959–1966 (2009)
15. Takahashi, Y.K., Kori, H., Masuda, N.: Self-Organization of Feed-Forward Structure and Entrainment in Excitatory Neural Networks with Spike-Timing-Dependent Plasticity. Phys. Rev. E 79, 051904 (2009)
16. Ren, Q., Kolwankar, K.M., Samal, A., Jost, J.: STDP-Driven Networks and The C. elegans Neuronal Network. Physica A 389, 3900–3914 (2010)
17. Strong, S.P., Koberle, R., van Steveninck, R.R.R., Bialek, W.: Entropy and Information in Neural Spike Trains. Phys. Rev. Lett. 80, 197–200 (1998)
18. Hennequin, G., Gerstner, W., Pfister, J.-P.: STDP in Adaptive Neurons Gives Close-To-Optimal Information Transmission. Front. Comput. Neurosci. 4, 143–158 (2010)
19. Borst, A., Theunissen, F.E.: Information Theory and Neural Coding. Nat. Neurosci. 2, 947–957 (1999)
20. Kennel, M.B., Shlens, J., Abarbanel, H.D.I., Chichilnisky, E.J.: Estimating Entropy Rates with Bayesian Confidence Intervals. Neural Comput. 7, 1531–1576 (2005)
21. Song, S., Miller, K.D., Abbott, L.F.: Competitive Hebbian Learning Through Spike-Timing-Dependent Synaptic Plasticity. Nat. Neurosci. 3, 919–954 (2000)
22. Brunel, N., Hakim, V.: Fast Global Oscillations in Networks of Integrate-and-Fire Neurons with Low Firing Rates. Neural Comput. 11, 1621–1671 (1999)
23. Bi, G., Poo, M.: Synaptic Modification by Correlated Activity: Hebb's Postulate Revisited. Annu. Rev. Neurosci. 24, 139–166 (2001)

# Generation and Analysis of 3D Virtual Neurons Using Genetic Regulatory Network Model*

Xianghong Lin and Zhiqiang Li

School of Computer Science and Engineering, Northwest Normal University,
Lanzhou 730070, China
linxh@nwnu.edu.cn

**Abstract.** Neuronal morphology is significant for understanding structure-function relationships and brain information processing in computational neuroscience. So it is very important to simulate neuronal morphology completely and accurately. In this paper, we present a novel approach for efficient generation of 3D virtual neurons using genetic regulatory network model. This approach describes dendritic geometry and topology by locally inter-correlating morphological variables which can be represented by the dynamics of gene expression. The experimental results show that the generating virtual neurons that are anatomically indistinguishable and accurate from experimentally traced real neurons.

**Keywords:** virtual neuron, neuronal morphology, genetic regulatory network.

## 1  Introduction

The brain, with its billions of interconnected neurons, is without any doubt the most complex organ in the body and it will be a long time before we understand all its mysteries. The Human Brain Project [1] proposes a completely new approach and it builds on the work of the Blue Brain Project [2], which has already taken an essential first towards simulation of the complete brain. The project is integrating everything we know about the brain into computer models and using these models to simulate the actual working of the brain. The remarkable progress of computer processing power in the past few decades has enabled the construction of greatly sophisticated models of neuronal function and behavior. Computational modeling of neuronal morphology is a powerful tool for understanding developmental processes and structure-function relationships. Dendrites and axons define the connectivity of the brain [3] and play a large role in information processing at the single cell level [4].

The research of neuronal activity and network connectivity in computational neuroanatomy has recently encouraged neuroscientists to develop and characterize computer

---

* This research is supported by the National Natural Science Foundation of China under Grant No. 61165002, and the Natural Science Foundation of Gansu Province of China under Grant No. 1010RJZA019.

C. Guo, Z.-G. Hou, and Z. Zeng (Eds.): ISNN 2013, Part I, LNCS 7951, pp. 9–18, 2013.

algorithms for the simulation of neuronal morphology. There are two main computer algorithms exist to generate virtual neurons with similar shape properties as their empirically observed counterparts: reconstruction algorithms and growth algorithms. Reconstruction algorithms focusing on both topological and metrical aspects, use the empirical distribution functions for geometrical properties, and provide models for randomly generating dendritic morphologies by a repeated process of random sampling of these distributions [5,6,7,8]. Growth algorithms, in contrast, aim at modeling dendritic morphologies from principles of dendritic development, use the hypothetical growth rules for branching and elongation in the generation of random dendritic trees. Growth algorithms explain topological variation by assuming that the branching probability depends on the type of segment (intermediate or terminal) as well as on the centrifugal order of the segment [9,10]. For this reason, growth algorithms are mechanistic models and 'grow' models in an abstract but biologically plausible manner. In addition, recent advances in computer graphics have become possible to use these algorithms for generation and display of 3D models of neuronal structures that are visually and statistically indistinguishable from the traced real neurons.

In this paper, using the recurrent genetic regulatory network model, the dynamics of gene expression can be treated as a model for dendritic development. We propose a novel growth algorithm for the 3D simulation of neuronal dendritic trees. This method generates dendrites in single neuron by local algorithms, in which each modeled dendrite grows depending on intrinsic influences, and independent of other dendritic trees or neurons. The aim of this paper is to generate virtual neurons that are anatomically accurate and realistic.

## 2    Morphological Data of Real Neurons

The morphological structure of a neuron can be represented in a SWC format [11], which extracted from 3D reconstructions of intracellularly stained cells. The reconstructions are publicly available at the online NeuroMorpho archive [12,13]. The motoneuron files corresponded to six alpha motor neurons from cat spinal cord. These typical motor neurons obtain from Burke's laboratory archive [6]. We can read SWC morphology file by the publicly available tools CVAPP [14] and Neuromantic [15]. Additionally, our method output the generated virtual neuron which is interpreted to SWC format file.

In this paper, we adopt some basic variables and emergent variables [16]. The basic variables can be usually measured from digital files of traced neurons. Such measurements result in distributions of values, which are fitted with a statistical function. The basic morphological variables include number of stems, diameter of stem, azimuth angle and elevation angle of stem, the length of compartment, taper rate, amplitude angle and torque angle at a bifurcation, Rall's power at a bifurcation, azimuth angle and elevation angle at an elongate point. A list of basic variables and their value ranges for motoneurons is shown in Table 1. On the other hand, the emergent variables emerge from the interaction between basic variables. An example is the overall length of a dendritic branch that emerges from several basic variables such as the number of stems, the stem length and individual compartment length. There are some emergent variables are used in this study: total number of bifurcations, total number of branches, branching order, surface, volume, total dendritic length, contraction, fragmentation, partition asymmetry, significant width, significant height and significant depth.

Table 1. List of basic morphological variables for motoneurons

| Variable | Description | Value range |
|---|---|---|
| $N_{stem}$ | Number of stems | (8, 16) |
| $D_{stem}$ | Diameter of stem | (3.0, 5.0) |
| $S_{az}$ | Azimuth angle of stem | (0, 360) |
| $S_{elev}$ | Elevation angle of stem | (0, 180) |
| $L_{com}$ | The length of compartment | (60, 100) |
| $R_{taper}$ | Taper rate | (0.1, 0.2) |
| $B_{amp}$ | Amplitude angle at a bifurcation | (10, 170) |
| $B_{tor}$ | Torque angle at a bifurcation | (0, 180) |
| $B_{rall}$ | Rall's power at a bifurcation | (1, 4.45) |
| $E_{az}$ | Azimuth angle at an elongate point | (−10, 10) |
| $E_{elev}$ | Elevation angle at an elongate point | (−10, 10) |

# 3    Generation of 3D Virtual Neurons

## 3.1    Genetic Regulatory Network Model

The genetic regulatory network can be viewed as expressed genes influencing the expression of other genes. It can be represented as a directed graph, in which each node represents a gene and edges represent direct transcriptional interactions. Each edge is directed from a gene that encodes a transcription factor to a gene that is regulated by that transcription factor. A genetic system is defined as a network of $N$ interacting nodes. The genetic regulatory network is structured in three layers, consisting $N_I$ input nodes, $N_R$ regulatory nodes, and $N_O$ output nodes respectively. $N_I$ and $N_O$ are determined by the developmental model described in the below section. The regulatory nodes that recurrently connect in the network represent genes that play a regulatory role only.

The activation of node $i$ at time $t$ during development is represented by a variable $A_i(t)$, $i=1, 2, \ldots, N$. The state of the network is updated synchronously in discrete time steps, with the activation $A_i(t+1)$ of node $i$ at time $t+1$, given by

$$A_i(t+1) = \sigma(\sum_{j=1}^{N} w_{ij} A_j(t) - \theta_i) . \tag{1}$$

where $w_{ij}$ is the level of the interaction from node $j$ to node $i$, which is a real-valued weight in the range [0, 5.0]. $\theta_i$ is the activation threshold of node $i$, given by

$$\theta_i = \sum_{j=1}^{N} w_{ij} \delta . \tag{2}$$

where $\delta$ is weight bias (by default, $\delta = 0.5$). $\sigma(x)$ is the sigmoid function, given by

$$\sigma(x) = \frac{1}{1+e^{-x}}. \tag{3}$$

Therefore, the activation of gene is continuous change in the interval [0, 1], where 0 represents gene inactivation and 1 represents complete gene expression.

## 3.2    Developmental Method of Virtual Neurons

In this paper, the recurrent genetic regulatory network control the dendritic development process, which includes 2 nodes in input layer, 10 nodes in regulatory layer, and 10 nodes in output layer. The initial genetic regulatory network can express as a soma, which activation is random initialization within the interval [0, 1]. The input nodes are used to specify the relative position of the compartment in the dendritic trees; the regulatory nodes are used to regulate other regulatory nodes and output nodes; the output nodes are used to represent the development process and the morphological variables.

Considering the convenience of description, the activation of a regulatory node at time $t$ during development is represented by $R_i(t)$, $i=1, 2, ..., N_R$. The activation of an input or output node at time $t$ is represented by $I_i(t)$, $i=1, 2, ..., N_I$ or $O_i(t)$, $i=1, 2, ..., N_O$, respectively. The activation of the output nodes express qualitative features, which include bifurcation probability, symmetric or asymmetric bifurcation mode, termination probability, and the others convert into specific parameter values within the given range $[P_{min}, P_{max}]$, which is represented as $P_{min}+(P_{max}-P_{min})O_i(t)$. These parameters correspond to the basic morphological variables as shown in Table 1.

The morphogenetic algorithm for generation of neuronal morphologies is shown in Fig. 1. The algorithm generates each dendritic tree as an independent process. For generation of the motoneurons, one of the basic variables is $N_{stem}$, which is the number of stems. Once a value $N_{stem}$ is sampled for this variable, the algorithm is repeated $N_{stem}$ times to generate the appropriate number of stems. Each stem originates from the soma with a certain initial diameter $D_{stem}$ and an orientation specified by $S_{az}$ and $S_{elev}$. Then in the spherical coordinate system, the dendrite is defined by the spherical coordinates of its end point taking the starting point as the origin (Fig. 2).

If the activation of the first output node, $O_1(t)$, is above a certain bifurcation threshold $\theta_{bif}$, a dendritic compartment bifurcates, the two daughter compartments will create two copies of the genetic regulatory network with identical interactions of nodes. The threshold of bifurcation dynamic changes in the developmental processes, it is assigned a value which depends exponentially on the diameter of current compartment. So it can be represented as $\theta_{bif} = k_1 e^{\lambda_1 d}$, where $k_1=0.06$, $d$ is the diameter, $\lambda_1$ represents development scale and uses for controlling the scale of neuronal development. The relative directions of the two daughters are determined with two variables, a bifurcation amplitude $B_{amp}$ and a bifurcation torque $B_{tor}$. Both daughters are determined by the activation of asymmetric output node ($O_2(t)$) and their diameters are calculated from the parent diameter based on the Rall's ratio [17]. After bifurcation, the input nodes $I_1(t)$ and $I_2(t)$ are set to {0, 1} in the left daughter and {1, 0} in the right daughter.

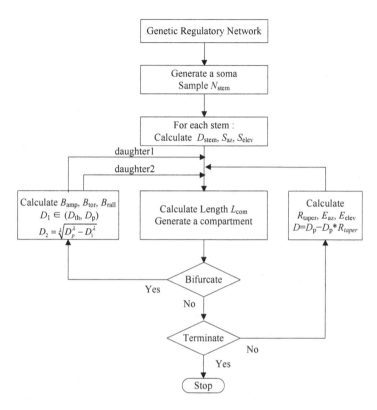

**Fig. 1.** Flow diagram of the morphogenetic algorithm. The algorithm begins at a soma and attaches dendritic variables. Initially, an initial azimuth and elevation angle of stem is calculated. Then, the algorithm decides recursively whether to bifurcate, terminate or stretch the current compartment. For motoneurons, a dendritic branch is also terminated if it grows in length larger than 2200 μm or in diameter less than 0.13 μm.

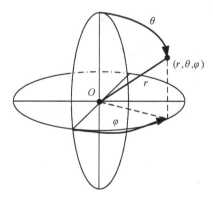

**Fig. 2.** The spherical coordinate system where a dendritic compartment is defined by the coordinate $(r, \theta, \varphi)$ of its end point taking the starting point as the origin: $r$ is the Euclidean distance between the two points, $\theta$ is the elevation angel and $\varphi$ is the azimuth angle

If the compartment does not bifurcate, the compartment may terminate or elongate with a new compartment. The activation of the third node, $O_3(t)$, is used to specify the termination probability. Similarly, the threshold of termination also dynamic changes, it depends exponentially on diameter of current compartment. It can be represented as $\theta_{ter} = k_2 e^{\lambda_2 d}$, where $k_2$=0.65, $d$ is the diameter, and $\lambda_2$ is the development scale. Additionally, the dendritic compartment is always terminated for the motoneurons when the current diameter is less than 0.13 μm or the path length to the soma is larger than 2200 μm, these limitations are motivated by biological knowledge. If the compartment does not terminate either, the current compartment will elongate with a new compartment. The growth direction is updated with an azimuth angle $E_{az}$ and elevation angle $E_{elev}$. The compartment diameter shrinks according to a taper rate $R_{taper}$ that is relative to the diameter of parent compartment, but it is independent of the compartment length. The algorithm continues recursively until all dendritic compartments have terminated.

## 4     Experimental Results

In the results reported here, since the experimentally traced datasets consisted of a limited number of neurons, we generated a lot of virtual neurons to allow the statistical analysis of their emergent variables. An example of the morphological structures of real and simulated virtual neurons is shown in Fig. 3.

We mainly investigated the relationship between emergent variables and developmental scales in the generation of 3D virtual neurons. In the simulations, we carried out a sensitivity analysis for each parameter. For each combination of parameter settings, averages were taken over 100 virtual neurons. Because there were so many possible combinations, we kept the values of every parameter at a fixed base value except for the parameter being tested. The base values of the developmental scales $\lambda_1$ and $\lambda_2$ are 0.5 and −3.0, respectively. The emergent variables were measured from each real and virtual neuron. Every group was statistically characterized with average and standard deviation values for each of the emergent variables. Virtual data were then analyzed in terms of mean and standard deviation of each value, and compared to the corresponding values of the experimental group.

Fig. 4(a), (c), (e) show the results of the each emergent variable for the motoneurons as the value of development scale $\lambda_1$ is 0.4, 0.45, 0.5 and 0.55 respectively. Accordingly, as the value of development scale $\lambda_2$ is −2.0, −3.0, −4.0 and −5.0 respectively, the corresponding results are shown in Fig. 4(b), (d), (f). The mean value of the surface and the total dendritic length is lower than corresponding value in real neurons, while the volume of the virtual neurons is larger than the real cells in the two different development scales. The Surface, volume and total dendritic length are relatively sensitive to $\lambda_1$, but little sensitive to $\lambda_2$. In particular, the contraction, fragmentation and partition asymmetry of real cells are all quantitatively reproduced in virtual neurons. In short, $\lambda_1$ and $\lambda_2$ have little effect on the contraction, fragmentation and partition asymmetry. However, this algorithm fails to reproduce the real values related to width, height and depth. It is shown that the generated virtual neurons have little

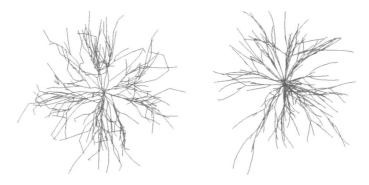

**Fig. 3.** The real and virtual neurons displayed with Neuromantic tool. This figure shows 2D projections of real (left) and virtual (right) morphological structures of the motoneurons

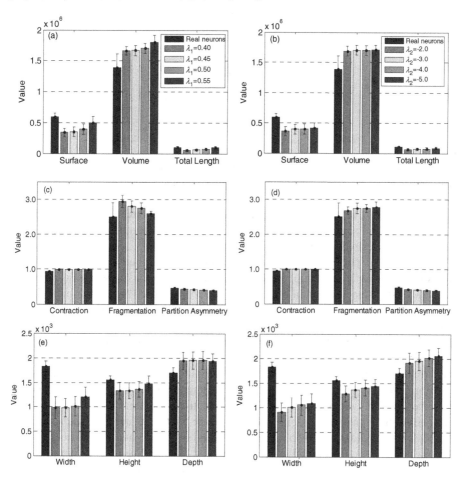

**Fig. 4.** Emergent morphological variables for the motoneurons. 6 real motoneurons and 4 groups of 100 virtual motoneurons were analyzed in different development scale

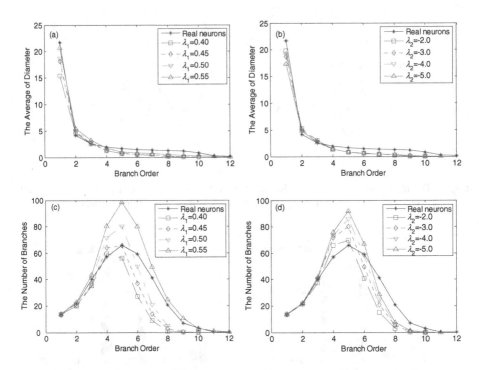

**Fig. 5.** Distributions of emergent variables for the motoneurons. (a) and (b) Branch diameter versus branch order, (c) and (d) Number of branches versus branch order

difference with real neurons in spatial structure at a certain extent. With the variation of the development scale, the value will get closer to the experimental value and may be achieved the desired results.

The analysis of virtual motoneurons is extended to distributions of emergent variables (Fig. 5). As an example, we reported the dependence of the average branch diameter versus branch order in different development scale (Fig. 5(a), (b)). Virtual neurons are very similar to the real ones and show typical decay function that conforms to biological detail. We concluded that the developmental scales $\lambda_1$ and $\lambda_2$ have little effect on this distribution. Sholl-like plots of the bifurcations or terminations versus branch order or path distance yielded classical bell shaped curves for all real and virtual neurons. In the same way, the distributions of the number of branches versus branch order are shown in Fig. 5(c), (d). Here, the number of branches is equal to number of bifurcations plus number of terminations. The real motoneurons group has a peek at 66 branches at order 5, while there are 56, 65, 80 and 98 branches in Fig. 5 (c) ($\lambda_1$ is 0.4, 0.45, 0.5 and 0.55 respectively), and 69, 80, 85 and 91 branches in Fig. 5 (d) ($\lambda_2$ is −2.0, −3.0, −4.0 and −5.0 respectively). The real neurons group decrease gradually as branch order increases, but virtual neurons decrease quickly. In conclusion, the generation of virtual neurons depends on the values of development scale $\lambda_1$ and $\lambda_2$, and some basic morphological variables.

# 5     Conclusions

We conclude that the genetic regulatory network model is useful techniques for efficient generation of 3D dendritic trees. The algorithm is able to generate virtual neurons that can be used in structure-function relationship studies, as building blocks for analysis and modeling of neurons and neural networks, as tools to search for the most efficient description of neuroanatomical data, or to aid researchers develop scientific intuitions and hypotheses [18]. It should be stressed that this algorithm does not describe a single neuron, but rather a morphological class. It will allow the same descriptive rule to generate neurons as diverse as Purkinje, Granule and Stellate neurons just changes the statistical distributions of morphological variables and the developmental control parameters. Since an arbitrarily huge amounts of virtual neurons can be generated from a finite data set, so it naturally amplifies the data. These features may have a great influence on the development of neuromorphological databases.

# References

1. Shepherd, G.M., Mirsky, J.S., Healy, M.D., Singer, M.S., Skoufos, E., Hines, M.S., Nadkarni, P.M., Miller, P.L.: The Human Brain Project: Neuroinformatics tools for integrating, searching and modeling multidisciplinary neuroscience data. Trends in Neuroscience 21(11), 460–468 (1998)
2. Markram, H.: The Blue Brain Project. Nature Reviews Neuroscience 7(2), 153–160 (2006)
3. Buckmaster, P.S., Alonso, A., Canfield, D.R., Amaral, D.G.: Dendritic morphology, local circuitry, and intrinsic electrophysiology of principal neurons in the entorhinal cortex of macaque monkeys. Journal of Comparative Neurology 470(3), 317–329 (2004)
4. Koch, C., Segev, I.: The role of single neurons in information processing. Nature Neuroscience 3(1), 1171–1177 (2000)
5. Hillman, D.: Neuronal shape parameters and substructures as a basis of neuronal form. In: Schmitt, F.O., Worden, F.G. (eds.) The Neurosciences, 4th Study Program, pp. 477–498. MIT Press, Cambridge (1979)
6. Burke, R., Marks, W., Ulfhake, B.: A parsimonious description of motorneuron dendritc morphology using computer simulation. Journal of Neuroscience 12(6), 2403–2416 (1992)
7. Ascoli, G.A., Krichmar, J.L.: L-Neuron: A modeling tool for the efficient generation and parsimonious description of dendritic morphology. Neurocomputing 32(33), 1003–1011 (2000)
8. López-Cruz, P.L., Bielza, C., Larrañaga, P., Benavides-Piccione, R., DeFelipe, J.: Models and simulation of 3D neuronal dendritic trees using bayesian networks. Neuroinformatics 9(4), 347–369 (2011)
9. Berry, M., Bradley, P.M.: The application of network analysis to the study of branching patterns of large dendritic fields. Brain Research 109(1), 111–132 (1976)
10. Van Pelt, J., Van Ooyen, A., Uylings, H.B.M.: Modeling dendritic geometry and the development of nerve connections. In: DeSchutter, E., Cannon, R.C. (eds.) Computational Neuroscience: Realistic Modeling for Experimentalist, pp. 179–208. CRC Press, Boca Raton (2001)
11. Cannon, R., Turner, D., Pyapali, G., Wheal, H.: An online archive of reconstructed hippocampal neurons. Journal of Neuroscience Methods 84(1), 49–54 (1998)

12. Ascoli, G.A., Donohue, D.E., Halavi, M.: NeuroMorpho.org: A central resource for neuronal morphologies. Journal of Neuroscience 27(35), 9247–9251 (2007)
13. NeuroMorpho.Org, http://neuromorpho.org/neuroMorpho/index.jsp
14. Hippocampal Neuronal Morphology, http://neuron.duke.edu/cells/cellArchive.html
15. Myatt, D.R., Hadlington, T., Ascoli, G.A., Nasuto, S.J.: Neuromantic–from semi-manual to semi-automatic reconstruction of neuron morphology. Frontiers in Neuroinformatics 6(4), 1–14 (2012)
16. Ascoli, G.A., Krichmar, J.L., Scorcioni, R., Nasuo, S.J., Senft, S.L.: Computer generation and quantitative morphometric analysis of virtual neurons. Anatomy and Embryology 204(4), 283–301 (2001)
17. Rall, W.: Branching dendritic trees and motoneuron membrane resistivity. Experimental Neurology 1(5), 491–527 (1959)
18. Ascoli, G.A.: Progress and perspectives in computational neuroanatomy. Anatomical Record 257(6), 195–207 (1999)

# A Finite-Time Convergent Recurrent Neural Network Based Algorithm for the $L$ Smallest $k$-Subsets Sum Problem

Shenshen Gu[*]

School of Mechatronic Engineering and Automation
Shanghai University
149 Yanchang Road, Shanghai, 200072, P.R. China
gushenshen@shu.edu.cn
http://www.gushenshen.com

**Abstract.** For a given set $S$ of $n$ real numbers, a $k$-subset means a subset of $k$ distinct elements of $S$. It is obvious that there are totally $C_n^k$ different combinations. The $L$ smallest $k$-subsets sum problem is defined as finding $L$ $k$-subsets whose summation of subset elements are the $L$ smallest among all possible combinations. This problem has many applications in research and the real world. However the problem is very computationally challenging. In this paper, a novel algorithm is proposed to solve this problem. By expressing all the $C_n^k$ $k$-subsets with a network, the problem is converted to finding the $L$ shortest loopless paths in this network. By combining the $L$ shortest paths algorithm and the finite-time convergent recurrent neural network, a new algorithm for the $L$ smallest $k$-subsets problem is developed. And experimental results show that the proposed algorithm is very effective and efficient.

**Keywords:** $L$ smallest $k$-subsets sum problem, $L$ shortest paths, convergence in finite time, recurrent neural network.

## 1 Introduction

For a given set $S$ of $n$ real numbers, a $k$-subset means a subset of $S$ containing $k$ distinct elements, where $k < n$ [1]. The number of $k$-subsets on $n$ elements is therefore given by the binomial coefficient $\binom{n}{k}$. For example, there are $\binom{4}{2} = 6$ 2-subsets of $\{2, 4, 6, 9\}$, namely $\{2, 4\}$, $\{2, 6\}$, $\{2, 9\}$, $\{4, 6\}$, $\{4, 9\}$ and $\{6, 9\}$. The values of summation of elements for these 2-subsets are 6, 8, 11, 10, 13 and 15 respectively. The $L$ smallest $k$-subsets sum problem is defined as finding $L$ $k$-subsetss whose summation of subset elements are the $L$ smallest among all possible combinations. It is obvious that the 3 smallest 2-subsets of $\{2, 4, 6, 9\}$ is $\{2, 4\}$, $\{2, 6\}$ and $\{4, 6\}$.

It is obvious that the total number of distinct $k$-subset on set $S$ of $n$ elements is given by $\sum_{k=1}^{n} \binom{n}{k} = 2^n - 1 \ (0 < k \leq n)$. For the previous example of $\{2, 4, 6, 9\}$,

[*] This work was supported by the Specialized Research Fund for the Doctoral Program of Higher Education (SRFDP) under Grant 20113108120010.

C. Guo, Z.-G. Hou, and Z. Zeng (Eds.): ISNN 2013, Part I, LNCS 7951, pp. 19–27, 2013.

these subsets are $\{2\}$, $\{4\}$, $\{6\}$, $\{9\}$, $\{2,4\}$, $\{2,6\}$, $\{2,9\}$, $\{4,6\}$, $\{4,9\}$, $\{6,9\}$, $\{2,4,6\}$, $\{2,4,9\}$, $\{2,6,9\}$, $\{4,6,9\}$ and $\{2,4,6,9\}$. Then finding the $L$ smallest subsets or finding some subsets satisfying certain specific conditions among these $2^n - 1$ subsets is known as the subset sum problem, which is proved to be NP-complete.

The subset sum problem has numerous applications in research and the real world. For example, in computer science, it is widely applied to the optimal memory management in multiple programming[2]. In the field of telecommunication, it is used in allocating wireless resources to support multiple scalable video sequences[3]. For the application in embedded system, it is used in generating application specific instructions for DSP applications to reduce the required code size and increasing performance in embedded DSP systems[4]. And in optimization, the subset sum problem can also be studied as a special case of the Knapsack problem[5].

Due to the importance of the subset sum problem, many algorithms have been proposed. Lagarias and Odlyzko proposed a polynomial time algorithm for this problem in 1983[6]. However, the algorithm can hardly find a solution only when the density of the problem is less than $1/n$. In 1990, Lobstein proved that there is no polynomial-time algorithm solving the general subset sum problem[7]. Several heuristic algorithm have been proposed such as quantum computation method[8], space-time tradeoff method [9] and penalty function method [2]. However, these algorithm do not always find a solution when one exists.

Since the subset sum problem can be thought of as the extension of the $L$ smallest $k$-subsets sum problem. Then, it is clear that the subset sum problem can be solve very efficiently if there is a fast and exact algorithm for the $L$ smallest $k$-subsets sum problem. For this reason, the $L$ smallest $k$-subsets sum problem is studied in detail in this paper. A specified structure network was creatively proposed to express all the $\binom{n}{k}$ $k$-subsets. Then, based on the relationship between the $\binom{n}{k}$ $k$-subsets and the proposed network, finding the $L$ smallest $k$-subsets sum is equivalent to searching the $L$ shortest loopless paths in this network. Furthermore, by combining the $L$ shortest paths algorithm and the finite-time convergent recurrent neural network, a fast and exact algorithm for the $L$ smallest $k$-subsets problem is developed.

The remainder of this paper is organized as follows. In Section 2, the problem formulation is presented and the expression of the problem with a network is illustrated. Then the procedure for finding the $L$ shortest loopless paths in this network is described in Section 3. In Section 4, by combining the $L$ shortest loopless paths algorithm with the finite-time convergent recurrent neural network, a new algorithm is developed. Next, in Section 5, experimental results are given to verify the efficiency and effectiveness of the proposed algorithm. Finally, Section 6 concludes this paper.

## 2 Problem Formulation and Model Description

To solve the $L$ smallest $k$-subsets problem effectively, an appropriate mathematic model is needed. It is obvious that finding the $L$ smallest $k$-subsets is equivalent to determine the $1^{st}$, $2^{nd}$, $3^{rd}$, ..., $(L-1)^{th}$ and $L^{th}$ smallest $k$-subsets step by step. Mathematically, the $l^{th}$ $(0 < l \leq L)$ smallest $k$-subset problem can be formulated as a function

$$x_i = \begin{cases} 1, & \text{if } v_i \in \{\text{the } l\text{th smallest } k\text{-subset}\}; \\ 0, & \text{otherwise}; \end{cases} \qquad (1)$$

for $i = 1, \ldots, n$; where $v \in R^n$ and $k \in \{1, \ldots, n-1\}$. Fig.1 shows the operation graphically.

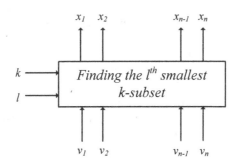

**Fig. 1.** Diagram of finding the $l^{th}$ $(0 < l \leq L)$ smallest $k$-subset operation

When $l = 1$ and $k$ is a nonnegative integer less than $n$, the above operation is almost the same with the $k$-Winners-Take-All ($k$WTA) operation[10]. The only difference is that $k$WTA operation finds $k$ largest elements from $n$ candidates instead of the smallest ones. However, solving the $L$ smallest $k$-subsets sum problem is much more complicated than solving the $k$WTA problem. On one hand, when the $(l-1)^{th}$ smallest $k$-subset is already known as $\hat{x}_{l-1}$, the $(l)^{th}$ smallest $k$-subset can be obtained by solving the following integer optimization(2).

$$\begin{aligned} \text{minimize} \quad & v^T x, \\ \text{subject to} \quad & e^T x = k \\ & x \neq \hat{x}_{l-1} \qquad \text{(C1)} \\ & v^T x \geq v^T \hat{x}_{l-1} \qquad \text{(C2)} \\ & x_i \in \{0, 1\}, \quad i = 1, 2, \ldots, n. \end{aligned} \qquad (2)$$

where $v = [v_1, \ldots, v_n]^T$, $e = [1, \ldots, 1]^T \in R^n$, $x = [x_1, \ldots, x_n]^T \in R^n$, $l$ is an integer greater than one and $k$ is a nonnegative integer less than $n$. Compared with the optimization formulation of $k$WTA problem[11], conditions (C1) and (C2) are added in finding the $(l)^{th}$ smallest $k$-subset, which increases the difficulty greatly. In addition, it is proved in [11] that condition $x_i \in \{0, 1\}$ in solving $k$WTA can be relaxed to $x_i \in [0, 1]$ . However, this integer condition in optimization problem(2) can not be relaxed. On the other hand, for the $k$WTA problem, only one round of optimization should be solved. However, for the $L$ smallest $k$-subsets sum problem, totally $L$ round of optimization(2) should be solved one by one from $l = 1$ to $l = L$.

Since integer optimization(2) is difficult to solve due to the inequality condition (C1) and the integer restriction for $x$, it is not wise to solve the original problem by integer programming. Here, a specific network structure is created to represent all $k$-subsets of set $S$. As shown in Fig.2, this network has one source node $n_{start}$ (i.e. start point), one sink node $n_{end}$ (i.e. ending point) and several intermediate nodes such as $n_{1.1}$ and $n_{2.1}$. There are some links connecting different nodes, and for each link one weight is assigned. For example, the weight for the link connecting nodes $n_{1.1}$ and $n_{2.1}$ is four. For this network, it can be easily enumerated that there are totally six different paths from the source node to the sink node. Each path represents a 2-subset by the weights of the links constituting this path. For instance, path $n_{start} \to n_{1.1} \to n_{2.1} \to n_{end}$ represents subset $\{2,4\}$. And all these six paths represent all 2-subsets of set $\{2,4,6,9\}$. When constructing a network for representing $k$-subsets of $n$ elements, the network is composed of a source node, a sink node and $k$ layers of intermediate nodes. For the first layer ($i = 1$), it has $(n-k+1)$ intermediate nodes. For other layers ($i = 2,\dots,k$), the $i^{th}$ layer has $(n-i+1)$ intermediate nodes. It can be proved that the upper bound for the number of nodes is $nk$. Then link the nodes in adjacent layers and assign the weights properly, a network representing all $k$-subsets of $n$ elements can be constructed. Therefore, it is clear that finding the $L$ smallest $k$-subsets sum is equivalent to searching $L$ shortest paths in a network.

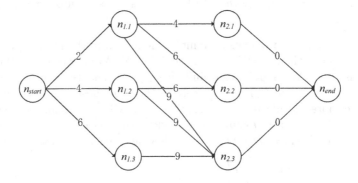

**Fig. 2.** A network representing all 2-subsets of $\{2,4,6,9\}$

## 3   Finding $L$ Shortest Paths in a Network

There are several algorithms available for finding $L$ shortest paths in a network, among which the algorithm proposed by Yen is one of the most efficient in term of the number of operations and the number of memory addresses[12]. Here the algorithm is adapted to find $L$ shortest paths in the specific structure network. The first step is to convert the network to a matrix form $M_{n,k}$. For the case in Fig.2, the matrix form is given in (3) where $M(i,j) = -1$ means there is no link from node $i$ to node $j$. And when $M(i,j)$ is a nonnegative number, it means there is a link node $i$ to node $j$ and the number is the weight on this link.

$$M_{8,2} = \begin{array}{c} n_{start} \\ n_{1.1} \\ n_{1.2} \\ n_{1.3} \\ n_{2.1} \\ n_{2.2} \\ n_{2.3} \\ n_{end} \end{array} \begin{pmatrix} \begin{array}{cccccccc} n_{start} & n_{1.1} & n_{1.2} & n_{1.3} & n_{2.1} & n_{2.2} & n_{2.3} & n_{end} \\ -1 & 2 & 4 & 6 & -1 & -1 & -1 & -1 \\ -1 & -1 & -1 & -1 & 4 & 6 & 9 & -1 \\ -1 & -1 & -1 & -1 & -1 & 6 & 9 & -1 \\ -1 & -1 & -1 & -1 & -1 & -1 & 9 & -1 \\ -1 & -1 & -1 & -1 & -1 & -1 & -1 & 0 \\ -1 & -1 & -1 & -1 & -1 & -1 & -1 & 0 \\ -1 & -1 & -1 & -1 & -1 & -1 & -1 & 0 \\ -1 & -1 & -1 & -1 & -1 & -1 & -1 & -1 \end{array} \end{pmatrix} \quad (3)$$

The flowchart of the algorithm is given in Fig.3. It is obvious that procedures P1 and P2 play main roles in the algorithm. Both these two procedures find the shortest path from a given network. Therefore, the efficiency of finding the shortest path completely determine the performance of the algorithm.

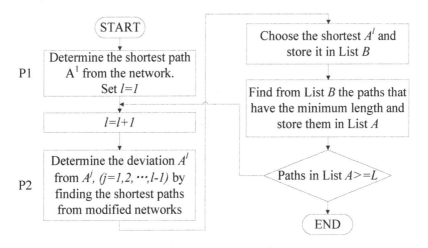

**Fig. 3.** Flowchart of the algorithm for finding $L$ shortest paths in a network

## 4 Finite-Time Convergent Recurrent Neural Network Based Algorithm

There are many algorithms proposed for finding the shortest path such as the famous dynamic programming method. However, for large-scale and real-time applications most of these series algorithms may not be efficient due to the drawback of sequential processing in computational time. Therefore, parallel computational models are more desirable. In [14], several neural networks were proposed for finding the shortest path. The shortest path problem was converted to the following linear program:

$$\text{minimize} \quad \sum_{i=1}^{n}\sum_{j=1}^{n} c_{ij}x_{ij}$$

$$\text{subject to} \quad \sum_{k=1}^{n} x_{ik} + \sum_{l=1}^{n} x_{li} = \delta_{il} - \delta_{in}$$

$$x_{ij} \geq 0, \quad i,j = 1,2,\ldots,n$$

where $c_{ij}$ is the weight of the link from node $i$ to node $j$, and $\delta_{ij}$ is the Kronecker delta function defined as $\delta_{ij} = 1(i = j)$ and $\delta_{ij} = 0(i \neq j)$. For the previous problem, the dual problem is

$$\text{maximize} \quad y_n - y_1$$

$$\text{subject to} \quad y_j - y_i \leq c_{ij}, \quad i,j = 1,2,\ldots,n$$

where $y_i$ is the dual decision variable associated with node $i$. The dual problem can be further simplified as follows by defining $z_i = y_i - y_1$ for $i = 1,2,\ldots,n$.

$$\begin{aligned} &\text{minimize} \quad z_n, \\ &\text{subject to} \quad z_j - z_i \leq c_{ij}, \quad i \neq j,\ i,j = 1,2,\ldots,n \end{aligned} \qquad (4)$$

where $z_1 = 0$. Based on the above formulation, several recurrent neural networks for shortest path routing have been proposed in succession, among which the finite-time convergent recurrent neural network is proved to be effective and efficient for its simple structure and the flexibility in choosing parameter for global convergence[13]. By combining (4) and the finite-time convergent recurrent neural network, the specific neural network model can be tailored as follows:

$$\epsilon\frac{dz}{dt} = -\sigma A^T g_{[0,1]}(Az - b) - c, \qquad (5)$$

where $\epsilon$ is a positive scaling constant, $\sigma$ is a gain parameter greater than one, $z = (z_2, z_3, \ldots, z_n)^T$, $b = (c_{12}, \ldots, c_{1n}, c_{21}, c_{23}, \ldots, c_{2n}, \ldots, c_{n1}, \ldots, c_{n,n-1})^T \in \mathbb{R}^{n(n-1)}$, $A$ is an $n(n-1) \times n$ matrix of 0, 1 and $-1$ to construct $n(n-1)$ inequality constraints, $g_{[0,1]}(v) = (g_{[0,1]}(v_1), \ldots, g_{[0,1]}(v_{n-1}))$, and its components are defined as

$$g_{[0,1]}(v_i) = \begin{cases} 1, & \text{if } v > 0, \\ [0,1], & \text{if } v = 0, \quad (i = 1,2,\ldots,n-1) \\ 0, & \text{if } v < 0. \end{cases} \qquad (6)$$

The block diagram of the specific neural network is shown in Fig.4. It consists of three main parts. The preprocessing part converts a certain network structure to a linear programming in the form of (4). Then the neural network processing part calculates the optimal solution. And finally, the optimal solution is decoded to the solution in terms of links by the postprocessing part.

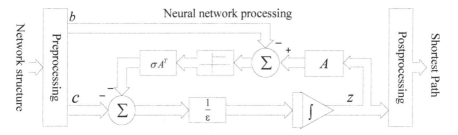

**Fig. 4.** Block diagram of the neural network for finding the shortest path

## 5   Experimental Results

Due to the efficiency of the finite-time convergent recurrent neural network in finding the shortest path, the neural network model is integrated with the $L$ shortest path algorithm to solve the $L$ smallest $k$-subsets problem.

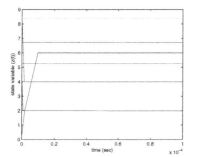
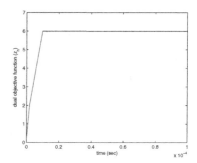

(a) Transient behaviors of the state variables.

(b) Global convergence of the dual objective function.

**Fig. 5.** Neural network dynamics for finding the shortest path in solving the $L$ smallest $k$-subsets problem

When $L = 4$, $k = 2$ and $S = \{2, 4, 6, 9\}$, the $L$ smallest $k$-subsets means finding 4 smallest 2-subsets from $S$. A specific network is constructed as in Fig.2 to represent all 2-subsets. Then the proposed algorithm is applied to this network. Fig.5 shows the dynamics of the finite-time convergent recurrent neural network in finding the shortest in the network, where Fig.5(a) and 5(b) presents the transient behaviors of the state variables and global convergence of the dual objective function respectively. It is obvious that the network converge quickly to the exact optimal value. For this case, the optimal value is six (i.e., $2 + 4 = 6$). The optimal solution is decoded to the solution in terms of links. After this postprocessing, the path in bold in Fig.6(a) is the first shortest path responding to subset $\{2, 4\}$. Then for the next three rounds, the second shortest path (i.e., Fig.6(b)), the third shortest path (i.e., Fig.6(c)) and the fourth shortest path

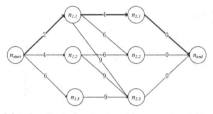

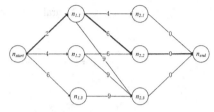

(a) The first shortest path corresponding to subset $\{2, 4\}$.

(b) The second shortest path corresponding to subset $\{2, 6\}$.

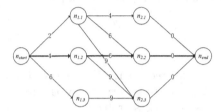

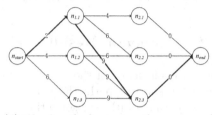

(c) The third shortest path corresponding to subset $\{4, 6\}$.

(d) The fourth shortest path corresponding to subset $\{2, 9\}$.

**Fig. 6.** Finding four smallest 2-subsets from $\{2, 4, 6, 9\}$

(i.e., Fig.6(d)) are determined in succession, which correspond to subsets $\{2, 6\}$, $\{4, 6\}$ and $\{2, 9\}$ respectively.

## 6   Conclusion

In this paper, based on a creative network structure as well as a finite-time convergent recurrent neural network, a new algorithm has been proposed for the $L$ smallest $k$-subsets sum problem. Experimental results show that this novel algorithm is efficient and is capable of finding the exact solutions.

## References

1. Nijenhuis, A., Wilf, H.: Combinatorial Algorithms for Computers and Calculators, 2nd ed. Academic Press, New York (1978)
2. Wang, H., Ma, Z., Nakayama, I.: Effectiveness of penalty function in solving the subset sum problem. In: Proceedings of IEEE International Conference on Evolutionary Computation, pp. 422–425 (1996)
3. Bocus, M.Z., Coon, J.P., Canagarajah, C.N., Armour, S.M.D., Doufexi, A., McGeehan, J.P.: Per-Subcarrier Antenna Selection for H.264 MGS/CGS Video Transmission Over Cognitive Radio Networks. IEEE Transactions on Vehicular Technology 61(3), 1060–1073 (2012)
4. Choi, H., Kim, J., Yoon, C.W., Park, I.C., Hwang, S.H., Kyung, C.H.: Synthesis of application specific instructions for embedded DSP software. IEEE Transactions on Computers 48(6), 603–614 (1999)

5. Martello, S., Toth, P.: 4 Subset-sum problem. In: Knapsack Problems: Algorithms and Computer Interpretations, pp. 105–136. Wiley-Interscience (1990)
6. Lagarias, J.C., Odlyzko, A.M.: Solving low density subset sum problems. In: 24th Annual Symposium on Foundations of Computer Science, pp. 1–10 (1983)
7. Lobstein, A.: The hardness of solving subset sum with preprocessing. IEEE Transactions on Information Theory 36(4), 943–946 (1990)
8. Chang, W.L., et al.: Quantum Algorithms of the Subset-Sum Problem on a Quantum Computer. In: WASE International Conference on Information Engineering, ICIE 2009, pp. 54–57 (2009)
9. Mine, T., Murakami, Y.: An implementation of space-time tradeoff method for subset sum problem. In: 6th International Conference on Computer Sciences and Convergence Information Technology (ICCIT), pp. 618–621 (2011)
10. Liu, S., Wang, J.: A simplified dual neural network for quadratic programming with its KWTA application. IEEE Transactions on Neural Networks 17(6), 1500–1510 (2006)
11. Gu, S.S., Wang, J.: A K-Winners-Take-All Neural Network Based on Linear Programming Formulation. In: International Joint Conference on Neural Networks, IJCNN 2007, pp. 37–40 (2007)
12. Yen, J.Y.: Finding the K shortest Loopless Paths in a Network. Management Science 17(11), 712–716 (1971)
13. Liu, Q.S., Wang, J.: Finite-time convergent recurrent neural network with a hard-limiting activation function for constrained optimization with piecewise-linear objective functions. IEEE Transactions on Neural Networks 22(4), 601–613 (2011)
14. Wang, J.: Primal and Dual Neural Networks for Shortest-Path Routing, Systems, Man and Cybernetics. IEEE Transactions on Part A: Systems and Humans 28(6) (1998)

# Spike Train Pattern and Firing Synchronization in a Model of the Olfactory Mitral Cell

Ying Du[1], Rubin Wang[1], and Jingyi Qu[2]

[1]Institute for Cognitive Neurodynamics, School of Science,
East China University of Science and Technology,
Shanghai, China, 200237
[2] School of Information Science and Technology, Civil Aviation University of China,
Tianjin, China, 300300
http://www.springer.com/lncs

**Abstract.** In the olfactory system, both the temporal spike structure and spatial distribution of neuronal activity are important for processing odor information. In this paper, a biophysically-detailed, spiking neuronal model is used to simulate the activity of olfactory bulb. It is shown that by varying some key parameters such as maximal conductances of $Ks$, $Nap$ and $Na$, the spike train of single neuron can exhibit various firing patterns. In the olfactory bulb, synchronization in coupled neurons is also investigated as the coupling strength gets increased. Synchronization process can be identified by correlation coefficient and phase plot. It is illustrated that the coupled neurons can exhibit different types of synchronization when the coupling strength increases. These results may be instructive to understand information transmission in olfactory system.

**Keywords:** olfactory, spike train, maximal conductance, synchronization.

## 1 Introduction

The study of the neural basis of olfactory system is important both for understanding the sense of smell and for understanding the mechanisms of neural computation. The olfactory bulb plays a central role in processing and relaying olfactory information(Laurent et al. 2001; Lledo et al. 2005; Shepherd et al. 2004). The main bulb receives signals from the population of olfactory receptor neurons(ORN) and transmits signals to the olfactory cortex and other brain regions.

Information about the environment is generally encoded into spike sequences by neurons in animal sensory nervous systems(Rieke.F et al. 1997). There is a lot of evidence that(Hudspeth AJ et al.2000; Laurent et al. 2001), in some systems, the representation of information comes through temporal encoding: each stimulus is characterized by a specific and reproducible sequence of firing spike train. In the olfactory bulb, the temporal structure of neuronal activity appear to be especially important for processing odor information(Hutcheon B et al. 2000). To build a reasonable dynamical theory of such an encoding, the

C. Guo, Z.-G. Hou, and Z. Zeng (Eds.): ISNN 2013, Part I, LNCS 7951, pp. 28–35, 2013.

rules on which the neuron model is based should be understood, and advantages that such stimulus representation has for further processing are also predicted. So it is necessary to explore the effect of some key parameters such as maximal conductances of the olfactory model on temporal spike train pattern.

Except the contribution of a single neuron on information transmitting, synchronization of a set of interacting individuals or units has been intensively studied because of its ubiquity in the natural world(Pikovsky et al. 2001; Davison et al. 2001 ). The synchronization of neuronal signals has been proposed as one of the mechanisms to transmit the information(Singer W. 1994; Eckhorn R. 1999). It is suggested that theoretical studies of such synchronized behaviors in neuronal assemblies play an important role in our understanding of information processing in olfactory systems. Hence, the synchronous firing of neurons has been extensively investigated by means of the theory of nonlinear dynamical systems.

Synchronization of action-potential firing has been demonstrated in the mitral and tufted cells of the rabbit olfactory bulb(Kashiwadani H et al.1999), which show that odor-evoked synchronization is functionally relevant for olfactory discrimination. Hence, the exploring for synchronization of two coupled neurons may be helpful to the understanding of olfactory system.

This paper is organized as follows: The Mitral cell model of olfactory bulb is introduced in Section 2. Single-neuron simulation and effects of maximal conductances to the spike train patterns are presented in Section 3, Section 4 describes synchronization in two coupled neurons, and a conclusion is given in Section 5.

## 2   Mitral Cell Model of Olfactory Bulb

A single compartment that includes voltage-dependent currents described by Hodgkin-Huxley kinetics was used to model mitral cells. The membrane capacitance for all cells is $1\mu F/cm^2$ and therefore not mentioned in the voltage equations. The following units were used: conductance per unit area in $[mS]/[cm^2]$, current per unit area in $[\mu A]/cm^2$, voltage in $[mV]$ and time(t) in $[ms]$. Their membrane potentials were calculated with the equations:

$$C_m \frac{dV}{dt} = -\frac{1}{R_m}(V - E_l) - I_{Na} - I_{Kfast} - I_{Ka} - I_{Ks} - I_{Nap} - I_e, \quad (1)$$

where $V$ is the membrane potential, $C_m$ is the membrane capacitance, $R_m$ is the input membrane resistance, and $E_l$ is the leak reversal potential.$I_e$ is external currents. The mitral cells have two sodium currents $I_{Na}$, $I_{Nap}$ and three potassium currents: $I_{Kfast}$, $I_{Ka}$ and $I_{Ks}$. All these intrinsic currents are described by the equation:

$$I_i = g_i m^M h^H (v - E_i), (i = Na, Kfast, Ka, Ks, Nap) \quad (2)$$

where $g_i$ is the maximal conductance and $E_i$ the reversal potential. The activation and inactivation variable $m$ and $h$ raised to the power $M$ and $H$ respectively follow the kinetic equation:

$$dm/dt = (m_\infty - m)/\tau_m \tag{3}$$

$$dh/dt = (h_\infty - h)/\tau_h \tag{4}$$

Values for all currents are taken from Bhalla and Bower(Bhalla et al 1993) and the Senselab databank(http://senselab.med.yale.edu).

## 3   Single-Neuron Simulations

### 3.1   Effect of Single Maximal Conductance to the Spike Train Pattern of Olfactory Model

In order to reproduce the characteristic properties of mitral cell, a set of parameters is selected to reflect the different behaviors of the model. Spike train patterns of olfactory bulb are sensitive to the persistent current $Nap$, $Ks$ and $Na$.

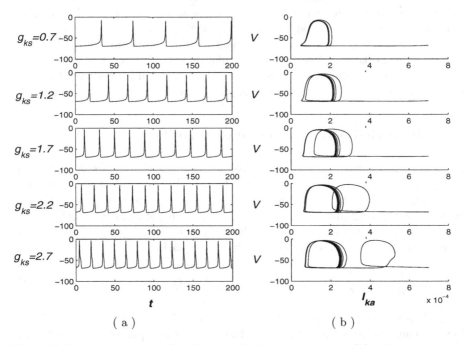

**Fig. 1.** Spike train patterns(a) and corresponding phase plots (b) with respect to variation of the maximal conductance of $Ks$

At first, the responses of spike train patterns at different maximal conductance of $Ks$ are considered. Fig. 1 presents the spike train patterns and corresponding phase plots when $g_{Ks}$ changes from 0.7 to 2.7, respectively. As shown in the Fig.2, with the $g_{Ks}$ increasing, the frequency of spike train firing is increasing. Then the situation of changing $g_{Nap}$ is considered while fixing other parameters. On

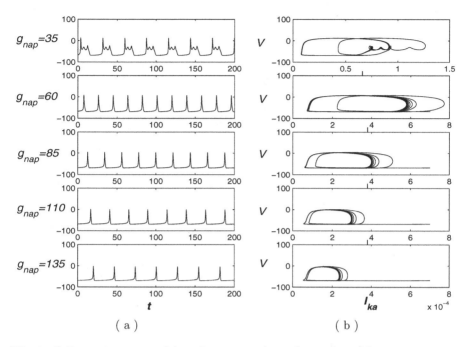

**Fig. 2.** Spike train patterns(a) and corresponding phase plots (b) with respect to variation of the maximal conductance of $Nap$

the contrary to the case of $g_{Ks}$, the frequency of spike train firing is decreasing when the $g_{Nap}$ is increased(Fig.2).

while varying the maximal conductance of $Na$ current in the model, the frequency of spike firing do not have large change. However, the top value of membrane potential increases from $10mv$ to $25mv$ and then to $40mv$ with the $g_{Na}$ taking the $10\mu S/m^2$, $20\mu S/m^2$ and $50\mu S/m^2$ respectively shown as (a), (b) and (c) of the Fig.3. The value of voltage will be stabilized at $40mv$ when the $g_{Na}$ is larger than $50\mu S/cm^2$. (d) of the Fig.3 shows the result taking $g_{Na}$ as linear function of time $t$. So it is supposed to simulate the experimental phenomenon through controlling the maximal conductance of the model.

## 3.2 Effect of Multi-Maximal Conductances to the Spike Train Pattern of Olfactory Model

If changing the maximal conductances of the $Nap$ and $Ks$ currents simultaneously, the variation tendency of spike train patterns are different to the case that only change one parameter. As shown in the Fig.4, when two maximal conductances decrease simultaneously, the frequencies of olfactory spike firing are increasing, which is opposite to the situation when $g_{Ks}$ decreases only. This state is similar to the case of changing single $g_{Nap}$, but if the value of $g_{Nap}$ is smaller

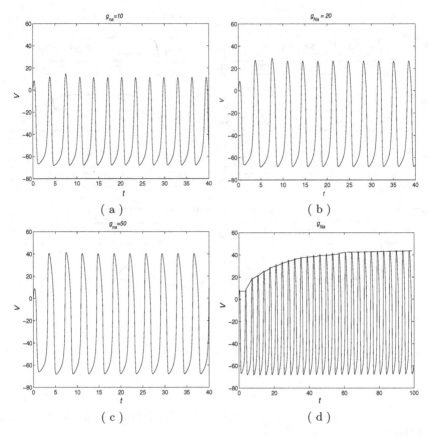

**Fig. 3.** The top value of membrane potential is increasing with the maximal conductance $g_{na}$ increasing

than $30\mu S/cm^2$, cell do not fire at all, only if the $g_{Ks}$ decrease with the $g_{Nap}$ simultaneously, the neuron can still generate membrane potential.

## 4   Firing Synchronization of Two Coupled Olfactory Neurons

In this section, synchronization of two identical coupled neurons is studied. It is shown that two coupled neurons can exhibit more complicated dynamical behaviors due to the effect of the coupling strength.

Dynamics of two coupled olfactory neurons are controlled as shown by the following differential equations:

$$C_m \frac{dV_i}{dt} = -\frac{1}{R_m}(V_i - E_l) - \sum_k I_k - I_e + \sum_j g_{syn}(V_i - V_j),  \qquad (5)$$

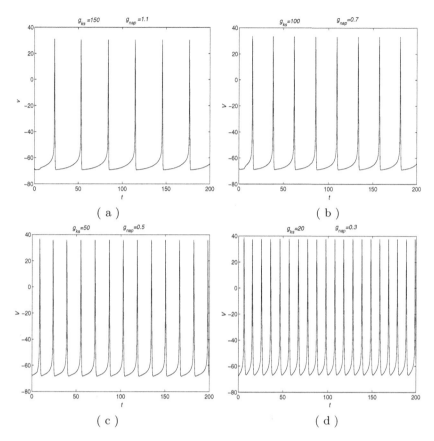

**Fig. 4.** Different spiking frequencies with the changing of $g_{Nap}$ and $g_{Ks}$. (a)$g_{Nap}=150S/m^2$,$g_{Ks}=1.1S/m^2$, (b)$g_{Nap}=100S/m^2$,$g_{Ks}=0.7S/m^2$, (c)$g_{Nap}=50S/m^2$,$g_{Ks}=0.5S/m^2$, (d)$g_{Nap}=20S/m^2$,$g_{Ks}=0.3S/m^2$,

where $I_k$ is ionic current($k = Na, Kfast, Ka, Ks, Nap$). $g_{syn}$ is coupling strength, $V_i$ and $V_j$ are voltages between adjacent neuron $i$ and $j$. Here, we take $i$ and $j$ as 1 and 2. All parameters are the same to the ones in section 2.

A correlation coefficient is introduced to measure the synchronization degree of the coupled neurons, and it is defined as follows:

$$C = \frac{\sum_{i=1}^{n} \mid (V_i^1 - \langle V_i^1 \rangle) \mid\mid (V_i^2 - \langle V_i^2 \rangle) \mid}{\sqrt{\sum_{i=1}^{n}(V_i^1 - \langle V_i^1 \rangle)^2 (V_i^2 - \langle V_i^2 \rangle)^2}} \tag{6}$$

where $V_i^1$ (or $V_i^2$) represents the samplings of the membrane potential $V^1(t)$ (or $V^2(t)$). $\langle \cdot \rangle$ denotes the average over the number of the sampling. It is easy to see that the more synchronous the coupled neurons are, the larger the correlation coefficient $C$ is, and the complete synchronization state o f the coupled neurons is achieved when $C$ is equal to 1.

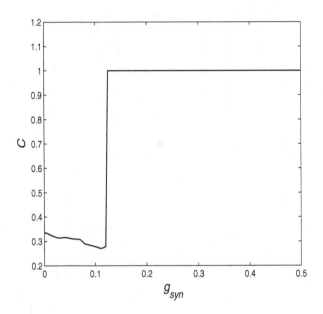

**Fig. 5.** Variation of the correlation coefficient $C$ of the membrane potential $V^1$ and $V^2$ with respect to $g_{syn}$

The correlation coefficient $C$ is calculated as illustrated in Fig.5. It is shown that synchronization exhibits a rout of process as: non-synchronization$\longrightarrow$ nearly synchronization $\longrightarrow$ spiking synchronization. Complete synchronization cannot occur when $g_{syn}$ is less than 0.14, while complete synchronization can be observed as shown in Fig.5, in which $C$ is equal to 1.

## 5 Conclusion

Based on the olfactory model, the analysis of this paper shows that maximal conductances have obviously effects on the spike train pattern: maximal conductance of $Na$ decides the top value of the membrane potential, however, the frequencies of firing can be controlled by the maximal conductances of $Nap$ and $Ks$. The frequency of spike train firing is increasing with the $g_{Ks}$ increasing, while the frequency of spike train firing is decreasing when the $g_{Ks}$ is increased; if the two maximal conductances decrease simultaneously, the frequency of olfactory spike firing is increased, which is opposite to the situation when $g_{Ks}$ decreases only. Synchronization of two coupled olfactory neurons has also been investigated in this paper. It was shown that the coupled neurons could achieve synchronization with the variation of coupling strength. It has been suggested that the variation of spike train pattern and neural synchronization can provide some important guidelines to understanding the process of neural information transmission. In the olfactory system particular, these analysis maybe helpful to understand the integration of the many factors influencing the construction and transformation of odor representations.

**Acknowledgement.** This work was supported by the National Natural Science Foundation of China (No.11002055, 11232005) and Young Teacher Fund of ECUST; Jingyi Qu is supported by the Special Fund of Civil Aviation University of China No.2012QD09X, No.ZXH2012C004.

# References

1. Rieke, F., et al.: Spikes: Exploring the Neural code. MIT Press, Cambridge (1997)
2. Hudspeth, A.J., Legothetis, N.K.: Sensory systems. Curr. Opin. Neurobiol. 10, 631 (2000)
3. Bhalla, U.S., Bower, J.M.: Exploring parameter space in detailed single cell models: simulations of the mitral and granule cells of the olfactory bulb. J. Neurophysiol. 69, 1948–1965 (1993)
4. Davison, A.P., Feng, J., Brown, D.: Spike synchronization in a biophysically-detailed model of the olfactory bulb. Neurocomputing 40, 515–521 (2001)
5. Hutcheon, B., Yarom, Y.: Resonance, oscillation and the intrinsic frequency preferences of neurons. Trends Neurosci. 23, 216–222 (2000)
6. Laurent, G., Stopfer, M., Friedrich, R.W., Rabinovich, M.I., Volkovskii, A., Abarbanel, H.D.I.: Odor encoding as an active, dynamical process: experiments, computation, and theory. Annu. Rev. Neurosci. 24, 263–297 (2001)
7. Lledo, P.-M., Gheusi, G., Vincent, J.D.: Information processing in the mammalian olfactory system. Physiol. Rev. 85, 281–317 (2005)
8. Shepherd, G.M., Wei, R.C., Greer, C.A.: Olfactory bulb. In: Shepherd, G.M. (ed.) The Synaptic Organization of the Brain, 5th edn., pp. 165–216. Oxford Univ. Press, Oxfrod (2004)
9. Pikovsky, A., Rosenblum, M., Kurths, J.: Synchronization, a universal concept in nonlinear sciences. Cambridge University Press, New York (2001)
10. Singer, W.: Time as coding space in Neocortical processing. Springer, Berlin (1994)
11. Eckhorn, R.: Neural mechanisms of scene segmentation: recording from the visual cortex suggest basic circuits or linking fieldmodels. IEEE Trans. Neural Netw. 10, 464–479 (1999)
12. Kashiwadani, H., Sasaki, Y.F., Uchida, N., Mori, K.: Synchronized oscillatory discharges of mitral/tufted cells with different molecular receptive ranges in the rabbit olfactory bulb. J. Neurophysiol. 82, 1786–1792 (1999)
13. Senselab databank, `http://senselab.med.yale.edu`

# Efficiency Improvements
# for Fuzzy Associative Memory

Nong Thi Hoa, The Duy Bui, and Trung Kien Dang

Human Machine Interaction Laboratory
University of Engineering and Technology
Vietnam National University, Hanoi

**Abstract.** FAM is an Associative Memory that uses operators of Fuzzy Logic and Mathematical Morphology (MM). FAMs possess important advantages including noise tolerance, unlimited storage, and one pass convergence. An important property, deciding FAM performance, is the ability to capture contents of each pattern, and associations of patterns. Standard FAMs capture either contents or associations of patterns well, but not both of them. In this paper, we propose a novel FAM that effectively stores both contents and associations of patterns. We improve both learning and recalling processes of FAM. In learning process, the associations and contents are stored by mean of input and output patterns and they are generalised by erosion operator. In recalling process, a new threshold is added to output function to improve outputs. Experiments show that noise tolerance of the proposed FAM is better than standard FAMs with different types of noise.

**Keywords:** Fuzzy Associative Memory, Noise Tolerance, Pattern Associations.

## 1 Introduction

Bidirectional Associative Memory (BAM) models store pattern associations and can retrieve desired output patterns from noisy input patterns. FAM is an Associative Memory that uses operators of Fuzzy Logic and Mathematical Morphology (MM). FAMs have three important advantages over traditional BAMs, which are noise tolerance, unlimited storage, and one pass convergence. Thanks to those advantages, FAMs have been widely applied in many fields such as image processing and optimization. Some standard FAMs [9,12,14] effectively store pattern associations by using the ratio of input pattern to output pattern. As a result, they do not store the content of patterns. Others [7,5,2,14] store the content of output patterns or some representative values, which means the associations of pattern pairs are not included.

In this paper, we propose a new standard FAM that can store both the content of patterns as well as pattern associations. Our FAM is improved in both learning and recalling process. In learning process, the associations and the contents are stored by the mean of input and output patterns, and they are generalised

C. Guo, Z.-G. Hou, and Z. Zeng (Eds.): ISNN 2013, Part I, LNCS 7951, pp. 36–43, 2013.

by erosion operator of Mathematical Morphology. In recalling process, a new threshold is added to the output function to improve recalled results. We have conducted experiments in face recognition and pattern recognition with three types of noise to confirm the effectiveness of our model.

The rest of the paper is organized as follows. Section 2 summarizes related work. In the section 3, we describe our novel FAM. Section 4 presents our experiments to show the advantages of the proposed FAM.

## 2   Related Work

Studies of FAMs can be divided into two categories: developing new models, and applying them into applications. In the first category, researchers mainly apply operators of Fuzzy Logic and MM to store pattern associations. The input and the association matrix are used to compute the output.

Kosko [7] used the minimum of input and output pattern to store the association and generalized them by dilation operator. A fuzzy implication operator was used to present associations by Junbo et al. [5]. Generalizing patterns was performed by erosion operator. The FAM set of Fulai [2] and some of Sussner [14] were similar to Junbo's FAM, in which the difference was only the output function. After that, Fulai and Tong proposed a way to add/delete a pattern pair [3]. These FAM, however, weakly presented the associations of patterns as they stored only input or output pattern for showing the association of each pattern pair.

Ping Xiao et al. [9] designed a model that applied the ratio of input to output patterns for the associations. Erosion operator was used for generalizing the associations. Wang and Lu [12] proposed a set of FAM that used division operator to describe the associations and erosion/dilations for generalizing the associations. Some FAMs of Sussner [14] used fuzzy implication to show the association and used s-norm operator in the output function. A threshold was added to the output function to improve weak outputs. Because of using the difference between input and output pattern for storing, the content of patterns was not presented in these FAMs.

An intuitive FAM that based on Junbo's model was proposed by Long et al. [8]. This FAM was added a complement value of each element of patterns and associative matrices. Valle and Sussner modified implicative FAMs tomake them be able to work with integer values by replacing values in [0,1] with values in [0,1,...,L] [15]. Other researchers focused on the stability of FAMs, the conditions for perfectly recalling stored patterns, and how to transform a given FAM to new FAMs [18,10,17,1].

In the second category, working with uncertain data is the reason why novel FAMs has been used in many fields such as pattern recognition, control, estimation, inference, and prediction. There are some typical examples of each field. Sussner and Valle used the implicative FAMs for face recognition [14]. Kim et al. predicted Korea stock price index [6]. Shahir and Chen inspected the quality of soaps on-line [11]. Wang and Valle detected pedestrian abnormal behaviour [16]. Sussner and Valle predicted the Furnas reservoir from 1991 to 1998 [13].

## 3   Our Approach

### 3.1   Design of the Proposed FAM

Because previous FAMs only effectively store the content or the associations, so some useful information from patterns is lost. Thus, the ability of recall is limited. We propose a novel FAM that stores both the content and the associations of patterns better. Furthermore, we propose a new threshold for the output function which can improve the noise tolerance.

Assuming that our FAM stores **p** pattern pairs, $(A_1, B_1), (A_2, B_2), ..., (A_p, B_p)$ in the general weight matrix **W**. The $k^{th}$ pair is represented by the vectors $A^k = (A_1^k, ..., A_m^k)$ and $B^k = (B_1^k, ..., B_n^k)$.

The design of our FAM is presented in the following processes:

**Learning Process Consists of Two Steps:**

*Step 1:* Learn the association of patterns $(A^k, B^k)$ by their mean values to store both the contents and the associations of patterns more clearly.

$$W_{ij}^k = \frac{1}{2}(A_i^k + B_j^k) \tag{1}$$

*Step 2:* Generalize the associations of patterns and store in the general weight matrix $W$.

$$W_{ij} = \bigwedge_{k=1}^{p} W_{ij}^k \tag{2}$$

**Recalling Process is Executed as Follows**

We use a new threshold for the output function. The threshold is used in case current output is much different from training output. That means the current output is equal to threshold when it is smaller than the minimum of training outputs. Models of Sussner [14] used a threshold that is the minimum of the training output patterns while our threshold is an arithmetic mean. Therefore, the ability of output correcting of Sussner's FAMs is lower than our model. The reason is that the ratio of minimum/maximum to the mean is smaller than the ratio of minimum/maximum to the minimum.

Our threshold is formulated as:

$$\theta_j = \frac{1}{p} \sum_{k=1}^{p} B_j^k \tag{3}$$

The summary of the current input and the general weight matrix is formulated by dilation operator. Then it is compared to the threshold. If it is smaller than the threshold then the current output is equal to the threshold, otherwise the current output is equal to it. Therefore, output Y is recalled from an input X by the equation:

$$Y_j = \bigvee_{i=1}^{m} X_i.W_{ij} \vee \theta_j \tag{4}$$

## 3.2  Discussion

To improve efficiency, our FAM employs arithmetic mean in both input and output functions. Thus, we name it MIOFAM (Mean Input and Output FAM).

MIOFAM has three important advantages over standard FAMs. First, it has unlimited capacity because of storing patterns in a single matrix. Second, recalling process performs in an iteration, which reduces computation and converges in only one pass. Finally, we expect to increase the noise tolerance ability because of the improvement in both learning and recalling process. In addition to the known advantages of FAM, MIOFAM is easy to understand and implement. As a result, we expect that our MIOFAM will perform well in different applications under harsh conditions.

## 4  Experiments

We have conducted three experiments with four image sets. FAMs are tested in the hetero-association mode since it is more general than the auto-association mode. The first experiment is face recognition from distorted inputs. The second and last experiment are pattern recognition from in-complete inputs and "salt & pepper" noise.

To prove the effectiveness, our novel FAM is compared to standard FAMs. Standard FAMs which are selected for comparison, are models of Kosko [7], Junbo et al. [4], Fulai and Tong [2], Ping Xiao et al. [9], Wang and Lu [12], and Valle and Sussner [14]. We choose best model of each set of FAM to compare. These FAMs and proposed FAM are similar to both learning and recalling process. Therefore, we only compare the noise tolerance of FAMs.

We use the peak signal-to-noise ratio (PSNR) to measure quality between the training and an output image. The higher the PSNR, the better the quality of the output image. PSNR is computed by the following equation:

$$PSNR = 40log_{10}\frac{R^2}{MSE} \qquad (5)$$

where R is the maximum fluctuation in the input image data type. Working with grey-scale images, value of R is 255. MSE represents the cumulative squared error between the training and an output image. MSE is formulated by the following equation:

$$MSE = \frac{\sum_{M,N}(I_1(m,n) - I_2(m,n))^2}{M * N} \qquad (6)$$

where M and N are the number of rows and columns in the input images. $I_1, I_2$ are the output and the training image.

## 4.1  Experiment 1: Face Recognition from Distorted Inputs

We choose the faces database of AT & T Laboratories Cambridge[1] including 40 people. Figure 1 shows some typical patterns in this experiment.

---

[1] Avaliable at: http://www.uk.research.att.com/facedatabase.htm

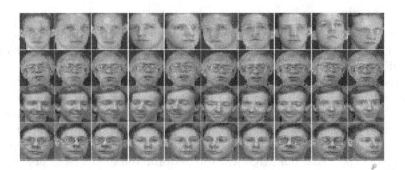

**Fig. 1.** Typical patterns for face recognition

There are 10 images for every person in the database, including one normal image and nine distorted images. Normal images are used to train and the distorted images are noisy inputs for experiments. Noisy inputs are made from the training images by rotating faces, changing position of light source, wearing glasses,...The size of each image is 112x92 pixels. We rescale the original images to 23x19 pixels. Figure 2 shows PSNR of models in Experiment 1. The results show that our FAM improves noise tolerance from 4% to 36 % comparing to standard FAMs.

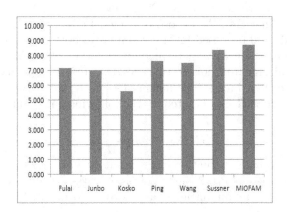

**Fig. 2.** PSNR of FAMs for face recognition from distorted inputs

## 4.2   Experiment 2: Recognition Applications from Incomplete Inputs

We select three image sets which include many groups of images, namely, human, animal, house, radar image, car, and thing. The first set contains 48 images of the grey-scale image database (CVG) of the Computer Vision Group, University of Granada, Spain[2] with six groups of images. The second set has 50 animal images

---

[2] Avaliable at: http://decsai.ugr.es/cvg/wellcome.html

**Fig. 3.** Some typical images of data sets.
(a), (b), (c) show CVG, CAR, ANIMAL datasets.

(ANIMAL) with many species from 50 Amazing Animals in the shared database of Torrentz in EU. The last set includes 48 images (CAR) of car features, which are selected from Corel database. Images are rescaled from 512x512, 1920x1080, 384x256 to 21x21, 23x19, 24x16 respectively. Figure 3 shows some typical images of the three datasets.

Normal images are used to create the training set and the test set is made from the training images by deleting parts of images. Figure 4 shows PSNR of FAMs in Experiment 2. This experiment shows that MIOFAM is better than standard FAMs in all datasets. Especially, our FAM achieves significant improvement in ANIMAL set (22.3% comparing to FAMs of Fulai, the second best FAM).

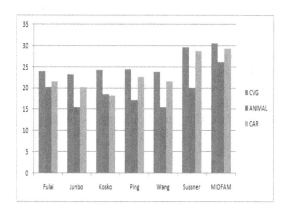

**Fig. 4.** PSNR of FAMs for pattern recognition from incomplete inputs

### 4.3 Experiment 3: Pattern Recognition from "Salt & Pepper" Noise

We use the same three image sets in Experiment 2. Noisy images are made from the training images by adding "salt & pepper". Figure 5 shows PSNR of FAMs

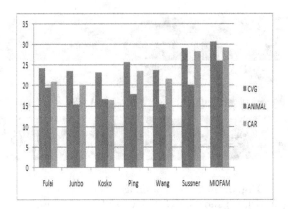

**Fig. 5.** PSNR of FAMs for pattern recognition from "salt & pepper" noise

in Experiment 3. Again, MIOFAM performs better than all other FAMs. In the ANIMAL dataset, our FAM tolerates noise 23.2% better than the second best FAM, FAMs of Sussner.

## 5 Conclusion

In this paper, we proposed a new FAM - the MIOFAM - that captures both content and associations of patterns. Our FAM improves both learning and recalling process by using arithmetic mean. While still possessing vital advantages of standard FAMs, the MIOFAM has better noise tolerance and is easy to construct. We have conducted three experiments in face recognition and pattern recognition to prove the efficiency of the proposed FAM. The obtained results show that MIOFAM is better than standard FAMs in experiments with three types of noise. Especially, our FAM performs much better than standards FAMs in one dataset, the ANIMAL dataset. This hints that our improvement in capturing pattern content and associations can be extremely effective. Our further work would investigate further into this direction, such as measuring the diffusion in pattern content and association to confirm that hypothesis.

**Acknowledgements.** This work is supported by Nafosted research project No. 102.02-2011.13.

## References

1. Cheng, Q., Fan, Z.T.: The stability problem for fuzzy bidirectional associative memories. Elsevier Science, Fuzzy Sets and Systems 132(1), 83–90 (2002)
2. Fulai, C., Tong, L.: Towards a High Capacity Fuzzy Associative Memory Model. In: IEEE World Congress on Computational Intelligence, 1994 IEEE International Conference on Neural Networks (1994)

3. Fulai, C., Tong, L.: On fuzzy associative memories with multiple-rule storage capacity. IEEE Transactions on Fuzzy System 4(3) (1996)
4. Junbo, F., Fan, J., Yan, S.: An Encoding Rule of FAM. In: Singapore ICCS/ISITA 1992, pp. 1415–1418 (1992)
5. Junbo, F., Fan, J., Yan, S.: A learning rule for FAM. In: 1994 IEEE International Conference on Neural Networks, pp. 4273–4277 (1994)
6. Kim, M.: Fuzzy Associative Memory-Driven Approach to Knowledge Integration. In: 1999 IEEE International Fuzzy Systems Conference Proceedings, pp. 298–303 (1999)
7. Kosko, B.: NeuralNetworks and Fuzzy Systems: A Dynamical Systems Approach to Machine Intelligence. Prentice Hall, Englewood Cliffs (1992)
8. Li, L., Yang, J., Wu, W., Wu, T.: An intuitionistic fuzzy associative memory network and its learning rule. In: GRC 2009, IEEE International Conference on Granular Computing, pp. 2–5 (2009)
9. Ping Xiao, F.Y., Yu, Y.: Max-Min Encoding Learning Algorithm for Fuzzy Max-Multiplication Associative Memory Networks. In: 1997 IEEE International Conference on Systems, Man, and Cybernetics (1997)
10. Ritter, G.X., Sussner, P., Diza-de Leon, J.L.: Morphological associative memories. IEEE Transactions on Neural Networks 9(2), 281–293 (1998)
11. Shahir, S., Chen, X.: Adaptive fuzzy associative memory for on-line quality control. In: Proceedings of the 35th South-eastern Symposium on System Theory, pp. 357–361 (2003)
12. Wang, S., Lu, H.J.: On New Fuzzy Morphological Associative Memories. IEEE Transactions on Fuzzy Systems 12(3), 316–323 (2004)
13. Sussner, P., Valle, M.E.: Implicative Fuzzy Associative Memories. IEEE Transactions on Fuzzy System 14(6), 793–807 (2006)
14. Sussner, P., Valle, M.E.: Fuzzy Associative Memories and Their Relationship to Mathematical Morphology. In: Handbook of Granular Computing, pp. 1–41 (2008)
15. Valle, M.E., Sussner, P.: Fuzzy associative memories from the perspective of mathematical morphology. IEEE Transactions on Fuzzy System, Mm (2007)
16. Wang, Z., Zhang, J.: Detecting Pedestrian Abnormal Behavior Based on Fuzzy Associative Memory. In: Fourth International Conference on Natural Computation, pp. 143–147 (2008)
17. Zeng, S., Xu, W., Yang, J.: Research on Properties of Max-Product Fuzzy Associative Memory Networks. In: Eighth International Conference on Intelligent Systems Design and Applications, pp. 438–443 (2008)
18. Zhang, Z., Zhou, W., Yang, D.: Global exponential stability of fuzzy logical BAM neural networks with Markovian jumping parameters. In: 2011 Seventh International Conference on Natural Computation, pp. 411–415 (2011)

# A Study of Neural Mechanism in Emotion Regulation by Simultaneous Recording of EEG and fMRI Based on ICA

Tiantong Zhou[1,2], Hailing Wang[1,2], Ling Zou[1,2,*], Renlai Zhou[4], and Nong Qian[2,4]

[1] State Key Laboratory of Robotics and System (HIT), Harbin Institute of Technology,
Harbin, Heilongjiang, 150080, China
[2] Changzhou Key Laboratory of Biomedical Information Technology, Changzhou University,
Changzhou, Jiangsu, 213164, China
[3] Beijing Key Laboratory of Applied Experimental Psychology, School of Psychology, Beijing
Normal University, Beijing, 100875, China
[4] Changzhou NO.2 People's Hospital attached to Nanjing Medical University
zoulingme@yahoo.cn

**Abstract.** The combination of electroencephalogram (EEG) and functional magnetic resonance imaging (fMRI) is a very attractive aim in neuroscience, in order to achieve both high temporal and spatial resolution for the non-invasive study of cognitive brain function. In this paper, we record simultaneous EEG–fMRI of the same subject in emotional processing experiment in order to explore the characteristics of different emotional picture processing, and try to find the difference of the subject' brain hemisphere when viewing different valence emotional pictures. For fMRI data, we study the participant's brain active region, and examine related blood oxygen level—dependent(BOLD) response. For EEG data, we focus on the amplitude of the late positive potential (LPP). We find that the amplitude of the LPP correlated significantly with BOLD intensity in visual cortex and amygdala, prefrontal is also modulated by different picture categories.

**Keywords:** emotion regulation, blood oxygen level-dependent, late positive potential, ICA.

## 1    Introduction

As an important psychological phenomenon, emotion plays a key role in the regulation of human social behavior. Emotion is a physically and mentally excited state generated by the individuals who are stimulated, it is a complicated higher nervous activity. Emotional processing, it is an emotional stimuli perception or evaluation that possible to wake up the emotional experience. Because of the importance and complexity of emotional itself, many scholars have made a series of research of emotion

---

* Address correspondence to Ling Zou, Faculty of Information Science & Engineering,
Changzhou university, Changzhou, Jiangsu, 213164, China.

C. Guo, Z.-G. Hou, and Z. Zeng (Eds.): ISNN 2013, Part I, LNCS 7951, pp. 44–51, 2013.

mechanism, and make some achievements. The main research methods are based on functional magnetic resonance imaging (fMRI) and electroencephalograph (EEG).

A key feature observed in event-related potential (ERP),which was derived from EEG, evoked by emotionally engaging stimuli is the late positive potential (LPP), which is characterized by an amplitude enhancement for positive and negative stimuli, relative to neutral stimuli. For affective picture viewing, LPP starts ~300 – 400 ms after picture onset and is often sustained throughout the duration of picture presentation. In parallel, fMRI has found that viewing of affective pictures is associated with increased blood oxygen level-dependent (BOLD) activity in widespread brain regions, including occipital, parietal, inferotemporal cortices, and amygdala: [1], [2], [3], [4]. So, if enhanced LPP and BOLD reflect a common underlying mechanism, one might expect a coupling between LPP amplitude and BOLD activity in the above reported regions. Consistent with prior work, the voltage strength of a positive slow wave recorded from parietal regions of the scalp (the LPP) reflects the rated emotional intensity of picture stimuli[5]. The strength of this scalp-positive waveform correlate significantly with BOLD signal in lateral occipital, inferior temporal, and medial parietal cortex. Thus, despite wide differences in signal origin and latency, the 2 measures of cortical reactivity show comparable modulation by emotional pictures. Past source-space modeling of LPP has only been able to identify generators in the visual system, including occipito-temporal, parietal, and inferior temporal cortices[6], despite the fact that the amplitude of LPP is closely related to the rated intensity of emotion. The contribution to scalp-recorded potentials by the emotional processing areas may be modulatory and mediated by the visual cortex. It has been hypothesized that, when observers view emotionally engaging scenes, cortical and deep sub-cortical structures modulate visual cortex in a reentrant fashion: [7],[8].

In this paper, we design the emotional processing experiment by ourself, and record simultaneous EEG–fMRI while one subject passively view three categories picture, neutral, negative and positive pictures. What's more, these pictures are all random present. We use Netstation software which come from America company EGI to preprocess EEG data, and use the 8th Statistical Parametric Mapping (SPM8) to preprocess fMRI data, then apply ICA algorithm for further study, and get some result.

## 2  Independent Component Analysis

Before estimating the independent components, the observed data can be whitened, that is, transformed to be uncorrelated and have unit variances. Whitening can be done using a linear transformation and does not constrain the estimation in any way, since independence implies uncorrelation. Additionally, whitening simplifies the following component estimation by restricting the structure of the mixing. If the whitening is done using principal component analysis (PCA), the number of free parameters can be reduced by taking only the K strongest principal components, and leaving out the weakest. Assuming that the weakest components contain mainly noise, the dimension of the data is reduced in an optimal manner to improve the signal-to-noise ratio.

As mentioned before, ICA is based only on the assumption that the source signals are statistically independent. This seems reasonable in many applications and in fact does not have to hold exactly for ICA to be applicable in practice. The generative model used in ICA is an instantaneous linear mixture of random variables. The original signals can be considered as a source matrix S, where each row of the matrix contains one of the K signals. Respectively, the observed mixed signals are denoted as matrix X . Again, each row of the matrix contains one of the N observed signals. Assuming a noiseless environment, the mixing model can be expressed in matrix form as:

$$X = AS \tag{1}$$

Each column $a_k$ of the full rank $N \times K$ mixing matrix A holds the mixing weights corresponding to source k. The problem of jointly solving both the mixing and the original sources is not only considerably difficult, but also ambiguous. Since both A and S are unknown, it immediately follows that the signs and scaling of the sources cannot be uniquely defined. One can multiply the mixing weights $a_k$ and divide the corresponding source respectively with any given coefficient. Additionally, the sources can appear in any order: [9], [10].

## 3    Experience

### 3.1    Participants and Experimental Material

One healthy adult, participate in simultaneous fMRI and EEG experiment, 22 years old, right handedness, normal vision, no nerve system disease and claustrophobia, and access to subjects' informed consent.

The stimuli consisted of 30 positive, 30 neutral, and 30 negative pictures select from the International Affective Picture System (IAPS) based on their normative valence and arousal levels. Positive picture average arousal is 5.32, neutral pictures average arousal is 3.26, negative picture average arousal is 6.41; the positive valence is 7.30, neutral valence is 4.30, negative valence is 3.46. Between two stimuli appears gray screen, and a red "+" at the central as fixation point.

### 3.2    Task

The experimental paradigm is implemented in an event-related fMRI design. Each picture is centrally displayed on a projector for 6 s followed by a 12 s inter-stimulus interval. Subject complete three experimental sessions in which the pictures are presented in different random orders. A red cross is displayed at the center of the screen between two stimuli appears as a fixation point, as shown in Fig.1. Stimuli are presented on an MR-compatible monitor using E-Prime software (Psychology Software Tools). Subject view the task presentation in the screen via a reflective mirror. During the experiment, subject should just view the picture, and don't have to feedback any information of the picture content. After one session is over, participants can close their eyes and have a rest for a period of time.

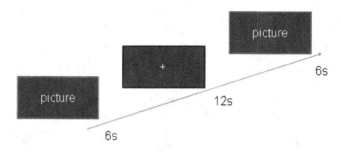

**Fig. 1.** Emotional picture stimulate experimental paradigm

# 4    Data Acquisition

FMRI data are acquired using a 3.0-Tesla superconducting type nuclear magnetic resonance imaging system from Philips company at Changzhou Second People's Hospital, and the fMRI time series of whole brain images are acquired with single excitation gradient echo planar imaging using a T2-weighted BOLD sequence. The sequence parameters are: TR=2000 ms, TE=35 ms, FA=90°, FOV=230 mm×125 mm, matrix=96×52, 20 continuous slices with a thickness of 3.5 mm.

EEG data is recorded using a 64-channel system (Electrical Geodesics, EGI) and all signals recorded are referenced to a Cz electrode, sampled at 250 Hz with an on-line bandpass filter of 0.3 to 30Hz using Netstation acquisition software and an EGI Net-Amps 300 amplifier. Impedance is checked online before recording and accepted when below 50 kΩ.

# 5    Data Analysis

## 5.1    fMRI Analysis

After acquiring the fMRI data, we perform using the software package of SPM8, including slice timing, realigning, spatial normalizing and smoothing.

Next, in order to get every condition activation model, we analyze the fMRI data with the generalized linear model, convolute each subject's sequential functional volumes and the hemodynamic response function(HRF), there will be a curve function, we take it as a reference function. Define experiment conditions as task conditions and rest conditions, then watch subject's activated brain regions when he passively accept visual task, correct threshold level P<0.05(family wise error, FWE). When analyse the data in following steps, we use the default parameter settings . At last, we get subject's fMRI activated information in MNI coordinate, which reflects the cerebral cortex BOLD activity on different picture stimulus conditions.

Fig.2 shows the subject's brain activate region on different picture stimulus conditions, as a 3D form. We can see that visual stimulation of the main active area is front

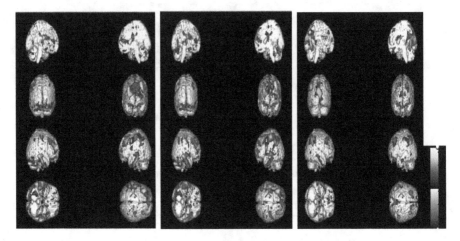

**Fig. 2.** 3D form brain active region among three categories pictures (positive, negative, neutral) stimulus, letf is the brain active region under positive picture stimulus, medial is the brain active region under negative picture stimulus, right is the brain active region under neutral picture stimulus.

occipital lobe, rear parietal lobe, inferior temporal gyrus, prefrontal and postcentral gyrus. We can also see that among the brain active regions, the frontal area and the temporal's activation strength under the emotional picture stimulus is greater than under the neutral picture stimulus, it suggests that subject's BOLD signal in the emotional stimulus is improved comparing with neutral picture stimulus.

## 5.2      EEG Analysis

The acquired continuous EEG data is firstly preprocessed by Netstation software, including the follow steps, magnetic resonance (MR) Artifact, QRS Detection, optimal basis set (OBS), a band pass digital filter between 0.01Hz and 40Hz, Segment from 200ms before stimulus to 1000ms after stimulus. Fig.3 shows the preprocessed EEG signals using the Netstation software. From the EEG signals we know that many artifacts like ECG and EOG are interfused into EEG signals. We remove all of the artifacts after segmenting with temporal ICA analysis, then we get the temporal and spatial independent components, shown as Fig.4. We can see that the channel of 7, 28, 32 are artifacts. We reconstruct the signals after removing the artifacts, which are showed in Fig.5. At last, the features of the de-noised EEG signals are extracted by using wavelet transform method, the characteristic signal are showed in Fig.6, it is the picture onset ERP, with the epoch ranging form 200 to 700 ms representing the LPP. The red circle line represents positive stimulus, blue star represents negative stimulus, green line represents neutral stimulus, the amplitude of positive and negative stimulus LPP are obviously higher than neutral stimulus, this suggests that the amplitude of the LPP is related to emotional arousal.

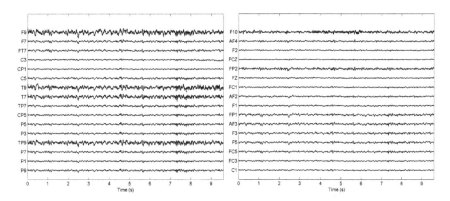

**Fig. 3.** The above 32 channels EEG signals after preprocessing. Left: EEG signals from 1 to 16 electrodes; Right: EEG signals from 17 to 32 electrodes.

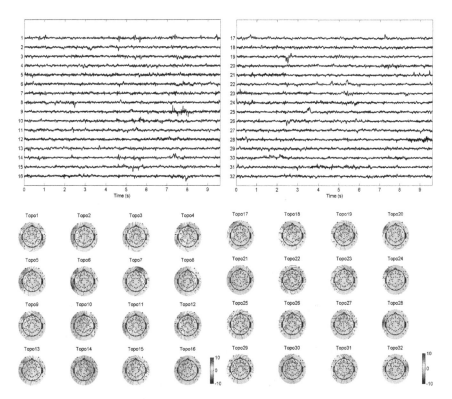

**Fig. 4.** Corresponding ICA component activations and scalp maps of the above 32 channel data., upper row is corresponding ICA component activations, lower row is corresponding scalp maps

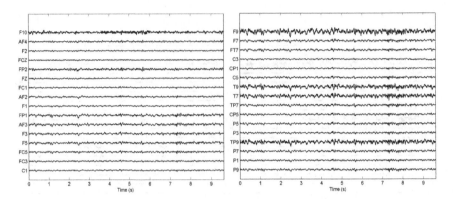

**Fig. 5.** ICA cleaned EEG signals from 1 to 32 channels

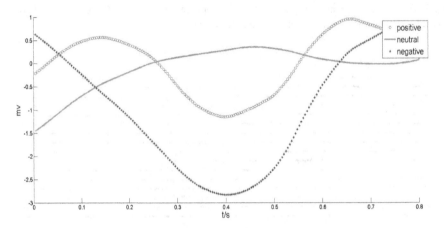

**Fig. 6.** The characteristic waveform of positive, negative, and neutral stimulus

## 6    Discussion

Currently, the most common methods of assessing emotional processing in the human brain are ERP, derived from the EEG, and BOLD contrast, assessed with fMRI. Here, we make a preliminary study of emotion regulation based on simultaneous recording of EEG and BOLD responses, we use temporal ICA to analysis the EEG signals, extracting LPP, the LPP and visual cortical blood oxygen level--dependent (BOLD) signals are both modulated by the rated intensity of picture arousal, the overall LPP amplitude variability across three picture categories(positive, negative and neutral, show as Fig. 6 is found to be mainly correlated with BOLD responses in visual cortex and amygdala, this is consistent with prior work, we also find temporal area and prefrontal are modulated by different picture categories.

**Acknowledgments.** This work has been partially supported by the open project of the State Key Laboratory of Robotics and System at Harbin Institute of Technology (SKLRS-2010-2D-09, SKLRS-2010-MS-10), National Natural Science Foundation of China (61201096), Natural Science Foundation of Changzhou City(CJ20110023) and Changzhou High-tech Reasearch Key Laboratory Project (CM20123006).

# References

1. Bradley, M.M., Sabatinelli, D., Lang, P.J., Fitzsimmons, J.R., King, W., Desai, P.: Activation of the visual cortex in motivated attention. Behav. Neurosci. 117, 369–380 (2003)
2. Norris, C.J., Chen, E.E., Zhu, D.C., Small, S.L., Cacioppo, J.T.: The interaction of social and emotional processes in the brain. J. Cogn. Neurosci. 16, 1818–1829 (2004)
3. Sabatinelli, D., Bradley, M.M., Fitzsimmons, J.R., Lang, P.J.: Parallel amygdala and inferotemporal activation reflect emotional intensity and fear relevance. Neuroimage 24, 1265–1270 (2005)
4. Sabatinelli, D., Lang, P.J., Bradley, M.M., Costa, V.D., Keil, A.: The timing of emotional discrimination in human amygdala, inferotemporal, and occipital cortex. J. Neurosci. 29, 14864–14868 (2009)
5. Schupp, H.T., Cuthbert, B.N., Bradley, M.M., Hillman, C.H., Hamm, A., Lang, P.J.: Brain processes in emotional perception: motivated attention. Cogn. Emotion 18, 593–611 (2004)
6. Sabatinelli, D., Lang, P.J., Keil, A., Bradley, M.M.: Emotional perception: correlation of functional MRI and event-related potentials. Cereb. Cortex 17, 1085–1091 (2007a)
7. Keil, A., Sabatinelli, D., Ding, M., Lang, P.J., Ihssen, N., Heim, S.: Re-entrant projections modulate visual cortex in affective perception: directional evidence from Granger causality analysis. Hum. Brain Mapp. 30, 532–540 (2009)
8. Pessoa, L., Adolphs, R.: Emotion processing and the amygdala: from a "low road " to "many roads " of evaluating biological significance. Nat. Rev. Neurosci. 11, 773–783 (2010)
9. Zou, L., Pu, H., Sun, Q., Su, W.: Analysis of Attention Deficit Hyperactivity Disorder and Control Participants in EEG Using ICA and PCA. In: Wang, J., Yen, G.G., Polycarpou, M.M. (eds.) ISNN 2012, Part I. LNCS, vol. 7367, pp. 403–410. Springer, Heidelberg (2012)
10. Zou, L., Zhang, Y., Yang, L.T., Zhou, R.: Single Trial Evoked Potentials Study by Combining Wavelet Denoising and Principal Component Analysis Method. Journal of Clinical Neurophysiology 27(1), 17–24 (2010)

# Emotion Cognitive Reappraisal Research Based on Simultaneous Recording of EEG and BOLD Responses

Ling Zou[1,2,*], Yi Zhang[1,2], Lin Yuan[3], Nong Qian[2,4], and Renlai Zhou[3]

[1] State Key Laboratory of Robotics and System (HIT), Harbin Institute of Technology, Harbin, Heilongjiang, 150080, China
[2] Changzhou Key Laboratory of Biomedical Information Technology, Changzhou University, Changzhou, Jiangsu, 213164, China
[3] Beijing Key Laboratory of Applied Experimental Psychology, School of Psychology, Beijing Normal University, Beijing, 100875, China
[4] Changzhou NO.2 People's Hospital attached to Nanjing Medical University
zoulingme@ yahoo.cn

**Abstract.** The combination of electroencephalogram (EEG) and functional magnetic resonance imaging (fMRI) is a very attractive aim in neuroscience, in order to achieve both high temporal and spatial resolution for the non-invasive study of cognitive brain function. In this paper, we record simultaneous EEG–fMRI of three subjects, study the participants' brain active regions, examine related blood oxygen level—dependent(BOLD) response for fMRI data, and focus on the effects of reappraisal instructions on the amplitude of the late positive potential(LPP) for EEG data. We find that emotion cognitive reappraisal result in early prefrontal cortex responses, decrease negative emotion experience and amygdala response. Besides, the study indicates that reappraisal decrease the magnitude of the LPP.

**Keywords:** emotional processing, late positive potential, Functional magnetic resonance imaging, blood oxygen level—dependent.

## 1 Introduction

Although the term "emotion regulation" has different meanings and refers to several modes of processing, researchers have focused particularly on better understanding a conscious, cognitive strategy for regulating emotions known as cognitive reappraisal:[1], [2]. Cognitive reappraisal severs to change the emotional meaning and significance of an event or stimulus[3].

The specific neural systems associated with antecedent-focused emotion regulation have been the target of numerous functional magnetic resonance imaging (fMRI) studies. Schaefer et al. first reported that the increased amygdala activity in response

---

* Address correspondence to Ling Zou, Faculty of Information Science & Engineering, Changzhou university, Changzhou, Jiangsu, 213164, China.

to unpleasant stimuli could be prolonged if subjects were given an explicit instruction to maintain their emotional response. Since then, multiple studies that utilize the specific strategies of detachment[4] and reappraisal have demonstrated that it is possible to consciously inhibit amygdala activation when instructed to decrease emotional response.

In addition to fMRI, event-related brain potentials (ERPs) have been used to investigate the processes of emotion regulation in recent studies. The excellent temporal resolution of ERPs has allowed for greater insight into the time course of emotion regulation processes, and these studies suggest that reappraisal modulates ERPs following unpleasant pictures[5]. These studies have focused on the late positive potential (LPP) in particular, which has been shown to be reliably increased in magnitude following both pleasant and unpleasant compared to neutral stimuli[6]. In one combined fMRI/ERP study, the LPP is found to be correlated with neural activity in the lateral occipital, inferotemporal, and parietal visual areas, supporting the notion that it reflects facilitated perceptual processing of motivationally relevant, emotional stimuli[7]. Instructions to reappraise unpleasant stimuli have been shown to decrease the magnitude of the LPP and, importantly, reappraisal modulates the LPP just 300 ms after stimulus onset.

In this paper, we record simultaneous EEG-fMRI while subjects passively view the neutral and unpleasant stimuli pictures which is preceded by a instruction of look or decrease. What's more, these pictures are all random present. We study the subjects' brain active regions, examine related BOLD response for fMRI data, and focus on the effects of reappraisal instructions on the amplitude of LPP for EEG data. We use Netstation software which come from America company EGI to preprocess EEG data, and use the 8th Statistical Parametric Mapping (SPM8) to preprocess fMRI data, then apply ICA algorithm for further study. From the scalp-recorded LPP and fMRI BOLD signal, we predict that reappraisal will decrease negative affect.

## 2     Independent Component Analysis (ICA)

Before estimating the independent components (ICs), the observed data can be whitened, that is, transformed to be uncorrelated and have unit variances. Whitening can be done using a linear transformation and does not constrain the estimation in any way, since independence implies uncorrelation. Additionally, whitening simplifies the following component estimation by restricting the structure of the mixing. If the whitening is done using principal component analysis (PCA), the number of free parameters can be reduced by taking only the K strongest principal components, and leaving out the weakest. Assuming that the weakest components contain mainly noise, the dimension of the data is reduced in an optimal manner to improve the signal-to-noise ratio.

As mentioned before, ICA is based only on the assumption that the source signals are statistically independent. This seems reasonable in many applications and in fact does not have to hold exactly for ICA to be applicable in practice. The generative model used in ICA is an instantaneous linear mixture of random variables. The original signals can be considered as a source matrix S, where each row of the matrix contains one of the K signals. Respectively, the observed mixed signals are denoted as matrix X. Again, each row of the matrix contains one of the N observed signals. Assuming a noiseless environment, the mixing model can be expressed in matrix form as:

$$X = AS \tag{1}$$

Each column $a_k$ of the full rank $N \times K$ mixing matrix A holds the mixing weights corresponding to source k. The problem of jointly solving both the mixing and the original sources is not only considerably difficult, but also ambiguous. Since both A and S are unknown, it immediately follows that the signs and scaling of the sources cannot be uniquely defined. One can multiply the mixing weights $a_k$ and divide the corresponding source respectively with any given coefficient. Additionally, the sources can appear in any order [8], [9].

## 3    Experimental Material and Design

### 3.1    Participants and Experimental Material

Three undergraduate right-hand students (1 female 2 male, mean age = 22.8 years, standard deviation 1.25 years) who reported no history of psychiatric or medical disorders or medication use, all participate in simultaneous EEG-fMRI experiment, normal vision, no nerve system disease and claustrophobia, and access to subjects' informed consent.

A total of 60 picture are taken from the International Affective Picture System; of these, 20 depicted neutral scenes (e.g., neutral faces, household objects) and 40 depicted unpleasant scenes (e.g., sad faces, violent images). The two categories differ on normative ratings of valence (M = 5.05, SD = 1.21, for neutral picture content; M = 2.82, SD = 1.64, for unpleasant picture content); additionally, the emotional pictures are reliably higher on normative arousal ratings (M = 5.71, SD = 2.16, for unpleasant picture content; and M = 2.91, SD = 1.93, for neutral picture content).

### 3.2    Emotion Cognitive Reappraisal Strategies and the Task

Prior to magnetic resonance imaging (MRI), participants are trained in specific reappraisal strategies while viewing four practice pictures. Reappraisal instructions encourage thinking objectively to decrease emotional reactivity to pictures (e.g., for decrease instruction, "I try to detach myself from the situation as much as possible. I imagine the scene is from a movie, or that the person was in no pain, or that the person got away from the bad situation").

The trial structure is identical to previous investigations of cognitive reappraisal. As shown in Fig.1, at the start of each trial, an instruction word is presented in the middle of the screen ('decrease' or 'look'; 4 seconds), a black blank screen is presented 2 seconds, a picture is presented (negative if instruction is decrease (regulation instruction), negative or neutral if instruction is look (non-regulation instruction); 4 seconds), follow by a rating period (scale from 1-9; 4 seconds)and then the word 'relax' (4 seconds). The comparisons from the 4-second picture presentation period are the only trial periods reported here. Following presentation of each picture, participants are prompted to answer the question 'How negative do you feel?' on a scale from 1-9 (where 1 is labeled 'weak' and 9 is labeled 'strong'). Responses are made on

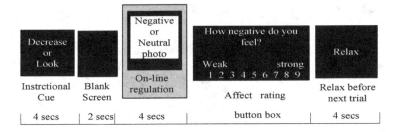

**Fig. 1.** Event-related emotion regulation task

a 9-button button box using the participant's dominant (right) hand. A total of 60 trials (20 of each trial type) are administered in 6 runs of 10 trials each.

# 4    Data Acquisition

fMRI is acquired using a 3-Tesla superconducting type nuclear magnetic resonance imaging system from Philips company at Changzhou Second People's Hospital, and the fMRI time series of whole brain images are acquired with single excitation gradient echo planar imaging using a T2-weighted BOLD sequence. The sequence parameters are: TR=2000ms, TE=35ms, FA=90°, FOV=230mm×180mm, matrix=96×74, 24 continuous slices with a thickness of 4mm. Subjects are instructed to lie on their back, stay awake, try to avoid a specific thinking activity, and lie inside the scanner with foam pads so as to prevent from head movement.

Continuous EEG data is recorded with a 64-channel Electrical Geodesics Inc. Net Station MR amplifier, via a dense array 64 electrode Geodesics Sensor Net. The Vertex (Cz) is chosen as the reference, and impedances are kept below 50 kΩ as recommended for the EGI high input impedance amplifier. Sampling rate is 250 Hz with an on-line bandpass filter of 0.3 to 30Hz.

# 5    Data Analysis

## 5.1    fMRI Analysis

After acquiring the fMRI data, we use the software package of SPM8 to preprocess the data, including slice timing, realigning, spatial normalizing and smoothing. Then we use ICA to calculate the average independent component. We sort all the space ICs into task related ICs, head moving ICs, instantaneous ICs, artifact ICs, and similar periodic ICs. Fig.2 and Fig.3 represent three different task related average ICs of the three subjects, and the corresponding time process curves.

To examine emotional reactivity, we choose two kinds of tasks related ICs, show in Fig.2. We examine reactivity by contrasting responses during the look negative condition with responses during look neutral condition. We observe greater amygdala activity during the look negative than the look neutral condition. At the same time, we find greater BOLD responses for viewing negative than viewing neutral stimulus.

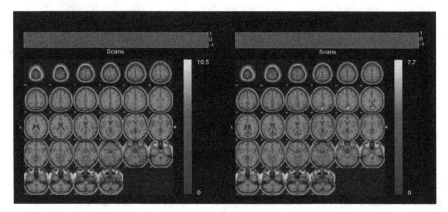

**Fig. 2.** Three subjects' average brain active area and the corresponding time process curves. left under neutral picture stimulus, right under negative picture stimulus

To examine emotion regulation, we consider first the a priori region of interest associated with emotional reactivity, the amygdala. We find the amygdala that are more active when participants are responding naturally to negative pictures than when they are actively regulating. The activities in the lateral occipital, inferotemporal and parietal visual areas are smaller than when subject view negative stimulus. The results are showed in Fig.3.

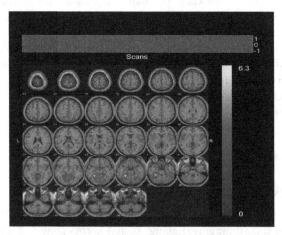

**Fig. 3.** Three subjects' brain active areas when cognitive reappraisal and corresponding time process curves

## 5.2    EEG Analysis

The acquired continuous EEG data of each subject is firstly processed by Netstation software, include the following steps, (MR) Artifact Removal, QRS Detection, Optimal Basis Set (OBS),  band pass digital filtering between 0.01Hz and 40Hz and

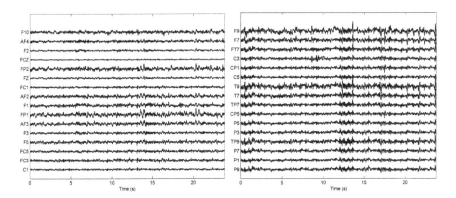

**Fig. 4.** One of the subject's above 32 channels source EEG signal. Left: raw data from 1 to 16 electrodes; Right: raw data from 17 to 32 electrodes.

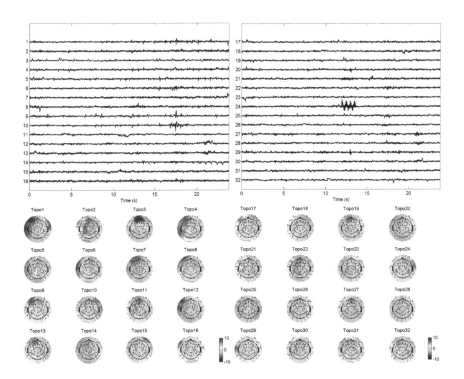

**Fig. 5.** One of the subjects' corresponding ICA component activations and scalp maps of the above 32 channel data., upper row is corresponding ICA component activations, lower row is corresponding scalp maps

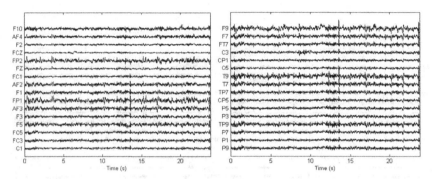

**Fig. 6.** One of the subjects' ICA cleaned EEG signals from 1 to 32 channels by removing artifacts

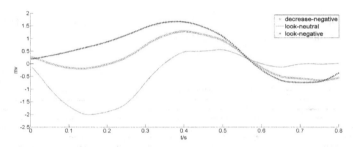

**Fig. 7.** The average characteristic waveform of the three subjects

Segmenting from 200ms before stimulus to 1000ms after stimulus. Fig.4 shows the first 32 channels preprocessed EEG signals for one subject. We can see that some artifacts like ECG and EOG are interfused into EEG signals. We remove these artifacts with ICA method, and get the temporal and spatial ICs, showed in Fig.5. From Fig.5, we can see that the channel of 3, 8, 9, 15, 16, 24, 29 are artifacts. We reconstruct the signals after removing the artifacts which are showed in Fig.6, these signals are clean and artifacts are removed. At last, the features of three subjects' de-noised EEG signals are extracted using wavelet packet method, and are averaged, the curves of characteristic signal are showed in Fig.7, it is the picture onset ERP, with the epoch ranging form 100ms to 600ms representing the LPP. Analyses conducted on the sustained LPP time windows confirm that the magnitude of the LPP will follow the direction of emotion regulation to negative pictures—relative to a passive viewing condition—such that it will be larger during viewing negative pictures and reappraisal. What's more, the amplitude of viewing negative pictures is higher than reappraisal.

## 6    Discussion

Currently, the most common methods of assessing emotional processing in the human brain are event-related potentials (ERP), derived from the electroencephalograph (EEG), and blood oxygen level-dependent (BOLD) contrast, assessed with functional magnetic resonance imaging (fMRI). Here, we make a preliminary study of emotion cognitive reappraisal based on simultaneous recording of EEG and BOLD responses. The current study built on previous findings and theory of emotion regulation by using ERPs elicited during negative and neutral processing to examine the time course of emotion generation and regulation. From Fig.7, we can see that decrease-negative and look-negative trail LPPs remain enhanced compared to look-neutral trail LPPs during the early picture onset. But this does not sustain all the time when picture onset. At the same time, from the results of fMRI data, we find that the activity of amygdala is different under different picture stimulus, it is more active when viewing negative pictures than viewing neutral, but the activity is smaller during reappraisal. The results are same with ERPs. Since the brain activation area is caused by the BOLD signal, we hypothesis that there are some kind of relationship between BOLD signal and LPP.

**Acknowledgments.** This work has been partially supported by the open project of the State Key Laboratory of Robotics and System at Harbin Institute of Technology (SKLRS-2010-2D-09, SKLRS-2010-MS-10), National Natural Science Foundation of China (61201096), Natural Science Foundation of Changzhou City(CJ20110023) and Changzhou High-tech Reasearch Key Laboratory Project (CM20123006).

# References

1. Gross, J.J., Thompson, R.A.: : Emotion regulation: Conceptual foundations. In: Gross, J.J. (ed.) Handbook of Emotion Regulation, vol. 3–26), pp. 3–26. Guilford Press, New York (2007)
2. Ochsner, K.N., Gross, J.J.: The neural architecture of emotion regulation. In: Gross, J.J. (ed.) Handbook of Emotion Regulation, New York, pp. 87–109 (2007)
3. Hajcak, G., Nieuwenhuis, S.: Reappraisal modulates the electrocortical response to unpleasant pictures. Cognitive, Affective and Behavioral Neuroscience 6, 291–297 (2006)
4. Kalisch, R., Wiech, K., Critchley, H.D., Dolan, R.J.: Levels of appraisal: A medial prefrontal role in high-level appraisal of emotional material. Neuroimage 30, 1458–1466 (2006)
5. Hajcak, G., Nieuwenhuis, S.: Reappraisal modulates the electrocortical response to unpleasant pictures. Cognitive, Affective, and Behavioral Neuroscience 6, 291–297 (2006)
6. Schupp, H.T., Junghofer, M., Weike, A.I., Hamm, A.O.: The selective processing of briefly presented affective pictures: An ERP analysis. Psychophysiology 41, 441–449 (2004)
7. Sabatinelli, D., Lang, P.J., Keil, A., Bradley, M.M.: Emotional perception: Correlation of functional MRI andevent-related potentials. Cerebral Cortex 17, 1085–1091 (2007)
8. Zou, L., Pu, H., Sun, Q., Su, W.: Analysis of Attention Deficit Hyperactivity Disorder and Control Participants in EEG Using ICA and PCA. In: Wang, J., Yen, G.G., Polycarpou, M.M. (eds.) ISNN 2012, Part I. LNCS, vol. 7367, pp. 403–410. Springer, Heidelberg (2012)
9. Zou, L., Zhang, Y., Yang, L.T., Zhou, R.: Single Trial Evoked Potentials Study by Combining Wavelet Denoising and Principal Component Analysis Method. Journal of Clinical Neurophysiology 27(1), 17–24 (2010)

# Convergence of Chaos Injection-Based Batch Backpropagation Algorithm For Feedforward Neural Networks

Huisheng Zhang[1,2,*], Xiaodong Liu[1,2], and Dongpo Xu[3]

[1] Research Center of Information and Control, Dalian University of Technology,
Dalian 116024, China
[2] Department of Mathematics, Dalian Maritime University, Dalian 116026, China
[3] College of Science, Harbin Engineering University, Harbin 150001, China
zhhuisheng@163.com

**Abstract.** This paper considers the convergence of chaos injection-based backpropagation algorithm. Both the weak convergence and strong convergence results are theoretically established.

**Keywords:** Convergence, Backpropagation algorithm, Chaos.

## 1   Introduction

Backpropagation (BP) is a popular training algorithm for feedforward neural networks. BP algorithm (BPA) can be implemented by two practical ways: the batch learning and the online learning [1]. The batch learning approach accumulates the weight correction over all the training samples before actually performing the update, nevertheless the online learning approach updates the network weights immediately after each training sample is fed. Though BPA is widely used, it also receives criticisms because of its slow convergence and the problem of being easily trapped into the local minimums. To overcome those problems, many improvements have been proposed , such as adding a penalty term to the error function [2], adding a momentum to the weight updation, injecting noise into the learning procedure, etc. The convergence theories for BPA and its various improvements have been well established by many authors [3,4,5,6,7,8,9,10].

Recently, a new approach is proposed by injecting chaos into the BP learning procedure and its effectiveness has been experimentally verified [11]. However, the convergence for this chaos injection-based BPA has not yet been theoretically established. The purpose of this paper is to investigate the convergence

---

\* Corresponding author. This work is partly supported by the National Natural Science Foundation of China (No.61101228), the China Postdoctoral Science Foundation (No.2012M520623), the Research Fund for the Doctoral Program of Higher Education of China (No.20122304120028), and the Fundamental Research Funds for the Central Universities of China.

C. Guo, Z.-G. Hou, and Z. Zeng (Eds.): ISNN 2013, Part I, LNCS 7951, pp. 60–66, 2013.

of the injection-based batch BPA. The corresponding convergence analysis for injection-based online BPA will be given in our future work.

The rest of this paper is organized as follows. The network structure and the learning algorithm are described in Section 3 and Section 4, respectively. Section 5 presents some assumptions and our main theorem. The detailed proof of the theorem is given in Section 6.

## 2   Network Structure

Consider a three-layer network consisting of $p$ input nodes, $q$ hidden nodes, and 1 output node. Let $\mathbf{w_0} = (w_{01}, w_{02}, \cdots, w_{0q})^T \in \mathbb{R}^q$ be the weight vector between all the hidden units and the output unit, and $\mathbf{w}_i = (w_{i1}, w_{i2}, \cdots, w_{ip})^T \in \mathbb{R}^p$ be the weight vector between all the input units and the hidden unit $i$   ($i = 1, 2, \cdots, q$). To simplify the presentation, we write all the weight parameters in a compact form, i.e., $\mathbf{w} = (\mathbf{w_0}^T, \mathbf{w}_1^T, \cdots, \mathbf{w}_q^T)^T \in \mathbb{R}^{q+pq}$ and we define a matrix $\mathbf{V} = (\mathbf{w}_1, \mathbf{w}_2, \cdots, \mathbf{w}_q)^T \in \mathbb{R}^{q \times p}$.

Given activation functions $f, g : \mathbb{R} \to \mathbb{R}$ for the hidden layer and output layer, respectively, we define a vector function $\mathbf{F}(\mathbf{x}) = (f(x_1), f(x_2), \cdots, f(x_q))^T$ for $\mathbf{x} = (x_1, x_2, \cdots, x_q)^T \in \mathbb{R}^q$. For an input $\boldsymbol{\xi} \in \mathbb{R}^p$, the output vector of the hidden layer can be written as $\mathbf{F}(\mathbf{V}\boldsymbol{\xi})$ and the final output of the network can be written as

$$\zeta = g(\mathbf{w_0} \cdot \mathbf{F}(\mathbf{V}\boldsymbol{\xi})), \tag{1}$$

where $\mathbf{w_0} \cdot \mathbf{F}(\mathbf{V}\boldsymbol{\xi})$ represents the inner product between the two vectors $\mathbf{w_0}$ and $\mathbf{F}(\mathbf{V}\boldsymbol{\xi})$.

## 3   Chaos Injection-Based Backpropagation Algorithm

Suppose that $\{\boldsymbol{\xi}^j, O^j\}_{j=1}^J \subset \mathbb{R}^p \times \mathbb{R}$ is a given set of training samples. The aim of the network training is to find the appropriate network weights $\mathbf{w}^*$ that can minimize the error function

$$E(\mathbf{w}) = \frac{1}{2} \sum_{j=1}^J \left( O^j - g(\mathbf{w_0} \cdot \mathbf{F}(\mathbf{V}\boldsymbol{\xi}^j)) \right)^2$$

$$= \sum_{j=1}^J g_j \left( \mathbf{w_0} \cdot \mathbf{F}(\mathbf{V}\boldsymbol{\xi}^j) \right), \tag{2}$$

where $g_j(t) := \frac{1}{2} \left( O^j - g(t) \right)^2$.

The gradient of the error function is given by

$$E_{\mathbf{w}}(\mathbf{w}) = \left( E_{\mathbf{w_0}}^T(\mathbf{w}), E_{\mathbf{w}_1}^T(\mathbf{w}), \cdots, E_{\mathbf{w}_q}^T(\mathbf{w}) \right)^T \tag{3}$$

with

$$E_{\mathbf{w}_0}(\mathbf{w}) = \sum_{j=1}^{J} g_j'(\mathbf{w}_0 \cdot \mathbf{F}(\mathbf{V}\boldsymbol{\xi}^j))\mathbf{F}(\mathbf{V}\boldsymbol{\xi}^j), \tag{4a}$$

$$E_{\mathbf{w}_i}(\mathbf{w}) = \sum_{j=1}^{J} g_j'(\mathbf{w}_0 \cdot \mathbf{F}(\mathbf{V}\boldsymbol{\xi}^j))w_{0i}f'(\mathbf{w}_i \cdot \boldsymbol{\xi}^j)\boldsymbol{\xi}^j, \quad i = 1, 2, \cdots, q. \tag{4b}$$

Starting from an arbitrary initial value $\mathbf{w}^0$, the chaos injection-based backpropagation algorithm updates the weights $\{\mathbf{w}^n\}$ iteratively by (cf. [11])

$$\mathbf{w}^{n+1} = \mathbf{w}^n - \eta_n(E_{\mathbf{w}}(\mathbf{w}^n) + \eta_n A\alpha v(t)(1 - v(t))\mathbf{I}), \quad n = 0, 1, 2, \cdots \tag{5}$$

where $\eta_n > 0$ is the learning rate, $A$ and $\alpha$ are positive constants, $\mathbf{I} = (1, \cdots, 1)^T \in \mathbb{R}^{pq+q}$, and $v(t+1) = \alpha v(t)(1 - v(t))$ is the logistic map/Verhust equation which is highly sensitive to the initial value $v(0)$ and the parameter $\alpha$. For specific values of $v(0)$(e.g., $0 < v(0) < 1$) and $\alpha$(e.g., $3.6 < \alpha < 4$), the logistic map produces a chaotic time series.

## 4    Main Results

Let $\Phi = \{\mathbf{w} : E_{\mathbf{w}}(\mathbf{w}) = 0\}$ be the stationary point set of the error function $E(\mathbf{w})$, and $\Phi_s = \{w_{ij} : \mathbf{w} = (w_{01}, \cdots, w_{ij}, \cdots, w_{qp}) \in \Phi, s = (i-1)p + j + q(\text{ if } i > 0) \text{ or } j(\text{ if } i = 0)\}$ be the projection of $\Phi$ onto the sth coordinate axis, for $s = 1, \cdots, pq+q$. The following assumptions are needed for our boundedness and convergence results.

($A1$)    The functions $f$ and $g$ are differentiable on $\mathbb{R}$. Moreover, $f, g, f'$, and $g'$ are uniformly bounded on $\mathbb{R}$.

($A2$)    $\eta_n > 0, \sum_{n=0}^{\infty} \eta_n = \infty, \sum_{n=0}^{\infty} \eta_n^2 < \infty.$

($A3$)    $\{\mathbf{w}^n\}$ is bounded over $\mathbb{R}^{pq+q}$.

($A4$)    The set $\Phi_s$ does not contain any interior point for every $s = 1, \cdots, pq+q$. Now we present our convergence result.

**Theorem 1.** *Suppose that the error function is given by* (2) *and that the weight sequence* $\{\mathbf{w}^n\}$ *is generated by the algorithm* (5) *for any initial value* $\mathbf{w}^0$. *Assume the conditions* $(A1) - (A3)$ *are valid. Then we have the weak convergence results*

(a)    *There is* $E^* > 0$ *such that* $\lim_{n \to \infty} E(\mathbf{w}^n) = E^*$; \hfill (6)

(b)    $\lim_{n \to \infty} \|E_{\mathbf{w}}(\mathbf{w}^n)\| = 0.$ \hfill (7)

*Moreover, if Assumption* $(A4)$ *is valid, then we have the strong convergence, i.e., there exists a point* $\mathbf{w}^* \in \Phi$ *such that*

(c)    $\lim_{n \to \infty} \mathbf{w}^n = \mathbf{w}^*.$ \hfill (8)

## 5    Proofs

We first list several lemmas which are crucial to our convergence analysis.

**Lemma 1.** *Let $Y_n, W_n$ and $Z_n$ be three sequences such that $W_n$ is nonnegative for all t. Assume that*

$$Y_{n+1} \leq Y_n - W_n + Z_n, n = 0, 1 \cdots, n$$

*and that the series $\sum_{n=0}^{\infty} Z_n$ is convergent. Then either $Y_n \to -\infty$ or else $Y_n$ converges to a finite value and $\sum_{n=0}^{\infty} W_n \leq \infty$.*

**Proof.** This lemma is directly from [12].                                                  □

**Lemma 2.** *Suppose the conditions (A1) and (A3) are valid, then $E_{\mathbf{w}}(\mathbf{w})$ satisfies Lipschitz conditon, that is, there exists a positive constant $L$, such that*

$$\|E_{\mathbf{w}}(\mathbf{w}^{n+1}) - E_{\mathbf{w}}(\mathbf{w}^n)\| \leq L\|\mathbf{w}^{n+1} - \mathbf{w}^n\|. \tag{9}$$

*Specially, we have*

$$\|E_{\mathbf{w}}(\mathbf{w}^n + \theta(\mathbf{w}^{n+1} - \mathbf{w}^n)) - E_{\mathbf{w}}(\mathbf{w}^n)\| \leq L\theta\|\mathbf{w}^{n+1} - \mathbf{w}^n\|. \tag{10}$$

**Proof.** The proof of this lemma is similar to Lemma 2 of [8] and thus omitted.                                                                                      □

**Lemma 3.** *(see Lemma 4.2 in [5]) Suppose that the learning rate $\eta_n$ satisfies (A2) and that the sequence $\{a_n\}(n \in \mathbb{N})$ satisfies $a_n \geq 0$, $\sum_{n=0}^{\infty} \eta_n a_n^{\beta} < \infty$ and $|a_{n+1} - a_n| \leq \mu\eta_n$ for some positive constants $\beta$ and $\mu$. Then we have $\lim_{n\to\infty} a_n = 0$.*

**Lemma 4.** *(see Lemma 5.3 in [6]) Let $F : \Phi \subset \mathbb{R}^k \to \mathbb{R}, (k \geq 1)$ be continuous for a bounded closed region $\Phi$, and $\Phi_0 = \{z \in \Phi : F(\mathbf{z}) = 0\}$. The projection of $\Phi_0$ on each coordinate axis does not contain any interior point. Let the sequence $\{\mathbf{z}^n\}$ satisfy:*
*(i) $\lim_{n\to\infty} F(\mathbf{z}^n) = 0$;*
*(ii) $\lim_{n\to\infty} \|\mathbf{z}^{n+1} - \mathbf{z}^n\| = 0$.*
*Then, there exists a unique $\mathbf{z}^* \in \Phi_0$ such that $\lim_{n\to\infty} \mathbf{z}^n = \mathbf{z}^*$.*

**Proof of (6).** Given that $0 < v(0) < 1$ and $3.6 < \alpha < 4$, it is easy to see

$$0 < v(t+1) = \alpha v(t)(1 - v(t)) \leq \alpha \frac{(v(t) + 1 - v(t))^2}{4} = \frac{\alpha}{4} < 1. \tag{11}$$

By the differential mean value theorem, there exists a constant $\theta \in [0, 1]$, such that

$$
\begin{aligned}
E(\mathbf{w}^{n+1}) &- E(\mathbf{w}^n) \\
&= (E_{\mathbf{w}}(\mathbf{w}^n + \theta(\mathbf{w}^{n+1} - \mathbf{w}^n)))^T(\mathbf{w}^{n+1} - \mathbf{w}^n) \\
&= (E_{\mathbf{w}}(\mathbf{w}^n))^T(\mathbf{w}^{n+1} - \mathbf{w}^n) \\
&\quad + (E_{\mathbf{w}}(\mathbf{w}^n + \theta(\mathbf{w}^{n+1} - \mathbf{w}^n)) - (E_{\mathbf{w}}(\mathbf{w}^n)))^T(\mathbf{w}^{n+1} - \mathbf{w}^n) \\
&\leq (E_{\mathbf{w}}(\mathbf{w}^n))^T(\mathbf{w}^{n+1} - \mathbf{w}^n) + L\theta\|\mathbf{w}^{n+1} - \mathbf{w}^n\|^2.
\end{aligned}
\tag{12}
$$

Considering (5) and (12), we have

$$
\begin{aligned}
E(\mathbf{w}^{n+1}) \leq E(\mathbf{w}^n) &+ \eta_n(E_{\mathbf{w}}(\mathbf{w}^n))^T[-E_{\mathbf{w}}(\mathbf{w}^n) - \eta_n A\alpha v(t)(1 - v(t))\mathbf{I}] \\
&+ L\theta\eta_n\|E_{\mathbf{w}}(\mathbf{w}^n) + A\eta_n\alpha v(t)(1 - v(t))\mathbf{I}\|^2.
\end{aligned}
\tag{13}
$$

Using (11) and the inequality $\|E_{\mathbf{w}}(\mathbf{w}^n)\| \leq \frac{(1+\|E_{\mathbf{w}}(\mathbf{w}^n)\|^2)}{2}$, we have

$$
\begin{aligned}
(E_{\mathbf{w}}(\mathbf{w}^n))^T&[-E_{\mathbf{w}}(\mathbf{w}^n) - \eta_n A\alpha v(t)(1 - v(t))\mathbf{I}] \\
&\leq -\|(E_{\mathbf{w}}(\mathbf{w}^n))\|^2 + \eta_n\|E_{\mathbf{w}}(\mathbf{w}^n)\|A\sqrt{pq + q}\alpha v(t)(1 - v(t)) \\
&\leq -\|(E_{\mathbf{w}}(\mathbf{w}^n))\|^2 + \eta_n A\sqrt{pq + q}\|E_{\mathbf{w}}(\mathbf{w}^n)\| \\
&\leq -\|(E_{\mathbf{w}}(\mathbf{w}^n))\|^2 + \eta_n\frac{A}{2}\sqrt{pq + q}(1 + \|E_{\mathbf{w}}(\mathbf{w}^n)\|^2)
\end{aligned}
\tag{14}
$$

Using inequality $(a + b)^2 \leq 2(a^2 + b^2)$, we have

$$
\begin{aligned}
\|\eta_n E_{\mathbf{w}}(\mathbf{w}^n) &+ A\eta_n^2\alpha v(t)(1 - v(t))\mathbf{I}\|^2 \\
&\leq 2\eta_n^2\|E_{\mathbf{w}}(\mathbf{w}^n)\|^2 + 2\eta_n^4\|A\alpha v(t)(1 - v(t))\mathbf{I}\|^2 \\
&\leq 2\eta_n^2\|E_{\mathbf{w}}(\mathbf{w}^n)\|^2 + 2A^2(pq + q)\eta_n^4.
\end{aligned}
\tag{15}
$$

Combing (13)-(15), we have

$$
\begin{aligned}
E(\mathbf{w}^{n+1}) &\leq E(\mathbf{w}^n) - \eta_n\|E_{\mathbf{w}}(\mathbf{w}^n)\|^2 + \eta_n^2\frac{A}{2}\sqrt{pq + q}(1 + \|E_{\mathbf{w}}(\mathbf{w}^n)\|^2) \\
&\quad + 2L\theta\eta_n^2\|E_{\mathbf{w}}(\mathbf{w}^n)\|^2 + 2L\theta A^2(pq + q)\eta_n^4 \\
&= E(\mathbf{w}^n) - \eta_n(1 - 2L\theta\eta_n - \frac{A}{2}\sqrt{pq + q}\eta_n)\|E_{\mathbf{w}}(\mathbf{w}^n)\|^2 \\
&\quad + \eta_n^2(\frac{A}{2}\sqrt{pq + q} + 2L\theta A^2\eta_n^2(pq + q)).
\end{aligned}
\tag{16}
$$

As $\eta_n \to 0$, for sufficiently large $n$, there exist positive constants $C_1$ and $C_2$, such that

$$
E(\mathbf{w}^{n+1}) \leq E(\mathbf{w}^n) - \eta_n C_1\|E_{\mathbf{w}}(\mathbf{w}^n)\|^2 + \eta_n^2 C_2.
\tag{17}
$$

Using $\sum\limits_{n=1}^{\infty} \eta_n^2 C_2 < \infty$, $E(\mathbf{w}^n) > 0$, and Lemma 1, we have

$$\lim_{n\to\infty} E(\mathbf{w}^n) = E^*, \tag{18}$$

$$\sum_{n=1}^{\infty} \|E_\mathbf{w}(\mathbf{w}^n)\|^2 \eta_n < \infty. \tag{19}$$

This completes the proof of (6). □

**Proof of** (7). By Assumptions $(A1)$ and $(A3)$, there is a constant $C_3 > 0$ such that for all $n = 0, 1, \cdots$

$$\|E_\mathbf{w}(\mathbf{w}^n)\| \leq C_3. \tag{20}$$

Using (5), (20), and Lemma 2, we have

$$\begin{aligned}
|\|E_\mathbf{w}(\mathbf{w}^{n+1})\| - \|E_\mathbf{w}(\mathbf{w}^n)\|| &\leq \|E_\mathbf{w}(\mathbf{w}^{n+1}) - E_\mathbf{w}(\mathbf{w}^n)\| \\
&\leq L\theta\|\mathbf{w}^{n+1} - \mathbf{w}^n\| \\
&\leq \eta_n L\theta(\|E_\mathbf{w}(\mathbf{w}^n)\| + \|A\eta_n\alpha v(t)(1 - v(t))\mathbf{I}\|) \\
&\leq C_4\eta_n,
\end{aligned} \tag{21}$$

where $C_4 = L\theta(C_3 + A\sqrt{pq + q}\sup_{n\in\mathbb{N}} \eta_n)$. Thus, by (19), (21), and Lemma 3, we conclude

$$\lim_{n\to\infty} E_\mathbf{w}(\mathbf{w}^n) = 0.$$

□

**Proof of** (8). Obviously $E_\mathbf{w}(\mathbf{w})$ is a continuous function under the Assumption (A1). Using (5) and (7), we have

$$\lim_{n\to\infty} \|\mathbf{w}^{n+1} - \mathbf{w}^n\| = \lim_{n\to\infty} \eta_n\|E_\mathbf{w}(\mathbf{w}^n) + A\eta_n\alpha v(t)(1 - v(t))\mathbf{I}\| = 0. \tag{22}$$

Furthermore, the Assumption (A4) is valid. Thus, applying Lemma 4, there exists a unique $\mathbf{w}^* \in \Phi$ such that $\lim\limits_{n\to\infty} \mathbf{w}^n = \mathbf{w}^*$. □

## References

1. Haykin, S.: Neural Networks and Learning Machines. Prentice Hall (2008)
2. Karnin, E.D.: A simple procedure for pruning back-propagation trained neural networks. IEEE Trans. Neural Netw. 1, 239–242 (1990)
3. Fine, T.L., Mukherjee, S.: Parameter convergence and learning curves for neural networks. Neural Computat. 11, 747–769 (1999)
4. Wu, W., Feng, G., Li, Z., Xu, Y.: Deterministic convergence of an online gradient method for bp neural networks. IEEE Trans. Neural Netw. 16, 533–540 (2005)
5. Wu, W., Wang, J., Chen, M.S., Li, Z.X.: Convergence analysis on online gradient method for BP neural networks. Neural Networks 24(1), 91–98 (2011)

6. Wang, J., Wu, W., Zurada, J.M.: Deterministic convergence of conjugate gradient mehtod for feedforward neural networks. Neurocomputing 74, 2368–2376 (2011)
7. Zhang, H.S., Wu, W., Liu, F., Yao, M.: Boundedness and convergence of online gadient method with penalty for feedforward neural networks. IEEE Trans. Neural Netw. 20(6), 1050–1054 (2009)
8. Zhang, H.S., Wu, W., Yao, M.C.: Boundedness and convergence of batch back-propagation algorithm with penalty for feedforward neural networks. Neurocomputing 89, 141–146 (2012)
9. Shao, H.M., Zheng, G.F.: Boundedness and convergence of online gradient method with penalty and momentum. Neurocomputing 74, 765–770 (2011)
10. Sum, J.P., Leung, C.S., Ho, K.I.: On-line node fault injection training algorithm for mlp networks: objective function and convergence analysis. IEEE Transactions on Neural Networks and Learning Systems 23(2), 211–222 (2012)
11. Ahmed, S.U., Shahjahan, M., Murase, K.: Injecting chaos in feedforward neetworks. Neural Process. Lett. 34, 87–100 (2011)
12. Bertsekas, D.P., Tsitsiklis, J.N.: Gradient convergence in gradient methods with errors. SIAM J. Optim. 3, 627–642 (2000)

# Discovering the Multi-neuronal Firing Patterns Based on a New Binless Spike Trains Measure

Hu Lu[1] and Hui Wei[1,2]

[1] Laboratory of Cognitive Model Algorithm, School of Computer Science, Fudan University, Shanghai 200433, China

[2] Shanghai Key Laboratory of Intelligent Information Processing, Shanghai 200433, China

**Abstract.** In this paper, we proposed a method which presented a new definition of different multi-step interval ISI-distance distribution of single neuronal spike trains and formed a new feature vector to represent the original spike trains. It is a binless spike train's measure method. We used spectral clustering algorithm on new multi-dimensional feature vectors to detect the multiple neuronal firing patterns. We tested this method on standard data set in machine learning, neuronal surrogate data set and in vivo multi-electrode recordings respectively. Results shown that the method proposed in this paper can effectively improve the clustering accuracy in standard data set and detect the firing patterns in neuronal spike trains.

**Keywords:** Spike trains, Spectral clustering, Firing patterns.

## 1 Introduction

How to compare the similarity between pairs of spike trains and discover the spatiotemporal firing patterns among spike trains from multi-electrode recordings have been an important part of research in computational neuroscience. With the development of multi-electrode recordings technology, dozens of individual neurons can be recorded simultaneously [1]. Currently, there are no good solutions because neuronal firing patterns are complexity. Many neuroscientists carried out relevant research. Fellous found the timing of the spikes is highly precise between many trials when a cortical neuron was repeatedly injected with the same current stimulus [2]. But the researches of firing patterns analysis between multiple neural spike trains data are still rare, also existing many challenges. In the other hand, how to determine the degree of similarity or dissimilarity between two spike trains is the key question to analyze the level of synchronization of neuronal firing. The general method calculated cross-correlation to measure the similarity of two spike trains [3,4]. This method required binning the spike trains using a moving-window. The choice of bin size directly impact on the analysis results. To avoid the difficulties associated with binning, several binless spike trains dissimilarity measures have been proposed. These metric-spaces are binless spike train measures without binning the time window to define the similarity between two spike

C. Guo, Z.-G. Hou, and Z. Zeng (Eds.): ISNN 2013, Part I, LNCS 7951, pp. 67–73, 2013.

trains. But these measures need to calculate all the dimensions of action potentials, the number of dimensions grows very large as the length of the spike trains grows. Paiva compared some binless spike trains measures [5].

In this paper, we proposed a definition of multi-step interval ISI-distance, a new binless spike trains measure to detect the similarity or synchrony of spike trains, by defining the different step interval ISI-distance distribution to represent the original spike trains. Regardless of the number of action potentials in a neuronal spike trains, spike trains all can be mapped onto a lower dimension. These values of the different step interval ISI-distance formed a new vector matrix. A new distance between two spike trains was also defined. Then, spectral clustering proposed by Andrew Y.Ng et al was applied to the new vector matrix to discover the neuronal firing patterns [6].

One of the main contributions of this work shows a new method which combines the new spike trains similarly distance with spectral clustering analysis. To test the effectiveness of the method, we tested this method on different data sets respectively. Firstly, the standard machine learning data set IRIS was used to examine the effectiveness of the method. Secondly, this method was tested on surrogate spike trains data set, for which the pattern structure was known in advance. The result shows that it is effective. Finally, we applied it to real spike trains in vivo. The spike trains were recorded from prefrontal cortex of rat when rat learned an ordered task in a Y-maze, for which the firing patterns were previously undetected. From the experimental results, we can see that the method can effectively find implicit patterns in the spike trains.

## 2       Methods

Spike distribution of neurons is often quite irregular, and certain information may be encoded in this irregular time distribution of the spikes.

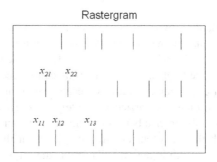

**Fig. 1.** Spike trains of three neurons, a row represent single neuronal spikes

Fig. 1 shows the spike trains of three neurons and a vertical line indicates an action potential. $x_{ij}$ represents the time of $j$th action potential of the $i$th neuron. Using n to represent the number of single neuronal action potentials, the value of n of each neuron is not necessarily the same. Each time interval between the $x_{i1}$ and $x_{i2}$ is called ISI ($x_{ISI} = x_{i2} - x_{i1}$).

This paper proposed the multi-step interval ISI-distance of single neuronal spike trains and the degree of similarity between two spike trains.

**Definition 1.** The p-step interval ISI-distance of $q$th neuron is defined as follows:

$$h_{qp} = \frac{\sqrt{\sum_{i=p+1}^{n}(x_{qi} - x_{q(i-p)})^2}}{\sqrt{\sum_{i=1}^{n}x_{qi}^2}} \tag{1}$$

According to definition 1, we obtained the 1-step interval ISI-distance of $q$th neuron,

$$h_{q1} = \frac{\sqrt{\sum_{i=2}^{n}(x_{qi} - x_{q(i-1)})^2}}{\sqrt{\sum_{i=1}^{n}x_{qi}^2}} \tag{2}$$

Other different multi-step interval ISI-distances are calculated in the same manner. Therefore, n neuronal spike trains were converted to a new multi-dimensional matrix V through the conversion of multi-step interval ISI-distance.

$$V = \begin{bmatrix} h_{11}, \dots, h_{1p} \\ h_{21}, \dots, h_{2p} \\ . \\ . \\ . \\ h_{n1}, \dots, h_{np} \end{bmatrix} \tag{3}$$

The new matrix V represents the original multi-neuronal spike trains and a line of matrix V represents a single neuronal spike trains. Same to principal component analysis (PCA), this method realized a dimensionality reduction of spike trains. But different to the PCA, the numbers of spikes in each neuronal spike trains can be differ.

The correlation coefficient between two neurons was given by the functional distance of two neuronal spike trains. Based on the matrix V, the correlation coefficient was defined using a Gaussian kernel, which has been widely used in the graph analysis methods.

$$s_{ij} = e^{-\|h_i - h_j\|^2 / 2\sigma^2} \tag{4}$$

Where $\left\| h_i - h_j \right\| = \sqrt{\sum_{k=1}^{p} (h_{ik} - h_{jk})^2}$ , it is the Euclidean distance between two

vectors. $\sigma$ is a scale parameter which controls the decay of the Gaussian kernel. The value of s is between 0 and 1.The closer the value of s is to its maximum 1, the stronger the synchrony.

The matrix V from Eq.(3) was a projection of the original spike trains. It realized nonlinear dimensionality reduction. Based on matrix V, we can detect the firing patterns of neurons using spectral clustering algorithm.

The matrix S based on Eq.(4) was a symmetrical matrix. Based on this matrix, we used the NJW spectral clustering algorithm to cluster the data set.

The details of NJW method are given as follows:

(1) Construct the affinity matrix A. The simility between a pair of spike trains is equal to the synchrony $A_{ij} = S_{ij}$ .

(2) Compute the degree matrix D and the normalized affinity matrix $L = D^{-1/2} A D^{-1/2}$ .

(3) Let $\lambda_1 \geq \lambda_2 \geq \cdots \lambda_k$ be the k largest eigenvalues of L and $v^1, v^2, \ldots, v^k$ be the corresponding eigenvectors. Construct the new matrix X=[$v^1, v^2, \ldots, v^k$], and here eigenvectors are the column vector.

(4) Let Y is normalized from affinity matrix X, $Y_{ij} = X_{ij} / (\sum X_{ij}^2)^{1/2}$ .

(5) Based on matrix Y, cluster them into k clusters by using k-means method.

NJW method can obtain satisfying clustering results through embedding the data points into a new space. Compared to the k-means algorithm, spectral clustering algorithm has the advantage to identify any shape data sets.

## 3     Experimental Results and Analysis

In this paper, in order to test the proposed method was effective, we firstly tested the performance of the above mentioned algorithm by applying it to a standard real data set IRIS and a surrogate spike trains data set. Finally, we applied the method to the analysis of multi-electrode recordings neuronal firing patterns.

### 3.1     Standard Data Set

First, we used a data set of IRIS to test the performance of proposed method in this paper. This data set was taken from the UCI repository. The IRIS data set is widely used as a relatively standard test data in many clustering algorithm testing. This data set has 150 samples. Each sample has four characteristics, containing a total of three classes, setosa, versicolor, and virginica. Each class has 50 samples.

In IRIS data set, ISI-distance represents the distance between the two properties, we selected the first three step interval ISI-distances for spectral clustering. Shown in experimental results, we can see the accuracy rate has been greatly improved after

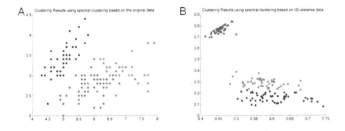

**Fig. 2.** Results of IRIS data set. (A) Results of spectral clustering on original IRIS data set. (B) Results of spectral clustering based on multi-step interval ISI-distances of IRIS data set.

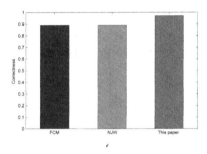

**Fig. 3.** The comparison of correctness of several different algorithms

ISI-distance conversion than other clustering algorithm. The results indicate that the testing of proposed method in this paper is effective in this standard data set.

## 3.2    Surrogate Spike Trains

We can not know what kinds of pattern exist in the real spike trains data set in advance. We need to use the surrogate data set to test the performance of the proposed method. There are many ways to generate the surrogate spike trains data [7]. In these data sets, neuronal firing patterns are known in advance. In this paper, a surrogate data set containing 20 neuronal spike trains was generated.

Shown in Fig. 4, the two clusters were identified according to similar spike firing. We can see from the experimental result of the surrogate data, combing the multi-step interval ISI-distance and spectral clustering proposed in this paper can correctly distinguish the firing patterns of neurons.

## 3.3    Real Spike Trains

In this section, we applied our method to the analysis of neuronal spike trains which structures were unknown in advance. We cannot give the standard partitioning results.

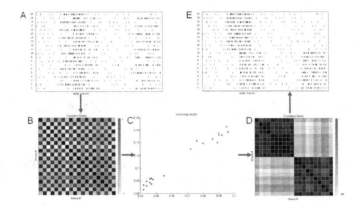

**Fig. 4.** Results of the surrogate data set. (A) The original generated surrogate data set. (B) The similarity matrix of 20 neurons. (C) The divided results by spectral clustering. (D) The sorted similarity matrix. (E) Two firing patterns were divided by spectral clustering.

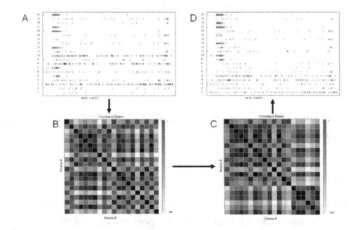

**Fig. 5.** Results of the Y-maze data set. (A) Spike trains of a trial process. (B) The similarity matrix. (C) the sorted similarity matrix after Spike trains were divided into two firing patterns. (D) Spike trains correspond to two firing patterns, with different colors.

Data was obtained from a trained rat. The rat had to perform a Y-maze task. Spike trains recordings were gotten from the medial prefrontal cortex (mPFC) of the rat. The experimental spike trains data was selected from a trial process. The time of spike trains was from 3082s to 3124s, lasted for 42s, containing 20 neurons.

The results showed that the firing patterns of 20 neurons can be roughly divided into two clusters, although the distinction between two clusters was not obvious. We used the method proposed in this paper can   identify the two different firing patterns.

# 4　Discussion and Conclusions

Understanding of firing patterns of neurons will play a very important role for the analysis of neural coding. This paper proposed the concept of multi-step interval ISI-distance. It realized a nonlinear mapping of spike trains, without binning the neuronal spike trains. This paper proposed a new similarity measure of pairs of neurons. Based on the similarity matrix, we used the spectral clustering algorithm to detect the firing patterns in the multi-electrode recordings. We tested this method in different experimental data sets, results shown that this method can clearly find the firing patterns of spike trains. We applied this method to analyze real multi-neuron spike trains from multi-electrode recordings, in which the rat performed the working memory task. The result shows there are similar firing patterns between neurons.

Of course, the present method also has some limitations. We can not know the number of clusters in advance in real spike trains. Existing spectral clustering algorithms typically require specification of the number of clusters at the outset. How to select the value of number is always need to be solved in clustering algorithms. In fact, there is a certain regularity of neuronal firing patterns. It is worthwhile further propose new methods which can identify the number of clusters automatically.

**Acknowledgements.** Multiple electrode recordings in vivo were provided by Dr. Ji-yun Peng (Institutes of Brain Science, Fudan University).

# References

1. Quiroga, R.Q., Panzeri, S.: Extracting information from neuronal populations: information theory and decoding approaches. Nature Reviews Neuroscience 10(3), 173–185 (2009)
2. Fellous, J.M., Tiesinga, P.H.E., Thomas, P.J., et al.: Discovering spike patterns in neuronal responses. The Journal of Neuroscience 24(12), 2989–3001 (2004)
3. Toups, J.V., Tiesinga, P.H.E.: Methods for finding and validating neural spike patterns. Neurocomputing 69(10), 1362–1365 (2006)
4. Schreiber, S., Fellous, J.M., Whitmer, D., et al.: A new correlation-based measure of spike timing reliability. Neurocomputing 52, 925–931 (2003)
5. Paiva, A.R.C., Park, I., Príncipe, J.C.: A comparison of binless spike train measures. Neural Computing and Applications 19(3), 405–419 (2010)
6. Paiva, A.R.C., Rao, S., Park, I., et al.: Spectral clustering of synchronous spike trains. In: IEEE International Joint Conference on Neural Networks, pp. 1831–1835 (2007)
7. Macke, J.II., Berens, P., Ecker, A.S., et al.: Generating spike trains with specified correlation coefficients. Neural Computation 21(2), 397–423 (2009)

# A Study on Dynamic Characteristics of the Hippocampal Two-Dimension Reduced Neuron Model under Current Conductance Changes

Yueping Peng[1,2], Xinxu Wang[1], and Xiaohua Qiu[1]

[1] Engineering University of Chinese Armed Police Force, 710086, Xi'an, China
[2] Xi'an Jiaotong University, 710049, Xi'an, China
Percy001@163.com

**Abstract.** In the paper, based on the computer simulation, the hippocampal two-dimension reduced neuron model is taken as the object, and its dynamic bifurcation characteristics are analyzed and discussed in detail by the neurodynamic analysis methods. When the maximum conductance of the instantaneous sodium channel and the maximum conductance of the delay-rectified potassium channel are changed, the neuron model undergoes the supercritical Andronov-Hopf bifurcation from the rest state to the continuous discharge state. The neuron model is a resonator with the monostable state and has the common dynamics of the resonator. This investigation is helpful to know and investigate deeply the dynamic characteristics and the bifurcation mechanism of the hippocampal neuron by the computer simulation.

**Keywords:** Neurodynamics, The two-dimension reduced neuron model, Bifurcation, Simulation.

## 1 Introduction

In the 1990s, neuron modeling in hippocampal region has been becoming a research hotspot in the field of neural science. Depended on electrophysiological experiments and new technologies such as optics imaging, some models of hippocampal pyramidal Neuron based on ionic conductance have been successfully constructed[1-14]; The work of R.D. Traub's research team is the most excellent[7-9].

The hippocampal CA1 pyramid neuron[10] has plenty of discharge actions. Yue found that bursting behavior persists in adult CA1 pyramidal cells after almost complete truncation of the apical dendrites. The mechanism of bursting is different from the "ping-pong" mechanism, which depends on the integrity of apical dendrites [11-13]. Based on the neuron's membrane ionic channel theory and the CA1 pyramidal neuron's electrophysiological experimental data, David had developed one-compartment model of CA1 Pyramidal Neuron by neurodynamic theory [14]. The model not only can simulate many electrophysiological features and experimental results of the hippocampal CA1 pyramid neuron, but also can spontaneously generate regular firing, tonic firing, rhythmic bursting, and so on.

C. Guo, Z.-G. Hou, and Z. Zeng (Eds.): ISNN 2013, Part I, LNCS 7951, pp. 74–82, 2013.

The nine-dimension one-compartment model of CA1 pyramid neuron developed by David [14] is different from former multi-compartment cable models of the hippocampal pyramid neuron. This model omits the effects of the apical dendrites and the complexity is much reduced. However, the model is still a set of complex high dimensional differential equations, and is very inconvenient to deeply analyze the neuron model's local dynamic characteristics. So this complex high dimensional neuron model is needed to be reduced by dimension reduction. In recent years, the reduction work of the nine-dimension one-compartment complex model of CA1 pyramid neuron developed by David is done by Yueping Peng, and et al[15-18]. They have reduced the nine-dimension complex neuron model to the four minimal models and the two-dimension reduced model by the neurodynamics theory and methods, and analyzed and discussed the dynamic characteristics of the normal model and these reduced models.

In the paper, based on computer stimulation, the two-dimension reduced neuron model[18] is taken as the object, and Its dynamic characteristics are analyzed and discussed in detail by the neurodynamic analysis methods such as the nullcline analysis method, the linear equalization analysis method, and so on.

## 2     The Hippocampal Two-dimension Reduced Neuron Model

The Hippocampal two-dimension reduced neuron model[18] is:

$$C\frac{dV}{dt} = -g_L(V - V_L) - g_{Na}m_\infty^3(V)(0.7334 - 0.8834n)(V - V_{Na}) - g_{Kdr}n^4(V - V_K) + I_{APP}$$

$$\frac{dn}{dt} = \frac{n_\infty(V) - n}{\tau_n(V)} \tag{1}$$

Where the physical meanings of parameters related to the model equation are showed in references[14].

The two-dimension reduced neuron model described by formula (1) has only two time variables: the membrane potential V and the delayed rectified $K^+$ current activation variables n. At numerical calculation, the values of model parameters are as follows:

$$C = 1uF/cm^2 \; ; \; g_L = 0.05mS/cm^2 \; ; \; V_L = -70mV \; ; \; g_{Na} = 35mS/cm^2 \; ;$$

$$g_{Kdr} = 6mS/cm^2 \; ; \; V_{Na} = 55mV \; ; \; V_K = -90mV \; ;$$

$$m_\infty(V) = \frac{1}{1 + \exp(\dfrac{-30 - V}{9.5})} \; ;$$

$$n_\infty(V) = \cfrac{1}{1+\exp(\cfrac{-35-V}{10})} \; ; \quad \tau_n(V) = 0.1 + 0.5 \times \cfrac{1}{1+\exp(\cfrac{-27-V}{-15})} \; ;$$

In addition, the state variable of the model is (V, n), and the initial state is (-65, 0.8).

The two-dimension reduced neuron model described by formula (1) has many dynamic characteristics. Under the different current stimulation, the two-dimension reduced model can also generate many discharge patterns such as period discharge pattern, bursting pattern, the chaos discharge pattern, and so on. Fig.1 shows the several common discharge patterns of the two-dimension reduced neuron model, where the model's simulation time is 0~2000ms.

In Fig.1, Figure (a) and Figure (d) are the period 1 discharge pattern, where the stimulation current is the direct current, and the current amplitude is 2nA. Figure (b) and Figure (e) are the bursting pattern, where the stimulation current is the half wave sine current, and the current amplitude and the current period are respectively 5nA and 60ms. Figure (c) and Figure (f) are the chaos discharge pattern, where the stimulation current is the half wave sine current, and the current amplitude and the current period are respectively 8nA and 10ms.

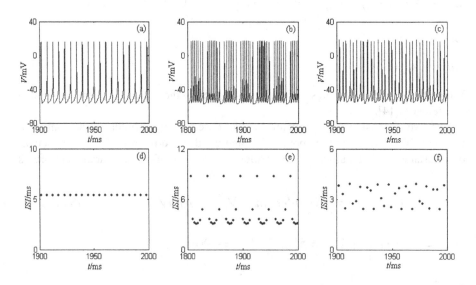

**Fig. 1.** The several common discharge patterns of the two-dimension reduced neuron model

# 3    The Dynamic Characteristics of the Two-dimension Reduced Neuron Model

In the following, we take the two-dimension reduced neuron model described by formula (1) as the object, and discuss its dynamic characteristics under the current conductance's changing.

## 3.1    The Dynamic Bifurcation under the Transient Sodium Channel's Maximum Conductance's Changing

The transient Na+ current is indispensable in the two-dimension reduced neuron model described by formula (1). Under the certain amplitude current's stimulation, the neuron model can generate the discharge process (change from the resting state to the continuous discharge state) only when the maximum conductance of the transient sodium channel ($g_{Na}$) reaches a certain threshold. During the changing process, the neuron also undergoes the dynamic bifurcation. Fig.2 shows the dynamic bifurcation process of the neuron in the V-n phase plane under the transient sodium channel's maximum conductance's changing, where the simulation time is 0~1000ms, and the stimulation current IApp is 5nA.

From Fig.2 (a), the neuron model has only one equilibrium point. In the following, the stability of the equilibrium point is analyzed by the neurodynamics theory.

In Fig.2 (a), the coordinate of the equilibrium point is (-41.8981, 0.3341) in the V-n phase plane. Jacobian matrix $J_{Na}$ and Its eigenvalues can be solved:

$$J_{Na} = \begin{bmatrix} -0.1247 & -43.0530 \\ 0.0479 & -2.1512 \end{bmatrix};$$

$$\lambda_{Na1} = -1.1380 + 1.0167i \, ; \quad \lambda_{Na2} = -1.1380 - 1.0167i$$

Because these two eigenvalues' real parts are both less than zero, the equilibrium point is stable, and is corresponding to the resting state of the neuron model.

From Fig.2, during the process of the transient sodium channel's maximum conductance's increasing gradually, the stable equilibrium point (Its coordinate is (-41.8981, 0.3341)) in the V-n phase plane gradually loses stability and gives birth to a small-amplitude subliminal oscillating phase trajectory around the stable equilibrium point. The oscillating amplitude is damped according to exponential form. Moreover, these subliminal oscillating phase trajectories finally return to the stable equilibrium point (the resting state of the neuron model). When $g_{Na}$ reaches some a value, the subliminal oscillating phase trajectories don't return to the stable equilibrium point, and give birth to a small-amplitude limit cycle attractor. At the same time, the stable equilibrium point loses stability. As $g_{Na}$ keeps on increasing, the amplitude of the

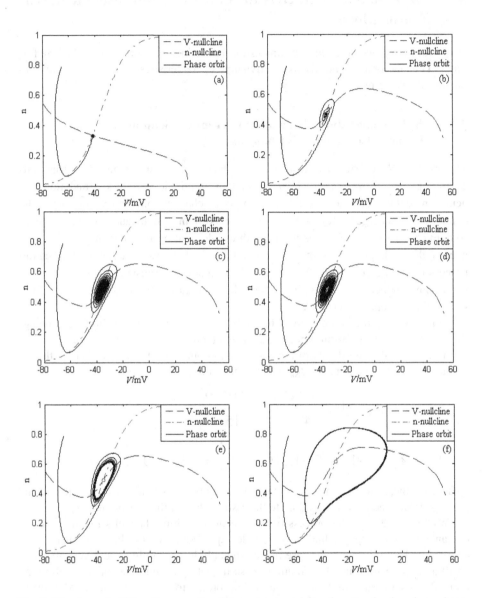

**Fig. 2.** The dynamic bifurcation process of the neuron model under $g_{Na}$'s changing, where the stimulation current $I_{App}$ is 5nA. (a) $g_{Na} = 0.0001mS/cm^2$. (b) $g_{Na} = 10\,mS/cm^2$. (c) $g_{Na} = 11.5mS/cm^2$. (d) $g_{Na} = 11.55mS/cm^2$. (e) $g_{Na} = 12mS/cm^2$. (f) $g_{Na} = 25mS/cm^2$.

limit cycle also increases and becomes full-size spiking limit cycle. From the neuro-dynamics theory point of view, the neuron model undergoes the supercritical Andronov-Hopf bifurcation process.

From the property of the supercritical Andronov-Hopf bifurcation, The neuron model is equivalent to a monostable resonator, and has the common dynamic properties of the resonator such as the subliminal discharge behaviour, class 2 neural excitability, and so on. When $g_{Na}$ is near 11.55mS/cm2 (showed in Fig.2 (d)), the stable limit cycle attractor begins to arise, and the neuron model goes into the stable "all or none" discharge state. So the neuron model undergoes the supercritical Andronov-Hopf bifurcation near 11.55mS/cm2. When the stimulation current IApp is 5nA, the transient sodium channel's maximum conductance's threshold of the neuron model which can produce the stable "all or none" discharge behaviour is near 11.55mS/cm2. In addition, because the neuron model has the monostable characteristics, the stable limit cycle (the continuous discharge state) and the stable equilibrium point can't coexist.

## 3.2    The Dynamic Bifurcation under the Delay-Rectified Potassium Channel's Maximum Conductance's Changing

The delay rectification K+ current is also indispensable in the two-dimension reduced neuron model described by formula (1). Under the certain amplitude current's stimulation, the neuron model can generate the discharge process (change from the resting state to the continuous discharge state) only when the maximum conductance of the delay-rectified potassium channel ( $g_{Kdr}$ ) reaches a certain threshold. During the changing process, the neuron also undergoes the dynamic bifurcation. Fig.3 shows the dynamic bifurcation process of the neuron in the V-n phase plane under the delay-rectified potassium channel's maximum conductance's changing, where the simulation time is 0~1000ms, and the stimulation current IApp is 5nA. The Values of related parameters are showed in the third section.

From Fig.3 (a), the neuron model has only one equilibrium point. In the following, the stability of the equilibrium point is analyzed by the neurodynamics theory.

In Fig.3 (a), the coordinate of the equilibrium point is (-18.9569, 0.8326) in the V-n phase plane. Jacobian matrix $J_{Kdr}$ and Its eigenvalues can be solved:

$$J_{Kdr} = \begin{bmatrix} -0.2000 & -1010.9000 \\ 0.0000 & -3.5000 \end{bmatrix}$$

$$\lambda_{Kdr1} = -1.8571 + 6.8386i \quad ; \quad \lambda_{Kdr2} = -1.8571 - 6.8386i$$

Because these two eigenvalues' real parts are both less than zero, the equilibrium point is stable, and is corresponding to the resting state of the neuron model.

From Fig.3, during the process of the delay-rectified potassium channel's maximum conductance's increasing gradually, the stable equilibrium point (Its coordinate is (-18.9569, 0.8326)) in the V-n phase plane gradually loses stability and gives birth

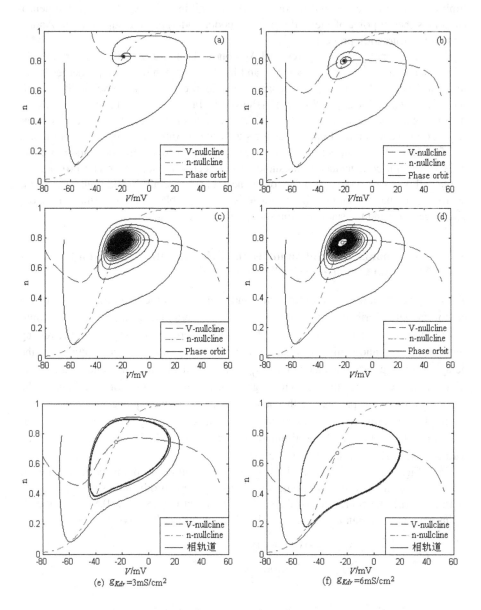

**Fig. 3.** The dynamic bifurcation process of the neuron model under $g_{Kdr}$'s changing, where the stimulation current $I_{App}$ is 5nA. (a) $g_{Kdr} = 0.0001 mS / cm^2$. (b) $g_{Kdr} = 1 mS / cm^2$ (c) $g_{Kdr} = 1.98 mS / cm^2$. (d) $g_{Kdr} = 1.99 mS / cm^2$. (e) $g_{Kdr} = 3 mS / cm^2$. (f) $g_{Kdr} = 6 mS / cm^2$.

to a small-amplitude subliminal oscillating phase trajectory around the stable equilibrium point. The oscillating amplitude is damped according to exponential form. Moreover, these subliminal oscillating phase trajectories finally return to the stable equilibrium point (the resting state of the neuron model). When $g_{Kdr}$ reaches some a value, the subliminal oscillating phase trajectories don't return to the stable equilibrium point, and gives birth to a small-amplitude limit cycle attractor. At the same time, the stable equilibrium point loses stability. As $g_{Kdr}$ keeps on increasing, the amplitude of the limit cycle also increases and it becomes full-size spiking limit cycle. From the neurodynamics theory point of view, the neuron model also undergoes the supercritical Andronov-Hopf  bifurcation process.

From the property of the supercritical Andronov-Hopf bifurcation, the neuron model is equivalent to a monostable resonator, and has the common dynamic properties of the resonator such as the subliminal discharge behaviour, class 2 neural excitability, and so on. When $g_{Kdr}$ is near 1.98mS/cm2 (showed in Fig.2 (c)), the stable limit cycle attractor begins to arise, and the neuron model goes into the stable "all or none" discharge state. So the neuron model undergoes the supercritical Andronov-Hopf bifurcation near 1.98mS/cm2. When the stimulation current IApp is 5nA, the delay-rectified potassium channel's maximum conductance's threshold of the neuron model which can produce the stable "all or none" discharge behaviour is near 1.98mS/cm2. In addition, because the neuron model has the monostable characteristics, the stable limit cycle (the continuous discharge state) and the stable equilibrium point can't coexist.

## 4    Conclusion

In the paper, the two-dimension reduced neuron model is taken as the object, and Its dynamic bifurcation characteristics are analyzed and discussed in detail by the neurodynamic analysis methods such as the nullcline analysis method, the phase space analysis method, the linear equalization analysis method, and so on. When the maximum conductance of the instantaneous sodium channel ($I_{Na}$) and the maximum conductance of the delay-rectified potassium channel ($I_{Kdr}$) are changed, the neuron model undergoes the supercritical Andronov-Hopf bifurcation from the rest state to the continuous discharge state. The neuron is a resonator with the monostable state and has the common dynamics of the resonator. This investigation is helpful to know and investigate deeply the dynamic characteristics and the bifurcation mechanism of the hippocampal neuron by the computer simulation.

## References

1. Urbani, A., Belluzzi, O.: Riluzole inhibits the persistent sodium current in mammalian CNS neurons. Eur. J. Neurosci. 12, 3567–3574 (2000)
2. Vasilyev, D.V., Barish, M.E.: Postnatal development of the hyperpolarization-activated excitatory current Ih in mouse hippocampal pyramidal neurons. J. Neurosci. 22, 8992–9004 (2002)

3. Vervaeke, K., Hu, H., Graham, L.J., Storm, J.F.: Contrasting effects of the persistent Na_current on neuronal excitability and spike timing. Neuron 49, 257–270 (2006)
4. Shuai, J., Bikson, M., Hahn, P.J., Lian, J., Durand, D.M.: Ionic mechanisms underlying spontaneous CA1 neuronal firing in $Ca2^+$-free solution. Biophys. J. 84, 2099–2111 (2003)
5. Chen, S., Yue, C., Yaari, Y.: A transitional period of calcium-dependent bursting triggered by spike backpropagation into apical dendrites in developing hippocampal neurons. J. Physiol. 567, 79–93 (2005)
6. Metz, A.E., Jarsky, T., Martina, M., Spruston, N.: R-type calcium channels contribute to afterdepolarization and bursting in hippocampal CA1 pyramidal neurons. J. Neurosci. 25, 5763–5773 (2005)
7. Traub, R.D., Wong, R.K., Miles, R., Michelson, H.: A model of a CA3 hippocampal pyramidal neuron incorporating voltage-clamp data on intrinsic conductances. J. Neurophysiol. 66, 635–649 (1991)
8. Traub, R.D., Miles, R.: Neuronal networks of the hippocampus, New York, Cambridge (1991)
9. Traub, R.D., Jefferys, J.G.R., Miles, R., Whittington, M.A., To´th, K.: A branching dendritic model of a rodent CA3 pyramidal neurone. J. Physiol. 481, 79–95 (1994)
10. Warman, E.N., Durand, D.M., Yuen, G.L.: Reconstruction of hippocampal CA1 pyramidal cell electrophysiology by computer simulations. J. Neurophysiol. 71, 2033–2045 (1994)
11. Yue, C., Remy, S., Su, H., Beck, H., Yaari, Y.: Proximal persistent Na+ channels drive spike afterdepolarizations and associated bursting in adult CA1 pyramidal cells. J. Neurosci. 25, 9704–9720 (2005)
12. Yue, C., Yaari, Y.: KCNQ/M channels control spike afterdepolarization and burst generation in hippocampal neurons. J. Neurosci. 24, 4614–4624 (2004)
13. Yue, C., Yaari, Y.: Axo-somatic and apical dendritic Kv7/M channels differentially regulate the intrinsic excitability of adult rat CA1 pyramidal cells. J. Neurophysiol. 95, 3480–3495 (2006)
14. Golomb, D., Yue, C., Yaari, Y.: Contribution of Persistent Na+ Current and M-Type $K^+$ Current to Somatic Bursting in CA1 Pyramidal Cells: Combined Experimental and Modeling Study. J. Neurophysiol, 96, 1912–1926 (2006)
15. Peng, Y.: Study on Dynamic Characteristics of the Hippocampal Neuron under Current Stimulation. Advanced Materials Research 341-342, 350–354 (2011)
16. Peng, Y., Zou, N., Wu, H.: Dynamic Characteristics of the Hippocampal Neuron under Conductance's Changing. Applied Mechanics and Materials 195-196, 868–873 (2012)
17. Peng, Y., Wu, H., Zou, N.: Study on the Hippocampal Neuron's Minimal Models' Discharge Patterns. I.J. Image, Graphics and Signal Processing 4, 32–38 (2011)
18. He, X., Peng, Y.: Study on Reduction Algorithm in Hippocampal Neuron Complex Model. In: The 2012 8th International Conference on Natural Computation, pp. 178–182 (2012)

# Overcoming the Local-Minimum Problem in Training Multilayer Perceptrons with the NRAE-MSE Training Method[*]

James Ting-Ho Lo[1], Yichuan Gui[2], and Yun Peng[2]

[1] Department of Mathematics and Statistics
[2] Department of Computer Science and Electrical Engineering
University of Maryland, Baltimore County
Baltimore, Maryland 21250, USA
{jameslo,yichgui1,ypeng}@umbc.edu

**Abstract.** The normalized risk-averting error (NRAE) training method presented in ISNN 2012 is capable of overcoming the local-minimum problem in training neural networks. However, the overall success rate is unsatisfactory. Motivated by this problem, a modification, called the NRAE-MSE training method is herein proposed. The new method trains neural networks with respect to NRAE with a fixed $\lambda$ in the range of $10^6$-$10^{11}$, and takes excursions to train with the standard mean squared error (MSE) from time to time. Once an excursion produces a satisfactory MSE with cross-validation, the entire NRAE-MSE training stops. Numerical experiments show that the NRAE-MSE training method has a success rate of 100% in all the testing examples each starting with a large number of randomly selected initial weights.

**Keywords:** Neural network, Training, Normalized risk-averting error, Global optimization, Local-minimum, Mean squared error.

## 1 Introduction

A standard formulation of training a multilayer perceptron (MLP) under supervision follows: A set of pairs, $(x_k, y_k)$, $k = 1, ..., K$, of which the vectors $x_k$ and the vectors $y_k$ are related by an unknown function $f$

$$y_k = f(x_k) + \xi_k$$

where $\xi_k$ are random noises. Find the weight vector $w$ of a MLP $\hat{f}(x, w)$ such that the mean squared error (MSE) criterion,

$$Q(w) = \frac{1}{K} \sum_{k=1}^{K} \left\| y_k - \hat{f}(x_k, w) \right\|^2 \tag{1}$$

[*] This material is based upon work supported in part by the National Science Foundation under Grant ECCS1028048, but does not necessarily reflect the position or policy of the Government.

C. Guo, Z.-G. Hou, and Z. Zeng (Eds.): ISNN 2013, Part I, LNCS 7951, pp. 83–90, 2013.

is minimized. If the MLP $\hat{f}(x_k, w)$ is nonlinear in $w$, the MSE criterion $Q(w)$ is usually nonconvex and has nonglobal local minima [1,2,3,4,5,6,7,8].

It is proven in [9] that the convexity region of $J_\lambda(w)/\lambda$ where the risk-averting error (RAE) $J_\lambda(w)$ is defined by

$$J_\lambda(w) := \sum_{k=1}^{K} \exp\left(\lambda \left\| y_k - \hat{f}(x_k, w) \right\|^2\right) \tag{2}$$

expands strictly as $\lambda$ increases, and that $\lim_{\lambda \to 0} \frac{1}{\lambda} \ln\left[\frac{1}{K} J_\lambda(w)\right] = Q(w)$. These properties confirmed the effectiveness of the adaptive training method reported in [10] for avoiding nonglobal local minima. However, $J_\lambda(w)$ is an exponential function of $\lambda \left\| y_k - \hat{f}(x_k, w) \right\|^2$ and is plagued with computer register overflow if $\lambda$ is large. This motivated the use of the normalized RAE (NRAE)

$$C_\lambda(w) := \frac{1}{\lambda} \ln\left[\frac{1}{K} J_\lambda(w)\right] \tag{3}$$

leading to the NRAE training method presented in ISNN 2012 [11]. However, its success rate over all the numerical experiments is 50% for the risk-sensitivity index $\lambda$ in the range $10^6$-$10^8$, 100% in the range $10^8$-$10^9$, and 75% in the range $10^9$-$10^{11}$, but fails to work for $\lambda > 10^{11}$.

Although the NRAE training method cannot reach a global minimum with a 100% success rate, it is able to bring $C_\lambda(w)$ and the corresponding $Q(w)$ significantly down for $10^6 \le \lambda \le 10^{11}$. Experiments show that if, after the NRAE training is performed for a reasonable number of iterations, the training criterion is switched from $C_\lambda(w)$ to $Q(w)$, a global minimum can be obtained each time for $\lambda$ in the range $10^6$ - $10^{11}$ in our numerical experiments. This method is called the NRAE-MSE method.

In this paper, numerical results of testing the NRAE-MSE method are presented that show the efficacy of the method. In the examples, cross-validation is performed to ensure the MLP trained with the new method has a good generalization capability.

## 2   The NRAE-MSE Training Method

The NRAE-MSE training method first select a value of $\lambda$ in the range $10^6$ - $10^{11}$, select positive integers $L$ and $M$, and randomly select an initial weight vector as the current weight vector for the MLP under training. The method then repeats the following two steps:

1. Use $C_\lambda(w)$ with the selected $\lambda$ to train the MLP for $L$ iterations. Each iteration replaces the current weight vector at the beginning of the iteration with the resultant weight vector as the current weight vector.
2. Starting with the current weight vector, use $Q(w)$ to train the MLP until the number of iterations exceeds $M$. After that, if $Q(w) < \varepsilon$, or if

a cross-validation test shows that overfitting of the training data occurs, stop the entire NRAE-MSE training. If a cross-validation test is needed, an early-stopping method is chosen to detect overfitting of the training data.

# 3   Numerical Experiments

Four function approximation experiments, where each function is intended to have nonglobal local minima, are provided to demonstrate the effectiveness of the proposed NRAE-MSE training method. For each task, ten different initial weight vectors are randomly chosen to start ten training groups. In each group, a NRAE-MSE training and a standard MSE training with the backpropagation (BP) and BFGS algorithm are performed by choosing the same initial weight vector. After ten training groups are completed, we analyze the obtained fitting plots and training errors for all function approximation tasks, and measure different performances between the MSE and NRAE-MSE training. For all NRAE-MSE training sessions, we record their corresponding MSE values as training errors and compare them to the results of the MSE training.

In order to test the capability of noise tolerance for the MSE and NRAE-MSE training, another ten training groups are initiated for these four function approximation tasks with measurement noises. Particularly, measurement noises are defined by the signal-to-noise ratio (SNR), which is generally used to indicate the proportion between the power of the target output $O$ and the power of the measurement noises $E$, and generated by the Gaussian white noise model. In our experiments, the SNR is chosen as $10 \log_{10} 2^2 = 6\text{dB}$. For the purpose of comparing different training criterions with noisy data involved, same training strategies performed to noiseless data are used to perform the training with noisy data for each function approximation task. In addition, we use the cross-validation to test generalization abilities and detect the overfitting of the MSE and NRAE-MSE training in both the noiseless and noisy experiments. The size of cross-validation data is chosen as half of the training data, and each cross-validation data is randomly selected from the target function domain without overlapping to training data. In order to clearly show the performance of the function approximation with measurement noises involved, we choose the trained MLP weights in noisy experiment and the target function data without adding measurement noises to compute the MSE as the training error for each noisy experiment.

Before each training session starts, several parameters in the MLP are chosen based on the suggestions in [8]: each synaptic weight in a weight vector is randomly selected from a uniform distribution between $-2.4/F_i$ and $2.4/F_i$, where $F_i$ is the number of input neurons of the connected unit; all input and output values defined in the training data are normalized into $[-1, 1]$; the activation function in each training neuron is chosen as the hyperbolic tangent function $\varphi(v) = a \tanh(bv)$, where $a = 1.7159$ and $b = 2/3$. For all MSE training sessions, the maximum number of training epochs is set to $10^6$. For all NRAE-MSE training sessions, we set $L = 1 \times 10^4$ for the NRAE training, $M = 1 \times 10^4$

for the MSE excursion, and $\lambda = 10^6$ as a sufficiently large value to achieve the condition $\lambda \gg 1$. In addition, the maximum number of iterations of the NRAE-MSE training is limited to 50, which indicates the maximum number of training epochs for the combination of the NRAE training sessions and the MSE excursions is $10^6$. A large number of numerical experiments are not shown in this paper because of the page limit. In those experiments, the NRAE-MSE method delivered a 100% success rate for $\lambda$ in the range $10^6$ - $10^{11}$.

## 3.1   Function Approximation

**Three-notch** A function with three notches is defined by

$$y = f(x) = \begin{cases} 0 & \text{if } x \in [0, 1.0] \cup [2.2, 2.3] \cup [3.5, 4.5] \\ 0.25 & \text{if } x \in [2.8, 3.0] \\ 0.5 & \text{if } x \in [1.5, 1.7] \\ 1 & \text{otherwise} \end{cases} \tag{4}$$

where $x \in X = [0, 4.5]$. For the training data, input values $x_k$ are obtained by random sampling 2000 non-repeatable numbers from $X$ with a uniform distribution, and corresponding output values $y_k$ are computed by (4). Then, a training data set of 2000 $(x_k, y_k)$ pairs is obtained. By the same way, a cross-validation data set of 1000 $(x_k, y_k)$ pairs is also obtained such that it and the training data set are disjoint. MLPs with 1:16:1 architecture are initiated to all training sessions with noiseless or noisy data. Approximated function plots of the MSE and NRAE-MSE training are presented from Fig. 1(a) to Fig. 1(d). Training errors obtained by ten initial weight vectors are shown in Fig. 2(a) for the MSE and NRAE-MSE trained MLPs with or without noisy data involved.

**Fine Features** A smooth function with two fine features as spikes is defined by

$$y = f(x) = g\left(x, \frac{1}{6}, \frac{1}{2}, \frac{1}{6}\right) + g\left(x, \frac{1}{64}, \frac{1}{4}, \frac{1}{128}\right)$$
$$+ g\left(x, \frac{1}{64}, \frac{11}{20}, \frac{1}{128}\right) \tag{5}$$

where $x \in X = [0, 1]$, and $g$ is defined as

$$g(x, \alpha, \mu, \sigma) = \frac{\alpha}{\sqrt{2\pi}\sigma} \cos\left(\frac{(x - \mu)\pi}{\sigma}\right) \exp\left(-\frac{(x - \mu)^2}{2\sigma^2}\right). \tag{6}$$

For the training data, input values $x_k$ are selected by sampling 2000 numbers from a uniform distributed grid on $X$, and corresponding output values $y_k$ are computed by (5). Then, a training data set of 2000 $(x_k, y_k)$ pairs is obtained. By the same way, a cross-validation data set of 1000 $(x_k, y_k)$ pairs is also obtained such that it and the training data set are disjoint. MLPs with 1:14:1 architecture are initiated to all training sessions with noiseless or noisy data. Approximated function plots of the MSE and NRAE-MSE training are presented from Fig. 1(e) to Fig. 1(h). Training errors obtained by ten initial weight vectors are shown in Fig. 2(b) for the MSE and NRAE-MSE trained MLPs with or without noisy data involved.

**Unevenly-Sampled Segments.** A smooth function with two unevenly-sampled segments is defined by

$$y = f(x) = g\left(x, \frac{1}{5}, \frac{1}{4}, \frac{1}{12}\right) + g\left(x, \frac{1}{5}, \frac{3}{4}, \frac{1}{12}\right)$$
$$+ g\left(x, \frac{1}{64}, \frac{5}{4}, \frac{1}{12}\right) \tag{7}$$

where $x \in X = [0, 1.5]$ and $g$ is defined in (6). For the training data, input values $x_k$ are collected by sampling 50 numbers from a uniform distributed grid on $[0, 0.5]$, 50 numbers from a uniform distributed grid on $[1.0, 1.5]$, and 2000 numbers from a uniform distributed grid on $(0.5, 1.0)$. Corresponding output values $y_k$ are computed by (7). Then, a training data set of 2100 $(x_k, y_k)$ pairs is obtained. By the same way, a cross-validation data set of 1050 $(x_k, y_k)$ pairs is also obtained such that it and the training data set are disjoint. MLPs with 1:12:1 architecture are initiated to all training sessions with noiseless or noisy data. Approximated function plots of the MSE and NRAE-MSE training are presented from Fig. 1(i) to Fig. 1(l). Training errors obtained by ten initial weight vectors are shown in Fig. 2(c) for the MSE and NRAE-MSE trained MLPs with or without noisy data involved.

**Unevenly-Sampled Square.** A three-dimensional function, which has a letter 'L' shape and an unevenly-sampled square raised from a plane, is defined by

$$z = f(x, y) = \begin{cases} 1 \text{ if } x \in [1.0, 5.5] \text{ and } y \in [1.0, 2.0] \\ 1 \text{ if } x \in [1.0, 2.0] \text{ and } y \in [2.0, 5.5] \\ 1 \text{ if } x \in [3.0, 5.5] \text{ and } y \in [3.0, 5.5] \\ 0 \text{ otherwise} \end{cases} \tag{8}$$

where $x \in X = [0, 6]$ and $y \in Y = [0, 6]$. For the training data, input values $x_k$ and $y_k$ are collected by sampling 289 numbers from an uniform distributed grid on $X' \times Y' = (2.5, 6] \times (2.5, 6]$ and 2522 numbers from an uniform distributed grid on $(X - X') \times (Y - Y')$. Corresponding output values $z_k$ are computed by (8). Then, a training data set of 2811 $(x_k, y_k)$ pairs is obtained. By the same way, a cross-validation data set of 1405 $(x_k, y_k)$ pairs is also obtained such that it and the training data set are disjoint. MLPs with 2:6:3:1 architecture are initiated to all training sessions with noiseless and noisy data. Approximated function plots of the MSE and NRAE-MSE training are presented from Fig. 1(m) to Fig. 1(p). Training errors obtained by ten initial weight vectors are shown in Fig. 2(d) for the MSE and NRAE-MSE trained MLPs with or without noisy data involved.

## 3.2  Discussion

Experimental results illustrated in Fig. 2 indicates that the NRAE-MSE training consistently leads all trained MLPs to achieve satisfactory training errors, which are lower than the MSE training. Fig. 1 shows the NRAE-MSE training captures all significant features located on the target function for both noiseless and noisy tests, but the MSE training can only find partial features. Moreover, based on the noisy test results presented in Fig. 1, all the NRAE-MSE training sessions are superior to the MSE training sessions in achieving low training errors with high generalization levels. These mentioned results illustrate that: with the combination of two steps in the NRAE-MSE training method, and the

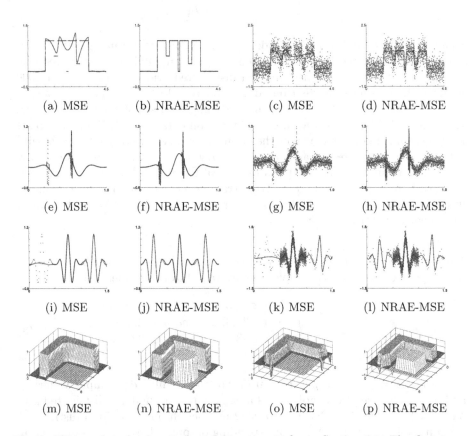

**Fig. 1.** Fitting plots for function approximation tasks in Section 3.1. The first two columns describe functions trained with noiseless data, and the last two columns show functions trained with noisy data. Numbers on the horizontal and vertical axes in each subfigure represent the input and output of the function, respectively. From Fig. 1(a) to Fig. 1(l), red dots denote target training samples, and blue dash lines are MLP approximated function plots. From Fig. 1(m) to Fig. 1(p), only MLP approximated function plots are shown by using blue and red colors to distinguish different function values on vertical axes.

following MSE excursion can successfully lead to a satisfactory training error; on the contrary, without the elimination of local minima, the MSE training error is difficult to converge to a lower one.

In Fig. 2, another fact which can be observed is that trying to use different initial weight vectors never makes the MSE training becoming better than the NRAE-MSE training in any of our experiments. The explanation is that one proper initial guess of weight vectors for the MSE training can only provide a good starting point to the local-searching method, but it cannot eliminate or avoid local minima existing in the searching path of the optimal weight vector. In contrast, the NRAE-MSE training with a large $\lambda$ has the capability to avoid nonglobal local minima, thus it is more insensitive to different selections of initial weight vectors than the MSE training.

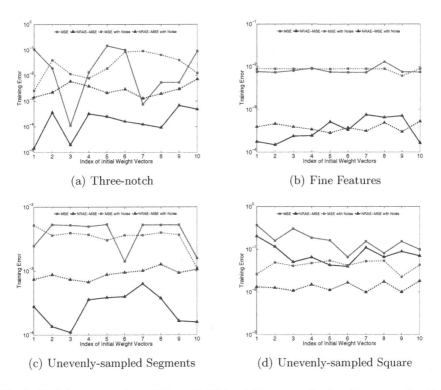

(a) Three-notch

(b) Fine Features

(c) Unevenly-sampled Segments

(d) Unevenly-sampled Square

**Fig. 2.** Training errors of ten different initial weight vectors for function approximation tasks in Section 3.1. Colors and symbols denote different training methods of the MSE (red square) and NRAE-MSE (blue triangle). Solid and dash lines represent different training sessions with noiseless and noisy data, respectively. In order to clearly show differences between MSE values obtained by the MSE and NRAE-MSE training, actual numbers in all vertical axes are converted to logarithmic numbers with respect to base 10.

# 4    Conclusion

The unsatisfactory success rate of the NRAE training method proposed in [11] motivated the development of the NRAE-MSE training method herein reported. The NRAE-MSE training method consists of 2 phases - training with respect to the NRAE criterion at a fixed $\lambda$ for a number of iterations and training with respect to the MSE criterion until convergence. Our numerical experiments show that the method is not sensitive to said number of iterations or the risk-sensitivity index $\lambda$ as long as $\lambda$ lies in the range $10^6$-$10^{11}$. The success rate over a large number of numerical examples that we have worked out is 100%.

It can be argued that the numerical examples are toy examples, which are not representative of more complex real-world applications. However, the target functions in these examples contain fine features or unevently-sampled functions are designed to create nonglobal local minima that are hard for a training method to escape from. Most real-world applications are not expected to be so "vicious".

More importantly, the NRAE-MSE training method roots in the convexfication idea in [9]. This idea is valid regardless of the complexity of the target function or training data. The new training method is therefore expected to do well in real-world applications. As soon as test results are obtained on real-world applications of the NRAE-MSE method, we will report the test results.

## References

1. Aarts, E., Korst, J.: The Neuron. Oxford University Press (1989)
2. Zurada, J.M.: Introduction to Artificial Neural Networks. West Publishing Company, St. Paul (1992)
3. Hassoun, M.H.: Fundamentals of Artificial Neural Networks. MIT Press, Cambridge (1995)
4. Principe, J.C., Euliano, N.R., Lefebvre, W.C.: Neural and Adaptive Systems: Fundamentals through Simulations. John Wiley and Sons, Inc., New York (2000)
5. Bishop, C.M.: Pattern Recognition and Machine Learning. Springer, New York (2006)
6. Du, K.L., Swamy, M.: Neural Networks in a Softcomputing Framework. Springer, New York (2006)
7. Press, W.H., Teukolsky, S.A., Vetterling, W.T., Flannery, B.P.: Numerical Recipes in C: The Art of Scientific Computing, 3rd edn. Cambridge University Press, New York (2007)
8. Haykin, S.: Neural Networks and Learning Machines, 3rd edn. Prentice Hall, Upper Saddle River (2008)
9. Lo, J.T.-H.: Convexification for data fitting. Journal of Global Optimization 46(2), 307–315 (2010)
10. Lo, J.T.-H., Bassu, D.: An adaptive method of training multilayer perceptrons. In: Proceedings of the International Joint Conference on Neural Networks (IJCNN 2001), vol. 3, pp. 2013–2018 (July 2001)
11. Lo, J.T.-H., Gui, Y., Peng, Y.: Overcoming the local-minimum problem in training multilayer perceptrons with the NRAE training method. In: Proceedings of the 9th International Conference on Advances in Neural Networks (ISNN 2012), vol. Part I, pp. 440–447 (July 2012)

# Generalized Single-Hidden Layer Feedforward Networks

Ning Wang[1,*], Min Han[2], Guifeng Yu[1], Meng Joo Er[1,3], Fanchao Meng[1], and Shulei Sun[1]

[1] Marine Engineering College, Dalian Maritime University, Dalian 116026, China
n.wang.dmu.cn@gmail.com
[2] Faculty of EIEE, Dalian University of Technology, Dalian 116024, China
minhan@dlut.edu.cn
[3] School of EEE, Nanyang Technological University, Singapore 639798
emjer@ntu.edu.sg

**Abstract.** In this paper, we propose a novel generalized single-hidden layer feedforward network (GSLFN) by employing polynomial functions of inputs as output weights connecting randomly generated hidden units with corresponding output nodes. The main contributions are as follows. For arbitrary $N$ distinct observations with $n$-dimensional inputs, the augmented hidden node output matrix of the GSLFN with $L$ hidden nodes using any infinitely differentiable activation functions consists of $L$ sub-matrix blocks where each includes $n+1$ column vectors. The rank of the augmented hidden output matrix is proved to be no less than that of the SLFN, and thereby contributing to higher approximation performance. Furthermore, under minor constraints on input observations, we rigorously prove that the GLSFN with $L$ hidden nodes can exactly learn $L(n+1)$ arbitrary distinct observations which is $n+1$ times what the SLFN can learn. If the approximation error is allowed, by means of the optimization of output weight coefficients, the GSLFN may require less than $N/(n+1)$ random hidden nodes to estimate targets with high accuracy. Theoretical results of the GSLFN evidently perform significant superiority to that of SLFNs.

**Keywords:** single-hidden layer feedforward networks, polynomial functions, output weights, hidden node numbers, approximation capability.

## 1   Introduction

In the field of neural networks, in addition to various fuzzy neural networks [1,2,3], single-hidden layer feedforward networks (SLFNs) have been investigated thoroughly in the past two decades. In the 1990's, it has been shown that SLFNs

* This work is supported by the National Natural Science Foundation of China (under Grant 51009017), Applied Basic Research Funds from Ministry of Transport of P. R. China (under Grant 2012-329-225-060), China Postdoctoral Science Foundation (under Grant 2012M520629), and Fundamental Research Funds for the Central Universities of China (under Grant 2009QN025, 2011JC002 and 3132013025).

C. Guo, Z.-G. Hou, and Z. Zeng (Eds.): ISNN 2013, Part I, LNCS 7951, pp. 91–98, 2013.

with $N$ hidden nodes can exactly learn $N$ distinct observations. Tamura *et al.* [4] proved that a SLFN with $N$ hidden units using sigmoid functions can give any $N$ input-target relations exactly. The further improvement proposed by Huang [5] revealed that if input weights and hidden biases are tunable the SLFN with at most $N$ hidden neurons using any bounded nonlinear activation function which has a limit at one infinity can learn $N$ distinct samples with zero error. In contrast to previous SLFNs which adjust all the parameters of hidden layers, some researchers suggested incremental SLFNs allowing only newly added hidden nodes to be tuned. In this case, parameters of hidden nodes need to be updated only once based on training data. Nevertheless, the computation burden would also be heavy. Alternatively, Huang *et al.* [6] developed an innovative learning scheme termed as extreme learning machine (ELM) for SLFNs with randomly generated hidden units using infinitely differentiable activation functions. Corresponding results [7] indicated that SLFNs with $N$ hidden nodes using any infinitely differentiable activation functions can learn $N$ distinct samples exactly and SLFNs may require less than $N$ hidden nodes if learning error is allowed. Similar to [8], Ferrari *et al.* showed that SLFNs with $N$ sigmoidal hidden nodes and with input weights randomly generated but hidden biases appropriately tuned can exactly learn $N$ distinct observations. Besides, several interesting investigations on compact structure of SLFNs were implemented by using singular value decomposition (SVD) [9] and regularized least-squares (RLS) [10] methods, *etc.* However, all the previous works focused on the SLFNs using constant output weights whether hidden node parameters are adjusted or not. Rationally, we refer to the abovementioned SLFNs as standard SLFNs since all the output weights are confined to be constants independent on inputs. In this case, the constant output weights would impose much deficiency on the capability of approximation and generalization.

In this paper, we propose a novel kind of generalized single-hidden layer feedforward networks (GSLFNs) which extend the standard SLFNs by using polynomial functions of inputs instead of constants as the output weights. To be specific, for arbitrary $N$ distinct observations $(\mathbf{x}_k, \mathbf{t}_k) \in \mathbf{R}^n \times \mathbf{R}^m$, $L$ hidden nodes using any infinitely differentiable activation functions are randomly generated and output weight coefficients are allowed to be adjustable for desired performance of approximation and generalization. In this case, the augmented hidden node output matrix consists of $L$ sub-matrix blocks whereby each one includes $n+1$ column vectors containing $N$ entities. Each column vector in the $i$th sub-matrix block is defined by the Hadamard product of the input vector in the $j$-dimension (i.e., $\mathbf{x}^j = [x_{1j}, \cdots, x_{Nj}]^T$) and the $i$th hidden node output vector with respect to the $k$th input observation. Accordingly, preliminary results reveal that the rank of augmented hidden node output matrix in the GSLFN would be no less than that of hidden node output matrix in SLFN, and thereby contributing to higher potentials for approximation capability. Furthermore, we rigorously prove that under minor constraints on input observations the GSLFN with any $L$ randomly generated hidden nodes can exactly learn $L(n+1)$ arbitrary distinct observations which are $n+1$ times what the SLFN can learn.

## 2  Preliminaries

Given $N$ arbitrary distinct samples $(\mathbf{x}_k, \mathbf{t}_k)$ where $\mathbf{x}_k = [x_{k1}, x_{k2}, \cdots, x_{kn}]^\mathrm{T} \in \mathbf{R}^n$ and $\mathbf{t}_k = [t_{k1}, t_{k2}, \cdots, t_{km}]^\mathrm{T} \in \mathbf{R}^m$, the standard single-hidden layer feedforward networks (SLFNs) with $L$ hidden nodes and activation function $g(\mathbf{x})$ can be mathematically modeled as,

$$\mathbf{y}_k = \sum_{i=1}^{L} \boldsymbol{\beta}_i g(\mathbf{a}_i \cdot \mathbf{x}_k + b_i), \ k = 1, 2, \cdots, N \tag{1}$$

where $\mathbf{a}_i = [a_{i1}, a_{i2}, \cdots, a_{in}]^\mathrm{T} \in \mathbf{R}^m$ is the weight vector connecting the $i$th hidden node and the input nodes, $\boldsymbol{\beta}_i = [\beta_{i1}, \beta_{i2}, \cdots, \beta_{im}]^\mathrm{T} \in \mathbf{R}^m$ is the weight vector connecting the $i$th hidden node and the output nodes, and $b_i$ is the threshold of the $i$th hidden node. $\mathbf{a}_i \cdot \mathbf{x}_k$ denotes the inner product of $\mathbf{a}_i$ and $\mathbf{x}_k$.

If the outputs of the SLFN are equal to the targets, we have the compact formulation as follows:

$$\mathbf{H}\boldsymbol{\beta} = \mathbf{T} \tag{2}$$

where,

$$\mathbf{H}(\mathbf{a}_1, \ldots, \mathbf{a}_L, b_1, \ldots, b_L, \mathbf{x}_1, \ldots, \mathbf{x}_N)$$
$$= \begin{bmatrix} g(\mathbf{a}_1, b_1, \mathbf{x}_1) & \cdots & g(\mathbf{a}_L, b_L, \mathbf{x}_1) \\ \vdots & \ddots & \vdots \\ g(\mathbf{a}_1, b_1, \mathbf{x}_N) & \cdots & g(\mathbf{a}_L, b_L, \mathbf{x}_N) \end{bmatrix}_{N \times L} \tag{3}$$

$$\boldsymbol{\beta} = \begin{bmatrix} \boldsymbol{\beta}_1^T \\ \vdots \\ \boldsymbol{\beta}_L^T \end{bmatrix}_{L \times m} \quad \text{and} \quad \mathbf{T} = \begin{bmatrix} \mathbf{t}_1^T \\ \vdots \\ \mathbf{t}_N^T \end{bmatrix}_{N \times m} \tag{4}$$

Here, $\mathbf{H}$ is called the hidden-layer output matrix of the SLFN, whereby the $i$th column is the $i$th hidden node's output vector with respect to inputs $\mathbf{x}_1, \ldots, \mathbf{x}_N$ and the $j$th row is the output vector of the hidden layer with respect to input $\mathbf{x}_j$. $\boldsymbol{\beta}$ and $\mathbf{T}$ are corresponding matrices of output weights and targets, respectively.

It has been proved that standard SLFNs with a wide type of random computational hidden nodes possess the universal approximation capability as follows.

**Lemma 1.** [6] Given a standard SLFN with $N$ hidden nodes and activation function $g : \mathbf{R}^n \to \mathbf{R}$ which is infinitely differentiable in any interval, for $N$ arbitrary distinct samples $(\mathbf{x}_k, \mathbf{t}_k)$, where $\mathbf{x}_k \in \mathbf{R}^n$ and $\mathbf{t}_k \in \mathbf{R}^m$, for any $\mathbf{a}_i$ and $b_i$ randomly chosen from any intervals of $\mathbf{R}^n$ and $\mathbf{R}$, respectively, according to any continuous probability distribution, then with probability one, the hidden layer output matrix $\mathbf{H}$ of the SLFN is invertible and $\|\mathbf{H}\boldsymbol{\beta} - \mathbf{T}\| = 0$. □

**Lemma 2.** [6] Given any small positive value $\varepsilon > 0$ and activation function $g : \mathbf{R}^n \to \mathbf{R}$ which is infinitely differentiable in any interval, there exists $L \le N$ such that for $N$ arbitrary distinct samples $(\mathbf{x}_k, \mathbf{t}_k)$, where $\mathbf{x}_k \in \mathbf{R}^n$ and $\mathbf{t}_k \in \mathbf{R}^m$, for any $\mathbf{a}_i$ and $b_i$ randomly chosen from any intervals of $\mathbf{R}^n$ and $\mathbf{R}$, respectively, according to any continuous probability distribution, then with probability one, $\|\mathbf{H}_{N \times L} \boldsymbol{\beta}_{L \times m} - \mathbf{T}_{N \times m}\| < \varepsilon.$ □

## 3 Generalized Single-Hidden Layer Feedforward Networks

We are now in a position to extend standard SLFNs to generalized SLFNs (GSLFNs) by defining the output weights as polynomial functions of input variables (i.e., $\boldsymbol{\beta} \triangleq \boldsymbol{\beta}(\mathbf{x})$) as follows:

$$\beta_{ij}(\mathbf{x}) = w_{ij}^{(0)} + w_{ij}^{(1)} x_1 + \cdots + w_{ij}^{(n)} x_n, \ i = 1, 2, \cdots, L, \ j = 1, 2, \cdots, m \quad (5)$$

where $w_{ij}^{(0)}, w_{ij}^{(1)}, \cdots, w_{ij}^{(n)}$ are corresponding weights for input variables. Accordingly, if the outputs of the GSLFN estimate the targets with zero errors, we obtain the following compact formulation,

$$\mathbf{GW} = \mathbf{T} \quad (6)$$

where,

$$\mathbf{G}(\mathbf{a}_1, \ldots, \mathbf{a}_L, b_1, \ldots, b_L, \mathbf{x}_1, \ldots, \mathbf{x}_N)$$
$$= \begin{bmatrix} g(\mathbf{a}_1, b_1, \mathbf{x}_1)\bar{\mathbf{x}}_1^{\mathrm{T}} & \cdots & g(\mathbf{a}_L, b_L, \mathbf{x}_1)\bar{\mathbf{x}}_1^{\mathrm{T}} \\ \vdots & \ddots & \vdots \\ g(\mathbf{a}_1, b_1, \mathbf{x}_N)\bar{\mathbf{x}}_N^{\mathrm{T}} & \cdots & g(\mathbf{a}_L, b_L, \mathbf{x}_N)\bar{\mathbf{x}}_N^{\mathrm{T}} \end{bmatrix}_{N \times L(n+1)} \quad (7)$$

$$\mathbf{W} = \begin{bmatrix} \mathbf{w}_{11} & \mathbf{w}_{12} & \cdots & \mathbf{w}_{1m} \\ \mathbf{w}_{21} & \mathbf{w}_{22} & \cdots & \mathbf{w}_{2m} \\ \vdots & \cdots & \ddots & \vdots \\ \mathbf{w}_{L1} & \mathbf{w}_{L2} & \cdots & \mathbf{w}_{Lm} \end{bmatrix}_{L(n+1) \times m} \quad (8)$$

$$\bar{\mathbf{x}}_k = [1, x_{k1}, x_{k2}, \cdots, x_{kn}]^{\mathrm{T}}, \ \mathbf{w}_{ij} = \left[ w_{ij}^{(0)}, w_{ij}^{(1)}, \cdots, w_{ij}^{(n)} \right]^{\mathrm{T}} \quad (9)$$

Here, $\mathbf{G}$ is referred to be the augmented hidden-layer output matrix of the GSLFN consisting of $N \times L$ blocks, whereby the $k$th block $\mathbf{G}_{ki}$ is the product of the $i$th hidden node output with respect to the $k$th input vector, i.e., $g_i(\mathbf{x}_k)$, and the corresponding augmented input vector $\bar{\mathbf{x}}_k^{\mathrm{T}}$, and thereby constituting a $[N \times L(n+1)]$-dimension matrix. Accordingly, $\mathbf{W}$ is the output coefficient matrix consisting of $L \times m$ blocks, whereby the block $\mathbf{w}_{ij}$ in the $i$th-row-$j$th-column position corresponds to the coefficient vector of the output weight connecting the $i$th hidden node to the $j$th output node, and thereby contributing a $[L(n+1) \times m]$-dimension matrix.

## 4   Main Results

Furthermore, one can obtain the main results on the approximation capabilities of the GSLFN as follows.

**Theorem 1.** Given a GSLFN with $N$ hidden nodes and activation function $g : \mathbf{R}^n \to \mathbf{R}$ which is infinitely differentiable in any interval, for $N$ arbitrary distinct samples $(\mathbf{x}_k, \mathbf{t}_k)$, where $\mathbf{x}_k \in \mathbf{R}^n$ and $\mathbf{t}_k \in \mathbf{R}^m$, for any $\mathbf{a}_i$ and $b_i$ randomly chosen from any intervals of $\mathbf{R}^n$ and $\mathbf{R}$, respectively, according to any continuous probability distribution, then with probability one, the rank of the augmented hidden-layer output matrix $\mathbf{G}$ for the GSLFN satisfies $N \leq rank(\mathbf{G}) \leq N(n+1)$, and there exists at least one coefficient matrix $\mathbf{W}$ such that $\|\mathbf{GW} - \mathbf{T}\| = 0$.                                      □

*Proof.* By Lemma 1, for any $\mathbf{a}_i$ and $b_i$ randomly chosen from any intervals of $\mathbf{R}^n$ and $\mathbf{R}$, respectively, according to any continuous probability distribution, then with probability one, the hidden-layer output matrix $\mathbf{H}$ satisfies $rank(\mathbf{H}) = N$. In addition, the augmented hidden-layer output matrix $\mathbf{G}$ can be represented by,

$$
\mathbf{G} = \mathbf{P} \begin{bmatrix} g(\mathbf{a}_1, b_1, \mathbf{x}_1) & \cdots & g(\mathbf{a}_L, b_L, \mathbf{x}_1) & g(\mathbf{a}_1, b_1, \mathbf{x}_1)\mathbf{x}_1^{\mathrm{T}} & \cdots & g(\mathbf{a}_L, b_L, \mathbf{x}_1)\mathbf{x}_1^{\mathrm{T}} \\ \vdots & \ddots & \vdots & \vdots & \ddots & \vdots \\ g(\mathbf{a}_1, b_1, \mathbf{x}_N) & \cdots & g(\mathbf{a}_L, b_L, \mathbf{x}_N) & g(\mathbf{a}_1, b_1, \mathbf{x}_N)\mathbf{x}_N^{\mathrm{T}} & \cdots & g(\mathbf{a}_L, b_L, \mathbf{x}_N)\mathbf{x}_N^{\mathrm{T}} \end{bmatrix} \mathbf{Q}
$$

where, $\mathbf{P} \in \mathbf{R}^{N \times N}$ and $\mathbf{Q} \in \mathbf{R}^{N(n+1) \times N(n+1)}$ are elementary transformation matrices. It follows, with probability one, that $N \leq rank(\mathbf{G}) \leq N(n+1)$. In this case, eqn. (6) becomes an under-determined problem with $N$ independent equations since the number of equations is larger than that of unknown parameters. As a consequence, there exists at least one solution for the coefficient matrix $\mathbf{W}$ in (6). This concludes the proof.                                      □

Similar to Lemma 2, for the GSLFN, we can straightforward obtain the following result.

**Theorem 2.** Given any small positive value $\varepsilon > 0$ and activation function $g : \mathbf{R}^n \to \mathbf{R}$ which is infinitely differentiable in any interval, there exists $L \leq N$ such that for $N$ arbitrary distinct samples $(\mathbf{x}_k, \mathbf{t}_k)$, where $\mathbf{x}_k \in \mathbf{R}^n$ and $\mathbf{t}_k \in \mathbf{R}^m$, for any $\mathbf{a}_i$ and $b_i$ randomly chosen from any intervals of $\mathbf{R}^n$ and $\mathbf{R}$, respectively, according to any continuous probability distribution, then with probability one, $\|\mathbf{G}_{N \times L(n+1)}\mathbf{W}_{L(n+1) \times m} - \mathbf{T}_{N \times m}\| < \varepsilon$.                                      □

*Proof.* Following the proof of Theorem 1, the rank of $\mathbf{G}$ would be less than $L$ with high probability. Accordingly, the columns of $\mathbf{G}$ might belong to a subspace of dimension no more than $N$. In other words, the independent equations of eqn. (6) would be no more than $N$, and thereby resulting in an under-determined equation with partial targets not be exactly estimated. Fortunately, the tuning of coefficient matrix $\mathbf{W}$ can make the approximation error infinitely small especially when $L = N$. The proof is completed.                                      □

Furthermore, the GSLFN using polynomial functions as output weights features significant characteristics as follows.

**Theorem 3.** Given a GSLFN with $L$ hidden nodes and activation function $g : \mathbf{R}^n \to \mathbf{R}$ which is infinitely differentiable in any interval, for $N \leq L(n + 1)$ arbitrary distinct samples $(\mathbf{x}_k, \mathbf{t}_k)$, where $\mathbf{x}_k \in \mathbf{R}^n$, $\mathbf{t}_k \in \mathbf{R}^m$ and $x_{kj} \neq x_{k'j'}, \exists k \neq k', j \neq j'$, for any $\mathbf{a}_i$ and $b_i$ randomly chosen from any intervals of $\mathbf{R}^n$ and $\mathbf{R}$, respectively, according to any continuous probability distribution, then with probability one, the rank of the augmented hidden-layer output matrix $\mathbf{G}$ satisfies,

$$rank(\mathbf{G}) = N \tag{10}$$

and there exists at least one coefficient matrix $\mathbf{W}$ such that $\|\mathbf{GW} - \mathbf{T}\| = 0$. □

*Proof.* The augmented hidden-layer output matrix $\mathbf{G}$ consists of $L$ sub-matrices $\mathbf{G}_i, i = 1, 2, \cdots, L$, and thereby totally contributing to $L(n + 1)$ column vectors given by,

$$\mathbf{G} = [\mathbf{G}_1, \mathbf{G}_2, \cdots, \mathbf{G}_L]_{N \times L(n+1)}$$

$$\mathbf{G}_i = \begin{bmatrix} g(\mathbf{a}_i, b_i, \mathbf{x}_1) & g(\mathbf{a}_i, b_i, \mathbf{x}_1)x_{11} & \cdots & g(\mathbf{a}_i, b_i, \mathbf{x}_1)x_{1n} \\ \vdots & \vdots & \ddots & \vdots \\ g(\mathbf{a}_i, b_i, \mathbf{x}_N) & g(\mathbf{a}_i, b_i, \mathbf{x}_N)x_{N1} & \cdots & g(\mathbf{a}_i, b_i, \mathbf{x}_N)x_{Nn} \end{bmatrix}$$

Note that $\mathbf{a}_i$ are randomly generated based on a continuous probability distribution, we can assume that $\mathbf{a}_i \cdot \mathbf{x}_k \neq \mathbf{a}_i \cdot \mathbf{x}_{k'}$ for all $k \neq k'$. Consider the $j$th column of the $i$th matrix block $\mathbf{G}_i$, i.e.,

$$\mathbf{g}(b_i, \mathbf{x}^j) = \mathbf{c}(b_i) \odot \mathbf{x}^j \tag{11}$$

where $\odot$ denotes the Hadamard product, and

$$\mathbf{c}(b_i) = [g(b_i + d_{i1}), \cdots, g(b_i + d_{iN})]^\mathrm{T}, \quad \mathbf{x}^j = [x_{1j}, \cdots, x_{Nj}]^\mathrm{T}, \quad j = 0, 1, \cdots, n$$

where $d_{ik} = \mathbf{a}_i \cdot \mathbf{x}_k$, $b_i \in (a, b) \subset \mathbf{R}$ and $x_{k0} = 1, k = 1, 2, \cdots, N$.

It can be proved by contradiction that vectors $\mathbf{g}$ does not belong to any subspace whose dimension is less than $N$. Suppose that $\mathbf{g}$ belongs to a subspace of dimension $N - 1$. Then there exists a vector $\boldsymbol{\alpha} \neq \mathbf{0}$ which is orthogonal to this subspace, i.e.,

$$\langle \boldsymbol{\alpha}, \mathbf{g}(b_i, \mathbf{x}^j) - \mathbf{g}(a, \mathbf{x}^{j'}) \rangle = 0 \tag{12}$$

Note that, for $N \leq L(n + 1)$ arbitrary distinct samples $(\mathbf{x}_k, \mathbf{t}_k) \in \mathbf{R}^n \times \mathbf{R}^m$, $x_{kj} \neq x_{k'j}, \exists k \neq k', j \neq j'$. For the cases $j = j' = 0$ and $j \neq j'$ in (12), we can simply set $b_i + d_{ij} \neq a + d_{ij'}$ and $b_i + d_{ij} = a + d_{ij'}$, respectively. As a consequence, it holds that

$$\mathbf{g}(b_i, \mathbf{x}^j) - \mathbf{g}(a, \mathbf{x}^{j'}) = \begin{cases} (\mathbf{c}(b_i) - \mathbf{c}(a)) \odot \mathbf{x}^j, & j = j' = 0 \\ \mathbf{c}(b_i) \odot \left( \mathbf{x}^j - \mathbf{x}^{j'} \right), & j \neq j' \end{cases}$$

It follows that $\mathbf{g}(b_i, \mathbf{x}^j) - \mathbf{g}(a, \mathbf{x}^{j'}) \neq \mathbf{0}$ is guaranteed since $\mathbf{x}^j \neq \mathbf{x}^{j'}, \forall j, j' \neq 0$ and $\mathbf{x}^0 = 1$.

Using (11), eqn. (12) can be further written as,

$$\alpha_1 g(b_i + d_{i1})x_{1j} + \alpha_2 g(b_i + d_{i2})x_{2j} + \cdots + \alpha_N g(b_i + d_{iN})x_{Nj} - \boldsymbol{\alpha} \cdot \mathbf{g}(a, \mathbf{x}^{j'}) = 0$$

Without loss of generality, we assume $\alpha_N \neq 0$ and obtain,

$$g(b_i + d_{iN})x_{Nj} = -\sum_{k=1}^{N-1} \gamma_k g(b_i + d_{ik})x_{kj} + \boldsymbol{\alpha}' \cdot \left( \mathbf{c}(a) \odot \mathbf{x}^{j'} \right)$$

where $\gamma_k = \alpha_k/\alpha_N$, $k = 1, 2, \cdots, N-1$ and $\boldsymbol{\alpha}' = \boldsymbol{\alpha}/\alpha_N$. Since the activation function $g(.)$ is infinitely differentiable in any interval, we have

$$g^{(l)}(b_i + d_{iN})x_{Nj} = -\sum_{k=1}^{N-1} \gamma_k g^{(l)}(b_i + d_{ik})x_{kj}, \ l = 1, 2, \cdots, N, N+1, \cdots$$

where $g^{(l)}$ is the $l$th derivative of function $g$ of $b_i$. However, there are only $N-1$ free coefficients: $\gamma_1, \cdots, \gamma_{N-1}$ for the derived more than $N-1$ linear equations, this is contradictory. Thus, vector $g$ does not belong to any subspace whose dimension is less than $N$.

As a consequence, from any interval $(a, b)$ it is possible to randomly choose $L' = ceil(N/(n+1))$ bias values $b_1, \cdots, b_{L'}$ for the $L'$ hidden nodes such that $N$ vectors $\mathbf{g}(b_i, \mathbf{x}^j), i = 1, 2, \cdots, L', j = 0, 1, \cdots, n$ span $\mathbf{R}^N$. This means that for any input weight vectors $\mathbf{a}_i$ and bias values $b_i$ chosen from any intervals of $\mathbf{R}^n$ and $\mathbf{R}$, respectively, according to any continuous probability distribution, then with probability one, the column vectors of $\mathbf{G}$ can be made row full-rank, i.e., $rank(\mathbf{G}) = N$ if $N \leq L(n+1)$.

Accordingly, the number of independent equations in eqn. (6) is $N$, and thereby resulting in a well- or under-determined problem. Hence, there exists at least one solution for $\mathbf{W}$ in (6). This concludes the proof.    □

**Theorem 4.** Given any small positive value $\varepsilon > 0$ and activation function $g : \mathbf{R}^n \rightarrow \mathbf{R}$ which is infinitely differentiable in any interval, there exists $L \leq N/(n+1)$ such that for $N$ arbitrary distinct samples $(\mathbf{x}_k, \mathbf{t}_k) \in \mathbf{R}^n \times \mathbf{R}^m$ where $x_{kj} \neq x_{k'j'}, \exists k \neq k', j \neq j'$, for any $\mathbf{a}_i$ and $b_i$ randomly chosen from any intervals of $\mathbf{R}^n$ and $\mathbf{R}$, respectively, according to any continuous probability distribution, then with probability one, $\|\mathbf{G}_{N \times L(n+1)} \mathbf{W}_{L(n+1) \times m} - \mathbf{T}_{N \times m}\| < \varepsilon$.    □

*Proof.* Following the proof of Theorem 3, the rank of $\mathbf{G}$ would no more than $N$ with high probability. In this case, (6) would be an over-determined equation. It means there might not exist exact solutions for $\mathbf{W}$ in (6). Alternatively, given any small error $\varepsilon > 0$ and the GSLFN with $L \leq N(n+1)$ hidden nodes, fine tuning of coefficient matrix $\mathbf{W}$ can make the estimation error less than $\varepsilon$.    □

## 5   Conclusions

This paper extends the standard single-hidden layer feedforward networks (SLFNs) to generalized SLFNs (GSLFNs) by employing the polynomial functions of inputs as output weights. Accordingly, we have rigorously proved the significant characteristics of the GSLFN as follows. On the one hand, similar to the SLFN, the GSLFN with at most $N$ hidden nodes using any infinitely differentiable activation functions can exactly learn $N$ distinct observations. On the other hand, for distinct $n$-input $m$-output observations with different data in each input dimension, the GSLFN features much higher approximation capability such that the GSLFN with only $N/(n+1)$ hidden nodes using any infinitely differentiable activation functions can exactly learn $N$ distinct observations. The number of hidden nodes in the GSLFN can be dramatically reduced by using polynomials as output weights, especially for high-dimension regressions and classifications.

## References

1. Wang, N., Er, M.J., Meng, X.: A Fast and Accurate Online Self-Organizing Scheme for Parsimonious Fuzzy Neural Networks. Neurocomput. 72, 3818–3829 (2009)
2. Wang, N., Er, M.J., Meng, X.Y., Li, X.: An Online Self-Organizing Scheme for Parsimonious and Accurate Fuzzy Neural Networks. Int. J. Neural Syst. 20, 389–403 (2010)
3. Wang, N.: A Generalized Ellipsoidal Basis Function Based Online Self-Constructing Fuzzy Neural Network. Neural Process. Lett. 34, 13–37 (2011)
4. Tamura, S., Tateishi, M.: Capabilities of a Four-Layered Feedforward Neural Network: Four Layers Versus Three. IEEE Trans. Neural Netw. 8, 251–255 (1997)
5. Huang, G.-B., Babri, H.A.: Upper Bounds on the Number of Hidden Neurons in Feedforward Networks with Arbitrary Bounded Nonlinear Activation Functions. IEEE Trans. Neural Netw. 9, 224–229 (1998)
6. Huang, G.-B., Zhu, Q.-Y., Siew, C.-K.: Extreme Learning Machine: Theory and Applications. Neurocomput. 70, 489–501 (2006)
7. Huang, G.-B., Chen, L., Siew, C.K.: Universal Approximation Using Incremental Constructive Feedforward Networks with Random Hidden Nodes. IEEE Trans. Neural Netw. 17, 879–892 (2006)
8. Ferrari, S., Stengel, R.F.: Smooth Function Approximation Using Neural Networks. IEEE Trans. Neural Netw. 16, 24–38 (2005)
9. Teoh, E.J., Tan, K.C., Xiang, C.: Estimating the Number of Hidden Neurons in a Feedforward Network Using the Singular Value Decomposition. IEEE Trans. Neural Netw. 17, 1623–1629 (2006)
10. Huynh, H.T., Won, Y., Kim, J.J.: An Improvement of Extreme Learning Machine for Compact Single-Hidden-Layer Feedforward Neural Networks. Int. J. Neural Syst. 18, 433–441 (2008)

# An Approach for Designing Neural Cryptography

Nankun Mu and Xiaofeng Liao

College of Computer Science, Chongqing University, Chongqing 400044, China
xfliao@cqu.edu.cn

**Abstract.** Neural cryptography is widely considered as a novel method of exchanging secret key between two neural networks through mutual learning. This paper puts forward a generalized architecture to provide an approach to designing novel neural cryptography. Meanwhile, by taking an in-depth investigation on the security of neural cryptography, a heuristic rule is proposed. These results can effectively guide us to designing secure neural cryptography. Finally, an example is given to demonstrate the effectiveness of the proposed structure and the heuristic rule.

**Keywords:** Neural synchronization, Neural cryptography, Generalized architecture, Heuristic rule.

## 1 Introduction

Public key exchange protocols (PKEPs) have played an important role in modern cryptography since initially introduced by Diffie and Hellman [1]. Actually, PKEPs enable two parties, named A and B, to share a common secret key on public channel, while an attacker E cannot retrieve the key even equipped with access to the communication channel. Then the key can be utilized to deal with some cryptographic problems, such as privacy, authentication, data integrity to name a few. In particular, PKEPs based on number theory have been extensively studied [1–4]. However, it has been recognized that neural synchronization is able to achieve the same objective bringing about what is known as *neural cryptography* [5]. The mechanism behind neural cryptography is similar to secret key agreement though public discussion [6]. In particular, benefited from the absence of large-scale computation, which is highly expected by the small-scale embedded systems [7,8], neural cryptography has gained considerable attention and has also been an increasingly important research. Besides, by substituting the neural networks, other synchronization systems, such as chaotic maps and coupled lasers [9–11], can also can be exploited in constructing the similar PKEPs.

So far, several models of neural cryptography have been proposed, typically permutation parity machine (PPM) [14,15], TPM [5] and TCM [16]. Meanwhile a probabilistic attack algorithm for PPM with a high success rate has also been presented [17] and the TCM does not work effectively [16]. In addition, the TPM containing one or two hidden units ($K = 1, 2$) has been also attacked successfully [18,19]. Besides, only TPM with the fixed structure ($K = 3$) can resist kinds

C. Guo, Z.-G. Hou, and Z. Zeng (Eds.): ISNN 2013, Part I, LNCS 7951, pp. 99–108, 2013.

of attacks depending on increasing the synaptic depth of its networks [20–22]. Then applicability and security of TPM have been extensively studied [23–29]. In [23], three most common learning rules are analyzed in detail, including Hebbian learning rule, Anti-Hebbian learning rule, and Random-walk learning rule. The dynamic process of synchronization in TPM has been carefully profiled [24]. Moreover, the model of the classical ruin problem is used to examine the average synchronization time of TPM [25]. A feedback mechanism to guarantee relevant input vectors partially unavailable to the attacker [26], and a queries mechanism is further introduced [27], which speeds up the synchronization through generating input vectors by means of queries instead of randomizer. Recently, an error prediction mechanism based on the algorithm called "Do not Trust My Partner" (DTMP) is presented [28]. It largely relies on that one party sends some erroneous bits while the other is capable of predicting and correcting corresponding errors. Meanwhile, Four kinds of attack algorithms are also experimentally investigated in detail, i.e., simple attack [5], geometric attack [29], majority attack [12], and genetic attack [23]. In spite of numerous studies on neural cryptography, there has been no secure neural cryptography other than TPM ($K = 3$).

Motivated by the situation in this area, the objective of this paper is to establish an approach to addressing this problem, with which we can formulate some better instances. Meanwhile, an example is used to demonstrate the usefulness of the proposed design methods.

## 2    Generalized Architecture and Mutual Learning

Shown in Fig.1 is principle graph of the generic structure named TSCM. It can be regarded as a tree-connected networks consisting of three layers. More precisely, a TSCM has $K$ hidden units, $N$ input neurons for each hidden unit, and a unique output neuron connecting all hidden units (we refer to the output neuron as state

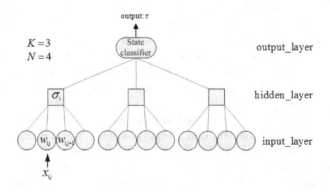

**Fig. 1.** The architecture with $K = 3, N = 4$

classifier). Each hidden unit works like an independent perceptron and elements of the weighted vector are integral numbers:

$$w_{i,j} \in \{-L, \ldots, 0, \ldots, +L\}. \tag{1}$$

Here, $L$ represents synaptic depth of the networks; the index $i = 1, \ldots, K$ denotes the $i$-th hidden unit of the networks and $j = 1, \ldots, N$ the $j$-th input neuron for each hidden unit. Meanwhile input vector is indicated by $\mathbf{x}$ and its elements are binary, i.e.,

$$x_{i,j} \in \{-1, +1\}. \tag{2}$$

When TSCM receives $\mathbf{x}$, the value of the $i$-th hidden unit is defined by as:

$$y = \begin{cases} \sigma_i = \text{sgn}(h_i); \\ h_i = \frac{1}{\sqrt{N}} \mathbf{w}_i \mathbf{x}_i = \frac{1}{\sqrt{N}} \sum_{i=1}^{N} w_{i,j} x_{i,j}. \end{cases} \tag{3}$$

In order to ensure $\sigma_i$ binary, $h_i = 0$ is mapped to $\sigma_i = -1$. Afterwards, the state classifier generates an output, denoted by $\tau$, according to the state of vector $\sigma$, i.e.,

$$\tau = \text{StateClassifier}(\sigma). \tag{4}$$

**Table 1.** Classification of TPM($K = 3$) and TCM($K = 3$)

| Name | $\tau$ | state vector $\sigma$ | Name | $\tau$ | state vector $\sigma$ |
|------|--------|------------------------|------|--------|------------------------|
| TPM | $+1$ | $(+1, +1, +1); (+1, -1, -1);$ $(-1, +1, -1); (-1, -1, +1)$ | TCM | $+1$ | $(+1, +1, +1); (-1, +1, +1);$ $(+1, -1, +1); (+1, +1, -1)$ |
| | $-1$ | $(-1, -1, -1); (-1, +1, +1);$ $(+1, -1, +1); (+1, +1, -1)$ | | $-1$ | $(-1, -1, -1); (+1, -1, -1);$ $(+1, -1, +1); (+1, +1, -1)$ |

Significantly, TSCM generalizes and unifies the existing structures, i.e., TPM and TCM. When the state classifier is defined as Table 1, it is observed that TCM and TPM are two special cases of TSCM.

The mutual learning algorithm of TSCM is illustrated as follows:

1. The two parties A and B start with a uniform TSCM-based networks and randomly choose weighted vectors $\mathbf{w}^A$ and $\mathbf{w}^B$, which are kept secret. This can effectively guarantee the uncorrelation between $\mathbf{w}^A$ and $\mathbf{w}^B$ at the beginning.
2. In each learning step, the two parties receive a common input vector $\mathbf{x}$ at the same time. Upon receiving $\mathbf{x}$, A and B work out $\sigma^A$ and $\sigma^B$, respectively. And then $\tau^A$ and $\tau^B$ can also be calculated by using $\sigma^A$ and $\sigma^B$, respectively. This computation process is well defined in the above subsection. Afterwards, $\tau^A$ and $\tau^B$ are exchanged with each other on the public channel while $\sigma^A$ and $\sigma^B$ are secretly kept.

3. All weights are iteratively adjusted using one of the following learning rules:
   (a) *Hebbian learning rule:*

$$w_{i,j}^{A/B} = g\big(w_{i,j}^{A/B} + x_{i,j}\tau^{A/B}\Theta(\sigma_i\tau^{A/B})\Theta(\tau^A\tau^B)\big). \qquad (5)$$

   (b) *Anti-Hebbian learning rule:*

$$w_{i,j}^{A/B} = g\big(w_{i,j}^{A/B} - x_{i,j}\tau^{A/B}\Theta(\sigma_i\tau^{A/B})\Theta(\tau^A\tau^B)\big). \qquad (6)$$

   (c) *Random-walk learning rule:*

$$w_{i,j}^{A/B} = g\big(w_{i,j}^{A/B} + x_{i,j}\Theta(\sigma_i\tau^{A/B})\Theta(\tau^A\tau^B)\big). \qquad (7)$$

Here $g(w)$ is introduced to ensure that each component of the weighted vector $\mathbf{w}^{A/B}$ resides within the range $[-L, +L]$. It is determined by

$$g(w) = \begin{cases} \mathrm{sgn}(w)L, & \text{for}\,|w| > L; \\ w, & \text{otherwise.} \end{cases} \qquad (8)$$

4. Repeating procedure 2 and 3 until synchronization ($\mathbf{w}^A = \mathbf{w}^B$) is achieved. The final identical weighted vector $\mathbf{w}^{A/B}$ can be used as the common secret key between A and B.

In whole synchronization process, the state vector $\sigma^{A/B}$ is consistently inaccessible to any others. Then two possible real update behavior can be defined:

- *An attractive step* ($\tau^A = \tau^B = \sigma_i^{A/B}$): The weighted vectors of both corresponding hidden units increase or decrease in the same direction. And a series of such steps leads to synchronization eventually.
- *A repulsive step* ($\tau^A = \tau^B, \sigma_i^A \neq \sigma_i^B$): Only one weighted vector of the corresponding hidden units in A or B update while B or A remains unchanged. A sequence of these repulsive steps may lower synchronization speed.

## 3    Analysis on Security

In this section, we analyze the security of neural cryptography by investigating the neural synchronization process. Meanwhile, TPM and TCM are illustrated to propose the heuristic rule. It is worth paying special attention that TPM($K = 3$) is a security instance and TCM is insecure.

Synchronization in neural cryptography is a Markovian process composed of finite learning steps. The level of synchronization is indicated by the normalized overlap between the two corresponding hidden units:

$$\rho = \frac{\mathbf{w}_i^A\mathbf{w}_i^B}{\sqrt{\mathbf{w}_i^A\mathbf{w}_i^A}\sqrt{\mathbf{w}_i^B\mathbf{w}_i^B}}, \rho \in [0, +1]. \qquad (9)$$

At the beginning of synchronization, $\rho$ locates approximately at $\rho = 0$, because of the random initial weighted vectors. Through a series of learning steps, synchronization is achieved and $\rho$ is stable at $\rho = +1$. In these learning steps, it is possible that the two corresponding hidden units have different $\sigma$. The probability of this event is defined by $\varepsilon$, which is known as the generation error

$$\varepsilon = \frac{1}{\pi} \arccos(\rho), \varepsilon \in [0, +1]. \tag{10}$$

For discussing the dynamics of $\rho$ in the synchronization process, it is very necessary to introduce

$$\langle \Delta\rho(\rho) \rangle = P_a(\rho)\langle \Delta\rho_a(\rho) \rangle + P_r(\rho)\langle \Delta\rho_r(\rho) \rangle. \tag{11}$$

Here, $\langle \Delta\rho(\rho) \rangle$ denotes the average change size of $\rho$ in each learning step; $\langle \Delta\rho_a(\rho) \rangle$ ($\langle \Delta\rho_r(\rho) \rangle$) represents the average change size of $\rho$ in each attractive (repulsive) step; $P_a(\rho)$ ($P_r(\rho)$) indicates the probability of the event that an attractive (repulsive) step occurs between one pair of corresponding hidden units.

In particular, a secure neural cryptography need to satisfy: as the synaptic depth $L$ grows, the averaged synchronization time by mutual learning for A and B grows at a polynomial rate; meanwhile the averaged synchronization time by unidirectional learning for E grows at an exponential rate. Remarkably, $\langle \Delta\rho(\rho) \rangle > 0$ in $(0,1)$ enables synchronization time grows at a polynomial rate with increasing $L$. So we can take two essential conditions for a secure neural cryptography: i) in synchronization process for A and B,

$$P_a(\rho)\langle \Delta\rho_a(\rho) \rangle + P_r(\rho)\langle \Delta\rho_r(\rho) \rangle > 0, \rho \in (0,1) \tag{12}$$

is true; ii) for E, there exists a region $G$ in (0,1), such that

$$P_a(\rho)\langle \Delta\rho_a(\rho) \rangle + P_r(\rho)\langle \Delta\rho_r(\rho) \rangle < 0, \rho \in G. \tag{13}$$

However, the following analysis mainly focuses on condition i). Clearly, the inequality (12) can be reduced to:

$$-\frac{\langle \Delta\rho_a(\rho) \rangle}{\langle \Delta\rho_r(\rho) \rangle} > \frac{P_r(\rho)}{P_a(\rho)}, \rho \in (0,1). \tag{14}$$

For convenience of analysis, the inequality (14) can be rewritten as:

$$\begin{cases} U(\rho) > R(\rho), \\ U(\rho) = -\frac{\langle \Delta\rho_a(\rho) \rangle}{\langle \Delta\rho_r(\rho) \rangle}, \\ R(\rho) = \frac{P_r(\rho)}{P_a(\rho)}, \\ \rho \in (0,1). \end{cases} \tag{15}$$

Substituting (10) to $R(\rho)$, we have

$$\begin{cases} U(\rho) > R(\varepsilon), \\ U(\rho) = -\frac{\langle \Delta\rho_a(\rho) \rangle}{\langle \Delta\rho_r(\rho) \rangle}, \\ R(\varepsilon) = \frac{P_r(\varepsilon)}{P_a(\varepsilon)}, \\ \rho \in (0,1), \varepsilon \in (0,0.5). \end{cases} \tag{16}$$

In [19], the authors have noted that $U(\rho)$ is only dependent on the equations of motion, which means $U(\rho)$ is constant. On the other hand, we can easily obtain $R(\rho)$ on the basis of $P_a(\varepsilon)$ and $P_r(\varepsilon)$. While $P_a(\varepsilon)$ and $P_r(\varepsilon)$ are described by formula as:

$$\begin{cases} P_a(\varepsilon) = P(\tau^{A/B} = \sigma^A = \sigma^A | \tau^A = \tau^B); \\ P_r(\varepsilon) = P(\sigma^A \neq \sigma^A | \tau^A = \tau^B); \\ P_u(\varepsilon) = P(\tau^A = \tau^B). \end{cases} \tag{17}$$

Here, $P_u(\varepsilon)$ denotes the probability of an agreement on outputs. Then we describe the computational procedure of $P_u(\varepsilon)$ in $\text{TPM}(K = 3)$. The following equalities always hold:

$$P_u^{\text{TPM}}(k, \varepsilon) \begin{cases} P_u^{\text{TPM}}(0, \varepsilon) = \binom{3}{0}\varepsilon^0(1-\varepsilon)^3, \\ P_u^{\text{TPM}}(1, \varepsilon) = 0\binom{3}{1}\varepsilon^1(1-\varepsilon)^2, \\ P_u^{\text{TPM}}(2, \varepsilon) = \binom{3}{2}\varepsilon^2(1-\varepsilon)^1, \\ P_u^{\text{TPM}}(3, \varepsilon) = 0\binom{3}{3}\varepsilon^3(1-\varepsilon)^0. \end{cases} \tag{18}$$

Here, $P_u^{\text{TPM}}(k, \varepsilon)$ denotes the probability of an update step in TPM as $k$ pairs of corresponding hidden units disagree. According to (17) and (18), we can get:

$$P_a^{\text{TPM}}(k, \varepsilon) \begin{cases} P_a^{\text{TPM}}(0, \varepsilon) = \frac{\frac{1}{2}\binom{3}{0}\varepsilon^0(1-\varepsilon)^3}{P_u^{\text{TPM}}(k,\varepsilon)}, \\ P_a^{\text{TPM}}(1, \varepsilon) = 0, \\ P_a^{\text{TPM}}(2, \varepsilon) = \frac{\frac{1}{6}\binom{3}{2}\varepsilon^2(1-\varepsilon)^1}{P_u^{\text{TPM}}(k,\varepsilon)}, \\ P_a^{\text{TPM}}(3, \varepsilon) = 0. \end{cases} \tag{19}$$

and

$$P_r^{\text{TPM}}(k, \varepsilon) \begin{cases} P_r^{\text{TPM}}(0, \varepsilon) = \frac{0\binom{3}{0}\varepsilon^0(1-\varepsilon)^3}{P_u^{\text{TPM}}(k,\varepsilon)}, \\ P_r^{\text{TPM}}(1, \varepsilon) = 0, \\ P_r^{\text{TPM}}(2, \varepsilon) = \frac{\frac{2}{3}\binom{3}{2}\varepsilon^2(1-\varepsilon)^1}{P_u^{\text{TPM}}(k,\varepsilon)}, \\ P_r^{\text{TPM}}(3, \varepsilon) = 0. \end{cases} \tag{20}$$

Sum them up to derive the $P_a^{\text{TPM}}(\varepsilon)$ and $P_r^{\text{TPM}}(\varepsilon)$:

$$\begin{cases} P_a^{\text{TPM}}(\varepsilon) = \frac{\frac{1}{2}(1-\varepsilon)^3 + \frac{1}{2}\varepsilon^2(1-\varepsilon)}{P_u^{\text{TPM}}(\varepsilon)}; \\ P_r^{\text{TPM}}(\varepsilon) = \frac{2\varepsilon^2(1-\varepsilon)}{P_u^{\text{TPM}}(\varepsilon)}. \end{cases} \tag{21}$$

Then, it follows that

$$R^{\text{TPM}}(\varepsilon) = \frac{P_r^{\text{TPM}}(\varepsilon)}{P_a^{\text{TPM}}(\varepsilon)} = \frac{4\varepsilon^2}{(1-\varepsilon)^2 + \varepsilon^2}. \tag{22}$$

Similarly, we also can obtain:

$$R^{\text{TCM}}(\varepsilon) = \frac{\frac{1}{2}\varepsilon(1-\varepsilon) + \varepsilon^2}{\frac{3}{4}(1-\varepsilon)^2 + \varepsilon(1-\varepsilon) + \frac{1}{2}\varepsilon^2}. \tag{23}$$

However, the equations of motion in TPM and TCM is the same, and this means $U^{\mathrm{TPM}}(\rho) = U^{\mathrm{TCM}}(\rho)$. So the difference in security between TPM and TCM is due to the $R(\varepsilon)$. Simulation experiment shows that $\langle \Delta\rho(\rho) \rangle < 0$ is most likely to occur nearly synchronization ($\rho \to 1, \varepsilon \to 0$). Closing to synchronization, $R^{\mathrm{TPM}}(\varepsilon)$ and $R^{\mathrm{TCM}}(\varepsilon)$ can be represented by two approximate value:

$$R^{\mathrm{TPM}}(\varepsilon) \sim 4\varepsilon^2 \tag{24}$$

and

$$R^{\mathrm{TCM}}(\varepsilon) \sim \frac{2}{3}\varepsilon. \tag{25}$$

Then, we can derive the following:

$$R^{\mathrm{TPM}}(\varepsilon) \sim 4\varepsilon^2 < U^{\mathrm{TCM/TPM}}(\rho) < \frac{2}{3}\varepsilon \sim R^{\mathrm{TCM}}(\varepsilon), \rho \to 1, \varepsilon \to 0. \tag{26}$$

Because $R(\varepsilon)$ is closely related to the state classifier, we make an assumption that some factor of the state classifier can impact the security of TSCM.

**Definition 1:** For two different state vectors in one class, i.e., $\sigma$ and $\sigma'$ in $c_1$, the Hamming Distance (HD) between them is defined by: $\mathrm{HD}^{c_1} = \mathrm{HD}(\sigma, \sigma') = \sum_{i=1}^{K}(\sigma_i \oplus \sigma_i')$. And the Smallest Hamming Distance (SHD) in class $c_1$ : $\mathrm{SHD}^{c_1} = \min\{\mathrm{HD}_1^{c_1}, \mathrm{HD}_2^{c_1}, \ldots\}$. And Minimum Hamming Distance ($d$) in the classifier: $d = \min\{\mathrm{SHD}^{c_1}, \mathrm{SHD}^{c_2}, \ldots\}$.
Note that (24) and (25) can be described by a uniform formula:

$$R(\varepsilon) \sim \lambda \varepsilon^d, \varepsilon \to 0^+, \lambda \in \mathbb{R}^+. \tag{27}$$

Here $\lambda$ is a real number related to $K$. The following illustrates that (27) is always true:

1. In the synchronization process, the probability of the occurrence of an attractive step between one pair of corresponding hidden units is calculated by

$$\mathrm{P_a}(\varepsilon) = \frac{\sum_{i=0}^{n} a_i \binom{n}{i}\varepsilon^i(1-\varepsilon)^{n-i}}{\mathrm{P_u}(\varepsilon)}. \tag{28}$$

In fact, it is certainly possible that an attractive step happens as all pairs of corresponding hidden units agree. So $a_0$, and the item $\frac{a_0\binom{n}{0}\varepsilon^0(1-\varepsilon)^n}{\mathrm{P_u}(\varepsilon)}$ constantly exist.

2. Similarly, we can obtain

$$\mathrm{P_r}(\varepsilon) = \frac{\sum_{i=0}^{n} r_i \binom{n}{i}\varepsilon^i(1-\varepsilon)^{n-i}}{\mathrm{P_u}(\varepsilon)}. \tag{29}$$

By Definition 1, it is impossible that $\tau^A = \tau^B$ occurs as $i < d$, and this also means the repulsive step can not occur as $i < d$. Therefore, the existence of the minimum $i$ equals $d$ in (29). Meanwhile, the item $\frac{r_d\binom{n}{d}\varepsilon^d(1-\varepsilon)^{n-d}}{\mathrm{P_u}(\varepsilon)}$ constantly exists.

3. Combining (28) and (29), it can be easily obtained

$$R(\varepsilon) = \frac{P_r(\varepsilon)}{P_a(\varepsilon)} = \frac{\sum_{i=d}^{n} r_i \binom{n}{i} \varepsilon^i (1-\varepsilon)^{n-i}}{\sum_{i=0}^{n} a_i \binom{n}{i} \varepsilon^i (1-\varepsilon)^{n-i}}. \tag{30}$$

As $\varepsilon \to 0$, it can be derived

$$R(\varepsilon) \sim \frac{r_d \binom{n}{d} \varepsilon^d (1-\varepsilon)^{n-d}}{a_0 \binom{n}{0} \varepsilon^0 (1-\varepsilon)^n} = \frac{r_d \binom{n}{d} \varepsilon^d}{a_0}. \tag{31}$$

Simplifying (31), (27) is hold.

According to (12),(26) and (27), a Heuristic Rule about security can be found.

**Heuristic Rule:** Without other behavior, by means of modifying the state classifier to enlarge $d$ appropriately, we can improve the security of TSCM.

## 4    Example

In this section, an example has been illustrated the benefits and effectiveness of our results for designing novel secure neural cryptography.

**Table 2.** Classification of Modified TCM($K = 3$)

| Name | $\tau$ | state vector $\sigma$ |
|---|---|---|
| MTCM | +1 | $(+1,+1,-1),(+1,-1,+1),(-1,+1,+1)$ |
|  | -1 | $(-1,-1,+1),(-1,+1,-1),(+1,-1,-1)$ |

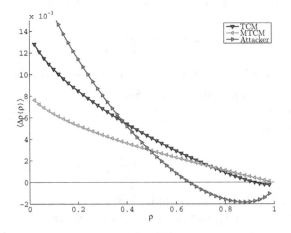

**Fig. 2.** Three types of dynamics of $\langle \Delta\rho(\rho) \rangle$

For simplicity, let us modify the insecure case, namely TCM. The state classifier of the modified TCM(MTCM) is shown in Table.2. And the $d$ of MTCM equals two. We can also obtain

$$R^{\mathrm{MTCM}}(\varepsilon) \sim 2\varepsilon^2 < U^{\mathrm{MTCM}}(\rho), \rho \to 1, \varepsilon \to 0. \tag{32}$$

The inequality (32) indicates that the modified TCM can meet $\langle \Delta\rho(\rho) \rangle > 0$ nearly synchronization. Besides, simulation experiments shown in Fig.2 can further prove the security of the modified TCM in whole synchronization.

## 5    Conclusion

The factor impacting on the security of neural cryptography is carefully investigated in this paper. The approach for designing novel neural cryptography is established by presenting the generalized architecture and the heuristic rule. In addition, main results obtained are efficient which have been demonstrated by an example.

**Acknowledgements.** This work was supported in part by the National Natural Science Foundation of China (No. 60973114 and No. 61170249), the Natural Science Foundation project of CQCSTC (2009BA2024), in part by the State Key Laboratory of Power Transmission Equipment & System Security and New Technology, Chongqing University (2007DA10512711206), in part by the program for Changjiang scholars, and supported by Specialized Research Fund for priority areas for the Doctoral Program of Higher Education.

## References

1. Diffie, W., Hellman, M.: New directions in cryptography. IEEE Trans. Inform. Theory 22(6), 644–654 (1976)
2. Menezes, A.J., Vanstone, S.A., Van Oorschot, P.C.: Handbook of Applied Cryptography. In: Applied Cryptography, CRC Press, Boca Raton (1996)
3. Schneier, B.: Applied Cryptography: Protocols, Algorithms, and Source Code in C, 2nd edn. Wiley, New York (1995)
4. Stallings, W.: Cryptography and Network Security: Principles and Practice. Pearson, Upper Saddle River (2002)
5. Kanter, I., Kanter, W., Kanter, E.: Secure exchange of information by synchronization of neural networks. Europhys. Lett. 57(1), 141–147 (2002)
6. Maurer, U.: Secert key agreement by public discussion from common information. IEEE Trans. Inform. Theory 39(3), 733–742 (1993)
7. Volkmer, M., Wallner, S.: Tree parity machine rekeying architectures. IEEE Trans. Comput. 54(4), 421–427 (2005)
8. Volkmer, M., Wallner, S.: Tree parity machine rekeying architectures for embedded security. "Cryptology ePrint Archive," Inst. Comput. Technol., Hamburg Univ. Technol., Hamburg, Germany, Rep. 2005/235 (2005)

9. Gross, N., Klein, E., Rosenbluh, M., Kinzel, W., Khaykovich, L., Kanter, I.: A framework for public-channel cryptography using chaotic lasers. Phys. Rev. E 73(6), 066 214-1–066 214-4 (2005)

10. Klein, E., Kanter, R.M.I., Kinzel, W.: Public-channel cryptography using chaos synchronization. Phys. Rev. E 72(1), 016 214-1–016 214-4 (2005)

11. Klein, E., Gross, N., Rosenbluh, M., Kinzel, W., Khaykovich, L., Kanter, I.: Stable isochronal synchronization of mutually coupled chaotic lasers. Phys. Rev. E 73(6), 066 214-1–066 214-4 (2006)

12. Shacham, L.N., Klein, E., Mislovaty, R., Kanter, I., Kinzel, W.: Cooperating attackers in neural cryptography. Phys. Rev. E 69(6), 066 137-1–066 137-4 (2004)

13. Saad, D.: On-line learning in neural networks. Cambridge University Press, Cambridge (1998)

14. Reyes, O.M., Kopitzke, I., Zimmermann, K.-H.: Permutation parity machines for neural synchronization. Phys. A: Math. Gen. 42, 195002 (2009)

15. Reyes, O.M., Zimmermann, K.-H.: Permutation parity machines for neural cryptography. J. Phys. A: Math. Theor. 42, 195 002-1–195 002-20 (2009)

16. Rosen-Zvi, M., Klein, E., Kanter, I., Kinzel, W.: Mutual learning in a tree parity machine and its application to cryptography. Phys. Rev. E 66(6), 066 135-1–066 135-13 (2002)

17. Seoane, L.F., Ruttor, A.: Successful attack on permutation-parity-machine-based neural cryptography. Phys. Rev. E 85(2), 025 101-1–025 101-4 (2012)

18. Metzler, R., Kinzel, W., Kanter, I.: Interacting neural networks. Phys. Rev. E 62(3), 2555–2565 (2000)

19. Kinzel, W., Metzler, R., Kanter, I.: Dynamics of interacting neural networks. J. Phys. A: Math. Gen. 33(14), L141–L147 (2000)

20. Kinzel, W., Kanter, I.: Disorder generated by interacting neural networks: application to econophysics and cryptography. J. Phys. A: Math. Gen. 36(43), 11 173–11 186 (2003)

21. Kinzel, W., Kanter, I.: Interacting neural networks and cryptography. Advances in Solid State Physics (2002)

22. Bornholdt, S., Schuster, H.: Theory of Interacting Neural Networks. Wiley VCH (2003)

23. Ruttor, A., Kinzel, W., Naeh, R., Kanter, I.: Genetic attack on neural cryptography. Phys. Rev. E 73(3), 036 121-1–036 121-8 (2006)

24. Ruttor, A., Kinzel, W., Kanter, I.: Dynamics of neural cryptographys. Phys. Rev. E 75(5), 056 104-1–056 104-4 (2007)

25. Ruttor, A., Reents, G., Kinzel, W.: Sychronizition of random walks with reflecting boundaries. J. Phys. A: Math. Gen. 37, 8609–8618 (2004)

26. Ruttor, A., Kinzel, W., Shacham, L., Kanter, I.: Neural cryptography with feedback. Phys. Rev. E 69(4), 046 110-1–046 110-7 (2004)

27. Ruttor, A., Kinzel, W., Kanter, I.: Neural cryptography with queries. J. Stat. Mech. 2005(1), 1–14 (2005)

28. Allam, A.M., Abbas, H.M.: On the improvement of neural cryptography using erroneous transmitted information with error prediction. IEEE Trans. Neural Networks 21(12), 1915–1924

29. Klimov, A., Mityaguine, A., Shamir, A.: Analysis of neural cryptography. In: Proc. 8th Int. Conf. Theory Appl. Cryptology Inform. Secur., pp. 288–298 (2002)

# Bifurcation of a Discrete-Time Cohen-Grossberg-Type BAM Neural Network with Delays

Qiming Liu

Institute of Applied Mathematics, Shijiazhuang Mechanical Engineering College,
Shijiazhuang 050003, China
lqmmath@163.com

**Abstract.** A tri-neuron discrete-time Cohen-Grossberg BAM neural network with delays is investigated in this paper. By analyzing the corresponding characteristic equations, the asymptotical stability of the null solution and the existence of Neimark-Sacker bifurcations are discussed. By applying the normal form theory and the center manifold theorem, the direction of the Neimark-Sacker bifurcation and the stability of bifurcating periodic solutions are obtained. Numerical simulations are given to illustrate the obtained results.

**Keywords:** Cohen-Grossberg neural network, discrete time, delay, stability, Neimark-Sacker bifurcation.

## 1 Introduction

Hopfield [1] proposed Hopfield neural networks in 1982, and in 1983, Cohen-Grossberg [2] proposed Cohen-Grossberg neural networks, which include Hopfield neural networks. These network models have been successfully applied to signal processing, pattern recognition, optimization and associative memories. The analysis of the dynamical behaviors is a necessary step for practical design of neural networks because their applications heavily depend on the dynamical behaviors, researchers have focused on the studying of simple systems in order to obtain a deep and clear understanding of the dynamics of complicated neural networks with time delays [3-15]. This is indeed very useful since the complexity found may be carried over to large networks.

In applications of continuous-time neural networks with or without delays to some practical problems, such as computer simulation, experimental or computational purposes, it is usual to formulate a discrete-time system which is a discrete version of the continuous-time system. Recently, bifurcation analysis for some discrete-time neural networks have been undertaken [11-15].

The authors discussed the Neimark-Sacker bifurcation for a tri-neuron discrete-time Hopfield-type BAM neural network [11], we discussed the bifurcation for a two-neuron discrete-time Cohen-Grossberg neural network with discrete delays

C. Guo, Z.-G. Hou, and Z. Zeng (Eds.): ISNN 2013, Part I, LNCS 7951, pp. 109–116, 2013.

[15]. Motivated by the work in [11], we study the following tri-neuron discrete-time Cohen-Grossberg-type BAM neural network with discrete delays further:

$$x_1(n+1)=x_1(n)-a(x_1(n))[b(x_1(n))-p_{11}f_1(y_1(n-k_2))-p_{12}f_1(y_2(n-k_2))],$$
$$y_1(n+1)=y_1(n)-c(y_1(n))[d(y_1(n))-q_{11}f_2(x_1(n-k_1))],$$
$$y_2(n+1)=y_2(n)-c(y_2(n))[d(y_2(n))-q_{21}f_3(x_1(n-k_1))],$$

$$(1)$$

where $x_1(n)$ denotes the state variable of the $i$th neuron from the neural field $F_X$ and and $y_i(n), i = 1,2$ denote the state variables of neurons from the neural field $F_Y$, $a(\cdot)$ and $c(\cdot)$ represent amplification functions which are positive for $R$; $f_i(\cdot)$ denote the signal functions of neurons; $b(\cdot)$ and $d(\cdot)$ are appropriately behaved functions; $p_{1i}(\cdot), i = 1,2$ and $q_{i1}(\cdot), i = 1,2$ are connection weights of the neural networks; discrete delays $k_1$ and $k_2$ correspond to the finite speed of the axonal signal transmission.

The neural network (1) may be regarded as a discrete version of the following continuous-time Cohen-Grossberg BAM neural network by Euler method.

$$\dot{x}_1(t) = -\tilde{a}(x_1(t))[b(x_1(t)) - p_{11}f_1(y_1(t-\tau_2)) - p_{12}f_1(y_2(t-\tau_2))],$$
$$\dot{y}_1(t) = -\tilde{c}(y_1(t))[d(y_1(t)) - q_{11}f_2(x_1(t-\tau_1))],$$
$$\dot{y}_2(t) = -\tilde{c}(y_2(t))[d(y_2(t)) - q_{21}f_3(x_1(t-\tau_1))].$$

$$(2)$$

The rest of this paper is organized as follows: The asymptotical stability and bifurcation are analyzed for the system in Section 2. The formula for determining the properties of Neimark-Sacker bifurcation of the model such as the direction of Neimark-Sacker bifurcation, stability of bifuricating periodic solutions are derived in Section 3. An example is given in Section 4 to demonstrate the main results. Conclusions are finally drawn in Section 5.

## 2    Stability Analysis and Existence of Bifurcations

Throughout this paper, we assume that

(H$_1$) $b(0) = d(0) = 0$, $f_i(0) = 0, i = 1,2,3$;
(H$_2$) $0 < a(0)b'(0) < 1, 0 < c(0)d'(0) < 1$.

It is clear that$(0,0,0)$ is a fixed point of system (1). We transform system (1) into the following system of $k_1 + 2k_2 + 3$ difference equations without delays:

$$x_1^{(0)}(n+1)=x_1^{(0)}(n)-a(x_1^{(0)}(n))[b(x_1^{(0)}(n))-p_{11}f_1(y_1^{(k_2)}(n))-p_{12}f_1(y_2^{(k_2)}(n))],$$
$$y_1^{(0)}(n+1)=y_1^{(0)}(n)-c(y_1^{(0)}(n))[d(y_1^{(0)}(n))-q_{11}f_2(x_1^{(k_1)}(n))],$$
$$y_2^{(0)}(n+1)=y_2^{(0)}(n)-c(y_2^{(0)}(n))[d(y_2^{(0)}(n))-q_{21}f_3(x_1^{(k_1)}(n))],$$
$$x_1^{(i)}(n+1)=x_1^{(i-1)}(n), i = 1,2,\ldots,k_1,$$
$$y_1^{(j)}(n+1)=y_1^{(j-1)}(n), y_2^{(l)}(n+1)=y_2^{(l-1)}(n), j,l = 1,2,\ldots,k_2.$$

$$(3)$$

The Jacobian matrix of system (3) at the fixed point $(0, \cdots, 0)$ as follows:

$$
\hat{A} =
\begin{pmatrix}
\alpha_1 & 0 & \cdots & 0 & 0 & 0 & 0 & \cdots & 0 & \gamma_1 & 0 & 0 & \cdots & 0 & \gamma_2 \\
1 & 0 & \cdots & 0 & 0 & 0 & 0 & \cdots & 0 & 0 & 0 & 0 & \cdots & 0 & 0 \\
0 & 1 & \cdots & 0 & 0 & 0 & 0 & \cdots & 0 & 0 & 0 & 0 & \cdots & 0 & 0 \\
\multicolumn{15}{c}{\cdots\cdots\cdots\cdots\cdots\cdots} \\
0 & 0 & \cdots & 1 & 0 & 0 & 0 & \cdots & 0 & 0 & 0 & 0 & \cdots & 0 & 0 \\
0 & 0 & \cdots & 0 & \beta_1 & \alpha_2 & 0 & \cdots & 0 & 0 & 0 & 0 & \cdots & 0 & 0 \\
0 & 0 & \cdots & 0 & 0 & 1 & 0 & \cdots & 0 & 0 & 0 & 0 & \cdots & 0 & 0 \\
0 & 0 & \cdots & 0 & 0 & 0 & 1 & \cdots & 0 & 0 & 0 & 0 & \cdots & 0 & 0 \\
\multicolumn{14}{c}{\cdots\cdots\cdots\cdots\cdots\cdots} & 0 & \cdots \\
0 & 0 & \cdots & 0 & 0 & 0 & 0 & \cdots & 1 & 0 & 0 & 0 & \cdots & 0 & 0 \\
0 & 0 & \cdots & 0 & \beta_2 & 0 & 0 & \cdots & 0 & 0 & \alpha_2 & 0 & \cdots & 0 & 0 \\
0 & 0 & \cdots & 0 & 0 & 0 & 0 & \cdots & 0 & 0 & 1 & 0 & \cdots & 0 & 0 \\
0 & 0 & \cdots & 0 & 0 & 0 & 0 & \cdots & 0 & 0 & 0 & 1 & \cdots & 0 & 0 \\
\multicolumn{15}{c}{\cdots\cdots\cdots\cdots\cdots\cdots} \\
0 & 0 & \cdots & 0 & 0 & 0 & 0 & \cdots & 0 & 0 & 0 & 0 & \cdots & 1 & 0
\end{pmatrix}
\tag{4}
$$

in which

$$
\begin{aligned}
&\alpha_1 = 1 - a(0)b'(0), \alpha_2 = 1 - c(0)d'(0), \\
&\beta_1 = a(0)p_{11}f_1'(0), \beta_2 = a(0)p_{12}f_1'(0), \gamma_1 = c(0)q_{11}f_2'(0), \gamma_2 = c(0)q_{21}f_3'(0).
\end{aligned}
\tag{5}
$$

The associated characteristic equation of system (3) is

$$
\lambda^{k_2}(\lambda - \alpha_2)[(\lambda - \alpha_1)(\lambda - \alpha_2)\lambda^k - b] = 0.
\tag{6}
$$

where $b = \beta_1\gamma_1 + \beta_2\gamma_2$ and $k = k_1 + k_2$.

Obviously, (6) has $k_2$ zero roots and the root $\alpha_2$, the other roots satisfy the following equation

$$
(\lambda - \alpha_1)(\lambda - \alpha_2)\lambda^k = b.
\tag{7}
$$

Denote

$$
b_0 = (1 - \alpha_1)(1 - \alpha_2),
$$

$$
b_j = (-1)^j \sqrt{(1 + \alpha_1^2 - 2\alpha_1 \cos\theta_j)(1 + \alpha_2^2 - 2\alpha_2 \cos\theta_j)}, j = 1, 2, \cdots, k+1,
\tag{8}
$$

$$
b_{k+2} = (-1)^{k+2}(1 + \alpha_1)(1 + \alpha_2),
$$

where $\theta_j = h^{-1}(j\pi)$, $j = 1, 2, \cdots, (k+1)$ and $h(\theta) = \cot^{-1}\left(\dfrac{\cos\theta - \alpha_1}{\sin\theta}\right)$

$+\cot^{-1}\left(\dfrac{\cos\theta - \alpha_2}{\sin\theta}\right) + k\theta$ in which $\cot^{-1}$ denotes the inverse of the cotangent function restricted to the interval $(0, \pi)$.

According to Theorem 2.1 in [15], together with $0 < \alpha_2 < 1$, we obtain the following results.

**Theorem 1.** *Under assumptions* $(H_1)$-$(H_2)$, *we have*

1. *If* $b \in (b_1, b_0)$, *the null solution of system* (1) *is asymptotically stable.*

2. *If* $b = b_0$, *a Fold bifurcation occurs at the origin in system* (1).

3. *If* $b = b_1$, *a Neimark-Sacker bifurcation occurs at the origin in system* (1), *i.e. a unique closed invariant curve bifurcates from the origin near* $b = b_1$. *Where*

$$b_0 = (1 - \alpha_1)(1 - \alpha_2), \ b_1 = -\sqrt{(1 + \alpha_1^2 - 2\alpha_1 \cos \theta_1)(1 + \alpha_2^2 - 2\alpha_2 \cos \theta_1)}$$

*in which* $\theta_1$ *is the unique solution in* $(0, \pi/(k+2))$ *of the equation* $\sin(k + 2)\theta - (\alpha_1 + \alpha_2)\sin(k+1)\theta + \alpha_1\alpha_2 \sin k\theta = 0$ *and* $k = k_1 + k_2$.

**Remark 1.** *If* $b = b_j, j = 2, 3, \cdots, k + 1$, *a Neimark-Sacker bifurcation also occurs at the origin in system* (1) *near* $b = b_j$ *except* $\theta_j = 2\pi/3$ *and* $\pi/2$, *and the value* $b_j$ *satisfy*

$$\cdots < b_5 < b_3 < b_1 < 0 < b_0 < b_2 < b_4 \cdots$$

*and* $|b_j| \le (1 + \alpha_1)(1 + \alpha_2)$ [15].

## 3    Direction and Stability of Neimark-Sacker Bifurcation for the Model

In this section, we investigate the direction of Neimark-Sacker bifurcation and the stability of periodic solutions bifurcating from the origin of system (1) by applying the normal form theory and the center manifold theorem for discrete time system developed by Kuznetsov [16].

We still discuss system (3). We know that if $b = b_1$, system (3) undergoes a Neimark-Sacker bifurcation at the origin and the associated characteristic equation of system (3) has a pair simple imaginary roots $e^{\pm i\theta_1}$.

Denote $\lambda_1 = e^{i\theta_1}$. Let $q \in C^{k_1 + 2k_2 + 3}$ be an eigenvector of $\hat{A}$ corresponding to eigenvalue $\lambda_1$, then $\hat{A}q = \lambda_1 q$. Again, let $p \in C^{k_1 + 2k_2 + 3}$ be an eigenvector of $\hat{A}^T$ corresponding to its eigenvalue $\bar{\lambda}_1$. By direct calculation we obtain

$$q = \left( \lambda_1^{k_1} q_1, \lambda_1^{k_1 - 1} q_1, \cdots, \lambda_1 q_1, q_1, \lambda_1^{k_2} q_2, \lambda_1^{k_2 - 1} q_2, \cdots, \lambda_1 q_2, q_2, \right.$$
$$\left. \lambda_1^{k_2} q_3, \lambda_1^{k_2 - 1} q_3, \cdots, \lambda_1 q_3, q_3 \right)^T,$$

$$p = \left( p_1, (\bar{\lambda}_1 - \alpha_1)p_1, \bar{\lambda}_1(\bar{\lambda}_1 - \alpha_1)p_1, \cdots, \bar{\lambda}_1^{k_1 - 1}(\bar{\lambda}_1 - \alpha_1)p_1, \right. \tag{9}$$
$$p_2, (\bar{\lambda}_1 - \alpha_2)p_2, \bar{\lambda}_1(\bar{\lambda}_1 - \alpha_2)p_2, \cdots, \bar{\lambda}_1^{k_2 - 1}(\bar{\lambda}_1 - \alpha_2)p_2,$$
$$\left. p_3, (\bar{\lambda}_1 - \alpha_3)p_3, \bar{\lambda}_1(\bar{\lambda}_1 - \alpha_3)p_3, \cdots, \bar{\lambda}_1^{k_2 - 1}(\bar{\lambda}_1 - \alpha_3)p_3 \right)^T$$

in which

$$q_1 = \lambda_1^{k_2}(\lambda_1 - \alpha_2), q_2 = \gamma_1, q_3 = \gamma_2,$$

$$\bar{p}_1 = \frac{1}{\lambda_1^k(\lambda_1 - \alpha_2)[\lambda_1 + k_1(\lambda_1 - \alpha_1)]},$$

$$\bar{p}_2 = \frac{\gamma_2}{\lambda_1^{k_2} + k_2\lambda_1^{k_2-1}(\lambda_1 - \alpha_2)}, \bar{p}_3 = \frac{-\gamma_1}{\lambda_1^{k_2} + k_2\lambda_1^{k_2-1}(\lambda_1 - \alpha_2)}$$

and

$$< p, q >= \lambda_1^k(\lambda_1-\alpha_2)[\lambda_1+k_1(\lambda_1-\alpha_1)]\bar{p}_1+[\lambda_1^{k_2}+k_2\lambda_1^{k_2-1}(\lambda_1-\alpha_2))](\gamma_1\bar{p}_2+\gamma_2\bar{p}_2) = 1.$$

Let function $F : R^{k_1+2k_2+3} \to R^{k_1+2k_2+3}$ be given by the right-hand side of system (3), and denote the operators

$$\hat{B} = D^2F(0), \hat{C} = D^3F(0), \tag{10}$$

when $b = b_1$, the restriction of system (3) to its two-dimensional center manifold at the critical parameter value can be transformed into the normal form written in complex coordinates [16]:

$$w \to \lambda_1 w \left(1 + \frac{1}{2}d|w|^2\right) + O(|w|^4), \ w \in C \tag{11}$$

in which

$$d = \bar{\lambda}_1 < p, \hat{C}(q, q, \bar{q}) + 2\hat{B}(q, (1 - \hat{A})^{-1}\hat{B}(q, q)) + \hat{B}(\bar{q}, (\lambda_1^2 I - \hat{A})^{-1}\hat{B}(q, q)) >,$$

where $p$ and $q$ are defined by (9), $\hat{A}$ and $\hat{B}$, $\hat{C}$ are defined by (4) and (10), respectively.

**Theorem 2.** ([16]) *The direction and stability of the Neimark-Sacker bifurcation for system (1) are determined by the sign of $\mathrm{Re}(d)$. If $\mathrm{Re}(d) < 0$, then the bifurcation is supercritical, i.e. the closed invariant curve bifurcating from the origin is asymptotically stable. If $\mathrm{Re}(d) > 0$, then the bifurcation is subcritical, i.e. the closed invariant curve bifurcating from the origin is unstable.*

**Remark 2.** If $a(\cdot) = 1, c(\cdot) = 1, b(x(n)) = \tilde{b}x_1(n), d(y_i(n)) = \tilde{d}y_i(n), i = 1, 2$, and $\tilde{b} = \tilde{d}$, system (1) reduces to the following Hopfield-type BAM neural network in [11].

$$\begin{aligned}
x_1(n+1) &= ax_1(n) + p_{11}f_1(y_1(n-k_2)) + p_{12}f_2(y_2(n-k_2)), \\
y_1(n+1) &= ay_1(n) + q_{11}f_2(x_1(n-k_1)), \\
y_2(n+1) &= ay_2(n) + q_{21}f_3(x_1(n-k_1))
\end{aligned} \tag{12}$$

where $a = 1 - \tilde{b}$, and $0 < a < 1$ according to the condition $(H_2)$. In this case, $b_j = (-1)^j|1 + a^2 - 2a\cos\theta_j|$ due to $\alpha_1 = \alpha_2 = a$. The Neimark-Sacker bifurcation for system (12) has been discussed in [11]. Note that $|b| = (a^4 + 4a^2 + 1 - (4a^3 + 4a)\cos\theta + 2a^2\cos 2\theta)^{\frac{1}{2}}$ in (2.1) in [11] is not a concise results, actually, $|b| = |1 + a^2 - 2a\cos\theta|$. Hence Theorem 3.1 and Theorem 4.1 in [11] are Corollary of Theorem 2.1 and Theorem 3.1 in this paper. Furthermore, the results in [11] can be simplified.

## 4    A Numerical Example

Consider the following discrete-time Cohen-Grossberg BAM neural network with discrete delays:

$$x_1(n+1) = x_1(n) - [0.7x_1(n) - \tanh(y_1(n-2)) - p_{12}\tanh(y_2(n-2))],$$
$$y_1(n+1) = y_1(n) - 0.2(2 + \cos(y_1(n)))[y_1(n) - 0.6\sin(x_1(n-3))], \qquad (13)$$
$$y_2(n+1) = y_2(n) - 0.2(2 + \cos(y_2(n)))[y_2(n) - 1.2\sin(x_1(n-3))].$$

We can obtain that $\theta_1 = 0.3823$ and furthermore we obtain that $b_1 = -0.4747$ in view of bisection method by using MATLAB. It is easy to know $b_0 = 0.42$. According to Theorem 1, the null solution of system (13) is asymptotically stable when $b = \beta_1\gamma_1 + \beta_2\gamma_2 = 0.36 + 0.72p_{12} \in (-0.4747, 0.42)$, and when $b = b_1$, the Neimark-Sacker bifurcation occurs at the origin.

Case 1: Let $p_{12} = -1.1$, $b = -0.2844 \in (-0.4747, 0.42)$, then the null solution of system (13) is asymptotically stable ( as shown in Fig. 1).

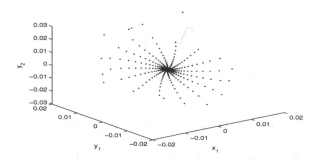

**Fig. 1.** Phase plot for system (13) with initial condition $(0.02, 0.02)$ and $p_{12} = -1.1$

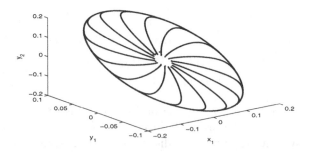

**Fig. 2.** Phase plot for system (13) with initial condition $(0.02, 0.02)$ and $p_{12} = -1.17$

Case 2: Let $p_{12} = -1.17$, $b = -0.4827$, and $b < b_1 = -0.4747$, thus Neimark-Sacker bifurcation occur at the origin and the bifurcating periodic solutions is stable as $\text{Re}(d) = -0.2137$ ( as shown in Fig. 2).

Case 3: Let $p_{12} = -3.27$, although Neimark-Sacker bifurcation does not occur due to $|b| > (1 + \alpha_1)(1 + \alpha_2)$, Fig. 3-4 show the chaotic dynamic behaviors of system (13).

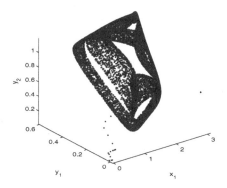

**Fig. 3.** Phase plot for system (13) with initial condition $(0.02, 0.02)$ and $p_{12} = -3.27$

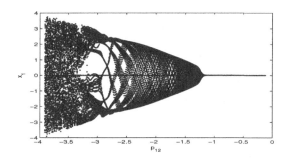

**Fig. 4.** Bifurcation diagram in space $(p_{12}, x_1)$ for system (13) with step size 0.01 for $p_{12}$ and initial condition $(0.02, 0.02)$, in which there are 100 points for each $p_{12}$

## 5    Conclusions

A discrete-time Cohen-Grossberg BAM neural network with discrete delays is analyzed. By using $b = \beta_1\gamma_1 + \beta_2\gamma_2 = a(0)c(0)f_1'(0)[p_{11}q_{11}f_2'(0) + p_{12}q_{21}f_3'(0)]$ as a bifurcation parameter, we show that this system undergoes Neimark-Sacker bifurcations at the critical parameter The direction of Neimark-Sacker bifurcation and the stability of the bifurcating periodic solutions are investigated for the system. In addition, system may be chaotic when $|b|$ is large enough.

**Acknowledgements.** This research was supported by the Hebei Provincial Natural Science Foundation of China under Grant No.A2012205028.

# References

1. Hopfield, J.J.: Neural networks and physical systems with emergent collective computational abilities. Proc. Nat. Acad. Sci. USA. 79, 2554–2558 (1982)
2. Cohen, M., Grossberg, S.: Absolute stability and global pattern formation and parallel memory storage by competitive neural networks. IEEE Trans. Syst. Man Cybernet. Smc-13, 815–826 (1982)
3. Wei, J., Ruan, S.: Stability and bifurcation in a neural network model with two delays. Physica D 130, 255–272 (1999)
4. Cao, J., Xiao, M.: Stability and Hopf bifurcation in a simplified BAM neural network with two time delays. IEEE Trans Neural Networks 18, 416–430 (2007)
5. Guo, S., Huang, L., Wang, L.: Lineartability and Hopf bifurcation in a two-neuron network with three delays. Int. J. Bifurcat. Chaos 14, 2799–2810 (2004)
6. Huang, C., Huang, L., Feng, J., Nai, M., He, Y.: Hopf bifurcation analysis for a two-neuron network with four delays. Chaos, Solitons and Fractals 34, 795–812 (2007)
7. Zhou, X., Wu, Y., Li, Y., Yao, X.: Stability and bifurcation analysis on a two-neuron networks with discrete and distributed delays. Chaos, Solitons and Fractals 40, 1493–1505 (2009)
8. Yang, Y., Ye, J.: Stability and bifurcation in a simplified five-neuron BAM neural networks with delays. Chaos, Solitons Fract 42, 2357–2363 (2009)
9. Zhao, H., Wang, L.: Hopf bifurcation in Cohen-Grossberg neural network with distributed delays. Nonlinear Anal.: Real World Appl. 8, 73–89 (2007)
10. Liu, Q., Xu, R.: Stability and bifurcation of a Cohen-Grossberg neural network with discrete delays. Applied Mathematics and Computation 218, 2850–2862 (2011)
11. Gan, Q., Xu, R., Hu, W., Yang, P.: Bifurcation analysis for a tri-neuron discrete-time BAM neural network with delays. Chaos, Solitons and Fractals 42, 2502–2511 (2009)
12. Kaslik, E., Balint, S.: Bifurcation analysis for a discrete-time Hopfield neural network of two neurons with two delays and self-connections. Chaos, Solitons and Fractals 39, 83–91 (2009)
13. Kaslik, E., Balint, S.: Complex and chaotic dynamics in a discrete-time-delayed Hopfield neural network with ring architecture. Neural Networks 22, 1411–1418 (2009)
14. Zhao, H., Wang, L.: Stability and bifurcation for discrete-time Cohen-Grossberg neural networks. Appl. Math. Comput. 179, 787–798 (2006)
15. Liu, Q., Xu, R., Wang, Z.: Stability and bifurcation of a class of discrete-time Cohen-Grossberg neural networks with delays. Discrete Dynamics in Nature and Society 2011, Article ID 403873 (2011)
16. Kuznetsov, A.: Elements of Applied Bifurcation Theory. Springer, New York (1998)

# Stability Criteria for Uncertain Linear Systems with Time-Varying Delay

Huimin Liao, Manchun Tan\*, and Shuping Xu

Department of Mathematics, Jinan University, Guangzhou 510632, China
tanmc@jnu.edu.cn

**Abstract.** This paper is concerned with the stability criteria for uncertain systems with time-varying delay. The parameter uncertainties are supposed to be norm-bounded. By using Lyapunov functional and integral inequality, some delay-dependent stability criteria are obtained. Numerical examples are given to demonstrate the effectiveness of proposed method.

**Keywords:** uncertain linear systems, Lyapunov functional, time-varying delay, stability.

## 1 Introduction

It is known that the phenomena of time-delay occur in many systems, such as communication systems, neural networks, automatic systems, and so on. The existence of time-delay has an adverse effect on system performance and stability. Hence, much attention is received in stability analysis of the systems with time-delay [1]-[22].

To derive the delay-dependent stability criteria, many methods have been proposed such as augmented functional approach [1], free-weighting matrix approach [2][3], integral inequality [4][5], convex combination method [6], delay partition approach [7][8], triple-integral approach [9][10], mathematical analysis [11], LMI [12]-[14], and so on. The free-weighting matrix approach is most widely used among these approaches.

In this paper, we investigate the delay-dependent stability of uncertain linear systems. By choosing an appropriate Lyapunov functional and using integral inequality, some delay-dependent stability criteria are obtained. In the end, numerical examples are given to show the advantages of the proposed stability criteria.

The notations used throughout this paper are as follows. Matrix $I$ refers to an identity matrix. The superscript $T$ stands for matrix transposition. The symmetric term in a matrix is denoted by $*$. $R^n$ represents $n$ dimensional Euclidean, $R^{n \times m}$ is the set of all $n \times m$ real matrices. For real symmetric matrices $A$ and $B$, the notion $A > B$ means the matrix $A - B$ is positive definite.

---

\* Correspondding author.

C. Guo, Z.-G. Hou, and Z. Zeng (Eds.): ISNN 2013, Part I, LNCS 7951, pp. 117–124, 2013.

## 2    Problem Formulation

Consider the following uncertain linear system with time-varying delay:

$$\dot{x}(t) = (A + \triangle A(t))x(t) + (A_1 + \triangle A_1(t))x(t - h(t)), \quad \forall\, t \geq 0,$$
$$x(t) = \Phi(t), \quad \forall\, t \in [-h_2, 0], \tag{1}$$

where $x(t)$ is the state vector, $A$ and $A_1$ are constant matrices of appropriate dimensions, $\Phi(t)$ is a given continuous vector valued function. The delay $h(t)$ is a bounded differentiable function that satisfies

$$0 \leq h_1 \leq h(t) \leq h_2, \quad \dot{h}(t) \leq d < \infty, \quad \forall\, t \geq 0, \tag{2}$$

where $h_1, h_2$ and $d$ are constants. Unknown real matrices $\triangle A(t)$ and $\triangle A_1(t)$ represent time-varying parametric uncertainties satisfying

$$[\triangle A(t) \quad \triangle A_1(t)] = DF(t)[E \quad E_1], \tag{3}$$

where $D, E$ and $E_1$ are constant matrices of appropriate dimensions, $F(t)$ is unknown matrix that satisfies

$$F^T(t)F(t) \leq I. \tag{4}$$

We first introduce the following lemmas which will play a great role on our derivation.

**Lemma 1.** [16] For any constant matrix $M \in R^{n \times n}$, there exist positive scalars $h_1, h_2$ such that $h_1 \leq h(t) \leq h_2$ and vector function $\dot{x} : [-h_2, -h_1] \to R^n$ such that the integration $\int_{t-h_2}^{t-h_1} \dot{x}^T(s)M\dot{x}(s)ds$ is well defined, then

$$-(h_2 - h_1)\int_{t-h_2}^{t-h_1} \dot{x}^T(s)M\dot{x}(s)ds \leq \xi^T(t)\Lambda\xi(t),$$

where $\xi(t) = \begin{bmatrix} x(t-h_1) \\ x(t-h(t)) \\ x(t-h_2) \end{bmatrix}, \Lambda = \begin{bmatrix} -M & M & 0 \\ * & -2M & M \\ * & * & -M \end{bmatrix}.$

**Lemma 2.** [17] For a given matrix $S = \begin{bmatrix} S_{11} & S_{12} \\ * & S_{22} \end{bmatrix}$ with $S_{11} = S_{11}^T, S_{22} = S_{22}^T$, then the following conditions are equivalent:

(1)$S < 0$;                (2)$S_{22} < 0, S_{11} - S_{12}S_{22}^{-1}S_{12}^T < 0$.

**Lemma 3.** [17] Let $U, F, W$ and $Q$ be real matrices of appropriate dimensions with $Q$ satisfying $Q = Q^T$, then

$$Q + UFW + W^TF^TU^T < 0 \quad \text{for all } F^TF \leq I$$

if and only if there exists a scalar $\varepsilon > 0$ such that

$$Q + \varepsilon^{-1}UU^T + \varepsilon W^TW < 0.$$

## 3  Main Results

**Theorem 1.** Given scalars $0 \leq h_1 < h_2$ and $d \geq 0$, system (1) with (3) is asymptotically stable if there exist positive definite matrices $P, M, R_{11}, R_{22}, R_{33}, Q_i (i = 1, 2, 3, 4)$, appropriately dimensioned matrices $R_{12}, R_{13}, R_{23}$ and a positive constant $\varepsilon$ such that the following LMI holds:

$$\varphi = \begin{bmatrix} \varphi_{11} & \varphi_{12} & 0 & 0 & A^T \Omega & PD & \varepsilon E^T \\ * & \varphi_{22} & M & M & A_1^T \Omega & 0 & \varepsilon E_1^T \\ * & * & \varphi_{33} & 0 & 0 & 0 & 0 \\ * & * & * & \varphi_{44} & 0 & 0 & 0 \\ * & * & * & * & -\Omega & \Omega D & 0 \\ * & * & * & * & * & -\varepsilon I & 0 \\ * & * & * & * & * & * & -\varepsilon I \end{bmatrix} < 0, \tag{5}$$

where

$$R = \begin{bmatrix} R_{11} & R_{12} & R_{13} \\ * & R_{22} & R_{23} \\ * & * & R_{33} \end{bmatrix} > 0,$$

$$\varphi_{11} = PA + A^T P + Q_1 + Q_2 + Q_3 + h_2 R_{11} + R_{13} + R_{13}^T,$$
$$\varphi_{12} = PA_1 + h_2 R_{12} - R_{13} + R_{23}^T,$$
$$\varphi_{22} = -(1 - d)Q_3 + h_2 R_{22} - R_{23} - R_{23}^T - M - M^T,$$
$$\varphi_{33} = -Q_1 + Q_4 - M,$$
$$\varphi_{44} = -Q_2 - Q_4 - M,$$
$$\Omega = h_2 R_{33} + (h_2 - h_1)^2 M.$$

**Proof.** Consider the Lyapunov function

$$V(t) = \sum_{i=1}^{4} V_i(t), \tag{6}$$

where

$$V_1(t) = x^T(t)Px(t),$$

$$V_2(t) = \int_{t-h_1}^{t} x^T(s)Q_1 x(s)ds + \int_{t-h_2}^{t} x(s)^T Q_2 x(s)ds + \int_{t-h(t)}^{t} x(s)^T Q_3 x(s)ds$$
$$+ \int_{t-h_2}^{t-h_1} x(s)^T Q_4 x(s)ds,$$

$$V_3(t) = \int_{-h_2}^{0} \int_{t+\theta}^{t} \dot{x}^T(s)R_{33}\dot{x}(s)dsd\theta + (h_2 - h_1) \int_{-h_2}^{-h_1} \int_{t+\theta}^{t} \dot{x}^T(s)M\dot{x}(s)dsd\theta,$$

$$V_4(t) = \int_{0}^{t} \int_{\theta-h(\theta)}^{\theta} \zeta^T(\theta, s)R\zeta(\theta, s)dsd\theta,$$

where

$$\zeta(\theta, s) = [x^T(\theta) \quad x^T(\theta - h(\theta)) \quad \dot{x}^T(s)]^T.$$

Taking the derivative of $V(t)$ along any trajectory of system (1), we have

$$\dot{V}_1(t) = 2x^T(t)P\dot{x}(t), \tag{7}$$

$$\begin{aligned}
\dot{V}_2(t) &\leq x^T(t)(Q_1 + Q_2 + Q_3)x(t) - x^T(t - h_1)Q_1 x(t - h_1) \\
&\quad - x^T(t - h_2)Q_2 x(t - h_2) - (1 - d)x^T(t - h(t))Q_3 x(t - h(t)) \\
&\quad + x^T(t - h_1)Q_4 x(t - h_1) - x^T(t - h_2)Q_4 x(t - h_2), \tag{8}
\end{aligned}$$

$$\begin{aligned}
\dot{V}_3(t) &= h_2 \dot{x}^T(t)R_{33}\dot{x}(t) - \int_{t-h_2}^{t} \dot{x}^T(s)R_{33}\dot{x}(s)ds + (h_2 - h_1)^2 \dot{x}^T(t)M\dot{x}(t) \\
&\quad - (h_2 - h_1)\int_{t-h_2}^{t-h_1} \dot{x}^T(s)M\dot{x}(s)ds \\
&\leq \dot{x}^T(t)[h_2 R_{33} + (h_2 - h_1)^2 M]\dot{x}(t) - \int_{t-h(t)}^{t} \dot{x}^T(s)R_{33}\dot{x}(s)ds \\
&\quad - (h_2 - h_1)\int_{t-h_2}^{t-h_1} \dot{x}^T(s)M\dot{x}(s)ds, \tag{9}
\end{aligned}$$

$$\begin{aligned}
\dot{V}_4(t) &= h(t)[x^T(t) \quad x^T(t - h(t))] \begin{bmatrix} R_{11} & R_{12} \\ * & R_{22} \end{bmatrix} [x^T(t) \quad x^T(t - h(t))]^T + 2x^T(t)R_{13}x(t) \\
&\quad - 2x^T(t)R_{13}x(t - h(t)) + 2x^T(t - h(t))R_{23}x(t) - 2x^T(t - h(t))R_{23}x(t - h(t)) \\
&\quad + \int_{t-h(t)}^{t} \dot{x}^T(s)R_{33}\dot{x}(s)ds \\
&\leq h_2 x^T(t)R_{11}x(t) + 2h_2 x^T(t)R_{12}x(t - h(t)) + h_2 x^T(t - h(t))R_{22}x(t - h(t)) \\
&\quad + 2x^T(t)R_{13}x(t) - 2x^T(t)R_{13}x(t - h(t)) + 2x^T(t - h(t))R_{23}x(t) \\
&\quad - 2x^T(t - h(t))R_{23}x(t - h(t)) + \int_{t-h(t)}^{t} \dot{x}^T(s)R_{33}\dot{x}(s)ds, \tag{10}
\end{aligned}$$

By Lemma 1, we have

$$- (h_2 - h_1)\int_{t-h_2}^{t-h_1} \dot{x}(s)M\dot{x}(s)ds \leq \phi^T(t) \begin{bmatrix} -2M & M & M \\ * & -M & 0 \\ * & * & -M \end{bmatrix} \phi(t), \tag{11}$$

where $\phi^T(t) = [x^T(t - h(t)) \quad x^T(t - h_1) \quad x^T(t - h_2)]^T$.

By adding (7)-(11), we obtain the following result:

$$\dot{V}(t) \leq \eta^T(t)[\psi + \Pi^T \Omega \Pi]\eta(t), \tag{12}$$

where $\varphi_{11}, \varphi_{22}, \varphi_{33}, \varphi_{44}$ are defined in (5) and

$$\eta(t) = [x^T(t) \quad x^T(t - h(t)) \quad x^T(t - h_1) \quad x^T(t - h_2)]^T,$$

$$\psi = \begin{bmatrix} \varphi_{11} + \Delta_1 & \varphi_{12} + \Delta_2 & 0 & 0 \\ * & \varphi_{22} & M & M \\ * & * & \varphi_{33} & 0 \\ * & * & * & \varphi_{44} \end{bmatrix},$$

$$\Delta_1 = PDF(t)E + E^T F^T(t)D^T P,$$
$$\Delta_2 = PDF(t)E_1,$$
$$\Pi = [A + DF(t)E \quad A_1 + DF(t)E_1 \quad 0 \quad 0],$$
$$\Omega = h_2 R_{33} + (h_2 - h_1)^2 M.$$

By Lemma 2, if

$$\psi + \Pi^T \Omega \Pi < 0, \tag{13}$$

then

$$\begin{bmatrix} \psi & \Pi^T \Omega \\ * & -\Omega \end{bmatrix} < 0. \tag{14}$$

Inequality (14) is equivalent to

$$\Xi + UF(t)W + W^T F^T(t)U^T < 0, \tag{15}$$

where

$$\Xi = \begin{bmatrix} \varphi_{11} & \varphi_{12} & 0 & 0 & A^T \Omega \\ * & \varphi_{22} & M & M & A_1^T \Omega \\ * & * & \varphi_{33} & 0 & 0 \\ * & * & * & \varphi_{44} & 0 \\ * & * & * & * & -\Omega \end{bmatrix},$$
$$U = [D^T P \quad 0 \quad 0 \quad 0 \quad D^T \Omega]^T,$$
$$W = [E \quad E_1 \quad 0 \quad 0 \quad 0].$$

Applying Lemma 2 and Lemma 3, inequality (12) can be expressed as follows:

$$\dot{V}(t) \le \eta^T(t)\varphi\eta(t). \tag{16}$$

If $\varphi < 0$, then $\dot{V}(t) < 0$. Hence, systems (1) is asymptotically stable if LMI (5) is satisfied. This completes the proof.

If the function $h(t)$ is not differentiable or $d$ is unknown, by eliminating $Q_3$, we have the following theorem.

**Theorem 2.** Given scalars $0 \le h_1 < h_2$, system (1) with (3) is asymptotically stable if there exist positive definite matrices $P, M, R_{11}, R_{22}, R_{33}, Q_1, Q_2, Q_4$, appropriately dimensioned matrices $R_{12}, R_{13}, R_{23}$ and a positive constant $\varepsilon$ such that the following LMI condition holds:

$$\hat{\varphi} = \begin{bmatrix} \hat{\varphi}_{11} & \varphi_{12} & 0 & 0 & A^T\Omega & PD & \varepsilon E^T \\ * & \hat{\varphi}_{22} & M & M & A_1^T\Omega & 0 & \varepsilon E_1^T \\ * & * & \varphi_{33} & 0 & 0 & 0 & 0 \\ * & * & * & \varphi_{44} & 0 & 0 & 0 \\ * & * & * & * & -\Omega & \Omega D & 0 \\ * & * & * & * & * & -\varepsilon I & 0 \\ * & * & * & * & * & * & -\varepsilon I \end{bmatrix} < 0, \tag{17}$$

where $\varphi_{33}, \varphi_{44}, \Omega$ are defined in (5) and

$$R = \begin{bmatrix} R_{11} & R_{12} & R_{13} \\ * & R_{22} & R_{23} \\ * & * & R_{33} \end{bmatrix} > 0,$$

$$\hat{\varphi}_{11} = PA + A^T P + Q_1 + Q_2 + h_2 R_{11} + R_{13} + R_{13}^T,$$

$$\hat{\varphi}_{22} = h_2 R_{22} - R_{23} - R_{23}^T - M - M^T.$$

## 4    Numerical Example

In this section, we consider three examples to show the reduced conservatism of the proposed stability criteria.

**Table 1.** Maximum allowable delay bound $h_2$ for various $d$ and $h_1 = 0$

| method | $d$ | 0 | 0.5 | 0.9 |
|--------|-----|---|-----|-----|
| Fridman [18] | $h_2$ | 1.1490 | 0.9247 | 0.6710 |
| Wu et al. [19] | $h_2$ | 1.1490 | 0.9247 | 0.6954 |
| Theorem 1 | $h_2$ | 1.2060 | 1.0103 | 0.8165 |

**Table 2.** Maximum allowable delay bound $h_2$ for various $h_1$ and unknown $d$

| method | $h_1$ | 0.5 | 0.8 | 1 | 2 |
|--------|-------|-----|-----|---|---|
| Jiang et al. [22] | $h_2$ | 1.07 | 1.33 | 1.50 | 2.39 |
| He et al. [20] | $h_2$ | 1.0991 | 1.3476 | 1.5187 | 2.4000 |
| Shao [21] | $h_2$ | 1.2191 | 1.4539 | 1.6169 | 2.4798 |
| Theorem 2 | $h_2$ | 1.2201 | 1.4892 | 1.6676 | 2.5436 |

**Example 1.** Consider the system (1) with

$$A = \begin{bmatrix} -2 & 0 \\ 0 & -0.9 \end{bmatrix}, A_1 = \begin{bmatrix} -1 & 0 \\ -1 & -1 \end{bmatrix}, D = \begin{bmatrix} 0.1 & 0 \\ 0 & 0.5 \end{bmatrix},$$

$$E = \begin{bmatrix} 16 & 0 \\ 0 & 0.1 \end{bmatrix}, E_1 = \begin{bmatrix} 1 & 0 \\ 0 & 0.6 \end{bmatrix}.$$

When the lower bound $h_1 = 0$, by using Theorem 1, Table 1 lists the maximum allowable delay bound for various delay derivative values. It clearly shows that the proposed criterion is less conservative than those obtained in [18][19].

**Example 2.** Consider the system (1) with

$$A = \begin{bmatrix} 0 & 1 \\ -1 & -2 \end{bmatrix}, A_1 = \begin{bmatrix} 0 & 0 \\ -1 & 1 \end{bmatrix}, D = E = E_1 = 0.$$

When $d$ is unknown, Table 2 gives the comparisons of the maximum allowable delay $h_2$ for various $h_1$ by Theorem 2.

## 5    Conclusion

In this paper, we discuss the delay-dependent stability criteria with uncertain systems. By using Lyapunov functional and integral inequality, some improved stability criteria are obtained. Examples are given to demonstrate the advantages of our method.

**Acknowledgments.** The research is supported by grants from the Natural Science Foundation of Guangdong Province in China (No.S2012010010356), the Foundation of Science and Technology of Guangdong Province in China (No. 2009B011400046), and the Fundamental Research Funds for the Central Universities (No. 21612443).

## References

1. Jafarov, E.M.: Comparative analysis of simple improved delay-dependent stability criterions for linear time-delay systems: an augmented functional approach. In: American Control Conference, pp. 3389–3394 (2001)
2. Lin, C., Wang, Q.G., Lee, T.H.: A less conservative robust stability test for linear uncertain time-delay systems. IEEE Transactions on Automatic Control 51(1), 87–91 (2006)
3. He, Y., Wang, Q.G., Xie, L.H., Lin, C.: Further improvement of free-weighting matrices technique for systems with time-varying delay. IEEE Transactions on Automatic Control 52(2), 293–299 (2007)

4. Sarabi, F.E., Momeini, H.R.: Less Conservative delay-dependent robust stability criteria for linear time-delay systems. In: International Conference on Electronics Computer Telecommunications and Information Technology, pp. 738–741 (2010)
5. Ramakrishnan, K., Ray, G.: Robust stability criteria for uncertain linear systems with interval time-varying delay. J. Control Theory Appl. 9(4), 559–566 (2011)
6. Tang, M., Wang, Y.W., Wen, C.: Improved delay-range-dependent stability criteria for linear systems with interval time-varying delays. IET Control Theory Appl. 6(6), 868–873 (2002)
7. Teng, X.: Improved stability criterion for linear systems with-varying delay. In: 4th International Conference on Intelligent Human-Machine Systems and Cybernetics, pp. 108–111 (2012)
8. Tian, E.A., Zhao, X., Wang, W.N.: Robust $H_\infty$ control for linear systems with delay in state or input. In: Chinese Control and Decision Conference, pp. 2046–2051 (2010)
9. Sun, J., Liu, G.P., Chen, J., Rees, D.: Improved stability criteria for linear systems with time-varying delay. IET Control Theory Appl. 4(4), 683–689 (2010)
10. Sun, J., Chen, J., Liu, G.P., Rees, D.: Delay-range-dependent and rate-range-dependent stability criteria for linear systems with time-varying delays. In: Joint 48th IEEE Conference on Decision and Control and 28th Control Conference, Shanghai, China, pp. 251–256 (2009)
11. Tan, M.C.: Exponential convergence behavior of fuzzy cellular neural networks with distributed delays and time-varying coefficients. International Journal of Bifurcation and Chaos 19(7), 2455–2462 (2009)
12. Tan, M.C.: Global asymptotic stability of fuzzy cellular neural networks with unbounded distributed delays. Neural Processing Letters 31(2), 147–157 (2010)
13. Tan, M.C., Zhang, Y., Su, W.L.: Exponential stability analysis of neural networks with variable delays. International Journal of Bifurcation and Chaos 20(5), 1541–1549 (2010)
14. Zhang, X.M., Tang, X.H., Xiao, S.P.: New criteria on delay-dependent stability for linear systems with time delays. In: 27th Chinese Control Conference, pp. 43–47 (2008)
15. Lien, C.H.: Stability and stabilization criteria for a class of uncertain neutral systems with time-varying delays. J. Optimization Theory Appl. 124(3), 637–657 (2005)
16. Peng, C., Tian, Y.C.: Improved delay-dependent robust stability criteria for uncertain systems with interval time-varying delay. IET Control Theory and Appl. 2(9), 752–761 (2008)
17. Zhang, A., Xu, Z.D., Liu, D.: New stability criteria for neutral systems with interval time-varying delays and nonlinear perturbations. In: Chinese Control and Decision Conference, pp. 2995–2999 (2011)
18. Fridman, E., Shaked, U.: A descriptor systems approach to $H_\infty$ control of linear time-delay systems. IEEE Transaction on Automatic Control 47(2), 253–270 (2002)
19. Wu, M., He, Y., She, J., Liu, G.: Delay-dependent criteria for robust stability of time-varying delay systems. Automatica 40(8), 1435–1439 (2004)
20. He, Y., Wang, Q., Lin, C., Wu, M.: Delay-range dependent stability for systems with time-varying delay. Automatica 43(2), 371–376 (2007)
21. Shao, H.Y.: New delay-dependent stability criteria for systems with interval delay. Automatica 45(3), 744–749 (2009)
22. Jiang, X.F., Liu, Q.L.: On $H_\infty$ control for linear systems with interval time-varying delay. Automatica 41(12), 2099–2106 (2005)

# Generalized Function Projective Lag Synchronization between Two Different Neural Networks

Guoliang Cai[1,*], Hao Ma[1], Xiangqian Gao[1], and Xianbin Wu[2]

[1] Nonlinear Scientific Research center, Jiangsu University, Zhenjiang, Jiangsu 212013, China
glcai@ujs.edu.cn, {597795513,745524677}@qq.com
[2] Junior College, Zhejiang Wanli University, Ningbo, Zhejiang, 315100, China
wxb3210@zwu.edu.cn

**Abstract.** The generalized function projective lag synchronization (GFPLS) is proposed in this paper. The scaling functions which we have investigated are not only depending on time, but also depending on the networks. Based on Lyapunov stability theory, a feedback controller and several sufficient conditions are designed such that the response networks can realize lag-synchronize with the drive networks. Finally, the corresponding numerical simulations are performed to demonstrate the validity of the presented synchronization method.

**Keywords:** generalized function projective lag synchronization, Lyapunov stability theorem, neural networks, feedback control.

## 1 Introduction

In recent years, neural networks (NNs) have drawn the attention of many researchers from different areas since they have been fruitfully applied in signal and image processing, associative memories, automatic control , secure communication, and so on[1-4]. Synchronization is a ubiquitous phenomenon in nature, roughly speaking, if two or more networks have something in common, a synchronization phenomenon will occur between them which are either chaotic or periodic share a common dynamical behavior. As a research hot spot, synchronization in neural networks has received a great deal of interests. Since the pioneering work of Pecora and Carroll [5], in which proposed a successful method to synchronize two identical chaotic systems with different initial conditions, chaos synchronization has received a great deal of interest among scientists from various fields. In the past decades, some new types of synchronization have appeared in the literatures. Such as complete synchronization [6], function projective synchronization [7], stochastic synchronization [8], impulsive synchronization [9] and projective synchronization [10], etc.

The problem of synchronization between two neural networks with time delay has been extensively studied [11-14]. Recently, considering a time delay will affect the synchronization of neural networks received a lot of attentions of researchers. Namely, the response networks' output lags behind the output of the driver system

---

* Corresponding author.

C. Guo, Z.-G. Hou, and Z. Zeng (Eds.): ISNN 2013, Part I, LNCS 7951, pp. 125–132, 2013.
© Springer-Verlag Berlin Heidelberg 2013

proportionally. Such as projective lag synchronization [15], exponential lag synchronization [16-17], function projective lag synchronization [18], and so on. In Ref.[15], the author considered adaptive lag synchronization in unknown stochastic chaotic neural networks with discrete and distributed time-varying delays, by the adaptive feedback technique and several sufficient conditions have been derived to ensure the synchronization of stochastic chaotic neural networks. In this paper, we investigate a special kind of function projective lag synchronization which is different from other literatures. A controller and sufficient conditions are designed such that the response networks can lag-synchronize with the drive networks.

The rest of this paper is organized as follows: Section 2 gives neural networks model and some preliminaries. In section 3, the function projective lag synchronization is presented for neural networks based on Lyapunov stability theory. In section 4, we give numerical simulation to verify the result. The conclusion is finally drawn in section 5.

## 2    Neural Networks Model and Preliminaries

In this paper, we consider the following neural networks as drive neural networks:

$$\dot{x}_i(t) = A_i x_i(t) + f_i(x_i(t)) + \sum_{j=1}^{N} c_{ij} P x_j(t) + I_i \qquad i = 1, 2, ..., N \qquad (1)$$

where $x_i(t) = (x_{i1}(t), x_{i2}(t), ..., x_{in}(t))^T \in R^n$ is the state vector of the $i$th neuron at time $t$, $N$ corresponds to the number of neurons, $A_i \in R^{n \times n}$ is a constant matrix, $f_i : R^n \to R^n$ denote the activation functions of the neurons, $I_i \in R^N, i = 1, 2, ..., N$ is a constant external input vector. $P \in R^{n \times n}$ is an inner coupling matrix, $C = (c_{ij})_{N \times N} \in R^{N \times N}$ is the coupling configuration matrix meaning the coupling strength and the topological structure of the networks, if there is a connection from node $i$ to node $j (i \neq j)$, then $c_{ij} \neq 0$, otherwise, $c_{ij} = 0$, the diagonal elements of matrix $C$ is defined by $c_{ii} = -\sum_{j \neq i, j=1}^{N} c_{ij} \; i = 1, 2, ..., N$.

We take the response neural networks is given by:

$$\dot{y}_i(t) = B_i y_i(t) + g_i(y_i(t)) + \sum_{j=1}^{N} d_{ij} Q y_j(t) + I_i + u_i(t) \qquad i = 1, 2, ..., N \qquad (2)$$

where the $y_i(t) = (y_{i1}(t), y_{i2}(t), ..., y_{in}(t))^T \in R^n$ denote the response state vector of the $i$th neuron at time $t$, respectively, $B_i, g_i, Q, I_i$ have the same meanings as $A_i, f_i, P, I_i$ of Eq.(1). $D = (d_{ij})_{N \times N} \in R^{N \times N}$ is the same meaning as $C$, $u_i \in R^n \; (i = 1, 2, ..., N)$ are the controllers to be designed.

Let the error term $e_i(t) = y_i(t) - \alpha_i(x_i(t - \tau(t)))$, $i = 1, 2, ..., N$. $\alpha_i(x, t)$ are nonzero scaling functions and continuously differentiable functions, $\tau(t)$ is the time lag.

**Definition 1** (GFPLS). For the drive neural networks (1) and the response neural networks (2), it is said that they are generalized function projective lag synchronized, if there exist the continuous function $\alpha_i(x,t)$ such that

$$\lim_{t\to\infty}\|e_i(t)\| = \lim_{t\to\infty}\|y_i(t) - \alpha_i(x_i(t-\tau(t)))\| = 0 .$$

**Definition 2** [19]. The Kronecker product of matrices A and B is defined as:

$$A \otimes B = \begin{pmatrix} a_{11}B & \cdots & a_{1m}B \\ \vdots & \ddots & \vdots \\ a_{n1}B & \cdots & a_{nm}B \end{pmatrix} \tag{3}$$

where if A is an $n\times m$ matrix and B is a $p\times q$ matrix, then $A \otimes B$ is an $np\times mq$ matrix.

**Remark 1.** If $\tau(t)=0$, we know that the generalized function projective lag synchronization reduced to generalized function projective synchronization.

**Remark 2.** If $\alpha_i(x,t)$ are taken as nonzero constants then the generalized function projective lag synchronization reduced to projective lag synchronization.

## 3    Main Result

In this section, we consider the generalized function projective lag synchronization of neural networks (1) and (2). For simple we regard $\alpha_i(x_i(t-\tau(t)))$ as a linear function $s_i x_i(t-\tau(t)) + \lambda_i$, $s_i$, $\lambda_i$ are constant $i=1,2,...,N$, then we have $e_i(t) = y_i(t) - \alpha_i(x_i(t-\tau(t))) = y_i(t) - s_i x_i(t-\tau(t)) - \lambda_i$. So we can get error networks as following:

$$\dot{e}_i(t) = \dot{y}_i(t) - s_i \dot{x}_i(t-\tau(t))(1-\dot{\tau}(t))$$

$$= B_i y_i(t) + g_i(y_i(t)) + \sum_{j=1}^{N} d_{ij}Qy_j(t) + u_i(t) - s_i A_i x_i(t-\tau(t))(1-\dot{\tau}(t))$$

$$-s_i f_i x_i(t-\tau(t))(1-\dot{\tau}(t)) - s_i \sum_{j=1}^{N} c_{ij} P x_j(t-\tau(t))(1-\dot{\tau}(t))$$

$$= B_i e_i(t) + (B_i - A_i) s_i x_i(t-\tau(t)) + s_i A_i x_i(t-\tau(t))\dot{\tau}(t) + g_i(y_i(t))$$

$$- s_i f_i(x_i(t-\tau(t))) + s_i f_i(x_i(t-\tau(t)))\dot{\tau}(t) + \sum_{j=1}^{N} d_{ij}Qy_j(t)$$

$$- s_i \sum_{j=1}^{N} c_{ij} P x_j(t-\tau(t)) + s_i \sum_{j=1}^{N} c_{ij} P x_j(t-\tau(t))\dot{\tau}(t) + B_i \lambda_i + u_i(t)$$

$$= B_i e_i(t) + (B_i - A_i) s_i x_i(t-\tau(t)) + g_i(y_i(t)) - s_i f_i(x_i(t-\tau(t)))$$

$$+ s_i \dot{\tau}(t)\dot{x}_i(t-\tau(t)) + \sum_{j=1}^{N} d_{ij}Qe_j(t) + s_i \sum_{j=1}^{N} d_{ij}Qx_i(t-\tau(t)) \tag{4}$$

$$+ \lambda_i \sum_{j=1}^{N} d_{ij}Q - s_i \sum_{j=1}^{N} c_{ij} P x_j(t-\tau(t)) + B_i \lambda_i + u_i(t)$$

$$= B_i e_i(t) + (B_i - A_i) s_i x_i(t - \tau(t)) + g_i(y_i(t)) - s_i f_i(x_i(t - \tau(t)))$$

$$+ s_i \dot{\tau}(t) \dot{x}_i(t - \tau(t)) + \sum_{j=1}^{N} d_{ij} Q e_j(t) + \lambda_i \sum_{j=1}^{N} d_{ij} Q - s_i \sum_{j=1}^{N} (c_{ij} P - d_{ij} Q) x_j(t - \tau(t))$$

$$+ B_i \lambda_i + u_i(t)$$

**Theorem 1.** The neural networks (2) and (1) can realize generalized function projective lag synchronization, if we choose the adaptive controller and update law such that:

$$u_i(t) = -(B_i - A_i) s_i x_i(t - \tau(t)) - g_i(y_i(t)) + s_i f_i(x_i(t - \tau(t)))$$

$$- s_i \dot{\tau}(t) \dot{x}_i(t - \tau(t)) - \lambda_i \sum_{j=1}^{N} d_{ij} Q + s_i \sum_{j=1}^{N} (c_{ij} P - d_{ij} Q) x_j(t - \tau(t)) \quad i = 1, 2, ..., N \quad (5)$$

$$- B_i \lambda_i - k^* e_i(t)$$

where $k^*$ is a positive constant to be determined.

**Proof.** Choose the following Lyapunov function:

$$V(t) = \frac{1}{2} \sum_{i=1}^{N} e_i^T(t) e_i(t) \qquad i = 1, 2, ..., N \qquad (6)$$

Substituting (4) into (6), with the controllers (5), we obtain:

$$\dot{V}(t) = \sum_{i=1}^{N} e_i^T(t) \left[ B_i e_i(t) + \sum_{j=1}^{N} d_{ij} Q e_j(t) - k^* e_i(t) \right]$$

$$= \sum_{i=1}^{N} e_i^T(t) B_i e_i(t) + \sum_{i=1}^{N} \sum_{j=1}^{N} d_{ij} e_i^T(t) Q e_j(t) - k^* \sum_{i=1}^{N} e_i^T(t) e_i(t)$$

$$= e^T(t) B e(t) + e^T(t) W e(t) - k^* e^T(t) e(t)$$

$$\leq \left( \Lambda_{\max}(\frac{B + B^T}{2})) + \Lambda_{\max}(\frac{W + W^T}{2}) - k^* \right) e^T(t) e(t)$$

where $B = \text{diag}(B_1, B_2, ..., B_N) \in R^{nN \times nN}$, $W = D \otimes Q \in R^{nN \times nN}$.

Taking $k^* \geq \Lambda_{\max}(\frac{B + B^T}{2})) + \Lambda_{\max}(\frac{W + W^T}{2}) + 1$, we can get $\dot{V} \leq -e^T(t) e(t) \leq 0$

Based on Lyapunov stability theory, we know that the neural networks which we have discussed achieved function projective lag synchronization. So, the proof is completed.

**Corollary 1.** If $\tau(t) = 0$ and other situations constant, the neural networks which we have discussed can achieve generalized function projective synchronization by the following controllers:

$$u_i(t) = -(B_i - A_i) s_i x_i(t - \tau(t)) - g_i(y_i(t)) + s_i f_i(x_i(t - \tau(t)))$$

$$- \lambda_i \sum_{j=1}^{N} d_{ij} Q + s_i \sum_{j=1}^{N} (c_{ij} P - d_{ij} Q) x_j(t - \tau(t)) - B_i \lambda_i - k^* e_i(t) \qquad \begin{matrix} i = 1, 2, ..., N \quad (7) \end{matrix}$$

where $k^* \geq \Lambda_{\max}(\frac{B + B^T}{2})) + \Lambda_{\max}(\frac{W + W^T}{2}) + 1$.

**Corollary 2.** If $\alpha_i(x,t)$ are taken as nonzero constants $s_i$ and other situations constant then the neural networks which we have discussed can achieve projective lag synchronization by the controllers as following:

$$u_i(t) = -\left(B_i - A_i\right)s_i x_i\left(t - \tau(t)\right) - g_i\left(y_i(t)\right) + s_i f_i\left(x_i(t - \tau(t))\right)$$

$$- s_i \dot{\tau}(t)\dot{x}_i\left(t - \tau(t)\right) + s_i \sum_{j=1}^{N}(c_{ij}P - d_{ij}Q)x_j(t - \tau(t)) - k^* e_i(t) \qquad i = 1,2,...,N \quad (8)$$

where $k^* \geq \Lambda_{\max}(\dfrac{B+B^T}{2})) + \Lambda_{\max}(\dfrac{W+W^T}{2}) + 1$.

**Corollary 3.** If $A_i = B_i, f_i = g_i, c_{ij} = d_{ij}, P = Q$ and other situations constant then we can make two neural networks which have the same structure to achieve generalized function projective lag synchronization by the following controllers:

$$u_i(t) = s_i x_i\left(t - \tau(t)\right) - g_i\left(y_i(t)\right) + s_i f_i\left(x_i(t - \tau(t))\right)$$

$$- s_i \dot{\tau}(t)\dot{x}_i\left(t - \tau(t)\right) - \lambda_i \sum_{j=1}^{N} d_{ij}Q - B_i\lambda_i - k^* e_i(t) \qquad i = 1,2,...,N \quad (9)$$

where $k^* \geq \Lambda_{\max}(\dfrac{B+B^T}{2})) + \Lambda_{\max}(\dfrac{W+W^T}{2}) + 1$.

## 4    Simulation

In this section, the simulation example is performed to verify the effectiveness of the proposed synchronization scheme in the previous section. We consider six Lorenz systems and six Chen systems as the drive neural network and response networks, respectively. Comparing Eq. (1), and Eq. (2), we can get:

$$\begin{cases} \dot{x}_1 = a(y_1 - x_1) \\ \dot{y}_1 = (b - z_1)x_1 - y_1 = \\ \dot{z}_1 = x_1 y_1 - cz_1 \end{cases} \begin{bmatrix} -a & a & 0 \\ b & -1 & 0 \\ 0 & 0 & -c \end{bmatrix}\begin{bmatrix} x_1 \\ y_1 \\ z_1 \end{bmatrix} + \begin{bmatrix} 0 \\ -x_1 z_1 \\ x_1 y_1 \end{bmatrix}, \qquad (10)$$

$$\begin{cases} \dot{x}_2 = \alpha(y_2 - x_2) \\ \dot{y}_2 = (\gamma - \alpha)x_2 - x_2 z_2 + \gamma y_2 = \\ \dot{z}_2 = x_2 y_2 - \beta z_2 \end{cases} \begin{bmatrix} -\alpha & \alpha & 0 \\ \gamma - \alpha & \gamma & 0 \\ 0 & 0 & -\beta \end{bmatrix}\begin{bmatrix} x_2 \\ y_2 \\ z_2 \end{bmatrix} + \begin{bmatrix} 0 \\ -x_2 z_2 \\ x_2 y_2 \end{bmatrix}. \quad (11)$$

where $x_1$, $y_1$, $z_1$ are the drive state variables, when the three real parameters $a = 10, b = 28, c = \dfrac{8}{3}$, the system shows chaotic behavior, $x_2$, $y_2$, $z_2$ are response state variables, when three real parameters $\alpha = 35$, $\beta = 3$, $\gamma = 28$, the system shows chaotic behavior. They are shown in Figs 1-2, respectively. Comparing Eq.(1) and (2), we have:

$$A = \begin{bmatrix} -10 & 10 & 0 \\ 28 & -1 & 0 \\ 0 & 0 & -\frac{8}{3} \end{bmatrix} \quad f = \begin{bmatrix} 0 \\ -x_1 z_1 \\ x_1 y_1 \end{bmatrix} \quad B = \begin{bmatrix} -35 & 35 & 0 \\ -7 & 28 & 0 \\ 0 & 0 & -3 \end{bmatrix} \quad g = \begin{bmatrix} 0 \\ -x_2 z_2 \\ x_2 y_2 \end{bmatrix} \quad (12)$$

We rewrite the drive neural networks and response networks as following:

$$\dot{x}_i(t) = Ax_i(t) + f(x_i(t)) + \sum_{j=1}^{6} c_{ij} P x_j(t) + I_i \quad i = 1, 2, ..., 6 \quad (13)$$

$$\dot{y}_i(t) = By_i(t) + g(y_i(t)) + \sum_{j=1}^{6} d_{ij} Q y_j(t) + I_i + u_i(t) \quad i = 1, 2, ..., 6 \quad (14)$$

For simply, we take the input vector $I_i$ as $(0,0,0)$, $\lambda_i = 0$, $s_i = 1$, $u_i$ is designed as Eq.(4), the inner coupling matrices of Eq.(13) and Eq.(14) $P = \text{diag}(1,1,1)$ and $Q = \text{diag}(1,1,1)$, the coupling configuration matrices $C = (c_{ij})_{6\times6}$, $D = (d_{ij})_{6\times6}$ are given respectively as follows:

$$C = \begin{pmatrix} -3 & 1 & 0 & 0 & 2 & 0 \\ 1 & -5 & 2 & 0 & 0 & 2 \\ 1 & 2 & -4 & 0 & 1 & 0 \\ 0 & 1 & 1 & -2 & 0 & 0 \\ 0 & 1 & 0 & 2 & -5 & 2 \\ 1 & 0 & 1 & 0 & 2 & -4 \end{pmatrix} \quad D = \begin{pmatrix} -5 & 2 & 0 & 2 & 0 & 1 \\ 0 & -4 & 1 & 0 & 2 & 1 \\ 2 & 0 & -3 & 0 & 1 & 0 \\ 1 & 2 & 0 & -4 & 0 & 1 \\ 1 & 0 & 2 & 2 & -5 & 0 \\ 1 & 0 & 0 & 0 & 2 & -3 \end{pmatrix}$$

Let $x_i(0) = (0.1 + 0.3i, 0.1 + 0.3i, 0.1 + 0.3i)^T$, $y_i(0) = (2 + 0.5i, 2 + 0.5i, 2 + 0.5i)^T$ and $\tau(t) = 1$. Based on theorem 1, the neural networks we have discussed can achieve function projective lag synchronization, the simulation results as shown in Fig 3.

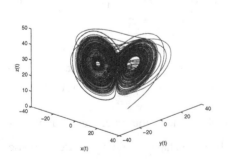

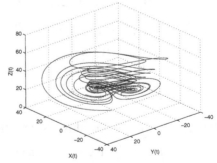

**Fig. 1.** Chaotic attractor of drive system Eq. (9)

**Fig. 2.** Chaotic attractor of response system Eq. (10)

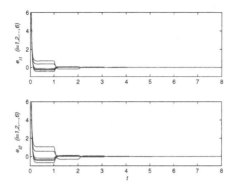

**Fig. 3.** The function projective lag synchronization errors of the drive networks(1) and response networks(2)

## 5    Conclusion

In this paper, we investigated the function projective lag synchronization between two different neural networks. Based on Lyapunov stability theory and feedback control method, the two neural networks can realize FPLS under the designed controller and updated law. Moreover the function projective lag synchronization which we have investigated is special and different from the other literatures, the scaling functions are not only depending on time, but also depending on the networks. We also discussed the FPLS of the same structure of the networks in Corollary 3. Finally, numerical simulations are provided to show the effectiveness of the main result.

**Acknowledgments.** This work was supported by the National Nature Science foundation of China (Nos. 11171135, 51276081), the Society Science Foundation from Ministry of Education of China (Nos. 12YJAZH002, 08JA790057), the Project Funded by the Priority Academic Program Development of Jiangsu Higher Education Institutions, the Advanced Talents' Foundation of Jiangsu University (Nos.07JDG054, 10JDG140), and the Students' Research Foundation of Jiangsu University (Nos Y11A079, 11A179). Especially, thanks for the support of Jiangsu University.

## References

1. Akhmet, U., Arugaslan, D., Yilmaz, E.: Stability in Cellular Neural Networks with a Piecewise Constant Argument. J. Comp. Appl. Math. 233, 2365–2373 (2010)
2. Pan, L.J., Cao, J.D.: Robust Stability for Uncertain Stochastic Neural Network with Delay and Impulses. Neurocomputing 94, 102–110 (2012)
3. Cai, G., Ma, H., Li, Y.: Adaptive Projective Synchronization and Function Projective Synchronization of a Chaotic Neural Network with Delayed and Non-Delayed Coupling. In: Wang, J., Yen, G.G., Polycarpou, M.M. (eds.) ISNN 2012, Part I. LNCS, vol. 7367, pp. 293–301. Springer, Heidelberg (2012)

4. Huang, W.Z., Huang, Y.: Chaos of a New Class of Hopfield Neural Networks. Appl. Math. Comp. 206, 1–11 (2008)
5. Pecora, L.M., Carroll, T.L.: Synchronization in Chaotic Systems. Phys. Rev. Lett. 64, 821–824 (1990)
6. Cai, G.L., Shao, H.J., Yao, Q.: A Linear Matrix Inequality Approach to Global Synchronization of Multi-Delay Hopfield Neural Networks with Parameter Perturbations. Chin. J. Phys. 50, 86–99 (2012)
7. Cai, G.L., Wang, H.X., Zheng, S.: Adaptive Function Projective Synchronization of Two Different Hyperchaotic Systems with Unknown Parameters. Chin. J. Phys. 47, 662–669 (2009)
8. Yang, X.S., Cao, J.D.: Stochastic Synchronization of Coupled Neural Networks with Intermittent Control. Phys. Lett. A 373, 3259–3272 (2009)
9. Zheng, S., Dong, G.G., Bi, Q.S.: Impulsive Synchronization of Complex Networks with Non-Delayed Coupling. Phys. Lett. A 373, 4255–4259 (2009)
10. Meng, J., Wang, X.Y.: Generalized Projective Synchronization of a Class of Delayed Neural Networks. Mod. Phys. Lett. B 22, 181–190 (2008)
11. Zhang, D., Xu, J.: Projective Synchronization of Different Chaotic Time-Delayed Neural Networks Based on Integral Sliding Mode Controller. Appl. Math. Comp. 217, 164–174 (2010)
12. Wang, Z.D., Wang, Y., Liu, Y.R.: Global Synchronization for Discrete-Time Stochastic Complex Networks with Randomly Occurred Nonlinearities and Mixed Time Delays. IEEE Trans. Neur. Netw. 21, 11–25 (2010)
13. Wang, K., Teng, Z.D., Jiang, H.J.: Adaptive Synchronization of Neural Networks with Time-Varying Delay and Distributed Delay. Phys. A 387, 631–642 (2008)
14. Cai, G.L., Yao, Q., Shao, H.J.: Global Synchronization of Weighted Cellular Neural Networks with Time-Varying Coupling Delay. Comm. Nonlin. Sci. Numer. Simul. 17, 3843–3847 (2012)
15. Tang, Y., Qiu, R.H., Fang, J.A., Miao, Q.Y., Xia, M.: Adaptive Lag Synchronization in Unknown Stochastic Caotic Neural Networks with Discrete and Distributed Time-Varying Delays. Phys. Lett. A 372, 4425–4433 (2008)
16. Hu, C., Yu, J., Jiang, H.J.: Exponential Lag Synchronization for Neural Networks with Mixed Delays via Periodically Intermittent Control. Chaos 20, 1–8 (2010)
17. Ding, W., Han, M.A., Li, M.L.: Exponential Lag Synchronization of Delayed Fuzzy Cellular Neural Networks with Impulses. Phys. Lett. A 373, 832–837 (2009)
18. Wu, X.J., Lu, H.T.: Adaptive Generalized Function Projective Lag Synchronization of Different Chaotic Systems with Fully Uncertain Parameters. Chaos, Solitons & Fractals 44, 802–810 (2011)
19. Amy, L., William, S.: The Kronecker Product and Stochastic Automata Networks. J.Comp. Appl. Math. 167, 429–447 (2004)

# Application of Local Activity Theory of CNN to the Coupled Autocatalator Model

Guangwu Wen, Yan Meng, Lequan Min, and Jing Zhang

Schools of Mathematics and Physics
University of Science and Technology Beijing
Beijing 100083, P.R. China
wenguangwu@yeah.net/mengyan1220@sohu.com

**Abstract.** The study of chemical reactions with oscillating kinetics has drawn increasing interest over the last few decades. However the dynamical properties of the coupled nonlinear dynamic system are difficult to deal with. The local activity principle of the Cellular Nonlinear Network (CNN) introduced by Chua has provided a powerful tool for studying the emergence of complex behaviors in a homogeneous lattice formed by coupled cells. Based on the Autocatalator Model introduced by Peng.B, this paper establishes a two dimensional coupled Autocatalator CNN system. Using the analytical criteria for the local activity calculates the chaos edge of the Autocatalator CNN system. The numerical simulations show that the emergence may exist if the selected cell parameters are nearby the edge of chaos domain. The Autocatalator CNN can exhibit periodicity and chaos.

**Keywords:** Cellular Nonlinear Network, edge of chaos, reaction diffusion equation, chaos.

## 1   Introduction

The study of autocatalator model for isothermal reaction in a thermodynamically closed system has drawn increasing interest over the last few decades, because it also contributes towards a deeper understanding of the complex phenomena of temporal and spatial organizations in biological systems. The role of diffusion in the modelling of many physical, chemical and biological processes has been extensively studied. However the dynamical properties of the coupled nonlinear dynamic system are difficult to deal with.

The cellular nonlinear network (CNN), first introduced by Chua and Yang [1], have been widely studied for image processing and biological versions([2],[3]). The theory of CNN local activity principle introduced by Chua([4],[5],[6]) has provided an effective analytical tool for determining whether a lattice dynamical system made of coupled cells can exhibit emergent and complex behaviors. It asserts that a wide spectrum of complex behaviors may exist if the corresponding cell parameters of the CNN's are chosen in or nearby the edge of chaos domain[7]. In particular, some analytical criteria for local activity of CNN's have

C. Guo, Z.-G. Hou, and Z. Zeng (Eds.): ISNN 2013, Part I, LNCS 7951, pp. 133–140, 2013.

been established and applied to the study of the dynamics of the CNN's related to the FitzHugh-Nagumo equation [7], the Yang-Epstein model [8], the Oregonator model[9] and smoothed Chua's circuit CNNs [10], respectively. Based on the Autocatalator Model introduced by Peng.B[11], this paper establishes a two dimensional coupled Autocatalator CNN system. Using the analytical criteria for the local activity calculates the bifurcation diagrams of the edge of chaos. Numerical simulations show that the emergence may appear if the chosen cell parameters are located nearby the edge of chaos domain.

The rest of this paper is organized as follows: Section 2 states the basic local activity theory developed by Chua. Bifurcation diagrams and numerical simulation of the Autocatalator CNN's are given in section 3. Concluding remarks are given in section 4.

## 2     Main Theorem of Local Activity

Generally speaking, in a reaction-diffusion CNN, every $C(i, j, k)$ has $n$ state variables but only $m(\leq n)$ state variables coupled directly to their nearest neighbors. Consequently, each cell $C(i, j, k)$ has the following state equations:

$$\dot{V}_1(j, k, l) = f_1(V_1, V_2, \ldots, V_n) + D_1\nabla^2 V_1$$

$$\vdots$$

$$\dot{V}_m(j, k, l) = f_m(V_1, V_2, \ldots, V_n) + D_m\nabla^2 V_m$$
$$\dot{V}_{m+1}(j, k, l) = f_{m+1}(V_1, V_2, \ldots, V_n) \tag{1}$$

$$\vdots$$

$$\dot{V}_n(j, k, l) = f_n(V_1, V_2, \ldots, V_n)$$
$$j, k, l = 1, 2, \ldots, N_x$$

In vector form, Eq.(1) becomes

$$\dot{\mathbf{V}}_{\mathbf{a}} = \mathbf{f}_{\mathbf{a}}(\mathbf{V}_{\mathbf{a}}, \mathbf{V}_{\mathbf{b}}) + \mathbf{I}_{\mathbf{a}}, \dot{\mathbf{V}}_{\mathbf{b}} = \mathbf{f}_{\mathbf{b}}(\mathbf{V}_{\mathbf{a}}, \mathbf{V}_{\mathbf{b}})$$

The cell equilibrium point $Q_i = (\overline{\mathbf{V}}_{\mathbf{a}}; \overline{\mathbf{V}}_{\mathbf{b}})(\in \mathbb{R}^n)$ of Eq.(2) for the restricted local activity domain [7] can be determined numerically or analytically, via

$$\mathbf{f}_{\mathbf{a}}(\mathbf{V}_{\mathbf{a}}, \mathbf{V}_{\mathbf{b}}) = 0, \mathbf{f}_{\mathbf{b}}(\mathbf{V}_{\mathbf{a}}, \mathbf{V}_{\mathbf{b}}) = 0.$$

The local linearized state equations at the cell equilibrium point $Q_i$ are defined

$$\dot{\mathbf{V}}_{\mathbf{a}} = \mathbf{A}_{\mathbf{aa}}\mathbf{V}_{\mathbf{a}} + \mathbf{A}_{\mathbf{ab}}\mathbf{V}_{\mathbf{b}} + \mathbf{I}_{\mathbf{a}},$$
$$\dot{\mathbf{V}}_{\mathbf{b}} = \mathbf{A}_{\mathbf{ba}}\mathbf{V}_{\mathbf{a}} + \mathbf{A}_{\mathbf{bb}}\mathbf{V}_{\mathbf{b}}. \tag{2}$$

where the Jacobian matrix at equilibrium point $Q_i$, for the restricted local activity domain, has the following form:

$$\mathbf{J}(Q_i) \overset{\triangle}{=} [a_{lk}(Q_i)] \overset{\triangle}{=} \begin{bmatrix} \mathbf{A}_{\mathbf{aa}}(Q_i) & \mathbf{A}_{\mathbf{ab}}(Q_i) \\ \mathbf{A}_{\mathbf{ba}}(Q_i) & \mathbf{A}_{\mathbf{bb}}(Q_i) \end{bmatrix} \tag{3}$$

The admittance matrix at equilibrium point $Q$ ([4]) is

$$Y_Q(s) \triangleq (s\mathbf{I} - \mathbf{A_{aa}}) - \mathbf{A_{ab}}(s\mathbf{I} - \mathbf{A_{bb}})^{-1}\mathbf{A_{ba}} \tag{4}$$

**Definition 1. (Local activity at cell equilibrium point $Q$)** ([4], [5], [6]) *A cell is said to be locally active at a cell equilibrium point $Q$ iff there exists a continuous function of time $i_a(t) \in \mathbf{R}^m, t \geq 0$, such that $0 < T < \infty$,*

$$\int_0^T \langle v_a(t), i_a(t) \rangle \, dt < 0 \tag{5}$$

*where $\langle \cdot, \cdot \rangle$ denotes the vector dot product, and $v_a(t)$ is a solution of the linearized state equation(2) about equilibrium point $Q_i$ with zero initial state $v_a(0) = 0$ and $v_b(0) = 0$, otherwise it is said to be locally passive at equilibrium point $Q_i$.*

**Theorem 1. Main Theorem on the Local Activity of CNN** ([4], [5]) *A three-port Reaction Diffusion CNN cell is locally active at a cell equilibrium point $Q \triangleq (\bar{V}_a, \bar{V}_b, \bar{I}_a)$ if and only if, its cell admittance matrix $Y_Q(s)$ satisfies at least one of the following four conditions:*
*1. $Y_Q(s)$ has a pole in $Re[s] > 0$.*
*2. $Y_Q^H(i\omega) = Y_Q^*(i\omega) + Y_Q(i\omega)$ is not a positive semi-definite matrix at some $\omega = \omega_0$, where $\omega_0$ is an arbitrary real number, $Y_Q^*(s)$ is constructed by first taking the transpose of $Y_Q(s)$, and then by taking the complex conjugate operation .*
*3. $Y_Q(s)$ has a simple pole $s = i\omega_\rho$ on the imaginary axis, where its associate residue matrix*

$$k_1 \triangleq \lim_{s \to i\omega_\rho} (s - i\omega_\rho)Y_Q(s)$$

*is either not a Hermitian matrix, or else not a positive semi-definite Hemitian matrix.*
*4. $Y_Q(s)$ has a multiple pole on the imaginary axis.*

**Theorem 2.** ([4], [6]) *When the number of nonzero diffusion coefficients is equal to the number of the state variables. RD CNN cell is locally active at a cell equilibrium point $Q$, if and only if the $J^s = \mathbf{A} + \mathbf{A}'$ is a nonpositive-define matrix.*

**Definition 2. (Edge of chaos with respect to equilibrium point $Q_i$)** ([4], [5]) *A "Reaction-Diffusion" CNN with one "diffusion coefficient" $D_1$ (resp. two diffusion coefficients $D_1$ and $D_2$; or three diffusion coefficients $D_1$, $D_2$, and $D_3$) is said to be operating on the edge of chaos with respect to an equilibrium point $Q_i$ if, and only if, $Q_i$ is both locally active and stable when $\bar{I}_1 = 0$ (resp. $\bar{I}_2 = 0$ and $\bar{I}_3 = 0$; or $\bar{I}_1 = 0$, $\bar{I}_2 = 0$, and $\bar{I}_3 = 0$).*

## 3    Autocatalator CNN Model and Numerical Simulation

In [11], Peng.B, Scott.S.K, and Showalte.K introduced the a three variable Autocatalator model

$$\begin{cases} \dot{x} = \mu + rz - xy^2 - x \\ \dot{y} = \gamma(x + xy^2 - y) \\ \dot{z} = S(y - z) \end{cases} \tag{6}$$

The equilibrium points of (6) are

$$Q_0 = (\frac{\mu(1-r)}{(1-r)^2 + \mu^2}, \frac{\mu}{1-r}, \frac{\mu}{1-r})$$

Now the prototype Autocatalator equations (6) can be mapped to a two-dimensional $N \times N$ Autocatalator CNN model.

$$\begin{cases} \dot{x}_{i,j} = \mu + rz_{i,j} - x_{i,j}y_{i,j}^2 - x_{i,j} + D_1[x_{i+1,j} + x_{i-1,j} \\ \qquad + x_{i,j+1} + x_{i,j-1} - 4x_{i,j}] \\ \dot{y}_{i,j} = \gamma(x_{i,j} + x_{i,j}y_{i,j}^2 - y_{i,j}) + D_2[y_{i+1,j} + y_{i-1,j} \\ \qquad + y_{i,j+1} + y_{i,j-1} - 4y_{i,j}] \\ \dot{z}_{i,j} = S(y_{i,j} - z_{i,j}) + D_3[z_{i+1,j} + z_{i-1,j} + z_{i,j+1} \\ \qquad + z_{i,j-1} - 4z_{i,j}] \\ \qquad i, j = 1, 2, \cdots, N. \end{cases} \tag{7}$$

periodic boundary condition: $x_{i,0} = x_{i,N}, y_{i,0} = y_{i,N}, z_{i,0} = z_{i,N}, x_{i,N+1} = x_{i,1}, y_{i,N+1} = y_{i,1}, z_{i,N+1} = z_{i,1}, x_{0,j} = x_{N,j}, y_{0,j} = y_{N,j}, z_{0,j} = z_{N,j}, x_{N+1,j} = x_{1,j}, y_{N+1,j} = y_{1,j}, z_{N+1,j} = z_{1,j}.$

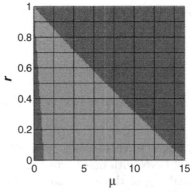

**Fig. 1.** Bifurcation diagram of the Autocatalator CNN model at $\mu \in [0, 15], r \in [0, 1], \gamma = 230, S = 40$, with respect to the equilibrium point $Q_0$. The domains are coded as follows: edge of chaos (red), locally active unstable domain (green).

Using Theorems 1 ∼ 2, the locally active domains and edges of chaos with respect to $Q_0$ of Autocatalator CNN can be numerically calculated via computer programs, respectively. The bifurcation diagram(Fig.1) with respect to the equilibrium point $Q_0$ are only locally active domains. The calculated results based on the bifurcation diagram are shown in Table 1.

**Table 1.** Cell parameters and corresponding dynamic properties of the Autocatalator CNN EQs. where $\gamma = 230, S = 40$. Symbols ⇓, ⇑, $np$ and ⊕ indicate that convergent patterns, divergent patterns, n-period patterns, and chaotic patterns are observed near to or far from the corresponding cell equilibrium points $Q_0$, respectively. The numbers which are marked by * indicate that the corresponding cell parameters lie on the edge of chaos domain. $M(\mathbf{A})$ represents the eigenvalue with the maximum real part at $Q_0$.

| No. | $\mu$ | $r$ | Equilibrium points $Q$ | $M(\mathbf{A})$ | Pattern |
|-----|-----|-----|-----|-----|-----|
| 1* | 12 | 0.19 | 0.0671, 14.8331, 14.8331 | -0.0028 ± 2.248i | 1p |
| 2* | 11 | 0.254 | 0.0675, 14.7453, 14.7453 | -0.0023 ± 2.2326i | ⇓ |
| 3* | 10 | 0.3172 | 0.0680, 14.6456, 14.6456 | -0.0002 ± 2.2151i | 1p |
| 4 | 10 | 0.1995 | 0.0795, 12.4922, 12.4922 | 0.3136 ± 1.8544i | 3p |
| 5 | 10 | 0.195 | 0.0800, 12.4224, 12.4224 | 0.3230 ± 1.8421i | 4p |
| 6* | 10 | 0.32 | 0.0677, 14.7059, 14.7059 | -0.0096 ± 2.2247i | ⇓ |
| 7 | 10 | 0.1965 | 0.0798, 12.4446, 12.4446 | 0.3199 ± 1.8462i | ⊕ |
| 8 | 8 | 0.446 | 0.0689, 14.4404, 14.4404 | 0.0039 ± 2.1789i | 1p |
| 9 | 8 | 0.445 | 0.0690, 14.4144, 14.4144 | 0.0078 ± 2.1747i | ⇑ |
| 10* | 6 | 0.55 | 0.0746, 13.3333, 13.3333 | 0.1365 ± 1.9874i | 1p |
| 11 | 6 | 0.515 | 0.0803, 12.3711, 12.3711 | 0.2671 ± 0.9923i | 2p |
| 12 | 5 | 0.540 | 0.0912, 12.8696, 12.8696 | 0.4362 ± 1.5162i | 4p |
| 13 | 5 | 0.51 | 0.0971, 10.2041, 10.2041 | 0.5122 ± 1.3777i | ⊕ |
| 14* | 4 | 0.716 | 0.0706, 14.0845, 14.0845 | -0.0002 ± 2.1148i | ⇓ |
| 15 | 4 | 0.715 | 0.0709, 14.0351, 14.0351 | 0.0070 ± 2.1063i | 1p |
| 16 | 4 | 0.600 | 0.0990, 10.0000, 10.0000 | 0.5154 ± 1.3230i | 3p |

Now let us choose parameters $\mu = 10, r = 0.1965, \gamma = 230, S = 40$, which are lie in the locally active unstable domains nearby the edges of chaos. The initial conditions are given by following.

$$\begin{cases} [x_{i,j}(0)] = 0.0464 + 0.001 rand(21, 21) \\ [y_{i,j}(0)] = 14.7669 + 0.01 rand(21, 21) \quad i, j = 1, 2, \cdots, 21 \\ [z_{i,j}(0)] = 15.4041 + 0.01 rand(21, 21). \end{cases}$$

where $rand(21, 21)'s$ are $21 \times 21$ matrices with uniformly distributed random generated, and we select $D_1 = 0.01, D_2 = 0.1, D_3 = 0.1$. The chaotic trajectories of cell located in $\{12, 12\}$ over the time interval $[0, 0.5]$ are shown in Fig.2.

We furthermore choose parameters $\mu = 10, r = 0.3172, \gamma = 230, S = 40$, which are lie in the edges of chaos domains. The periodic trajectories of cell located in $\{12, 12\}$ are shown in Fig.3. Three-dimensional views of the state variables are shown in Fig.6. Observe that three state variables exhibit oscillations and a new type of limit cycle spiral waves has emerged. Some of simulation results based on the bifurcation diagram are shown in Figs.4-5.

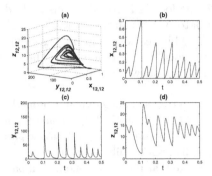

**Fig. 2.** Chaotic trajectories: $\mu = 10, r = 0.1965, \gamma = 230, S = 40$ (a) $x_{12,12}, y_{12,12}, z_{12,12}$. (b) $x_{12,12}$. (c) $y_{12,12}$. (d)$z_{12,12}$.

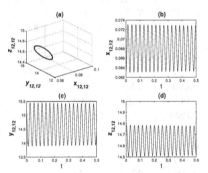

**Fig. 3.** Periodic trajectories: $\mu = 8, r = 0.446, \gamma = 230, S = 40$.(a) $x_{12,12}, y_{12,12}, z_{12,12}$. (b) $x_{12,12}$. (c) $y_{12,12}$. (d) $z_{12,12}$

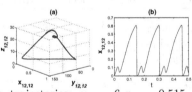

**Fig. 4.** Double periodic trajectories: $\mu = 6, r = 0.515, \gamma = 230, S = 40$. (a) $x_{12,12}, y_{12,12}, z_{12,12}$. (b) $x_{12,12}$.

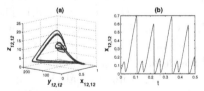

**Fig. 5.** Multiperiodic trajectories: $\mu = 5, r = 0.54, \gamma = 230, S = 40$. (a) $x_{12,12}, y_{12,12}, z_{12,12}$. (b) $x_{12,12}$.

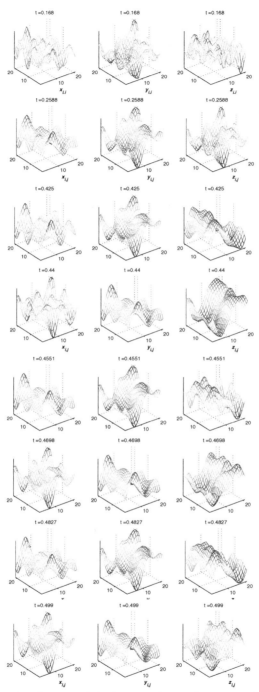

**Fig. 6.** Three-dimensional views of the limit cycle spiral waves at different time $t$. The vertical axes represent the state variables $x'_{i,j}s$, $y'_{i,j}s$ and $z'_{i,j}s$. The horizontal axes define the spatial coordinates $(i, j)$.

# 4    Concluding Remarks

The local activity criteria of CNN's provide a new tool to the research on the Coupled nonlinear dynamical cell models.This paper uses the criteria of the local activity for the CNN with three state variables and three ports to study the coupled Autocatalator model with three coupling terms. It has been found that for the parameters $\mu = 10, r = 0.1965, \gamma = 230, S = 40$, the corresponding RD CNN can exhibit chaos behaviors. When $\mu = 10, r = 0.3, \gamma = 230, S = 40$,it can exhibit a periodic behaviors. Roughly speaking, if the parameter group of the Autocatalator CNN are selected nearby the edge of chaos domain, the dynamics of the correspondingly Autocatalator CNN always exhibit chaos or period trajectory. If the selected cell parameters are located on the edge of chaos domain with respect to $Q_0$, then the trajectories of the Autocatalator equations will always converge to equilibrium point $Q_0$ or exhibit period trajectory. The effectiveness of the local activity principle in the study for the emergence of complex behaviors in a homogeneous lattice formed by coupled cells is confirmed.

**Acknowledgments.** This project is jointly supported by the National Natural Science Foundations of China (Grant No. 61074192).

# References

1. Chua, L.O., Yang, L.: Cellular neural networks: Theory and Applications. IEEE Trans. Circuits Syst. 35, 1257–1290 (1988)
2. Li, H., Liao, H., Li, C., Huang, H., Li, C.: Edge detection of noisy images based on cellular neural networks. Communications in Nonlinear Science and Numerical Simulation 16(9), 3746–3759 (2011)
3. Shaojiang, D., Yuan, T., Xipeng, H., Pengcheng, W., Mingfu, Q.: Application of new advanced CNN structure with adaptive thresholds to color edge detection. Int. CSNS 17, 1637–1648 (2012)
4. Chua, L.O.: CNN: Visions of complexity. Int. J. Bifur. and Chaos 7, 2219–2425 (1997)
5. Chua, L.O.: Passivity and Complexity. IEEE Trans. Circuits Syst. I. 46, 71–82 (1999)
6. Chua, L.O.: Local Activity is the origin of complexity. Int. J. Bifur. Chaos 15, 3435–3456 (2005)
7. Dogaru, R., Chua, L.O.: Edge of chaos and local activity domain of Fitzhugh-Nagumo equation. Int. J. Bifur. and Chaos 8, 211–257 (1998)
8. Min, L., Meng, Y., Chua, L.O.: Applications of Local Activity Theory of CNN to Controlled Coupled Oregonator Systems. International Journal of Bifurcation and Chaos 18(11), 3233–3297 (2008)
9. Min, L., Crounse, K.R., Chua, L.O.: Analytical criteria for local activity and applications to the Oregonator CNN. Int. J. Bifur. and Chaos 10, 25–71 (2000)
10. Dong, X., Min, L.: Analytical Criteria for Local Activity of One-Port CNN with Five State Variables and Application. Dynamics of Continuous, Discrete and Impulsive Systems 11, Series B. Supplyment Issue. 78–93 (2004)
11. Peng, B., Scott, S.K., Showalte, K.: Period Doubling and Chaos in a Three-Variable Autocatalator. Int. J. Phys. Chem. 94, 5243–5246 (1990)

# Passivity Criterion of Stochastic T-S Fuzzy Systems with Time-Varying Delays

Zhenjiang Zhao[1] and Qiankun Song[2]

[1] Department of Mathematics, Huzhou Teachers College, Huzhou 313000, China
[2] Department of Mathematics, Chongqing Jiaotong University,
Chongqing 400074, China
qiankunsong@163.com

**Abstract.** In this paper, the passivity for stochastic Takagi-Sugeno (T-S) fuzzy systems with time-varying delays is investigated without assuming the differentiability of the time-varying delays. By utilizing the Lyapunov functional method, the Itô differential rule and the matrix inequality techniques, a delay-dependent criterion to ensure the passivity for T-S fuzzy systems with time-varying delays is established in terms of linear matrix inequalities (LMIs) that can be easily checked by using the standard numerical software.

**Keywords:** T-S fuzzy systems, Passivity, Time-varying delays, Stochastic disturbance.

## 1 Introduction

The Takagi-Sugeno (T-S) fuzzy system, first proposed and studied by Takagi and Sugeno[1], have been widely applied within various engineering and scientific fields since it provides a general framework to represent a nonlinear plant by using a set of local linear models which are smoothly connected through nonlinear fuzzy membership functions [2]. In practice, time-delays often occur in many dynamic systems such as chemical processes, metallurgical processes, biological systems, and neural networks [3]. The existence of time-delays is usually a source of instability and poor performance [4]. Besides, stochastic disturbance is probably another source leading to undesirable behaviors of T-S fuzzy systems [5]. As a result, stability analysis for T-S fuzzy systems with time-delay has not only important theoretical interest but also practical value [6]. Recently, the stability analysis for T-S fuzzy systems with delays and stochastic T-S fuzzy systems with delays have been extensively studied, for example, see [2–6], and references therein.

On the other hand, the passivity theory is another effective tool to the stability analysis of system [7]. The main idea of passivity theory is that the passive properties of system can keep the system internal stability [8]. For these reasons, the passivity and passification problems have been an active area of research recently. The passification problem, which is also called the passive control problem, is formulated as the one of finding a suitable controller such that the resulting

C. Guo, Z.-G. Hou, and Z. Zeng (Eds.): ISNN 2013, Part I, LNCS 7951, pp. 141–148, 2013.
© Springer-Verlag Berlin Heidelberg 2013

closed-loop system is passive. Recently, some authors have studied the passivity of some systems and obtained sufficient conditions for checking the passivity of the systems that include linear systems with delays [7], delayed neural networks [8], and T-S fuzzy systems with delays [9]-[14]. In [9–11], authors considered the continuous-time T-S fuzzy system with constant delays, and presented several criteria for checking the passivity and feedback passification of the system. In [12], discrete-time T-S fuzzy systems with delays were considered, several sufficient conditions for checking passivity and passification were obtained. In [11, 13, 14], the contiguous-time T-S fuzzy systems with time-varying delays were investigated, several criteria to ensure the passivity and feedback passification were given. It is worth noting that it has been assumed in [11, 13] that the time-varying delays are *differentiable*. However, when the time-varying delays are not differentiable, the presented methods in [11, 13] are difficult to be applied to investigate the passivity and passification of T-S fuzzy systems with time-varying. Therefore, there is a need to further improve and generalize the passivity results reported in [11, 13].

Motivated by the above discussions, the objective of this paper is to study the passivity of stochastic T-S fuzzy system with time-varying delays by employing Lyapunov-Krasovskii functionals.

## 2    Problem Formulation and Preliminaries

Consider a continuous time stochastic T-S fuzzy system with time-varying delays, and the $i$th rule of the model is of the following form:

**Plant Rule $i$**

IF $z_1(t)$ is $M_{i1}$ and ... and $z_p(t)$ is $M_{ip}$, THEN

$$\begin{cases} dx(t) = [A_i x(t) + B_i x(t - \tau(t)) + U_i w(t)]dt + \sigma_i(t, x(t), x(t - \tau(t)))d\omega(t), \\ y(t) = C_i x(t) + D_i x(t - \tau(t)) + V_i w(t), \\ x(s) = \phi(s), \quad s \in [\tau, 0], \end{cases} \quad (1)$$

where $t \geq 0$, $i = 1, 2, \ldots, r$ and $r$ is the number of IF-THEN rules; $z_1(t), \ldots, z_p(t)$ are the premise variables, each $M_{ij}$ $(j = 1, 2, \ldots, p)$ is a fuzzy set; $x(t) = (x_1(t), x_2(t), \ldots, x_n(t))^T \in \mathbb{R}^n$ is the state vector of the system at time $t$; $w(t) = (w_1(t), w_2(t), \ldots, w_q(t))^T \in \mathbb{R}^q$ is the square integrable exogenous input; $y(t) = (y_1(t), y_2(t), \ldots, y_q(t))^T \in \mathbb{R}^q$ is the output vector of the system; $\tau(t)$ denote the time-varying delay, and satisfies $0 \leq \tau(t) \leq \tau$, where $\tau$ is constant; $\phi(s)$ is bounded and continuously differentiable on $[-\tau, 0]$; $A_i$, $B_i$, $U_i$, $C_i$, $D_i$ and $V_i$ are some given constant matrices with appropriate dimensions. $\omega(t) \in \mathbb{R}^m$ is a scalar Brownian motion defined on $(\Omega, \mathcal{F}, \mathcal{P})$; $\sigma_i : \mathbb{R} \times \mathbb{R}^n \times \mathbb{R}^n \to \mathbb{R}^{n \times m}$ is the noise intensity function.

Let $\mu_i(t)$ be the normalized membership function of the inferred fuzzy set $\gamma_i(t)$, that is

$$\mu_i(t) = \frac{\gamma_i(t)}{\sum\limits_{i=1}^{r} \gamma_i(t)},$$

where $\gamma_i(t) = \prod\limits_{j=1}^{p} M_{ij}(z_j(t))$ with $M_{ij}(z_j(t))$ being the grade of membership function of $z_j(t)$ in $M_{ij}(t)$. It is assumed that $\gamma_i(t) \geq 0$ $(i = 1, 2, \ldots, r)$ and $\sum\limits_{i=1}^{r} \gamma_i(t) > 0$ for all $t$. Thus $\mu_i(t) \geq 0$ and $\sum\limits_{i=1}^{r} \mu_i(t) = 1$ for all $t$. And the T-S fuzzy model (1) can be defuzzied as

$$
\begin{cases}
dx(t) = \sum\limits_{i=1}^{r} \mu_i(t)[A_i x(t) + B_i x(t - \tau(t)) + U_i w(t)]dt \\
\quad + \sum\limits_{i=1}^{r} \mu_i(t)\sigma_i(t, x(t), x(t - \tau(t)))d\omega(t), \\
y(t) = \sum\limits_{i=1}^{r} \mu_i(t)[C_i x(t) + D_i x(t - \tau(t)) + V_i w(t)], \\
x(s) = \phi(s), \quad s \in [-\tau, 0].
\end{cases}
\tag{2}
$$

Let $f(t) = \sum\limits_{i=1}^{r} \mu_i(t)[A_i x(t) + B_i x(t - \tau(t)) + U_i w(t)]$, $\alpha(t) = \sum\limits_{i=1}^{r} \mu_i(t)\sigma_i(t, x(t),$
$x(t - \tau(t)))$, then model (2) is rewritten as

$$
\begin{cases}
dx(t) = f(t)dt + \alpha(t)d\omega(t), \\
y(t) = \sum\limits_{i=1}^{r} \mu_i(t)[C_i x(t) + D_i x(t - \tau(t)) + V_i w(t)], \\
x(s) = \phi(s), \quad s \in [-\tau, 0].
\end{cases}
\tag{3}
$$

Throughout this paper, we make the following assumptions:
(**H**). There exist constant matrices $R_{1i}$ and $R_{2i}$ of appropriate dimensions such that the following inequality

$$
\text{trace}(\sigma_i^T(t, \alpha, \beta)\sigma_i(t, \alpha, \beta)) \leq \|R_{1i}\alpha\|^2 + \|R_{2i}\beta\|^2
\tag{4}
$$

holds for all $i = 1, 2, \ldots, r$ and $(t, \alpha, \beta) \in \mathbb{R} \times \mathbb{R}^n \times \mathbb{R}^n$.

**Definition 1.** *System (1) is called passive if there exists a scalar $\gamma > 0$ such that*

$$
2\mathbf{E}\left\{ \int_0^{t_p} y^T(s)w(s)ds \right\} \geq -\gamma\mathbf{E}\left\{ \int_0^{t_p} w^T(s)w(s)ds \right\}
$$

*for all $t_p \geq 0$ and for the solution of (1) with $\phi(\cdot) \equiv 0$.*

Let $C^{1,2}(\mathbb{R} \times \mathbb{R}^n, \mathbb{R}^+)$ denote the family of all nonnegative function $V(t, x(t))$ on $\mathbb{R} \times \mathbb{R}^n$ which are continuously once differentiable in $t$ and twice differentiable in $x$. For each $V \in C^{1,2}(\mathbb{R} \times \mathbb{R}^n, \mathbb{R}^+)$, By Itô's differential formula, the stochastic derivative of $V(t, x(t))$ along (3) can be obtained as:

$$
dV(t, x(t)) = \mathcal{L}V(t, x(t))dt + V_x(t, x(t))\alpha(t)d\omega(t),
\tag{5}
$$

where $\mathcal{L}$ is the weak infinitesimal operator of the stochastic process $\{x_t = x(t + s)|t \geq 0, -\rho \leq s \leq 0\}$, and the mathematical expectation of $\mathcal{L}V(t, x(t))$ is given by

$$\mathbf{E}\{\mathcal{L}V(t, x(t))\} = \mathbf{E}\Big\{V_t(t, x(t)) + V_x(t, x(t))f(t)$$
$$+ \frac{1}{2}\text{trace}(\alpha^T(t)V_{xx}(t, x(t))\alpha(t))\Big\}, \tag{6}$$

in which

$$V_t(t, x(t)) = \frac{\partial V(t, x(t))}{\partial t}, V_x(t, x(t)) = \Big(\frac{\partial V(t, x(t))}{\partial x_1}, \dots, \frac{\partial V(t, x(t))}{\partial x_1}\Big),$$

$$V_{xx}(t, x(t)) = \Big(\frac{\partial^2 V(t, x(t))}{\partial x_i \partial x_j}\Big)_{n \times n}.$$

To prove our results, the following lemmas that can be found in [15] are necessary.

**Lemma 1.** *For any constant matrix $W \in \mathbb{R}^{m \times m}$, $W > 0$, scalar $0 < h(t) < h$, vector function $\omega(\cdot) : [0, h] \to \mathbb{R}^m$ such that the integrations concerned are well defined, then*

$$\Big(\int_0^{h(t)} \omega(s)ds\Big)^T W\Big(\int_0^{h(t)} \omega(s)ds\Big) \leq h(t)\int_0^{h(t)} \omega^T(s)W\omega(s)ds.$$

**Lemma 2.** *Let $a, b \in R^n$, $P$ be a positive definite matrix, then $2a^T b \leq a^T P^{-1} a + b^T Pb$.*

**Lemma 3.** *Given constant matrices $P$, $Q$ and $R$, where $P^T = P$, $Q^T = Q$, then*

$$\begin{bmatrix} P & R \\ R^T & -Q \end{bmatrix} < 0$$

*is equivalent to the following conditions*

$$Q > 0 \quad \text{and} \quad P + RQ^{-1}R^T < 0.$$

## 3    Main Result

**Theorem 1.** *Model (1) is passive if there exist four symmetric positive definite matrices $P_i$ ($i = 1, 2, 3, 4$), three matrices $Q_i$ ($i = 1, 2, 3$), three scalars $\lambda_i > 0$ ($i = 1, 2$) and $\gamma > 0$, such that the following LMIs hold for $i = 1, 2, \dots, r$:*

$$P_1 < \lambda_1 I, \tag{7}$$

$$P_3 < \lambda_2 I, \tag{8}$$

$$\Omega_i = \begin{bmatrix} \Omega_{11,i} & Q_2 & 0 & A_i^T Q_1 + P_1 & -C_i^T & Q_2 & Q_2 & 0 & 0 \\ * & \Omega_{22,i} & Q_3 & B_i^T Q_1 & -D_i^T & 0 & 0 & Q_3 & Q_3 \\ * & * & -P_2 & 0 & 0 & 0 & 0 & 0 & 0 \\ * & * & * & \Omega_{44} & Q_1^T U_i & 0 & 0 & 0 & 0 \\ * & * & * & * & \Omega_{55,i} & 0 & 0 & 0 & 0 \\ * & * & * & * & * & -\frac{1}{\tau}P_4 & 0 & 0 & 0 \\ * & * & * & * & * & * & -P_3 & 0 & 0 \\ * & * & * & * & * & * & * & -\frac{1}{\tau}P_4 & 0 \\ * & * & * & * & * & * & * & * & -P_3 \end{bmatrix} < 0, \quad (9)$$

where $\Omega_{11,i} = (\lambda_1 + \tau\lambda_2)R_{1i}R_{1i}^T + P_2 - Q_2 - Q_2^T$, $\Omega_{22,i} = (\lambda_1 + \tau\lambda_2)R_{2i}R_{2i}^T - Q_3 - Q_3^T$, $\Omega_{44} = -Q_1 - Q_1^T + \tau P_4$, $\Omega_{55,i} = -V_i - V_i^T - \gamma I$.

*Proof.* Consider the following Lyapunov-Krasovskii functional as

$$V(t) = x^T(t)P_1 x(t) + \int_{t-\tau}^{t} x^T(s)P_2 x(s)ds$$

$$+ \int_{-\tau}^{0}\int_{t+\xi}^{t}\left(\text{trace}(\alpha^T(s)P_3\alpha(s)) + f^T(s)P_4 f(s)\right)dsd\xi. \quad (10)$$

By Itô differential rule, the mathematical expectation of the stochastic derivative of $V(t)$ along the trajectory of system (3) can be obtained as

$$\mathbf{E}\{dV(t)\} = \mathbf{E}\Big\{\Big[2x^T(t)P_1 f(t) + \text{trace}(\alpha^T(t)P_1\alpha(t)) + x^T(t)P_2 x(t)$$

$$-x^T(t-\tau)P_2 x(t-\tau) + \tau\text{trace}(\alpha^T(t)P_3\alpha(t)) + \tau f^T(t)P_4 f(t)$$

$$-\int_{t-\tau}^{t}\text{trace}(\alpha^T(s)P_3\alpha(s))ds - \int_{t-\tau}^{t}f^T(s)P_4 f(s)ds\Big]dt$$

$$+2x^T(t)P_1\alpha(t)dw(t)\Big\}$$

$$\leq \mathbf{E}\Big\{\Big[2x^T(t)P_1 f(t) + (\lambda_1 + \tau\lambda_2)\text{trace}(\alpha^T(t)\alpha(t))$$

$$+x^T(t)P_2 x(t) - x^T(t-\tau)P_2 x(t-\tau) + \tau f^T(t)P_4 f(t)$$

$$-\int_{t-\tau}^{t}\text{trace}(\alpha^T(s)P_3\alpha(s))ds - \int_{t-\tau}^{t}f^T(s)P_4 f(s)ds\Big]dt\Big\}. \quad (11)$$

In deriving inequality (11), we have utilized $\mathbf{E}\{2x^T(t)P_1\alpha(t)dw(t)\} = 0$ and conditions (7) and (8).

From the definition of $f(t)$, we have

$$0 = 2\Big(-f(t) + \sum_{i=1}^{r}\mu_i(t)[A_i x(t) + B_i x(t-\tau(t)) + U_i w(t)]\Big)^T Q_1 f(t)$$

$$= \sum_{i=1}^{r}\mu_i(t)\Big[2x^T(t)A_i^T Q_1 f(t) + 2x^T(t-\tau(t))B_i^T Q_1 f(t)$$

$$+2w^T(t)U_i^T Q_1 f(t) + f^T(t)(-Q_1 - Q_1^T)f(t)\Big]. \quad (12)$$

Integrating both sides of first equation in (3) from $t - \tau(t)$ to $t$, we get

$$x(t) - x(t - \tau(t)) = \int_{t-\tau(t)}^{t} f(s)ds + \int_{t-\tau(t)}^{t} \alpha(s)d\omega(s).$$

By using Lemma 1 and Lemma 2 and noting $0 \leq \tau(t) \leq \tau$, we have

$$0 = 2x^T(t)Q_2\Big[ - x(t) + x(t - \tau(t)) + \int_{t-\tau(t)}^{t} f(s)ds + \int_{t-\tau(t)}^{t} \alpha(s)d\omega(s)\Big]$$

$$\leq x^T(t)\Big( - Q_2 - Q_2^T + \tau Q_2 P_4^{-1} Q_2^T + Q_2 P_3^{-1} Q_2^T \Big)x(t) + 2x^T(t)Q_2 x(t - \tau(t))$$

$$+ \int_{t-\tau(t)}^{t} f^T(s)P_4 f(s)ds + \Big( \int_{t-\tau(t)}^{t} \alpha(s)d\omega(s) \Big)^T P_3 \Big( \int_{t-\tau(t)}^{t} \alpha(s)d\omega(s) \Big) \tag{13}$$

Similarly, we can get that

$$0 = 2x^T(t - \tau(t))Q_3\Big[ - x(t - \tau(t)) + x(t - \tau)$$

$$+ \int_{t-\tau}^{t-\tau(t)} f(s)ds + \int_{t-\tau}^{t-\tau(t)} \alpha(s)d\omega(s)\Big]$$

$$\leq x^T(t - \tau(t))\Big( - Q_3 - Q_3^T + \tau Q_3 P_4^{-1} Q_3^T + Q_3 P_3^{-1} Q_3^T \Big)x(t - \tau(t))$$

$$+ 2x^T(t - \tau(t))Q_3 x(t - \tau) + \int_{t-\tau}^{t-\tau(t)} f^T(s)P_4 f(s)ds$$

$$+ \Big( \int_{t-\tau}^{t-\tau(t)} \alpha(s)d\omega(s) \Big)^T P_3 \Big( \int_{t-\tau}^{t-\tau(t)} \alpha(s)d\omega(s) \Big), \tag{14}$$

From the proof of [16], we can get that

$$\mathbf{E}\Big\{ \Big( \int_{t-\tau}^{t-\tau(t)} \alpha(s)d\omega(s) \Big)^T P_3 \Big( \int_{t-\tau}^{t-\tau(t)} \alpha(s)d\omega(s) \Big) \Big\}$$

$$= \mathbf{E}\Big\{ \int_{t-\tau}^{t-\tau(t)} \text{trace}[\alpha^T(s)P_3\alpha(s)]ds \Big\} \tag{15}$$

and

$$\mathbf{E}\Big\{ \Big( \int_{t-\tau(t)}^{t} \alpha(s)d\omega(s) \Big)^T P_3 \Big( \int_{t-\tau(t)}^{t} \alpha(s)d\omega(s) \Big) \Big\}$$

$$= \mathbf{E}\Big\{ \int_{t-\tau(t)}^{t} \text{trace}[\alpha^T(s)P_3\alpha(s)]ds \Big\}. \tag{16}$$

From definition of $\alpha(t)$ and assumption (**H**), we have

$$\text{trace}(\alpha^T(t)\alpha(t)) = \text{trace}\Big( \sum_{i=1}^{r} \sum_{j=1}^{r} \mu_i(t)\mu_j(t)\sigma_i^T(t)\sigma_j(t) \Big)$$

$$\leq \text{trace}\left(\frac{1}{2}\sum_{i=1}^{r}\sum_{j=1}^{r}\mu_i(t)\mu_j(t)(\sigma_i^T(t)\sigma_i(t)+\sigma_j^T(t)\sigma_j(t))\right)$$

$$=\sum_{i=1}^{r}\mu_i(t)\text{trace}(\sigma_i^T(t)\sigma_i(t))$$

$$\leq \sum_{i=1}^{r}\mu_i(t)\left(\|R_{1i}x(t)\|^2+\|R_{2i}x(t-\tau(t))\|^2\right). \tag{17}$$

It follows from (11)-(17) that

$$\mathbf{E}\{dV(t)-2y^T(t)w(t)dt-\gamma w^T(t)w(t)dt\}\leq\mathbf{E}\left\{\sum_{i=1}^{r}\mu_i(t)z^T(t)\Pi_i z(t)dt\right\},\tag{18}$$

where $z(t)=\left(x^T(t),x^T(t-\tau(t)),x^T(t-\tau),f^T(t),w^T(t)\right)^T$, and

$$\Pi_i=\begin{bmatrix}\Pi_{11,i} & Q_2 & 0 & A_i^TQ_1+P_1 & -C_i^T\\ * & \Pi_{22,i} & Q_3 & B_i^TQ_1 & -D_i^T\\ * & * & -P_2 & 0 & 0\\ * & * & * & \Pi_{44} & Q_1^TU_i\\ * & * & * & * & \Pi_{55,i}\end{bmatrix}$$

with $\Pi_{11,i}=(\lambda_1+\tau\lambda_2)R_{1i}R_{1i}^T+P_2-Q_2-Q_2^T+\tau Q_2P_4^{-1}Q_2^T+Q_2P_3^{-1}Q_2^T$, $\Pi_{22,i}=(\lambda_1+\tau\lambda_2)R_{2i}R_{2i}^T-Q_3-Q_3^T+\tau Q_3P_4^{-1}Q_3^T+Q_3P_3^{-1}Q_3^T$, $\Pi_{44}=-Q_1-Q_1^T+\tau P_4$, $\Pi_{55,i}=-V_i-V_i^T-\gamma I$.

It is easy to verify the equivalence of $\Pi_i<0$ and $\Omega_i<0$ by using Lemma 3. Thus, one can derive from (9) and (19) that

$$\frac{\mathbf{E}\{dV(t)\}}{dt}-\mathbf{E}\{2y^T(x(t))w(t)+\gamma w^T(t)w(t)\}\leq 0. \tag{19}$$

It follows from (19) and the definition of $V(t)$ that

$$2\mathbf{E}\{\int_0^{t_p}f^T(x(s))u(s)ds\}\geq-\gamma\mathbf{E}\{\int_0^{t_p}u^T(s)u(s)ds\}.$$

From Definition 1, we know that the stochastic T-S fuzzy system (1) is globally passive in the sense of expectation. The proof is completed.

## 4  Conclusions

In this paper, the passivity for stochastic Takagi-Sugeno (T-S) fuzzy systems with time-varying delays has been investigated without assuming the differentiability of the time-varying delays. By utilizing the Lyapunov functional method, the Itô differential rule and the matrix inequality techniques, a delay-dependent criterion to ensure the passivity for T-S fuzzy systems with time-varying delays has been established in terms of linear matrix inequalities (LMIs) that can be easily checked by using the standard numerical software.

**Acknowledgments.** This work was supported by the National Natural Science Foundation of China under Grants 61273021, 60974132 and 11172247.

# References

1. Takagi, T., Sugeno, M.: Fuzzy Identification of Systems and Its Applications to Modeling and Control. IEEE Transactions on Systems, Man and Cybernetics 15, 116–132 (1985)
2. Cao, Y., Frank, P.M.: Stability Analysis and Synthesis of Nonlinear Time-Delay Systems Via Linear Takagi-Sugeno Fuzzy Models. Fuzzy Sets and Systems 124, 213–229 (2001)
3. Liu, F., Wu, M., He, Y., Yokoyama, R.: New Delay-Dependent Stability Criteria for T-S Fuzzy Systems with Time-Varying Delay. Fuzzy Sets and Systems 161, 2033–2042 (2010)
4. Kwon, O., Park, M., Lee, S., Park, J.: Augmented Lyapunov-Krasovskii Functional Approaches to Robust Stability Criteria for Uncertain Takagi-Sugeno Fuzzy Systems with Time-Varying Delays. Fuzzy Sets and Systems 201, 1–19 (2012)
5. Wang, Z., Ho, D.W.C., Liu, X.: A Note on The Robust Stability of Uncertain Stochastic Fuzzy Systems with Time-Delays. IEEE Transactions on Systems, Man, and Cybernetics, Part A 34, 570–576 (2004)
6. Senthilkumar, T., Balasubramaniam, P.: Delay-Dependent Robust $H_\infty$ Control for Uncertain Stochastic T-S Fuzzy Systems with Time-Varying State and Input Delays. International Journal of Systems Science 42(5), 877–887 (2011)
7. Mahmoud, M.S., Ismail, A.: Passivity and Passification of Time-Delay Systems. Journal of Mathematical Analysis and Applications 292, 247–258 (2004)
8. Park, J.H.: Further Results on Passivity Analysis of Delayed Cellular Neural Networks. Chaos, Solitons and Fractals 34, 1546–1551 (2007)
9. Calcev, G.: Passivity Approach to Fuzzy Control Systems. Automatica 33(3), 339–344 (1998)
10. Li, C., Zhang, H., Liao, X.: Passivity and Passification of Fuzzy Systems with Time Delays. Computers and Mathematics with Applications 52, 1067–1078 (2006)
11. Chang, W., Ku, C., Huang, P.: Robust Fuzzy Control for Uncertain Stochastic Time-Delay Takagi-Sugeno Fuzzy Models for Achieving Passivity. Fuzzy Sets and Systems 161, 2012–2032 (2010)
12. Liang, J., Wang, Z., Liu, X.: On Passivity and Passification of Stochastic Fuzzy Systems with Delays: The Discrete-Time Case. IEEE Transactions on Systems, Man, and Cybernetics, Part B 40, 964–969 (2010)
13. Zhang, B., Zheng, W., Xu, S.: Passivity Analysis and Passive Control of Fuzzy Systems with Time-Varying Delays. Fuzzy Sets and Systems 174, 83–98 (2011)
14. Song, Q., Wang, Z., Liang, J.: Analysis on Passivity and Passification of T-S Fuzzy Systems with Time-Varying Delays. Journal of Intelligent & Fuzzy Systems 24, 21–30 (2013)
15. Song, Q., Cao, J.: Global Dissipativity Analysis on Uncertain Neural Networks with Mixed Time-varying Delays. Chaos 18, 043126 (2008)
16. Yu, J., Zhang, K., Fei, S.: Further Results on Mean Square Exponential Stability of Uncertain Stochastic Delayed Neural Networks. Communications in Nonlinear Science and Numerical Simulation 14, 1582–1589 (2009)

# Parallel Computation of a New Data Driven Algorithm for Training Neural Networks

Daiyuan Zhang

College of Computer, Nanjing University of Posts and Telecommunications, Nanjing, China
dyzhang@njupt.edu.cn, zhangdaiyuan2011@sina.com

**Abstract.** Different from some early learning algorithms such as backpropagation (BP) or radial basis function (RBF) algorithms, a new data driven algorithm for training neural networks is proposed. The new data driven methodology for training feedforward neural networks means that the system modeling are performed directly using the input-output data collected from real processes, To improve the efficiency, the parallel computation method is introduced and the performance of parallel computing for the new data driven algorithm is analyzed. The results show that, by using the parallel computing mechanisms, the training speed can be much higher.

**Keywords:** artificial intelligence, neural networks, weight function, B-spline function, algorithm, data driven methodology, parallel computation.

## 1 Introduction

It is well know that there are many drawbacks in BP algorithm, for example, local minima, the architecture of neural networks are difficult to be settled, as the networks are at least three-layer structures and the hidden layers need to be extended or modified again and again, which usually take a very long training time. To improve the learning performance of BP algorithm, many studies have been made ([1] [2] [3] [4]), but those studies can not overcome the drawbacks mentioned above. In addition, the trained weights by BP or RBF algorithms are constant data (constant weights), which have no information about relations between input and output patterns. Therefore, the constant weights can not be used as useful information for understanding any relations, regularities or structure inherent in some source of trained data, or we say that the constant weights serve no useful purpose.

In recent years, to overcome the drawbacks of BP or RBF algorithms completely, the new training algorithms using weight functions have been proposed [5] [6], in which the architecture of networks is simple and the learning processes of the algorithms using weight functions are to find the weight functions by interpolation and approximate theories.

In this paper, the weight functions are obtained directly by the input-output data. The forms of weight functions are B-spline functions defined on the sets of the given input variables and can be established easily, which are linear combination of some

C. Guo, Z.-G. Hou, and Z. Zeng (Eds.): ISNN 2013, Part I, LNCS 7951, pp. 149–156, 2013.
© Springer-Verlag Berlin Heidelberg 2013

values associated with the given output patterns. From this point of view, we say that the new algorithm can get very good approximate results directly without training. The approximately analytic expression can be used to analysis some interested performance.

## 2    Network's Architecture and Fundamentals of the Algorithm

In this paper, the new network's architecture is different from that used in BP or RBF networks. The network has two layers, one is input layer, and the other is output layer. There are $m$ points in the input layer, and denoted by $x_i$, $i = 1, 2, \cdots, m$, or we say, the input dimension is $m$. There are $n$ points in the output layer, and denoted by $N_j$, $j = 1, 2, \cdots, n$, or we say, the output dimension is $n$. The neuron $N_j$ is used as an adder. Note that each of the $m$ inputs is connected to each of the neurons (adders). The mapping relations between output layer and input layer are

$$y_j = \sum_{i=1}^{m} y_{ji} = \sum_{i=1}^{m} g_{ji}(x_i) \tag{1}$$

The one-variable function $g_{ji}(x_i)$ is called weight function between the $j$th output point (neuron) and the $i$th input point (variable).

The $y_j$ denotes the output values of the network's $j$th neuron, and the $z_j$ indicates the target patterns of the $j$th point of the network.

Given the knot sequence in the following

$$a = t_0 < t_1 \cdots < t_{N+1} = b \tag{2}$$

Now, let's begin with a system of knots on the real line, named $t_i$. Usually, only a finite set of knots is ever needed for practical purposes (see (1)), but for the theoretical development it is much easier to suppose that the knots form an infinite set from (1) extending to $+\infty$ on the right and to $-\infty$ on the left in the following:

$$\cdots < t_{-2} < t_{-1} < t_0 < t_1 < t_2 \cdots \tag{3}$$

We suppose that a function $f_{[a,b]}(x)$ is given on $[a,b]$. Now let us extend function $f_{[a,b]}(x)$ in the following.

$$f(x) = \begin{cases} f_{[a,b]}(t_0) & \text{if } x \le t_0 \\ f_{[a,b]}(x) & \text{if } x \in [t_0, t_{N+1}] = [a,b] \\ f_{[a,b]}(t_{N+1}) & \text{if } x \ge t_{N+1} \end{cases} \tag{4}$$

The set of B-splines $\left\{B_{-r}^{r}, B_{-r+1}^{r}, \cdots, B_{N-1}^{r}\right\}$ is linearly independent on $(t_0, t_N)$. Therefore, any spline function of degree $r$ defined on interval $(t_0, t_N)$ can be expressed as

$$s(x) = \sum_{p=-\infty}^{\infty} c_p B_p^r(x) = \sum_{p=-r}^{N} c_p B_p^r(x) \tag{5}$$

The next step is to introduce a spline function that approximates function $f(x)$. For this purpose we choose:

$$g(x) = \sum_{p=-\infty}^{\infty} f\left(t_{p+\lceil r/2\rceil}\right) B_p^r(x) \tag{6}$$

Obviously, $g(x)$ is a special form of (4). With the aid of function (5) we have that, If $f(x)$ is a function on $[t_0, t_n]$, then the spline function $g(x)$ in   (5) satisfies

$$\max_{t_0 \le x \le t_n} \left| f(x) - g(x) \right| \le (1 + r/2) \omega\left(f_{[a,b]}; \delta\right) \tag{7}$$

where
$$\delta = \max_{-k \le p \le n+1} \left(t_p - t_{p-1}\right) \tag{8}$$

and $\omega\left(f_{[a,b]}, \delta\right)$ is the modulus of continuity of function $f(x)$ on $[a,b]$.

The preceding result can be stated in terms of the distance from $f(x)$ to the space $S_{N+1}^r$. The distance from a function $f(x)$ to a subspace $G$ in a normed space is defined by

$$\text{dist}(f, G) = \inf_{g \in G} \|f - g\|_\infty \tag{9}$$

Let us use the norm defined by

$$\|f\|_\infty = \max_{a \le x \le b} \left| f(x) \right| \tag{10}$$

We have

$$\text{dist}\left(f, S_{N+1}^r\right) \le \lambda \cdot \omega\left(f_{[a,b]}; \delta\right) \tag{11}$$

If $f$ is continuous on $[a,b]$ , then

$$\lim_{\delta \to 0} \omega\left(f_{[a,b]}; \delta\right) = 0 \tag{12}$$

Hence, by increasing the density of the knots, the upper bound in (10) can be made to approach zero.

For functions possessing some derivatives, more can be said in the following: let $\alpha < r < N$. If $f \in C^\alpha(t_0, t_{N+1})$, then

$$\|f(x) - g(x)\|_\infty \leq (1 + r/2)^\alpha \, \delta^\alpha \, \|f^{(\alpha)}(x)\|_\infty \tag{13}$$

Now, we introduce the notation including networks' parameters (the number of input and output layers) to the neural network's architecture, and from (5), we can construct the weight function between the $j$th output neuron and the $i$th input point (variable), i.e.,

$$g_{ji}(x_i) = \sum_{p=-\infty}^{\infty} f_{ji}\left(t_{p+\lceil r/2 \rceil}\right) B_{ip}^r(x_i) \tag{14}$$

The learning algorithm can be expressed as the form:

$$\begin{cases} g_{ji}(x_i) = \sum_{p=-\infty}^{\infty} f_{ji}\left(t_{p+\lceil r/2 \rceil}\right) B_{ip}^r(x_i) \\ y_j = \sum_{i=1}^{m} g_{ji}(x_i) \end{cases} \tag{15}$$

The most important advantage in (14) is that we can get the forms of weight functions by the given patterns directly and immediately. This important advantage may have some applications in data-centric technologies or approaches.

For the finitely discrete data (patterns), the final expression of learning algorithm in this paper can be expressed in the following

$$\begin{cases} g_{ji}(x_i) = \sum_{p=-r}^{N} \eta_{ji} z_{j(p+\lceil r/2 \rceil)} B_{ip}^r(x_i) \\ y_j = \sum_{i=1}^{m} g_{ji}(x_i) \end{cases} \tag{16}$$

Now, the next step is to investigate whether continuous functions can be approximated to arbitrary precision by the algorithm proposed in this paper.

**Theorem:** Let $\alpha < r < N$, $z \in C^\alpha(t_0, t_{N+1})$, then the following holds:

$$\|z_j - y_j\|_\infty \leq m(1 + r/2)^\alpha \, \delta_j \, \|z_j^{(\alpha)}\|_\infty \tag{17}$$

*Proof* From (12) and (15), we have

$$\left\| z_j - y_j \right\|_\infty = \left\| \sum_{i=1}^{m} \left( z_{ji} - g_{ji}(x_i) \right) \right\|_\infty$$

$$\leq \sum_{i=1}^{m} \left\| z_{ji}(x_i) - g_{ji}(x_i) \right\|_\infty \leq \sum_{i=1}^{m} (1+r/2)^\alpha \, \delta_{ji}^\alpha \left\| z_{ji}^{(\alpha)}(x_i) \right\|_\infty \qquad (18)$$

$$= (1+r/2)^\alpha \sum_{i=1}^{m} \delta_{ji}^\alpha \left\| \eta_{ji} z_j^{(\alpha)} \right\|_\infty \leq (1+r/2)^\alpha \sum_{i=1}^{m} \delta_{ji}^\alpha \left\| z_j^{(\alpha)} \right\|_\infty$$

$$\leq m(1+r/2)^\alpha \, \delta_j \left\| z_j^{(\alpha)} \right\|_\infty$$

Where $z_{ji}(x_i)$ is the theoretical weight function between the $j$th output neuron and the $i$th input point. In many practical applications, we do not know the form of function $z_{ji}(x_i)$. Function $g_{ji}(x_i)$ is the weight function constructed by B-splines between the $j$th output neuron and the $i$th input point. $\delta_{ji}$ is the maximum step of knots between the $j$th output neuron and the $i$th input point described as that in Theorem, and $\delta_j = \max_{1 \leq i \leq m} \delta_{ji}$.

The theorem states that, if $\left\| z_j^{(\alpha)} \right\|_\infty < \infty$ and $r$ is a constant, any desired precision can be reached by increasing the density of knots.

Obviously, in order to obtain higher precision, we can increase the number of knots (or decrease the values of step $\delta_j$).

The learning procedure proposed in this paper is to construct a system of (15) for getting the forms of weight functions $g_{ji}(x_i)$. Equation (15) indicates that each of weight functions takes one correspond input neuron (input point) as an argument. The forms of weight functions are linear combination of some values associated with the given output patterns and the B-spline functions defined on the sets of given input variables (input knots or input patterns).

Obviously, (15) is an analytic form of mathematic expression, which can be obtained directly by knots (or extended patterns). Analytic expression is very useful for getting some performance in the interval $X = \left[ t_0, t_{N+1} \right]$ or some interested sub-intervals like $X_q = \left[ t_q, t_{q+1} \right] \subset X$ and so on.

## 3    Parallel Mechanisms

In order to accelerate the training speed, parallel mechanisms must be introduced. Some devices must be used to implement the overhead of preparing the data packet at the sender or processing it when it arrives at the receiver. Obviously, networks interconnecting more than two devices require mechanisms to physically connect the source to its destination in order to transport the data packet and deliver it to the correct destination. These mechanisms can be implemented in different ways and the types of network structure and functions performed by those mechanisms are very much the same, regardless of the domain.

In addition to the devices, the algorithm can be divided into some independent parts are the premises on which the parallel computation is based. Eq.(15) shows that the weight functions can be found independently, and the number of independent parts is $mn$., which means that the algorithm proposed in this paper can be implemented by parallel computation.

Parallel mechanisms indicate that the devices (more than two) can process the data at the same time. The time can be found by the parallel mechanisms in the following

$$
\begin{cases}
T_\mathrm{P} = \min_i \max \left\{ \lambda_i \left( \Delta t_i + \Delta u_i \right) \middle| i = 1, 2, \cdots, N_\mathrm{P} \right\} \\
\text{s.t.} \sum_{i=1}^{N_\mathrm{P}} \lambda_i = mn
\end{cases}
\tag{19}
$$

where $\lambda_i \in \{1, 2, \cdots\}$, $\Delta t_i$ is the overhead of processing of the $i$th device (processor), $\Delta u_i$ denotes the overhead of preparing the data packet of the $i$th device. $N_\mathrm{P}$ is the number of devices working at the same time.

If there is only one device, we have

$$
T = mn \left( \Delta t + \Delta u \right)
\tag{20}
$$

The speedup is

$$
S = \frac{T}{T_\mathrm{P}} = \frac{mn \left( \Delta t + \Delta u \right)}{\min_i \max \left\{ \lambda_i \left( \Delta t_i + \Delta u_i \right) \middle| i = 1, 2, \cdots, N_\mathrm{P} \right\}}
\tag{21}
$$

Suppose each device has the same performance, i.e. $\Delta t_i = \Delta t$, $\Delta u_i = \Delta u$, we have

$$
T_\mathrm{P} = \left\lceil \frac{mn}{N_\mathrm{P}} \right\rceil \left( \Delta t + \Delta u \right)
\tag{22}
$$

And the speedup will be

$$
S = \frac{T}{T_\mathrm{P}} \approx N_\mathrm{P}
\tag{23}
$$

In general, $N_\mathrm{P} \gg 1$, it shows that the parallel computation has high training speedup.

## 4    Simulations

The most important issue developed in this paper is that the weight functions can be obtained before network training.

*Example*: We choose $r=3$ in (15) and the coefficients $\eta_{ji} = 1/3$. The patterns are obtained by

$$\begin{cases} z_1 = \sin\left(x_1^2 + x_2^2 + x_3^2\right) \\ z_2 = e^{-(x_1+x_2+x_3)} \sin\left(x_1^2 + x_2^2 + x_3^2\right) \\ z_3 = \left(x_1^2 + x_2^2 + x_3^2\right)\sin\left(x_1^2 + x_2^2 + x_3^2\right) \end{cases} \tag{24}$$

And the learning curve [5] is

$$x_1 = t, \ x_2 = 1.8t, \ x_3 = t+3 \tag{25}$$

where $t \in [0, 1]$. The 10 input patterns are equally spaced knots, using the values of $t$:

$$t = 0 + (1-0)\, p/(10-1) = p/9, \ p = 0, 1, 2, \cdots, N+1, N = 8 \tag{26}$$

Obviously, the learning procedure of the new algorithm developed in this paper is to construct the linear combination of B-spline weight functions (15) by the given knots and output patterns, which states that if the patterns are obtained, then the weight functions can be given directly by (15). For example, if the patterns are the same as that from example 1, we can write the analytic form of weight functions directly in the following without network training:

$$\begin{cases} g_{ji}\left(x_i\right) = \dfrac{1}{3} \displaystyle\sum_{p=-3}^{8} z_{j(p+2)} B_{ip}^3\left(x_i\right) \\ y_j = \displaystyle\sum_{i=1}^{3} g_{ji}\left(x_i\right) \end{cases} \tag{27}$$

The extended knots (input patterns) are equally spaced outside [0, 1].

The architecture of the neural network in example is 3-3, which indicates that the input and output nodes are 3 and 3 respectively. Or we say that the number of weight functions needed in this example is 3×3=9. Using parallel mechanisms proposed in this paper, the speedup can be found by (20)

$$S = \frac{T}{T_p} = \frac{9\left(\Delta t + \Delta u\right)}{\min\limits_{i} \max \left\{\lambda_i\left(\Delta t_i + \Delta u_i\right)\big| i = 1, 2, \cdots, N_p\right\}} \tag{28}$$

Suppose each device has the same performance, from (21), we have

$$T_p = \left\lceil \frac{9}{N_p} \right\rceil \left(\Delta t + \Delta u\right) \tag{29}$$

The overhead of processing and preparing is $\Delta t + \Delta u$ if there is only one device in the system, which means the overhead of processing and preparing for one weight function is $\Delta t + \Delta u$. The speedup is about $N_P$.

## 5    Conclusions

The new algorithm proposed in this paper inherits many advantages of cubic spline weight functions, i.e. the neural network's architecture is very simple and without problems such as local minima, slow convergence, and dependent on initialized values arising from the steepest descent-like algorithms (BP and RBF algorithms). The weights obtained, instead of constant weights found by BP and RBF algorithms, are weight functions. When relations between input and output patterns are established, the weight functions, at least to some extent, can be used to understand some relations, regularities or structure inherent in some trained patterns. The new algorithm has good property of generalization.

The analytic forms of mathematical expressions can be easily found immediately as long as the expended patterns are given, which is a new data driven methodology for feedforward neural networks. This advantage may be important in some data-centric technologies, which can deals with the automatic detection of patterns in data, and plays a central role in many modern artificial intelligence and computer science problems. The parallel computing mechanisms are introduced to the network training, the results show that, by using the parallel computing mechanisms, the parallel algorithm has high speedup.

## References

1. Ampazis, N., Perantonis, S.J.: Two highly efficient second-order algorithms for training feedforward networks. IEEE Trans. Neural Netw. 13, 1064–1073 (2002)
2. Khashman, A.: A Modified Backpropagation Learning Algorithm With Added Emotional Coefficients. IEEE Trans. Neural Netw. 19, 1896–1909 (2008)
3. Bortman, M., Aladjem, M.: A Growing and Pruning Method for Radial Basis Function Networks. IEEE Trans. Neural Netw. 20, 1039–1045 (2009)
4. Wedge, D., Ingram, D., McLean, D., Mingham, C., Bandar, Z.: On global-local artificial neural networks for function approximation. IEEE Trans. Neural Netw. 17, 942–952 (2006)
5. Zhang, D.Y.: New Theories and Methods on Neural Networks. Tsinghua University Press, Beijing (2006) (in Chinese)
6. Zhang, D.Y.: New Algorithm for Training Feedforward Neural Networks with Cubic Spline weight functions. Systems Engineering and Electronics 28, 1434–1437 (2006) (in Chinese)

# Stability Analysis of a Class of High Order Fuzzy Cohen-Grossberg Neural Networks with Mixed Delays and Reaction-Diffusion Terms

Weifan Zheng[1,2], Jiye Zhang[1,*], and Mingwen Wang[2]

[1] Traction Power State Key Laboratory, Southwest Jiaotong University (SWJTU)
[2] Information Research Institute, SWJTU, Chengdu 610031, People's Republic of China
{wfzheng,jyzhang,wangmw}@home.swjtu.edu.cn

**Abstract.** In this paper, we investigate a class of high order fuzzy Cohen-Grossberg neural networks (HOFCGNN) with mixed delays which include time variable dalay and unbounded delays. Based on the properties of M-matrix, by constructing vector Lyapunov functions and applying differential inequalities, the sufficient conditions ensuring existence, uniqueness, and global exponential stability of the equilibrium point of HOFCGNN with mixed delays and reaction-diffusion terms are obtained.

**Keywords:** Stability, HOFCGNN, mixed Delay, Reaction-diffusion.

## 1    Introduction

Recently, some kinds of Cohen-Grossberg[1], fuzzy cellular[2] and high order neural networks[6] have attracted the attention of the scientific community due to their promising potential for tasks of associative memory, parallel computation and their ability to solve difficult optimization problems. In these applications it is required that there is a well-defined computable solution for all possible initial states. This means that the neural network should have a unique equilibrium point that is globally stable. Thus, the qualitative analysis of dynamic behaviors is a prerequisite step for the practical design and application of neural networks. There are many papers discuss the qualitative properties for neural networks [3]. In hardware implementation, time delays are unavoidable, and may lead to an oscillation and instability of networks [4]. In most situations, the time delays are variable, and in fact unbounded. Therefore, the study of stability of neural networks with mixed delay is practically important. Stability of neural networks with constant and variable time delays stand by differential equation has been studied in [4~5]. The global exponential stability of Cohen-Grossberg neural networks and fuzzy logic with reaction-diffusion terms were obtained in [7~10, 12]. Stability of high-order neural networks was studied in [6, 8-12]. But they do not consider the unbounded delays.

In this paper, we study a class of HOFCGNN, which contain both variable time delays and unbounded delay. We relax some conditions on activation functions and

---

[*] Corresponding Author.

C. Guo, Z.-G. Hou, and Z. Zeng (Eds.): ISNN 2013, Part I, LNCS 7951, pp. 157–165, 2013.

diffusion functions of systems similar to that discussed in [8-9], [11-12], by using M-matrix theory and nonlinear integro-differential inequalities, even type Lyapunov functions were constructed to analyze the conditions ensuring the existence, uniqueness and global exponential stability of the equilibrium point of the models.

## 2     Model Description and Preliminaries

In this paper, we analyze the stability of reaction-diffusion HOFCGNN with both variable delays and unbounded delay described by the following differential equations

$$\frac{\partial u_i(t)}{\partial t} = \sum_{k=1}^{m} \frac{\partial}{\partial x_k} [ D_{ik}(t,x,u_i) \frac{\partial u_i}{\partial x_k} ] - d_i(u_i(t))[ \rho_i(u_i(t)) - \sum_{j=1}^{n} a_{ij} f_j(u_j(t))$$

$$- \wedge_{j,k=1}^{n} b_{ijk}^1 f_j(u_j(t-\tau_{ij}(t)))f_k(u_k(t-\tau_{ik}(t))) - \vee_{j,k=1}^{n} b_{ijk}^2 f_j(u_j(t-\tau_{ij}(t)))f_k(u_k(t-\tau_{ik}(t)))$$

$$- \wedge_{j,k=1}^{n} c_{ijk}^1 \int_{-\infty}^{t} k_{ij}(t-s)f_j(u_j(s))f_k(u_k(s))ds - \vee_{j,k=1}^{n} c_{ijk}^2 \int_{-\infty}^{t} k_{ij}(t-s)f_j(u_j(s))f_k(u_k(s))ds + J_i ] \quad (1)$$

$$\frac{\partial u_i}{\partial \tilde{n}} = col(\frac{\partial u_i}{\partial x_1}, ..., \frac{\partial u_i}{\partial x_m}) = 0 \quad t \in I, \quad x \in \partial\Omega, \quad (i=1,2,\cdots,n) \quad (2)$$

where $u_i$ is the state of neuron $i$, $(i=1,2,\cdots,n)$ and $n$ is the number of neurons; $D_i(t,x,u_i)$ is smooth reaction-diffusion function, $a_{ij}$ are the first order connection weights, $b_{ijk}^1$, $b_{ijk}^2$, $c_{ijk}^1$ and $c_{ijk}^2$ are second order connection weights of fuzzy feedback MIN template and MAX template, respectively, $J = (J_1, \cdots, J_n)^T$ is the constant input vector. $f(u) = (f_1(u_1),...,f_n(u_n))^T$, is the activation function of the neurons; $d_i(u_i)$ represents an amplification function; $\rho_i(u_i)$ is an appropriately behaved function such that the solutions of model (1) remain bounded. The variable delays $\tau_{ij}(t)$, $\tau_{ik}(t)$ ($i,j,k=1,2,\Lambda,n$) are bounded functions, i.e. $0 \le \tau_{ij}(t), \tau_{ik}(t) \le \tau$, and $k_{ij}:[0,\infty) \to [0,\infty)$, $(i,j=1,2,\cdots,n)$ are piecewise continuous on $[0,\infty)$. Let $k = (k_{ij})_{n \times n}$. $\wedge$ and $\vee$ denote the fuzzy AND and fuzzy OR operation, respectively.

The conventional conditions for kernel functions of Eq. (1) meet the following assumptions:

**Assumption A:** $\int_0^\infty k_{ij}(s)ds = 1$, $\int_0^\infty sk_{ij}(s)ds < +\infty$, $(i,j=1,2,\cdots,n)$.

**Assumption B:** $\int_0^\infty k_{ij}(s)ds = 1$, $\int_0^\infty e^{\beta s} k_{ij}(s)ds = K_{ij} < +\infty$, $(i,j=1,2,\cdots,n)$.

In order to study the exponential stability of neural networks (1) conveniently, we inquire kernel functions meet the following assumption:

**Assumption   C:** $\int_0^\infty e^{\beta s} k_{ij}(s)ds = N_{ij}(\beta)$, $(i,j=1,2,\cdots,n)$, where $N_{ij}(\beta)$ are continuous functions in $[0,\delta)$, $\delta > 0$, and $N_{ij}(0) = 1$.

It is easy to prove that the Assumption C includes Assumption A and B [5].

The initial conditions of equation (1) are of the form $u_i(s) = \phi_i(s)$, $s \le 0$, where $\phi_i$ is bounded and continuous on $(-\infty, 0]$. Equation (2) is the boundary condition of equation (1), in which $x \in \Omega \subset R^m$, $\Omega$ is a compact set with smooth boundary and $mes\Omega > 0$, $\partial\Omega$ is the boundary of $\Omega$, $t \in I = [0,+\infty]$.

For convenience, we introduce some notations. The express $u = (u_1, u_2, ..., u_n)^T \in R^n$ represents a column vector (the symbol $(\ )^T$ denotes transpose). For matrix $A = (a_{ij})_{n \times n}$, $|A|$ denotes absolute value matrix given by $|A| = (|a_{ij}|)_{n \times n}$, $i, j = 1, 2, ..., n$; $[A]^S$ is defined as $(A^T + A)/2$. For $x \in R^n$, $|x| = (|x_1|, ..., |x_n|)^T$, $\|x\|$ denotes a vector norm defined by $\|x\| = \max\limits_{1 \le i \le n} \{|x_i|\}$. $\bar{p} = diag(\rho_1, ..., \rho_n)$, $\bar{d} = diag(d_1, ..., d_n)$. And

$$D_i(t, x, u_i) = \sum_{k=1}^{m} \frac{\partial}{\partial x_k}(D_{ij}(t, x, u_i)\frac{\partial u_i}{\partial x_k}),$$ so model (1) becomes the following system:

$$\dot{u}_i(t) = D_i(t, x, u_i(t)) - d_i(u_i(t))[\rho_i(u_i(t)) - \sum_{j=1}^{n} a_{ij}f_j(u_i(t))$$

$$-\wedge_{j,k=1}^{n} b_{ijk}^1 f_j(u_j(t-\tau_{ij}(t)))f_k(u_k(t-\tau_{ik}(t))) - \vee_{j,k=1}^{n} b_{ijk}^2 f_j(u_j(t-\tau_{ij}(t)))f_k(u_k(t-\tau_{ik}(t)))$$

$$-\wedge_{j,k=1}^{n} c_{ijk}^1 \int_{-\infty}^{t} k_{ij}(t-s)f_j(u_j(s))f_k(u_k(s))ds - \vee_{j,k=1}^{n} c_{ijk}^2 \int_{-\infty}^{t} k_{ij}(t-s)f_j(u_j(s))f_k(u_k(s))ds + J_i]. \quad (3)$$

Therefore, system (3) and (1) has the same properties of stability.

Now we consider the activation functions of the neurons, amplification function and behaved function satisfying the following assumption:

**Assumption D.** For each $i \in \{1,2,\cdots,n\}$, $f_i : R \to R$, there exist numbers $M_i > 0$, such that $|f_i(y_i)| \le M_i$, and there exist the real number $p_i > 0$, such that

$$p_i = \sup_{y \ne z} |\frac{f_i(y) - f_i(z)}{y - z}|,$$

for every $y \ne z$. Let $P = diag(p_1, ..., p_n)$, $M = diag(M_1, ..., M_n)$.

**Remark 1:** Assumption D introduced the supremum of Global/Local Lipschitz constants, and expanded the scope of system application. So, the activation functions such as sigmoid type and piecewise linear type are the special case of that satisfying it.

Let $B_{ij} = \sum_{k=1}^{n} M_k(|b_{ijk}^1| + |b_{ikj}^1| + |b_{ijk}^2| + |b_{ikj}^2|)$, $C_{ij} = \sum_{k=1}^{n} M_k(|c_{ijk}^1| + |c_{ikj}^1| + |c_{ijk}^2| + |c_{ikj}^2|)$

and $B = (B_{ij})_{n \times n}$, $C = (C_{ij})_{n \times n}$ in which the $M_k$ is defined as the Assumption D.

**Assumption E.** For each $i \in \{1,2,\cdots,n\}$, $e_i : R \to R$ is strictly monotone increasing, i.e. there exists a positive diagonal matrix $\bar{p} = diag(\rho_1, \rho_2, \cdots, \rho_n) > 0$ such that

$$\frac{\rho_i(u) - \rho_i(v)}{u - v} \ge \rho_i, (u \ne v).$$

**Assumption F.** For each $i \in \{1,2,\cdots,n\}$, $d_i : R \to R$ is continuous function and $0 < \sigma_i \le d_i$, where $\sigma_i$ is a constant.

## 3    Existence and Uniqueness of the Equilibrium Point

In this section, we shall study the condition which ensures the existence and uniqueness of the equilibrium point of system (1).

For convenience, we introduce some definitions and lemmas as follows.

**Definition 1 [13].** A real matrix $A = (a_{ij})_{n \times n}$ is said to be an M-matrix if $a_{ij} \leq 0$ $i, j = 1, 2, \ldots, n$, $i \neq j$, $a_{ii} > 0$ and all successive principal minors of $A$ are positive.

**Definition 2.** The equilibrium point $u*$ of (1) is said to be globally exponentially stable, if there exist constant $\lambda > 0$ and $\beta > 0$ such that $\| u(t) - u* \| \leq \beta \| \phi - u* \| e^{-\lambda t}$ ($t \geq 0$), where $\| \phi - u* \| = \max\limits_{1 \leq i \leq n} \sup\limits_{s \in [-\tau, 0]} | \phi_i(s) - u_i^* |$.

If there is a constant $u_0 = u_0^* = const$ (const denotes invariable constant) which is the solution of the following equations:

$$-\rho_i(u_i(t)) + \sum_{j=1}^{n} a_{ij} f_j(u_i(t)) + \wedge_{j,k=1}^{n} b_{ijk}^1 f_j(u_j(t - \tau_{ij}(t))) f_k(u_k(t - \tau_{ik}(t)))$$

$$+ \vee_{j,k=1}^{n} b_{ijk}^2 f_j(u_j(t - \tau_{ij}(t))) f_k(u_k(t - \tau_{ik}(t))) - \wedge_{j,k=1}^{n} c_{ijk}^1 \int_{-\infty}^{t} k_{ij}(t-s) f_j(u_j(s)) f_k(u_k(s)) ds$$

$$- \vee_{j,k=1}^{n} c_{ijk}^2 \int_{-\infty}^{t} k_{ij}(t-s) f_j(u_j(s)) f_k(u_k(s)) ds + J_i = 0 \tag{4}$$

then $\dfrac{\partial u_i^*}{\partial x} = 0$. That is to say equations (4) and (3) have the same equilibrium point, and so, system (4) has the same equilibrium point as that of system (1).

We firstly study the solutions of the nonlinear map associated with (1) as follows:

$$H_i(u_i) = -\rho_i(u_i) + \sum_{j=1}^{n} a_{ij} f_j(u_i) + \wedge_{j,k=1}^{n} b_{ijk}^1 f_j(u_j) f_k(u_k)$$

$$+ \vee_{j,k=1}^{n} b_{ijk}^2 f_j(u_j) f_k(u_k) + \wedge_{j,k=1}^{n} c_{ijk}^1 f_j(u_j) f_k(u_k) + \vee_{j,k=1}^{n} c_{ijk}^2 f_j(u_j) f_k(u_k) + J_i \tag{5}$$

Let $H(x) = (H_1(u_1), H_2(u_2), \ldots, H_n(u_n))^T$. It is well known that the solutions of $H(u) = 0$ are equilibriums in (1). If map $H(u) = 0$ is a homeomorphism on $R^n$, then system (1) has a unique equilibrium $u*$ (see [3]). In the following, we will give condition ensuring $H(u) = 0$ is a homeomorphism.

**Lemma 1[5].** If $H(u) \in C^0$ satisfies the following conditions: (i) $H(u)$ is injective on $R^n$; (ii) $\| H(u) \| \to \infty$ as $\| u \| \to \infty$; then $H(u)$ is a homeomorphism of $R^n$.

**Lemma 2 [2].** Suppose $x$ and $y$ are two states of system (1), then

$$| \wedge_{j=1}^{n} c_{ij}^1 f_j(x_j) - \wedge_{j=1}^{n} c_{ij}^1 f_j(y_j) | \leq \sum_{j=1}^{n} c_{ij}^1 \| f_j(x_j) - f_j(y_j) \|, \quad | \vee_{j=1}^{n} c_{ij}^2 f_j(x_j) - \vee_{j=1}^{n} c_{ij}^2 f_j(y_j) | \leq \sum_{j=1}^{n} c_{ij}^2 \| f_j(x_j) - f_j(y_j) \|.$$

**Lemma 3 [8]** For $u* = (u_1^*, u_2^*, \ldots, u_n^*)^T$ $i \in \{1, 2, \cdots, n\}$, and $f_i : R \to R$ are continuously differentiable, then we have

$$\sum_{j=1}^{n} \sum_{k=1}^{n} b_{ijk} [f_j(u_j) f_k(u_k) - f_j(u_j^*) f_k(u_k^*)] = \sum_{j=1}^{n} \sum_{k=1}^{n} (b_{ijk} + b_{ikj}) \frac{\partial f_j(\xi_j)}{\partial u_j} (u_j - u_j^*) f_k(\xi_k)$$

$$\sum_{j=1}^{n} \sum_{k=1}^{n} b_{ijk} [f_j(u_j) f_k(u_k) - f_j(u_j^*) f_k(u_k^*)] = \sum_{j=1}^{n} \sum_{k=1}^{n} (b_{ijk} + b_{ikj}) [f_j(u_j) - f_j(u_j^*)] f_k(\xi_k)$$

Where $\xi_j$ lies between $u_j$ and $u_j^*$ and $\xi_k$ lies between $u_k$ and $u_k^*$.

So, from Lemma 2 and Lemma 3, we have the following lemma.

**Lemma 4.** Suppose $x$ and $y$ are two states of system (1), from Lemma 3, then

$$|\wedge_{j,k=1}^{n} b_{ijk}^1 f_j(x_j)f_k(x_k) - \wedge_{j,k=1}^{n} b_{ijk}^1 f_j(y_j)f_k(y_k)| \leqslant \sum_{j=1}^{n}\sum_{k=1}^{n}(|b_{ijk}^1|+|b_{ikj}^1|)|f_j(x_j)-f_j(y_j)||f_k(\xi_k)|$$

$$|\vee_{j,k=1}^{n} b_{ijk}^2 f_j(x_j)f_k(x_k) - \vee_{j,k=1}^{n} b_{ijk}^2 f_j(y_j)f_k(y_k)| \leqslant \sum_{j=1}^{n}\sum_{k=1}^{n}(|b_{ijk}^2|+|b_{ikj}^2|)|f_j(x_j)-f_j(y_j)||f_k(\xi_k)|$$

$$|\wedge_{j,k=1}^{n} c_{ij}^1 f_j(x_j)f_k(x_k) - \wedge_{j,k=1}^{n} c_{ijk}^1 f_j(y_j)f_k(y_k)| \leqslant \sum_{j=1}^{n}\sum_{k=1}^{n}(|c_{ijk}^1|+|c_{ikj}^1|)|f_j(x_j)-f_j(y_j)||f_k(\xi_k)|$$

$$|\vee_{j,k=1}^{n} c_{ijk}^2 f_j(x_j)f_k(x_k) - \vee_{j,k=1}^{n} c_{ijk}^2 f_j(y_j)f_k(y_k)| \leqslant \sum_{j=1}^{n}\sum_{k=1}^{n}(|c_{ijk}^2|+|c_{ikj}^2|)|f_j(x_j)-f_j(y_j)||f_k(\xi_k)|.$$

**Theorem 1.** If Assumption D, E, and F are satisfied, and $\alpha = \overline{\rho}-(|A|+|B|+|C|)P$ is an M-matrix, then, for every input $J$, system (1) has a unique equilibrium $u^*$.

**Proof:** In order to prove that for every input $J$, system (1) has a unique equilibrium point $u^*$, it is only to prove that $H(u)$ is a homeomorphism on $R^n$. In following, we shall prove it in two steps.

*Step* 1, we will prove that condition (i) in Lemma 1 is satisfied. Suppose, for purposes of contradiction, that there exist $x, y \in R^n$ with $x \neq y$ such that $H(x) = H(y)$.

From Assumption E, we know that there exists matrix $\overline{\beta} = \mathrm{diag}(\beta_1, \beta_2, \cdots, \beta_n)$, $(\beta_i \geq \rho_i)$ such that $\rho_i(x_i) - \rho_i(y_i) = \beta_i(x_i - y_i)$, for $i = 1,2,...,n$.

Form (5), we get

$$-[\rho_i(x_i)-\rho_i(y_i)]+\sum_{j=1}^{n}a_{ij}(f_j(x_j)-f_j(y_j)) +\wedge_{j,k=1}^{n} b_{ijk}^1 f_j(x_j)f_k(x_k) -\wedge_{j,k=1}^{n} b_{ijk}^1 f_j(y_j)f_k(y_k)$$

$$+\vee_{j,k=1}^{n} b_{ijk}^2 f_j(x_j)f_k(x_k) -\vee_{j,k=1}^{n} b_{ijk}^2 f_j(y_j)f_k(y_k) +\wedge_{j,k=1}^{n} c_{ijk}^1 f_j(x_j)f_k(x_k) -\wedge_{j,k=1}^{n} c_{ijk}^1 f_j(y_j)f_k(y_k)$$

$$+\vee_{j,k=1}^{n} c_{ijk}^2 f_j(x_j)f_k(x_k) -\vee_{j,k=1}^{n} c_{ijk}^2 f_j(y_j)f_k(y_k) = 0. \tag{6}$$

We have $|\rho_i(x_i)-\rho_i(y_i)| \leq |\sum_{j=1}^{n} a_{ij}(f_j(x_j)-f_j(y_j))|+$

$$|\wedge_{j,k=1}^{n} b_{ijk}^1 f_j(x_j)f_k(x_k)-\wedge_{j,k=1}^{n} b_{ijk}^1 f_j(y_j)f_k(y_k)|+|\vee_{j,k=1}^{n} b_{ijk}^2 f_j(x_j)f_k(x_k)-\vee_{j,k=1}^{n} b_{ijk}^2 f_j(y_j)f_k(y_k)|$$

$$+|\wedge_{j,k=1}^{n} c_{ij}^1 f_j(x_j)f_k(x_k)-\wedge_{j,k=1}^{n} c_{ijk}^1 f_j(y_j)f_k(y_k)|+|\vee_{j,k=1}^{n} c_{ijk}^2 f_j(x_j)f_k(x_k)-\vee_{j,k=1}^{n} c_{ijk}^2 f_j(y_j)f_k(y_k)|$$

$$\leq \sum_{j=1}^{n}[|a_{ij}|+\sum_{k=1}^{n}(|b_{ijk}^1|+|b_{ikj}^1|)M_k +\sum_{k=1}^{n}(|b_{ijk}^2|+|b_{ikj}^2|)M_k]|f_j(x_j)-f_j(y_j)|$$

$$+\sum_{j=1}^{n}[\sum_{k=1}^{n}(|c_{ijk}^1|+|c_{ikj}^1|)M_k +\sum_{k=1}^{n}(|c_{ijk}^2|+|c_{ikj}^2|)M_k]|f_j(x_j)-f_j(y_j)|.$$

From Assumption D, E and Lemma 2 and 4, we get $[\overline{\beta}-(|A|+|B|+|C|)P]|x-y|\leq 0$ . (7)

Because of $\alpha$ being an M-matrix, from Definiton 1, we know that all elements of $(\overline{\beta}-(|A|+|B|+|C|)P)^{-1}$ are nonnegative. Therefore $|x-y|\leq 0$ , i.e., $x=y$ . From the supposition $x \neq y$, thus this is a contradiction. So $H(u)$ is injective.

*Step*2. We now prove that condition (ii) in Lemma 1 is satisfied. Let $\overline{H}(u) = H(u) - H(0))$. From (5), we get $\overline{H}_i(u_i) = -\beta_i u_i + \sum_{j=1}^{n} a_{ij}(f_j(u_i) - f_j(0))$

$+ \wedge_{j,k=1}^{n} b_{ijk}^1 f_j(u_j) f_k(u_k) - \wedge_{j,k=1}^{n} b_{ijk}^1 f_j(0) f_k(0) + \vee_{j,k=1}^{n} b_{ijk}^2 f_j(u_j) f_k(u_k) - \vee_{j,k=1}^{n} b_{ijk}^2 f_j(0) f_k(0)$

$+ \wedge_{j,k=1}^{n} c_{ij}^1 k f_j(u_j) f_k(u_k) - \wedge_{j,k=1}^{n} c_{ijk}^1 f_j(0) f_k(0) + \vee_{j,k=1}^{n} c_{ijk}^2 f_j(u_j) f_k(u_k) - \vee_{j,k=1}^{n} c_{ijk}^2 f_j(0) f_k(0)$.

To show that $H(u)$ is homeomorphism, it suffices to show that $\overline{H}(u)$ is homeomorphism. According to Assumption D, we get $| f_i(u) - f_i(0) | \lessgtr p_i | u |$.

Since $\alpha = \overline{p} - (|A| + |B| + |C|)P$ is an M-matrix, so $\overline{\alpha} = \overline{\beta} - (|A| + |B| + |C|)P$ is an M-matrix. From the property of M-matrix[13], there exists a matrix $T = \text{diag}(T_1, \cdots, T_n) > 0$ such that $[T(-\overline{\beta} + (|A| + |B| + |C|)P)]^s \le -\varepsilon E_n < 0$     (8)

for sufficiently small $\varepsilon > 0$, where $E_n$ is the identity matrix. Calculating

$$[Tu]^{\mathrm{T}} \overline{H}(u) = \sum_{i=1}^{n} u_i T_i [-d_i u_i + \sum_{j=1}^{n} a_{ij}(f_j(u_i) - f_j(0)) +$$

$\wedge_{j,k=1}^{n} b_{ijk}^1 f_j(u_j) f_k(u_k) - \wedge_{j,k=1}^{n} b_{ijk}^1 f_j(0) f_k(0) + \vee_{j,k=1}^{n} b_{ijk}^2 f_j(u_j) f_k(u_k) - \vee_{j,k=1}^{n} b_{ijk}^2 f_j(0) f_k(0)$

$+ \wedge_{j,k=1}^{n} c_{ij}^1 k f_j(u_j) f_k(u_k) - \wedge_{j,k=1}^{n} c_{ijk}^1 f_j(0) f_k(0) + \vee_{j,k=1}^{n} c_{ijk}^2 f_j(u_j) f_k(u_k) - \vee_{j,k=1}^{n} c_{ijk}^2 f_j(0) f_k(0)]$

$\lessgtr u |^{\mathrm{T}} [T(-\overline{\beta} + (|A| + |B| + |C|)P)]^s | u | \le -\varepsilon \| u \|^2$.     (9)

From (9) and using Schwartz inequality, we get $\varepsilon \| u \|^2 \lessgtr \| T \| \| u \| \| \overline{H}(u) \|$, namely,

$$\varepsilon \| u \| \lessgtr \| T \| \cdot \| \overline{H}(u) \| .$$     (10)

So, $\| \overline{H}(u) \| \to +\infty$, i.e., $\| H(u) \| \to +\infty$ as $\| u \| \to +\infty$. From Lemma 1, we know that for every input $J$, map $H(u)$ is homeomorphism on $R^n$. So systems (1) have a unique equilibrium point $u^*$. The proof is completed.

## 4     Global Exponential Stability of Equilibrium Point

In this section, we shall apply the ideal of vector Lyapunov method to analyze global exponential stability of model (1).

**Theorem 2.** If Assumption D, E, and F are satisfied and $\alpha = \overline{p} - (|A| + |B| + |C|)P$ is an M-matrix, then for each input $J$, systems (1) have a unique equilibrium point, which is globally exponentially stable.

**Proof:** Since $\alpha$ is an M-matrix, from Theorem 1, system (1) has a unique equilibrium point $u^*$. Let $z(t) = u(t) - u^*$, model (1) can be written as

$$\frac{\partial z_i(t)}{\partial t} = \sum_{k=1}^{m} \frac{\partial}{\partial x_k} [D_{ik}(t, x, (z_i(t) + u^*)) \frac{\partial z_i}{\partial x_k}] - d_i^{\wedge}(z_i(t))[\rho_i^{\wedge}(z_i(t)) - \sum_{j=1}^{n} a_{ij} F_j(z_j(t))$$

$+ \wedge_{j,k=1}^{n} b_{ijk}^1 F_j(z_j(t - \tau_{ij}(t))) F_k(z_k(t - \tau_{ik}(t))) - \wedge_{j,k=1}^{n} b_{ijk}^1 f_j(u_j^*) f_k(u_k^*)$

$+ \vee_{j,k=1}^{n} b_{ijk}^2 F_j(z_j(t - \tau_{ij}(t))) F_k(z_k(t - \tau_{ik}(t))) - \vee_{j,k=1}^{n} b_{ijk}^2 f_j(u_j^*) f_k(u_k^*)$

$$- \wedge_{j,k=1}^{n} c_{ijk}^{1} \int_{-\infty}^{t} k_{ij}(t-s)F_{j}(z_{j}(s))F_{k}(z_{k}(s))ds + \wedge_{j,k=1}^{n} c_{ijk}^{1} f_{j}(u_{j}^{*})f_{k}(u_{k}^{*})$$

$$- \vee_{j,k=1}^{n} c_{ijk}^{2} \int_{-\infty}^{t} k_{ij}(t-s)F_{j}(z_{j}(s))F_{k}(z_{k}(s))ds + \vee_{j,k=1}^{n} c_{ijk}^{2} f_{j}(u_{j}^{*})f_{k}(u_{k}^{*})] \quad (i=1,2,...,n) , (11)$$

where $F_{j}(z_{j}(t)) = f_{j}(z_{j}(t)+u_{j}*)-f_{j}(u_{j}*)$,

$$F_{j}(z_{j}(t-\tau_{ij}(t))) = f_{j}(z_{j}(t-\tau_{ij}(t)+u_{j}^{*}))-f_{j}(u_{j}^{*})$$
$$d_{i}^{\wedge}(z_{i}(t)) = d_{i}(z_{i}(t)+u_{i}^{*}), \quad \rho_{i}^{\wedge}(z_{i}(t)) = \rho_{i}(z_{i}(t)+u_{i}^{*})-\rho_{i}(u_{i}^{*}).$$

The initial condition of Eqs. (11) is $\psi(s) = \phi(s)-u^{*}$, $s \leq 0$, and Eqs. (11) have a unique equilibrium at $z = 0$. So according to the Assumption D, we get

$$|F_{j}(z_{j}(t))| \leq p_{j} |z_{j}(t)|.$$

Due to $\alpha$ being an M-matrix, so $\overline{\alpha} = \overline{\beta} - (|A|+|B|+|C|)P$ is an M-matrix. Using property of M-matrix [13], there exist $\xi_{i} > 0$ $(i=1,2,...,n)$ satisfy

$$-\xi_{i}\beta_{i} + \sum_{j=1}^{n} \xi_{j}(|a_{ij}|+|B_{ij}|+|C_{ij}|)p_{j} < 0 \quad (i=1,2,...,n). \quad (12)$$

From Assumption F, we know that $0 \leq \sigma_{i} \leq d_{i}(z_{i}(t)+u_{i}^{*})$, so $d_{i}(z_{i}(t)+u_{i}^{*})/\sigma_{i} \geq 1$.

So there exist a constant $\lambda > 0$ such that

$$-\xi_{i}(\beta_{i}-\frac{\lambda}{\sigma_{i}})+\sum_{j=1}^{n}\xi_{j}(|a_{ij}|+e^{\lambda\tau}|B_{ij}|+|C_{ij}|N_{ij}(\lambda))p_{j} < 0. \quad (13)$$

Here, $\tau$ is a fixed number according to assumption of neural networks (1). Let $V_{i}(t) = e^{\lambda t} z_{i}(t)$ $(i=1,2,...,n)$, and Lyapunov function $\overline{V_{i}}(t) = \int_{\Omega} |V_{i}(t)| dx$, calculating the upper right derivative $D^{+}\overline{V_{i}}(t)$ of $\overline{V_{i}}(t)$ along the solutions of (11), we get

$$D^{+}\overline{V_{i}}(t) = \int_{\Omega} D^{+}|V_{i}(t)|dx = \int_{\Omega}(e^{\lambda t}\,\text{sgn}\,z_{i}\dot{z}_{i}+\lambda e^{\lambda t}|z_{i}|)dx . \quad (14)$$

From (11) and (14), according to Assumption D, Lemma 4, boundary condition (2), and $mes\Omega > 0$, we get

$$D^{+}\overline{V_{i}}(t) = \int_{\Omega}[e^{\lambda t}\,\text{sgn}\,z_{i}(t)\sum_{k=1}^{m}\frac{\partial}{\partial x_{k}}[D_{ik}(t,x,(z_{i}(t)+u^{*}))\frac{\partial u_{i}}{\partial x_{k}}]dx + \int_{\Omega}\lambda e^{\lambda t}|z_{i}(t)|dx$$

$$- \int_{\Omega} d_{i}^{\wedge}(z_{i}(t))e^{\lambda t}\{sgn\,z_{i}(t)[\rho_{i}^{\wedge}(z_{i}(t))-\sum_{j=1}^{n}a_{ij}F_{j}(z_{j}(t))$$

$$-\wedge_{j,k=1}^{n}b_{ijk}^{1}F_{j}(z_{j}(t-\tau_{ij}(t)))F_{k}(z_{k}(t-\tau_{ik}(t)))+\wedge_{j,k=1}^{n}b_{ijk}^{1}f_{j}(u_{j}^{*})f_{k}(u_{k}^{*})$$

$$-\vee_{j,k=1}^{n}b_{ijk}^{2}F_{j}(z_{j}(t-\tau_{ij}(t)))F_{k}(z_{k}(t-\tau_{ik}(t)))+\vee_{j,k=1}^{n}b_{ijk}^{2}f_{j}(u_{j}^{*})f_{k}(u_{k}^{*})$$

$$+\wedge_{j,k=1}^{n}c_{ijk}^{1}\int_{-\infty}^{t}k_{ij}(t-s)F_{j}(z_{j}(s))F_{k}(z_{k}(s))ds-\wedge_{j,k=1}^{n}c_{ijk}^{1}f_{j}(u_{j}^{*})f_{k}(u_{k}^{*})$$

$$+\vee_{j,k=1}^{n}c_{ijk}^{2}\int_{-\infty}^{t}k_{ijk}(t-s)F_{j}(z_{j}(s))F_{k}(z_{k}(s))ds-\vee_{j,k=1}^{n}c_{ijk}^{2}f_{j}(u_{j}^{*})f_{k}(u_{k}^{*})]\}dx$$

$$\leq -\int_{\Omega}d_{i}^{\wedge}(z_{i}(t))[(\beta_{i}-\frac{\lambda}{\sigma_{i}})|V_{i}(t)|-\sum_{j=1}^{n}|a_{ij}|p_{j}|V_{j}(t)|-e^{\lambda\tau_{ij}(t)}|B_{ij}|p_{j}|V_{j}(t-\tau_{ij}(t))|]dx$$

$$+\int_{\Omega}d_{i}^{\wedge}(z_{i}(t))[\sum_{j=1}^{n}|C_{ij}|p_{j}\times\int_{-\infty}^{t}k_{ij}(t-s)e^{\lambda(t-s)}|V_{j}(s)|ds]dx \quad (15)$$

Defining the curve $\gamma = \{y(l): y_i = \xi_i l, l > 0, i = 1,2,...,n\}$ , and the set $\Omega(y) = \{u : 0 \le u \le y, y \in \gamma\}$ . Let $\xi_{\min} = \min_{1 \le i \le n}\{\xi_i\}$ , $\xi_{\max} = \max_{1 \le i \le n}\{\xi_i\}$ . Taking $l_0 = (1+\delta)\|\psi\|/\xi_{\min}$ , where $\delta > 0$ is a constant number, then $\{|V|:|V| = e^{\lambda s}|\psi(s)|, -\tau \le s \le 0\} \subset \Omega(z_0(l_0))$ , namely $|V_i(s)| = e^{\lambda s}|\psi_i(s)| < \xi_i l_0$ , $-\tau \le s \le 0$ , $i = 1,2,...,n$ .

We claim that $|V_i(t)| < \xi_i l_0$ , for $t \in [0,+\infty)$ , $i = 1, 2, ... , n$ . If it is not true, then there exist some index $i$ and $t_1$ ($t_1 > 0$) such that $|V_i(t_1)| = \xi_i l_0$ , $D^+|V_i(t_1)| \ge 0$ , and $|V_j(t)| \le \xi_j l_0$ , for $-\tau < t \le t_1$ , $j = 1, 2, ... , n$ . So we could get $D^+\overline{V_i}(t_1) = \int_\Omega D^+|V_i(t_1)|dx \ge 0$ . However, from (13) and (15), we get

$$D^+\overline{V_i}(t_1) = \int_\Omega D^+|V_i(t_1)|dx \le -d_i^\wedge(z_i(t))[\xi_i(\beta_i - \frac{\lambda}{\sigma_i}) - \sum_{j=1}^n \xi_j(|a_{ij}|p_j$$

$$+ e^{\lambda\tau}|B_{ij}|p_j + |C_{ij}|N_{ij}(\lambda)p_j)]l_0 < 0$$

There is a contradiction. So $|V_i(t)| < \xi_i l_0$ , for $t \in [0,+\infty)$ , therefore,

$$|z_i(t)| < \xi_i l_0 e^{-\lambda t} \le (1+\delta)\|\psi\|\xi_{\max}/\xi_{\min} e^{-\lambda t} = \beta\|\psi\|e^{-\lambda t} ,$$

where $\beta = (1+\delta)\xi_{\max}/\xi_{\min}$ . From definition 2, the zero solution of systems (11) is globally exponentially stable, i.e., the equilibrium point of systems (1) is globally exponentially stable. The proof is completed.

**Remarks 2**: If the activation function of second order satisfy $f_k(u_k(t - \tau_{ik}(t))) = 1$ , $f_k(u_k(s)) = 1$ , then the system (1) is become the normal fuzzy Cohen-Grossberg neural networks discussed in [7]. And if $D_i(t,x,u_i) = 0$ or $d_i(u_i) = 0$ , then the system (1) is become the generally studied neural networks in many literatures such as [2~5].

**Remarks 3**: If $c_{ijk}^1 = 0$ and $c_{ijk}^2 = 0$ , then the high order system is the same as [8,10,11,12], which not include the unbounded delays. In other words, the system in these literatures is the special case of this paper.

# 5    Conclusions

In the paper, a thorough analysis of existence, uniqueness, and global exponential stability of the equilibrium point for a class of HOFCGNN with reaction-diffusion terms and both variable delays and unbounded delay have been presented. The conditions ensuring the existence and uniqueness of the equilibrium are obtained. By constructing proper Lyapunov functionals, using M-matrix theory and qualitative property of the differential inequalities, the sufficient conditions for global exponential stability of the equilibrium point for HOFCGNN are obtained. Some restrictions on HOFCGNN are removed, and the results in this paper are explicit and convenient to verify in practice.

**Acknowledgments.** This work is supported by the Fundamental Research Funds for the Central Universities (SWJTU11BR091), Natural Science Foundation of China (No. 10772152, 60974132, 11172247, 61003142, 61100118).

# References

[1] Cohen, M.A., Grossberg, S.: Absolute stability and global pattern formation and parallel memory storage by competitive neural networks. IEEE Trans. Syst. Man Cybern. 13(5), 815–825 (1983)

[2] Yang, T., Yang, L.B.: The global stability of fuzzy cellular network. IEEE Trans. Circuits. System-I 43, 880–883 (1996)

[3] Forti, M., Tesi, A.: New conditions for global stability of neural networks with application to linear and quadratic programming problems. IEEE Trans. Circuits System-I 42(7), 354–366 (1995)

[4] Civalleri, P.P., Gilli, M., Pandolfi, L.: On stability of cellular neural networks with delay. IEEE Transactions on Circuits and Systems-I 40(3), 157–164 (1993)

[5] Zhang, J.Y., Suda, Y., Iwasa, T.: Absolutely exponential stability of a class of neural networks with unbounded delay. Neural Networks 17, 391–397 (2004)

[6] Milenkovic, S., Obradovic, Z., Litovski, V.: Annealing Based Dynamic Learning in Second-Order Neural Networks (1996)

[7] Zhang, J., Ren, D., Zhang, W.: Global Exponential Stability of Fuzzy Cohen-Grossberg Neural Networks with Variable Delays and Distributed Delays. In: Huang, D.-S., Heutte, L., Loog, M. (eds.) ICIC 2007. LNCS (LNAI), vol. 4682, pp. 66–74. Springer, Heidelberg (2007)

[8] Ke, Y., Miao, C.: Periodic Solutions for High-Order Cohen-Grossberg-Type BAM Neural Networks with Time-Delays. In: Liu, D., Zhang, H., Polycarpou, M., Alippi, C., He, H. (eds.) ISNN 2011, Part I. LNCS, vol. 6675, pp. 375–384. Springer, Heidelberg (2011)

[9] Wang, Z.D., Fang, J.A., Liu, X.H.: Global stability of stochastic high-order neural networks with discrete and distributed delays. Chaos, Solitons and Fractals 36, 388–396 (2008)

[10] Zheng, C.D., Zhang, H.G., Wang, Z.S.: Novel Exponential Stability Criteria of High-Order Neural Networks With Time-Varying Delays. IEEE Trans. Syst. Man Cybern-B: Cybernetics 41(2) (April 2011)

[11] Ren, F.L., Cao, J.D.: Periodic solutions for a class of higher-order Cohen–Grossberg type neural networks with delays. Computers and Mathematics with Applications 54, 826–839 (2007)

[12] Yan, P., Lv, T., Lei, J.S., He, W.M.: Synchronization of Fuzzy High-order Cohen-Grossberg Neural Networks with Reaction-diffusion Term and Time-varying. Journal of Computational Information Systems 7(11), 4145–4152 (2011)

[13] Siljak, D.D.: Large-scale Dynamic Systems — Stability and Structure. Elsevier North-Holland, Inc., New York (1978)

# A Study on the Randomness Reduction Effect of Extreme Learning Machine with Ridge Regression

Meng Joo Er[12,*], Zhifei Shao[2], and Ning Wang[1]

[1] Dalian Maritime University, 1 Linghai Road, Dalian, China, 116026
[2] Nanyang Technological University, 50 Nanyang Ave, Singapore, 639798
{emjer,zshao1}@ntu.edu.sg,
n.wang.dmu.cn@gmail.com

**Abstract.** In recent years, Extreme Learning Machine (ELM) has attracted comprehensive attentions as a universal function approximator. Comparing to other single layer feedforward neural networks, its input parameters of hidden neurons can be randomly generated rather than tuned, and thereby saving a huge amount of computational power. However, it has been pointed out that the randomness of ELM parameters would result in fluctuating performances. In this paper, we intensively investigate the randomness reduction effect by using a regularized version of ELM, named Ridge ELM (RELM). Previously, RELM has been shown to achieve generally better generalization than the original ELM. Furthermore, we try to demonstrate that RELM can also greatly reduce the fluctuating performance with 12 real world regression tasks. An insight into this randomness reduction effect is also given.

**Keywords:** Extreme Learning Machine, Ridge Regression, Randomness Reduction.

## 1 Introduction

Extreme Learning Machine (ELM) achieves its extremely fast learning speed through random generalization of input parameters. However, one of the biggest concerns towards ELM is also the reason of its popularity: randomness nature. Zhu *et al* pointed out that the random assignment of parameters could introduce non-optimal solutions [1]. In [2], fluctuating performance of ELM was reported with different initial parameters. Various approaches have been applied to reduce the randomness effect of ELM, such as evolutionary algorithms [1,3]. However,

---

* Corresponding Author. This work is supported by the National Natural Science Foundation of China (under Grant 51009017), Applied Basic Research Funds from Ministry of Transport of P. R. China (under Grant 2012-329-225-060), China Postdoctoral Science Foundation (under Grant 2012M520629), and Fundamental Research Funds for the Central Universities of China (under Grant 2009QN025, 2011JC002).

C. Guo, Z.-G. Hou, and Z. Zeng (Eds.): ISNN 2013, Part I, LNCS 7951, pp. 166–173, 2013.
© Springer-Verlag Berlin Heidelberg 2013

their slow learning speed becomes the bottleneck. Another approach is to use an ensemble of ELMs, which gives an average output of each individual one [2,4,5,6]. The third approach is to first create a large pool of neurons, and a subset of more significant neurons are selected using various ranking algorithms [7,8,9]. Ridge regression [10] was also applied to improve the performance of ELM [11]. It has been found out that the generalization ability of ridge ELM (RELM) is less sensitive to the choice of ridge parameter $C$ and the number of neurons than traditional ELM [12]. And for some activation function, sigmoid for instance, it appears that the generalization performance reaches a plateau rather than deteriorating when the number of neurons exceeds some value [12].

In this paper, we attempt to demonstrate an additional feature of RELM, the randomness reduction effect. Different to previous works, we design our tests solely to investigate the ELM fluctuating performance caused by different initial parameters, excluding the influences such as random data partition and target value approximation (caused by classification tasks). The results are compared with the original ELM and ELM ensemble since their computational time are at relatively same level. The remaining of this paper is organized as follows: Section 2 gives the preliminaries of ELM, RELM and ensemble method. The randomness effect comparison of the above methods is presented in Section 3. An attempt to explain the reason why RELM is less affected by random weights than ELM is given in Section 4. Conclusions are drawn in Section 5.

## 2 Preliminaries on Original ELM, ELM Ensemble and RELM

The structure of the original ELM is a generalized single layer feed forward Neural Network, and its output $y$ with $L$ hidden nodes can be represented by:

$$y = \sum_{i=1}^{L} \beta_i g_i(\mathbf{x}) = \sum_{i=1}^{L} \beta_i G(\boldsymbol{\omega}_i, b_i, \mathbf{x}) = \mathbf{H}\boldsymbol{\beta} \tag{1}$$

where $\mathbf{x}, \boldsymbol{\omega}_i \in \mathbb{R}^d$ and $g_i$ denotes the $i^{th}$ hidden node output function $G(\boldsymbol{\omega}_i, b_i, \mathbf{x})$; $\mathbf{H}$ and $\boldsymbol{\beta}$ are the hidden layer output matrix and output weight matrix respectively. For $N$ distinct samples $(\mathbf{x_j}, t_j)$, $j = 1, \ldots N$, Eqn. 1 can be written as:

$$\mathbf{H}\boldsymbol{\beta} = \mathbf{T} \tag{2}$$

$$\mathbf{H} = \begin{bmatrix} \mathbf{h}_1 \\ \vdots \\ \mathbf{h}_N \end{bmatrix} = \begin{bmatrix} \mathbf{h(x_1)} \\ \vdots \\ \mathbf{h(x_N)} \end{bmatrix} = \begin{bmatrix} G(\boldsymbol{\omega}_1, b_1, \mathbf{x_1}) & \cdots & G(\boldsymbol{\omega}_L, b_L, \mathbf{x_1}) \\ \vdots & \cdots & \vdots \\ G(\boldsymbol{\omega}_1, b_1, \mathbf{x_N}) & \cdots & G(\boldsymbol{\omega}_L, b_L, \mathbf{x_N}) \end{bmatrix}_{N \times L} \tag{3}$$

Since the input weights of its hidden neurons $(\boldsymbol{\omega}_i, b_i)$ can be randomly generated instead of tuned [12], the only parameters that need to be calculated in ELM is the output weight matrix $\boldsymbol{\beta}$, which can be easily done through Least Squares Estimate (LSE):

$$\boldsymbol{\beta} = \mathbf{H}^\dagger \mathbf{T} \tag{4}$$

where $\mathbf{H}^{\dagger}$ is the *Moore-Penrose generalized inverse* of matrix $\mathbf{H}$ [11], which can be calculated through orthogonal projection, where $\mathbf{H}^{\dagger} = (\mathbf{H^TH})^{-1}\mathbf{H^T}$.

The idea of neural network ensembles was first introduced by Hansen and Salamon [13]. Its structure consists of $P$ individual neural networks. In the common implementation, the final output of the ensemble is the average of each individual one's result. By using the ensemble, it has been shown that the the overall network performance can be expected to be improve, and it can be applied to ELM to reduce its fluctuating performance [2,4,6].

According to ridge regression theory [10], more stable and better generalization performance can be achieved by adding a positive value $1/C$ to the diagonal elements of $\mathbf{H^TH}$ when calculating the output weight $\boldsymbol{\beta}$ [12,14]. Therefore, the corresponding ELM with ridge regression becomes:

$$\mathbf{H}^{\dagger} = (\mathbf{H^T H} + \frac{\mathbf{I}}{C})^{-1}\mathbf{H^T} \tag{5}$$

## 3 Randomness Effect Comparison

To better explain the fluctuating performance of ELM, Auto MPG data (datasets used in this paper are taken from UCI Machine Learning Repository [15] and Statlib [16]) is selected for demonstration. To be fair, $L$ in ELM and $C$ in RELM ($L$ in RELM is uniformly selected as 1000) are chosen using 10-fold-cross-validation. The specific data configuration is shown in Table 1. Totally 10 trials are carried out with random permutation. Within each trial, 50 runs are done to study the randomness effect with data unchanged, therefore it eliminates the performance fluctuation caused by different data partitions.

From Figure 1, it can be seen that the Standard Deviation (STD) caused by the random generalization of input parameters are reduced to about 1/3 to the original level, and therefore ELM ensemble can indeed achieve more stable performance than using a single ELM. For the case of RELM, not only the testing results are generally improved, the STD has also been reduced to about half of

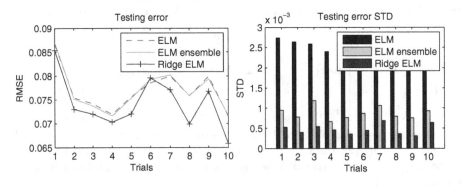

**Fig. 1.** Random effect comparison of original ELM and ELM ensemble with Auto MPG

ELM ensemble and 1/6 of original ELM. For the Auto MPG problem, the STD is only around half a thousandth of the whol e data range, pretty insignificant.

To thoroughly study the randomness reduction effect of RELM, 12 regression datasets are selected for comparison. Regression tasks are preferred in this study, since the correction rate in classification may be affected by the output approximation, therefore not able to truly reflect the randomness effect[1].

**Table 1.** Specification of datasets and results. 1:Number of features; 2: Number of training data; 3: Number of testing data

| Datasets | L | C | 1 | 2 | 3 | ELM | | ELM Ensemble | | RELM | |
|---|---|---|---|---|---|---|---|---|---|---|---|
| | | | | | | RMSE | STD | RMSE | STD | RMSE | STD |
| Basketball | 5 | $2^0$ | 5 | 64 | 32 | 0.2980 | 0.0034 | 0.2975 | 0.0011 | **0.2923** | **2.064e-4** |
| Autoprice | 19 | $2^{-1}$ | 16 | 106 | 53 | 0.1847 | 0.0119 | 0.1841 | 0.0036 | **0.1767** | **6.412e-4** |
| Bodyfat | 24 | $2^0$ | 15 | 168 | 84 | 0.0326 | 0.0034 | 0.0324 | 0.0010 | **0.0291** | **2.983e-4** |
| Auto MPG | 28 | $2^{10}$ | 8 | 261 | 131 | 0.0766 | 0.0026 | 0.0766 | 8.563e-4 | **0.0738** | **4.507e-4** |
| Housing | 30 | $2^5$ | 14 | 337 | 169 | 0.0673 | 0.0022 | 0.0674 | 7.245e-4 | **0.0605** | **4.748e-4** |
| Forest Fire | 1 | $2^0$ | 13 | 345 | 172 | 0.0945 | 0.0031 | 0.0947 | 0.0010 | **0.0669** | **2.344e-4** |
| Strike | 19 | $2^{-5}$ | 7 | 416 | 209 | 0.2987 | 0.0095 | 0.2981 | 0.0029 | **0.2959** | **6.043e-5** |
| Concrete | 139 | $2^{13}$ | 9 | 687 | 343 | 0.0900 | 0.0076 | 0.0897 | 0.0024 | **0.0766** | **0.0011** |
| Balloon | 16 | $2^{20}$ | 3 | 1334 | 667 | 0.0544 | 0.0026 | **0.0543** | 8.155e-4 | 0.0577 | **1.607e-4** |
| Quake | 18 | $2^4$ | 4 | 1452 | 726 | 0.1764 | 2.927e-4 | **0.1764** | 9.116e-5 | 0.1774 | **5.193e-5** |
| Space-ga | 69 | $2^4$ | 7 | 2071 | 1036 | **0.0341** | 8.104e-4 | 0.0342 | 2.732e-4 | 0.0357 | **2.743e-5** |
| Abalone | 30 | $2^0$ | 9 | 2784 | 1393 | 0.0775 | 0.0011 | **0.0775** | 3.617e-4 | 0.0777 | **5.478e-5** |

From the results in Table 1 (best results shown in bold letters), it can be seen that the fluctuating performance of ELM caused by random generalization of input weights can be greatly reduced by using RELM. Generality speaking, RELM offers better test performance, which is consistent with the results reported in [12]. Although by using an ensemble of ELMs, the random effect of input parameters can also be reduced, but still not enough to beat RELM in our tests. In summary, we believe RELM is a better choice compared to ELM and ELM ensemble because it offers the following benefits (first two suggested in [12]):

- Better generalization performance may be achieved.
- Only one ridge parameter $C$ needs to be defined by the user, which can be selected easily and efficiently.
- The fluctuating performance of ELM caused by random generalization of input parameters can be greatly reduced.

---

[1] The final STD is the average of STD derived from each trial rather than all trials.

## 4   An Attempt to Explain the Randomness Reduction Effect of RELM

Although the simulation results strongly demonstrate the randomness reduction effect of RELM, it is preferred that this phenomenon can be explained. From Eqn. 2 and 3 and , the target estimation matrix in Eqn. 2 can be rewritten as:

$$\begin{aligned}
\mathbf{H}\boldsymbol{\beta} &= [\,G(\mathbf{X}\boldsymbol{\omega}_1 + \mathbf{b}_1)\, G(\mathbf{X}\boldsymbol{\omega}_2 + \mathbf{b}_2))\, \cdots \, G(\mathbf{X}\boldsymbol{\omega}_L + \mathbf{b}_L))\,]\boldsymbol{\beta} \\
&= \beta_1 G(\mathbf{X}\boldsymbol{\omega}_1 + \mathbf{b}_1) + \beta_2 G(\mathbf{X}\boldsymbol{\omega}_2 + \mathbf{b}_2) + \cdots + \beta_L G(\mathbf{X}\boldsymbol{\omega}_L + \mathbf{b}_L) \\
&= \beta_1 \mathbf{G}_1 + \beta_2 \mathbf{G}_2 + \cdots + \beta_L \mathbf{G}_L = \mathbf{T}
\end{aligned} \tag{6}$$

Therefore the learning process can be considered as finding the best linear combination of $[\mathbf{G}_1 \, \mathbf{G}_2 \, \cdots \, \mathbf{G}_L]$ to approximate the target vector $\mathbf{T}$. According to [12], to solve Eqn. 6 for any $\mathbf{T}$ with zero error, the linear combination of $\mathbf{G}_i$, should be able to cover the whole space of $\mathbb{R}^N$. Given the fact that when $L \geq N$, serious overfitting problem can appear in ELM, $L$ is usually much smaller than $N$ for the optimal structure in ELM. Consequently, $\mathbf{H}\boldsymbol{\beta}$ can only approximate $T$ to a certain degree. In the simulation tests carried out in Section 3, $\mathbf{X}$ and $\mathbf{T}$ are hold constant in Eqn. 6. However, because of the effect of randomly generalized $\boldsymbol{\omega}$ and $\mathbf{b}$ ($\boldsymbol{\omega} \in [-1,1]$ and $b \in [0,1]$), each $\mathbf{G}_i$ will be oscillating in the range of $[-G(-\mathbf{X}), G(\mathbf{X}+1)]$ (with sigmoid function). For each run, the linear combination of $\mathbf{G}_i$, can only approximate $\mathbf{T}$ with varying performances, and this is where the problem of random performance stems from.

   The output regularization of RELM can suppress the overfitting problem and therefore more neurons can be used in the network structure. In [12], $L$ is uniformly selected as 1000 for all tasks, which results in $L > N$ for most cases. Intuitively, the linear combination of $\mathbf{G}_i$ should be able to approximate $\left\| \mathbf{H}\boldsymbol{\beta} - \tilde{\mathbf{T}} \right\|^2$, with $\left\| \mathbf{T} - \tilde{\mathbf{T}} \right\| = \varepsilon$, much easier than ELM. Hence the fluctuating training performance can be reduced. And with properly selected $C$, the fluctuating testing performance can also be reduced. However, to thoroughly explain this phenomenon, we propose the following lemmas and a remark.

**Lemma 1.** *In RELM, let $L \geq N$, the fluctuating training performance will decrease towards zero when $C \to \infty$*

*Proof.* When $C \to \infty$, $\frac{1}{C} \to 0$, therefore the RELM will approach to the normal ELM. ELM can approximate $\mathbf{T}$ with zero error if at least $N$ $\mathbf{G}_i$ are linearly independent Consequently, the STD is close to zero because all training errors are close to zero.

**Lemma 2.** *In RELM, the fluctuating performance will decrease towards zero when $C \to 0$*

*Proof.* When $C \to 0$, $\frac{1}{C} \to \infty$, then $(\mathbf{H}^T\mathbf{H} + \frac{I}{C})$ can be considered as a square matrix with diagonal elements approaching to $\infty$ and others approaching to 0, and this results in its inverse $(\mathbf{H}^T\mathbf{H} + \frac{I}{C})^{-1}$ approach zero matrix. Therefore the estimated output $\mathbf{H}\boldsymbol{\beta} = \mathbf{H}(\mathbf{H}^T\mathbf{H} + \frac{I}{C})^{-1}\mathbf{H}^T\mathbf{T}$ is overwhelmed by the zero matrix and produces nearly zero. Thus the STD of outputs approaches zero.

The two extreme cases mentioned above rarely happens in practical implementations since the scenario in Lemma 1 will lead to serious overfitting problem and the RELM in Lemma 2 has outputs approaching zero. Consequently, it is more important to analyse the scenarios where $C$ and $L$ enable RELM to have optimal generalization ability, which has the most practical significance. Figure 2 and Figure 3 shows the randomness reduction effect with varying $C$ and $L$, the STD comparison of ELM and RELM across a wide range of $C$ and $L$.

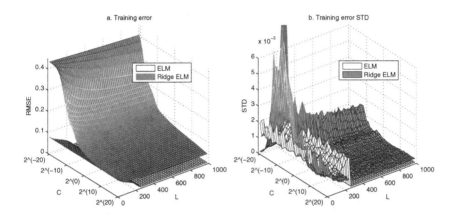

**Fig. 2.** Training error and STD comparison between ELM and RELMfor Auto MPG task

The above two Lemmas have been verified in Figure 2. From Figure 2.a, It can be seen that the STD of RELM is pretty low when $C$ takes two extreme values. However, as mentioned in Lemma 2, this is caused by two different scenarios, where RELM is close to ELM when $C$ approaches $\infty$, while when $C$ approaches 0, it almost loses approximation ability and its output error is very high. When $L$ is small, the available $\mathbf{G}_i$ has difficulties approximating the target $T$, thus the STD of ELM oscillates at a relatively high value. At the same time, adding $C$ can stabilize the output, and this explains why the STD of RELM is smaller than one of ELM when $L$ is small, but with slightly higher training error.

From Figure 2.b, it can be seen that implementing $C$ generally has some negative effect on controlling the training STD, since the ELM has lower STD than RELM when $L > 200$. However, in real applications, optimal ELM usually takes a small $L$ (28 in Auto MPG), then RELM with much more hidden neurons (e.g., 1000), has more advantage in controlling STD, as long as $C$ is not too small (rarely happens since normally $\frac{1}{C}$ takes the value in $[0, 1]$ [17]). But of course, it is more important to analyse generalization performance, shown in Figure 3.

Generally speaking, by using a small amount of $1/C$, the ELM generalization ability can be improved, given the same $L$ is used in RELM, except when $L$ is small (probably before the overfitting effect appears in ELM). With increasing number of hidden neurons used, RELM can achieve better generalization ability

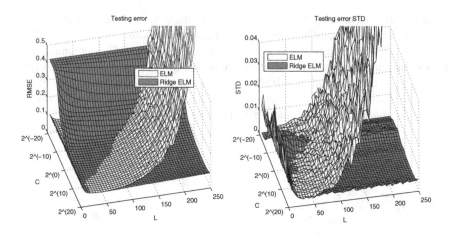

**Fig. 3.** Testing error and STD comparison between ELM and RELM for Auto MPG task

with bigger amount of $1/C$. This demonstrates that RELM is more resistant to overfitting problem. Although whether optimal RELM performs better than optimal ELM is still remain to be determined, the results in Table 1 seem to give an affirmative answer. From Figure 3.b, given the same $L$, RELM has more stable performance, and only a very small region in the plot shows otherwise. In this specific Auto MPG task, RELM already has lower STD than ELM when $L = 28$, $C = 2^{10}$. Since more neurons tends to offer better approximation ability, and the $C$ is selected so that overfitting effect is limited in RELM, we can almost certain to say that RELM offers more stable results with carefully chosen parameters, and experiments in Table 1 also shows similar results. Therefore we propose the following remark:

*Remark 1.* Given the same approximation task, and optimal $L$ in ELM and $C$ in RELM ($L$ is large enough) are selected, RELM can achieve more stable performance.

## 5    Conclusions

In this paper, the randomness reduction effect on the ELM performance caused by the implementation of ridge regression is investigated. According to the results of the 12 real world regression task experiments, it has been shown that the fluctuating performance can be greatly reduced by using RELM comparing to normal ELM, even an ensemble of 10 ELMs. Furthermore, we investigated the reason of the randomness effect through a dimensional viewpoint and proposed two Lemmas to discuss two extreme scenarios in RELM. After examining the performances across a wide range of $C$ and $L$, together with the results from experiments, we conclude that RELM can achieve more stable performance than normal ELM, given appropriate parameters are selected.

RELM is generally deemed to achieve better generalization performance and the parameter $C$ is easier to select. We demonstrated that in addition to these two benefits, it can also offer more stable performances. We believe our work can make ELM to attract more researchers, considering the fact that the random performance is one of the biggest concerns in ELM. In future works, we also plan to develop an algorithm to automatically derive the parameter $C$, replacing the tedious manual tuning procedure.

# References

1. Zhu, Q.-Y., Qin, A.K., Suganthan, P.N., Huang, G.-B.: Evolutionary extreme learning machine. Pattern Recognition 38, 1759–1763 (2005)
2. Lan, Y., Soh, Y., Huang, G.-B.: Ensemble of online sequential extreme learning machine. Neurocomputing 72(13-15), 3391–3395 (2009)
3. Suresh, S., Saraswathi, S., Sundararajan, N.: Performance enhancement of extreme learning machine for multi-category sparse data classification problems. Engineering Applications of Artificial Intelligence 23(7), 1149–1157 (2010)
4. Liu, N., Wang, H.: Ensemble based extreme learning machine. IEEE Signal Processing Letters 17(8), 754–757 (2010)
5. Miche, Y., Sorjamaa, A., Bas, P., Simula, O., Jutten, C., Lendasse, A.: Op-elm: optimally pruned extreme learning machine. IEEE Transactions on Neural Networks 21(1), 158–162 (2010)
6. Sun, Z., Choi, T., Au, K., Yu, Y.: Sales forecasting using extreme learning machine with applications in fashion retailing. Decision Support Systems 46(1), 411–419 (2008)
7. Feng, G., Huang, G.-B., Lin, Q., Gay, R.: Error minimized extreme learning machine with growth of hidden nodes and incremental learning. IEEE Transactions on Neural Networks 20(8), 1352–1357 (2009)
8. Lan, Y., Soh, Y., Huang, G.-B.: Constructive hidden nodes selection of extreme learning machine for regression. Neurocomputing 73, 3191–3199 (2010)
9. Lan, Y., Soh, Y., Huang, G.-B.: Two-stage extreme learning machine for regression. Neurocomputing 73(16-18), 3028–3038 (2010)
10. Hoerl, A., Kennard, R.: Ridge regression: Biased estimation for nonorthogonal problems. Technometrics, 55–67 (1970)
11. Huang, G.-B., Wang, D., Lan, Y.: Extreme learning machines: a survey. International Journal of Machine Learning and Cybernetics (2011)
12. Huang, G.-B., Zhou, H., Ding, X., Zhang, R.: Extreme learning machine for regression and multiclass classification. IEEE Transactions on Systems, Man, and Cybernetics, Part B: Cybernetics (99), 1–17 (2010)
13. Hansen, L., Salamon, P.: Neural network ensembles. IEEE Transactions on Pattern Analysis and Machine Intelligence 12(10), 993–1001 (1990)
14. Toh, K.: Deterministic neural classification. Neural Computation 20(6), 1565–1595 (2008)
15. Newman, D., Asuncion, A.: UCI machine learning repository (2007)
16. Vlachos, P.: Statlib project repository. Carnegie Mellon University (2000)
17. Mardikyan, S., Cetin, E.: Efficient choice of biasing constant for ridge regression. Int. J. Contemp. Math. Sciences 3(11), 527–536 (2008)

# Stability of Nonnegative Periodic Solutions of High-Ordered Neural Networks

Lili Wang[1] and Tianping Chen[2]

[1] Department of Applied Mathematics, Shanghai University of Finance and Economics,
Shanghai, P.R. China
[2] School of Computer Sciences, Key Laboratory of Nonlinear Mathematics Science,
School of Mathematical Sciences, Fudan University, Shanghai, P.R. China
wang.lili@mail.shufe.edu.cn,
tchen@fudan.edu.cn

**Abstract.** In this paper, a class of high-ordered neural networks are investigated. By rigorous analysis, a set of sufficient conditions ensuring the existence of a nonnegative periodic solution and its $R_+^n$-asymptotical stability are established. The results obtained can also be applied to the first-ordered neural networks.

**Keywords:** High-ordered neural networks, nonnegative periodic solutions, global stability, asymptotical stability.

## 1 Introduction

Recently, the neural networks with high-ordered interactions have attracted considerable attentions due to the fact that they have stronger approximation property, faster convergence rate, greater storage capacity, and higher fault tolerance than first-ordered neural networks. It is of great importance to study the dynamics underlying these systems in both theory and applications. Consider a class of high-ordered Cohen-Grossberg neural networks described by

$$\frac{du_i(t)}{dt} = -a_i(u_i(t))\left[b_i(u_i(t)) + \sum_{j=1}^{n} c_{ij}(t)f_j(u_j(t))\right.$$

$$\left. + \sum_{j_1=1}^{n}\sum_{j_2=1}^{n} d_{ij_1j_2}(t)g_{j_1}(u_{j_1}(t-\tau_{ij_1}))g_{j_2}(u_{j_2}(t-\tau_{ij_2})) + I_i(t)\right],$$

$$i = 1, \cdots, n, \quad (1)$$

where $u_i(t)$ represents the state of the $i$th unit at time $t$, $a_i(u) > 0$ is the amplification function, $c_{ij}(t)$ and $d_{ij_1j_2}(t)$ are the first-order and high-order connection weights, respectively, $\tau_{ij} \geq 0$ stands for the transmission delay, and $\tau = \max_{i,j} \tau_{ij}$, $I_i(t)$ denotes the external input at time $t$, $f_j$, $g_j$ are the activation functions, $i, j = 1, \cdots, n$. The initial condition is

$$u_i(s) = \phi_i(s) \quad for \quad s \in [-\tau, 0], \quad (2)$$

where $\phi_i \in C[-\tau, 0]$, $i = 1, \cdots, n$.

C. Guo, Z.-G. Hou, and Z. Zeng (Eds.): ISNN 2013, Part I, LNCS 7951, pp. 174–180, 2013.
© Springer-Verlag Berlin Heidelberg 2013

In the competition models of biological species, $u_i(t)$, which represents the amount of the $i$-th species, must be nonnegative. In [2], [4]-[6], theoretical analysis have been provided on the stability of positive solutions of neural networks. The dynamics of high-ordered neural networks were also studied in [7]-[9]. In this paper, we are concerned with the Cohen-Grossberg neural networks (1) with high-ordered connections, and to investigate the existence and stability of nonnegative periodic solutions.

## 2  Preliminaries

First of all, we present some assumptions and definitions required throughout the paper.

**Assumption 1.**    The amplification function $a_i(\rho)$ is continuous with $a_i(0) = 0$ and $a_i(\rho) > 0$ when $\rho > 0$. And for any $\epsilon > 0$, it holds that $\int_0^\epsilon \frac{d\rho}{a_i(\rho)} = +\infty$, $i = 1, 2, \cdots, n$.

**Assumption 2.**    $c_{ij}(t), d_{ij_1 j_2}(t), I_i(t)$ are continuous and periodic functions with period $\omega$. $b_i(x)$ is continuous and satisfies $\frac{b_i(x) - b_i(y)}{x - y} \geq \gamma_i > 0$, $i = 1, 2, \cdots, n$.

**Assumption 3.**    There exist positive constants $F_j > 0, G_j, M_j > 0$, such that,

$$|f_j(x) - f_j(y)| \leq F_j|x - y|, \quad |g_j(x) - g_j(y)| \leq G_j|x - y|, \quad |g_j(x)| \leq M_j$$

for any $x, y \in R, j = 1, 2, \cdots, n$.

**Definition 1.**    $\{\xi, \infty\}$-norm: $\|u(t)\|_{\{\xi, \infty\}} = \max_{i=1,\dots,n} \xi_i^{-1}|u_i(t)|$, where $\xi_i > 0, i = 1, \dots, n$.

**Definition 2.**    A nonnegative periodic solution $u^*(t)$ of the system (1) is said to be $R_+^n$-asymptotically stable if for any trajectory solution $u(t)$ of (1) with initial condition $\phi_i(s) > 0$, $s \in [-\tau, 0]$, $i = 1, 2, \cdots, n$, there holds that $\lim_{t \to +\infty} [u(t) - u^*(t)] = 0$.

## 3  Main Results

**Lemma 1.**    *Suppose that Assumption 1 is satisfied. If the initial condition $\phi_i(s) > 0$, then, the corresponding trajectory $u(t) = [u_1(t), u_2(t), \cdots, u_n(t)]^T$ satisfies $u_i(t) > 0$ for $t > 0$.*

**Proof:**    If for some $t_0 > 0$ and some index $i_0$, $u_{i_0}(t_0) = 0$, then, by the Assumption 1, we have

$$\int_0^{t_0} \left[ b_{i_0}(u_{i_0}(t)) + \sum_{j=1}^n c_{i_0 j}(t) f_j(u_j(t)) \right.$$

$$\left. + \sum_{j_1=1}^n \sum_{j_2=1}^n d_{i_0 j_1 j_2}(t) g_{j_1}(u_{j_1}(t - \tau_{i_0 j_1})) g_{j_2}(u_{j_2}(t - \tau_{i_0 j_2})) + I_{i_0}(t) \right] dt$$

$$= -\int_0^{t_0} \frac{\dot{u}_{i_0}(t)}{a_{i_0}(u_{i_0}(t))} dt = \int_0^{\phi_{i_0}(0)} \frac{d\rho}{a_{i_0}(\rho)} = +\infty. \tag{3}$$

Because of the continuity of $u_i(t)$, the left side of (3) is finite, which contradicts with the infinity of right side. Hence, $u_i(t) \neq 0$ for all $t > 0$, which implies that $u_i(t) > 0$ for all $t > 0$, $i = 1, 2, \cdots, n$. Lemma 1 is proved.

**Lemma 2.**    *Suppose that Assumption 1 $\sim$ Assumption 3 are satisfied. If there are positive constants $\xi_1, \ldots, \xi_n$ such that*

$$-\gamma_i \xi_i + \sum_{j=1}^{n} |c_{ij}(t)| \xi_j F_j + \sum_{j_1=1}^{n} \sum_{j_2=1}^{n} |d_{ij_1j_2}(t)| (\xi_{j_1} G_{j_1} M_{j_2} + \xi_{j_2} G_{j_2} M_{j_1}) < 0, \quad (4)$$

*for all $0 \leq t < \omega$, $i = 1, \cdots, n$, then any solution $u(t)$ of the high-ordered Cohen-Grossberg neural networks (1) is bounded.*

**Proof:**    By (4), we can find a small positive constant $\eta$ such that

$$-\gamma_i \xi_i + \sum_{j=1}^{n} |c_{ij}(t)| \xi_j F_j + \sum_{j_1=1}^{n} \sum_{j_2=1}^{n} |d_{ij_1j_2}(t)| (\xi_{j_1} G_{j_1} M_{j_2} + \xi_{j_2} G_{j_2} M_{j_1}) < -\eta < 0,$$

for all $t > 0$.

Define $M(t) = \sup_{t-\tau \leq s \leq t} \|u(s)\|_{\{\xi, \infty\}}$. For any fixed $t_0$, there are two possibilities:

(i) $\|u(t_0)\|_{\{\xi, \infty\}} < M(t_0)$.

In this case, there exists $\delta > 0$, $\|u(t)\|_{\{\xi, \infty\}} < M(t_0)$ for $t \in (t_0, t_0 + \delta)$. Thus $M(t) = M(t_0)$ for $t \in (t_0, t_0 + \delta)$.

(ii) $\|u(t_0)\|_{\{\xi, \infty\}} = M(t_0)$.

In this case, let $i_{t_0}$ be an index such that $\xi_{i_{t_0}}^{-1} |u_{i_{t_0}}(t_0)| = \|u(t_0)\|_{\{\xi, \infty\}}$. Notice that

$$|f_j(x)| \leq F_j |x| + |f_j(0)|, \quad |g_{j_1}(x) g_{j_2}(y)| \leq G_{j_1} M_{j_2} |x| + G_{j_2} M_{j_1} |y| + |g_{j_1}(0) g_{j_2}(0)|.$$

Denote

$$\tilde{C} = \max_i \left\{ \sum_{j=1}^{n} |b_i(0)| + c_{ij}^* |f_j(0)| + \sum_{j_1=1}^{n} \sum_{j_2=1}^{n} d_{ij_1j_2}^* |g_{j_1}(0) g_{j_2}(0)| + I_i^* \right\}, \quad (5)$$

where $c_{ij}^* = \sup_t |c_{ij}(t)|$, $d_{ij_1j_2}^* = \sup_t |d_{ij_1j_2}(t)|$, $I_i^* = \sup_t |I_i(t)|$.

Then we have

$$\left. \frac{d|u_{i_{t_0}}(t)|}{dt} \right|_{t=t_0}$$

$$= -\operatorname{sign}(u_{i_{t_0}}(t_0)) a_{i_{t_0}}(u_{i_{t_0}}(t_0)) \left\{ b_{i_{t_0}}(u_{i_{t_0}}(t_0)) + \sum_{j=1}^{n} c_{i_{t_0}j}(t_0) f_j(u_j(t_0)) \right.$$

$$\left. + \sum_{j_1=1}^{n} \sum_{j_2=1}^{n} d_{i_{t_0},j_1,j_2}(t_0) g_{j_1}(u_{j_1}(t_0 - \tau_{i_{t_0}j_1})) g_{j_2}(u_{j_2}(t_0 - \tau_{i_{t_0}j_2})) + I_{i_{t_0}}(t_0) \right\}$$

$$\leq a_{i_{t_0}}(u_{i_{t_0}}(t_0)) \left\{ -\gamma_{i_{t_0}} |u_{i_{t_0}}(t_0)| + \sum_{j=1}^{n} |c_{i_{t_0}j}(t_0)| F_j |u_j(t_0)| \right.$$

$$\left. + \sum_{j_1=1}^{n} \sum_{j_2=1}^{n} |d_{i_{t_0},j_1,j_2}(t_0)| (G_{j_1} M_{j_2} |u_{j_1}(t_0 - \tau_{i_{t_0}j_1})| + G_{j_2} M_{j_1} |u_{j_2}(t_0 - \tau_{i_{t_0}j_2})|) + \tilde{C} \right\}$$

$$\leq a_{i_{t_0}}(u_{i_{t_0}}(t_0))\left\{\left[-\gamma_{i_{t_0}}\xi_{i_{t_0}} + \sum_{j=1}^{n}|c_{i_{t_0}j}(t_0)|\xi_j F_j\right.\right.$$

$$\left.\left. + \sum_{j_1=1}^{n}\sum_{j_2=1}^{n}|d_{i_{t_0},j_1,j_2}(t_0)|(\xi_{j_1}G_{j_1}M_{j_2} + \xi_{j_2}G_{j_2}M_{j_1})\right]M(t_0) + \widetilde{C}\right\}$$

$$< a_{i_{t_0}}(u_{i_{t_0}}(t_0))\left\{-\eta M(t_0) + \widetilde{C}\right\}. \tag{6}$$

If $M(t_0) \geq \widetilde{C}/\eta$, then, $\left.\dfrac{d|u_{i_{t_0}}(t)|}{dt}\right|_{t=t_0} \leq 0$, $M(t)$ is non-increasing nearby $t_0$, that is, there exists $\delta_1 > 0$, such that $M(t) = M(t_0)$ for $t \in (t_0, t_0 + \delta_1)$. On the other hand, if $M(t_0) < \widetilde{C}/\eta$, then there exist $\delta_2 > 0$, such that $M(t) < \widetilde{C}/\eta$ for $t \in (t_0, t_0 + \delta_2)$. Let $\delta = \min\{\delta_1, \delta_2\}$, then, $M(t) \leq \max\{M(t_0), \widetilde{C}/\eta\}$ holds for every $t \in (t_0, t_0 + \delta)$.

In summary, $\|u(t)\|_{\{\xi,\infty\}} \leq M(t) \leq \max\{M(0), \widetilde{C}/\eta\}$. Lemma 2 is proved.

**Theorem 1.**    *Suppose that Assumption 1 $\sim$ Assumption 3 are satisfied. If there are positive constants $\xi_1, \ldots, \xi_n, \zeta_1, \cdots, \zeta_n$, such that for every $0 \leq t < \omega$, there hold (4) and*

$$-\gamma_i\zeta_i + \sum_{j=1}^{n}|c_{ji}(t)|\zeta_j F_i + \sum_{j_1=1}^{n}\sum_{j_2=1}^{n}(\zeta_{j_1}d_{j_1ij_2}^*M_{j_2} + \zeta_{j_2}d_{j_2j_1i}^*M_{j_1})G_i < 0, \tag{7}$$

*for $i = 1, \ldots, n$. Then the system (1) has a nonnegative periodic solution with periodic $\omega$, which is $R_+^n$-asymptotically stable.*

**Proof:** In fact, for any positive solution $u(t)$ of system (1), let $x_i(t) = u_i(t + \omega) - u_i(t)$, $y_i(t) = \int_{u_i(t)}^{u_i(t+\omega)}\frac{d\rho}{a_i(\rho)}$. It is obvious that $sign(x_i(t)) = sign(y_i(t))$. Then, by direct calculation, we have

$$\frac{d|y_i(t)|}{dt} = sign(y_i(t))\left\{-b_i(u_i(t+\omega)) + b_i(u_i(t))\right.$$

$$-\sum_{j=1}^{n}c_{ij}(t+\omega)f_j(u_j(t+\omega)) + \sum_{j=1}^{n}c_{ij}(t)f_j(u_j(t))$$

$$-\sum_{j_1=1}^{n}\sum_{j_2=1}^{n}d_{ij_1j_2}(t+\omega)g_{j_1}(u_{j_1}(t+\omega-\tau_{ij_1}))g_{j_2}(u_{j_2}(t+\omega-\tau_{ij_2}))$$

$$+\sum_{j_1=1}^{n}\sum_{j_2=1}^{n}d_{ij_1j_2}(t)g_{j_1}(u_{j_1}(t-\tau_{ij_1}))g_{j_2}(u_{j_2}(t-\tau_{ij_2}))$$

$$\left.-I_i(t+\omega) + I_i(t)\right\}$$

$$\leq -\gamma_i|x_i(t)| + \sum_{j=1}^{n}|c_{ij}(t)|F_j|x_j(t)|$$

$$+\sum_{j_1=1}^{n}\sum_{j_2=1}^{n}|d_{ij_1j_2}(t)|\left(G_{j_1}M_{j_2}|x_{j_1}(t-\tau_{ij_1})| + G_{j_2}M_{j_1}|x_{j_2}(t-\tau_{ij_2})|\right).$$

Define

$$V(t) = \sum_{i=1}^{n} \zeta_i |y_i(t)| + \sum_{i=1}^{n} \sum_{j_1=1}^{n} \sum_{j_2=1}^{n} \zeta_i d^*_{ij_1j_2} \left( G_{j_1} M_{j_2} \int_{t-\tau_{ij_1}}^{t} |x_{j_1}(s)| ds + G_{j_2} M_{j_1} \int_{t-\tau_{ij_2}}^{t} |x_{j_2}(s)| ds \right).$$

(8)

Differentiating it along the trajectory $u(t)$, it gets

$$\frac{dV(t)}{dt} \leq \sum_{i=1}^{n} \zeta_i \left\{ -\gamma_i |x_i(t)| + \sum_{j=1}^{n} |c_{ij}(t)| F_j |x_j(t)| \right.$$

$$\left. + \sum_{j_1=1}^{n} \sum_{j_2=1}^{n} |d_{ij_1j_2}(t)| \left( G_{j_1} M_{j_2} |x_{j_1}(t-\tau_{ij_1})| + G_{j_2} M_{j_1} |x_{j_2}(t-\tau_{ij_2})| \right) \right\}$$

$$+ \sum_{i=1}^{n} \sum_{j_1=1}^{n} \sum_{j_2=1}^{n} \zeta_i d^*_{ij_1j_2} \left( G_{j_1} M_{j_2} |x_{j_1}(t)| + G_{j_2} M_{j_1} |x_{j_2}(t)| \right.$$

$$\left. -G_{j_1} M_{j_2} |x_{j_1}(t-\tau_{ij_1})| + G_{j_2} M_{j_1} |x_{j_2}(t-\tau_{ij_2})| \right)$$

$$\leq \sum_{i=1}^{n} (-\gamma_i \zeta_i) |x_i(t)| + \sum_{i=1}^{n} \sum_{j=1}^{n} |c_{ij}(t)| \zeta_i F_j |x_j(t)|$$

$$+ \sum_{i=1}^{n} \sum_{j_1=1}^{n} \sum_{j_2=1}^{n} \zeta_i d^*_{ij_1j_2} (G_{j_1} M_{j_2} |x_{j_1}(t)| + G_{j_2} M_{j_1} |x_{j_2}(t)|)$$

$$= \sum_{i=1}^{n} \left\{ -\gamma_i \zeta_i + \sum_{j=1}^{n} |c_{ji}(t)| \zeta_j F_i + \sum_{j_1=1}^{n} \sum_{j_2=1}^{n} (\zeta_{j_1} d^*_{j_1ij_2} M_{j_2} + \zeta_{j_2} d^*_{j_2j_1i} M_{j_1}) G_i \right\} |x_i(t)|$$

$$\leq -\lambda \sum_{i=1}^{n} |x_i(t)| = -\lambda \|x(t)\|_1,$$

(9)

where $\lambda$ is defined by

$$-\lambda = \max_i \sup_t \left\{ -\gamma_i \zeta_i + \sum_{j=1}^{n} |c_{ji}(t)| \zeta_j F_i + \sum_{j_1=1}^{n} \sum_{j_2=1}^{n} (\zeta_{j_1} d^*_{j_1ij_2} M_{j_2} + \zeta_{j_2} d^*_{j_2j_1i} M_{j_1}) G_i \right\} < 0.$$

Integrating both sides of (9), we have

$$\int_0^\infty \|x(t)\|_1 dt \leq \frac{1}{\lambda} V(0) < +\infty,$$

(10)

that is,

$$\sum_{n=1}^{\infty} \int_0^\omega \|u(t+n\omega) - u(t+(n-1)\omega)\|_1 dt \leq \frac{1}{\lambda} V(0) < +\infty.$$

(11)

By Cauchy convergence principle, we have that $u(t+n\omega)$ converges in $L^1[0,\omega]$ as $n \to +\infty$. On the other hand, from Lemma 2, we know that $u_i(t)$ is bounded, so that $a_i(u_i(t))$ is bounded and $u(t)$ is uniformly continuous correspondingly. Then the sequence $u(t+n\omega)$ is uniformly bounded and equicontinuous. By $Arzéla - Ascoli$ theorem, there exists a subsequence $u(t+n_k\omega)$ converging on any compact set of $R$. Denote the limit as $u^*(t)$.

It is easy to see that $u^*(t)$ is also the limit of $u(t + n\omega)$ in $L^1[0, \omega]$, i.e.,

$$\lim_{n \to +\infty} \int_0^\omega \|u(t + n\omega) - u^*(t)\|_1 dt = 0.$$

Then, we have that $u(t + n\omega)$ converges to $u^*(t)$ uniformly on $[0, \omega]$. Similarly, $u(t + n\omega)$ converges to $u^*(t)$ uniformly on any compact set of $R$.

Now we will prove that $u^*(t)$ is a nonnegative periodic solution of (1). Clearly, $u^*(t)$ is nonnegative. And, $u^*(t+\omega) = \lim_{n \to +\infty} u(t+(n+1)\omega) = \lim_{n \to +\infty} u(t+n\omega) = x^*(t)$, so that $u^*(t)$ is periodic with period $\omega$. Then, replace $u(t)$ with $u(t + n_k\omega)$ in system (1), let $k \to \infty$, and it gets

$$\frac{du_i^*(t)}{dt} = -a_i(u_i^*(t))\Big[b_i(u_i^*(t)) + \sum_{j=1}^n c_{ij}(t)f_j(u_j^*(t))$$

$$+ \sum_{j_1=1}^n \sum_{j_2=1}^n d_{ij_1 j_2}(t)g_{j_1}(u_{j_1}^*(t - \tau_{ij_1}))g_{j_2}(u_{j_2}^*(t - \tau_{ij_2})) + I_i(t)\Big]. \quad (12)$$

Hence $u^*(t)$ is a solution of system (1).

Let $t = t_1 + n\omega$, where $t_1 \in [0, \omega]$. Then, $\|u(t) - u^*(t)\| = \|u(t_1 + n\omega) - u^*(t_1)\|$. And the uniform convergence of $\{u(t + n\omega)\}$ on $[0, \omega]$ leads to

$$\lim_{t \to +\infty} \|u(t) - u^*(t)\| = 0. \quad (13)$$

Finally, we prove that any positive solution $v(t)$ of system (1) converges to $u^*(t)$. In fact, redefine $x_i(t) = v_i(t) - u_i(t)$, $y_i(t) = \int_{u_i(t)}^{v_i(t)} \frac{d\rho}{a_i(\rho)}$. Using the same method above, it is easy to get that $\lim_{t \to +\infty} \|v(t) - u(t)\| = 0$. Combining with (13), we get $\lim_{t \to +\infty} \|v(t) - u^*(t)\| = 0$. Theorem 1 is proved.

## 4   Conclusions

In this paper, the high-ordered Cohen-Grossberg neural networks are addressed. Under some mild conditions, the existence of a nonnegative periodic solution and its $R_+^n$-asymptotical stability are presented.

**Acknowledgments.** This work is jointly supported by the National Natural Sciences Foundation of China under Grant No. 61101203, Shanghai Young College Teachers Program scd11011, and Shanghai Municipal International Visiting Scholar Project for College Teachers.

## References

1. Cohen, M.A., Grossberg, S.: Absolute Stability and Global Pattern Formation and Parallel Memory Storage by Competitive Neural Networks. IEEE Trans. Syst. Man Cybern. B. 13, 815–821 (1983)

2. Lin, W., Chen, T.: Positive Periodic Solutions of Delayed Periodic Lotka-Volterra Systems. Physics Letters A 334(4), 273–287 (2005)
3. Chen, T., Wang, L., Ren, C.: Existence and Global Stability Analysis of Almost Periodic Solutions for Cohen- Grossberg Neural Networks. In: Wang, J., Yi, Z., Żurada, J.M., Lu, B.-L., Yin, H. (eds.) ISNN 2006. LNCS, vol. 3971, pp. 204–210. Springer, Heidelberg (2006)
4. Chen, T., Bai, Y.: Stability of Cohen-Grossberg Neural Networks with Nonnegative Periodic Solutions. In: Proceedings of International Joint Conference on Neural Networks, Orlando, Florida, USA, August 12–17 (2007)
5. Lu, W., Chen, T.: $R_+^n$-global Stability of A Cohen-Grossberg Neural Network System with Nonnegative Equilibria. Neural Networks 20(6), 714–722 (2007)
6. Fan, Q., Shao, J.: Positive Almost Periodic Solutions for Shunting Inhibitory Cellular Neural Networks with Time-Varying and Continuously Distributed Delays. Communications in Nonlinear Science And Numerical Simulation 15(6), 1655–1663 (2010)
7. Wang, L.: Existence and Global Attractivity of Almost Periodic Solutions for Delayed High-Ordered Neural Networks. Neurocomputing 73, 802–808 (2010)
8. Wang, Y., Lin, P., Wang, L.: Exponential Stability of Reaction-Diffusion High-Order Markovian Jump Hopfield Neural Networks with Time-Varying Delays. Nonlinear Analysis-Real World Applications 13(3), 1353–1361 (2012)
9. Li, D.: Dynamical Analysis for High-Order Delayed Hopfield Neural Networks with Impulses. Abstract And Applied Analysis (2012), doi:10.1155/2012/825643

# Existence of Periodic Solution for Competitive Neural Networks with Time-Varying and Distributed Delays on Time Scales

Yang Liu, Yongqing Yang, Tian Liang, and Xianyun Xu

Key Laboratory of Advanced Process Control for Light Industry
(Ministry of Education), School of Science,
Jiangnan University, Wuxi 214122, P.R. China
yyq640613@gmail.com,
634196363@qq.com

**Abstract.** In this paper, under the condition without assuming the boundedness of the activation functions, the competitive neural networks with time-varying and distributed delays are studied. By means of contraction mapping principle, the existence and uniqueness of periodic solution are investigated on time scales.

## 1 Introduction

Over past few decades, competitive neural networks have been extensively investigated as is apparent from a large number of publications [1–11]. Generally, a competitive neural network model contains two types of state variable, the short-term memory (STM), and the long-term memory (LTM). The STM describes rapid changes in neuronal dynamics, and the LTM describes the unsupervised neural cell synaptic slow behavior. For the detailed hardware implementation of competitive neural networks, please refer to the reference [2].

So far, a large number of papers involved in the existence of solutions of competitive neural networks had been published in various magazines [3–11]. Based on flow-invariance, the existence and uniqueness of the equilibrium of CNNs were discussed [3, 7]. In addition, the existence and uniqueness of the equilibrium point of CNNs were investigated by using the nonsmooth analysis techniques [4], Browers fixed-point theorem [5], the theory of uncertain singularly perturbed systems [6], the topological degree theory [8], the nonlinear Lipschitz measure (NLM) method [9] and the Leray-Schauder alternative theorem [10], respectively. But there is no papers on studying the existence of the solution of CNNs by adopting contraction mapping principle on time scales.

The theory of time scales, which unify the continuous-time and discrete-time demains, was initiated by Hilger in 1988 [16]. We recommend the interested readers to crack the reference [17] which summarized and organized the time scale calculous theory in detail.

Recently, many excellent results have been reported on existence of periodic solution of several types of neural networks on time scales [12–15]. However,

C. Guo, Z.-G. Hou, and Z. Zeng (Eds.): ISNN 2013, Part I, LNCS 7951, pp. 181–188, 2013.

there is still little work dedicated to studying the existence of periodic solutions for CNNs on time scales. Motivated by above mentioned, in this paper, we will discuss the existence of periodic solution for competitive neural networks with time-varying and distributed delays on time scales. Such a model is described by the following form:

$$
\begin{cases}
STM : x_i^{\Delta}(t) = & -\alpha_i(t)x_i(t) + \sum_{j=1}^{N} D_{ij}(t)f_j(x_j(t)) \\
& + \sum_{j=1}^{N} D_{ij}^{\tau}(t)f_j(x_j(t - \tau_{ij}(t))) \\
& + \sum_{j=1}^{N} \bar{D}_{ij}(t) \int_0^{+\infty} K_{ij}(u)f_j(x_j(t - u))\Delta u \qquad (1) \\
& + B_i(t)S_i(t) \\
LTM : S_i^{\Delta}(t) = & -c_i(t)S_i(t) + E_i(t)f_i(x_i(t))
\end{cases}
$$

with the initial values

$$
\begin{aligned}
x_i(s) &= \phi_i(s), & s \in (-\infty, 0]_T \\
S_i(s) &= \psi_i(s), & s \in (-\infty, 0]_T
\end{aligned}
$$

where $i, j = 1, ..., N$; $x_i(t)$ is the neuron current activity level, $\alpha_i(t), c_i(t)$ are the time variable of the neuron, $f_j(x_j(t))$ is the output of neurons, $D_{ij}(t)$ and $D_{ij}^{\tau}(t)$, $\bar{D}_{ij}(t)$ represent the connection weight and the synaptic weight of delayed feedback between the $i$th neuron and the $j$th neuron respectively, $B_i(t)$ is the strength of the external stimulus, $E_i(t)$ denotes disposable scale, transmission delays $\tau_{ij}(t)$ satisfies $0 < \tau_{ij}(t) \leqslant \tau$ $(\tau > 0)$.

$T$ is an $\omega$-periodic time scale, and $\phi_i(\cdot)$, $\psi_i(\cdot)$ are rd-continuous. Assume that $0 \in T$, $T$ is unbounded above, ie. $\sup T = \infty$ and $T^+ = \{t \in T, t \geq 0\}$. We denote

$$
\begin{aligned}
\bar{\mu} &= \max_{t \in [0,\omega]_T} |\mu(t)|, & D_{ij} &= \max_{t \in [0,\omega]_T} |D_{ij}(t)|, & D_{ij}^{\tau} &= \max_{t \in [0,\omega]_T} |D_{ij}^{\tau}(t)|, \\
\bar{D}_{ij} &= \max_{t \in [0,\omega]_T} |\bar{D}_{ij}(t)|, & B_i &= \max_{t \in [0,\omega]_T} |B_i(t)|, & E_i &= \max_{t \in [0,\omega]_T} |E_i(t)|.
\end{aligned}
$$

Through this paper, we make the following assumptions:

$(H_1)$: $\alpha_i(t), c_i(t)$, $D_{ij}(t)$, $D_{ij}^{\tau}(t)$, $\bar{D}_{ij}(t)$, $B_i(t)$, $E_i(t)$, $\tau_{ij}(t)$ are continuous $\omega$-periodic functions with $\omega > 0$, and there exist positive numbers $\underline{\alpha}_i$, $\bar{\alpha}_i$, $\underline{c}_i$, $\bar{c}_i$ such that $\underline{\alpha}_i \leq \alpha_i(\cdot) \leq \bar{\alpha}_i$, $\underline{c}_i \leq c_i(\cdot) \leq \bar{c}_i$, for $i, j = 1, ..., N$.

$(H_2)$: The delay kernels $K_{ij}(s) : [0, +\infty) \to [0, +\infty)$ are continuous integral functions, and satisfy

$$
\int_0^{+\infty} K_{ij}(s)\Delta s = 1, \qquad \int_0^{+\infty} K_{ij}(s)e_{\eta}(s, t)\Delta s < \infty, \qquad (2)
$$

for $i, j = 1, ..., N$.

$(H_3)$: The functions $f_i \in C(R, R)$ are Lipschitz functions, that is, there exists positive constants $k_i > 0$, such that for all $x, y \in R$

$$
|f_i(x) - f_i(y)| \leq k_i|x - y|. \qquad (3)
$$

## 2    Preliminary

**Definition 1.** *([17]) A time scale $T$ is an arbitrary nonempty closed subset of the real set $R$ with the topology and ordering inherited from $R$. The set of all right-dense continuous functions on $T$ is defined by $C_{rd} = C_{rd}(T) = C_{rd}(T, R)$. The graininess of the time scale $T$ and is determined by the formula $\mu(t) = \sigma(t) - t$, $\sigma(t) = \inf\{s \in T, s > t\}$.*

**Definition 2.** *([17]) For $f : T \to R$ and $t \in T^k$, we define the so-called $\Delta$-derivative of $f$, $f^{\Delta}(t)$, to be the number (if it exists) with the following property, for any $\varepsilon > 0$ there is a $N$-neighborhood of $t$ with*

$$|f(\sigma(t)) - f(s) - f^{\Delta}(t)(\sigma(t) - s)| \leq \varepsilon|\sigma(t) - s|, \ \ for \ all \ s \in N.$$

*We call $f^{\Delta}(t)$ the delta derivative of $f$ at $t$.*

**Definition 3.** *([17]) We say that a function $f : T \to R$ is called regressive if $1 + \mu(t)f(t) \neq 0$, for all $t \in T^k$. If $p, q \in R$, we define*
$(p \oplus q)(t) = p(t) + q(t) + \mu(t)p(t)q(t)$, $\quad p \ominus q = p \oplus (\ominus q)$, $\quad \ominus p(t) = -\frac{p(t)}{1+\mu(t)p(t)}$.

**Definition 4.** *([17]) For $s, t \in T$, if $p$ is a regressive function then we define the exponential function $e_p(t, s)$ by*

$$e_p(t, s) = exp\left( \int_s^t \xi_{\mu(\tau)}(p(\tau))\Delta\tau \right), \quad \xi_h(z) = \begin{cases} \frac{Log(1+zh)}{h}, \ h \neq 0 \\ z, \qquad\qquad h = 0 \end{cases}.$$

**Lemma 1.** *([17]) If $p, q \in R$, $t, r, s \in T$, then*
*(i) $e_0(t, s) \equiv 1$ and $e_p(t, t) \equiv 1$;*
*(ii) $e_p(\sigma(t), s) = (1 + \mu(t)p(t))e_p(t, s)$;*
*(iii) $e_p(t, s) = \frac{1}{e_p(s,t)} = e_{\ominus p}(s, t)$;*
*(iv) $e_p(t, s)e_p(s, r) = e_p(t, r)$;*
*(v) $e_p(t, s)e_q(t, s) = e_{p \oplus q}(t, s)$;*
*(vi) $e_p^{\Delta}(t, t_0) = p(t)e_p(t, t_0)$.*

**Lemma 2.** *Suppose $(H_1) - (H_3)$ hold, then $Z(t) \in X$ is an $\omega$-periodic solution of system (1), if and only if $Z(t)$ is an $\omega$-periodic solution of the following system*

$$
\begin{cases}
x_i(t) = \dfrac{1}{e_{\ominus(-\alpha_i)}(\omega, 0) - 1} \displaystyle\int_t^{t+\omega} \dfrac{e_{\ominus(-\alpha_i)}(s, t)}{1 - \alpha_i(s)\mu(s)} \times \left\{ \sum_{j=1}^N D_{ij}(s)f_j(x_j(s)) \right. \\
\qquad\qquad + \displaystyle\sum_{j=1}^N D_{ij}^{\tau}(s)f_j(x_j(s - \tau_{ij}(s))) + \sum_{j=1}^N \bar{D}_{ij}(s)\int_0^{+\infty} K_{ij}(u)f_j(x_j(s - u))\Delta u \\
\qquad\qquad + B_i(s)S_i(s) \Big\}\Delta s \\
S_i(t) = \dfrac{1}{e_{\ominus(-c_i)}(\omega, 0) - 1} \displaystyle\int_t^{t+\omega} \dfrac{e_{\ominus(-c_i)}(s, t)}{1 - c_i(s)\mu(s)} E_i(s)f_i(x_i(s))\Delta s
\end{cases}
$$

$$(4)$$

*for $t \in T^+$; $i = 1, 2, \cdots, N$.*

*Proof. The proof of the lemma is similar to that of the lemma 2.3 in [12].*

# 3    Existence and Uniqueness of Periodic Solution

**Theorem 1.** *Assume $(H_1) - (H_3)$ hold. Further, assume that $M = \max\{M_1, M_2\} < 1$ with $\bar{\mu} < \min\{\frac{1}{\bar{\alpha}_i}, \frac{1}{\bar{c}_i}\}$, for $i = 1, 2, \cdots, N$, where*

$$M_1 = \omega \max_{1 \leq i \leq N} \left( \frac{A(\alpha_i)}{(1 - \bar{\alpha}_i\bar{\mu})} W_i + \frac{A(c_i)}{(1 - \bar{c}_i\bar{\mu})} k_i E_i \right), \quad M_2 = \omega \max_{1 \leq i \leq N} \frac{A(\alpha_i)}{(1 - \bar{\alpha}_i\bar{\mu})} B_i \tag{5}$$

*and where*

$$A(\theta) = \frac{e_{\ominus(-\theta)}(\omega, 0)}{e_{\ominus(-\theta)}(\omega, 0) - 1}, \quad W_i = \sum_{j=1}^{N}(D_{ij} + D_{ij}^\tau + \bar{D}_{ij})k_j,$$

*Then system (1) has a unique $\omega$-periodic solution.*

*Proof.* For convenience, we consider system (1) as the following equations:

$$\begin{cases} x_i^\Delta(t) = -\alpha_i(t)x_i(t) + h_{ij}(t) \\ S_i^\Delta(t) = -c_i(t)S_i(t) + l_{ij}(t) \end{cases} \tag{6}$$

where

$$\begin{cases} h_{ij}(t) = \sum_{j=1}^{N} D_{ij}(t)f_j(x_j(t)) + \sum_{j=1}^{N} D_{ij}^\tau(t)f_j(x_j(t - \tau_{ij}(t))) \\ \qquad\quad + \sum_{j=1}^{N} \bar{D}_{ij}(t) \int_0^{+\infty} K_{ij}(u)f_j(x_j(t - u))\Delta u + B_i(t)S_i(t) \\ l_{ij}(t) = \quad E_i(t)f_i(x_i(t)) \end{cases} \tag{7}$$

Let $X = \{Z(t)|Z \in C_{rd}(T, R^{2N}), Z(t + \omega) = Z(t)\}$ with the norm $\|Z\| = \sum_{k=1}^{N}(\sup_{t \in T^+}|x_k(t)| + \sup_{t \in T^+}|S_k(t)|)$, where $Z(t) = (x_1(t), \cdots, x_N(t), S_1(t), \cdots, S_N(t))^T$, then $X$ is a Banach space. Let

$$\Omega = \left\{ Z(t)|Z \in X, \|Z\| \leq k, k > \frac{M_3}{1 - M} \right\}$$

where

$$M_3 = \omega \sum_{i=1}^{N} \frac{A(\alpha_i)}{(1 - \bar{\alpha}_i\bar{\mu})} \sum_{j=1}^{N}(D_{ij} + D_{ij}^\tau + \bar{D}_{ij})|f_j(0)| + \omega \sum_{i=1}^{N} \frac{A(c_i)}{(1 - \bar{c}_i\bar{\mu})} E_i|f_i(0)|.$$

Obviously, $\Omega$ is a closed nonempty subset of $X$.

Now, we define an operator $F$ on the Banach $X$.

$F : X \to X, (FZ)(t) = ((Fx)_1(t), \cdots, (Fx)_N(t), (FS)_1(t), \cdots, (FS)_N(t))$, $t \in T^+$ for any $Z(t) = (x_1(t), \cdots, x_N(t), S_1(t), \cdots, S_N(t)) \in X$, where

$$\begin{cases} (Fx)_i(t) = \frac{1}{e_{\ominus(-\alpha_i)}(\omega, 0) - 1} \int_t^{t+\omega} \frac{e_{\ominus(-\alpha_i)}(s, t)}{1 - \alpha_i(s)\mu(s)} h_{ij}(s)\Delta s \\ (FS)_i(t) = \frac{1}{e_{\ominus(-c_i)}(\omega, 0) - 1} \int_t^{t+\omega} \frac{e_{\ominus(-c_i)}(s, t)}{1 - c_i(s)\mu(s)} l_{ij}(s)\Delta s \end{cases}$$

for $t \in T^+, i = 1, \cdots, N$.

In order to obtain the result of theorem 1, the following three steps are made:

**Step 1.** Proving the following inequalities

$$e_{\ominus(-\alpha_i)}(s,t) \le e_{\ominus(-\alpha_i)}(\omega, 0), \quad e_{\ominus(-c_i)}(s,t) \le e_{\ominus(-c_i)}(\omega, 0)$$

Firstly, we prove the first inequality.

Considering two possible cases: (i) $\mu(t) \ne 0$; (ii) $\mu(t) = 0$;

Noting $\alpha_i(t) > 0, \mu(t) > 0, \alpha_i(t)\mu(t) < 1$, that is, $Log(1 - \alpha_i(t)\mu(t))^{-1} > 0$.

(i) When $\mu(t) \ne 0$, we have

$$e_{\ominus(-\alpha_i)}(s,t) \le \exp\left\{ \int_t^{t+\omega} \frac{Log(1 - \mu(\tau)\alpha_i(\tau))^{-1}}{\mu(\tau)} \Delta\tau \right\}$$

$$= \exp\left\{ \int_0^{\omega} \frac{Log(1 - \mu(\tau)\alpha_i(\tau))^{-1}}{\mu(\tau)} \Delta\tau \right\}$$

$$= e_{\ominus(-\alpha_i)}(\omega, 0).$$

(ii) When $\mu(t) = 0$, we have

$$e_{\ominus(-\alpha_i)}(s,t) = \exp\left\{ \int_t^s \alpha_i(\tau)\Delta\tau \right\}$$

$$\le \exp\left\{ \int_0^{\omega} \alpha_i(\tau)\Delta\tau \right\}$$

$$= e_{\ominus(-\alpha_i)}(\omega, 0).$$

So, $e_{\ominus(-\alpha_i)}(s,t) \le e_{\ominus(-\alpha_i)}(\omega, 0)$. Similarly, $e_{\ominus(-c_i)}(s,t) \le e_{\ominus(-c_i)}(\omega, 0)$.

**Step 2.** Proving $F$ maps $\Omega$ into itself.

Firstly, from inequality (3), we know

$$|f_i(x)| \le |f_i(x) - f_i(0)| + |f_i(0)| \le k_i|x| + |f_i(0)|.$$

Noting that,

$$|h_{ij}(s)| \le \sum_{j=1}^{N} |D_{ij}(s)||f_j(x_j(s))| + \sum_{j=1}^{N} |D_{ij}^\tau(s)||f_j(x_j(s - \tau_{ij}(s)))|$$

$$+ \sum_{j=1}^{N} |\bar{D}_{ij}(s)| \int_0^{+\infty} K_{ij}(u)|f_j(x_j(s - u))|\Delta u + |B_i(s)||S_i(s)|$$

$$\le \sum_{j=1}^{N} D_{ij}(k_j|(x_j(s))| + |f_j(0)|) + \sum_{j=1}^{N} D_{ij}^\tau(k_j|x_j(s - \tau_{ij}(s))| + |f_j(0)|)$$

$$+ \sum_{j=1}^{N} \bar{D}_{ij} \int_0^{+\infty} K_{ij}(u)(k_j|x_j(s - u)| + |f_j(0)|)\Delta u + B_i|S_i(s)|$$

$$\le W_i \sup_{s \in T^+} |x_i(s)| + B_i \sup_{s \in T^+} |S_i(t)| + \sum_{j=1}^{N} (D_{ij} + D_{ij}^\tau + \bar{D}_{ij})|f_j(0)|$$

$$(8)$$

$$|l_{ij}(s)| \leq |E_i(s)||f_i(x_i(s))| \leq E_i(k_i|x_i(s)| + |f_i(0)|)$$
$$\leq E_i k_i \sup_{s \in T^+} |x_i(s)| + E_i|f_i(0)|.$$

For any $Z(t) \in \Omega$, we have

$$\|(FZ)(t)\| = \sup_{t \in T^+} \left[ \sum_{i=1}^{N} |(Fx)_i(t)| + \sum_{i=1}^{N} |(FS)_i(t)| \right]$$

$$\leq \sup_{t \in T^+} \sum_{i=1}^{N} \frac{e_{\ominus(-\alpha_i)}(\omega, 0)}{(1 - \bar{\alpha}_i \bar{\mu}) \cdot (e_{\ominus(-\alpha_i)}(\omega, 0) - 1)} \int_t^{t+\omega} |h_{ij}(s)| \Delta s$$

$$+ \sup_{t \in T^+} \sum_{i=1}^{N} \frac{e_{\ominus(-c_i)}(\omega, 0)}{(1 - \bar{c}_i \bar{\mu}) \cdot (e_{\ominus(-c_i)}(\omega, 0) - 1)} \int_t^{t+\omega} |l_{ij}(s)| \Delta s$$

$$\leq \sum_{i=1}^{N} \omega \left[ \frac{A(\alpha_i)W_i}{(1 - \bar{\alpha}_i \bar{\mu})} + \frac{A(c_i)k_i E_i}{(1 - \bar{c}_i \bar{\mu})} \right] \sup_{t \in T^+} |x_i(t)|$$

$$+ \sum_{i=1}^{N} \omega \frac{A(\alpha_i)B_i}{(1 - \bar{\alpha}_i \bar{\mu})} \sup_{t \in T^+} |S_i(t)|$$

$$+ \sum_{i=1}^{N} \omega \frac{A(\alpha_i)}{(1 - \bar{\alpha}_i \bar{\mu})} \sum_{j=1}^{N} (D_{ij} + D_{ij}^\tau + \bar{D}_{ij})|f_j(0)| + \sum_{i=1}^{N} \omega \frac{A(c_i)E_i}{(1 - \bar{c}_i \bar{\mu})}|f_i(0)|$$

$$\leq M_1 \sum_{i=1}^{N} \sup_{t \in T^+} |x_i(t)| + M_2 \sum_{i=1}^{N} \sup_{t \in T^+} |S_i(t)| + M_3$$

$$\leq M\|Z\| + M_3$$

$$\leq k$$

So, $FZ \in \Omega$.

**Step 3.** Proving that $F$ is a contraction mapping. For any
$Z(t) = (x_1(t), \cdots, x_N(t), S_1(t), \cdots, S_N(t))^T \in \Omega$, $Z^*(t) = (x_1^*(t), \cdots, x_N^*(t), S_1^*(t), \cdots, S_N^*(t))^T \in \Omega$, we have

$$\|(FZ)(t) - (FZ^*)(t)\| = \sup_{t \in T^+} \sum_{i=1}^{N} \left[ |(Fx)_i(t) - (Fx^*)_i(t)| + |(FS)_i(t) - (FS^*)_i(t)| \right]$$

$$\leq \sum_{i=1}^{N} \omega \left[ \frac{A(\alpha_i)}{(1 - \bar{\alpha}_i \bar{\mu})}W_i + \frac{A(c_i)}{(1 - \bar{c}_i \bar{\mu})}k_i E_i \right] \sup_{t \in T^+} |x_i(t) - x_i^*(t)|$$

$$+ \sum_{i=1}^{N} \omega \frac{A(\alpha_i)B_i}{(1 - \bar{\alpha}_i \bar{\mu})} \sup_{t \in T^+} |S_i(t) - S_i^*(t)|$$

$$\leq M_1 \sum_{i=1}^{N} \sup_{t \in T^+} |x_i(t) - x_i^*(t)| + M_2 \sum_{i=1}^{N} \sup_{t \in T^+} |S_i(t) - S_i^*(t)|$$

$$\leq M\|Z - Z^*\|$$

$$(9)$$

Note that $M < 1$, thus $F$ is a contraction mapping.

According to the fixed point theorem, $F$ has and only one unique fixed point. Therefore, system (1) has a unique $\omega$-periodic solution.

**Corollary 1.** When $\mu(t) = 0$, suppose that $(H_1) - (H_3)$ hold. Further, suppose the following.

For $i = 1, 2, \cdots, N$

$$
\begin{aligned}
M_1 = \omega \max_{1 \leq i \leq N} &\left( \frac{\exp\left( \int_0^\omega \alpha_i(t)dt \right)}{\exp\left( \int_0^\omega \alpha_i(t)dt \right) - 1} \sum_{j=1}^N (D_{ij} + D_{ij}^\tau + \bar{D}_{ij}) k_j \right. \\
&\left. + \frac{\exp\left( \int_0^\omega c_i(t)dt \right) k_i E_i}{\exp\left( \int_0^\omega c_i(t)dt \right) - 1} \right) < 1,
\end{aligned}
\tag{10}
$$

$$
M_2 = \omega \max_{1 \leq i \leq N} \frac{\exp\left( \int_0^\omega \alpha_i(t)dt \right) B_i}{\exp\left( \int_0^\omega \alpha_i(t)dt \right) - 1} < 1.
$$

*Then the periodic solution of the system (1) is existent and unique.*

**Corollary 2.** When $\mu(t) = 1$ and $\bar{\alpha}_i, \bar{c}_i < 1$, suppose $(H_1) - (H_3)$ hold. Further, suppose the following. For $i = 1, 2, \cdots, N$

$$
\begin{aligned}
M_1 = \omega \max_{1 \leq i \leq N} &\left( \frac{\prod_{k=0}^{\omega-1} [1 - \alpha_i(k)]^{-1}}{\left( \prod_{k=0}^{\omega-1} [1 - \alpha_i(k)]^{-1} - 1 \right) \cdot \left( 1 - \bar{\alpha}_i \right)} \sum_{j=1}^N (D_{ij} + D_{ij}^\tau + \bar{D}_{ij}) k_j \right. \\
&\left. + \frac{\prod_{k=0}^{\omega-1} [1 - c_i(k)]^{-1} k_i E_i}{\left( \prod_{k=0}^{\omega-1} [1 - c_i(k)]^{-1} - 1 \right) \cdot \left( 1 - \bar{c}_i \right)} \right) < 1,
\end{aligned}
$$

$$
M_2 = \omega \max_{1 \leq i \leq N} \frac{\prod_{k=0}^{\omega-1} [1 - \alpha_i(k)]^{-1} B_i}{\left( \prod_{k=0}^{\omega-1} [1 - \alpha_i(k)]^{-1} - 1 \right) \cdot \left( 1 - \bar{\alpha}_i \right)} < 1.
\tag{11}
$$

*Then the periodic solution of the system (1) is existent and unique.*

## 4   Conclusion

Under the condition without assuming that the activation functions are zero at the zero [12] and the activation functions are bounded [13], some conditions are obtained to ensure the existence of periodic solution for competitive neural networks with time-varying and distributed delays on time scales by using contraction mapping principle. This is the first time applying the time scale calculus theory to discuss the existence of periodic solution of competitive neural networks, therefore, the results derived in this paper extend some previously existing results[1].

---

[1] This work was jointly supported by the Fundamental Research Funds for the Central Universities ( JUSRP51317B), the Foundation of Key Laboratory of Advanced Process Control for Light Industry (Jiangnan University), Ministry of Education of China.

# References

1. Sowmya, B., Sheela Rani, B.: Colour image segmentation using fuzzy clustering techniques and competitive neural network. Appl. Soft Comput. 11, 3170–3178 (2011)
2. Engel, P., Molz, R.: A new proposal for implementation of competitive neural networks in analog hardware. In: Proceedings of the 5th Brazilian Symposium on Neural Networks, pp. 186–191. IEEE Press, New York (1998)
3. Meyer-Baese, A., Pilyugin, S., Wismüller, A., Foo, S.: Local exponential stability of competitive neural networks with different time scales. Eng. Appl. Artif. Intell. 17, 227–232 (2004)
4. Lu, H., He, Z.: Global exponential stability of delayed competitive neural networks with different time scales. Neural Netw. 18, 243–250 (2005)
5. Meyer-Baese, A., Thümmler, V.: Local and global stability of an unsupervised competitive neural network. IEEE Trans. Neural Netw. 19, 346–351 (2008)
6. Meyer-Baese, A., Roberts, R., Yu, H.: Robust stability analysis of competitive neural networks with different time-scales under perturbations. Neurocomputing 71, 417–420 (2007)
7. Meyer-Baese, A., Pilyugin, S., Chen, Y.: Global exponential stability of competitive neural networks with different time scale. IEEE Trans. Neural Netw. 14, 716–719 (2003)
8. Kao, Y., Ming, Q.: Global robust stability of competitive neural networks with continuously distributed delays and different time scales. In: de Sá, J.M., Alexandre, L.A., Duch, W., Mandic, D.P. (eds.) ICANN 2007. LNCS, vol. 4668, pp. 569–578. Springer, Heidelberg (2007)
9. Gu, H., Jiang, H., Teng, Z.: Existence and global exponential stability of equilibrium of competitive neural networks with different time-scales and multiple delays. J. Franklin Inst. 347, 719–731 (2010)
10. Nie, X., Cao, J.: Existence and global stability of equilibrium point for delayed competitive neural networks with discontinuous activation functions. Int. J. Syst. Sci. 43, 459–474 (2012)
11. Liao, W., Wang, L.: Existence and global attractability of almost periodic solution for competitive neural networks with time-varying delays and different time scales. In: Wang, J., Yi, Z., Żurada, J.M., Lu, B.-L., Yin, H. (eds.) ISNN 2006. LNCS, vol. 3971, pp. 297–302. Springer, Heidelberg (2006)
12. Chen, A., Chen, F.: Periodic solution to BAM neural network with delays on time scales. Neurocomputing 73, 274–282 (2009)
13. Zhang, Z., Peng, G., Zhou, D.: Periodic solution to Cohen-Grossberg BAM neural networks with delays on time scales. J. Franklin Inst. 348, 2759–2781 (2011)
14. Kaufmann, E., Raffoul, Y.: Periodic solutions for a neutral nonlinear dynamical equation on a time scale. J. Math. Anal. Appl. 319, 315–325 (2006)
15. Li, Y., Zhao, L., Zhang, T.: Global exponential stability and existence of periodic solution of impulsive Cohen-Grossberg neural networks with distributed delays on time scales. Neural Proc. Lett. 33, 61–81 (2011)
16. Hilger, S.: Analysis on measure chains-a unified approach to continuous and discrete calculus. Results Math. 18, 18–56 (1990)
17. Bohner, M., Peterson, A.: Dynamic Equations on Time Scales: an Introduction with Applications. Birkhäuser, Boston (2001)

# Global Exponential Stability in the Mean Square of Stochastic Cohen-Grossberg Neural Networks with Time-Varying and Continuous Distributed Delays

Tian Liang, Yongqing Yang*, Manfeng Hu, Yang Liu, and Li Li

Key Laboratory of Advanced Process Control for Light Industry
(Ministry of Education), School of Science,
Jiangnan University, Wuxi 214122, P.R. China
liangtian2007@126.com, yyq640613@gmail.com

**Abstract.** In this paper, the global exponential stability in the mean square of stochastic Cohen-Grossberg neural networks (SCGNNS) with mixed delays is studied. By applying the Lyapunov function, stochastic analysis technique and inequality techniques, some sufficient conditions are obtained to ensure the exponential stability in the mean square of the SCGNNS. An example is given to illustrate the theoretical results.

## 1 Introduction

In 1983, Cohen and Grossberg constructed an important kind of simplified neural networks model which is now called Cohen-Grossberg neural networks (CGNNS) [1]. This kind of neural networks is very general and includes Hopfield neural networks, cellular neural networks and BAM neural networks as its special cases.

In general, the model of Cohen-Grossberg neural networks is described by the set of ordinary differential equations:

$$x_i^{'}(t) = -a_i(x_i(t))[b_i(x_i(t)) - \sum_{j-1}^{n} c_{ij}g_j(x_j(t))], \quad i = 1, 2, \cdots, n,$$

where $n$ is the number of neurons and $x_i(\cdot)$ is the state variable, $a_i(\cdot)$ is an amplification function and $b_i(\cdot)$ represents a behaved function, $C = (c_{ij})_{n \times n}$ is a connection matrix, which shows how the neurons are connected in the network, and the activation function $g_j(\cdot)$ tells how the $j_{th}$ neuron reacts to the input.

In recent years, the stability of the Cohen-Grossberg neural networks with or without delays has been widely studied. Many useful and interesting results have been obtained in [2]-[13]. Distributed delays were introduced into the neural networks in [3], and by using the theory of dissipative systems, several conditions were obtained to ensure the stability of the system. Without considering the global Lipschitz activation functions, Zhou [4] studied the stability of the

---

* Corresponding author.

C. Guo, Z.-G. Hou, and Z. Zeng (Eds.): ISNN 2013, Part I, LNCS 7951, pp. 189–196, 2013.
© Springer-Verlag Berlin Heidelberg 2013

almost periodic solutions for delayed neural networks. Recently, Balasubramaniam [7] investigated the Takagi-Sugeno fuzzy Cohen-Grossberg BAM neural
networks. In [10], by utilizing the Lyapunov-Krasovkii function and combining
with the linear matrix inequality (LMI) approach, some well conclusions were
obtained about the global exponential stability of neutral type neural networks
with distributed delays. The problem of robust global exponential stability for a
class of neutral-type neural networks was investigated in [9], where the interval
time-varying delays allowed for both fast and slow time-varying delays. However,
to the best of our knowledge, there have been few authors to study the global
exponential stability in the mean square of stochastic Cohen-Grossberg neural
networks with time-varying and continuous distributed delays. This motivates
our present research.

The rest organization of this paper is as follows: the model and some preliminaries are introduced in section 2. In section 3, some sufficient conditions are
given to guarantee the global exponential stability in the mean square of the
model. One example is given in section 4. And in the last section: we give some
conclusions.

## 2    Model Description and Preliminaries

In this paper, we consider a system of SCGNNS as follows:

$$
dx_i(t) = - a_i(x_i(t)) \Big[ b_i(x_i(t)) - \sum_{j=1}^{n} c_{ij}(t) f_j(x_j(t)) - \sum_{j=1}^{n} d_{ij}(t) f_j(x_j(t - \tau))
$$
$$
- \sum_{j=1}^{n} e_{ij}(t) \int_0^{+\infty} K_{ij}(s) f_j(x_j(t - s)) ds \Big] dt + \sum_{j=1}^{n} \sigma_{ij}(x_j(t)) d\omega_j(t),
\tag{1}
$$

$t \geq 0$, where $n$ is the number of neurons and $x_i(t)$ is the state variable, $a_i$ is
an amplification function and $b_i$ represents a behaved function, $c_{ij}(t)$, $d_{ij}(t)$,
$e_{ij}(t)$ present the strengths of connectivity between cells $i$ and $j$ at time $t$,
the activation function $f_j(\cdot)$ tells how the $j_{th}$ neuron reacts to the input, $\tau$
corresponds to the time delay. $K_{ij}(\cdot)$ is the delay kernel function and satis-
fies that $\int_0^{+\infty} |K_{ij}(s)| ds \leq \overline{K}_{ij}$, $\int_0^{+\infty} K_{ij}(s) e^{\lambda s} ds \leq k_{ij}$, $\int_0^{+\infty} K_{ij}(s) e^{\lambda s} s ds \leq$
$\overline{k}_{ij}$. The noise pertuibation $\sigma_{ij}$ is a Borel measurable function, and $\omega(t) =$
$(\omega_1(t), \cdots, \omega_n(t))^T$ is an n-dimensional Brownian motion defined on a complete
probability space $(\Omega, \mathcal{F}, \mathbf{P})$ with a natural filtration $\{\mathcal{F}_t\}_{t \geq 0}$.

Let $C((-\infty, 0]; R^n)$ be the Banach space of continuous functions, which from
$(-\infty, 0]$ to $R^n$. The initial condition associated with (1) is: $x_i(s) = \varphi_i(s)$, $s \in$
$(-\infty, 0]$, $\varphi \in \mathrm{L}^1_{\mathcal{F}_0}((-\infty, 0]; R^n)$, where $\mathrm{L}^1_{\mathcal{F}_0}((-\infty, 0]; R^n)$ is the family of all $\mathcal{F}_0$-
measurable, $C((-\infty, 0]; R^n)$-valued stochastic process. Moreover, $\int_{-\tau}^0 \mathrm{E}|\varphi(t)| dt$
$< \infty$, where $\mathrm{E}[\cdot]$ is the correspondent expectation operator with respect to the
given probability measure P.

Following are some assumptions and one definition which can be used later.

$A_1$: The weight functions $c_{ij}(t)$, $d_{ij}(t)$, and $e_{ij}(t)$ are continuous, and $\bar{c}_{ij} = \sup_{t\geq 0}\{|c_{ij}(t)|\}$, $\bar{d}_{ij} = \sup_{t\geq 0}\{|d_{ij}(t)|\}$, $\bar{e}_{ij} = \sup_{t\geq 0}\{|e_{ij}(t)|\}$.

$A_2$ : $0 < \underline{a}_i \leq a_i(u) \leq \bar{a}_i, \forall u \in R$, and $\forall x, y \in R$, there exists a positive constant $\gamma_i$, such that $(x - y)[b_i(x) - b_i(y)] \geq \gamma_i(x - y)^2$, $b_i(0) \equiv 0$.

$A_3$ : $\forall x, y \in R$, there exists a constant $M_j > 0$, such that $|f_j(x) - f_j(y)| \leq M_j|x - y|$, Moreover, $f_j(0) \equiv 0$.

$A_4$ : $\sigma_{ij}(0) \equiv 0$, and there exists a constant $\Gamma_{ij} > 0$, such that $|\sigma_{ij}(x) - \sigma_{ij}(y)| \leq \Gamma_{ij}|x - y|$, $\forall x, y \in R$.

**Definition 1.** *The solution of model (1) is said to be mean square exponential stability if* $\forall \varphi \in L^1_{\mathcal{F}_0}((-\infty, 0]; R^n)$, *there exist constants* $\mu, \lambda > 0$, *such that*

$$Ex^2(t) \leq \mu e^{-\lambda t} \sup_{s\leq 0} E\varphi^2(s), \quad t > 0.$$

## 3   Globally Exponentially Stable in the Mean Square

In this section, we give the global mean square exponential stability of SCGNNS.

**Theorem 1.** *Assume that $A_1 - A_4$ hold, if there exists a positive constant $p_i$, such that the following inequality holds:*

$$2p_i\underline{a}_i\gamma_i > \sum_{j=1}^{n} p_i\bar{a}_i M_j[\bar{c}_{ij} + \bar{d}_{ij} + \bar{e}_{ij}\overline{K}_{ij}]$$

$$+ \sum_{j=1}^{n} p_j\bar{a}_j M_i[\bar{c}_{ji} + \bar{d}_{ji} + \bar{e}_{ji}k_{ji}] + \sum_{j=1}^{n} p_j\Gamma_{ji}^2 \qquad (2)$$

*then the solution of system (1) is global exponential stability in the mean square.*

*Proof.* Define the following Lyapunov-Krasovskii functions:

$$V_1(t) = e^{\lambda t}\sum_{i=1}^{n} p_i x_i^2(t), \quad V_2(t) = \int_{t-\tau}^{t} e^{\lambda(s+\tau)}\sum_{i=1}^{n}\sum_{j=1}^{n} p_j\bar{a}_j\bar{d}_{ji}M_i x_i^2(s)ds,$$

$$V_3 = \sum_{i=1}^{n}\sum_{j=1}^{n} p_i\bar{a}_i\bar{e}_{ij}M_j\int_{0}^{+\infty} K_{ij}(s)e^{\lambda s}\Big[\int_{t-s}^{t} e^{\lambda u}x_j^2(u)du\Big]ds.$$

Then for $t > 0$, consider the Itô formula, we have:

$$\mathcal{L}V_1(t) = \lambda e^{\lambda t}\sum_{i=1}^{n} p_i x_i^2(t) + 2e^{\lambda t}\sum_{i=1}^{n} p_i x_i(t)\Big\{-a_i(x_i(t))[b_i(x_i(t))$$

$$- \sum_{j=1}^{n} c_{ij}(t)f_j(x_j(t)) - \sum_{j=1}^{n} d_{ij}(t)f_j(x_j(t-\tau))$$

$$- \sum_{j=1}^{n} e_{ij}(t)\int_{0}^{+\infty} K_{ij}(s)f_j(x_j(t-s))ds]\Big\}$$

$$+ e^{\lambda t}\sum_{i=1}^{n} p_i \sum_{j=1}^{n} \sigma_{ij}^2(x_j(t))$$

$$= \lambda e^{\lambda t} \sum_{i=1}^{n} p_i x_i^2(t) + 2e^{\lambda t} \sum_{i=1}^{n} p_i x_i(t) \big[ -a_i(x_i(t)) b_i(x_i(t)) \big]$$

$$+ 2e^{\lambda t} \sum_{i=1}^{n} p_i x_i(t) a_i(x_i(t)) \sum_{j=1}^{n} c_{ij}(t) f_j(x_j(t))$$

$$+ 2e^{\lambda t} \sum_{i=1}^{n} p_i x_i(t) a_i(x_i(t)) \sum_{j=1}^{n} d_{ij}(t) f_j(x_j(t - \tau))$$

$$+ 2e^{\lambda t} \sum_{i=1}^{n} p_i x_i(t) a_i(x_i(t)) \sum_{j=1}^{n} e_{ij}(t) \int_{0}^{+\infty} K_{ij}(s) f_j(x_j(t - s)) ds$$

$$+ e^{\lambda t} \sum_{i=1}^{n} p_i \sum_{j=1}^{n} \sigma_{ij}^2(x_j(t))$$

$$\leq \lambda e^{\lambda t} \sum_{i=1}^{n} p_i x_i^2(t) - 2e^{\lambda t} \sum_{i=1}^{n} p_i \underline{a}_i \gamma_i x_i^2(t)$$

$$+ 2e^{\lambda t} \sum_{i=1}^{n} p_i \overline{a}_i \sum_{j=1}^{n} \overline{c}_{ij} M_j |x_i(t) x_j(t)|$$

$$+ 2e^{\lambda t} \sum_{i=1}^{n} p_i \overline{a}_i \sum_{j=1}^{n} \overline{d}_{ij} M_j |x_i(t) x_j(t - \tau)|$$

$$+ 2e^{\lambda t} \sum_{i=1}^{n} p_i \overline{a}_i \sum_{j=1}^{n} \overline{e}_{ij} M_j \int_{0}^{+\infty} K_{ij}(s) |x_i(t) x_j(t - s)| ds$$  $$(3)$$

$$+ e^{\lambda t} \sum_{i=1}^{n} p_i \sum_{j=1}^{n} \Gamma_{ij}^2 x_j^2(t)$$

$$\leq \lambda e^{\lambda t} \sum_{i=1}^{n} p_i x_i^2(t) - 2e^{\lambda t} \sum_{i=1}^{n} p_i \underline{a}_i \gamma_i x_i^2(t) + e^{\lambda t} \sum_{i=1}^{n} p_i \overline{a}_i \sum_{j=1}^{n} \overline{c}_{ij} M_j [x_i^2(t)$$

$$+ x_j^2(t)] + e^{\lambda t} \sum_{i=1}^{n} p_i \overline{a}_i \sum_{j=1}^{n} \overline{d}_{ij} M_j [x_i^2(t) + x_j^2(t - \tau)]$$

$$+ e^{\lambda t} \sum_{i=1}^{n} p_i \overline{a}_i \sum_{j=1}^{n} \overline{e}_{ij} M_j \int_{0}^{+\infty} K_{ij}(s) [x_i^2(t) + x_j^2(t - s)] ds$$

$$+ e^{\lambda t} \sum_{i=1}^{n} p_i \sum_{j=1}^{n} \Gamma_{ij}^2 x_j^2(t).$$

$$\mathcal{L}V_2(t) = e^{\lambda(t+\tau)} \sum_{i=1}^{n} \sum_{j=1}^{n} p_j \overline{a}_j \overline{d}_{ji} M_i x_i^2(t) - e^{\lambda t} \sum_{i=1}^{n} \sum_{j=1}^{n} p_j \overline{a}_j \overline{d}_{ji} M_i x_i^2(t - \tau). \quad (4)$$

$$\mathcal{L}V_3(t) = \sum_{i=1}^{n}\sum_{j=1}^{n} p_i \bar{a}_i \bar{e}_{ij} M_j \int_0^{+\infty} K_{ij}(s) e^{\lambda s}\left[e^{\lambda t}x_j^2(t) - e^{\lambda(t-s)}x_j^2(t-s)\right]ds. \quad (5)$$

Then from (3) to (5), we have

$$\mathcal{L}V(t) \leq \lambda e^{\lambda t}\sum_{i=1}^{n}p_i x_i^2(t) - 2e^{\lambda t}\sum_{i=1}^{n}p_i \underline{a}_i \gamma_i x_i^2(t) + e^{\lambda t}\sum_{i=1}^{n}p_i \bar{a}_i \sum_{j=1}^{n}\bar{c}_{ij}M_j x_i^2(t)$$

$$+ e^{\lambda t}\sum_{i=1}^{n}p_i \bar{a}_i \sum_{j=1}^{n}\bar{c}_{ij}M_j x_j^2(t) + e^{\lambda t}\sum_{i=1}^{n}p_i \bar{a}_i \sum_{j=1}^{n}\bar{d}_{ij}M_j x_i^2(t)$$

$$+ e^{\lambda t}\sum_{i=1}^{n}p_i \bar{a}_i \sum_{j=1}^{n}\bar{e}_{ij}M_j \int_0^{+\infty} K_{ij}(s)x_i^2(t)ds + e^{\lambda t}\sum_{i=1}^{n}p_i \sum_{j=1}^{n}\Gamma_{ij}^2 x_j^2(t)$$

$$+ e^{\lambda(t+\tau)}\sum_{i=1}^{n}\sum_{j=1}^{n}p_j \bar{a}_j \bar{d}_{ji}M_i x_i^2(t)$$

$$+ \sum_{i=1}^{n}\sum_{j=1}^{n}p_i \bar{a}_i \bar{e}_{ij}M_j \int_0^{+\infty} K_{ij}(s)e^{\lambda s}e^{\lambda t}x_j^2(t)ds$$

$$\leq e^{\lambda t}\sum_{i=1}^{n}\left[\lambda p_i - 2p_i\underline{a}_i\gamma_i + \sum_{j=1}^{n}p_i\bar{a}_i\bar{c}_{ij}M_j + \sum_{j=1}^{n}p_j\bar{a}_j\bar{c}_{ji}M_i\right.$$

$$+ \sum_{j=1}^{n}p_i\bar{a}_i\bar{d}_{ij}M_j + \sum_{j=1}^{n}p_i\bar{a}_i\bar{e}_{ij}M_j\overline{K}_{ij} + \sum_{j=1}^{n}p_j\Gamma_{ji}^2 + e^{\lambda\tau}\sum_{j=1}^{n}p_j\bar{a}_j\bar{d}_{ji}M_i$$

$$\left.+ \sum_{j=1}^{n}p_j\bar{a}_j\bar{e}_{ji}M_ik_{ji}\right]x_i^2(t)$$

$$(6)$$

From (2), we have

$$-2p_i\underline{a}_i\gamma_i + \sum_{j=1}^{n}p_i\bar{a}_iM_j[\bar{c}_{ij} + \bar{d}_{ij} + \bar{e}_{ij}\overline{K}_{ij}]$$

$$+ \sum_{j=1}^{n}p_j\bar{a}_jM_i[\bar{c}_{ji} + \bar{d}_{ji} + \bar{e}_{ji}k_{ji}] + \sum_{j=1}^{n}p_j\Gamma_{ji}^2 < 0.$$

Then we can choose an appropriate constant $\lambda$, $0 < \lambda \ll 1$, such that

$$\lambda p_i - 2p_i\underline{a}_i\gamma_i + \sum_{j=1}^{n}p_i\bar{a}_i\bar{c}_{ij}M_j + \sum_{j=1}^{n}p_j\bar{a}_j\bar{c}_{ji}M_i + \sum_{j=1}^{n}p_i\bar{a}_i\bar{d}_{ij}M_j + \sum_{j=1}^{n}p\Gamma_{ji}^2$$

$$+ \sum_{j=1}^{n}p_i\bar{a}_i\bar{e}_{ij}M_j\overline{K}_{ij} + e^{\lambda\tau}\sum_{j=1}^{n}p_j\bar{a}_j\bar{d}_{ji}M_i + \sum_{j=1}^{n}p_j\bar{a}_j\bar{e}_{ji}M_ik_{ji} < 0. \quad (7)$$

Combining (6) and (7), we can obtain that $\mathcal{L}V(t) < 0$. Then by applying the Dynkin formula, we have the following inequality:

$$EV(t) - EV(0) = E \int_0^t \mathcal{L}V(u)du < 0.$$

It implies that

$$
\begin{aligned}
EV(t) < EV(0) &\le \sum_{i=1}^n p_i Ex_i^2(0) + e^{\lambda\tau} \sum_{i=1}^n \sum_{j=1}^n p_j \bar{a}_j \bar{d}_{ji} M_i \int_{-\tau}^0 Ex_i^2(s)ds \\
&+ \sum_{i=1}^n \sum_{j=1}^n p_i \bar{a}_i \bar{e}_{ij} M_j \int_0^{+\infty} K_{ij}(s)e^{\lambda s} \int_{-s}^0 Ex_j^2(u)duds \\
&\le \Big( \max_{1\le i\le n} p_i + \max_{1\le i,j\le n} \tau e^{\lambda\tau} p_j \bar{a}_j \bar{d}_{ji} M_i \\
&+ \max_{1\le i,j\le n} p_j \bar{a}_j \bar{e}_{ji} M_i \bar{k}_{ji} \Big) \sup_{s\le 0} E\varphi^2(s)
\end{aligned}
\tag{8}
$$

On the other hand, by the definition of the Lyapunov function $V(t)$, it is easy to get the following inequality:

$$EV(t) \ge Ee^{\lambda t} \sum_{i=1}^n p_i x_i^2(t) \ge e^{\lambda t} \min_{1\le i\le n} p_i Ex^2(t). \tag{9}$$

Combining (8) and (9), we can obtain: $Ex^2(t) \le \alpha e^{-\lambda t} E\varphi^2(s)$, where

$$\alpha = \frac{\max_{1\le i\le n} p_i + \max_{1\le i,j\le n} \tau e^{\lambda\tau} p_j \bar{a}_j \bar{d}_{ji} M_i + \max_{1\le i,j\le n} p_j \bar{a}_j \bar{e}_{ji} M_i \bar{k}_{ji}}{\min_{1\le i\le n} p_i} > 0.$$

Then by the Definition 1, the trivial solution of the model (1) is globally exponentially stable in the mean square. Thus completes the proof.

## 4   An Example

In this section, we will give one example to illustrate the results obtained.

*Example 1.* Consider the following two dimensional Cohen-Grossberg neural networks with the following parameters:

$$a(x(t)) = \begin{bmatrix} 3 - \cos(t) & 0 \\ 0 & 3 + \sin(t) \end{bmatrix}, \quad b(x(t)) = \begin{bmatrix} 6 & 0 \\ 0 & 5 \end{bmatrix} \begin{bmatrix} x_1(t) \\ x_2(t) \end{bmatrix},$$

$$(c_{ij})(t)_{2\times 2} = \sin(t) \begin{bmatrix} \frac{3}{5} & \frac{1}{5} \\ \frac{2}{5} & \frac{2}{5} \end{bmatrix}, \quad (d_{ij})(t)_{2\times 2} = \sin(t) \begin{bmatrix} \frac{1}{5} & \frac{1}{10} \\ \frac{1}{5} & \frac{1}{10} \end{bmatrix},$$

$$(e_{ij})(t)_{2\times 2} = \sin(t) \begin{bmatrix} \frac{1}{3} & \frac{1}{3} \\ \frac{1}{3} & \frac{1}{3} \end{bmatrix}, \quad f(x(t)) = \begin{bmatrix} \sin(x_1(t)) \\ \sin(x_2(t)) \end{bmatrix},$$

$$(\sigma_{ij}(x_j(t)))_{2\times 2} = \begin{bmatrix} 2 & 0 \\ 0 & 1 \end{bmatrix} \begin{bmatrix} x_1(t) & 0 \\ 0 & x_2(t) \end{bmatrix},$$

$$(K_{ij}(s))_{2\times2} = \begin{bmatrix} \frac{1}{2}e^{-s} & \frac{1}{2}e^{-s} \\ \frac{1}{2}e^{-s} & \frac{1}{2}e^{-s} \end{bmatrix}, 0 \le s \le 40, K_{ij}(s) = 0, \forall s \in (40, +\infty).$$

So $\underline{a}_1 = \underline{a}_2 = 2,\ \bar{a}_1 = \bar{a}_2 = 4,\ \gamma_1 = 6,\ \gamma_2 = 5, M_1 = M_2 = 1,\ \Gamma_{11} = 2,\ \Gamma_{22} = 1,\ \Gamma_{12} = \Gamma_{21} = 0$, select $K_{ij} = k_{ij} = 1, p_1 = p_2 = 1$, then

$$
\begin{aligned}
2p_1\underline{a}_1\gamma_1 =& 2 \times 1 \times 2 \times 6 = 24 \\
>& p_1\bar{a}_1 M_1(\bar{c}_{11} + \bar{d}_{11} + \bar{e}_{11}\overline{K}_{11}) + p_1\bar{a}_1 M_2(\bar{c}_{12} + \bar{d}_{12} + \bar{e}_{12}\overline{K}_{12}) \\
& + p_1\bar{a}_1 M_1(\bar{c}_{11} + \bar{d}_{11} + \bar{e}_{11}\overline{K}_{11}) + p_2\bar{a}_2 M_1(\bar{c}_{21} + \bar{d}_{21} + \bar{e}_{21}\overline{K}_{21}) \\
& + p_1\Gamma_{11}^2 + p_2\Gamma_{21}^2 = \frac{328}{15}, \\
2p_2\underline{a}_2\gamma_2 =& 2 \times 1 \times 2 \times 5 = 20 \\
>& p_2\bar{a}_2 M_1(\bar{c}_{21} + \bar{d}_{21} + \bar{e}_{21}\overline{K}_{21}) + p_2\bar{a}_2 M_2(\bar{c}_{22} + \bar{d}_{22} + \bar{e}_{22}\overline{K}_{22}) \\
& + p_1\bar{a}_1 M_2(\bar{c}_{12} + \bar{d}_{12} + \bar{e}_{12}\overline{K}_{12}) + p_2\bar{a}_2 M_2(\bar{c}_{22} + \bar{d}_{22} + \bar{e}_{22}\overline{K}_{22}) \\
& + p_1\Gamma_{12}^2 + p_2\Gamma_{22}^2 = \frac{47}{3}.
\end{aligned}
$$

From Theorem 1, we can check that the solution of the model is global mean square exponential stability. The results are shown in Fig.1 and Fig.2.

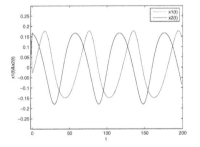
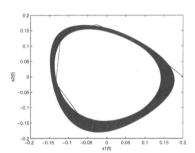

**Fig. 1.** The trajectory of $x_1(t)$ and $x_2(t)$     **Fig. 2.** The stability of the solutions

## 5   Conclusions

In this paper, we investigate the stochastic Cohen-Grossberg neural networks with mixed delays. Some sufficient conditions are obtained to ensure the global mean square exponential stability of SCGNNS. From Theorem 1, we know that if system (1) satisfies the conditions in the theorem, then the stability is independent of the delays and the noise. In [11] and [12], the authors also investigated the mean square stability of the SCGNNS. However, they did not consider the distributed delays, which is very important in a neural reaction process in the real world. Compared with them, the model in our paper is more general [1].

---

[1] This work was jointly supported by the National Natural Science Foundation of China under Grant 60875036, the Key Research Foundation of Science and Technology of the Ministry of Education of China under Grant 108067, and the Fundamental Research Funds for the Central Universities JUSRP51317B.

# References

1. Cohen, M., Grossberg, S.: Absolute stability of global pattern formation and parallel memory storage by competitive neural networks. IEEE Transactions on Systems Man and Cybernetics 3, 815–826 (1983)
2. Oliveira, J.: Global stability of a Cohen-Grossberg neural network with both time-varying and continuous distributed delays. Nonlinear Analysis: Real World Applications 12, 2861–2870 (2011)
3. Wang, L.: Stability of Cohen-Grossberg neural networks with distributed delays. Applied Mathematics and Computation 160, 93–110 (2005)
4. Zhou, J., Zhao, W., Lv, X., Zhu, H.: Stability analysis of almost periodic solutions for delayed neural networks without global Lipschitz activation functions. Mathematics and Computers in Simulation 81, 2440–2445 (2011)
5. Zhang, Z., Liu, W., Zhou, D.: Global asymptotic stability to a generalized Cohen-Grossberg neural networks of neutral type delays. Neural Networks 25, 94–105 (2012)
6. Tojtovska, B., Janković, S.: On a general decay stability of stochastic Cohen-Grossberg neural networks with time-varying delays. Applied Mathematics and Computation 219, 2289–2302 (2012)
7. Balasubramaniam, P., Ali, M.: Stability analysis of Takagi-Sugeno fuzzy Cohen-Grossberg BAM neural networks with discrete and distributed time-varying delays. Mathematical and Computer Modelling 53, 151–160 (2011)
8. Li, L., Fang, Z., Yang, Y.: A shunting inhibitory cellular neural network with continuously distributed delays of neutral type. Nonlinear Analysis: Real World Applications 13, 1186–1196 (2012)
9. Mahmoud, M., Ismail, A.: Improved results on robust exponential stability criteria for neutral-type delayed neural networks. Applied Mathematics and Computation 217, 3011–3019 (2010)
10. Rakkiyappan, R., Balasubramaniam, P.: New global exponential stability results for neutral type neural networks with distributed time delays. Neurocomputing 71, 1039–1045 (2008)
11. Zhu, Q., Li, X.: Exponential and almost sure exponential stability of stochastic fuzzy delayed Cohen-Grossberg neural networks. Fuzzy Sets and Systems 203, 74–79 (2012)
12. Li, D., He, D., Xu, D.: Mean square exponential stability of impulsive stochastic reaction-diffusion Cohen-Grossberg neural networks with delays. Mathematics and Computers in Simulation 82, 1531–1543 (2012)
13. Zhou, W., Wang, T., Mou, J., Fang, J.: Mean square exponential synchronization in Lagrange sense for uncertain complex dynamical networks. Journal of the Franklin Institute 349, 1267–1282 (2012)

# A Delay-Partitioning Approach to Stability Analysis of Discrete-Time Recurrent Neural Networks with Randomly Occurred Nonlinearities

Jianmin Duan[1], Manfeng Hu[1,2], and Yongqing Yang[1,2]

[1] School of Science, Jiangnan University, Wuxi 214122, China
`duanjianmin1988@yahoo.cn, humanfeng@jiangnan.edu.cn`
[2] Key Laboratory of Advanced Process Control for Light Industry
(Jiangnan University), Ministry of Education, Wuxi 214122, China
`yongqingyang@163.com`

**Abstract.** This paper considers the problem of stability analysis for discrete-time recurrent neural networks with randomly occurred non-linearities (RONs) and time-varying delay. By utilizing new Lyapunov-Krasovskii functions and delay-partitioning technique, the stability criteria are proposed in terms of linear matrix inequality (LMI). We have also shown that the conservatism of the conditions is a non-increasing function of the number of delay partitions. A numerical example is provided to demonstrate the effectiveness of the proposed approach.

## 1 Introduction

Due to the successful applications in pattern recognition, associative memories, signal processing and the other fields [1], the study of neural networks has received a great deal of attention during the past decades [1]-[8]. It is well known that the stability of neural networks is a prerequisite in modern control theories for these applications. However, time delays are often attributed as the major sources of instability. Therefore, how to find sufficient conditions to guarantee the stability of neural networks with time delays is an important research topic. A vast amount of effort has been devoted to this topic [2]-[8].

Recently, by introducing free-weighting matrices, LMI approach and adopting the concept of delay partitioning, criteria have been established [2,7,9]. Stability results for neural networks with time-varying delay have been proposed by employing complete delay-decomposing approach and LMI in [2]. Compared with continuous-time systems with time delay, discrete-time systems with state delay have a strong background in engineering applications. It has been well recognized that networked-based control is a typical example. However, little effort has been made for studying the problem of stability of discrete time-delay systems. On the other hand, it is worth mentioning that, a number of practical systems are influenced by additive randomly occurred nonlinear disturbances which are caused by environmental circumstances. The nonlinear disturbances

C. Guo, Z.-G. Hou, and Z. Zeng (Eds.): ISNN 2013, Part I, LNCS 7951, pp. 197–204, 2013.

may occur in a probabilistic way, what's more, they are randomly changeable in terms of their types and/or intensity. For example, in a networked environment, nonlinear disturbances may be subject to random abrupt changes, which may result from abrupt phenomena such as random failures and repairs of the components, environmental disturbance and so on. The stochastic nonlinearities, which are then named as randomly occurred nonlinearities (RONs), have recently attracted much attention [10,11,12]. However, to the best of the authors' knowledge, the stability analysis of discrete-time recurrent neural networks with RONs via delay-partitioning method has not been tackled in the previous literatures. This motivates our research.

## 2  Problem Formulation

Consider the following discrete recurrent neural networks with time-varying delay and RONs:

$$
\begin{aligned}
y(k+1) =& (A + \Delta A)y(k) + (B + \Delta B)g(y(k)) + (C + \Delta C)g(y(k - d(k))) \\
& + \xi(k)\tilde{E}f(y(k)), \\
y(k) =& \phi(k), \quad k = -d_2, -d_2 + 1, \cdots, 0,
\end{aligned}
\tag{1}
$$

where $y(k) \in R^n$ is the state vector. $d(k)$ is a positive integer representing time-varying delay with lower and upper bounds $1 \leq d_1 \leq d(k) \leq d_2$, $d_1$ and $d_2$ are known positive integers. Note that the lower bound of the delay can be always described by $d_1 = \tau m$, where $\tau$ and $m$ are integers, and $\tau$ presents the partition size. $\phi(k)$ is an initial value at time $k$. $A$, $B$, $C$ and $\tilde{E}$ are known real matrices with appropriate dimensions. $\Delta A$, $\Delta B$ and $\Delta C$ are unknown matrices representing parameter uncertainties, which are assumed to satisfy the following admissible condition:

$$
[\Delta A, \Delta B, \Delta C] = MF[N_1, N_2, N_3], \qquad FF^T \leq I,
\tag{2}
$$

where $M$, $N_1$, $N_2$ and $N_3$ are known constant matrices.
$g(y(\cdot)) = \left(g_1(y_1(\cdot))\ g_2(y_2(\cdot))\ \cdots\ g_n(y_n(\cdot))\right)^T$ denotes the neuron activation function. It satisfies the following assumptions:
$A_1$: For $i \in \{1, 2, \ldots, n\}$ , the function $g(y(\cdot))$ is continuous and bounded.
$A_2$: For any $t_1$, $t_2 \in R$, $t_1 \neq t_2$:

$$
0 \leq \frac{g_i(y_i)}{y_i} \leq s_i, \quad g_i(0) = 0, \quad \forall y_i \neq 0, \quad i = 1, 2, \ldots, n,
\tag{3}
$$

where $s_i$ are known constant scalars.
The nonlinear function $f(y)$ satisfies the following sector-bounded condition:

$$
[f(y) - K_1 y]^T[f(y) - K_2 y] \leq 0, \quad \forall y \in R^n,
\tag{4}
$$

where $K_1$ and $K_2$ are known real matrices and $K_1 - K_2 \geq 0$.

**Remark 1.** The nonlinear function $f(y)$ satisfying (4) is customarily said to belong to the sector $[K_1, K_2]$. Due to such a nonlinear condition is quite general that includes the usual Lipschitz condition as a special case, the systems with sector-bounded nonlinearities have been intensively studied in [10].

The stochastic variable $\xi(k) \in R$, which accounts for the phenomena of RONs, is a Bernoulli distributed white sequence taking values of 1 and 0 with

$$Prob\{\xi(k) = 1\} = \bar{\xi}, \quad Prob\{\xi(k) = 0\} = 1 - \bar{\xi}, \quad \bar{\xi} \in [0, 1].$$

**Remark 2.** According to the given hypothesis, we have $E\{\xi(k) - \bar{\xi}\} = 0, E\{(\xi(k) - \bar{\xi})^2\} = \bar{\xi}(1 - \bar{\xi})$. Besides, as emphasised in [13], $\xi(k)$ is a Markovian process and follows an unknown but exponential distribution of switchings.

Under $A_1$ and $A_2$, it is not difficult to ensure the existence of equilibrium point for (1) by employing the well-known Brouwer's fixed point theorem. To end this section, the lemmas are introduced as follows.

**Lemma 1.** [3] *For any symmetric positive-definite matrix $M \in R^{n \times n}$, two integers $r_1$ and $r_2$ satisfying $r_2 \geq r_1$, and vector function $\omega(i) \in R^n$, such that the sums in the following are well defined, then*

$$-(r_2 - r_1 + 1) \sum_{i=r_1}^{r_2} \omega^T(i) M \omega(i) \leq -(\sum_{i=r_1}^{r_2} \omega^T(i)) M (\sum_{i=r_1}^{r_2} \omega(i)).$$

**Lemma 2.** [14] *There exists a matrix $X$ such that*

$$\begin{pmatrix} P & Q + XE & X \\ (Q + XE)^T & R & V \\ X^T & V^T & S \end{pmatrix} > 0,$$

*if and only if*

$$\begin{pmatrix} P & Q \\ Q^T & R - VE - E^T V^T + E^T SE \end{pmatrix} > 0,$$

$$\begin{pmatrix} R - VE - E^T V^T + E^T SE & V - E^T S \\ (V - E^T S)^T & S \end{pmatrix} > 0.$$

## 3    Main Results

In this section, we give the LMI-based asymptotic stability conditions for the system (1). The main results are stated as follows.

**Theorem 1.** *For given positive integers $\tau$, $m$, and $d_2$, the system (1) is asymptotically stable if there exist real matrices $P > 0$, $Q_i > 0$ $(i = 1, 2, 3)$, $Z_i > 0$ $(i = 1, 2)$, $X$, $Y$, $W$, $Y - W > 0$, and a scalar $\mu > 0$, such that the following LMI holds:*

$$\begin{pmatrix} -Z_2 & Y^T & X \\ * & \Psi & W \\ * & * & -Z_2 \end{pmatrix} < 0, \tag{5}$$

*where*

$$\Psi = W_1^T P W_1 + W_2^T \Phi_1 W_2 + W_3^T \Phi_2 W_3 + W_4^T \Phi_3 W_4 + W_5^T \Phi_4 W_5 + W_6^T \Phi_5 W_6$$
$$+ W_7^T Q_2 W_7 - W_8^T Z_1 W_8 + W_3^T Z_1 W_8 + W_8^T Z_1 W_3 - \mu \tilde{K} + (\lambda_2 + 1)\tilde{Q}$$
$$+ \left(O\ Y\ -(Y-W)S^{-1}\ -W\ O\right) + \left(O\ Y\ -(Y-W)S^{-1}\ -W\ O\right)^T,$$

$$\lambda_1 = \sqrt{\bar{\xi}(1-\bar{\xi})}, \qquad \lambda_2 = d_2 - \tau m, \qquad \Phi_1 = P + \tau^2 Z_1 + \lambda_2^2 Z_2,$$

$$\Phi_2 = -P + Q_2 + (\lambda_2 + 1)Q_3 - Z_1, \qquad \Phi_3 = \tau^2 Z_1 + \lambda_2^2 Z_2,$$

$$\Phi_4 = \begin{pmatrix} Q_1 & O \\ O & -Q_1 \end{pmatrix}, \qquad \Phi_5 = -(S^{-1})^T Q_3 S^{-1},$$

$$W_1 = \left(A + \Delta A\ O_{n,mn}\ C + \Delta C\ O_n\ B + \Delta B\ \bar{\xi}\tilde{E}\right),$$

$$W_2 = \left(O_{n,(m+4)n}\ \lambda_1 \tilde{E}\right), \qquad W_3 = \left(I_n\ I_{n,(m+4)n}\right), \qquad W_4 = W_1 - W_3,$$

$$W_5 = \begin{pmatrix} I_{mn} & O_{mn,5n} \\ O_{mn,n} & I_{mn}\ O_{mn,4n} \end{pmatrix}, \qquad W_6 = \left(O_{n,(m+1)n}\ I_n\ O_{n,3n}\right),$$

$$W_7 = \left(O_{n,(m+2)n}\ I_n\ O_{n,2n}\right), \qquad W_8 = \left(O_n\ I_n\ O_{n,(m+3)n}\right),$$

$$\tilde{Q} = \{Q_1, O_{(n,3n)}, Q_2, O_n\},$$

$$\tilde{K} = \begin{pmatrix} \tilde{K}_1 & * & * & * \\ O_{(m+2)n,n} & O_{(m+2)n} & * & * \\ O_n & O_{n,(m+2)n} & O_n & * \\ \tilde{K}_2 & O_{n,(m+2)n} & O_n & I_n \end{pmatrix},$$

$$\tilde{K}_1 = (K_1^T K_2 + K_2^T K_1)/2, \qquad \tilde{K}_2 = -(K_1^T + K_2^T)/2.$$

*Proof.* Construct the following Lyapunov-Krasovskii function:

$$V(k) = V_1(k) + V_2(k) + V_3(k) + V_4(k), \tag{6}$$

where

$$V_1(k) = y^T(k) P y(k) + \sum_{j=-d_2+1}^{-d_1+1} \sum_{i=k-1+j}^{k-1} \zeta^T(i) \tilde{Q} \zeta(i),$$

$$V_2(k) = \sum_{i=k-\tau}^{k-1} \gamma^T(i) Q_1 \gamma(i) + \sum_{i=k-d_2}^{k-1} y^T(i) Q_2 y(i),$$

$$V_3(k) = \sum_{j=-d_2}^{-\tau m} \sum_{i=k+j}^{k-1} y^T(i) Q_3 y(i),$$

$$V_4(k) = \sum_{j=-\tau}^{-1} \sum_{i=k+j}^{k-1} \tau \Delta y^T(i) Z_1 \Delta y(i) + \sum_{j=-d_2}^{-\tau m-1} \sum_{i=k+j}^{k-1} \lambda_2 \Delta y^T(i) Z_2 \Delta y(i),$$

with

$$\gamma(i) = \left(y^T(i)\ y^T(i-\tau) \dots y^T(i-(m-1)\tau)\right)^T, \Delta y(i) = y(i+1) - y(i),$$

$$\zeta(k) = \left(\gamma^T(k)\ y^T(k-\tau m)\ g^T(y(k-d(k)))\ y^T(k-d_2)\ g^T(y(k))\ f^T(y(k))\right)^T.$$

Calculating the difference of $V(k)$ along the system (1):

$$E\{\Delta V_1(k)\} \leq E\{\zeta^T(k)W_1^T PW_1\zeta(k) + \zeta^T(k)W_2^T PW_2\zeta(k) + (\lambda_2 + 1)\zeta^T(k)\tilde{Q}\zeta(k)$$
$$- y^T(k)Py(k)\},$$
$$E\{\Delta V_2(k)\} = E\{y^T(k)Q_2y(k) - y^T(k - d_2)Q_2y(k - d_2) + \gamma^T(k)Q_1\gamma(k)$$
$$- \gamma^T(k - \tau)Q_1\gamma(k - \tau)\},$$
$$E\{\Delta V_3(k)\} = E\{(\lambda_2 + 1)y^T(k)Q_3y(k) - \sum_{i=k-d_2}^{k-\tau m} y^T(i)Q_3y(i)\},$$
$$E\{\Delta V_4(k)\} = E\{\Delta y^T(k)(\tau^2 Z_1 + \lambda_2^2 Z_2)\Delta y(k) - \sum_{i=k-\tau}^{k-1} \tau \Delta y^T(i)Z_1\Delta y(i)$$
$$- \sum_{i=k-d_2}^{k-\tau m-1} \lambda_2\Delta y^T(i)Z_2\Delta y(i)\}.$$

$$(7)$$

Using Lemma 1, we have

$$- \sum_{i=k-\tau}^{k-1} \tau \Delta y^T(i)Z_1\Delta y(i) \leq -(y(k) - y(k - \tau))^T Z_1(y(k) - y(k - \tau)). \quad (8)$$

Note that

$$- \sum_{i=k-h_2}^{k-\tau m} y^T(i)Q_3y(i) \leq -y^T(k - d(k))Q_3y(k - d(k)). \quad (9)$$

For any matrices $Y$ and $W$ ($Y - W > 0$), the following equations always hold:

$$2\zeta^T(k)Y[y(k - \tau m) - y(k - d(k)) - \sum_{j=k-d(k)}^{k-\tau m-1} \Delta y(j)] = 0,$$

$$2\zeta^T(k)W[y(k - d(k)) - y(k - d_2) - \sum_{j=k-d_2}^{k-d(k)-1} \Delta y(j)] = 0. \quad (10)$$

On the other hand, for $\mu > 0$, (4) is equivalent to

$$-\mu\zeta^T(k)\tilde{K}\zeta(k) \geq 0. \quad (11)$$

From (6) to (11), we obtain

$$E\{\Delta V(k)\} \leq E\left\{\frac{1}{\lambda_2} \sum_{i=k-d(k)}^{k-\tau m-1} \begin{pmatrix} \zeta(k) \\ -\lambda_2\Delta y(i) \end{pmatrix}^T \begin{pmatrix} \Psi & Y \\ * & -Z_2 \end{pmatrix} \begin{pmatrix} \zeta(k) \\ -\lambda_2\Delta y(i) \end{pmatrix}$$

$$+ \frac{1}{\lambda_2} \sum_{i=k-d_2}^{k-d(k)-1} \begin{pmatrix} \zeta(k) \\ -\lambda_2\Delta y(i) \end{pmatrix}^T \begin{pmatrix} \Psi & W \\ * & -Z_2 \end{pmatrix} \begin{pmatrix} \zeta(k) \\ -\lambda_2\Delta y(i) \end{pmatrix}\right\}. \quad (12)$$

According to Lemma 2, it can be shown that there exists a matrix $X$ of appropriate dimensions such that (5) holds if and only if

$$\begin{pmatrix} \Psi & Y \\ * & -Z_2 \end{pmatrix} < 0, \quad \begin{pmatrix} \Psi & W \\ * & -Z_2 \end{pmatrix} < 0. \tag{13}$$

Therefore, if the condition (5) is satisfied, the condition (13) is satisfied. By (12), there exists a scalar $\lambda > 0$ such that $E\{\Delta V(k)\} \leq -\lambda \|E\{y(k)\}\|^2 < 0$ for $y(k) \neq 0$. Then, from the Lyapunov stability theory, we can conclude that the system (1) is asymptotically stable.

**Remark 3.** The Lyapunov-Krasovskii functional candidate (6) introduces the term $\sum_{j=-d_2+1}^{-d_1+1} \sum_{i=k-1+j}^{k-1} \zeta^T(i)\tilde{Q}\zeta(i)$, which makes use of the information of $g_i(y_i(k))$ and the involved delay $d(k)$. Additionally, (11) is introduced, which makes our result take the full advantage of the information of $f(y)$. Thus, Theorem 1 in this investigation is expected to be less conservative.

**Remark 4.** Using similar steps as in the proof of [7], it is easy to establish that the conservatism is reduced as delay partitions grow, which largely benefits from the fact that the delay-partitioning approach is employed. Thus our results provide the flexibility that allows us to trade off between complexity and performance of the stability analysis.

In view of Theorem 1, we consider the case $d_1 \geq 1$, however, the $d_1$ could be zero in applications, thus we introduce the following criterion.

**Theorem 2.** *For given positive integer $d_2$, the system (1) is asymptotically stable if there exist real matrices $P > 0$, $\hat{Q}_i > 0$ ($i = 1, 2,$), $Z > 0$, $X$, $\hat{Y}$, $\hat{W}$, $\hat{Y} - \hat{W} > 0$, and a scalar $\mu > 0$, such that the following LMI holds:*

$$\begin{pmatrix} -Z & \hat{Y}^T & X \\ * & \hat{\psi} & \hat{W} \\ * & * & -Z \end{pmatrix} < 0, \tag{14}$$

*where*

$$\hat{\psi} = \hat{W}_1^T P \hat{W}_1 + \hat{W}_2^T \hat{\Phi}_1 \hat{W}_2 + \hat{W}_3^T \hat{\Phi}_2 \hat{W}_3 + d_2^2 \hat{W}_4^T Z \hat{W}_4 + \hat{W}_5^T \hat{\Phi}_3 \hat{W}_5 + (d_2 + 1)\hat{Q}$$
$$\quad - \mu\hat{T} + \left(\hat{Y} - (\hat{Y} - \hat{W})S^{-1} - \hat{W} \ O\right) + \left(\hat{Y} - (\hat{Y} - \hat{W})S^{-1} - \hat{W} \ O\right)^T,$$
$$\hat{\Phi}_1 = P + d_2^2 Z, \quad \hat{\Phi}_2 = -P + \hat{Q}_1 + (d_2 + 1)\hat{Q}_2, \quad \hat{\Phi}_3 = -(S^{-1})^T \hat{Q}_2 S^{-1},$$
$$\hat{W}_1 = \left(A + \Delta A \ C + \Delta C \ O_n \ B + \Delta B \ \bar{\xi}\tilde{E}\right), \quad \hat{W}_2 = \left(O_{n,4n} \ \lambda_1\tilde{E}\right),$$
$$\hat{W}_3 = \left(I_n \ I_{n,4n}\right), \quad \hat{W}_4 = \hat{W}_1 - \hat{W}_3, \quad \hat{W}_5 = \left(O_n \ I_n \ O_{n,3n}\right),$$
$$\hat{W}_6 = \left(O_{n,2n} \ I_n \ O_{n,2n}\right), \quad \hat{Q} = \{\hat{Q}_1, O_{(2n)}, \hat{Q}_2, O_n\},$$
$$\hat{K} = \begin{pmatrix} \tilde{K}_1 & * & * & * \\ O_{2n,n} & O_{2n} & * & * \\ O_n & O_{n,2n} & O_n & * \\ \tilde{K}_2 & O_{n,2n} & O_n & I_n \end{pmatrix}.$$

*Proof.* The proof can then be derived by following a similar line of arguments as that in Theorem 1.

## 4  An Simulation Example

To demonstrate the effectiveness of the proposed method, we now consider the system (1) with parameters as follows:

$$A = \begin{pmatrix} 0.3 & 0 \\ 0 & 0.6 \end{pmatrix}, \quad B = \begin{pmatrix} 0.2 & 0 \\ 0 & 0.4 \end{pmatrix}, \quad C = \begin{pmatrix} -0.1 & 0 \\ 0 & -0.2 \end{pmatrix}, \quad \tilde{E} = \begin{pmatrix} 0.1 & 0 \\ 0 & 0.1 \end{pmatrix},$$

$$M = \begin{pmatrix} 1 & 1 \end{pmatrix}^T, \quad N_1 = \begin{pmatrix} 0.001 & 0 \end{pmatrix}, \quad N_2 = \begin{pmatrix} 0.001 & 0 \end{pmatrix}, \quad N_3 = \begin{pmatrix} 0.001 & 0 \end{pmatrix},$$

$$\bar{\xi} = 0.8, \quad m = 2, \quad s_i = 1 \; (i = 1, 2), \quad d(k) = 10 + 6\sin\frac{k\pi}{2} \; (k \in Z),$$

$$\hat{g}_i(s) = \tanh(s) \; (i = 1, 2), \quad F = \sin(k) \; (k \in Z),$$

$$\hat{f}(x(k)) = \frac{1}{2} \left( \frac{0.3(x_1(k) + x_2(k))}{1 + x_1^2(k) + x_2^2(k)} + 0.1x_1(k) + 0.1x_2(k) \; 0.3x_1(k) + 0.3x_2(k) \right)^T,$$

$$K_1 = \begin{pmatrix} 0.2 & 0.1 \\ 0 & 0.2 \end{pmatrix}, \quad K_2 = \begin{pmatrix} -0.1 & 0 \\ -0.1 & 0.1 \end{pmatrix}.$$

$$(15)$$

By simple computation, we have $d_1 = 4$, $d_2 = 16$ and $\tau = 2$. The LMI (5) can be verified by using the Matlab LMI Toolbox. According to Theorem 1, the system (1) with parameters in (15) is asymptotically stable. The simulation result is shown in Fig. 1, which also confirms our method.

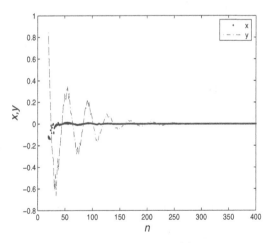

**Fig. 1.** State development of system (1) with parameters (15)

## 5  Conclusions

In this paper, we discuss the problem of asymptotic stability for discrete-time recurrent neural networks with time-varying delay and RONs. By using new

Lyapunov-Krasovskii functions and delay-partitioning technique, the LMI-based criteria for the asymptotic stability of such systems are established. The results reported in this paper are not only dependent on the delay but also dependent on the partitioning, which aims at reducing the conservatism. An illustrative example is exploited to show the usefulness of the results obtained[1].

# References

1. Arik, S.: Global asymptotic stability of a class of dynamical neural networks. IEEE Trans. Circuits Syst. 47, 568–571 (2000)
2. Zeng, H.B., He, Y.: Complete delay-decomposing approach to asymptotic stability for neural networks with time-varying delays. IEEE Trans. Neural Networks 22, 806–812 (2011)
3. Wu, Z.G., Su, H.Y., Chu, J., Zhou, W.N.: Improved delay-dependent stability condition of discrete recurrent neural networks with time-varying delays. IEEE Trans. Neural Networks 21, 692–697 (2010)
4. He, Y., Liu, G., Rees, D., Wu, M.: Stability analysis for neural networks with time-varying interval delay. IEEE Trans. Neural Networks 18, 1850–1854 (2007)
5. Arik, S.: Stability analysis of delayed neural networks. IEEE Trans. Circuits Syst.I 47, 1089–1092 (2000)
6. Hua, C., Long, C., Guan, X.: New results on stability analysis of neural networks with time-varying delays. Phys. Lett. A 36, 335–340 (2006)
7. Meng, X.Y., Lam, J., Du, B.Z., Gao, H.J.: A delay-partitioning approach to the stability analysis of discrete-time systems. Automatica 46, 610–614 (2010)
8. Chen, W., Lu, X., Liang, D.: Global exponential stability for discrete-time neural networks with variable delays. Phys. Lett. A 358, 186–198 (2006)
9. Li, Z.C., Liu, H.Y., Gao, H.J.: A delay partitioning approach to $H_\infty$ filtering for continuous time-delay systems. Circuits Syst. Signal Process. 30, 501–513 (2011)
10. Liu, Y.S., Wang, Z.D., Wang, W.: Reliable $H_\infty$ filtering for discrete time-delay systerms with randomly occurred nonlinearities via delay-partitioning method. Signal Process. 91, 713–727 (2011)
11. Shen, B., Wang, Z., Shu, H., Wei, G.: $H_\infty$ filtering for nonlinear discrete-time stochastic systerms with randomly varying sensor delays. Automatica 45, 1032–1037 (2009)
12. Shen, B., Wang, Z., Shu, H., Wei, G.: Robust $H_\infty$ finite-horizon filtering with randomly occurred nonlinearities and quantization effects. Automatica 46, 1743–1751 (2010)
13. Yue, D., Tian, E., Wang, Z., Lam, J.: Stabilization of systems with probabilistic interval input delays and its applications to networked control systems. IEEE Trans. Syst. 39, 939–945 (2009)
14. Han, Q.L., Gu, K.Q.: Stability of linear systems with time-varying delay: a generalized discretized lyapunov functional approach. Asian J. Contr. 3, 170–180 (2001)

---

[1] This work was supported by the Fundamental Research Funds for the Central Universities (JUSRP51317B) and the National Natural Science Foundation of China under Grant 10901073.

# The Universal Approximation Capabilities
# of Mellin Approximate Identity Neural Networks

Saeed Panahian Fard* and Zarita Zainuddin

School of Mathematical Sciences, Universiti Sains Malaysia
11800 USM, Pulau Pinang, Malaysia
saeedpanahian@yahoo.com,
zarita@cs.usm.my
http://math.usm.my/

**Abstract.** Universal approximation capability of feedforward neural networks with one hidden layer states that these networks are dense in the space of functions. In this paper, the concept of the Mellin approximate identity functions is proposed. By using this concept, It is shown that feedforward Mellin approximate identity neural networks with one hidden layer can approximate any positive real continuous function to any degree of accuracy. Moreover, universal approximation capability of these networks is extended to positive real Lebesgue spaces.

**Keywords:** Mellin approximate identity, Mellin approximate identity neural networks, Universal approximation, Electromyographic signals analysis, Lognormal distribution.

## 1   Introduction

The problem of function approximation by an artificial neural networks has been perused by many researchers. The generalization of this problem to some function spaces leads to the concept of universal approximation property. " Universal approximation property states that the set of approximation functions lies dense in the set of approximated function $C[X]$ w.r.t. a proper norm, where $X$ is a compact domain." [15].

The question of universal approximation capability by feedforward neural netwotks (FNNs) is reviewed as follows. Cybenko (1989) and Funahashi (1989) proved that any continuous function can be approximated by a one hidden layer feedforward neural networks using sigmoid functions as activation functions. Hornik, Stinchcombe, and Withe (1989) showed that a one hidden layer FNNs by using arbitrary squashing functions can approximate any continuous function to any degree of accuracy. Girosi and Poggio (1990) and Park and Sandberg (1991, 1993) proved the universal approximation capability of radial basis function FNNs. In addition, Leshno et al. (1993) showed that one hidden layer FNNs are universal approximators provided their activation functions are not polynomials. Moreover, Scarselli and Tosi (1998) reviewed universal approximation by using FNNs.

---

* Corresponding author.

C. Guo, Z.-G. Hou, and Z. Zeng (Eds.): ISNN 2013, Part I, LNCS 7951, pp. 205–213, 2013.

Furthermore, a few works have been done in the approximate identity neural networks (AINNs) in the last fifteen years. Turchetti et al. (1998) showed that feedforward approximate identity neural networks with one hidden layer are capable of providing the universal approximation property. By using approximate identity functions, Hahm and Hong (2004) proved that the universal approximation capability of FNNs with one hidden layer is with Sigmoidal activation functions and fixed weights. Then, Li (2008) discussed the universal approximation capability of a radial basis function FNNs with fixed weights by using approximate identity functions. Lately, Zainuddin and Panahian Fard (2012) showed that the double approximate identity neural networks are universal approximators in a real Lebesguse space. In addition, Panahian Fard and Zainuddin (2013a) proved that flexible approximate identity neural networks are universal approximators in the space of continuous functions. Moreover, these authors (2013b) proved that these networks are universal approximators in the real Lebesgue spaces.

On the other hand, Rupasov et al. (2012) showed that the analysis of electromyographic signals is well described by the lognormal distribution which is obviously superior to the normal distribution. The question that arises is whether lognormal radial basis functions belonging to some general class possess universal approximation capability. This question is the motivation of this paper.

The approach of this paper is to introduce Mellin approximate identity which is the "multiplicative" counterpart of approximate identity. A feedforward neural networks of Mellin approximate identity functions as activation functions with one hidden layer is considered. Then, the universal approximation capability of these networks are shown in two functions spaces. First, the universal approximation capability of feedforward Mellin approximate identity neural networks is proved in a space of positive continuous functions. Then, this property is extended to a positive real Lebesgue space.

The organization of this paper is as follows. In section 2, the definition of the Mellin approximate identity will be proposed. Besides, an example will be provided to clarify this definition. Then, two Theorems will be proven in a space of positive real continuous functions. In section 3, two Theorem will be obtained in a positive real Lebesgue space. In section 4, conclusions will be presented

## 2    Theoretical Results for Positive Real Continuous Functions

In this section, the definition of Mellin approximate identity functions is introduced. This definition will be used in further sections.

**Definition 1.** *Let* $\{\phi_n(x)\}_{n\in\mathbb{N}}$, $\phi_n : \mathbb{R}^+ := ]0,\infty[ \to [0,\infty[$ *is said to be Mellin approximate identity if the following properties hold:*
1) $\int_{\mathbb{R}^+} \phi_n(x)\frac{dx}{x} = 1$;
2) *Given* $\epsilon > 0$ *and* $1 < \delta < +\infty$, *there exists* $N$ *such that if* $n \geqslant N$ *then* $\int_{\mathbb{R}^+\setminus[\frac{1}{\delta},\delta]} \phi_n(x)\frac{dx}{x} \leq \epsilon$.

To present the clarification of the previous definition, the following example is given.

**Example 1.** [1] Let us consider the following sequence
$$\phi_n(x) = \tfrac{1}{\sqrt{4\pi}} exp\big( - (\tfrac{1}{2} \log nx)^2 \big), \; x \in \mathbb{R}^+.$$
It is obvious that this sequence is a Mellin approximate identity. By using Mellin approximate identity functions, the theoretical results will be obtained in the next section.

Now, the universal approximation capability of FMAINNs in a positive real continuous functions space $C[\tfrac{1}{\delta}, \delta]$ where $1 < \delta < \infty$, is described. First, Theorem 1 is presented. This Theorem states that any positive real continuous function $f$, convolved with another function $\phi_n$ where $\phi_n$ belongs to the Mellin approximate identity as $n \to \infty$, converges to itself.

**Theorem 1.** *Let $\{\phi_n(x)\}_{n\in\mathbb{N}}$, $\phi_n : \mathbb{R}^+ :=]0, \infty[\to [0, \infty[$, be a Mellin approximate identity. Let $f$ be a function in $C[\tfrac{1}{\delta}, \delta]$ where $1 < \delta < \infty$. Then, $\phi_n * f$ uniformly converges to $f$ on $[\tfrac{1}{\delta}, \delta]$.*

*Proof.* Let $x \in [\tfrac{1}{\delta}, \delta]$ and $\epsilon > 0$. There exists a $\eta > 0$ such that $|f(xy) - f(x)| < \tfrac{\epsilon}{2\|\phi\|_{L^1(\mathbb{R}^+)}}$ for all $y, |xy - x| < \eta$. Let us define $\{\phi_n * f\}_{n\in\mathbb{N}}$ by $\phi_n(x) = n\phi(x^n)$. Then,

$$\phi_n * f(x) - f(x) = \int_{\mathbb{R}^+} n\phi(y^n)\{f(xy) - f(x)\}\frac{dy}{y}$$

$$= \Big( \int_{[\tfrac{1}{\delta},\delta]} + \int_{\mathbb{R}^+\setminus[\tfrac{1}{\delta},\delta]} \Big) n\phi(y^n)\{f(xy) - f(x)\}\frac{dy}{y}$$

$$= I_1 + I_2,$$

where $I_1$ and $I_2$ are as follows:

$$|I_1| \leq \int_{[\tfrac{1}{\delta},\delta]} n\phi(y^n)\{f(xy) - f(x)\}\frac{dy}{y}$$

$$< \frac{\epsilon}{2\|\phi\|_{L^1(\mathbb{R}^+)}} \int_{[\tfrac{1}{\delta},\delta]} n\phi(y^n)\frac{dy}{y}$$

$$= \frac{\epsilon}{2\|\phi\|_{L^1(\mathbb{R}^+)}} \int_{[\tfrac{1}{\delta},\delta]} \phi(t)\frac{dt}{t}$$

$$\leq \frac{\epsilon}{2\|\phi\|_{L^1(\mathbb{R}^+)}} \int_{\mathbb{R}^+} \phi(t)\frac{dt}{t} = \frac{\epsilon}{2}.$$

For $I_2$, we have

$$|I_2| \leq 2\|f\|_{C[\tfrac{1}{\delta},\delta[} \int_{\mathbb{R}^+\setminus[\tfrac{1}{\delta},\delta]} n\phi(y^n)\frac{dy}{y}$$

$$= 2\|f\|_{C[\tfrac{1}{\delta},\delta]} \int_{\mathbb{R}^+\setminus[\tfrac{1}{\delta},\delta]} \phi(t)\frac{dt}{t}.$$

Since

$$\lim_{n\to\infty} \int_{\mathbb{R}^+\setminus[\frac{1}{\delta},\delta]} \phi(t)\frac{dt}{t} = 0,$$

there exists an $n_0 \in \mathbb{N}$ such that for all $n \geq n_0$,

$$\int_{\mathbb{R}^+\setminus[\frac{1}{\delta},\delta]} \phi(t)\frac{dt}{t} < \frac{\epsilon}{4\|f\|_{C[\frac{1}{\delta},\delta]}}.$$

Combining $I_1$ and $I_2$ for $n \geq n_0$, we have

$$\|\phi_n * f(x) - f(x)\|_{C[\frac{1}{\delta},\delta]} < \epsilon.$$

This Theorem provides the core of the following theorem.

**Theorem 2.** *Let $C[\frac{1}{\delta}, \delta]$ be a linear space of all continuous functions on the closed and bounded interval $[\frac{1}{\delta}, \delta]$ where $1 < \delta < \infty$, and $V \subset C[\frac{1}{\delta}, \delta]$ a compact set. Let $\{\phi_n(x)\}_{n\in\mathbb{N}}$, $\phi_n : \mathbb{R}^+ :=]0, \infty[ \to [0, \infty[$, be Mellin approximate identity functions. Let the family of functions $\{\sum_{j=1}^{M} \lambda_j \frac{\phi_j(x)}{x} | \lambda_j \in \mathbb{R}, x \in \mathbb{R}^+, M \in \mathbb{N}\}$, be dense in $C[\frac{1}{\delta}, \delta]$, and given $\epsilon > 0$. Then there exists $N \in \mathbb{N}$ which depends on $V$ and $\epsilon$ but not on $f$, such that for any $f \in V$, there exist weights $c_k = c_k(f, V, \epsilon)$ satisfying*

$$\left\| f(x) - \sum_{k=1}^{N} c_k \frac{\phi_k(x)}{x} \right\|_{C[\frac{1}{\delta},\delta]} < \epsilon$$

*Moreover, every $c_k$ is a continuous function of $f \in V$.*

*Proof.* The method of the proof is similar to the proof of Theorem 1 in [17]. Since $V$ is compact, for any $\epsilon > 0$, there is a finite $\frac{\epsilon}{2}$-net $\{f^1, ..., f^M\}$ for $V$. This implies that for any $f \in V$, there is an $f^j$, such that $\| f - f^j \|_{C[\frac{1}{\delta},\delta]} < \frac{\epsilon}{2}$. For any $f^j$, by assumption of the theorem, there are $\lambda_i^j \in \mathbb{R}, N_j \in \mathbb{N}$, and $\frac{\phi_i^j(x)}{x}$, such that

$$\left\| f^j(x) - \sum_{i=1}^{N_j} \lambda_i^j \frac{\phi_i^j(x)}{x} \right\|_{C[\frac{1}{\delta},\delta]} < \frac{\epsilon}{2}. \tag{1}$$

For any $f \in V$, we define

$$F_-(f) = \{j| \| f - f^j \|_{C[\frac{1}{\delta},\delta]} < \frac{\epsilon}{2}\},$$

$$F_0(f) = \{j| \| f - f^j \|_{C[\frac{1}{\delta},\delta]} = \frac{\epsilon}{2}\},$$

$$F_+(f) = \{j| \| f - f^j \|_{C[\frac{1}{\delta},\delta]} > \frac{\epsilon}{2}\}.$$

Therefore, $F_-(f)$ is not empty according to the definition of $\frac{\epsilon}{2}$-net. If $\tilde{f} \in V$ approaches $f$ such that $\| \tilde{f} - f \|_{C[\frac{1}{\delta},\delta]}$ is small enough, then we have $F_-(f) \subset$

$F_-(\tilde{f})$ and $F_+(f) \subset F_+(\tilde{f})$. Thus $F_-(\tilde{f}) \cap F_+(f) \subset F_-(\tilde{f}) \cap F_+(\tilde{f}) = \emptyset$, which implies $F_-(\tilde{f}) \subset F_-(f) \cup F_0(f)$. We finish with the following.

$$F_-(f) \subset F_-(\tilde{f}) \subset F_-(f) \cup F_0(f). \tag{2}$$

Define

$$d(f) = \left[ \sum_{j \in F_-(f)} \left( \frac{\epsilon}{2} - \| f - f^j \|_{C[\frac{1}{\delta}, \delta]} \right) \right]^{-1}$$

and

$$f_h = \sum_{j \in F_-(f)} \sum_{i=1}^{N_j} d(f) \left( \frac{\epsilon}{2} - \| f - f^j \|_{C[\frac{1}{\delta}, \delta]} \right) \lambda_i^j \frac{\phi_i^j(x)}{x} \tag{3}$$

then $f_h \in \left\{ \sum_{j=1}^{M} \lambda_j \frac{\phi_j(x)}{x} \right\}$ approximates $f$ with accuracy $\epsilon$ :

$$\| f - f_h \|_{C[\frac{1}{\delta}, \delta]}$$

$$= \left\| \sum_{j \in F_-(f)} d(f) \left( \frac{\epsilon}{2} - \| f - f^j \|_{C[\frac{1}{\delta}, \delta]} \right) \right.$$

$$\left. \left( f - \sum_{i=1}^{N_j} \lambda_i^j \frac{\phi_i^j(x)}{x} \right) \right\|_{C[\frac{1}{\delta}, \delta]}$$

$$= \left\| \sum_{j \in F_-(f)} d(f) \left( \frac{\epsilon}{2} - \| f - f^j \|_{C[\frac{1}{\delta}, \delta]} \right) \right.$$

$$\left. \left( f - f^j + f^j - \sum_{i=1}^{N_j} \lambda_i^j \frac{\phi_i^j(x)}{x} \right) \right\|_{C[\frac{1}{\delta}, \delta]}$$

$$\leq \sum_{j \in F_-(f)} d(f) \left( \frac{\epsilon}{2} - \| f - f^j \|_{C[\frac{1}{\delta}, \delta]} \right)$$

$$\left( \| f - f^j \|_{C[\frac{1}{\delta}, \delta]} + \left\| f_j - \sum_{i=1}^{N_j} \lambda_i^j \frac{\phi_i^j(x)}{x} \right\|_{C[\frac{1}{\delta}, \delta]} \right)$$

$$\leq \sum_{j \in F_-(f)} d(f) \left( \frac{\epsilon}{2} - \| f - f^j \|_{C[\frac{1}{\delta}, \delta]} \right) \left( \frac{\epsilon}{2} + \frac{\epsilon}{2} \right) = \epsilon. \tag{4}$$

Now, the continuity of $c_k$ is proved for the next step. For the proof, we use (2) to obtain

$$\sum_{j \in F_-(f)} \left( \frac{\epsilon}{2} - \| \tilde{f} - f^j \|_{C[\frac{1}{\delta}, \delta]} \right)$$

$$\leq \sum_{j \in F_-(\tilde{f})} \left( \frac{\epsilon}{2} - \| \tilde{f} - f^j \|_{C[\frac{1}{\delta}, \delta]} \right)$$

$$\leq \sum_{j\in F_-(\widetilde{f})} \left(\frac{\epsilon}{2} - \|\widetilde{f} - f^j\|_{C[\frac{1}{\delta},\delta]}\right) +$$

$$\sum_{j\in F_0(f)} \left(\frac{\epsilon}{2} - \|\widetilde{f} - f^j\|_{C[\frac{1}{\delta},\delta]}\right). \tag{5}$$

Let $\widetilde{f} \to f$ in (5), then we have

$$\sum_{j\in F_-(\widetilde{f})} \left(\frac{\epsilon}{2} - \|\widetilde{f} - f^j\|_{C[\frac{1}{\delta},\delta]}\right) \to \sum_{j\in F_-(f)} \left(\frac{\epsilon}{2} - \|f - f^j\|_{C[\frac{1}{\delta},\delta]}\right). \tag{6}$$

This obviously shows $d(\widetilde{f}) \to d(f)$. Thus, $\widetilde{f} \to f$ results

$$d(\widetilde{f})\left(\frac{\epsilon}{2} - \|\widetilde{f} - f^j\|_{C[\frac{1}{\delta},\delta]}\right)\lambda_i^j \to d(f)\left(\frac{\epsilon}{2} - \|f - f^j\|_{C[\frac{1}{\delta},\delta]}\right)\lambda_i^j. \tag{7}$$

Let $N = \sum_{j\in F_-(f)} N_j$ and define $c_k$ in terms of

$$f_h = \sum_{j\in F_-(f)} \sum_{i=1}^{N_j} d(f)\left(\frac{\epsilon}{2} - \|f - f^j\|_{C[\frac{1}{\delta},\delta]}\right)\lambda_i^j \frac{\phi_i^j(x)}{x}$$

$$\equiv \sum_{k=1}^{N} c_k \frac{\phi_k(x)}{x}$$

From (7), $c_k$ is a continuous functional of $f$. This completes the proof.

Now, the theoretical results for a positive real Lebesgue space will be extended in the following section.

## 3   Theoretical Results for Positive Real Lebesgue Functions

In this section, the universal approximation capability of FMAINNs in a positive real Lebesgue space Let $L^p[\frac{1}{\delta}, \delta]$ where $1 \leq p < \infty$, and $1 < \delta < \infty$, is considered. First, the following simple Lemma is presented. This Lemma will be used in the proof of Theorem 3.

**Lemma 1.** Let $L^p[\frac{1}{\delta}, \delta]$ where $1 \leq p < \infty$, and $1 < \delta < \infty$, be a linear space of all positive real Lebesgue integrable functions on any compact subset of the positive real space. Then, $\lim_{s\to 1}\|f(xs) - f(x)\|_{L^p[\frac{1}{\delta},\delta]} = 0$.

Theorem 3 described that by any positive real Lebesgue integrable function $f$, convolved with another function $\phi_n$ where $\phi_n$ belongs to the Mellin approximate identity as $n \to \infty$, converges to itself.

**Theorem 3.** *Let $L^p[\frac{1}{\delta}, \delta]$ where $1 \le p < \infty$, and $1 < \delta < \infty$, be a linear space of all positive real Lebesgue integrable functions on any compact subset of the positive real space. Let $\{\phi_n(x)\}_{n \in \mathbb{N}}$, $\phi_n : \mathbb{R}^+ :=]0, \infty[ \to [0, \infty[$, be Mellin approximate identity functions. Let $f$ be a function in $L^p[\frac{1}{\delta}, \delta]$. Then $\phi_n * f$ uniformly converges to $f$ in $L^p[\frac{1}{\delta}, \delta]$.*

*Proof.* Generalized Minkowski inequality implies that

$$\|\phi_n * f - f\|_{L^p[\frac{1}{\delta}, \delta]} \le \int_{\mathbb{R}^+} \|f(xs) - f(x)\|_{L^p[\frac{1}{\delta}, \delta]} |\phi_n(s)| \frac{ds}{s}. \tag{8}$$

Using Lemma 1, for any $\epsilon > 0$, there exists a $1 < \delta < +\infty$ such that if $\frac{1}{\delta} < s < \delta$,

$$\|f(xs) - f(x)\|_{L^p[\frac{1}{\delta}, \delta]} \le \frac{\epsilon}{4M}. \tag{9}$$

Also, the triangular inequality implies that

$$\|f(xs) - f(x)\|_{L^p[\frac{1}{\delta}, \delta]} \le 2\|f\|_{L^p[\frac{1}{\delta}, \delta]}. \tag{10}$$

By substituting the last two inequalities (10) and (9) in inequality (8), we obtain

$$\|\phi_n * f - f\|_{L^p[\frac{1}{\delta}, \delta]} \le \int_{[\frac{1}{\delta}, \delta]} \frac{\epsilon}{2M} |\phi_n(s)| \frac{ds}{s} + \int_{\mathbb{R}^+ \setminus [\frac{1}{\delta}, \delta]} 2\|f\|_{L^p[a, b]} |\phi_n(s)| \frac{ds}{s}$$
$$\le \frac{\epsilon}{2M} \int_{[\frac{1}{\delta}, \delta]} |\phi_n(s)| \frac{ds}{s} + 2\|f\|_{L^p[\frac{1}{\delta}, \delta]} \int_{\mathbb{R}^+ \setminus [\frac{1}{\delta}, \delta]} |\phi_n(s)| \frac{ds}{s} \tag{11}$$

By definition 1 there exists an $N$ such that for $n \ge N$

$$\int_{\mathbb{R}^+ \setminus [\frac{1}{\delta}, \delta]} |\phi_n(s)| \frac{ds}{s} \le \frac{\epsilon}{4\|f\|_{L^p[\frac{1}{\delta}, \delta]}} \tag{12}$$

Using inequality (12) in (11), it follows that for $n \ge N$

$$\|\phi_n * f - f\|_{L^p[\frac{1}{\delta}, \delta]} \le \frac{\epsilon}{2M}.M + 2\|f\|_{L^p[\frac{1}{\delta}, \delta]}.\frac{\epsilon}{4\|f\|_{L^p[\frac{1}{\delta}, \delta]}}$$
$$= \epsilon.$$

This Theorem constructs the core of the Theorem 4.

**Theorem 4.** *Let $L^p[\frac{1}{\delta}, \delta]$ where $1 \le p < \infty$, and $1 < \delta < \infty$, be a linear space of all positive real Lebesgue integrable functions on any compact subset of the positive real space, and $V \subset L^p[\frac{1}{\delta}, \delta]$ a compact set. Let $\{\phi_n(x)\}_{n \in \mathbb{N}}$, $\phi_n : \mathbb{R}^+ :=]0, \infty[ \to [0, \infty[$, be Mellin approximate identity functions. Let the family of functions $\{\sum_{j=1}^{M} \lambda_j \frac{\phi_j(x)}{x} | \lambda_j \in \mathbb{R}, x \in \mathbb{R}^+, M \in \mathbb{N}\}$, be dense in $L^p[\frac{1}{\delta}, \delta]$, and given $\epsilon > 0$. Then there exists an $N \in \mathbb{N}$ which depends on $V$ and $\epsilon$ but not* *on $f$, such that for any $f \in V$, there exist weights $c_k = c_k(f, V, \epsilon)$ satisfying*

$$\left\| f(x) - \sum_{k=1}^{N} c_k \frac{\phi_k(x)}{x} \right\|_{L^p[\frac{1}{\delta}, \delta]} < \epsilon$$

*Moreover, every $c_k$ is a continuous function of $f \in V$.*

*Proof.* The construction of the proof of this theorem is similar to the proof of Theorem 2, and hence will be omitted.

## 4   Conclusions

In this study, the definition of Mellin approximate identity functions has been proposed. Based on this definition, Theorem 1 proves that any positive real continuous function $f$, convolved with another function $\phi_n$ where $\phi_n$ belongs to Mellin approximate identity functions, converges to itself. By using this result, Theorem 2 has been presented as the first main result. This theorem states that the universal approximation capability of feedforward Mellin approximate identity neural networks with one hidden layer in a positive real continuous functions space. Again, making use of Mellin approximate identity functions, Theorem 3 proves that any positive real Lebesgue integrable function $f$, convolved with another function $\phi_n$, where $\phi_n$ belongs to the Mellin approximate identity functions, converges to itself. Based on this Theorem, Theorem 4 has been proved as the second main result. This Theorem shows the universal approximation capability of these networks in a positive real Lebesgue space.

**Acknowledgements.** This work was supported by Universiti Sains Malaysia (1001/PMATHS/811161).

## References

1. Arik, S.: Global asymptotic stability of a class of dynamical neural networks. IEEE Trans. Circuits Syst. 47, 568–571 (2000)
2. Cybenko, G.: Approximation by superpositions of a sigmoid function. Mathematics of Control, Signal, and Systems 3, 303–314 (1989)
3. Funuhashi, K.: On the approximate realization of continuous mapping by neural networks. Neural Networks 2, 359–366 (1989)
4. Hahm, N., Hong, B.I.: An approximation by neural networks with a fixed weight. Computers and Mathematics with Applications 47, 1897–1903 (2004)
5. Hornik, K., Stinchcombe, M., White, H.: Multilayer feedforward networks are universal approximators. Neural Networks 2, 359–366 (1989)
6. Leshno, M., Pinkus, A., Shocken, S.: Multilayer feedforward neural networks with a polynomial activation function can approximate any function. Neural Networks 6, 861–867 (1993)
7. Li, F.: Function approximation by neural networks. In: Proceedings 5th International Symposium on Neural Networks, pp. 348–390 (2008)

8. Panahian Fard, S., Zainuddin, Z.: Analyses for $L^p[a, b]$-norm approximation capability of flexible approximate identity neural networks. Accepted in Neural Computing and Applications 23: Special Issues on ICONIP 2012 (2013b)

9. Panahian Fard, S., Zainuddin, Z.: On the universal approximation capability of flexible approximate identity neural networks. In: Wong, W.E., Ma, T. (eds.) Emerging Technologies for Information Systems, Computing, and Management. LNEE, vol. 236 (2013a) ISBN 978-1-4614-7010-6, doi:10.1007/978-1-4614-7010-6-23

10. Park, J., Sandberg, I.W.: universal approximation using radial-basis-function networks. Neural Computation 3, 246–257 (1991)

11. Park, J., Sandberg, I.W.: Approximation and radial-basis-functions networks. Neural Computation 5, 305–316 (1993)

12. Poggio, T., Girosi, F.: A theory of networks for approximation and learning. Ai, Memo 1140. Artificial Intelligence Laboratory, MIT, Cambridge (1989)

13. Rupasov, V.I., Lebedev, M.A., Erlichman, J.S., Linderman, M.: Neuronal variability during handwriting: lognormal distribution. PLoS One 7, e34759 (2012)

14. Scarselli, F., Tsoi, A.C.: Universal approximation using feed forward neural networks: a survey of some existing methods, and some new results. Neural Networks 11, 15–37 (1998)

15. Tikk, D., Koczy, L.T., Gedeon, T.D.: A survey on universal approximation and its limits in soft computing techniques. International Journal of Approximate Reasoning 33, 185–202 (2003)

16. Turchetti, C., Conti, M., Crippa, P., Orcioni, S.: On the approximation of stochastic processes by approximate identity neural networks. IEEE Transactions on Neural Networks 9, 1069–1085 (1998)

17. Wu, W., Nan, D., Li, Z., Long, J., Wang, J.: Approximation to compact set of functions by feedforward neural networks. In: Proceedings 20th International Joint Conference on Neural Networks, pp. 1222–1225 (2007)

18. Zainuddin, Z., Fard, S.P.: Double approximate identity neural networks universal approximation in real lebesgue spaces. In: Huang, T., Zeng, Z., Li, C., Leung, C.S. (eds.) ICONIP 2012, Part I. LNCS, vol. 7663, pp. 409–415. Springer, Heidelberg (2012)

# $H_\infty$ Filtering of Markovian Jumping Neural Networks with Time Delays[*]

He Huang, Xiaoping Chen, and Qiang Hua

School of Electronics and Information Engineering,
Soochow University, Suzhou 215006, P.R. China
cshhuang@gmail.com

**Abstract.** This paper focuses on studying the filtering problem of Markovian jumping neural networks with time delays. Based on a stochastic Lyapunov functional, a delay-dependent design criterion is presented under which the resulting filtering error system is stochastically stable and a prescribed $H_\infty$ performance is guaranteed. It is shown that the gain matrices of the desired filter and the optimal performance index are simultaneously obtained by handing a convex optimization problem subject to some coupled linear matrix inequalities, which can be efficiently solved by some standard algorithms.

**Keywords:** Markovian jumping neural networks, filter design, time delays, stochastic stability.

## 1 Introduction

In recent years, the so-called Markovian jumping neural networks have been proposed and received considerable attention [7, 10]. It is known that time delays are frequently encountered in real systems including neural networks, chemical processes and communication systems, etc. One of the disadvantages of the existence of time delays is to lead to instability and poor performance of the underlying systems. Consequently, the study on the dynamical behaviors of delayed Markovian jumping neural networks has become an active research topic and a great number of interesting conditions have been published in the literature (see, e.g., [15–17]).

As mentioned in [4, 11], it is generally hard (or sometimes impossible) to completely acquire the state information of the neurons in a relatively large-scale neural network. However, in some neural-network-based applications, these information are utilized to achieve certain objectives. That is to say, in some situations, one needs to know these information in advance. Therefore, the state estimation problem of delayed neural networks is of great importance. It has been extensively investigated in the literature [4, 8, 9]. At the same time, the state

[*] This work was jointly supported by the National Natural Science Foundation of China under Grant Nos. 61005047 and 61273122, and the Natural Science Foundation of Jiangsu Province of China under Grant No. BK2010214.

C. Guo, Z.-G. Hou, and Z. Zeng (Eds.): ISNN 2013, Part I, LNCS 7951, pp. 214–221, 2013.

estimation problem was also discussed for Markovian jumping neural networks with time delays [2, 12, 14].

On the other hand, in the VLSI implementations of neural networks, noise is inevitable due to the tolerance of the used electronic elements. It means that the performance analysis is also of practical significance. It should be pointed out that, in [2, 12, 14], only the state estimator design problem was considered for Markovian jumping neural networks. Furthermore, although some results on the performance analysis of delayed neural networks without Markovian jumping parameters were reported in [5, 6], this issue has not yet been studied for Markovian jumping neural networks with time delays.

Motivated by the above discussion, this paper is concerned with the delay-dependent $H_\infty$ filtering problem of Markovian jumping neural networks with time delays. The activation function is assumed to satisfy a sector-bounded condition, which is less restrictive than the Lipschitz continuous condition. By constructing a stochastic Lyapunov functional, a delay-dependent criterion is derived under which the resulting filtering error system is stochastically stable and a prescribed performance is ensured. It is shown that the gain matrices of the designed filter and the optimal $H_\infty$ performance index are obtained by solving linear matrix inequalities (LMIs) [1], which is facilitated readily by standard algorithms.

## 2 Problem Formulation

Let $\{r_t, t \geq 0\}$ be a right-continuous Markov chain defined on a probability space and taking values in a finite set $\mathcal{M} = \{1, 2, \ldots, M\}$. It is assumed that its transition probability matrix $\Pi = [\pi_{ij}]_{M \times M}$ is given by

$$\Pr\{r_{t+h} = j | r_t = i\} = \begin{cases} \pi_{ij} h + o(h), & i \neq j \\ 1 + \pi_{ii} h + o(h), & i = j \end{cases}$$

where $h > 0$, $\lim_{h \to 0+} o(h)/h = 0$, $\pi_{ij} \geq 0$ for $j \neq i$ is the transition rate from mode $i$ at time $t$ to mode $j$ at time $t + h$, and for each $i \in \mathcal{M}$,

$$\pi_{ii} = - \sum_{j=1, j \neq i}^{M} \pi_{ij}. \tag{1}$$

Consider the following delayed Markovian jumping neural network subject to noise disturbances:

$$\dot{x}(t) = -A(r_t)x(t) + B_1(r_t)f(x(t))$$
$$+ B_2(r_t)f(x(t - \tau)) + J(r_t) + E_1(r_t)w(t) \tag{2}$$
$$y(t) = C(r_t)x(t) + D(r_t)x(t - \tau) + E_2(r_t)w(t) \tag{3}$$
$$z(t) = H(r_t)x(t) \tag{4}$$
$$x(t) = \phi(t) \quad \forall t \in [-\tau, 0] \tag{5}$$

where $x(t) = [x_1(t), x_2(t), \ldots, x_n(t)]^T \in \mathbb{R}^n$ is the state vector with $n$ being the number of the neurons, $w(t) \in \mathbb{R}^q$ is a noise disturbance belonging to $L_2[0, \infty)$, $y(t) \in \mathbb{R}^m$ is the available measurement, $z(t) \in \mathbb{R}^p$, to be estimated, is a linear combination of the neuron states, $f(x(t)) = [f_1(x_1(t)), f_2(x_2(t)), \ldots, f_n(x_n(t))]^T$ is the activation function, $\tau$ is a constant time delay, and $\phi(t)$ is an initial condition on $[-\tau, 0]$ with $\tau > 0$. For a fixed $r_t \in \mathcal{M}$, $A(r_t)$ is a diagonal matrix with positive entries, $B_1(r_t)$ and $B_2(r_t)$ are respectively the connection weight matrix and delayed connection weight matrix, $E_1(r_t), E_2(r_t), C(r_t), D(r_t)$ and $H(r_t)$ are real known constant matrices with appropriate dimensions, and $J(r_t)$ is an external input vector.

**Assumption 1.** *There exist scalars $L_i^-$ and $L_i^+$ ($i = 1, 2, \ldots, n$) such that for any $u \neq v \in \mathbb{R}$,*

$$L_i^- \leq \frac{f_i(u) - f_i(v)}{u - v} \leq L_i^+. \tag{6}$$

Denote $L^- = \mathrm{diag}(L_1^-, L_2^-, \ldots, L_n^-)$ and $L^+ = \mathrm{diag}(L_1^+, L_2^+, \ldots, L_n^+)$.

In sequel, for each $r_t = i \in \mathcal{M}$, $A(r_t), B_1(r_t), B_2(r_t), E_1(r_t), E_2(r_t), C(r_t), D(r_t), H(r_t)$ and $J(r_t)$ are respectively denoted by $A_i, B_{1i}, B_{2i}, E_{1i}, E_{2i}, C_i, D_i, H_i$ and $J_i$.

As stated before, this study aims to deal with the $H_\infty$ filtering problem of the neural network (2)-(5). For each $i \in \mathcal{M}$, a causal Markovian jumping filter is constructed as follows:

$$\dot{\hat{x}}(t) = -A_i \hat{x}(t) + B_{1i} f(\hat{x}(t)) + B_{2i} f(\hat{x}(t - \tau)) + J_i$$
$$+ K_i[y(t) - C_i \hat{x}(t) - D_i \hat{x}(t - \tau)], \tag{7}$$
$$\hat{z}(t) = H_i \hat{x}(t), \tag{8}$$
$$\hat{x}(t) = 0, \quad t \in [-\tau, 0], \tag{9}$$

where $\hat{x}(t) \in \mathbb{R}^n$, $\hat{z}(t) \in \mathbb{R}^p$, and $K_i (i \in \mathcal{M})$ are the gain matrices of the filter.

Let the error signals $e(t) = x(t) - \hat{x}(t)$ and $\bar{z}(t) = z(t) - \hat{z}(t)$. Then, the filtering error system is obtained from (2)-(4) and (7)-(8) and is described by

$$\dot{e}(t) = -(A_i + K_i C_i)e(t) - K_i D_i e(t - \tau)$$
$$+ B_{1i} g(t) + B_{2i} g(t - \tau) + (E_{1i} - K_i E_{2i})w(t), \tag{10}$$
$$\bar{z}(t) = H_i e(t), \tag{11}$$

with $g(t) = f(x(t)) - f(\hat{x}(t))$ and $g(t - \tau) = f(x(t - \tau)) - f(\hat{x}(t - \tau))$.

Let $e_t = e(t + s), -\tau \leq s \leq 0$. It is known from [13] that $\{(e_t, r_t), t \geq 0\}$ is a Markov process. Then, its weak infinitesimal operator $\mathcal{L}$ acting on a functional $V(e_t, i, t)$ is defined by

$$\mathcal{L}V(e_t, i, t) = \lim_{\Delta \to 0+} \frac{1}{\Delta} \left\{ \mathbb{E}\left[ V(e_{t+\Delta}, r_{t+\Delta}, t + \Delta) | e_t, r_t = i \right] - V(e_t, i, t) \right\}.$$

According to Dynkin's formula, one has

$$\mathbb{E}V(e_t, r_t, t) = V(e(0), r_0, 0) + \mathbb{E}\left\{ \int_0^t \mathcal{L}V(e_s, r_s, s)ds \right\}.$$

**Definition 1.** *The filtering error system* (10) *with $w(t) \equiv 0$ is said to be stochastically stable if $\lim_{t \to \infty} \mathbb{E}\{\int_0^t e^T(s)e(s)ds\} < \infty$ holds for any initial conditions $\phi(t)$ and $r_0 \in \mathcal{M}$.*

The objective of this study is to present a delay-dependent approach to addressing the $H_\infty$ filtering problem of the delayed Markovian jumping neural network (2)-(5) via the causal filter (7)-(9). Detailedly, for a prescribed level of noise attenuation $\rho > 0$, the filter (7)-(9) is designed such that (i) the resulting filtering error system (10) with $w(t) \equiv 0$ is stochastically stable, and (ii)

$$\|\bar{z}\|_{\mathbb{E}_2} < \rho \|w\|_2 \tag{12}$$

under the zero-initial conditions for all nonzero $w(t) \in L_2[0, \infty)$, where $\|\bar{z}\|_{\mathbb{E}_2} = \left(\mathbb{E}\{\int_0^\infty \bar{z}^T(t)\bar{z}(t)dt\}\right)^{\frac{1}{2}}$ and $\|w\|_2 = \sqrt{\int_0^\infty w^T(t)w(t)dt}$.

## 3   Main Result

**Theorem 1.** *Consider the delayed Markovian jumping neural network* (2)-(5), *and let $\rho > 0$ be a prescribed scalar. Then there exist $K_i(i = 1, 2, \ldots, M)$ such that the $H_\infty$ filtering problem is solvable if there are real matrices $P_i > 0, Q_i = \begin{bmatrix} Q_{1i} & Q_{2i} \\ * & Q_{3i} \end{bmatrix} > 0, Q = \begin{bmatrix} Q_1 & Q_2 \\ * & Q_3 \end{bmatrix} > 0, R > 0, \Lambda_i = diag(\lambda_{1i}, \lambda_{2i}, \ldots, \lambda_{ni}) > 0, \Gamma_i = diag(\gamma_{1i}, \gamma_{2i}, \ldots, \gamma_{ni}) > 0$ and $N_i$ $(i = 1, 2, \ldots, M)$ such that the following LMIs*

$$\begin{bmatrix} \Sigma_{1i} & \Sigma_{2i}^T \\ * & -2P_i + R \end{bmatrix} < 0, \tag{13}$$

$$\sum_{j=1}^M \pi_{ij} Q_j \leq Q, \tag{14}$$

*are satisfied for $i = 1, 2, \ldots, M$, where $*$ represents the symmetric block in a symmetric matrix, and*

$$\Sigma_{1i} = \begin{bmatrix} \Phi_{11i} & \Phi_{12i} & \Phi_{13i} & P_i B_{2i} & \Phi_{15i} \\ * & \Phi_{22i} & 0 & \Phi_{24i} & 0 \\ * & * & \Phi_{33i} & 0 & 0 \\ * & * & * & \Phi_{44i} & 0 \\ * & * & * & * & -\rho^2 I_q \end{bmatrix},$$

$$\Sigma_{2i} = \tau \begin{bmatrix} -P_i A_i - N_i C_i & -N_i D_i & P_i B_{1i} & P_i B_{2i} & P_i E_{1i} - N_i E_{2i} \end{bmatrix},$$

$$\Phi_{11i} = -P_i A_i - A_i^T P_i - N_i C_i - C_i^T N_i^T + \sum_{j=1}^M \pi_{ij} P_j$$

$$+ H_i^T H_i - 2L^+ \Lambda_i L^- + Q_{1i} + \tau Q_1 - R,$$

$$\Phi_{12i} = -N_i D_i + R, \quad \Phi_{13i} = P_i B_{1i} + Q_{2i} + \tau Q_2 + L^+ \Lambda_i + L^- \Lambda_i,$$

$$\Phi_{15i} = P_i E_{1i} - N_i E_{2i}, \quad \Phi_{22i} = -Q_{1i} - R - 2L^+ \Gamma_i L^-,$$
$$\Phi_{24i} = -Q_{2i} + L^+ \Gamma_i + L^- \Gamma_i, \quad \Phi_{33i} = Q_{3i} + \tau Q_3 - 2\Lambda_i, \quad \Phi_{44i} = -Q_{3i} - 2\Gamma_i.$$

*Furthermore, the gain matrices $K_i (i = 1, 2, \ldots, M)$ can be designed as*

$$K_i = P_i^{-1} N_i. \tag{15}$$

*Proof.* It is firstly shown that, under the zero-initial conditions, (12) holds for any nonzero $w(t)$. Choose a stochastic Lyapunov functional candidate for each $i \in \mathcal{M}$ as

$$V(e_t, i, t) = e^T(t) P_i e(t) + \int_{t-\tau}^t \xi^T(s) Q_i \xi(s) ds$$
$$+ \int_{-\tau}^0 \int_{t+\theta}^t \xi^T(s) Q \xi(s) ds d\theta + \tau \int_{-\tau}^0 \int_{t+\theta}^t \dot{e}^T(s) R \dot{e}(s) ds d\theta, \tag{16}$$

with $\xi(t) = [e^T(t), g^T(t)]^T$.

By taking the weak infinitesimal operator $\mathcal{L}$ on the stochastic Lyapunov functional $V(e_t, i, t)$, one has

$$\mathcal{L}V(e_t, i, t) = e^T(t)\Big[ -P_i(A_i + K_i C_i) - (A_i + K_i C_i)^T P_i \Big] e(t)$$
$$-2e^T(t) P_i K_i D_i e(t - \tau) + 2e^T(t) P_i B_{1i} g(t) + 2e^T(t) P_i B_{2i} g(t - \tau)$$
$$+2e^T(t) P_i (E_{1i} - K_i E_{2i}) w(t) + e^T(t) \Big( \sum_{j=1}^M \pi_{ij} P_j \Big) e(t) + \xi^T(t) Q_i \xi(t)$$
$$-\xi(t - \tau)^T Q_i \xi(t - \tau) + \int_{t-\tau}^t \xi^T(s) \Big( \sum_{j=1}^M \pi_{ij} Q_j \Big) \xi(s) ds$$
$$+\tau \xi^T(t) Q \xi(t) - \int_{t-\tau}^t \xi^T(s) Q \xi(s) ds + \tau^2 \dot{e}^T(t) R \dot{e}(t)$$
$$-\tau \int_{t-\tau}^t \dot{e}^T(s) R \dot{e}(s) ds. \tag{17}$$

It follows from (14) that

$$\int_{t-\tau}^t \xi^T(s) \Big( \sum_{j=1}^M \pi_{ij} Q_j \Big) \xi(s) ds \leq \int_{t-\tau}^t \xi^T(s) Q \xi(s) ds. \tag{18}$$

By Jensen's inequality [3], one has

$$-\tau \int_{t-\tau}^t \dot{e}^T(s) R \dot{e}(s) ds \leq -\Big[ e(t) - e(t - \tau) \Big]^T R \Big[ e(t) - e(t - \tau) \Big]. \tag{19}$$

Noting that $g(t) = f(x(t)) - f(\hat{x}(t))$ and (6), it is known that $L_i^- \leq \frac{g_i(t)}{e_i(t)} \leq L_i^+$. Then, for any positive diagonal matrices $\Lambda_i$ and $\Gamma_i$,

$$
0 \leq -2 \sum_{k=1}^{n} \lambda_{ki} \Big[ g_k(t) - L_k^- e_k(t) \Big] \Big[ g_k(t) - L_k^+ e_k(t) \Big]
$$
$$
= -2g^T(t)\Lambda_i g(t) + 2g^T(t)\Lambda_i L^+ e(t)
$$
$$
+2e^T(t)L^- \Lambda_i g(t) - 2e^T(t)L^+ \Lambda_i L^- e(t), \tag{20}
$$
$$
0 \leq -2g^T(t-\tau)\Gamma_i g(t-\tau) + 2g^T(t-\tau)\Gamma_i L^+ e(t-\tau)
$$
$$
+2e^T(t-\tau)L^- \Gamma_i g(t-\tau) - 2e^T(t-\tau)L^+ \Gamma_i L^- e(t-\tau). \tag{21}
$$

Define

$$
\mathcal{J}(T) = \mathbb{E}\left\{ \int_0^T [\bar{z}^T(t)\bar{z}(t) - \rho^2 w^T(t)w(t)]dt \right\} \tag{22}
$$

for $T > 0$. Under the zero-initial conditions, it is known from (16) that $V(e(0), r_0, 0) = 0$ and $\mathbb{E}V(e_t, r_t, t) \geq 0$, and thus $\mathbb{E}\{ \int_0^t \mathcal{L}V(e_s, r_s, s)ds \} \geq 0$. Then, by combining (17)-(21) together, for any $T > 0$, one has

$$
\mathcal{J}(T) \leq \mathbb{E}\left\{ \int_0^T [\bar{z}^T(t)\bar{z}(t) - \rho^2 w^T(t)w(t) + \mathcal{L}V(e_t, i, t)]dt \right\}
$$
$$
\leq \mathbb{E}\left\{ \int_0^T \eta^T(t)\Big( \Xi_{1i} + \tau^2 \Xi_{2i}^T R\Xi_{2i} \Big)\eta(t)dt \right\}, \tag{23}
$$

where

$$
\eta(t) = \begin{bmatrix} e^T(t) & e^T(t-\tau) & g^T(t) & g^T(t-\tau) & w^T(t) \end{bmatrix}^T,
$$
$$
\Xi_{1i} = \begin{bmatrix}
\bar{\Phi}_{11i} & \bar{\Phi}_{12i} & \Phi_{13i} & P_i B_{2i} & \bar{\Phi}_{15i} \\
* & \Phi_{22i} & 0 & \Phi_{24i} & 0 \\
* & * & \Phi_{33i} & 0 & 0 \\
* & * & * & \Phi_{44i} & 0 \\
* & * & * & * & -\rho^2 I_q
\end{bmatrix},
$$
$$
\Xi_{2i} = \begin{bmatrix} -A_i - K_i C_i & -K_i D_i & B_{1i} & B_{2i} & E_{1i} - K_i E_{2i} \end{bmatrix},
$$
$$
\bar{\Phi}_{11i} = -P_i A_i - A_i^T P_i - P_i K_i C_i - C_i^T K_i^T P_i + \sum_{j=1}^{M} \pi_{ij} P_j
$$
$$
+ H_i^T H_i - 2L^+ \Lambda_i L^- + Q_{1i} + \tau Q_1 - R,
$$
$$
\bar{\Phi}_{12i} = -P_i K_i D_i + R, \quad \bar{\Phi}_{15i} = P_i E_{1i} - P_i K_i E_{2i}.
$$

By Schur complement, $\Xi_{1i} + \tau^2 \Xi_{2i}^T R\Xi_{2i} < 0$ is equivalent to

$$
\begin{bmatrix} \Xi_{1i} & \tau \Xi_{2i}^T R \\ * & -R \end{bmatrix} < 0. \tag{24}
$$

By pre- and post- multiplying (24) by $\mathrm{diag}(I_{4n+q}, P_i R^{-1})$ and $\mathrm{diag}(I_{4n+q}, R^{-1}P_i)$ respectively and noting that $K_i = P_i^{-1}N_i$ and $-P_i R^{-1}P_i \leq -2P_i + R$, it yields that (24) is guaranteed by (13). Then, it has $\bar{\Xi}_{1i} + \tau^2 \bar{\Xi}_{2i}^T R \bar{\Xi}_{2i} < 0$. It thus implies that $\mathcal{J}(T) \leq 0$ for any $T > 0$. According to (22) that (12) holds under the zero-initial conditions for any nonzero $w(t)$.

Secondly, it is shown that the filtering error system (10) with $w(t) \equiv 0$ is stochastically stable. When $w(t) \equiv 0$, (10) is of the form:

$$\dot{e}(t) = -(A_i + K_i C_i)e(t) - K_i D_i e(t - \tau) + B_{1i}g(t) + B_{2i}g(t - \tau). \quad (25)$$

As proven above, (24) is true because of (13). Then, it follows from (24) that

$$\begin{bmatrix} \bar{\Xi}_{1i} & \tau \bar{\Xi}_{2i}^T R \\ * & -R \end{bmatrix} < 0, \quad (26)$$

where

$$\bar{\Xi}_{1i} = \begin{bmatrix} \bar{\Phi}_{11i} & \bar{\Phi}_{12i} & \Phi_{13i} & P_i B_{2i} \\ * & \Phi_{22i} & 0 & \Phi_{24i} \\ * & * & \Phi_{33i} & 0 \\ * & * & * & \Phi_{44i} \end{bmatrix},$$

$$\bar{\Xi}_{2i} = \begin{bmatrix} -A_i - K_i C_i & -K_i D_i & B_{1i} & B_{2i} \end{bmatrix}.$$

By Schur complement again, it immediately implies that $\bar{\Xi}_{1i} + \tau^2 \bar{\Xi}_{2i}^T R \bar{\Xi}_{2i} < 0$. We still consider the stochastic Lyapunov functional (16). Similar to the derivative of (23), it is not difficult to obtain that

$$\mathcal{L}V(e_t, i, t) \leq \zeta^T(t)(\bar{\Xi}_{1i} + \tau^2 \bar{\Xi}_{2i}^T R \bar{\Xi}_{2i})\zeta(t) < 0 \quad (27)$$

for any $\zeta(t) = \begin{bmatrix} e^T(t), e^T(t-\tau), g^T(t), g^T(t-\tau) \end{bmatrix}^T \neq 0$. Therefore, the filtering error system (10) with $w(t) \equiv 0$ is stochastically stable. This completes the proof. $\quad\square$

**Remark 1.** *Theorem 1 provides a delay-dependent criterion to the $H_\infty$ filter design to the Markovian jumping neural network with time delays (2)-(5). It is worth noting that the $H_\infty$ performance index $\rho$ can be optimized by a convex optimization algorithm:*

**Algorithm 1.** $\min_{P_i,Q_i,Q,R,\Lambda_i,\Gamma_i,N_i} \rho^2$ *subject to the LMIs (13)-(14).*

**Remark 2.** *The performance of Theorem 1 can be further improved by employing some recently-established techniques such as the free-weighting matrices based technique and the delay partitioning approach, etc.*

**Remark 3.** *Some illustrative examples, used to show the effectiveness of Theorem 1, are omitted due to page limit.*

# 4   Conclusion

The delay-dependent $H_\infty$ filtering problem has been studied in this paper for Markovian jumping neural networks with time delays. A delay-dependent design criterion has been derived such that the resulting filtering error system is stochastically stable and a prescribed $H_\infty$ performance is achieved. Its advantage is that the gain matrices and the optimal performance index can be obtained by solving a corresponding convex optimization problem with some LMIs constraints.

# References

1. Boyd, S.: Linear Matrix Inequalities in System and Control Theory. SIAM, Philadelphia (1994)
2. Chen, Y., Zheng, W.X.: Stochastic state estimation for neural networks with distributed delays and Markovian jump. Neural Netw. 25, 14–20 (2012)
3. Gu, K., Kharitonov, V.L., Chen, J.: Stability of Time-delay Systems. Birkhauser, Massachusetts (2003)
4. He, Y., Wang, Q.-G., Wu, M., Lin, C.: Delay-dependent state estimation for delayed neural networks. IEEE Trans. Neural Netw. 17, 1077–1081 (2006)
5. Huang, H., Feng, G.: Delay-dependent $H_\infty$ and generalized $H_2$ filtering for delayed neural networks. IEEE Trans. Circuits Syst. I 56, 846–857 (2009)
6. Huang, H., Feng, G., Cao, J.: Guaranteed performance state estimation of static neural networks with time-varying delay. Neurocomputing 74, 606–616 (2011)
7. Kovacic, M.: Timetable construction with Markovian neural network. Eur. J. Oper. Res. 69, 92–96 (1993)
8. Liu, Y., Wang, Z., Liu, X.: Design of exponential state estimators for neural networks with mixed time delays. Phys. Lett. A 364, 401–412 (2007)
9. Park, J.H., Kwon, O.M.: State estimation for neural networks of neutral-type with interval time-varying delays. Appl. Math. Comput. 203, 217–223 (2008)
10. Tino, P., Cernansky, M., Benuskova, L.: Markovian architectural bias of recurrent neural networks. IEEE Trans. Neural Netw. 15, 6–15 (2004)
11. Wang, Z., Ho, D.W.C., Liu, X.: State estimation for delayed neural networks. IEEE Trans. Neural Netw. 16, 279–284 (2005)
12. Wu, Z., Su, H., Chu, J.: State estimation for discrete Markovian jumping neural networks with time delay. Neurocomputing 73, 2247–2254 (2010)
13. Xu, S., Chen, T., Lam, J.: Robust $H^\infty$ filtering for uncertain Markovian jump systems with mode-dependent time delays. IEEE Trans. Autom. Control 48, 900–907 (2003)
14. Zhang, D., Yu, L.: Exponential state estimation for Markovian jumping neural networks with time-varying discrete and distributed delays. Neural Netw. 35, 103-111 (2012)
15. Zhang, H., Wang, Y.: Stability analysis of Markovian jumping stochastic Cohen-Grossberg neural networks with mixed time delays. IEEE Trans. Neural Netw. 19, 366–370 (2008)
16. Zhao, Y., Zhang, L., Shen, S., Gao, H.: Robust stability criterion for discrete-time uncertain Markovian jumping neural networks with defective statistics of mode transitions. IEEE Trans. Neural Netw. 22, 164–170 (2011)
17. Zhu, Q., Cao, J.: Exponential stability of stochastic neural networks with both Markovian jump parameters and mixed time delays. IEEE Trans. Syst. Man Cybern. 41, 341–353 (2011)

# Convergence Analysis for Feng's MCA Neural Network Learning Algorithm*

Zhengxue Li**, Lijia You, and Mingsong Cheng

School of Mathematical Sciences, Dalian University of Technology,
Dalian 116024, China
lizx@dlut.edu.cn

**Abstract.** The minor component analysis is widely used in many fields, such as signal processing and data analysis, so it has very important theoretical significance and practical values for the convergence analysis of these algorithms. In this paper we seek the convergence condition for Feng's MCA learning algorithm in deterministic discrete time system. Finally numerical experiments show the correctness of our theory.

**Keywords:** MCA learning algorithm, DDT system, eigenvalue, eigenvector, invariant sets.

## 1 Introduction

The minor component analysis (MCA) is a very important statistical method, which is used to search a space direction such that the signal data has the smallest variance in it. Through the analysis, it is found that the minor component is just the corresponding eigenvector of the smallest eigenvalue of the signal data correlation matrix. The minor component analysis has been widely used in curve/surface fitting, digital beamforming, total least squares (TLS), etc. So it is very important for the convergence problem of the minor component analysis.

There are many algorithms for MCA learning method, including Feng's MCA, EXIN MCA, OJAn MCA [1-8], etc. Most of the MCA learning algorithms are about stochastic discrete time system (SDT), which results in the difficulty of the direct convergence analysis [9, 10], and we can study them via some indirect methods. Later the deterministic continuous time system (DCT) is proposed [11]. However, the demanded conditions are very harsh, such as the learning rate of the algorithm tends to zero [12], which in practice are very difficult to get satisfied. To study the convergence of algorithms better, the deterministic discrete time (DDT) system is proposed in [13]. DDT system doesn't require the learning rate to approach zero, moreover, it preserves the discrete nature of the original algorithms. Furthermore, artificial neural networks have the properties

---

* This work was supported by the National Natural Science Foundation of China (No. 11171367) and by the Fundamental Research Funds for the Central Universities (No. DUT12JS04 and No. DUT13LK04).
** Corresponding author.

C. Guo, Z.-G. Hou, and Z. Zeng (Eds.): ISNN 2013, Part I, LNCS 7951, pp. 222–229, 2013.

of fault tolerance capability, self organization and parallel computation [16, 17], so it is successfully used in the study of MCA algorithms. Some scholars have deeply studied the learning algorithms such as Oja-Xu MCA, AMEX MCA as yet [14, 15]. A necessary condition for the convergence of Feng's MCA learning algorithm is given in [2]: $\eta < \min\{\lambda_n/\lambda_1, 0.5\}$. Based on that paper, we propose a more general condition: $\eta < \min\{2\lambda_n/(\lambda_1 - \lambda_n), 0.5\}$.

## 2 Feng's MCA Learning Algorithm

Consider a linear neuron with the following input/output relation

$$y(k) = w(k)^T x(k), \quad k = 0, 1, \cdots, \tag{1}$$

where the neuron input $\{x(k)|x(k) \in \mathbb{R}^n, k = 0, 1, \cdots\}$ is a stationary stochastic process with mean zero, $y(k)$ denotes the neuron output, $w(k)$ the weight vector. Feng's MCA learning algorithm is represented as follows (see [1])

$$w(k + 1) = w(k) + \eta \left(w(k) - \|w(k)\|^2 y(k) x(k)\right), \tag{2}$$

where $\eta$ is the learning rate.

Let $R = E[x(k)x(k)^T]$ $(R \neq 0)$ is the correlation matrix of the input vector $\{x(k)|\, x(k) \in \mathbb{R}^n, k = 0, 1, \cdots\}$, then $R$ is a symmetric nonnegative definite matrix. Suppose $\lambda_1 \geq \lambda_2 \geq \cdots \geq \lambda_n \geq 0$ are all the eigenvalues of $R$, $\{v_1, v_2, \cdots, v_n\}$ is an orthonormal basis of $\mathbb{R}^n$, where each $v_i$ $(i = 1, 2, \cdots, n)$ is the unit eigenvector of the correlation matrix $R$ associated with the eigenvalue $\lambda_i$. Obviously, for each $k \geq 0$, the weight vector can be represented

$$w(k) = \sum_{i=1}^{n} z_i(k) v_i, \tag{3}$$

where $z_i(k)$ $(i = 1, 2, \cdots, n)$ are some real constants.

Taking conditional expectations $E\{w(k+1)/w(0), x(i), i < k\}$ in both sides of the equation (2) via deterministic discrete time method, we can get the following DDT system

$$w(k + 1) = w(k) + \eta \left(w(k) - \|w(k)\|^2 Rw(k)\right). \tag{4}$$

By (3), we have

$$Rw(k) = \sum_{i=1}^{n} z_i(k) R v_i = \sum_{i=1}^{n} \lambda_i z_i(k) v_i. \tag{5}$$

Based on the equation (3), (4) and (5), we have

$$z_i(k + 1) = [1 + \eta(1 - \lambda_i \|w(k)\|^2)] z_i(k), \quad i = 1, 2, \cdots, n. \tag{6}$$

# 3    Convergence Analysis

## 3.1    Assumptions

In [2] Peng has proved that if $\lambda_n = 0$, then Feng's MCA learning algorithm is divergent. Based on the above discussion, we propose the following assumptions throughout this paper:

(A1)    $\lambda_1 > \lambda_2 > \cdots > \lambda_n > 0$.
(A2)    $\eta < \min\{2\lambda_n/(\lambda_1 - \lambda_n), 0.5\}$.

## 3.2    Invariant Sets

For preserving the boundness of the system (4), invariant sets is proposed in [2].

**Definition 1.** *A subset $S$ of $\mathbb{R}^n$ space is called an invariant set of the system (4), if every trajectory with any initial $w(0) \in S$ always lies in the interior of $S$.*

**Lemma 1.** *[2] Assume $\eta \leq 0.5$, then for all $\mu \in [0, 1/\sqrt{\lambda_n}]$, we have*

$$[1 + \eta(1 - \lambda_n\mu^2)]\mu \leq 1/\sqrt{\lambda_n}.$$

**Theorem 1.** *Let $S_0 = \{w(k) \,|\, w(k) \in \mathbb{R}^n, \|w(k)\| \leq 1/\sqrt{\lambda_n}\}$. If the assumptions (A1), (A2) are satisfied, then $S_0$ is an invariant set of the system (4).*

*Proof.* Let $k \geq 0$, and suppose $w(k) \in S_0$, then $1 + \eta(1 - \lambda_n\|w(k)\|^2) > 0$.
By the assumptions (A1), (A2) and the known, we have

$$\begin{aligned}
&1 + \eta(1 - \lambda_1\|w(k)\|^2) + 1 + \eta(1 - \lambda_n\|w(k)\|^2) \\
&= 2 + 2\eta - \eta(\lambda_1 + \lambda_n)\|w(k)\|^2 \geq 2 + 2\eta - \eta(\lambda_1 + \lambda_n)/\lambda_n \qquad (7) \\
&= 2 - \eta(\lambda_1 - \lambda_n)/\lambda_n > 0.
\end{aligned}$$

For $i = 1, 2, \cdots, n$, due to

$$1 + \eta(1 - \lambda_1\|w(k)\|^2) \leq 1 + \eta(1 - \lambda_i\|w(k)\|^2) \leq 1 + \eta(1 - \lambda_n\|w(k)\|^2).$$

Combined with (7), we get

$$\left|1 + \eta(1 - \lambda_i\|w(k)\|^2)\right| \leq 1 + \eta(1 - \lambda_n\|w(k)\|^2). \qquad (8)$$

By (3), (6), (8), it is easy to prove

$$\begin{aligned}
\|w(k+1)\|^2 = \sum_{i=1}^{n} z_i^2(k+1) &= \sum_{i=1}^{n} \left[1 + \eta(1 - \lambda_i\|w(k)\|^2)\right]^2 z_i^2(k) \\
&\leq \left[1 + \eta(1 - \lambda_n\|w(k)\|^2)\right]^2 \sum_{i=1}^{n} z_i^2(k) \qquad (9) \\
&= \left[1 + \eta(1 - \lambda_n\|w(k)\|^2)\right]^2 \|w(k)\|^2,
\end{aligned}$$

that is

$$\|w(k+1)\| \leq \left[1 + \eta(1 - \lambda_n \|w(k)\|^2)\right] \|w(k)\|.$$

By Lemma 1, we get $\|w(k+1)\| \leq 1/\sqrt{\lambda_n}$.

Finally, for any $k \geq 0$, if $w(k) \in S_0$, then it must be have $w(k+1) \in S_0$, and so $S_0$ is an invariant set of the system (4).

### 3.3   Convergence Analysis

**Lemma 2.** *Suppose* $(A1)$, $(A2)$ *are satisfied,* $w(0) \in S_0$ *and* $w(0)^T v_n \neq 0$, *then there exists a constant* $l > 0$ *such that for any* $k \geq 0$, $\|w(k)\|^2 \geq l$.

*Proof.* By the known and (3), we have $z_n(0) = w(0)^T v_n \neq 0$.
    Let $l = z_n^2(0) > 0$, then

$$\|w(0)\|^2 = \sum_{i=1}^{n} z_i^2(0) \geq z_n^2(0) = l.$$

Since $S_0$ is an invariant set of (4), then for any $k \geq 0$, $\|w(k)\|^2 \leq 1/\lambda_n$, and so we have

$$1 + \eta(1 - \lambda_n \|w(k)\|^2) \geq 1. \tag{10}$$

Combined with (6), it follows that the sequence $\{z_n^2(k)\}$ is monotonically increasing, hence

$$\|w(k+1)\|^2 = \sum_{i=1}^{n} \left[1 + \eta(1 - \lambda_i \|w(k)\|^2)\right]^2 z_i^2(k)$$

$$\geq \left[1 + \eta(1 - \lambda_n \|w(k)\|^2)\right]^2 z_n^2(k) \geq z_n^2(k) \geq z_n^2(0) = l.$$

This completes the proof.

**Lemma 3.** *Suppose* $(A1)$, $(A2)$ *are satisfied,* $w(0) \in S_0$ *and* $w(0)^T v_n \neq 0$, *then there exists constant numbers* $\alpha > 0$, $\beta > 0$, *such that* $\sum_{i=1}^{n-1} z_i^2(k) \leq \alpha e^{-\beta k}$ *for any* $k \geq 0$.

*Proof.* For $i = 1, 2, \ldots, n-1$, from $(A1)$, $(A2)$ and Lemma 2 we know

$$-1 < 1 + \eta\left(1 - \lambda_1/\lambda_n\right) \leq \frac{1 + \eta(1 - \lambda_i \|w(k)\|^2)}{1 + \eta(1 - \lambda_n \|w(k)\|^2)} \leq \frac{1 + \eta(1 - \lambda_{n-1}l)}{1 + \eta(1 - \lambda_n l)} < 1.$$

Let $\rho = \max\{[1 + \eta(1 - \lambda_1/\lambda_n)]^2, [\frac{1+\eta(1-\lambda_{n-1}l)}{1+\eta(1-\lambda_n l)}]^2\}$, then $0 < \rho < 1$.
    Combined with (6),

$$\frac{z_i^2(k+1)}{z_n^2(k+1)} = \left[\frac{1 + \eta(1 - \lambda_i \|w(k)\|^2)}{1 + \eta(1 - \lambda_n \|w(k)\|^2)}\right]^2 \frac{z_i^2(k)}{z_n^2(k)}$$

$$\leq \rho \frac{z_i^2(k)}{z_n^2(k)} \leq \rho^{k+1} \frac{z_i^2(0)}{z_n^2(0)} = \frac{z_i^2(0)}{l} e^{-\beta(k+1)},$$

where $\beta = -\ln\rho$.

From the definition of invariant set $S_0$ and (3), for any $k \geq 0$, $i = 1, 2, \ldots, n$, it follows that

$$z_i^2(k) \leq \|w(k)\|^2 \leq 1/\lambda_n, \tag{11}$$

so we have

$$\sum_{i=1}^{n-1} z_i^2(k) = \sum_{i=1}^{n-1}\left[\frac{z_i(k)}{z_n(k)}\right]^2 z_n^2(k) \leq \alpha e^{-\beta k},$$

where $\alpha = \frac{n-1}{l\lambda_n^2}$.

From Lemma 3 we immediately get

**Theorem 2.** *Suppose* (A1), (A2) *are satisfied,* $w(0) \in S_0$ *and* $w(0)^T v_n \neq 0$, *then* $\lim_{k\to\infty} z_i(k) = 0$, $i = 1, 2, \cdots, n-1$.

**Theorem 3.** *Suppose* (A1), (A2) *are satisfied,* $w(0) \in S_0$ *and* $w(0)^T v_n \neq 0$, *then* $\lim_{k\to\infty} w(k) = (\pm 1/\sqrt{\lambda_n})v_n$.

*Proof.* By the equation (6) and (10), it is easy to know that if $z_n(0) < 0$, then the sequence $\{z_n(k)\}$ is monotonically decreasing, and if $z_n(0) > 0$, then $\{z_n(k)\}$ is monotonically increasing. Combined with (11), we can get: $\lim_{k\to\infty} z_n(k)$ exists and is not zero. Denote $\lim_{k\to\infty} z_n(k) = c \neq 0$.

From (3) and Theorem 2, we can deduce that

$$\lim_{k\to\infty} \|w(k)\|^2 = \lim_{k\to\infty} z_n^2(k) = c^2.$$

Let $i = n$ in (6) and take limit on both sides we have

$$c = \pm 1/\sqrt{\lambda_n}.$$

Hence

$$\lim_{k\to\infty} w(k) = \lim_{k\to\infty} z_n(k)v_n = (\pm 1/\sqrt{\lambda_n})v_n.$$

This completes the proof.

## 4    Numerical Experiments

### 4.1    Experiments Conditions

Throughout the experiments, we use the following $5 \times 5$ random correlation matrix

$$R = \begin{pmatrix} 0.1944 & 0.0861 & 0.0556 & 0.1322 & 0.1710 \\ 0.0861 & 0.2059 & 0.1656 & 0.1944 & 0.1467 \\ 0.0556 & 0.1656 & 0.2358 & 0.1948 & 0.1717 \\ 0.1322 & 0.1944 & 0.1948 & 0.3135 & 0.2927 \\ 0.1710 & 0.1467 & 0.1717 & 0.2927 & 0.3241 \end{pmatrix},$$

all the eigenvalues are $\lambda_5 = 0.0059$, $\lambda_4 = 0.0526$, $\lambda_3 = 0.0887$, $\lambda_2 = 0.1838$, $\lambda_1 = 0.9428$, and the minor component is $v_5 = [-0.2470 \quad 0.3892 \quad -0.1422 \quad -0.6295 \quad 0.6091]^T$. Define the direction cosine between the weight vector $w(k)$ and the minor component $v_n$ as $\theta(k) = w(k)^T v_n/\|w(k)\|$, which used to measure the convergence rate of the weight vector $w(k)$ into the minor component $v_n$.

## 4.2    Verification of Theoretical Results

### 4.2.1    Invariant Set $S_0$

Generate three initial weight vector randomly as following

$$w_1(0) = \begin{pmatrix} 0.7851 & -1.2715 & -1.2967 & 0.5490 & -1.3122 \end{pmatrix}^T, \|w_1(0)\| = 2.4368.$$
$$w_2(0) = \begin{pmatrix} 3.2253 & -0.5377 & 1.1313 & -9.5904 & 8.0924 \end{pmatrix}^T, \quad \|w_2(0)\| = 13.0167.$$
$$w_3(0) = \begin{pmatrix} 2.3471 & 1.5690 & 4.7120 & -3.1450 & -2.6800 \end{pmatrix}^T, \quad \|w_3(0)\| = 6.8736.$$

It is easy to know that $\|w_i(0)\| < \sqrt{1/\lambda_5} \approx 13.0189$, $w_i(0)v_5 \neq 0$, $i = 1, 2, 3$. Taking the learning rate $\eta = \frac{1.99\lambda_5}{\lambda_1 - \lambda_5} \approx 0.0126$, and the total iteration number is 600. The change for the norm of the weight vector as different initial weight vector in the algorithm is given in figure 1.

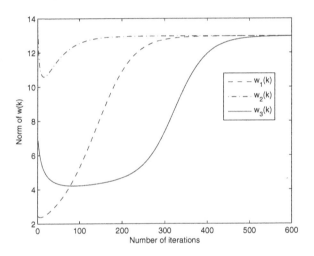

**Fig. 1.** The boundedness of $\|w(k)\|$

From figure 1 we can see that for any initial weight vector $w(0) \in S_0$, as long as the learning rate satisfies the assumption $(A2)$, then $w(k) \in S_0$, which verifies the accuracy of Theorem 1.

### 4.2.2    Convergence for Every Component $\{z_i(k)|\, i = 1, 2, \cdots, 5\}$ of the Weight Vector $w(k)$

Taking initial weight vector $w(0) = w_2(0)$, then $z_5(0) = w(0)^T v_n > 0$. Let the learning rate $\eta = 0.0126$, the total iteration number 100 and 400, respectively. The change of every component of the weight vector $w(k)$ is given in figure 2.

From figure 2(a) we can deduce that $z_i(k) \to 0$, $i = 1, 2, 3, 4$ as $k \to \infty$, and from figure 2(b) we can easily see that the sequence $\{z_5(k)\}$ is monotonically increasing and is convergent. The experiment results equate with Theorem 2 and Theorem 3.

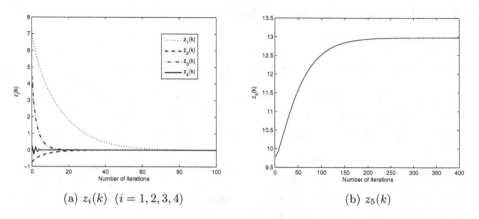

(a) $z_i(k)$ $(i = 1, 2, 3, 4)$                    (b) $z_5(k)$

**Fig. 2.** The convergence of the sequence $\{z_i(k)|\, i = 1, 2, \cdots, 5\}$

### 4.2.3    Comparison of Convergence Rate

The initial weight vector $w(0) = w_3(0)$, taking three different learning rate as $\eta_1 = 1.99\lambda_5/(\lambda_1 - \lambda_5) \approx 0.0126$, $\eta_2 = 1.5\lambda_5/(\lambda_1 - \lambda_5) \approx 0.0095$, $\eta_3 = \lambda_5/\lambda_1 \approx 0.0063$, where $\eta_1$ and $\eta_2$ satisfy the assumption $(A2)$, $\eta_3$ is the result in [2]. The total iteration number is 400. The convergence for the direction cosine $\theta_i(k)$ $(i = 1, 2, 3)$ as different learning rate is given in figure 3.

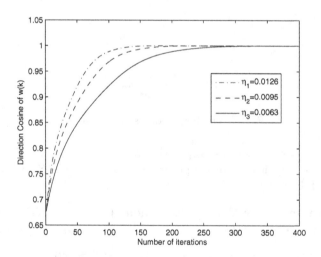

**Fig. 3.** Comparison of convergence rate for direction cosine

From figure 3 we can know the choice of learning rate has certain effect on the convergence rate of the algorithm. Since the learning rate $\eta_1$ and $\eta_2$ are all greater than the learning rate $\eta_3$ in [2], the convergence is faster.

## 4.3   Conclusion

Based on [2], we study the convergence of Feng's MCA learning algorithm deeply, and we enlarge the original learning rate twice by the premise of guarantee of the algorithm convergence, hence improve the convergence rate.

# References

[1] Feng, D., Bao, Z., Jiao, L.: Total least mean squares algorithm. IEEE Transactions on Signal Processing 46, 2122–2130 (1998)

[2] Peng, D., Yi, Z.: Convergence analysis of a deterministic discrete time system of Feng's MCA learning algorithm. IEEE Transactions on Signal Processing 54, 3626–3632 (2006)

[3] Cirrincione, G., Cirrinicione, M., Herault, J., et al.: The MCA EXIN neuron for the minor component analysis. IEEE Transactions on Neural Networks 13, 160–187 (2002)

[4] Luo, F., Unbehauen, R., Cichocki, A.: A minor component analysis algorithm. Neural Networks 10, 291–297 (1997)

[5] Xu, L., Oja, E., Suen, C.: Modified Hebbian learning for curve and surface fitting. Neural Networks 5, 441–457 (1992)

[6] Feng, D., Zheng, W., Jia, Y.: Neural network learning algorithms for tracking minor subspace in high-dimensional data stream. IEEE Transactions on Neural Networks 16, 513–521 (2005)

[7] Ouyang, S., Bao, Z., Liao, G., Ching, P.C.: Adaptive minor component extraction with modular structure. IEEE Transactions on Signal Processing 49, 2127–2137 (2001)

[8] Oja, E.: Principal components, minor components and linear neural networks. Neural Networks 5, 927–935 (1992)

[9] Zhang, Q.: On the discrete-time dynamics of a PCA learning algorithm. Neuro-computing 55, 761–769 (2003)

[10] Yi, Z., Ye, M., Lv, J., Tan, K.K.: Convergence analysis of a deterministic discrete time system of Oja's PCA learning algorithm. IEEE Transactions on Neural Networks 16, 1318–1328 (2005)

[11] Manton, J.H., Helmke, U., Merells, I.M.Y.: Dynamical systems for principal and minor component analysis. In: Proceedings of the IEEE Conference on Decision and Control, vol. 2, pp. 1863–1868 (2003)

[12] Ljung, L.: Analysis of recursive stochastic algorithms. IEEE Transactions on Automatic Control 22, 551–575 (1997)

[13] Zufiria, P.J.: On the discrete-time dynamics of the basic Hebbian neural network node. IEEE Transactions on Neural Networks 13, 1342–1352 (2002)

[14] Peng, D., Zhang, Y.: A modified Oja-Xu MCA learning algorithm and its convergence analysis. IEEE Transactions on Circuits and Systems II 54, 348–352 (2007)

[15] Li, X., Fan, Y., Peng, C.: Convergence analysis of adaptive total least square based on deterministic discrete time. Control and Decision 25, 1399–1402 (2010)

[16] Wu, W.: Neural networks computing, pp. 5–48. China Higher Education Press (2003)

[17] Jiang, Z.: Introduction to artificial neural networks, pp. 27–29. China Higher Education Press (2001)

# Anti-periodic Solutions
# for Cohen-Grossberg Neural Networks
# with Varying-Time Delays and Impulses

Abdujelil Abdurahman and Haijun Jiang*

College of Mathematics and System Sciences, Xinjiang University,
Urumqi, 830046, Xinjiang, P.R. China
jianghaixju@163.com

**Abstract.** In this paper, we discuss the existence and exponential stability of the anti-periodic solution for delayed Cohen-Grossberg neural networks with impulsive effects. First we give some sufficient conditions to ensure existence and stability of the anti-periodic solutions. Then we present an example with numerical simulations to illustrate our results.

**Keywords:** Cohen-Grossberg neural networks, Anti-periodic solution, Impulsive effects, Exponential stability.

## 1   Introduction and Preliminaries

Since Cohen-Grossberg neural networks (CGNNs) have been first introduced by Cohen and Grossberg in 1983 [1], they have been intensively studied due to their promising potential applications in classification, parallel computation, associative memory and optimization problems. In these applications, the dynamics of networks such as the existence, uniqueness, Hopf bifurcation and global asymptotic stability or global exponential stability of the equilibrium point, periodic and almost periodic solutions for networks plays a key role, see [2–11], and the references therein.

Over the past decades, the anti-periodic solution of Hopfield neural networks, recurrent neural networks and cellular neural networks have actively been investigated by a large number of scholars. For details, see [12–14] and references therein. However, till now, there are very few or even no results on the problems of anti-periodic solutions for delayed CGNNs with impulsive effects, while the existence of anti-periodic solutions plays a key role in characterizing the behavior of nonlinear differential equations (see [15]). Thus, it is worth investigating the existence and stability of anti-periodic solutions of CGNNs with both time-delays an impulsive effects.

---

* Corresponding author.

C. Guo, Z.-G. Hou, and Z. Zeng (Eds.): ISNN 2013, Part I, LNCS 7951, pp. 230–238, 2013.

Motivated by above analysis, in this paper, we consider the following impulsive Cohen-Grossberg neural networks model with time-varying delays

$$
\begin{cases}
x_i'(t) = a_i(x_i(t))\Big[ -b_i(t, x_i(t)) + \sum_{j=1}^{n} c_{ij}(t) f_j(x_j(t)) \\
\qquad + \sum_{j=1}^{n} d_{ij}(t) g_j\left(x_j(t - \tau_{ij}(t))\right) + I_i(t)\Big], \quad t \geq 0, \ t \neq t_k, \\
\Delta x_i(t_k) = x_i(t_k^+) - x_i(t_k^-) = \gamma_{ik} x_i(t_k), \quad k \in N,
\end{cases}
\tag{1}
$$

where $n$ denotes the number of neurons in the network, $x_i(t)$ corresponds to the state of the $i$th unit at time $t$, $a_i(x_i(t))$ represents an amplification function, $b_i(t, x_i(t))$ is an appropriate behaved function, $f_j(x_j(t))$ and $g_j(x_j(t - \tau_{ij}(t)))$ denote, respectively, the measures of activation to its incoming potentials of the unit $j$ at time $t$ and $t - \tau_{ij}(t)$, $\tau_{ij}(t)$ corresponds to the transmission delay along the axon of the $j$th unit and is non-negative function, and $I_i(t)$ denotes the external bias on the $i$th unit at time $t$. Concerning coefficients of differential system (1), $c_{ij}(t)$ denotes the synaptic connection weight of the unit $j$ on the unit $i$ at time $t$, $d_{ij}(t)$ denotes the synaptic connection weight of the unit $j$ on the unit $i$ at time $t - \tau_{ij}(t)$, where $\tau_{ij}(t) > 0$, $i, j \in \mathcal{I} \triangleq \{1, 2, ..., n\}$.

Throughout this paper, for the model (1), we introduce the following assumptions

(H$_1$) $a_i \in C(R, R^+)$ and there exist positive constants $\underline{a}_i$ and $\bar{a}_i$ such that

$$
\underline{a}_i \leq a_i(u) \leq \bar{a}_i, \quad \text{for all } u \in R, \quad i \in \mathcal{I}.
$$

(H$_2$) For each $u$, $b_i(\cdot, u)$ is continuous, $b_i(t, 0) \equiv 0$ and there exists a continuous and $\omega$-periodic function $\beta_i(t) > 0$ such that

$$
\frac{b_i(t, u) - b_i(t, v)}{u - v} \geq \beta_i(t), \quad u, v \in R, \ u \neq v, \quad i \in \mathcal{I}.
$$

(H$_3$) The activation functions $f_j$, $g_j$ are continuous, bounded and there exist Lipschitz constants $L_j^1, L_j^2 > 0$ such that, for each $j \in \mathcal{I}$

$$
|f_j(u) - f_j(v)| \leq L_j^1 |u - v|, \quad |g_j(u) - g_j(v)| \leq L_j^2 |u - v|, \quad u, v \in R.
$$

(H$_4$) $c_{ij}, d_{ij}, \tau_{ij}, I_i \in C(R, R)$, and there exists a constant $\omega > 0$ such that

$$
a_i(-u) = a_i(u), \quad b_i(t + \omega, u) = -b_i(t, -u), \quad c_{ij}(t + \omega) f_j(u) = -c_{ij}(t) f_j(-u),
$$
$$
I_i(t + \omega) = -I_i(t), \quad d_{ij}(t + \omega) g_j(u) = -d_{ij}(t) g_j(-u), \quad \tau_{ij}(t + \omega) = \tau_{ij}(t).
$$

(H$_5$) $|1 + \gamma_{ik}| \leq 1$, $t_{k+1} - t_k \geq \kappa$, $\lim_{k \to \infty} = \infty$ and there exists $q \in N$ such that

$$
[0, \omega] \bigcap t_{k \geq 1} = \{t_1, t_2 \cdots, t_q\}, \quad t_{k+q} = t_k + \omega, \ \gamma_{i(k+q)} = \gamma_{ik}, \quad i \in \mathcal{I}, \ k \in N.
$$

Let $\tau = \max_{i,j \in \mathcal{I}} \sup_{0 \leq t \leq \omega} \tau_{ij}(t)$, the initial conditions associated with system (1) are given by

$$
x_i(s) = \varphi_i(s), \quad s \in [-\tau, 0], \quad i \in \mathcal{I},
$$

where $\varphi_i(t)$ denotes a real-valued continuous function defined on $[-\tau, 0]$.

A solution $x(t) = (x_1(t), x_2(t), \cdots, x_n(t))^T$ of impulsive system (1) is a piecewise continuous vector function whose components belong to the space

$$PC([-\tau, +\infty), R) = \{\varphi(t) : [-\tau, +\infty) \longrightarrow R \text{ is continuous for } t \neq t_k,$$
$$\varphi(t_k^-), \varphi(t_k^+) \in R \text{ and } \varphi(t_k^-) = \varphi(t_k)\}.$$

A function $u(t) \in PC([-\tau, +\infty), R)$ is said to be $\omega$-anti-periodic, if

$$\begin{cases} u(t + \omega) = -u(t), & t \neq t_k, \\ u((t_k + \omega)^+) = -u(t_k^+), & k \in N. \end{cases}$$

For any $u = (u_1, u_2, \cdots, u_n) \in R^n$, its norm is defined by $\|u\|_1 = \max_{i \in \mathcal{I}} |u_i|$.

In addition, in the proof of the main results we shall need the following lemma.

**Lemma 1.** *Let hypotheses* $(H_1) - (H_3)$ *and* $(H_5)$ *are satisfied and suppose there exist $n$ positive constants* $p_1, p_2, \cdots, p_n$ *such that*

$$-\beta_i(t)p_i + \sum_{j=1}^n |c_{ij}(t)|L_j^1 p_j + \sum_{j=1}^n |d_{ij}(t)|L_j^2 p_j + D_i(t) < 0, \qquad (2)$$

*where*

$$D(t) = \sum_{j=1}^n |c_{ij}(t)||f_j(0)| + \sum_{j=1}^n |d_{ij}(t)||g_j(0)| + I_i(t).$$

*Then any solution* $x(t) = (x_1(t), \cdots, x_n(t))^T$ *of system (1) with initial conditions*

$$x_i(s) = \varphi_i(s), \quad |\varphi(s)| < p_i, \quad s \in [-\tau, 0], \quad i \in \mathcal{I}, \qquad (3)$$

*verifies*

$$|x(t)| < p_i, \quad \forall\, t > 0, \quad i \in \mathcal{I}. \qquad (4)$$

*Proof.* For any assigned initial condition, hypotheses $(H_1)$, $(H_2)$ and $(H_3)$ guarantee the existence and uniqueness of $x(t)$, the solution to (1) in $[0, +\infty)$.

Now, we first prove that, for each $i \in \mathcal{I}$,

$$|x_i(t)| < p_i, \; t \in [0, t_1]. \qquad (5)$$

In fact, if it dos not hold, then there exist some $i$ and time $\sigma \in [0, t_1]$ such that

$$|x_i(\sigma)| = p_i, \quad D^+|x_i(t)|_{t=\sigma} \geq 0 \text{ and } |x_j(t)| < p_i, \; \forall t \in [-\tau, \sigma), \; j \in \mathcal{I}. \qquad (6)$$

From $(H_1) - (H_3)$, (2) and (6), we get

$$0 \leq \; a_i(\sigma)\Big[ -\beta(\sigma)|x_i(\sigma)| + \sum_{j=1}^n |c_{ij}(\sigma)||f_j(x_j(\sigma))|$$
$$+ \sum_{j=1}^n |d_{ij}(\sigma)||g_j(x_j(\sigma - \tau_{ij}(\sigma)))| + |I_i(\sigma)|\Big]$$
$$\leq \; a_i(\sigma)\Big[ -\beta_i(\sigma)p_i + \sum_{j=1}^n |c_{ij}(\sigma)|L_j^1 p_j + \sum_{j=1}^n |d_{ij}(\sigma)|L_j^2 p_j + D_i(\sigma)\Big]$$
$$< 0$$

which is a contradiction and shows that (5) is true for $t \in [0, t_1]$.

Since $|x_i(t_1^+)| = |1 + \gamma_{i1}||x_i(t_1)| \le |x_i(t_1)|$, using the same method we can prove that

$$|x_i(t)| < p_i, \quad t \in [t_1, t_2], \quad i \in \mathcal{I},$$

and so on. The proof of Lemma 1 is now completed.    □

## 2    Main Results

In this section, we consider the existence and global exponentially stability of periodic solutions for system (1).

Suppose that $x^*(t) = (x_1^*(t), \cdots, x_n^*(t))^T$ is a solution of system (1) with initial conditions $x_i^*(s) = \varphi_i^*(s)$, $|\varphi_i^*(s)| < p_i$, $s \in [-\tau, 0]$, where $p_i$ are defined in Lemma 1. Then on the stability of system (1), we have a following result.

**Theorem 1.** *Let* $(H_1) - (H_3)$ *and* $(H_5)$ *hold. Assume that following inequality is satisfied*

$$(\lambda - \beta_i(t))p_i + \sum_{j=1}^{n} |c_{ij}(t)| L_j^1 p_j + \sum_{j=1}^{n} |d_{ij}(t)| L_j^2 p_j e^{\lambda \tau} < 0, \tag{7}$$

*where* $\lambda$ *is a positive constant such that* $\left(\lambda - \frac{\ln A}{\kappa}\right) > 0$, $A = \max_{i \in \mathcal{I}} \frac{\overline{a}_i}{\underline{a}_i}$. *Then* $x^*(t)$ *is globally exponentially stable.*

*Proof.* Let $x(t) = (x_1(t), x_2(t), \cdots, x_n(t))^T$ is an arbitrary solutions of system (1) with initial value $\varphi = (\varphi_1(s), \varphi_2(s), \cdots, \varphi_n(s))^T$ and set

$$u_i(t) = \mathrm{sign}(x_i(t) - x_i^*(t)) \int_{x_i^*(t)}^{x_i(t)} \frac{ds}{a_i(s)}. \tag{8}$$

From $(H_1)$, $(H_2)$ and $(H_3)$, for $t \ne t_k$, we have

$$u_i'(t) = \mathrm{sign}(x_i(t) - x_i^*(t)) \Big[ -(b_i(t, x_i(t)) - b_i(t, x_i^*(t))) + \sum_{j=1}^{n} c_{ij}(t) \big(f_j(x_j(t))$$

$$- f_j(x_j^*(t))\big) + \sum_{j=1}^{n} d_{ij}(t) \big(g_j(x_j(t - \tau_{ij}(t))) - g_j(x_j^*(t - \tau_{ij}(t)))\big) \Big]$$

$$\le -\beta_i(t)\underline{a}_i u_i(t) + \sum_{j=1}^{n} |c_{ij}(t)| L_j^1 \overline{a}_j u_j(t) + \sum_{j=1}^{n} |d_{ij}(t)| L_j^2 \overline{a}_j u_j(t - \tau_{ij}(t)).$$

$$\tag{9}$$

Also

$$u_i(t_k^+) \le \left| \int_{x_i^*(t_k^+)}^{x_i(t_k^+)} \frac{ds}{a_i(s)} \right| = \left| \int_{(1+\gamma_{ik})x_i^*(t_k)}^{(1+\gamma_{ik})x_i(t_k)} \frac{ds}{a_i(s)} \right| \le \frac{\overline{a}_i}{\underline{a}_i} |1 + \gamma_{ik}| u_i(t_k). \tag{10}$$

Considering the Lyapunov function

$$V_i(t) = u_i(t)e^{\lambda t}, \quad i \in \mathcal{I}, \tag{11}$$

and for $t \neq t_k$, calculating its upper right derivative along the trajectory of system (1) leads to

$$D^+ V_i(t) \leq \left( \lambda - \beta_i(t)\underline{a_i} \right) V_i(t) + \sum_{j=1}^{n} |c_{ij}(t)| L_j^1 \overline{a}_j V_j(t)$$

$$+ \sum_{j=1}^{n} |d_{ij}(t)| L_j^2 \overline{a}_j e^{\lambda \tau} V_j(t - \tau_{ij}(t)) . \tag{12}$$

Let $m > 1$ be an arbitrary real number such that

$$m p_i > \sup_{s \in [-\tau, 0]} \max_{i \in \mathcal{I}} \frac{|\varphi_i(s) - \varphi_i^*(s)|}{\underline{a_i}}, \quad i \in \mathcal{I} .$$

It follows from (8) and (11) that

$$V_i(t) = u_i(t)e^{\lambda t} < m p_i, \quad t \in [-\tau, 0], \quad i \in \mathcal{I} .$$

Now, we claim that

$$V_i(t) = u_i(t)e^{\lambda t} < A^k m p_i, \quad t \in [t_k, t_{k+1}], \quad i \in \mathcal{I} . \tag{13}$$

First, we prove that

$$V_i(t) = u_i(t)e^{\lambda t} < m p_i, \quad i \in \mathcal{I},$$

is true for $0 < t \leq t_1$. Otherwise, there exist $i \in \mathcal{I}$ and $\eta \in (0, t_1]$ such that

$$V_i(\eta) = m p_i \quad \text{and} \quad V_j(t) < m p_j, \quad t \in [-\tau, \eta), \quad j \in \mathcal{I} . \tag{14}$$

Combining (7), (12) and (14), we get

$$0 \leq D^+ V_i(t)\big|_{t=\eta} \leq m \left[ (\lambda - a_i(\eta)) p_i + \sum_{j=1}^{n} |b_{ij}(\eta)| L_j^1 p_j + \sum_{j=1}^{n} |c_{ij}(\eta)| L_j^2 p_j e^{\lambda \tau} \right] < 0,$$

which is a contradiction. Hence (13) holds for $t \in [-\tau, t_1]$.

From (10) and $|1 + \gamma_{ik}| \leq 1$, we know that

$$V_i(t_1^+) \leq \frac{\overline{a}_i}{\underline{a_i}} |1 + \gamma_{i1}(t_1)| V_i(t_1) \leq A V_i(t_1) \leq A m p_i .$$

Repeating above process, when $t \in [t_k, t_{k+1}]$, we get

$$V_i(t) = u_i(t)\, e^{\lambda t} < A^k m p_i, \quad t \in [t_k, t_{k+1}], \quad i \in \mathcal{I} .$$

Since $\triangle t_k = t_{k+1} - t_k \geq \kappa$ and $A > 1$, when $t \in [t_k, t_{k+1}]$, we get

$$A^k = e^{k \ln A} \leq e^{\left[1 + \frac{(t_2 - t_1) + (t_3 - t_2) + \ldots + (t_k - t_{k-1})}{\kappa}\right] \ln A} = e^{\left(1 + \frac{t_k - t_1}{\kappa}\right) \ln A} \leq M' e^{\frac{t}{\kappa} \ln A},$$

where $M' = e^{\left(1 + \frac{t_1}{\kappa}\right) \ln A}$. From (8) and above inequality, we have

$$|x_i(t) - x_i^*(t)| \leq \bar{a}_i u_i(t) \leq m \bar{a}_i M' p_i e^{-\left(\lambda - \frac{\ln A}{\kappa}\right)t}, \quad \forall t > 0, \ i \in \mathcal{I}. \tag{15}$$

Letting $M > 1$ such that

$$\max_{i \in \mathcal{I}} \{m \bar{a}_i M' p_i\} \leq M \|\varphi - \varphi^*\|_1, \quad i \in \mathcal{I}. \tag{16}$$

Together with (8), (15) and (16), we get

$$\|x(t) - x^*(t)\|_1 = \max_{i \in \mathcal{I}} |x_i(t) - x^*(t)| \leq M \|\varphi - \varphi^*\|_1 e^{-\delta t}, \quad \forall t > 0. \tag{17}$$

where $\delta = \left(\lambda - \frac{\ln A}{\kappa}\right) > 0$. Therefor, according to the definition 1 in [8], $x^*(t)$ is globally exponentially stable. The proof of Theorem 1 is completed. $\qquad \square$

**Theorem 2.** *Suppose that hypotheses* $(H_1) - (H_5)$ *hold. If the inequality (2) is satisfied, then system (1) admits an* $\omega - $ *anti-periodic solution.*

*Proof.* Let $x(t) = (x_1(t), x_2(t), \cdots, x_n(t))^T$ be a solution of system (1) with initial conditions (3). By Lemma 1, we have

$$|x_i(t)| \leq p_i, \quad \forall t \in [-\tau, \infty], \ i \in \mathcal{I}. \tag{18}$$

From hypotheses $(H_4)$, for any $h \in N$, we have

$$\begin{aligned}
\left[(-1)^{h+1} x_i(t + (h+1)\omega)\right]' &= (-1)^{h+1} x'_i(t + (h+1)\omega) \\
&= a_i \left((-1)^{h+1} x_i(t + (h+1)\omega)\right) \left[- b_i \left(t, (-1)^{h+1} x_i(t + (h+1)\omega)\right)\right. \\
&\quad + \sum_{j=1}^n c_{ij}(t) f_j \left((-1)^{h+1} x_j(t + (h+1)\omega)\right) + I_i(t) \\
&\quad + \left. \sum_{j=1}^n d_{ij}(t) g_j \left((-1)^{h+1} x_j \left(t + (h+1)\omega - \tau_{ij}(t)\right)\right)\right].
\end{aligned} \tag{19}$$

Moreover, by hypothesis of $(H_5)$, we have

$$(-1)^{h+1} x_i \left((t_k + (h+1)\omega)^+\right) = (1 + \gamma_{ik}) \left[(-1)^{h+1} x_i (t_k + (h+1)\omega)\right]. \tag{20}$$

Thus, for any natural number $h$, $x_i(t + (h+1)\omega)$ are the solutions of system (1). Therefore, by Theorem 1, there exist constants $M > 1$ and $\lambda > 0$ such that

$$|(-1)^{h+1} x_i (t + (h+1)\omega) - (-1)^h x_i (t + h\omega)| \leq 2M p_i e^{-\lambda(t + h\omega)} \tag{21}$$

and

$$|(-1)^{h+1}x_i\left((t_k + (h+1)\omega)^+\right) - (-1)^h x_i\left((t_k + h\omega)^+\right)| \le 2Mp_i e^{-\lambda(t_k + h\omega)}. \tag{22}$$

Thus, for any natural number m, we obtain

$$
\begin{cases}
|(-1)^{m+1}x_i\left(t + (m+1)\omega\right)| \\
\quad \le |x_i(t)| + \sum_{l=0}^{m}(-1)^{l+1}|x_i\left(t + (l+1)\omega\right) - (-1)^l x_i\left(t + l\omega\right)|, \quad t \ne t_k, \\
|(-1)^{m+1}x_i\left((t_k + (m+1)\omega)^+\right)| \le |(1 + \gamma_{ik})(-1)^{m+1}x_i\left(t_k + (m+1)\omega\right)| \\
\quad \le |x_i\left(t_k + (m+1)\omega\right)|.
\end{cases} \tag{23}
$$

In view of (21), we can choose a sufficiently large constant $N^0$ and a positive constant $\varepsilon$ such that

$$|x_i\left(t + (l+1)\omega\right) - (-1)^l x_i\left(t + l\omega\right)| \le \varepsilon(e^{-\lambda\omega})^l, \quad l > N^0, \tag{24}$$

on any compact set of $R^+$. Together with (23) and (24), it follows that fundamental sequence $\{(-1)^{m+1}x_i\left(t + (m+1)\omega\right)\}$ uniformly converges to a continuous function $x^*(t) = (x_1^*(t), x_2^*(t), \cdots, x_n^*(t))$ on any compact set of $R^+$.

Now we will show that $x^*(t)$ is $\omega$−anti-periodic solution of system (1). It is easily known that $x^*(t)$ is anti-periodic, since

$$
\begin{aligned}
x_i^*(t + \omega) &= \lim_{m \to +\infty}(-1)^m x_i\left(t + \omega + m\omega\right) \\
&= -\lim_{(m+1) \to +\infty}(-1)^{m+1}x_i\left(t + (m+1)\omega\right) = -x_i^*(t), \quad \text{for all} \quad t \ne t_k,
\end{aligned}
$$

and

$$
\begin{aligned}
x_i^*\left((t_k + \omega)^+\right) &= \lim_{m \to +\infty}(-1)^m x_i\left((t_k + \omega + m\omega)^+\right) \\
&= -\lim_{(m+1) \to +\infty}(-1)^{m+1}x_i\left((t + (m+1)\omega)^+\right) = -x_i^*(t_k^+), \quad \text{for any} \quad k \in N.
\end{aligned}
$$

In addition, together with the piece-wise continuity of the right side of (1), (19) and (20) imply that $x^*(t)$ converges uniformly to a piece-wise continuous function on any compact set of $R^+$. Therefore, letting $m \to +\infty$ on both side of (19) and (20), we know that $x^*(t)$ satisfies the impulsive system (1). Thus, $x^*(t)$ is a $\omega$-anti-periodic solution of system (1). The proof of Theorem 2 is now completed                                                                      □

Remark 1. If we let $d_{ij}(t) = 0$ or $g_j(x) = 0$, and $a_i(u) \equiv 1$, $b_i(t, u) \equiv b_i(t)$, then system (1) become

$$
\begin{cases}
x_i'(t) = -b_i(t)x_i(t) + \sum_{j=1}^{n}c_{ij}(t)f_j(x_j(t)) + I_i(t), \quad t \ge 0, \ t \ne t_k, \\
x_i(t_k^+) = (1 + \gamma_{ik})x_i(t_k), \ i \in \mathcal{I}, \quad k \in N,
\end{cases} \tag{25}
$$

which is studied in [12]. It is not difficult to see that Theorem 2 includes the main results in [12] as a special cases.

## 3   Numerical Simulations

For $n = 2$, consider the following delayed Cohen-Grossberg neural networks system

$$\begin{cases} x_i'(t) = a_i(x_i(t))\Big[ - b_i(t, x_i(t)) + \sum_{j=1}^{2} b_{ij}(t) f_j(x_j(t)) \\ \\ \qquad + \sum_{j=1}^{2} c_{ij}(t) g_j(x_j(t - \tau_{ij}(t))) + I_i(t)\Big], \ t \neq t_k, \end{cases} \tag{26}$$

with impulses $\triangle x_1(t_k) = 0.6 x_1(t_k)$, $\triangle x_2(t_k) = 0.4 x_2$, $k \in N$. Where

$$f_i(u) = \tanh(u/2), \quad g_i(u) = \frac{(|u+1|-|u-1|)}{2}, \quad \tau_{i1}(t) = \tau_{i2}(t) = 1, \ i = 1, 2$$

and

$$a_1(u) = 5 - \cos(u), \qquad a_2(u) = 5 + \cos(u), \ b_1(t, u) = (6 + \cos \pi t)u,$$
$$b_2(t, u) = (7 - \cos \pi t)u, \ I_1(t) = 3\sin(\pi t), \quad I_2(t) = 4\sin(\pi t),$$

$$(c_{ij})_{2\times 2} = \begin{pmatrix} 0.9|\sin(\pi t)| & 0.35|\cos(\pi t)| \\ |\cos(\pi t)| & |\cos(\pi t)| \end{pmatrix}, (d_{ij})_{2\times 2} = \begin{pmatrix} 0.7|\cos(\pi t)| & 0.5|\sin(\pi t)| \\ |\cos(\pi t)| & |\cos(\pi t)| \end{pmatrix}.$$

If we let $t_k = k - 0.5$, $\beta_1(t) = (3 + \cos \pi t)$ and $\beta_2(t) = (7 - \cos \pi t)$, then it is not difficult to check that hypothesis $(H_1) - (H_5)$ and inequalities (2) and (7) are all satisfied. Therefore, by Theorem 1 and Theorem 2, the impulsive system (26) has a unique 1-anti-periodic solution which is globally exponentially stable. The fact can be seen by simulation in Figure 1.

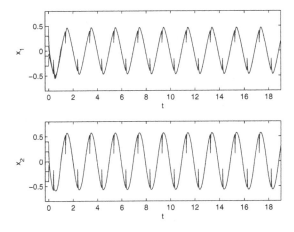

**Fig. 1.** Time response of state variables $x_i$ $(i = 1, 2)$ with impulsive effects

# 4   Conclusion

In this paper, we study the existence and exponential stability of anti-periodic solution of impulsive Cohen-Grossberg neural networks with periodic coefficients and time-varying delays. An examples with numerical simulations is presented to illustrate our results. The results obtained in this paper indicate that suitable impulsive effects can maintain the stability of system.

**Acknowledgments.** This work was supported by the National Natural Science Foundation of People's Republic of China (Grant No. 61164004). The authors are grateful to the Editor and anonymous reviewers for their kind help and constructive comments.

# References

1. Cohen, M.A., Grossberg, S.: Absoulute stability of global pattern formation and parallel memory storage by competetive neural networks. IEEE Trans. Syst. Man Cybern. 13, 815–826 (1983)
2. Arik, S., Orman, Z.: Global stability analysis of Cohen-Grossberg neural networks with time varying delays. Phys. Lett. A 341, 410–421 (2005)
3. Cao, J., Song, Q.: Stability in Cohen-Grossberg-type bidirectional associative memory neural networks with time-varying delays. Nonlinearity 19, 1601–1617 (2006)
4. Song, Q., Cao, J.: Stability analysis of Cohen-Grossberg neural network with both time-varying and continuously distributed delays. J. Comp. Appl. Math. 197, 188–203 (2006)
5. Zhao, H., Wang, L.: Stability and bifurcation for discrete-time Cohen-Grossberg neural network. Appl. Math. Comp. 179, 787–798 (2006)
6. Jiang, H., Cao, J., Teng, Z.: Dynamics of Cohen-Grossberg neural networks with time-varying delays. Phys. Lett. A 354, 414–422 (2006)
7. Xiang, H., Cao, J.: Almost periodic solution of Cohen-Grossberg neural networks with bounded and unbounded delays. Nonlinear Anal. RWA 10, 2407–2419 (2009)
8. Li, B., Xu, D.: Existence and exponential stability of periodic solution for impulsive Cohen-Grossberg neural networks with time-varying delays. Appl. Math. Comput. 219, 2506–2520 (2012)
9. Yang, X.: Existence and global exponential stability of periodic solution for Cohen-Grossberg shunting inhibitory cellular neural networks with delays and impulses. Neurocomputing 72, 2219–2226 (2009)
10. Li, K.: Stability analysis for impulsive Cohen-Grossberg neural networks with time-varying delays and distributed delays. Nonlinear Anal. RWA 10, 2784–2798 (2009)
11. Li, J., Yan, J., Jia, X.: Dynamical analysis of Cohen-Grossberg neural networks with time-delays and impulses. Comput. Math. Appl. 58, 1142–1151 (2009)
12. Shi, P., Dong, L.: Existence and exponential stability of anti-periodic solutions of Hopfield neural networks with impulses. Appl. Math. Comput. 216, 623–630 (2010)
13. Pan, L., Cao, J.: Anti-periodic solution for delayed cellular neural networks with impulsive effects. Nonlinear Anal. RWA 12, 3014–3027 (2011)
14. Shao, J.: An anti-periodic solution for a class of recurrent neural networks. J. Comput. Appl. Math. 228, 231–237 (2009)
15. Chen, Y., Nieto, J.J., Oregan, D.: Anti-periodic solutions for fully nonlinear first-order differential equations. Math. Comput. Model. 46, 1183–1190 (2007)

# Global Robust Exponential Stability in Lagrange Sense for Interval Delayed Neural Networks

Xiaohong Wang[1,*], Xingjun Chen[2], and Huan Qi[1]

[1] Department of Control Science and Engineering, Huazhong University of Science and Technology, Wuhan, 430074, China
{wxhong2006,xjchenmail}@163.com
[2] Department of Sciences Research, Dalian Naval Academy, Dalian, 116018, China

**Abstract.** The problem of global robust exponential stability in Lagrange sense for the interval delayed neural networks (IDNNs) with general activation functions is investigated. Based on the Lyapunov stability, a differential inequality and linear matrix inequalities (LMIs) technique, some conditions to guarantee the IDNNs global exponential stability in Lagrange sense are provided. Meanwhile, the specific estimation of globally exponentially attractive sets of the addressed system are also derived. Finally, a numerical example is provided to illustrate the effectiveness of the method proposed.

**Keywords:** Interval neural networks, Lagrange stability, Globally exponentially attractive set, Linear Matrix Inequality, Infinite distributed delays.

## 1 Introduction

In recent years, the problem of Lagrange stability has been widely paid attention because it refers to the stability of the total system, which is not only distinctly different from Lyapunov stability, but also include Lyapunov stability as a special case by regarding an equbrium point as an attractive set [1]. Generally speaking, the goal of study on global exponential Lagrange stability for neural networks is to determine globally exponentially attractive sets. Therefore, many initial findings on the global Lagrange stability analysis of neural networks have been reported [2-7]. For instance, in [2], the authors studied globally exponentially stability in Lagrange sense for continuous recurrent neural network with three different types of activation functions and constant time delays. Nextly, [3] and [4] continued to research the Lagrange stability for neutral type and periodic recurrent neural networks. Furthermore, [5] and [6] also investigated Lagrange stability for Cohen-Grossberg neural networks with mixed delays. [7] made use of the Linear Matrix inequality technique to further study the global exponential stability in Lagrange sense for recurrent neural networks. Even so, up to now, there is not existing work on interval neural networks with infinite

---

* Corresponding author.

C. Guo, Z.-G. Hou, and Z. Zeng (Eds.): ISNN 2013, Part I, LNCS 7951, pp. 239–249, 2013.

distributed delay, particularly made on it by means of LMIs [8]. Hence, this gives the motivation of our present investigation.

In this paper, we focus on the problem of global robust exponential stability in Lagrange sense for a class of interval neural networks with general activation functions and mixed delays. In the next section, some preliminaries, including some definitions, assumptions and significant lemmas will be described. Section 3 will state the main results. Section 4 will present a illustrative example to verify the main results and finally a summery will be given in Section 5.

***Notations.*** Throughout this paper, $I$ represents the unit matrix; $R^+ = [0, \infty)$, the symbols $R^n$ and $R^{n \times m}$ stand, respectively, for the $n$-dimensional Euclidean space and the set of all $n \times m$ real matrices. $A^T$ and $A^{-1}$ denote the matrix transpose and matrix inverse. $A > 0$ or $A < 0$ denotes that the matrix $A$ is a symmetric and positive definite or negative definite matrix. Meanwhile, $A < B$ indicates $A - B < 0$ and $\| * \|$ is the Euclidean vector norm. When $x$ is a variable, $\|x\| = \sum_{i=1}^{n} |x_i|$. $\Lambda = \{1, 2, \cdots, n\}$. Moreover, in symmetric block matrices, we use as an ellipsis for the terms that are introduced by symmetry. we use an asterisk " $*$ " to represent a term that is induced by symmetry and diag$\{\cdots\}$ stands for a block-diagonal matrix.

## 2   Problem Statement

The interval neural networks with infinite distributed delays is described by the following equation group:

$$\dot{x}(t) = -Dx(t) + Ag(x(t)) + Bg(x(t - \tau(t)))$$
$$+C \int_{-\infty}^{t} h(t - s)g(x(s))ds + U, \qquad (2.1)$$

where $x(t) = (x_1(t), \ldots, x_n(t))^T$ is the neuron state vector of the neural network; $U = (U_1, \cdots, U_n)^T$ is an external input; $\tau(t)$ is the transmission delay of the neural networks, which is time-varying and satisfies $0 \leq \tau(t) \leq \tau$, where $\tau$ is a positive constant; $g(x(\cdot)) = (g_1(x_1(\cdot)), \cdots, g_n(x_n(\cdot)))^T$ represents the neuron activation function and $h(\cdot) = \text{diag}\{h_1(\cdot), \cdots, h_n(\cdot)\}$ represents the delay kernel function. The matrices $D = \text{diag}\{d_1, \cdots, d_n\} > 0$ is a positive diagonal matrix, $A = (a_{ij})_{n \times n}, B = (b_{ij})_{n \times n}$, and $C = (c_{ij})_{n \times n}$ are the connection weight matrix, delayed weight matrix and distributively delayed connection weight matrix, respectively. In electronic implementation of neural networks, the values of the constant and weight coefficients depend on the resistance and capacitance values which are subject to uncertainties. This may lead some deviations in the values of $d_i, a_{ij}, b_{ij}$, and $c_{ij}$. Since these deviations are bounded in practice, the quantities $d_i, a_{ij}, b_{ij}$, and $c_{ij}$ may be described with interval numbers as follows:

$$D \in [\underline{D}, \overline{D}], A \in [\underline{A}, \overline{A}], B \in [\underline{B}, \overline{B}], C \in [\underline{C}, \overline{C}]$$

where $\underline{D} = \text{diag}\{\underline{d_1}, \ldots, \underline{d_n}\} > 0$, $\overline{D} = \text{diag}\{\overline{d_1}, \ldots, \overline{d_n}\}$, $\underline{A} = (\underline{a_{ij}})_{n \times n}$, $\overline{A} = (\overline{a_{ij}})_{n \times n}$, $\underline{B} = (\underline{a_{ij}})_{n \times n}$, $\overline{B} = (\overline{a_{ij}})_{n \times n}$, $\underline{C} = (\underline{a_{ij}})_{n \times n}$, $\overline{C} = (\overline{a_{ij}})_{n \times n}$. In addition, let $\Omega := \{\mathrm{diag}(\delta_{11},\cdots,\delta_{1n},\cdots,\delta_{n1},\cdots,\delta_{nn}) \in R^{n^2 \times n^2} || \ \delta_{ij}| \leq 1, i,j \in \Lambda\}$,

$D_0 = \frac{\overline{D}+\underline{D}}{2}, A_0 = \frac{\overline{A}+\underline{A}}{2}, B_0 = \frac{\overline{B}+\underline{B}}{2}, C_0 = \frac{\overline{C}+\underline{C}}{2}$,

$(\alpha_{ij})_{n\times n} = \frac{\overline{D}-\underline{D}}{2}, (\beta_{ij})_{n\times n} = \frac{\overline{A}-\underline{A}}{2}, (\gamma_{ij})_{n\times n} = \frac{\overline{B}-\underline{B}}{2}, (\vartheta_{ij})_{n\times n} = \frac{\overline{C}-\underline{C}}{2}$,

$M_1 = [\sqrt{\alpha_{11}}e_1,\cdots,\sqrt{\alpha_{1n}}e_1,\cdots,\sqrt{\alpha_{n1}}e_n,\cdots,\sqrt{\alpha_{nn}}e_n]_{n\times n^2}$,

$M_2 = [\sqrt{\beta_{11}}e_1,\cdots,\sqrt{\beta_{1n}}e_1,\cdots,\sqrt{\beta_{n1}}e_n,\cdots,\sqrt{\beta_{nn}}e_n]_{n\times n^2}$,

$M_3 = [\sqrt{\gamma_{11}}e_1,\cdots,\sqrt{\gamma_{1n}}e_1,\cdots,\sqrt{\gamma_{n1}}e_n,\cdots,\sqrt{\gamma_{nn}}e_n]_{n\times n^2}$,

$M_4 = [\sqrt{\vartheta_{11}}e_1,\cdots,\sqrt{\vartheta_{1n}}e_1,\cdots,\sqrt{\vartheta_{n1}}e_n,\cdots,\sqrt{\vartheta_{nn}}e_n]_{n\times n^2}$,

$J_1 = [\sqrt{\alpha_{11}}e_1,\cdots,\sqrt{\alpha_{1n}}e_n,\cdots,\sqrt{\alpha_{n1}}e_1,\cdots,\sqrt{\alpha_{nn}}e_n]_{n^2\times n}$,

$J_2 = [\sqrt{\beta_{11}}e_1,\cdots,\sqrt{\beta_{1n}}e_n,\cdots,\sqrt{\beta_{n1}}e_1,\cdots,\sqrt{\beta_{nn}}e_n]_{n^2\times n}$,

$J_3 = [\sqrt{\gamma_{11}}e_1,\cdots,\sqrt{\gamma_{1n}}e_n,\cdots,\sqrt{\gamma_{n1}}e_1,\cdots,\sqrt{\gamma_{nn}}e_n]_{n^2\times n}$,

$J_4 = [\sqrt{\vartheta_{11}}e_1,\cdots,\sqrt{\vartheta_{1n}}e_n,\cdots,\sqrt{\vartheta_{n1}}e_1,\cdots,\sqrt{\vartheta_{nn}}e_n]_{n^2\times n}$,

where $e_i \in R^n$ denotes the column vector with $i$th element to be 1 and others to be 0.

By some simple calculations, one can transform system (2.1) into the following form:

$$\dot{x}(t) = -[D_0 + M_1\Omega_1 J_1]x(t) + [A_0 + M_2\Omega_2 J_2]g(x(t))$$

$$+[B_0 + M_3\Omega_3 J_3]g(x(t-\tau(t))) + [C_0 + M_4\Omega_4 J_4]\int_{-\infty}^{t} h(t-s)g(x(s))ds + U,$$

or equivalently,

$$\dot{x}(t) = -D_0 x(t) + A_0 g(x(t)) + B_0 g(x(t-\tau(t)))$$

$$+C_0 \int_{-\infty}^{t} h(t-s)g(x(s))ds + \mathbb{M}\Psi(t) + U, \qquad (2.2)$$

where $\mathbb{M} = [M_1, M_2, M_3, M_4]_{n\times 4n^2}$, $\Omega_i \in \Omega, i = 1,2,3,4$,

$$\Psi(t) = \begin{pmatrix} \Omega_1 J_1 & 0 & 0 & 0 \\ 0 & \Omega_2 J_2 & 0 & 0 \\ 0 & 0 & \Omega_3 J_3 & 0 \\ 0 & 0 & 0 & \Omega_4 J_4 \end{pmatrix} \times \begin{pmatrix} x(t) \\ g(x(t)) \\ g(x(t-\tau(t))) \\ \int_{-\infty}^{t} h(t-s)g(x(s))ds \end{pmatrix}.$$

In this paper, the system (2.1) is supplemented the initial condition given by $x(s) = \phi(s), s \in (-\infty, 0], i = 1,\cdots,n$, where $\phi(\cdot) \in \mathcal{C}$, $\mathcal{C}$ denotes real-valued continuous functions defined on $(-\infty, 0]$. Here, it is assumed that for any initial condition $\phi(\cdot) \in \mathcal{C}$, there exists at least on solution of model (2.1). As usual, we will also assume that $g(0) = 0$ for all $t \in R^+$ in this paper.

For further discussion, the following assumptions and lemmas are needed:

**(A1).** The activation function $g$ satisfies $g(0) = 0$, and

$$l_j^- \leq \frac{g_j(x) - g_j(y)}{x - y} \leq l_j^+,$$

for all $x \neq y, x, y \in R$, where $l_j^+$ and $l_j^-, j \in \Lambda$, are some real constants.

**(A2).** The delay kernels $h_j(t)$, $j \in \Lambda$ are some real value nonnegative continuous function defined in $(-\infty, 0]$ and satisfy $h_j(t) \leq \tilde{h}(t)$, $j \in \Lambda$,

$$\int_0^\infty h_j(t)\mathrm{d}t = 1, \quad \int_0^\infty \tilde{h}(t)\mathrm{d}t = \tilde{h}, \quad \int_0^\infty \tilde{h}(t)e^{\varrho t}\mathrm{d}t \doteq \tilde{h}^\star < \infty,$$

in which $\tilde{h}(t)$ corresponds to some nonnegative function defined in $(-\infty, 0]$, constants $\varrho, \tilde{h}, \tilde{h}^\star$ are some positive numbers.

**Definition 1.** [2] The neural network defined by (2.1) or (2.2) is globally exponentially stable in Lagrange sense, if there exists a radially unbounded and positive definite Lyapunov function $V(x(t))$, which satisfies $V(x(t)) \geq \|x\|^\alpha$, where $\alpha > 0$ is a constant, and constants $\zeta > 0, \beta > 0$, such that for $V(x(t_0)) > \zeta, \bar{V}(x(t)) > \zeta, t \geq t_0$, the inequality $V(x(t)) - \zeta \leq (\bar{V}(x(t_0)) - \zeta) \exp\{-\beta(t - t_0)\}$ always holds. And $\{x | V(x(t)) \leq \zeta\}$ is said to be a globally exponentially attractive set of (2.1) or (2.2), where $\bar{V}(x(t_0)) \geq V(x(t_0))$ and $V(x(t_0))$ is a constant.

**Lemma 1.** [9] For any vectors $a, b \in R^n$, the inequality $\pm 2a^T b \leq a^T X a + b^T X^{-1} b$ holds, in which $X$ is any $n \times n$ matrix with $X > 0$.

**Lemma 2.** [10] For a given matrix $S = \begin{pmatrix} S_{11} & S_{12} \\ S_{12}^T & S_{22} \end{pmatrix}$, with $S_{11} = S_{11}^T$, $S_{22} = S_{22}^T$, then the following conditions are equivalent:

$(1) S < 0, (2) S_{22} < 0, S_{11} - S_{12}S_{22}^{-1}S_{12}^T < 0, (3) S_{11} < 0, S_{22} - S_{12}^T S_{11}^{-1} S_{12} < 0.$

**Lemma 3.** Let $p, q, r$ and $\tau$ denote nonnegative constants, and function $f \in C(R, R^+)$ satisfies the scalar differential inequality

$$D^+ f(t) \leq -pf(t) + q \sup_{t-\tau \leq s \leq t} f(s) + r \int_0^{+\infty} k(s) f(t - s)\mathrm{d}s, \quad t \geq t_0, \quad (2.3)$$

where $k(\cdot) \in C([0, +\infty), R^+)$ satisfies $\int_0^{+\infty} k(s)e^{\eta_0 s}\mathrm{d}s < \infty$ for some positive constant $\eta_0 > 0$. Assume that

$$p > q + r \int_0^{+\infty} k(s)\mathrm{d}s, \tag{2.4}$$

then $f(t) \leq \bar{f}(t_0)\exp(-\lambda(t - t_0))$ for all $t \geq t_0$, where $\bar{f}(t_0) = \sup_{-\infty \leq s \leq t_0} f(s)$, and $\lambda \in (0, \eta_0)$ satisfies the inequality

$$\lambda < p - qe^{\lambda \tau} - r \int_0^{+\infty} k(s)e^{\lambda s}\mathrm{d}s. \tag{2.5}$$

**Proof:** Let $\sigma = +\infty, a_k = 0, b_k = 0, M = 1, \eta = 0$ in [11, Lemma 2.1], then we can obtain above result easily.

## 3   Main Results

In this part, sufficient conditions for global robust exponential stability in Lagrange sense of (2.1) or (2.2) are got by using Lyapunov functions and inequality techniques.

**Theorem 1.**   Assume that assumptions (A1)-(A2) hold, if there exist three constants $\beta_i > 0, i = 1, 2, 3$, seven positive diagonal matrices $Q_1, Q_2, Q_3 \in R^{n \times n}$, $Q_4, Q_5, Q_6, Q_7 \in R^{n^2 \times n^2}$ and two positive definite matrices $P, Q_8 \in R^{n \times n}$ such that the following inequalities hold:

$$
\left.
\begin{pmatrix}
\Pi_1 & PA_0 & PB_0 & PC_0 & PM & P \\
* & Q_1 & 0 & 0 & 0 & 0 \\
* & * & Q_2 & 0 & 0 & 0 \\
* & * & * & Q_3 & 0 & 0 \\
* & * & * & * & \Pi_2 = \begin{pmatrix} Q_4 & 0 & 0 & 0 \\ 0 & Q_5 & 0 & 0 \\ 0 & 0 & Q_6 & 0 \\ 0 & 0 & 0 & Q_7 \end{pmatrix} & 0 \\
* & * & * & * & * & Q_8
\end{pmatrix} \geq 0 \\
\beta_2 P \geq WQ_2W + WJ_3^T Q_6 J_3 W \\
\beta_3 P \geq WQ_3W + WJ_4^T Q_7 J_4 W \\
\beta_1 > \beta_2 + \beta_3 \tilde{h}
\end{array}
\right\},(3.6)
$$

where $\Pi_1 = PD_0 + D_0 P - WQ_1W - J_1^T Q_4 J_1 - WJ_2^T Q_5 J_2 W - \beta_1 P$, $W = \text{diag}\{w_1, \cdots, w_n\}$, $w_j = \max\{|l_j^-|, |l_j^+|\}$. Then the IDNNs defined by (2.1) or (2.2) is globally robustly exponentially stable in Lagrange sense, and the set $\Phi = \{x \in R^n \mid x^T(t)Px(t) \leq \frac{U^T Q_8 U}{\beta_1 - \beta_2 - \beta_3 \tilde{h}}\}$ is a globally exponentially attractive set.

**Proof:** Now, we consider the following Lyapunov function

$$V(x(t)) = x^T(t)Px(t). \tag{3.7}$$

Calculating the derivative of $V(x(t))$ along the trajectories of (2.2), we can obtain

$$\frac{dV(x(t))}{dt}\Big|_{(2.2)} \leq 2x^T(t)P\left[-D_0 x(t) + A_0 g(x(t))\right.$$
$$\left. + B_0 g(x(t - \tau(t))) + C_0 \int_{-\infty}^{t} h(l - s)g(x(s))ds + M\Psi(t) + U\right]. \tag{3.8}$$

From Assumption (A1) and Lemma 1, we know that there exist three positive diagonal matrices $Q_1, Q_2, Q_3 \in R^{n \times n}$ and a positive definite matrix $Q_8 \in R^{n \times n}$ such that the following inequalities hold

$$2x^T(t)PA_0 g(x(t))) \leq x^T(t)PA_0 Q_1^{-1} A_0^T Px(t) + g^T(x(t))Q_1 g(x(t))$$
$$\leq x^T(t)PA_0 Q_1^{-1} A_0^T Px(t) + x^T(t)WQ_1 Wx(t), \tag{3.9}$$

$$2x^T(t)PB_0g(x(t-\tau(t))))$$
$$\leq x^T(t)PB_0Q_2^{-1}B_0^T Px(t) + g^T(x(t-\tau(t)))Q_2g(x(t-\tau(t)))$$
$$\leq x^T(t)PB_0Q_2^{-1}B_0^T Px(t) + x^T(t-\tau(t))WQ_2Wx(t-\tau(t)). \qquad (3.10)$$

$$2x^T(t)PC_0\int_{-\infty}^{t} h(t-s)g(x(s))ds \leq x^T(t)PC_0Q_3^{-1}C_0^T Px(t)$$
$$+\left(\int_{-\infty}^{t} h(t-s)g(x(s))ds\right)^T Q_3\left(\int_{-\infty}^{t} h(t-s)g(x(s))ds\right),$$

and by well-know Cauchy-Schwarz inequality and assumption (A2), we get

$$\left(\int_{-\infty}^{t} h(t-s)g(x(s))ds\right)^T Q_3\left(\int_{-\infty}^{t} h(t-s)g(x(s))ds\right)$$
$$=\sum_{j=1}^{n} q_j^{(3)}\left(\int_0^{+\infty} h_j(u)g_j(x_j(t-u))du\right)^2$$
$$\leq \sum_{j=1}^{n} q_j^{(3)}\int_0^{+\infty} h_j(u)du\int_0^{+\infty} h_j(u)g_j^2(x_j(t-u))du$$
$$\leq \sum_{j=1}^{n} q_j^{(3)}\int_0^{+\infty} \tilde{h}(u)w_j^2x_j^2(t-u)du = \int_0^{+\infty} \tilde{h}(u)\sum_{j=1}^{n} q_j^{(3)}w_j^2x_j^2(t-u)du$$
$$=\int_0^{+\infty} \tilde{h}(s)x^T(t-s)WQ_3Wx(t-s)ds,$$

which implies that

$$2x^T(t)PC_0\int_{-\infty}^{t} h(t-s)g(x(s))ds \leq x^T(t)PC_0Q_3^{-1}C_0^T Px(t) +$$
$$+\int_0^{+\infty} \tilde{h}(s)x^T(t-s)WQ_3Wx(t-s)ds. \qquad (3.11)$$

$$2x^T(t)PU \leq x^T(t)PQ_8^{-1}Px(t) + U^T Q_8 U. \qquad (3.12)$$

In view of the definition of $\Omega$, we have the following inequality:

$$\Psi^T(t)\Psi(t) \leq x^T(t)J_1^T J_1x(t) + g^T(x(t))J_2^T J_2g(x(t))$$
$$+g^T(x(t-\tau(t)))J_3^T J_3g(x(t-\tau(t)))$$
$$+\left(\int_{-\infty}^{t} h(t-s)g(x(s))ds\right)^T J_4^T J_4\left(\int_{-\infty}^{t} h(t-s)g(x(s))ds\right). \qquad (3.13)$$

Considering Lemmas 1 and (3.13), we derive

$$2x^T(t)PM\Psi(t) \leq x^T(t)PM\Pi_2^{-1}M^T Px(t) + \Psi^T(t)\Pi_2\Psi(t)$$

$$\leq x^T(t)\{PM\Pi_2^{-1}M^TP + J_1^TQ_4J_1 + WJ_2^TQ_5J_2W\}x(t)$$
$$+x^T(t-\tau(t))[WJ_3^TQ_6J_3W]x(t-\tau(t))$$
$$+\int_0^{+\infty}\tilde{h}(s)x^T(t-s)[WJ_4^TQ_7J_4W]x(t-s)ds. \tag{3.14}$$

Now, adding the terms on the right of (3.9)-(3.12) and (3.14) to (3.8), and considering conditions (3.7), also making use of Lemma 2, we can obtain that

$$\frac{dV(x(t))}{dt}\Big|_{(2.2)} \leq x^T(t)\{-PD_0 - D_0P + PA_0Q_1^{-1}A_0^TP + WQ_1W$$
$$+PB_0Q_2^{-1}B_0^TP + PC_0Q_3^{-1}C_0^TP + PQ_8^{-1}P + PM\Pi_2^{-1}M^TP + J_1^TQ_4J_1$$
$$+WJ_2^TQ_5J_2W\}x(t) + x^T(t-\tau(t))[WQ_2W + WJ_3^TQ_6J_3W]x(t-\tau(t))$$
$$+\int_0^{+\infty}\tilde{h}(s)x^T(t-s)[WQ_3W + WJ_4^TQ_7J_4W]x(t-s)ds + U^TQ_8U$$
$$\leq -\beta_1 x^T(t)Px(t) + \beta_2 x^T(t-\tau(t))Px(t-\tau(t))$$
$$+\beta_3\int_0^{+\infty}\tilde{h}(s)x^T(t-s)Px(t-s)ds + U^TQ_8U$$
$$\leq -\beta_1 V(x(t)) + \beta_2 V(x^T(t-\tau(t)))$$
$$+\beta_3\int_0^{+\infty}\tilde{h}(s)V(x(t-s))ds + U^TQ_8U. \tag{3.15}$$

Transforming (3.15) into the following inequality, we get

$$\frac{d(V(x(t))-\eta)}{dt}\Big|_{(2.2)} \leq -\beta_1(V(x(t))-\eta) + \beta_2\left(\sup_{t-\tau\leq s\leq t} V(x(s))-\eta\right)$$
$$+\beta_3\int_0^{+\infty}\tilde{h}(s)(V(x(t-s))-\eta)ds, \quad t\geq t_0, \tag{3.16}$$

where $\eta = \frac{U^T HU}{\beta_1 - \beta_2 - \beta_3\tilde{h}}$.

From the formula (3.16), we can know that it satisfies the (2.3) of Lemma 3. Meanwhile, noticed assumption (A2), it can be deduced that $\beta_1 > \beta_2 + \beta_3\tilde{h} \Leftrightarrow \beta_1 > \beta_2 + \beta_3\int_0^{+\infty}\tilde{h}(s)ds$. So the (2.4) of Lemma 3 is also satisfied. From this, when $V(x(t)) > \eta$, $\sup_{t-\tau\leq s\leq t} V(x(s)) > \eta$, and $\sup_{-\infty\leq s\leq t} V(x(s)) > \eta$, according to Lemma 3, we are able to derive

$$V(x(t)) - \eta \leq (\bar{V}(x(t))-\eta)\exp(-\lambda(t-t_0)),$$

where $\bar{V}(x(t)) = \sup_{-\infty\leq s\leq t} V(x(s))$, $\lambda \in (0,\varrho)$ satisfies

$$\lambda < \beta_1 - \beta_2 e^{\lambda\tau} - \beta_3\int_0^{+\infty}\tilde{h}(s)e^{\lambda s}ds. \tag{3.17}$$

Simultaneously, judging by [12], it is easy to prove that there exists a constant $\alpha$ such that $V(x(t)) \geq \|x\|^\alpha$. In terms of Definition 1, we know that the IDNNs defined by (2.1) or (2.2) is globally robustly exponentially stable in Lagrange sense and $\Phi = \{x \in R^n \mid x^T(t)Px(t) \le \frac{U^T HU}{\beta_1-\beta_2-\beta_3\tilde{h}}\}$ is a globally exponentially attractive set of (2.1) or (2.2). Hence, the proof of Theorem 1 is completed.

*Remark 1:* It should be noted that the exponential convergence rate $\lambda$ of IDNNs (2.1) or (2.2) is also derived in (3.17). Moreover, one may find that condition $\beta_1 - \beta_2 - \beta_3\tilde{h} > 0$ implies that there exists constant $\lambda \in (0,\varrho)$ such that (2.1) or (2.2) holds for any given $\tau > 0$.

In the IDNNs system (2.1) or (2.2), when getting rid of the term of infinite distributed delay $\int_{-\infty}^t h(t-s)g(x(s))\,ds$, we get the following Corollary 1.

**Corollary 1:**  Assume that assumptions (A1) holds, if there exist three constants $\beta_i > 0, i = 1,2$, five positive diagonal matrices $Q_1, Q_2 \in R^{n\times n}$, $Q_3, Q_4$, $Q_5 \in R^{n^2\times n^2}$ and two positive definite matrices $P, Q_6 \in R^{n\times n}$ such that the following inequalities hold:

$$\left.\begin{pmatrix} \Pi_1 & PA_0 & PB_0 & & PM & & P \\ * & Q_1 & 0 & & 0 & & 0 \\ * & * & Q_2 & & 0 & & 0 \\ * & * & * & \Pi_3 = \begin{pmatrix} Q_3 & 0 & 0 \\ 0 & Q_4 & 0 \\ 0 & 0 & Q_5 \end{pmatrix} & & 0 \\ * & * & * & & * & & Q_6 \\ & & & \beta_2 P \ge WQ_2W + WJ_3^T Q_5 J_3 W \\ & & & \beta_1 > \beta_2 \end{pmatrix} \ge 0\right\}, \qquad (3.18)$$

where $\Pi_1 = PD_0 + D_0P - WQ_1W - J_1^T Q_3 J_1 - WJ_2^T Q_4 J_2 W - \beta_1 P$, $W = \text{diag}\{w_1, \cdots, w_n\}$, $w_j = \max\{|l_j^-|, |l_j^+|\}$. Then the IDNNs defined by (2.1) or (2.2) is globally robustly exponentially stable in Lagrange sense and the set $\Phi = \{x \in R^n \mid x^T(t)Px(t) \le \frac{U^T Q_8 U}{\beta_1-\beta_2}\}$ is a globally exponentially attractive set.

**Proof:** In front of the course of proof is almost parallel to that of Theorem 1, except for the inequality (3.11) in Theorem 1, here no longer say. In the end, We can also obtain

$$\frac{d(V(x(t)) - \eta)}{dt} \le -\beta_1(V(x(t)) - \eta) + \beta_2(\bar{V}(x(t)) - \eta), \quad t \ge t_0,$$

where $\eta = \frac{U^T HU}{\beta_1-\beta_2}$, $\bar{V}(x(t)) = \sup_{t-\tau \le s \le t} V(x(s))$.
It is noticed that $\beta_1 > \beta_2$, hence according to the famous Halanay Inequality[13], when $V(x(t)) > \eta$ and $\sup_{t-\tau \le s \le t} V(x(s)) > \eta$, we are able to derive

$$V(x(t)) - \eta \le (\bar{V}(x(t)) - \eta)\exp(-\lambda(t - t_0)),$$

where $\lambda$ is the unique positive root of $\lambda = \beta_1 - \beta_2 e^{\lambda\tau}$. Similarly, it is obtained that $\Omega = \{x \in R^n \mid x^T(t)Px(t) \le \frac{U^T HU}{\beta_1-\beta_2}\}$ is a positive invariant and globally exponentially attractive set of system (2.1). Hence, the proof is gained.

*Remark 2:* For the Lagrange stability condition given in [2-7], the time delay are constant delays or time-varying delay that are differentiable such that their derivatives are not greater than one or finite. Note that in this paper we do not impose those restrictions on our time-varying delays, which means that our presented results have wider application range.

## 4  Simulation Example

In this section, we give a numerical example to illustrate the theoretical results.
*Example 1:* Consider the interval neural networks model (2.1) with the following parameters:

$$\tau(t) = 0.3 + 0.5|\sin(t)|, U = (1\ \ 1)^T, \overline{D} = \begin{pmatrix} 7.8 & 0 \\ 0 & 9.6 \end{pmatrix}, \underline{D} = \begin{pmatrix} 6.2 & 0 \\ 0 & 8.4 \end{pmatrix},$$

$$\overline{A} = \begin{pmatrix} 1.6 & 0.1 \\ 0.9 & 1.2 \end{pmatrix}, \underline{A} = \begin{pmatrix} 0 & -0.5 \\ 0.1 & 0 \end{pmatrix}, \overline{B} = \begin{pmatrix} 2 & 0.4 \\ -0.3 & 1.6 \end{pmatrix}, \underline{B} = \begin{pmatrix} 0.4 & -0.2 \\ -1.1 & 0.4 \end{pmatrix},$$

$$\overline{C} = \begin{pmatrix} 0.2 & 0.8 \\ 0 & 1.2 \end{pmatrix}, \underline{C} = \begin{pmatrix} -1.4 & 0.2 \\ -0.8 & 0 \end{pmatrix}.$$

And the delay kernel $h(s)$ is elected as $h_i(s) = e^{-s}$ for $s \in [0, +\infty), i = 1, 2$.
In this case, by simple calculation, it can be obtained that

$$D_0 = \frac{\overline{D}+\underline{D}}{2} = \begin{pmatrix} 7 & 0 \\ 0 & 9 \end{pmatrix}, A_0 = \frac{\overline{A}+\underline{A}}{2} = \begin{pmatrix} 0.8 & -0.2 \\ 0.5 & 0.6 \end{pmatrix}, B_0 = \frac{\overline{B}+\underline{B}}{2} = \begin{pmatrix} 1.2 & 0.1 \\ -0.7 & 1 \end{pmatrix},$$

$$C_0 = \frac{\overline{C}+\underline{C}}{2} = \begin{pmatrix} -0.6 & 0.5 \\ -0.4 & 0.6 \end{pmatrix}, (\alpha_{ij})_{2\times2} = \frac{\overline{D}-\underline{D}}{2} = \begin{pmatrix} 0.8 & 0 \\ 0 & 0.6 \end{pmatrix},$$

$$(\beta_{ij})_{2\times2} = \frac{\overline{A}-\underline{A}}{2} = \begin{pmatrix} 0.8 & 0.3 \\ 0.4 & 0.6 \end{pmatrix}, (\gamma_{ij})_{2\times2} = \frac{\overline{B}-\underline{B}}{2} = \begin{pmatrix} 0.8 & 0.3 \\ 0.4 & 0.6 \end{pmatrix},$$

$$(\upsilon_{ij})_{2\times2} = \frac{\overline{C}-\underline{C}}{2} = \begin{pmatrix} 0.8 & 0.3 \\ 0.4 & 0.6 \end{pmatrix}, M_1 = \begin{bmatrix} \sqrt{0.8} & 0 & 0 & 0 \\ 0 & 0 & 0 & \sqrt{0.6} \end{bmatrix},$$

$$M_2 = M_3 = M_4 = \begin{bmatrix} \sqrt{0.8} & \sqrt{0.3} & 0 & 0 \\ 0 & 0 & \sqrt{0.4} & \sqrt{0.6} \end{bmatrix},$$

$$J_1 = \begin{bmatrix} \sqrt{0.8} & 0 \\ 0 & 0 \\ 0 & 0 \\ 0 & \sqrt{0.6} \end{bmatrix}, J_2 = J_3 = J_4 = \begin{bmatrix} \sqrt{0.8} & 0 \\ \sqrt{0.3} & 0 \\ 0 & \sqrt{0.4} \\ 0 & \sqrt{0.6} \end{bmatrix}.$$

Clearly, $\tilde{h} = 1, \tau = 0.8$ and we choose $\varrho = 0.8 < 1$. In addition, let $g(x(t)) = \frac{1}{4}(|x+1| - |x-1|)$, the activation function satisfies the assumption (A1) with $L = \text{diag}\{-0.5, -0.5\}, F = W = \text{diag}\{0.5, 0.5\}$. In this case, we choose $\beta_1 = 1, \beta_2 = 0.5, \beta_3 = 0.4$. Obviously, it satisfies the condition $\beta_2 + \beta_3\tilde{h} < \beta_1$. Note that $\mathbb{M} = [M_1, M_2, M_3, M_4]$ and solving the LMIs in Theorem 1 using the Matlab LMI Control Toolbox, we obtain the feasible solutions as follows:

$$P = \begin{pmatrix} 3.907 & -0.044 \\ -0.044 & 5.761 \end{pmatrix}, Q_1 = \begin{pmatrix} 24.13 & 0 \\ 0 & 44.59 \end{pmatrix}, Q_2 = \begin{pmatrix} 1.601 & 0 \\ 0 & 1.293 \end{pmatrix},$$

$$Q_3 = \begin{pmatrix} 0.946 & 0 \\ 0 & 1.010 \end{pmatrix}, Q_4 = 14.672I_{4\times4}, Q_8 = 13.912I_{2\times2},$$

$$Q_5 = \begin{pmatrix} 29.5 & 0 & 0 & 0 \\ 0 & 76.3 & 0 & 0 \\ 0 & 0 & 111.2 & 0 \\ 0 & 0 & 0 & 74.8 \end{pmatrix}, Q_6 = \begin{pmatrix} 1.137 & 0 & 0 & 0 \\ 0 & 3.464 & 0 & 0 \\ 0 & 0 & 2.626 & 0 \\ 0 & 0 & 0 & 1.923 \end{pmatrix},$$

$$Q_7 = \begin{pmatrix} 1.501 & 0 & 0 & 0 \\ 0 & 3.464 & 0 & 0 \\ 0 & 0 & 2.626 & 0 \\ 0 & 0 & 0 & 1.923 \end{pmatrix}.$$

Thereby, by Theorem 1, we obtain that the system (2.1) is globally robustly exponentially stable in Lagrange sense. Moreover, by calculating the eigenvalues of $P$, we gain that the set $\Omega = \{x \in R^n \mid 3.906x_1^2 + 5.762x_2^2 \leq \frac{U^T R_8 U}{\beta_1 - \beta_2 - \beta_3 h} = \frac{13.912}{0.1} = 139.12\}$ is a globally exponentially attractive set of (2.1).

*Remark 3.* In order to imitate the dynamic behavior of system (2.1), we choose some parameters randomly as follows:

$$D = \begin{pmatrix} 6.4 & 0 \\ 0 & 8 \end{pmatrix}, A = \begin{pmatrix} 1 & 0.8 \\ 0.5 & 1 \end{pmatrix},$$
$$B = \begin{pmatrix} 1.5 & 0.1 \\ -0.8 & 1 \end{pmatrix}, C = \begin{pmatrix} -1 & 0.5 \\ -0.6 & 1 \end{pmatrix}. \tag{4.19}$$

Fig.1 depicts the phase plots of system (2.1) with parameters (4.19) in the case of $g(x(t)) = \frac{1}{4}(|x+1| - |x-1|)$. The numerical results show that system (2.1) with parameters (4.19) is globally robustly exponentially stable in Lagrange sense.

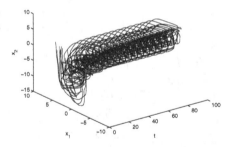

**Fig. 1.** The phase plots of system (3.1) with random initials

## 5   Conclusions

The paper has mainly studied the problem of global robust exponential stability in Lagrange sense for interval neural networks with general activation functions and infinite distributed delays. By employing the LMI techniques, some sufficient conditions to ensure the global robust exponential stability in Lagrange sense for IDNNs are derived. In addition, a series of globally exponentially attractive sets of IDNNs are also obtained. In the end, an example has been provided to demonstrate the validity of the proposed theoretical results.

**Acknowledgements.** The authors thank the editor and the anonymous reviewers for their careful reading of this paper and constructive comments.

# References

[1] Liao, X., Luo, Q., Zeng, Z.: Positive invariant and global exponential attractive sets of neural networks with time-varying delays. Neurocomputing 71, 513–518 (2008)

[2] Liao, X., Luo, Q., Zeng, Z., Guo, Y.: Global exponential stability in Lagrange sense for recurrent neural networks with time delays. Nonlinear Anal.: RWA 9, 1535–1557 (2008)

[3] Luo, Q., Zeng, Z., Liao, X.: Global exponential stability in Lagrange sense for neutral type recurrent neural networks. Neurocomputing 74, 638–645 (2011)

[4] Wu, A., Zeng, Z., Fu, C., Shen, W.: Global exponential stability in Lagrange sense for periodic neural networks with various activation functions. Neurocomputing 74, 831–837 (2011)

[5] Wang, X., Jiang, M., Fang, S.: Stability analysis in Lagrange sense for a non-autonomous Cohen-Grossberg neural network with mixed delays. Nonlinear Anal.: TMA 70, 4294–4306 (2009)

[6] Wang, B., Jian, J., Jiang, M.: Stability in Lagrange sense for Cohen-Grossberg neural networks with time-varying delays and finite distributed delays. Nonlinear Anal.: Hybrid Syst. 4, 65–78 (2010)

[7] Tu, Z., Jian, J., Wang, K.: Global exponential stability in Lagrange sense for recurrent neural networks with both time-varying delays and general activation functions via LMI approach. Nonlinear Anal.: RWA 12, 2174–2182 (2011)

[8] Muralisankar, S., Gopalakrishnan, N., Balasubramaniam, P.: An LMI approach for global robust dissipativity analysis of T-S fuzzy neural networks with interval time-varying delays. Expert Syst. Appl. 39, 3345–3355 (2012)

[9] Berman, A., Plemmons, R.: Nonnegative Matrices in Mathematical Sciences. Academic Press, New York (1979)

[10] Gahinet, P., Nemirovski, A., Laub, A., Chilali, M.: LMI control toolbox user's guide. Mathworks, Natick (1995)

[11] Li, X.: Existence and global exponential stability of periodic solution for impulsive Cohen-Grossberg-type BAM neural networks with continuously distributed delays. Applied Mathematics and Computation 215, 292–307 (2009)

[12] Liao, X., Fu, Y., Xie, S.: Globally exponential stability of Hopfied networks. Adv. Syst. Sci. Appl. 5, 533–545 (2005)

[13] Halanay, A.: Differential Equations: Stability, Oscillations, Time Lags. Academic Press, New York (1966)

# The Binary Output Units of Neural Network

Qilin Sun, Yan Liu, Zhengxue Li, Sibo Yang, Wei Wu, and Jiuwu Jin

School of Mathematical Sciences, Dalian University of Technology,
Dalian, 116024, P.R. China
wuweiw@dlut.edu.cn

**Abstract.** When solving a multi-classification problem with k kinds of samples, if we use a multiple linear perceptron, k output nodes will be widely-used. In this paper, we introduce binary output units of multiple linear perceptron by analyzing the classification problems of vertices of the regular hexahedron in the Three-dimensional Euclidean Space. And we define Binary Approach and One-for-Each Approach to the problem. Then we obtain a theorem with the help of which we can find a Binary Approach that requires more less classification planes than the One-for-Each Approach when solving a One-for-Each Separable Classification Problem. When we apply the Binary Approach to the design of output units of multiple linear perceptron, the output units required will decrease greatly and more problems could be solved.

**Keywords:** Multiple linear perceptron, Binary Approach, One-for-Each Approach.

## 1 Introduction

When we use a multiple linear perceptron to solve a multi-classification problem, every kind of samples will correspond to a specific output unit if we use existing method of designing output units of multiple linear perceptron[1-3]. If and only if we input the i-th kind of samples, the corresponding output unit will output '1'. And we need k output units to solve the classification problem with k kinds of samples. In this paper, we introduce Binary Approach by analyzing the classification problems of vertices of the regular hexahedron in the Three-dimensional Euclidean Space. When we apply this approach to the design of output units of multiple linear perceptron, the output units required will decrease greatly and more problems could be solved.

## 2 Approach to the Classification Problems

Assuming the vertices of the regular hexahedron in the Three-dimensional Euclidean Space, as shown in Fig.1, constitute the set of $A = \{P_1, P_P, P_3, P_4, P_5, P_6, P_7, P_8\}$, we can use a plane in the space to divide these vertices into two disjoint sets, and the plane is called the classification plane. For example, as shown in Fig.2, plane $S_1S_2S_3S_4$ divides A into two classes each of which has

C. Guo, Z.-G. Hou, and Z. Zeng (Eds.): ISNN 2013, Part I, LNCS 7951, pp. 250–257, 2013.

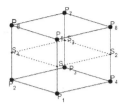

**Fig. 1.** The regular hexahedron in the Three-dimensional Euclidean Space

**Fig. 2.** Plane $S_1 S_2 S_3 S_4$ divides A into two classes

four vertices and the vertices of two classes constitute respectively the sets $A_1 = \{P_1, P_2, P_3, P_4\}$ and $A_2 = \{P_5, P_6, P_7, P_8\}$.

We define the classification problem of the set A as follows. Set A is divided into k ($2 \leq k \leq 8$) pairwise disjoint subsets, and the i-th subset has $n_i$ ($1 \leq n_i < 8$) vertices. Assuming that the vertices of the i-th subset constitute the set $A_i$, we can get $\bigcup_{i=1}^{k} A_i = A$, $A_i \bigcup A_j = \varnothing, (1 \leq i, j \leq k, i \neq j)$

**Definition 1.** *In terms of the classification problem described above, we will call it* linearly separable, *if there are some planes which create disjoint regions by incising the Three-dimensional Euclidean Space, and the vertices in subset $A_i$ located in the same region and the vertices in subset $A_i$ and the vertices in subset $A_j$ ($i \neq j$) are in different regions. We define the approach as* Binary Approach.

**Definition 2.** *A approach to solving a linearly separable classification problem is as follows. We find k planes: $S_1$, $S_2$ $\cdots$ $S_k$. And $S_i$ divides set A into two parts, making all the $n_i$ vertices in subset $A_i$ locate in the same side of $S_i$ but the remaining vertices locate in the other side. We define the approach above as* One-for-Each Approach. *If there are such planes in the classification problem, we will name it as* One-for-Each Separable.

Obviously, One-for-Each Approach is a special Binary Approach and $k$ planes divide set A in the process into k pairwise disjoint subsets. But k planes can divide the space into $2^k$ disjoint regions at most. So the regions created by the planes usually have no vertex in set A.

In terms of a classification problem which is linearly separable, we expect to use as few planes as possible to solve it. The $p$ planes represents a solution of the classification problem only if $p$ satisfies inequality $k \geq 2^p$, $p \in N$, because p planes can divide the space into $2^p$ disjoint regions at most,

So $p \geq \lceil log_2 k \rceil$ ($\lceil \bullet \rceil$ is the rounded up sign, such as $\lceil log_2 5 \rceil = 3$).

Can a linearly separable classification problem certainly be solved by $\lceil log_2 k \rceil$ planes? The answer is NO, as shown in Fig.3(a), $A_1 = \{P_1, P_2\}$, $A_2 = \{P_3, P_4\}$, $A_3 = \{P_5, P_8\}$, $A_4 = \{P_6, P_7\}$. We use bold lines to connect the vertices which are in the same subset. We can prove that to solve this classification problem, 3 planes are required at least. Fig.3(b) is a Binary Approach to it.

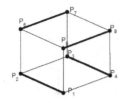

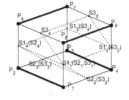

(a) The counter-example                    (b) The solution

**Fig. 3.** The counter-example and it's solution

But we can prove the theorem bellow:

**Theorem 1.** *When a classification problem with k classes is linearly separable, if it is One-for-Each Separable, we can find a Binary Approach which requires p planes to solve the same problem. And the correspondent relationship between p and k is shown in Table 1. Conversely, if we can find a Binary Approach to a linearly separable classification problem, we may not be able to find an One-for-Each Approach to the problem.*

**Table 1.** The relationship between p and k

| k | 2 | 3 | 4 | 5 | 6 | 7 | 8 |
|---|---|---|------|---|---|---|---|
| p | 1 | 2 | 2or3 | 3 | 3 | 3 | 3 |

## 3  The Application in the Neural Networks

Figure.4 shows the architectural graph of a multilayer perceptron with a hidden layer which has 3 neurons and an output layer which has n neurons. The network is fully connected and the activation function used is the threshold function $f(x)$:

$$f(x) = \begin{cases} 1 & x \geq 0 \\ 0 & x \leq 0 \end{cases} \tag{1}$$

The classification problem of the samples can be seen as the classification problem of corresponding outputs of the hidden layer, because outputs of the output layer is decided by corresponding outputs of the hidden layer. The output of the hidden layer must be a three-dimensional $\begin{bmatrix} 0 \ or \ 1 \\ 0 \ or \ 1 \\ 0 \ or \ 1 \end{bmatrix}$. So if we use the coordinate in the Three-dimensional Euclidean Space to denote outputs of the hidden layer,

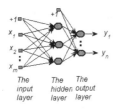

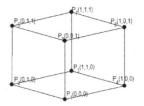

**Fig. 4.** A multiplayer perceptron

**Fig. 5.** The regular hexahedron of the Three-dimensional Euclidean Space

they will constitute the regular hexahedron in the space as shown in Fig.5. In this case, the multilayer perceptron can be simplified as Fig.6 and the input of the networks will be $\begin{bmatrix} 0\ or\ 1 \\ 0\ or\ 1 \\ 0\ or\ 1 \end{bmatrix}$, if we only consider about the design of the output layer. Obviously, the output can also be represented by a binary number, so the differences among binary numbers reflect the differences among Binary Approaches. For example, when we use the One-for-Each Approach, the output corresponding to the i-th subset of the networks is a k dimension vector with '1' at the i-th bit and '0' at others, namely $[0, \cdots, 0, 1, 0, \cdots, 0]^T$, which is widely used in the traditional design of the neural network. We can conclude from the Theorem:

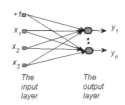

**Fig. 6.** The neural network simplified

A) As when we use this kind of network, outputs of the hidden layer will have 8 cases at most for any sample set , the sample set can be divided into 8 pairwise disjoint subsets at most. If we can divide the set into k pairwise disjoint subsets $(2 \leq k \leq 8)$ using One-for-Each Approach, it will require k neurons in the output layer but meanwhile we can find a Binary Approach which requires p neurons in the output layer. The correspondent relationship between p and k is shown in Table 1.

B) Conversely, if we can divide the set into k pairwise disjoint subsets ($2 \leq k \leq 8$) using a Binary Approach, we may not be able to divide the set into the same subsets using One-for-Each Approach no matter how much neurons in output layer.

From A), B), we can make conclusions: With appropriate Binary Approach, far less neurons are required in the output layer than the One-for-Each Approach and we can solve some classification problems that cannot be solved by One-for-Each Approach. Thus the complexity of the neural network is reduced and we can solve more classification problems.

## 4   Proof

We will prove the Theorem when $k \geq 4$, for k=2 or 3 is easy to solve. Firstly, we define some special planes.

1) As show in Fig.7, by connecting the midpoints of prism $P_4P_8$, $P_5P_8$, $P_7P_8$, namely $S1_1$, $S1_2$, $S1_3$, we can get a plane that separates $P_8$ from other vertices, so we call it the 1-plane of $\{P_8\}$. We can get the 1-plane of a vertex set composed of one vertex in a similar way.

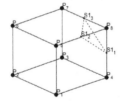   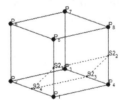

**Fig. 7.** The 1-plane of $\{P_8\}$          **Fig. 8.** The 2-plane of $\{P_1, P_4\}$

2) As show in Fig.8, by connecting the midpoints of prism $P_1P_5$, $P_4P_8$, $P_3P_4$, $P_1P_2$, namely $S2_1$, $S2_2$, $S2_3$, $S2_4$, we can get a plane that separate $\{P_1, P_4\}$ from other vertices, so we call it the 2-plane of $\{P_1, P_4\}$. We can get the 2-plane of a vertex set composed of two vertices in the same prisms in a similar way.

3) As shown in Fig.9, by connecting the midpoint of prism $P_5P_8$, a quarter point of prism $P_8P_7$ closest to $P_8$, a quarter point of prism $P_4P_3$ closest to $P_3$, a quarter point of prism $P_1P_2$ closest to $P_1$ and the midpoint of prism $P_1P_5$, namely $S3_1$, $S3_2$, $S3_3$, $S3_4$, $S3_5$, we can get a plane that separate $P_1$, $P_4$, $P_8$ from other vertices, so we call it the 3-plane of $\{P_1, P_4, P_8\}$. We can get the 3-plane of a vertex set composed of three vertices in the same face in a similar way.

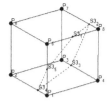 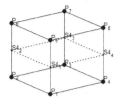

**Fig. 9.** The 3-plane of $\{P_1, P_4, P_8\}$      **Fig. 10.** The 4-1-plane of $\{P_1, P_2, P_3, P_4\}$

4) As shown in Fig.10, by connecting the midpoints of prism $P_1P_5$, $P_2P_6$, $P_3P_7$, $P_4P_8$, namely $S4_1$, $S4_2$, $S4_3$, $S4_4$, we can get a plane that divides A into $\{P_1, P_2, P_3, P_4\}$, $\{P_5, P_6, P_7, P_8\}$, so we call it the 4-1-plane of $\{P_1, P_2, P_3, P_4\}$. We can get the 4-1-plane of a vertex set composed of four vertices in the same face in a similar way.

5) As shown in Fig.11, by connecting the midpoints of prism $P_1P_2$, $P_1P_5$, $P_5P_8$, $P_7P_8$, $P_3P_7$, $P_2P_3$ namely $S42_1$, $S42_2$, $S42_3$, $S42_4$, $S42_5$, $S42_6$, we can get a plane that divides A into $\{P_1, P_3, P_4, P_8\}$, $\{P_2, P_5, P_6, P_7\}$, so we call it the 4-2-plane of $\{P_1, P_3, P_4, P_8\}$. We can get the 4-2-plane of a vertex set composed of four vertices that each three of them is in the same face in a similar way.

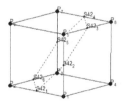

**Fig. 11.** The 4-2-plane of $\{P_1, P_3, P_4, P_8\}$

Secondly, we will list all the cases of classification problems which are One-for-Each Separable and satisfy $k \geq 4$. Please note that we will consider about the isomorphism among different cases. That is to say case $A_1 = \{P_1, P_2, P_3, P_4\}$, $A_2 = \{P_5, P_6, P_7, P_8\}$ is equivalent to case $A_1 = \{P_5, P_6, P_7, P_8\}$, $A_2 = \{P_1, P_2, P_3, P_4\}$, which we can get only by turning the regular hexahedron upside down.

1) $k = 4$.
   a) $A_1 = \{P_1, P_2, P_3, P_4, P_5\}$, $A_2 = \{P_6\}$, $A_3 = \{P_7\}$, $A_4 = \{P_8\}$.
   b) $A_1 = \{P_1, P_2, P_3\}$, $A_2 = \{P_4, P_5, P_8\}$, $A_3 = \{P_6\}$, $A_4 = \{P_7\}$.

c) $A_1 = \{P_1, P_2, P_3\}$, $A_2 = \{P_5, P_6, P_7\}$, $A_3 = \{P_4\}$, $A_4 = \{P_8\}$.
d) $A_1 = \{P_1, P_2, P_3\}$, $A_2 = \{P_5, P_7, P_8\}$, $A_3 = \{P_4\}$, $A_4 = \{P_6\}$.
e) $A_1 = \{P_1, P_2, P_3\}$, $A_2 = \{P_6, P_7, P_8\}$, $A_3 = \{P_4\}$, $A_4 = \{P_5\}$.
f) $A_1 = \{P_1, P_2, P_3, P_4\}$, $A_2 = \{P_5, P_6\}$, $A_3 = \{P_7\}$, $A_4 = \{P_8\}$.
g) $A_1 = \{P_1, P_2, P_3\}$, $A_2 = \{P_5, P_6\}$, $A_3 = \{P_7, P_8\}$, $A_4 = \{P_4\}$.
h) $A_1 = \{P_1, P_2, P_3\}$, $A_2 = \{P_4, P_8\}$, $A_3 = \{P_6, P_7\}$, $A_4 = \{P_5\}$.
i) $A_1 = \{P_1, P_2\}$, $A_2 = \{P_3, P_4\}$, $A_3 = \{P_5, P_6\}$, $A_4 = \{P_7, P_8\}$.
j) $A_1 = \{P_1, P_2\}$, $A_2 = \{P_3, P_4\}$, $A_3 = \{P_5, P_8\}$, $A_4 = \{P_6, P_7\}$.

2) $k = 5$.
   a) $A_1 = \{P_1, P_2, P_3, P_4\}$, $A_2 = \{P_5\}$, $A_3 = \{P_6\}$, $A_4 = \{P_7\}$, $A_5 = \{P_8\}$.
   b) $A_1 = \{P_1, P_2, P_3\}$, $A_2 = \{P_4, P_8\}$, $A_3 = \{P_5\}$, $A_4 = \{P_6\}$, $A_5 = \{P_7\}$.
   c) $A_1 = \{P_1, P_2, P_3\}$, $A_2 = \{P_5, P_6\}$, $A_3 = \{P_4\}$, $A_4 = \{P_7\}$, $A_5 = \{P_8\}$.
   d) $A_1 = \{P_1, P_2, P_3\}$, $A_2 = \{P_5, P_8\}$, $A_3 = \{P_4\}$, $A_4 = \{P_6\}$, $A_5 = \{P_7\}$.
   e) $A_1 = \{P_1, P_2\}$, $A_2 = \{P_3, P_4\}$, $A_3 = \{P_5, P_6\}$, $A_4 = \{P_7\}$, $A_5 = \{P_8\}$.
   f) $A_1 = \{P_1, P_2\}$, $A_2 = \{P_3, P_4\}$, $A_3 = \{P_5, P_8\}$, $A_4 = \{P_6\}$, $A_5 = \{P_7\}$.

3) $k = 6$.
   a) $A_1 = \{P_1, P_2, P_3\}$, $A_2 = \{P_4\}$, $A_3 = \{P_5\}$, $A_4 = \{P_6\}$, $A_5 = \{P_7\}$, $A_6 = \{P_8\}$.
   b) $A_1 = \{P_1, P_2\}$, $A_2 = \{P_3, P_4\}$, $A_3 = \{P_5\}$, $A_4 = \{P_6\}$, $A_5 = \{P_7\}$, $A_6 = \{P_8\}$.
   c) $A_1 = \{P_1, P_2\}$, $A_2 = \{P_7, P_8\}$, $A_3 = \{P_3\}$, $A_4 = \{P_4\}$, $A_5 = \{P_5\}$, $A_6 = \{P_6\}$.
   d) $A_1 = \{P_1, P_2\}$, $A_2 = \{P_5, P_8\}$, $A_3 = \{P_3\}$, $A_4 = \{P_4\}$, $A_5 = \{P_6\}$, $A_6 = \{P_7\}$.

4) $k = 7$.
   a) $A_1 = \{P_1, P_2\}$, $A_2 = \{P_3\}$, $A_3 = \{P_4\}$, $A_4 = \{P_5\}$, $A_5 = \{P_6\}$, $A_6 = \{P_7\}$, $A_7 = \{P_8\}$.

5) $k = 8$.
   a) $A_1 = \{P_1\}$, $A_2 = \{P_2\}$, $A_3 = \{P_3\}$, $A_4 = \{P_4\}$, $A_5 = \{P_5\}$, $A_6 = \{P_6\}$, $A_7 = \{P_7\}$, $A_8 = \{P_8\}$.

Third, we will solve the classification problems using planes defined in the first part of the Proof. The classification problems have many cases, but almost each of them is easy to find a correspondent Binary Approach which meets the demand. So we only discuss some of the complex cases.

1) As shown in Fig.12, $A_1 = \{P_1, P_2, P_3\}$, $A_2 = \{P_4, P_5, P_8\}$, $A_3 = \{P_6\}$, $A_4 = \{P_7\}$. We can use a 2-plane of $\{P_6, P_7\}$ and a 4-2-plane of $\{P_1, P_2, P_3, P_6\}$ to solve the problem, as shown in Fig.12. Namely, there is a Binary Approach requiring 2 planes to solve it.

2) As shown in Fig.13, $A_1 = \{P_1, P_2\}$, $A_2 = \{P_3, P_4\}$, $A_3 = \{P_5, P_8\}$, $A_4 = \{P_6\}$, $A_5 = \{P_7\}$. We can use a 4-1-plane of $\{P_1, P_2, P_3, P_4\}$, a 3-plane of $\{P_1, P_2, P_6\}$ and a 1-plane of $\{P_7\}$ to solve the problem, as shown in Fig.13. Namely, there is a Binary Approach requiring 3 planes to solve it.

3) As shown in Fig.14, $A_1 = \{P_1, P_2\}$, $A_2 = \{P_5, P_8\}$, $A_3 = \{P_3\}$, $A_4 = \{P_4\}$, $A_5 = \{P_6\}$, $A_6 = \{P_7\}$. We can use a 4-1-plane of $\{P_1, P_2, P_3, P_4\}$, a 3-plane of $\{P_1, P_2, P_6\}$ and a 2-plane of $\{P_3, P_7\}$ to solve the problem, as shown in Fig.14. Namely there is a Binary Approach requiring 3 planes to solve it.

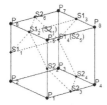

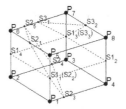

**Fig. 12.** Case b) when k=4 and it's solution

**Fig. 13.** Case f) when k=5 and it's solution

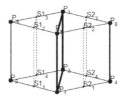

**Fig. 14.** Case d) when k=6 and it's solution

**Fig. 15.** The example for the second part of the proof

The second part of the theorem can be proved by a counter-example that $A_1 = \{P_1, P_3, P_5, P_7\}$, $A_2 = \{P_2, P_6\}$, $A_3 = \{P_4, P_8\}$, as shown in Fig.15. We can use a 2-plane of $\{P_2, P_6\}$ and a 2-plane of $\{P_4, P_8\}$ to solve the problem, as shown in Fig.15. Namely there is a Binary Approach requiring 2 planes to solve it. But we cannot find an One-for-Each Approach to solve it.

# References

1. Wu, W., Zhou, C.G., Liang, Y.C.: Intelligent computing. Higher Education Press, Beijing (2009)
2. Haykin, S.: Neural Networks and Learing Machines, 3rd edn. China Machine Press (2009)
3. SateeshBabu, G., Suresh, S.: Meta-cognitive Neural Network for classification problems in a sequential learning framework. Neurocomputing 81, 86–96 (2012)

# Support Vector Machine with Customized Kernel

Guangyi Chen[1], Tien Dai Bui[1], Adam Krzyzak[1], and Weihua Liu[2]

[1] Department of Computer Science and Software Engineering, Concordia University,
1455 de Maisonneuve West, Montreal, Quebec, Canada H3G 1M8
{guang_c,bui,krzyzak}@cse.concordia.ca
[2] State Key Lab. of Virtual Reality Technology and Systems, Beihang University,
ZipCode 100191, No 37, Xueyuan Rd., Haidian District, Beijing, P.R. China
liuwh_99@hotmail.com

**Abstract.** In the past two decades, Support Vector Machine (SVM) has become one of the most famous classification techniques. The optimal parameters in an SVM kernel are normally obtained by cross validation, which is a time-consuming process. In this paper, we propose to learn the parameters in an SVM kernel while solving the dual optimization problem. The new optimization problem can be solved iteratively as follows:

(a) Fix the parameters in an SVM kernel; solve the variables $\alpha_i$ in the dual optimization problem.
(b) Fix the variables $\alpha_i$; solve the parameters in an SVM kernel by using the Newton–Raphson method.

It can be shown that (a) can be optimized by using standard methods in training the SVM, while (b) can be solved iteratively by using the Newton-Raphson method. Experimental results conducted in this paper show that our proposed technique is feasible in practical pattern recognition applications.

**Keywords:** Support vector machine (SVM), feature extraction, SVM kernels, pattern recognition, pattern classification.

## 1 Introduction

Support vector machine (SVM) was developed by Vapnik et al. ([1], [2], [3]) for pattern recognition and function regression. The SVM assumes that all samples in the training set are independent and identically distributed. It uses an approximate implementation to the structure risk minimization principal in statistical learning theory, rather than the empirical risk minimization method. A kernel is utilized to map the input data to a higher dimensional feature space so that the problem becomes linearly separable. An SVM kernel plays a very important role in the performance of the SVM applications.

We briefly review recent advances in SVM applications. Chen and Dudek [4] developed the auto-correlation wavelet kernel for pattern recognition. It was shown that this kernel is better than the wavelet kernel [5] because the auto-correlation

C. Guo, Z.-G. Hou, and Z. Zeng (Eds.): ISNN 2013, Part I, LNCS 7951, pp. 258–264, 2013.

wavelet is shift-invariant whereas the wavelet is not. This shift-invariant property is very important in pattern recognition. Chen [6] also proposed the dual-tree complex wavelet (DTCWT) kernel for SVM classification. The DTCWT developed by Kingsbury [7] has the approximate shift invariant property and better orientation selectivity. These good properties have made the DTCWT a better candidate for pattern recognition.

In this paper, we propose to learn the parameters in an SVM kernel while solving the SVM optimization problem. We break the optimization problem into two smaller optimization problems: (a) Fix the parameters in an SVM kernel, and then solve the variables $\alpha_i$ in the dual optimization problem. (b) Fix the variables $\alpha_i$, and solve the parameters in an SVM kernel by using the *Newton–Raphson* method. We solve (a) and (b) iteratively for at most $\tau$ iterations in each round of optimization, respectively. We repeat the optimization of (a) and (b) in a loop manner until they converge or the maximum number of iterations is reached. Our simulation results show that our proposed method achieves higher classification rates than the standard SVM for recognizing traffic light signals and the vowel dataset ($\tau=100$).

The organization of this paper is as follows. Section 2 proposes to learn the parameters in an SVM kernel while training the SVM for pattern recognition. Section 3 conducts some experiments in order to show that by optimizing the parameters in an SVM kernel we can achieve higher classification rates. Finally, Section 4 draws the conclusions of this paper, and gives future research direction.

## 2    Proposed Method

An SVM can be used as a classifier for a pattern recognition problem with $n>2$ classes, which can be resolved by solving $n \times (n-1)/2$ two-class SVM problems. A two-class SVM problem can be summarized as follows. Let $(x_i, y_i)$ be a set of training samples, where $x_i$ is the feature vector and $y_i = +1$ or $-1$.

The primal form of an SVM problem is formulated as:

Min: $\frac{1}{2} \|w\|^2$
Subject to: $y_i (w^T x_i - b) \geq 1$ for all $i = 1, 2, ..., n$.

This is an optimization problem that can be solved by introducing a set of Lagrange multiplier $\alpha_i \geq 0$. We have to solve the following quadratic dual optimization problem:

$$\text{Max:} \sum_{i=1}^{n} \alpha_i - \frac{1}{2} \sum_{i=1}^{n} \sum_{j=1}^{n} \alpha_i \alpha_j y_i y_j k(x_i, x_j)$$

$$\text{Subject to:} \ 0 \leq \alpha_i \leq C \text{ and } \sum_{i=1}^{n} \alpha_i y_i = 0.$$

In this paper, we will restrict the kernel to be the radial basis function (RBF) kernel:

$$k(x_i, x_j) = \exp(-\gamma \| x_i - x_j \|^2)$$

or the exponential radial basis function (ERBF) kernel:

$$k(x_i, x_j) = \exp(-\gamma \| x_i - x_j \|),$$

where the parameter $\gamma \geq 0$. In the above dual optimization problem, the parameter $\gamma$ can be chosen by the user or it can be learned by cross-validation, which is a time-consuming process. We have decided to fix the parameter $C$ in this paper.

We propose to solve the above dual optimization problem in two steps:

(a) Fix the parameter $\gamma$ in an SVM kernel; solve the variables $\alpha_i$ in the dual optimization problem.
(b) Fix the variables $\alpha_i$, solve the parameters in an SVM kernel by using the *Newton–Raphson* method iteratively.

The first optimization problem (a) can be solved by the standard optimization method in training an SVM. We restrict the number of iteration in solving this optimization problem to be at most $\tau$ iterations, instead of looping for many iterations. We modify the C++ code of LIBSVM [8] to solve this optimization problem. After obtaining the approximate parameters $\alpha_i$, we will solve the second optimization problem (b) iteratively by using the *Newton–Raphson* method.

Let us derive the formula to solve the second optimization problem (b). Since

$$\sum_{i=1}^{n} \alpha_i y_i = 0,$$

We have

$$\sum_{i=1}^{n-1} \alpha_i y_i y_n = -\alpha_n.$$

By plugging $\alpha_n$ into the dual optimization problem, we obtain the following optimization problem without any constrains:

$$\text{Max:}\ W(\gamma) = \sum_{i=1}^{n-1} \alpha_i (1 - y_n y_i) - \tfrac{1}{2} \sum_{i=1}^{n-1} \sum_{j=1}^{n-1} \alpha_i \alpha_j y_i y_j$$

$$k(x_i,x_j)+ \left(\sum_{i=1}^{n-1} \alpha_i y_i \right)\left( \sum_{j=1}^{n-1} \alpha_j y_j\, k(x_n,x_j)\right)- \tfrac{1}{2}\left(\sum_{j=1}^{n-1} \alpha_i y_i\right)^2$$

In order to obtain the maximization, we need to set the first derivative $W'(\gamma)=0$. From the above equation, we can derive:

$$W'(\gamma)= \tfrac{1}{2}\sum_{i=1}^{n-1}\ \sum_{j=1}^{n-1}\ \alpha_i \alpha_j y_i y_j\, k(x_i,x_j)(-\|x_i-x_j\|^2)\, -$$
$$\alpha_n y_n \left(\sum_{j=1}^{n-1} \alpha_j y_j\, k(x_n,x_j)\right)(-\|x_n-x_j\|^2) \quad \text{for RBF.}$$

$$W'(\gamma)= \tfrac{1}{2}\sum_{i=1}^{n-1}\ \sum_{j=1}^{n-1}\ \alpha_i \alpha_j y_i y_j\, k(x_i,x_j)(-\|x_i-x_j\|)\, -$$
$$\alpha_n y_n \left(\sum_{j=1}^{n-1} \alpha_j y_j\, k(x_n,x_j)\right)(-\|x_n-x_j\|) \quad \text{for ERBF.}$$

and

$$W''(\gamma)= \tfrac{1}{2}\sum_{i=1}^{n-1}\ \sum_{j=1}^{n-1}\ \alpha_i \alpha_j y_i y_j\, k(x_i,x_j)(\|x_i-x_j\|^4)\, -$$
$$\alpha_n y_n \left(\sum_{j=1}^{n-1} \alpha_j y_j\, k(x_n,x_j)\right)(\|x_n-x_j\|^4) \quad \text{for RBF.}$$

$$W''(\gamma)= \tfrac{1}{2}\sum_{i=1}^{n-1}\ \sum_{j=1}^{n-1}\ \alpha_i \alpha_j y_i y_j\, k(x_i,x_j)(\|x_i-x_j\|^2)\, -$$
$$\alpha_n y_n \left(\sum_{j=1}^{n-1} \alpha_j y_j\, k(x_n,x_j)\right)(\|x_n-x_j\|^2) \quad \text{for ERBF.}$$

From the *Newton-Raphson* method, we obtain the following formula for the second optimization problem (b):

$$\gamma_{k+1} = \gamma_k - W'(\gamma_k)\, /\, W''(\gamma_k).$$

We would like to restrict the number of iterations for the second optimization problem (b) to be at most $\tau=100$ iterations, and then switch to the first optimization problem (a). We repeat to solve the two optimization problems (a) and (b) interchangeably until convergence or the maximum number of iterations is reached. The above solutions are for a two-class classification problem. Let $\Delta\gamma_k = \gamma_{k+1} - \gamma_k$ be for a two-class classification problem.  Since we have to solve $n\times(n-1)/2$ two-class SVM problems, we can take the mean of $\Delta\gamma_k$ over all these $n\times(n-1)/2$ two-class SVM problems. Therefore, the iterative formula for solving the parameter $\gamma_k$ can be given as

$$\gamma_{k+1} = \gamma_k + \varepsilon \times mean(\Delta\gamma_k).$$

It is expected that by solving the two optimization problems interchangeably, we can obtain better solutions for pattern recognition. The decision function of a two-class SVM problem is

$$f(x) = sign(\sum_{i=1}^{n} \alpha_i y_i k(x_i,x)+b)$$

where

$$b = y_r - \sum_{i=1}^{n} \alpha_i y_i k(x_i,x_r)$$

and $(x_r,y_r)$ is a training sample. For the one-versus-one SVM problem, classification is performed by a max-wins voting approach, in which every classifier assigns the instance to one of the two classes. The vote for the assigned class is increased by one, and the class with most votes determines the instance classification.

## 3    Experimental Results

We conducted some experiments by using the data sets *svmguide4* and *vowel* provided in [8]. The *svmguide4* data set is for traffic light signals, which has 6 classes with 300 training samples and 312 testing samples. The number of features was chosen to 10. The *vowel* data set has 11 classes with 528 training samples and 462 testing samples. The number of features was also chosen to 10. We used the following Code One and Two to train and test the standard LIBSVM and our proposed SVM, where the parameters $C$ and $g$ can be changed as desired. In our experiments, we choose $\tau=100$ and $\varepsilon=0.01$ for the traffic light dataset and for the vowel dataset.

```
-------------------------------------Code One-------------------------------------
svm-scale -l 0 -s range1 svmguide4 > svmguide4.scale
svm-scale -r range1 svmguide4.t > svmguide4.t.scale
svm-train -c 100 -g 0.2 svmguide4.scale
svm-predict svmguide4.t.scale svmguide4.scale.model svmguide4.t.predict

-------------------------------------Code Two-------------------------------------
svm-scale -l -1 -u 1 -s range3 vowel.scale > vowel.scale.scale
svm-scale -r range3 vowel.scale.t > vowel.scale.t.scale
svm-train -c 100 -g 0.2 -t 2 vowel.scale.scale
svm-predict vowel.scale.t.scale vowel.scale.scale.model vowel.scale.t.predict
-------------------------------------------------------------------------------------
```

Tables 1-2 tabulate the parameters $C$ and $g$, and the recognition rates for both the standard LIBSVM and our proposed SVM for the *traffic light* dataset and *vowel* dataset, respectively. From the two tables, it can be seen that our proposed SVM obtains higher classification rates than the standard LIBSVM due to the learning strategy introduced in our proposed SVM. Note that we only used the RBF in our experiments. We leave ERBF to our future research.

**Table 1.** A comparison between the standard LIBSVM and our proposed SVM for the traffic light dataset

| Parameter $C$ | Parameter $g$ | Classification rate (LIBSVM) | Classification rate (Proposed SVM) |
|---|---|---|---|
| 100 | 0.2 | 78.53% | **81.73%** |
| 10 | 0.2 | 54.17% | **66.67%** |
| 1 | 0.2 | 29.17% | **46.79%** |

**Table 2.** A comparison between the standard LIBSVM and our proposed SVM for the vowel dataset

| Parameter $C$ | Parameter $g$ | Classification rate (LIBSVM) | Classification rate (Proposed SVM) |
|---|---|---|---|
| 100 | 0.2 | 55.19% | **64.94%** |
| 10 | 0.2 | 53.03% | **63.20%** |
| 1 | 0.2 | 59.74% | **61.26%** |

# 4    Conclusions and Future Work

We have proposed a solution for solving an n-class classification problem by using SVM. We resolve the n-class SVM classification problem by solving $n \times (n-1)/2$ two-class SVM problems. Each two-class SVM classification problem can be resolved by (a) fixing the parameter $\gamma$ in an SVM kernel and then solve the variables $\alpha_i$ in the dual optimization problem, and by (b) fixing the variables $\alpha_i$ and solve the parameter $\gamma$ in an SVM kernel by using the *Newton–Raphson* method iteratively. We solve for (a) and (b) interchangeably until they converge or the maximum number of iterations $\tau=100$ is reached. Experimental results show that the proposed method in this paper is feasible in pattern recognition.

Further research needs to be done by learning the upper bound $C$ as well while solving the optimization problems. It is believed that, by optimizing both $C$ and $\gamma$, we can obtain higher classification rates for $n$-class pattern recognition problems. We may also apply our proposed SVM to the recognition of handwritten digits and handwritten characters. We are very interested in extracting the dual-tree complex wavelet features, the ridgelet features, the contourlet features, the curvelet features, etc. ([7], [9], [10]).

**Acknowledgments.** This research was supported by the research grant from the Natural Science and Engineering Research Council of Canada (NSERC) and Beijing Municipal Science and Technology Plan: Z111100074811001.

# References

1. Vapnik, V.N.: The nature of statistical learning. Springer, New York (1995)
2. Vapnik, V.N.: Statistical learning theory. Wiley, New York (1998)
3. Cortes, C., Vapnik, V.N.: Support vector networks. Machine Learning 20, 273–297 (1995)
4. Chen, G.Y., Dudek, G.: Auto-correlation wavelet support vector machine. Image and Vision Computing 27(8), 1040–1046 (2009)
5. Zhang, L., Zhou, W., Jiao, L.: Wavelet support vector machine. IEEE Transactions on Systems, Man, and Cybernetics - Part B 34(1), 34–39 (2004)
6. Chen, G.Y.: Dual-tree Complex Wavelet Support Vector Machine. International Journal of Tomography & Statistics 15(F10), 1–8 (2010)
7. Kingsbury, N.G.: Complex wavelets for shift invariant analysis and filtering of signals. Journal of Applied and Computational Harmonic Analysis 10(3), 234–253 (2001)
8. Chang, C.C., Lin, C.J.: LIBSVM: a library for support vector machines. ACM Transactions on Intelligent Systems and Technology 2(3) (2011), Software available at http://www.csie.ntu.edu.tw/~cjlin/libsvm
9. Candes, E.J.: Ridgelets and the representation of mutilated Sobolev functions. SIAM J. Math. Anal. 33(2), 2495–2509 (1999)
10. Do, M.N., Vetterli, M.: The contourlet transform: an efficient directional multiresolution image representation. IEEE Transactions on Image Processing 14(12), 2091–2106 (2005)

# Semi-supervised Kernel Minimum Squared Error Based on Manifold Structure

Haitao Gan*, Nong Sang, and Xi Chen

Science & Technology on Multi-spectral Information Processing Laboratory,
Institute for Pattern Recognition and Artificial Intelligence,
Huazhong University of Science and Technology, China, 430074

**Abstract.** Kernel Minimum Squared Error (KMSE) has been receiving much attention in data mining and pattern recognition in recent years. Generally speaking, training a KMSE classifier, which is a kind of supervised learning, needs sufficient labeled examples. However, there are usually a large amount of unlabeled examples and few labeled examples in real world applications. In this paper, we introduce a semi-supervised KMSE algorithm, called *Laplacian regularized KMSE* (LapKMSE), which explicitly exploits the manifold structure. We construct a $p$ nearest neighbor graph to model the manifold structure of labeled and unlabeled examples. Then, LapKMSE incorporates the structure information of labeled and unlabeled examples in the objective function of KMSE by adding a Laplacian regularized term. As a result, the labels of labeled and unlabeled examples vary smoothly along the geodesics on the manifold. Experimental results on several synthetic and real-world datasets illustrate the effectiveness of our algorithm.

**Keywords:** Kernel MSE, Semi-supervised learning, Manifold structure.

## 1  Introduction

In the last decades, kernel method has been receiving more and more attention in nonlinear classification and regression. A training example can be mapped into a high-dimensional feature space by using kernel trick [1,2] satisfying the Mercer condition and then a classifier can be trained in the new feature space. In most case, the kernel trick can achieve good generalization performance. Hence, many researchers have been studying the idea and various methods have been proposed, such as Kernel Minimum Squared Error (KMSE) [3], Support Vector Machine (SVM) [4], Least Squares SVM (LS-SVM) [5], Kernel Principal Component Analysis (KPCA) [6], Kernel Fisher Discriminant Analysis (KFDA) [7]. Among the above methods, KMSE has received many attention due to its higher computational efficiency in the training phase. However, it is difficult to control the generalization ability [8]. Xu *et al.* [9] proposed two versions of KMSE using different regularization terms and proved their relation to KFDA and LS-SVM.

* This work is supported by the National Natural Science Foundation of China under grant No. 61271328 and 61105014.

C. Guo, Z.-G. Hou, and Z. Zeng (Eds.): ISNN 2013, Part I, LNCS 7951, pp. 265–272, 2013.
© Springer-Verlag Berlin Heidelberg 2013

Nevertheless, the performance of KMSE, which is a kind of supervised learning, relies on sufficient labeled examples to train a good classifier [10]. In fact, labeled examples are usually insufficient while unlabeled data are *often* abundant in real world. Consequently, semi-supervised classification, which uses both labeled and unlabeled examples to train a classifier, has become a recent topic of interest. In semi-supervised learning, how to learn from unlabeled examples is still an open problem. One of the most used ways is manifold regularization. Belkin *et al.* [11] proposed Laplacian Regularized Least Squares (LapRLS) and Laplacian Support Vector Machines (LapSVM) which employ Laplacian regularization to learn from labeled and unlabeled examples. Cai [12] introduced a Semi-supervised Discriminant Analysis (SDA) where unlabeled examples are used to exploit the intrinsic manifold structure. We refer the readers to some excellent surveys [13,14] for more details.

In this paper, we propose a semi-supervised KMSE algorithm, called Laplacian regularized KMSE (LapKMSE), which explicitly reveals the local manifold structure. Naturally, if two examples are close on the manifold, they are likely to be drawn from the same class. In fact, the manifold is usually unknown. Hence, we construct a $p$ nearest neighbor graph to model the manifold and employ graph Laplacian to incorporate the Laplacian regularized term in the objective function of KMSE. Based on this, the information of labeled and unlabeled examples are exploited by Laplacian regularization which smooths the labels of labeled and unlabeled examples along the geodesics on the manifold.

The rest of the paper is organized as follows: In Sect.2, we briefly review the naïve KMSE. In Sect.3, we describe our algorithm in detail. Section 4 presents the experimental results on several synthetic and real-world datasets. Finally, we conclude the paper and discuss some future directions in Sect.5.

## 2    Naïve KMSE

Let $X = \{(x_1, y_1), \cdots, (x_l, y_l)\}$ be a training set of size $l$, where $x_i \in \mathbb{R}^D$ and $y_i \in \mathbb{R}$. For the binary classification problem, $y_i = -1$ if $x_i \in \omega_1$ or $y_i = 1$ if $x_i \in \omega_2$. By a nonlinear mapping $\Phi$, a training example is transformed into a new feature space $\Phi(x_i)$ from the original feature space. The task of KMSE is to build a linear model in the new feature. The outputs of the training examples obtained by the linear model are equal to the labels

$$\Phi W = Y \tag{1}$$

where

$$\Phi = \begin{bmatrix} 1 & \Phi(x_1)^T \\ \vdots & \vdots \\ 1 & \Phi(x_l)^T \end{bmatrix}, \quad W = \begin{bmatrix} \alpha_0 \\ w \end{bmatrix}, \quad \text{and} \quad Y = [y_1, \cdots, y_l]^T$$

According to the reproducing kernel theory [4,7], one can note that $w$ can be expressed as

$$w = \sum_{i=1}^{l} \alpha_i \Phi(x_i) \tag{2}$$

By substituting Eq.(2) into Eq.(1), we can get

$$K\alpha = Y \tag{3}$$

where

$$K = \begin{bmatrix} 1 & k(x_1,x_1) & \cdots & k(x_1,x_l) \\ \vdots & \vdots & \ddots & \vdots \\ 1 & k(x_l,x_1) & \cdots & k(x_l,x_l) \end{bmatrix} \quad \text{and} \quad \alpha = \begin{bmatrix} \alpha_0 \\ \vdots \\ \alpha_l \end{bmatrix}$$

here the matrix $K$ is kernel matrix whose entry $k(x_i, x_j) = (\Phi(x_i) \cdot \Phi(x_j))$.

The goal of KMSE is to find the optimal vector $\alpha$ by minimizing the objective function as follows:

$$\mathcal{J}_0(\alpha) = (Y - K\alpha)^T (Y - K\alpha) \tag{4}$$

By setting the derivation of $\mathcal{J}_0(\alpha)$ with respect to $\alpha$ to zero, we can obtain the solution:

$$\alpha^* = (K^T K)^{-1} K^T Y \tag{5}$$

From Eq.(5), we can find that the dimension of $\alpha^*$ is $l+1$ and $Rank(K^T K) \leq l$. In other words, $K^T K$ is always singular. Consequently, the solution $\alpha^*$ is not unique. In the last decades, the regularization approach [9] is often used to deal with the ill-posed problem. The corresponding regularized objective function can be described as

$$\mathcal{J}_1(\alpha) = (Y - K\alpha)^T (Y - K\alpha) + \mu \alpha^T \alpha \tag{6}$$

where $\mu$ is the coefficient of the regularization term.

By minimizing the above objective function (6), we can obtain

$$\alpha^* = (K^T K + \mu I)^{-1} K^T Y \tag{7}$$

where $I$ is an identity matrix of size $(l+1) \times (l+1)$.

When the optimal weight coefficients $\alpha^*$ is obtained, the linear model of KMSE can be presented as

$$f(x) = \sum_{i=1}^{l} \alpha_i^* k(x_i, x) + \alpha_0^* \tag{8}$$

In the testing phase, $x \in \omega_1$ if $f(x) < 0$ and $x \in \omega_2$ if $f(x) > 0$

## 3    Method

In this section, we will discuss how to learn from labeled and unlabeled examples in KMSE.

## 3.1   Manifold Regularization

Recall the standard learning framework. There is a probability distribution $P$ on $X \times \mathbb{R}$ according to which examples are generated for function learning. Labeled examples are $(x, y)$ pairs drawn according to $P$. Unlabeled examples are $x \in X$ generated according to the marginal distribution $\mathcal{P}_X$ of $P$. In many applications, the marginal distribution $\mathcal{P}_X$ is unknown. Related works show that there may be a relationship between the marginal and conditional distribution [11]. It is assumed that if two examples $x_1, x_2 \in X$ are *similar* in the intrinsic geometry of $\mathcal{P}_X$, then the conditional distribution $\mathcal{P}(y|x_1)$ and $\mathcal{P}(y|x_2)$ are *similar*. This is referred to as *manifold assumption* which is *often* used in semi-supervised learning [14].

Given a set $X = \{(x_1, y_1), \cdots, (x_l, y_l), x_{l+1}, \cdots, x_n\}$ with $l$ labeled examples and $u = n - l$ unlabeled examples. In order to exploit the manifold structure, Belkin *et al.* [11] introduced a Laplacian regularization by using graph Laplacian. The Laplacian regularization is defined as

$$\mathcal{R} = f^T L f \tag{9}$$

where $L$ is the graph Laplacian defined as $L = D - W$, and $f = [f(x_1), \cdots, f(x_n)]^T$ is the output of labeled and unlabeled examples. Here $D$ is a diagonal matrix whose entry $D_{ii} = \sum_j W_{ij}$ and the edge weight matrix $W = [W_{ij}]_{n \times n}$ can be defined as follows:

$$W_{ij} = \begin{cases} 1 & \text{if } x_i \in N_p(x_j) \text{ or } x_j \in N_p(x_i) \\ 0 & \text{otherwise} \end{cases}$$

where $N_p(x_i)$ denotes the data sets of $p$ nearest neighbors of $x_i$.

## 3.2   Laplacian Regularized KMSE (LapKMSE)

In this section, we introduce Laplacian regularized KMSE (LapKMSE) which is extended from KMSE by incorporating Laplacian regularizer into the objective function of KMSE.

By integrating regularization term (9) into Eq.(6), the objective function of LapKMSE can be given as

$$\mathcal{J}_r(\alpha) = (Y - GK\alpha)^T (Y - GK\alpha) + \gamma_A \alpha^T \alpha + \gamma_I \mathcal{R} \tag{10}$$

where

$$G = \begin{bmatrix} I_{l \times l} & \mathbf{0}_{l \times u} \\ \mathbf{0}_{u \times l} & \mathbf{0}_{u \times u} \end{bmatrix}, Y = [y_1, \cdots, y_l, 0, \cdots, 0]^T, \text{and } K = \begin{bmatrix} 1 & k(x_1, x_1) & \cdots & k(x_1, x_n) \\ \vdots & \vdots & \ddots & \vdots \\ 1 & k(x_n, x_1) & \cdots & k(x_n, x_n) \end{bmatrix}$$

According to the Representer Theorem, the solution can be given as [11]

$$f(x) = \sum_{i=1}^n \alpha_i^* k(x_i, x) + \alpha_0^* \tag{11}$$

Substituting Eq.(11) into Eq.(10), the modified objective function becomes

$$\mathcal{J}_r'(\alpha) = (Y - GK\alpha)^T(Y - GK\alpha) + \gamma_A \alpha^T \alpha + \gamma_I \alpha^T K^T LK\alpha \qquad (12)$$

The derivative of Eq.(12) with respect to $\alpha$ is

$$(-GK)^T(Y - GK\alpha) + \gamma_A \alpha + \gamma_I K^T LK\alpha = 0 \qquad (13)$$

By solving Eq.(13), we can get

$$\alpha^* = ((GK)^T GK + \gamma_A I + \gamma_I K^T LK)^{-1}(GK)^T Y \qquad (14)$$

As we can see, when $\gamma_I = 0$, the coefficients of unlabeled examples will be zeros and LapKMSE will be equivalent to the original KMSE (i.e., Eq.(6)).

# 4    Experimental Results

In this section, a series of experiments are conducted to evaluate our algorithm on several synthetic and real-world datasets. we compare the performance of our algorithm with KMSE, LapRLS and LapSVM[1].

We begin with one synthetic example to demonstrate how LapKMSE works.

## 4.1    A Synthetic Dataset

In this section, since our goal is to illustrate how to learn from unlabeled examples in LapKMSE, we mainly compare our algorithm with KMSE. Let us consider one synthetic example to evaluate the proposed algorithm.

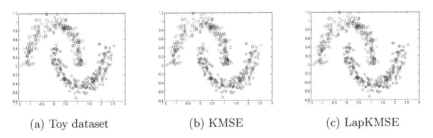

(a) Toy dataset          (b) KMSE          (c) LapKMSE

**Fig. 1.** Classification results on two moons dataset:(a)Toy dataset;(b)Results obtained by KMSE;(c)Results obtained by LapKMSE. Labeled points are shown in red color, the other points are unlabeled.

Figure 1 gives the plots of the two moons dataset and classification results of KMSE and LapKMSE. The dataset contains 400 points and 1 labeled point in each class. The parameter in KMSE is set as $\mu = 0.001$. And the parameters in

---

[1] The matlab codes are available at:
http://www.cs.uchicago.edu/~vikass/manifoldregularization.html

LapKMSE are set as $\gamma_A = 0.01, \gamma_I = 0.1$ and $p = 4$. As we can see, LapKMSE performs better than KMSE. LapKMSE achieves the desired classification result such that the two moons are well separated, while KMSE not. This is mainly because LapKMSE takes into account the manifold structure of the data set by incorporating a Laplacian regularizer.

## 4.2    Real-World Datasets

In this experiment, four real-world datasets are tested to evaluate our algorithm. The datasets are all from UC Irvine Machine Learning Repository [15]. The statistics of the four datasets are described in Table 1.

**Table 1.** Description of the experimental datasets

| Dataset | #Features | #Classes | #Traing examples | #Testing examples |
|---|---|---|---|---|
| Breast Cancer | 30 | 2 | 350 | 219 |
| Diabetes | 8 | 2 | 500 | 268 |
| Heart | 13 | 2 | 162 | 108 |
| IRIS | 4 | 3 | 60 | 90 |

Additionally, the parameters of the different methods are determined by using grid search. In each experiment, we employ the one-against-all strategy to train the multi-class classifier.

## 4.3    Results

To evaluate the performance of our algorithm, we carry out a series of experiments against KMSE, LapRLS and LapSVM. First, we randomly divide the training set into 10 subsets. Then we select each subset in order as labeled examples and the remaining subsets as unlabeled examples. We repeat this process 10 times to test our algorithm and the other three methods. The results are shown in Table 2.

**Table 2.** Classification results on each dataset

| Dataset | KMSE | LapRLS | LapSVM | LapKMSE |
|---|---|---|---|---|
| Breast Cancer | 94.01±3.68 | **95.89±2.48** | 95.70±3.07 | 95.47±3.36 |
| Diabetes | 71.56±3.33 | 77.35±2.78 | 77.01±2.38 | **77.72±2.43** |
| Heart | 53.42±9.17 | 75.46±4.43 | 75.46±4.43 | **77.03±3.54** |
| IRIS | 91.77±4.89 | 92.88±3.19 | 92.88±3.19 | **94.55±2.69** |

From Table 2, we have two observations. First, as can be seen, all three semi-supervised learning methods (i.e., LapRLS, LapSVM and our LapKMSE) make use of the manifold structure and achieve fairly good results. Therefore, it can be concluded that unlabeled data can help train a better classifier and improve the generalization ability of the classifier. And considering the manifold structure help learn a better discriminative function. Second, our LapKMSE algorithm outperforms LapRLS and LapSVM on the latter three datasets and gives comparable results on Breast Cancer. Especially, our algorithm has an obvious improvement on the Heart and IRIS datasets.

Besides the above comparison experiments, we next analysis the impact of the ratio of labeled examples on the performance of the different methods. Experiments are conducted to compare the performance of our algorithm with that of the other three methods while the ratio of labeled examples in the training set increases from 10% to 90%. The results on the four datasets are shown in Figure. 2.

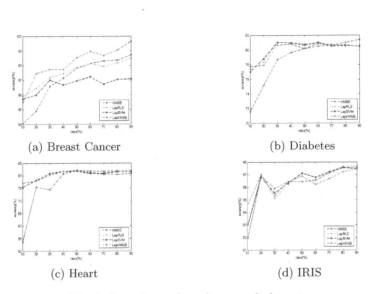

(a) Breast Cancer                    (b) Diabetes

(c) Heart                          (d) IRIS

**Fig. 2.** Experimental results on each dataset

Not surprisingly seen from these figures, the testing accuracy of all four methods increases overall as the ratio of labeled examples increases. On the whole, our LapKMSE algorithm gives comparable results with LapRLS and LapSVM. And the accuracy of semi-supervised learning methods are generally greater than KMSE, especially when the ratio is less than 30% on Diabetes and Heart datasets. Interestingly, we find that sometimes (e.g., the ratio of labeled examples is 90% in Figure. 2(b)) the performance of LapRLS, LapSVM and LapKMSE are worse than that of KMSE that is only trained by labeled examples. It may be because in those cases the labeled examples can cover the whole data space fairly well, while the locality of the manifold structure may increase the probability of overfitting.

## 5   Conclusion

In this paper, we introduce a semi-supervised learning algorithm, called Laplacian regularized KMSE. The method incorporates the manifold structure of labeled and unlabeled examples in the objective function of KMSE. We construct a $p$ nearest neighbor graph to exploit the manifold structure and use a Laplacian regularization term to smooth the labels of labeled and unlabeled examples along the geodesics on the manifold. A series of experiments are conducted on several synthetic and real-world datasets and the results show the effectiveness of our LapKMSE algorithm. In the future work, we will mainly focus on the analysis the impact of the parameters on the performance of LapKMSE.

## References

1. Muller, K.-R., Mika, S., Ratsch, G., Tsuda, S., Scholkopf, B.: An introduction to kernel-based learning algorithms. IEEE Transactions on Neural Networks 12(2), 181–202 (2001)
2. Scholkopf, B., Smola, A.J.: Learning with Kernels: Support Vector Machines, Regularization, Optimization, and Beyond. MIT Press, Cambridge (2001)
3. Ruiz, A., de Teruel, P.E.L.: Nonlinear kernel-based statistical pattern analysis. IEEE Transactions on Neural Networks 12(1), 16–32 (2001)
4. Cristianini, N., Shawe-Taylor, J.: An Introduction to Support Vector Machines and Other Kernel-based Learning Methods. Cambridge University Press (2000)
5. Suykens, J.A.K., Vandewalle, J.: Least squares support vector machine classifiers. Neural Processing Letters 9(3), 293–300 (1999)
6. Schlkopf, B., Smola, A.J., Müller, K.R.: Kernel principal component analysis. Advances in Kernel Methods: Support Vector Learning, 327–352 (1999)
7. Mika, S., Ratsch, G., Weston, J., Scholkopf, B., Mullers, K.R.: Fisher discriminant analysis with kernels. In: Proceedings of the 1999 IEEE Signal Processing Society Workshop on Neural Networks for Signal Processing IX, pp. 41–48 (1999)
8. Xu, J., Zhang, X., Li, Y.: Regularized kernel forms of minimum squared error method. Frontiers of Electrical and Electronic Engineering in China 1, 1–7 (2006)
9. Xu, J., Zhang, X., Li, Y.: Kernel mse algorithm: A unified framework for KFD, LS-SVM and KRR . In: Proceedings of International Joint Conference on Neural Networks, pp. 1486–1491 (2001)
10. Gan, H., Sang, N., Huang, R., Tong, X., Dan, Z.: Using clustering analysis to improve semi-supervised classification. Neurocomputing 101, 290–298 (2013)
11. Belkin, M., Niyogi, P., Sindhwani, V.: Manifold regularization: A geometric framework for learning from labeled and unlabeled examples. The Journal of Machine Learning Research 7, 2399–2434 (2006)
12. Cai, D., He, X., Han, J.: Semi-supervised discriminant analysis. In: Proceeding of the IEEE 11th International Conference on Computer Vision, ICCV (2007)
13. Chapelle, O., Scholkopf, B., Zien, A.: Semi-Supervised Learning. MIT Press, Cambridge (2006)
14. Zhu, X.: Semi-supervised learning literature survey. Technical Report 1530, Computer Sciences, University of Wisconsin-Madison (2005)
15. Frank, A., Asuncion, A.: UCI machine learning repository (2010)

# Noise Effects on Spatial Pattern Data Classification Using Wavelet Kernel PCA

## A Monte Carlo Simulation Study

Shengkun Xie[1], Anna T. Lawniczak[2], and Sridhar Krishnan[3]

[1] Global Management Studies, Ted Rogers School of Management,
Ryerson University
[2] Department of Math and Stat, University of Guelph, Guelph, ON N1G 2W1, CA
[3] Department of Electrical and Computer Engineering,
Ryerson University, Toronto, ON, Canada

**Abstract.** The kernel-based feature extraction method is of importance in applications of artificial intelligence techniques to real-world problems. It extends the original data space to a higher dimensional feature space and tends to perform better in many non-linear classification problems than a linear approach. This work makes use of our previous research outcomes on the construction of wavelet kernel for kernel principal component analysis (KPCA). Using Monte Carlo simulation approach, we study noise effects of the performance of wavelet kernel PCA in spatial pattern data classification. We investigate how the classification accuracy change when feature dimension is changed. We also compare the classification accuracy obtained from the single-scale and multi-scale wavelet kernels to demonstrate the advantage of using multi-scale wavelet kernel in KPCA. Our study show that multi-scale wavelet kernel performs better than single-scale wavelet kernel in classification of data that we consider. It also demonstrates the usefulness of multi-scale wavelet kernels in application of feature extraction in kernel PCA.

**Keywords:** Wavelet Kernel, Kernel Principal Component Analysis, Spatial Pattern Data, Non-linear Classification.

## 1 Introduction

Principal component analysis (PCA) has been broadly used for feature extraction of high dimensional data classification problems. The objective of PCA is to map the data attributes into a new feature space that contains better, i.e. more linearly separable, features than those in the original input space. However, as the standard PCA is linear in nature, the projections in the new feature space do not always yield meaningful results for classification purposes. For example, the linear classifiers do not perform well when groups are separated by a quadratic function. One possible solution to this problem is to introduce to the data new dimensions that combine the original data attributes in a non-linear fashion such as in a polynomial relation. In this way, the resulting number of new dimensions may be so

C. Guo, Z.-G. Hou, and Z. Zeng (Eds.): ISNN 2013, Part I, LNCS 7951, pp. 273–282, 2013.

large that the task of generating new features and calculating eigenvalue decompositions becomes computationally expensive. Various kernel methods have been applied successfully in statistical learning and data analysis, including data classification and regression, to address the question of non-linearity (e.g., [1,2,3,4,5]). The introduction of the kernel allows working implicitly in some extended feature space, while doing all computations in the original input space. Dot-product kernels in the extended feature space, that are expressed in terms of kernel functions in the original input space, are often used for non-linear classification and regression problems [6]. The major consequence of the use of dot-product kernels is that *"any algorithm which only uses scalar product can be turned to nonlinear version of it, using kernel methods "* [7].

Although kernel-based classification methods enable capturing of non-linearity of data attributes in the feature space, they are usually sensitive to the choices of kernel parameters [1]. Additionally, in KPCA ([8,9]), optimization of kernel parameters is difficult. The search of hyper-parameters via cross-validation methods can be computationally expensive because of many possible choices of parameters values [7]. This calls for the construction of a type of kernel [10,11] that performs well in KPCA. Besides the choice of a kernel and the determination of its parameters, another important issue is the feature dimensions. The classification accuracy for a given data set may highly depend on the choice of feature dimensions. This implies that investigation of classification accuracy related to the feature dimensions in data classification is important, particularly, for wavelet kernel. Because of this, a simulation study was conducted to illustrate the performance of wavelet kernel in kernel PCA and to see how classification results are related to feature dimensions.

In this paper, we first discuss the KPCA method and the wavelet kernels. Next, a simulation model is proposed to produce training and test data. Wavelet KPCA is then applied to extract data features and a linear classifier is used to determine their group memberships of test data. Finally, we discuss the results and summarize our findings.

## 2   Methods

### 2.1   Kernel PCA

PCA is a powerful tool in analyzing multivariate data and has become very useful for feature extraction of high dimensional data in the machine learning context. The underlying assumption of this statistical approach is the linearity among the data attributes, which may not be always the case in real-world applications. As an extension of PCA, KPCA spans the original variable space in terms of higher dimensional basis functions through a feature map. Mathematically, the feature map transforms the original observations $\mathbf{x}_1, \ldots, \mathbf{x}_n \in R^d$ into column vectors $\Phi(\mathbf{x}_1), \ldots, \Phi(\mathbf{x}_n)$, where $d$ represents the original data dimension and $\Phi(\cdot)$ is a kernel function. Finding principal components in KPCA can be done through the eigenvalues decomposition of kernel matrix $\mathbf{K}$, whose element are defined by:

$$k(\mathbf{x}_i, \mathbf{x}_j) = \Phi(\mathbf{x}_i) \cdot \Phi(\mathbf{x}_j)^\top, \ 1 \leq i, j \leq n. \tag{1}$$

The eigenvalues decomposition produces a set of eigenvectors, denoted by $V^l = (v_1^l, v_2^l, \ldots, v_n^l)^\top$, for $l = 1, 2, \ldots, n$, and the data features in the mapped feature space can be extracted by

$$\Phi(\mathbf{x}) \cdot V^l = \sum_{i=1}^{n} v_i^l k(\mathbf{x}_i, \mathbf{x}). \tag{2}$$

Due to the extension of the dimension of feature space from $d$ to $n$, the extracted features often appear to be more linear than the features extracted from PCA, particularly for a small $d$. However, in practice, it is important to retain a low dimensional feature vector in order to facilitate the classification process. This requires an investigation of how wavelet KPCA as a feature extraction method perform in data classification.

## 2.2   Wavelet Kernel

Besides the dot-product type kernel defined in (1), the translation invariant kernel is also popular, for example, the wavelet kernel. The advantage of using a dot-product type kernel is that it guarantees the required property of being Mercer kernel (i.e., non-negative definite kernel matrix) while the translation invariant kernel requires a proof of this property. Although the proof of being Mercer kernel is often difficult, the translation invariant kernel is still desirable as it has a reduced set of kernel parameters, which makes it more practical.

**Single-Scale Wavelet Kernel.** A signle-scale wavelet kernel is a function that uses a mother wavelet at a particular scale, but various translation factors are used for all variables within $\mathbf{x}_i$. A single-scale dot-product wavelet kernel (SDWK) is defined as

$$k(\mathbf{x}, \mathbf{y}) = \prod_{i=1}^{d} \psi(\frac{x_i - b_i}{a}) \psi(\frac{y_i - b_i}{a}), \tag{3}$$

where $\psi(x)$ is a mother wavelet function, $a \in R^+$ is the unique dilation factor and $b_i \in R$ are the translation factors, for each variable $x_i$ in $\mathbf{x}_i$ and $y_i$ in $\mathbf{y}_i$. A single-scale translation invariant wavelet kernel (STIWK) is defined as

$$k(\mathbf{x}, \mathbf{y}) = \prod_{i=1}^{d} \psi(\frac{x_i - y_i}{a}). \tag{4}$$

In practice, the parameter $a$ in both (3) and (4) needs to be estimated. This is possible for supervised problems, but it is not feasible when dealing with unsupervised problems as it requires a tedious cross-validation process in which all kinds of possible combinations of parameter values need to be considered.

**Multi-Scale Wavelet Kernel.** A multi-scale dot-product wavelet kernel (MDWK) ([10,11]) is defined as

$$k(\mathbf{x}, \mathbf{y}) = \prod_{i=1}^{d} \sum_{j=0}^{J-1} \sum_{k=0}^{2^{J-j}-1} \lambda_i \psi(\frac{x_i - b_k}{a_j}) \cdot \psi(\frac{y_i - b_k}{a_j}), \tag{5}$$

and a multi-scale translation invariant wavelet kernel (MTIWK) is given by

$$k(\mathbf{x}, \mathbf{y}) = \prod_{i=1}^{d} \sum_{j=0}^{J-1} \lambda_i \psi(\frac{x_i - y_i}{a_j}), \tag{6}$$

where, respectively, $\lambda_i = \frac{1}{a_j}$ in (5) and $\lambda_i = \frac{1}{J}$ in (6). The benefit of using the multi-scale wavelet kernel results from the fact that the kernel is a sum of multiple kernel functions under different wavelet scales. The kernel aims, potentially, to capture the multi-scale behavior of data and its application leads to a better performance of feature extraction, without a search of optimal parameter set within the parameter space. The values of $a_j$ are in powers of 2, that is $a_j \in \{1, 2^{0.25}, \ldots, 2^{0.25j}, \ldots, 2^{0.25(J-1)}\}$ for a given level $J$, which is 6 in this paper. For each $a_j$, the sequence $b_k$ is selected as $b_k = k u_0 a_j$, as suggested by Daubechies (1992). Here, $u_0$ controls the resolution of $b_k$ and is set to be 0.5. The range of $k$ is the set $\{0, 1, \ldots, 10\}$ which is determined by the border of the mother wavelet function. In this paper, the mother wavelet functions that we consider are Gaussian and Mexican hat wavelets.

## 2.3    Simulation Model

In this simulation study, we consider a two-class two dimensional heterogeneous clustered data, denoted by $\mathcal{D}=\{\mathbf{x}_i, \mathbf{y}_i: i=1, \ldots, n\}$, where each $\mathbf{x}_i=(x_i^1, x_i^2)$ represents the data of Cluster 1, each $\mathbf{y}_i=(y_i^1, y_i^2)$ is the data of Cluster 2, and $n$ is the total number of data points of each cluster. The simulation model is given by the following expressions

$$x_i^1 = x_0^1 + \sigma_x e_i; \quad x_i^2 = x_0^2 + \sigma_x e_i, \tag{7}$$
$$y_i^1 = y_0^1 + \sigma_y e_i; \quad y_i^2 = y_0^2 + \sigma_y e_i, \tag{8}$$

where $(x_0^1, x_0^2)$ and $(y_0^1, y_0^2)$ are the coordinates of the centers of Cluster 1 and Cluster 2, respectively; $\sigma_x$ and $\sigma_y$ are the standard deviations of the data on each dimension of Cluster 1 and Cluster 2, respectively, and $e_i \sim N(0,1)$. An example of the simulation data is presented in Fig. 1. From Fig. 1 we see that the type of data we investigate is related to multivariate non-stationary signals. Indeed, the spatial pattern data shown in Fig. 1 (a) and (c) can be treated as two-dimensional two-regime switching signals. This type of signals are often seen in biomedical and financial applications. As one increases the value of $\sigma_x$ or $\sigma_y$ or both, the separability of these two-dimensional data become lower and lower. It is of interest to see how classification accuracy is affected by the increase of noise level and which type of wavelet kernel, i.e. single-scale or multi-scale, performs more robustly to the added noises.

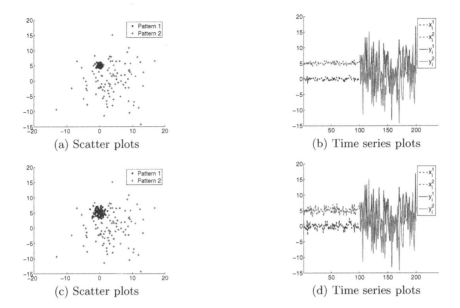

**Fig. 1.** Scatter plots and time series plots for heterogeneous clustered simulation data with $\sigma_x = 0.5$ & $\sigma_y = 5$ (a-b) and with $\sigma_x = 1$ & $\sigma_y = 5$ (c-d)

## 3    Results

We simulate the training and test data using the simulation model described in Section 2.3. We set the model parameters as $x_0^1=0$, $x_0^2=5$, $y_0^1=4$, $y_0^2=0$, $\sigma_x=1$, $\sigma_y=5$ and $n=100$, respectively, for simulating both sets, i.e. the training and test data. To simplify the problem, we fix the value of $\sigma_y$ to be 5 and take different values for $\sigma_x$. The values of $\sigma_x$ are taken as $\sigma_x^1=0.1$, $\sigma_x^2=0.2$, ..., and $\sigma_x^{30}=3$, respectively. We use each value of $\sigma_x$ for the simulation of the training and test data. We denote the training data and the test data by $\mathcal{D}_1^r$ and $\mathcal{D}_2^r$, respectively, for $r=1, 2, \ldots, 30$.

To each training data $\mathcal{D}_1^r$ and to each test data $\mathcal{D}_2^r$, first, we apply WKPCA, respectively, to map the original data into the feature space. We then investigate the effect of feature dimension to the classification accuracy with respect to different experimental setups, i.e. different values of $\sigma_x$ and different choices of wavelet kernel function. Fig. 2 and 3 report the results of classification accuracy. When $\sigma_x$ is small, i.e. $\sigma_x < 2.5$, the classification accuracy converges at a small value of feature dimension, i.e. 4. The increase of feature dimension does not lead to an improvement of classification performance of WKPCA. However, for larger $\sigma_x$, e.g. $\sigma_x > 2.5$, the classification accuracy increases with the increase of feature dimension. This implies that for highly noisy data, a larger value of feature dimension is required in order for obtaining a higher precision of classification. Also, generally speaking, both types of wavelet kernel end up with a similar pattern of classification results. With the increase of noise level the

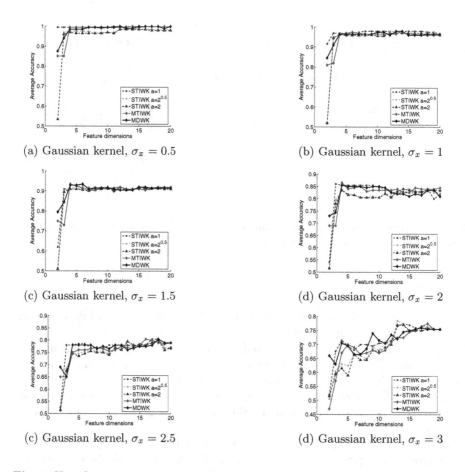

(a) Gaussian kernel, $\sigma_x = 0.5$

(b) Gaussian kernel, $\sigma_x = 1$

(c) Gaussian kernel, $\sigma_x = 1.5$

(d) Gaussian kernel, $\sigma_x = 2$

(c) Gaussian kernel, $\sigma_x = 2.5$

(d) Gaussian kernel, $\sigma_x = 3$

**Fig. 2.** Classification accuracy rate for heterogeneous clustered data classification with Gaussian kernel, various values of $\sigma_x$ and various feature extraction methods, namely STIWKPCA, MTIWKPCA and SDWKPCA, with respect to different numbers of retained feature dimensions

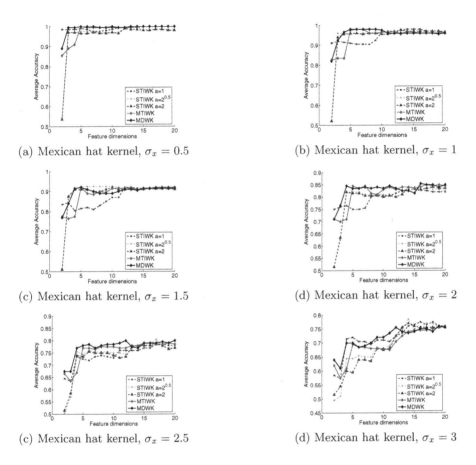

(a) Mexican hat kernel, $\sigma_x = 0.5$

(b) Mexican hat kernel, $\sigma_x = 1$

(c) Mexican hat kernel, $\sigma_x = 1.5$

(d) Mexican hat kernel, $\sigma_x = 2$

(c) Mexican hat kernel, $\sigma_x = 2.5$

(d) Mexican hat kernel, $\sigma_x = 3$

**Fig. 3.** Classification accuracy rate for heterogeneous clustered data classification with Mexican hat kernel, various values of $\sigma_x$ and various feature extraction methods, namely STIWKPCA, MTIWKPCA and SDWKPCA, with respect to different numbers of retained feature dimensions

**Table 1.** The table reports the classification accuracy for the simulation data with selected values of $\sigma_x$ and different types of wavelet kernel based on Gaussian mother wavelet function in data classification with feature dimensions 4, for 100 Monte Carlo simulations

| Gaussian | $\sigma_x$=.1 | $\sigma_x$=0.3 | $\sigma_x$=0.5 | $\sigma_x$=0.7 | $\sigma_x$=0.9 | $\sigma_x$=1.1 | $\sigma_x$=1.3 | $\sigma_x$=1.5 |
|---|---|---|---|---|---|---|---|---|
| STIWK | | | | | | | | |
| $a=1$ | 0.9441 | 0.9434 | 0.9417 | 0.9392 | 0.9345 | 0.9289 | 0.9212 | 0.9122 |
| $a=2^{0.25}$ | 0.9136 | 0.9128 | 0.9119 | 0.9100 | 0.9078 | 0.9033 | 0.8974 | 0.8912 |
| $a=2^{0.5}$ | 0.9002 | 0.9001 | 0.8991 | 0.8981 | 0.8962 | 0.8938 | 0.8892 | 0.8836 |
| $a=2^{0.75}$ | 0.8965 | 0.8962 | 0.8956 | 0.8943 | 0.8931 | 0.8904 | 0.8863 | 0.8809 |
| $a=2$ | 0.8953 | 0.8946 | 0.8936 | 0.8931 | 0.8920 | 0.8899 | 0.8857 | 0.8803 |
| $a=2^{1.25}$ | 0.8945 | 0.8942 | 0.8933 | 0.8927 | 0.8920 | 0.8899 | 0.8855 | 0.8802 |
| MTIWK | 0.9236 | 0.9228 | 0.9211 | 0.9188 | 0.9158 | 0.9126 | 0.9074 | 0.8989 |
| MDWK | 0.9194 | 0.9188 | 0.9183 | 0.9165 | 0.9135 | 0.9112 | 0.9078 | 0.9011 |

**Table 2.** The table reports the classification accuracy for the simulation data with selected values of $\sigma_x$ and different types of wavelet kernel based on Mexican hat mother wavelet function in data classification with 4 features, for 100 Monte Carlo simulations

| Mexican hat | $\sigma_x$=.1 | $\sigma_x$=0.3 | $\sigma_x$=0.5 | $\sigma_x$=0.7 | $\sigma_x$=0.9 | $\sigma_x$=1.1 | $\sigma_x$=1.3 | $\sigma_x$=1.5 |
|---|---|---|---|---|---|---|---|---|
| STIWK | | | | | | | | |
| $a=1$ | 0.9697 | 0.9692 | 0.9680 | 0.9663 | 0.9610 | 0.9488 | 0.9332 | 0.9112 |
| $a=2^{0.25}$ | 0.9238 | 0.9235 | 0.9225 | 0.9210 | 0.9192 | 0.9155 | 0.9109 | 0.9046 |
| $a=2^{0.5}$ | 0.9064 | 0.9062 | 0.9059 | 0.9052 | 0.9038 | 0.9012 | 0.8969 | 0.8912 |
| $a=2^{0.75}$ | 0.9035 | 0.9035 | 0.9029 | 0.9020 | 0.9011 | 0.8991 | 0.8943 | 0.8882 |
| $a=2$ | 0.9028 | 0.9027 | 0.9023 | 0.9016 | 0.9007 | 0.8987 | 0.8940 | 0.8876 |
| $a=2^{1.25}$ | 0.9027 | 0.9026 | 0.9020 | 0.9013 | 0.9003 | 0.8985 | 0.8938 | 0.8877 |
| MTIWK | 0.9409 | 0.9402 | 0.9379 | 0.9352 | 0.9317 | 0.9267 | 0.9204 | 0.9111 |
| MDWK | 0.9457 | 0.9455 | 0.9438 | 0.9418 | 0.9382 | 0.9337 | 0.9257 | 0.9143 |

performance of different types of wavelet kernel varies, and multi-scale wavelet kernel outperforms the single-scale wavelet kernel when a low dimensional feature vector is considered. However, this advantage disappears when the level of noise is high, i.e. $\sigma_x > 3$.

Tables 1-4 reports the results corresponding to some selected values of $\sigma_x$ and different choices of wavelet kernel, under the optimal feature dimension, i.e. 4 for small value of $\sigma_x$, and 20 when $\sigma_x$ is large. The obtained results are based on 100 runs of Monte Carlo simulation of different training and test data sets. From these tables, one can see that, under the optimal choice of feature dimension, a multi-scale kernel performs more robustly than a single-scale kernel for both the Gaussian kernel and the Mexican hat kernel. In the presence of a higher level of added noise, both single-scale and multi-scale wavelet kernels lead to similar results no matter what mother wavelet function is chosen. The improvement of classification accuracy for wavelet kernel multi-scale decreases as one increases the noise level. Although the dimensional feature vector can be extended when

**Table 3.** The table reports the classification accuracy for the simulation data with selected values of $\sigma_x$ and different types of wavelet kernels based on Gaussian mother wavelet function in data classification with feature dimensions 20, for 100 Monte Carlo simulations

| Gaussian | $\sigma_x=1.7$ | $\sigma_x=1.9$ | $\sigma_x=2.1$ | $\sigma_x=2.3$ | $\sigma_x=2.5$ | $\sigma_x=2.7$ | $\sigma_x=2.9$ |
|---|---|---|---|---|---|---|---|
| STIWK | | | | | | | |
| $a=1$ | 0.9054 | 0.8881 | 0.8738 | 0.8593 | 0.8462 | 0.8333 | 0.8217 |
| $a=2^{0.25}$ | 0.9026 | 0.8855 | 0.8691 | 0.8543 | 0.8423 | 0.8297 | 0.8179 |
| $a=2^{0.5}$ | 0.8995 | 0.8821 | 0.8664 | 0.8518 | 0.8387 | 0.8268 | 0.8154 |
| $a=2^{0.75}$ | 0.8978 | 0.8812 | 0.8651 | 0.8501 | 0.8374 | 0.8267 | 0.8148 |
| $a=2$ | 0.8970 | 0.8807 | 0.8645 | 0.8496 | 0.8372 | 0.8263 | 0.8145 |
| $a=2^{1.25}$ | 0.8945 | 0.8824 | 0.8683 | 0.8545 | 0.8412 | 0.8290 | 0.8165 |
| MTIWK | 0.9041 | 0.8875 | 0.8719 | 0.8592 | 0.8458 | 0.8326 | 0.8205 |
| MDWK | 0.9039 | 0.8859 | 0.8721 | 0.8582 | 0.8457 | 0.8325 | 0.8198 |

**Table 4.** The table reports the classification accuracy for the simulation data with selected values of $\sigma_x$ and different types of wavelet kernels based on Mexican hat mother wavelet function in data classification with feature dimensions 20, for 100 Monte Carlo simulations

| Mexican hat | $\sigma_x=1.7$ | $\sigma_x=1.9$ | $\sigma_x=2.1$ | $\sigma_x=2.3$ | $\sigma_x=2.5$ | $\sigma_x=2.7$ | $\sigma_x=2.9$ |
|---|---|---|---|---|---|---|---|
| STIWK | | | | | | | |
| $a=1$ | 0.9028 | 0.8851 | 0.8694 | 0.8550 | 0.8389 | 0.8238 | 0.8090 |
| $a=2^{0.25}$ | 0.9035 | 0.8867 | 0.8707 | 0.8555 | 0.8429 | 0.8310 | 0.8197 |
| $a=2^{0.5}$ | 0.8999 | 0.8827 | 0.8676 | 0.8515 | 0.8393 | 0.8277 | 0.8160 |
| $a=2^{0.75}$ | 0.8980 | 0.8816 | 0.8650 | 0.8506 | 0.8377 | 0.8270 | 0.8147 |
| $a=2$ | 0.8974 | 0.8806 | 0.8645 | 0.8498 | 0.8373 | 0.8263 | 0.8145 |
| $a=2^{1.25}$ | 0.8958 | 0.8805 | 0.8653 | 0.8505 | 0.8378 | 0.8255 | 0.8138 |
| MTIWK | 0.9032 | 0.8881 | 0.8724 | 0.8595 | 0.8474 | 0.8327 | 0.8197 |
| MDWK | 0.9030 | 0.8866 | 0.8724 | 0.8593 | 0.8470 | 0.8336 | 0.8201 |

a high level of noise is present, the classification accuracy cannot be improved even for multi-scale wavelet kernels.

## 4   Conclusion

In this paper, a new type of kernel, the multi-scale wavelet kernel, was proposed for KPCA in a feature extraction problem. Based on analysis of simulated data, we observed that application of multi-scale wavelet kernels in kernel PCA improves the classification accuracy rates when the kernel parameters are not cross-validated. As our examples show, the use of multi-scale wavelet kernels in kernel PCA is promising for improving the robustness of the classification performance and achieving a moderate level of accuracy without the search of the values of a kernel parameter of a given kernel. Our study suggests that the multi-scale WKPCA outperforms the single-scale WKPCA. Although multi-scale WKPCA may not lead to the highest classification accuracy rates, it is highly desirable as a feature extraction method for multi-scale data due to its ability of explaining a higher level of linearity in the feature space.

**Acknowledgements.** S. Xie acknowledges the financial support from MITACS and Ryerson University, under MITACS Elevate Strategic Post-doctoral Award. A. T. Lawniczak acknowledges support from NSERC (Natural Sciences and Engineering Research Council of Canada).

# References

1. Muller, K.R., Mika, S., Ratsch, G., Tsuda, K., Scholkopf, B.: An introduction to kernel-based learning alrorithms. IEEE Transactions on Netural Networks 12(2), 181–201 (2001)
2. Zhu, M.: Kernels and ensembles: perspectives on statistical learning. The American Statistician 62, 97–109 (2008)
3. Karatzoglou, A., Smola, A., Hornik, K., Zeileis, A.: Kernlab - an S4 package for kernel methods in R. Journal of Statistical Software 11(9), 1–20 (2004)
4. Vapnik, V.: The Nature of Statistical Learning Theory. Springer, NY (1995)
5. Vapnik, V.: Statistical Learning Theory. Wiley, New York (1998)
6. Chen, G.Y., Bhattacharya, P.: Function dot product kernels for support vector machine. In: 18th International Conference on Pattern Recognition (ICPR 2006), vol. 2, pp. 614–617 (2006)
7. Scholkopf, B., Smola, A.J., Muller, K.R.: Nonlinear component analysis as a kernel eigenvalue problem. Neur. Comput. 10, 1299–1319 (1998)
8. Takiguchi, T., Ariki, Y.: Robust Feature Extraction Using Kernel PCA. In: ICASSP 2006, pp. 509–512 (2006)
9. Scholkopf, B., Smola, A., Muller, K.R.: Nonlinear Component Analysis as a Kernel Eigenvalue Problem. Technical Report 44, Max-Planck-Institut fur biologische Kybernetik Arbeitsgruppe Bulthoff, Tubingen (1996)
10. Xie, S., Lawniczak, A.T., Krishnan, S., Liò, P.: Wavelet Kernel Principal Component Analysis in Noisy Multi-scale Data Classification. ISRN Computational Mathematics 2012, Article ID 197352, 13 Pages (2012), doi:10.5402/2012/197352
11. Xie, S., Lawniczak, A.T., Liò, P.: Feature Extraction Via Wavelet Kernel PCA for Data Classification. In: Proceedings of 2010 IEEE International Workshop on Machine Learning for Signal Processing, pp. 438–443 (2010)

# SVM-SVDD: A New Method to Solve Data Description Problem with Negative Examples

Zhigang Wang[1], Zeng-Shun Zhao[2], and Changshui Zhang[1,*]

[1] Department of Automation, Tsinghua University
State Key Lab of Intelligent Technologies and Systems
Tsinghua National Laboratory for Information Science and Technology(TNList),
Beijing, P.R. China
zcs@mail.tsinghua.edu.cn
[2] College of Information and Electrical Engineering
Shandong University of Science and Technology, Qingdao, P.R. China

**Abstract.** Support Vector Data Description(SVDD) is an important method to solve data description or one-class classification problem. In original data description problem, only positive examples are provided in training. The performance of SVDD can be improved when a few negative examples are availablewhich is known as SVDD_neg. Intuitively, these negative examples should cause an improvement on performance than SVDD. However, the performance of SVDD may become worse when some negative examples are available. In this paper, we propose a new approach "SVM-SVDD", in which Support Vector Machine(SVM) helps SVDD to solve data description problem with negative examples efficiently. SVM-SVDD obtains its solution by solving two convex optimization problems in two steps. We show experimentally that our method outperforms SVDD_neg in both training time and accuracy.

## 1 Introduction

Binary classification problem has been studied carefully in the field of machine learning. Given two classes of examples labeled +1 and -1, the binary classification task is to learn a classifier to predict whether the label of one unseen example is +1 or -1. Many classification algorithms have been developed such as SVM[1], Boosting[2]. However, examples from only one class are provided and no or only a few examples from non-given class in some applications. It is required to learn a model to detect whether one unseen example comes from given class or not. This problem is called data description or one-class classification[3]. Those much different from the given examples would be taken as non-given class. Here, the given class example is called "positive example" or "target" and non-given class example is called "negative example" or "outlier". Data description problem is usually caused by the fact that examples from one class can be collected easily while those from non-given class are very difficult to obtain. It is obvious that

---

* Corresponding author.

C. Guo, Z.-G. Hou, and Z. Zeng (Eds.): ISNN 2013, Part I, LNCS 7951, pp. 283–290, 2013.

the data description problem, which happens frequently in real life, cannot be solved directly by binary classification algorithms.

To solve data description problem, [4] adapted classical two-class SVM to one-class SVM, whose idea is to separate the given class examples from origin in feature space according to the principle of maximum margin. [5] proposed SVDD based on the idea of SVM. It makes the hypothesis that the examples from the given class should be inside a supersphere while non-given class examples should be outside. SVDD has become a popular method for its intuitive idea and good performance. It has been applied successfully to many real applications such as remote sensing[6][7], face detection and recognition[8][9], fault detection[10], document retrieval[11].

If a few outliers are available, they can be used to improve the performance with only targets. SVDD_neg[5], as an extension of SVDD, can solve the problem of data description with a few negative examples. But SVDD_neg often gets worse performance than SVDD[5]. Furthermore, the training of SVDD_neg is time consuming and the global optima is difficult to obtain for its non-convex formulation. This paper proposes an approach SVM-SVDD to data description problem with negative examples. The experimental results show that the proposed SVM-SVDD achieves better performances and less training time than SVDD_neg in benchmark data sets.

## 2 Our Method

### 2.1 SVDD and SVDD_neg

At the beginning, we present a brief review of SVDD and SVDD_neg[5]. Assume that we have a training set $\{x_i\}, i = 1, 2, ..., N$, which are targets for training in SVDD setting. The task is to learn a predictor that tests whether any new example is a target or not.

The hypothesis of SVDD is that targets should locate inside a close boundary in the feature space. The boundary is modeled as a supersphere. Two parameters can describe a supersphere: center $a$ and radius $R$. The goal is to obtain a supersphere that encloses nearly all training examples with the smallest radius $R$. This optimization problem is written as follows:

$$\min R^2 + C \sum_{i=1}^{N} \xi_i$$
$$\text{s.t.} \quad \|x_i - a\|^2 \leq R^2 + \xi_i, i = 1, 2, ..., N, \xi_i \geq 0,$$

(1)

where $\xi_i$ are the slack variables like those in SVM and $C$ penalizes the loss term $\sum \xi_i$. If error rate of SVDD is expected to be less than a given threshold, $C$ is set:

$$C = \frac{1}{N \cdot \text{threshold}} \geq 0.$$

(2)

The parameter $a$ is computed as:

$$a = \sum_{i=1}^{N} \alpha_i x_i, 0 \leq \alpha_i \leq C. \tag{3}$$

Similar to SVM, only a small fraction of $\alpha_i$ are non-zero.

To test an observation $v$, the distance from $v$ to center $a$ should be calculated:

$$\begin{aligned} \|v - a\|^2 \leq R^2 &\Rightarrow \text{v is target.} \\ \|v - a\|^2 > R^2 &\Rightarrow \text{v is outlier.} \end{aligned} \tag{4}$$

Based on SVDD, SVDD_neg is given below. The outliers available should be outside the supersphere. Thus, the distance from an outlier to the center $a$ should be larger than $R$. Assume that we have a training set containing $N$ targets $\{x_i\}$, $i = 1, 2, ..., N$ and $M$ outliers $\{x_j\}$, $j = N + 1, ..., N + M$. The optimization problem of SVDD_neg is written as follows:

$$\min R^2 + C_1 \sum_{i=1}^{N} \xi_i + C_2 \sum_{j=N+1}^{N+M} \xi_j,$$
$$\text{s.t.} \quad \|x_i - a\|^2 \leq R^2 + \xi_i, i = 1, 2, ..., N, \xi_i \geq 0,$$
$$\|x_j - a\|^2 \geq R^2 - \xi_j, j = N + 1, N + 2, ..., N + M, \xi_j \geq 0. \tag{5}$$

The same as (Eq. (2)), $C_1$ and $C_2$ are to control the error rates of targets and outliers respectively. For example, if we can accept 5% error rate for targets and 1% error rate for outliers, $C_1 = \frac{1}{0.01 \cdot N}$ and $C_2 = \frac{1}{0.05 \cdot M}$.

## 2.2  SVM-SVDD

It can be seen that SVDD_neg and SVM have the similar ideas and forms. If a few outliers is available to training, SVDD_neg is expected to achieve better performance than SVDD. However, performance of SVDD_neg may become worse especially when there are overlap areas between targets and outliers in feature space[5]. In our opinions, this problem is caused by the following reasons. First, SVDD_neg only puts targets inside supershere and outliers outside supersphere, which does not separate outliers far from targets. Compared with SVM, SVDD_neg does not use the concept of "margin", which acts an important role in separating two classes well. When there are overlap areas between targets and outliers in the feature space, SVDD_neg usually excludes a few outlier outside at the cost of more targets inside supershere. Second, the close boundary in SVDD_neg has two tasks: to enclose most or all targets inside and separate targets from outliers. It is difficult for SVDD_neg to fulfill these two tasks well at the same time. In addition, due to the existence of outliers, the optimization problem Eq. (5) of SVDD_neg becomes a non-convex programming problem from a convex programming Eq. (1) only using targets. Therefore, it becomes more difficult to solve SVDD_neg. Non-convex programming is likely to converge to a local minimum rather than a global minimum.

Based on the reasons mentioned above, it is clear why performance of SVDD_neg becomes worse with some outliers in some cases. Here, it is natural to improve SVDD by means of SVM to solve data description with negative examples efficiently.

SVM[1][12] is a classic binary classification approach. There are two classes of examples, labeled +1 and -1. The class +1 examples' number is $N$ and the class -1 examples' number is $M$. The set $\{x_i\}, i = 1, ..., N + M$ are training examples. $\{y_i\}, i = 1, ..., N+M$ are the labels of $x_i$, in which $y_i = +1, i = 1, ..., N$ and $y_i = -1, i = N + 1, ..., N + M$. The task is to learn a classifier which separates the +1 class and -1 class as well as possibly.If linear kernel is applied, the form of prediction hyperplane is $y = wx + b$, $x$ denotes the observation to test and sign of $y$ is the predicted label of $x$. The kerneled discriminant function is $\text{sign}(\sum\limits_{i=1}^{N+M} y_i \alpha_i \text{ker}(x_i \cdot x) + b)$. The optimization formulation of SVM is[1]:

$$\min \frac{\|w\|^2}{2} + C \sum_{i=1}^{N+M} \xi_i,$$
$$\text{s.t.} \quad wx_i + b \geq 1 - \xi_i, i = 1, ..., N,$$
$$wx_i + b \leq -1 + \xi_i, i = N + 1, ..., N + M. \tag{6}$$

$\xi_i$ are the introduced slack variables. $C$ is the same as the $C$ in Eq. (1), which controls the error rate.

Based on the analysis above, SVM-SVDD is proposed by the following formulation:

$$\min_{w,r} \frac{\|w\|^2}{2} + r^2 + C_0 \sum_{k=1}^{N} \zeta_k + C_1 \sum_{i=1}^{N} \xi_i + C_2 \sum_{j=N+1}^{N+M} \xi_j,$$
$$\text{s.t.} \quad wx_i + b \geq 1 - \xi_i, i = 1, ..., N$$
$$wx_j + b \leq -1 + \xi_j, j = N + 1, ..., N + M$$
$$\|x_k - a\|^2 \leq r^2 + \zeta_k, k = 1, 2, ..., N, \tag{7}$$
$$\xi_i \geq 0, \xi_j \geq 0, \zeta_k \geq 0.$$

Eq. (7) can replace Eq. (5) to solve the data description problem with negative examples. Eq. (7) is to find both a hyperplane $y = wx + b$ to separate targets and outliers and a supershpere $\|x - a\|^2 \leq r^2$ enclosing targets at the same time. $C_0$ is advised to be a large constant so that more targets can be enclosed inside supersphere. $C_1$ and $C_2$ are to control error rates of targets and outliers respectively. $C_1 < C_2$ is advised. $\xi$ and $\zeta$ are slack variables. The formulation Eq. (7) can be rewritten in dual form:

$$\min_{\alpha,\beta} \frac{1}{2} \sum_{i,j=1}^{N+M} y_i y_j \alpha_i \alpha_j ker_1(x_i, x_j) + \sum_{i,j=1}^{N} \beta_i \beta_j ker_2(x_i, x_j) - \sum_{i=1}^{N+M} \alpha_i - \sum_{i=1}^{N} \beta_i ker_2(x_i, x_i)$$

$$\text{s.t.} \sum_{i=1}^{M+N} y_i \alpha_i = 0, \quad 0 \le \alpha_i \le C_1 (i = 1, ..., N), \quad 0 \le \alpha_i \le C_2 (i = N+1, ..., N+M),$$

$$\sum_{i=1}^{N} \beta_i = 1, \quad 0 \le \beta_i \le C_0,$$

$$(8)$$

where $ker_1()$ and $ker_2()$ are two different kernel functions. $y_i = +1, i = 1, ..., N$ and $y_i = -1, i = N+1, ..., N+M$. These two kernel functions make SVM-SVDD more flexible. $\alpha = (\alpha_1, ..., \alpha_{N+M})^T$ and $\beta = (\beta_1, ..., \beta_N)^T$ can be solved separately in Eq. (8) for the independence between $\alpha$ and $\beta$. For example, we can solve $\alpha$ with $\beta$ fixed and then solve $\beta$ with $\alpha$. When $\alpha$ or $\beta$ is fixed, the formulation Eq. (8) becomes a convex quadratic programming problem. So the time complexity of Eq. (8) is $O((M + N)^2 + N^2) = O(M^2 + 2MN + 2N^2)$.

Only when both Eq. (9) and Eq. (10) hold at the same time, the observation $v$ is taken as a target. Otherwise it is an outlier.

$$f_1(v) = sign(\sum_{i=1}^{N+M} y_i \alpha_i ker_1(x_i, v) + b) = +1, \tag{9}$$

and

$$f_2(v) = sign(r^2 - ker_2(v, v) + 2\sum_{i=1}^{N} \alpha_i ker_2(v, x_i) - \sum_{i=1}^{N} \sum_{j=1}^{N} \alpha_i \alpha_j ker_2(x_i, x_j)) = +1. \tag{10}$$

## 3   Experiments

In this section, we compare SVM-SVDD with SVDD_neg on some benchmark data sets. We used "libsvm" [13] and "dd_tools" [14] in our experiments. These data sets are downloaded from UCI maching learning repository[15]. Table 1 gives the details on these data sets.

**Table 1.** Datasets description

| dataset | example        number (target/outlier) | dimension |
|---------|------------------|-----------|
| australian | 307/383 | 14 |
| breast-cancer | 239/444 | 10 |
| diabetes | 500/268 | 8 |
| haberman | 225/81 | 3 |
| german-credit | 700/300 | 24 |

Each of these data sets has targets and outliers with label +1 and -1. Because the data sets have no testing set, we use five-fold cross validation to train and test these data sets. The performances and speeds of both SVM-SVDD and SVDD_neg are mean values of 10 runs of training and testing in Table 2.

Table 2 shows the accuracies and training time on the data sets for comparison between SVM-SVDD and SVDD_neg. The parameters in optimization problems of both SVM-SVDD and SVDD_neg are fine-tuned by grid search. In the second column (training time(seconds)) of Table 2, the training time is those of SVM-SVDD. Here, the accuracy results include three indexes: target error (first error), outlier error(second error) and total error. These three indexes are defined according to four abbreviations: TT(True Target), TO(True Outlier), FT(False Target), FO(False Outlier). These three definitions are listed as follows[12]:

$$
\begin{aligned}
target \quad error &= \frac{FO}{TT + FO}, \\
outlier \quad error &= \frac{FT}{TO + FT}, \\
total \quad error &= \frac{FO + FT}{TT + TO + FT + FO}.
\end{aligned}
\tag{11}
$$

**Table 2.** The comparison of training time and accuracy(percentage) between SVM-SVDD and SVDD_neg is shown. In each blank of the Table 2, the number before "/" is the result of SVM-SVDD and after '/' is of SVDD_neg.

| data sets | training time (seconds) | target error (percentage) | outlier error (percentage) | total error (percentage) |
|---|---|---|---|---|
| australian | 0.144/1.399 | 17.32/31.58 | 11.21/11.74 | 13.92/20.57 |
| breast-cancer | 0.118/0.522 | 1.26/12.52 | 2.93/2.71 | 2.2/6.15 |
| diabetes | 0.165/2.219 | 13.2/27.8 | 41.38/40.68 | 23.05/32.30 |
| haberman | 0.102/0.498 | 8.44/26.22 | 66.54/60.29 | 23.86/35.29 |
| german-credit | 0.3131/1.8513 | 10.43/27.29 | 55.33/48.67 | 23.90/33.70 |

By comparison, the proposed SVM-SVDD achieves the less training time than SVDD_neg. On the other hand, SVM-SVDD outperforms SVDD_neg with the higher accuracies. In Table 2, SVM-SVDD achieves higher enhancement on target error at the cost of lower reduction on outlier error than SVDD_neg. The advantage of SVM-SVDD is illustrated by comparing the ROC curves of SVM-SVDD with SVDD_neg on these data sets in Figure 1. For each of data sets, ROC curve of SVM-SVDD locates above that of SVDD_neg. That means that at the same outlier acceptance rate, SVM-SVDD achieves higher target acceptance rate.

## 4    Conclusion

The task of SVDD_neg is to find a supersphere with targets inside and outliers outside. But when some outliers are available for training, the SVDD_neg can not improve the performance of training with only targets. The non-convex optimization problem of SVDD_neg is difficult to solve and time consuming. We

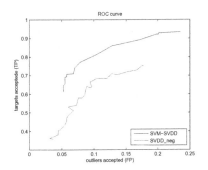

(a) The data set "australian" ROC curves comparison.

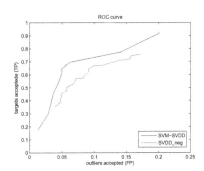

(b) The data set "breast-cancer" ROC curves comparison.

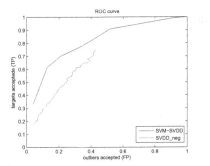

(c) The data set "diabetes" ROC curves comparison.

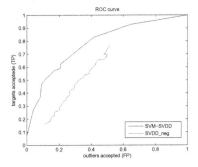

(d) The data set "german-credit" ROC curves comparison.

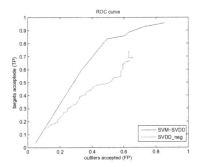

(e) The dataset "haberman" ROC curves comparison.

**Fig. 1.** The roc curves comparison between SVDD_neg and SVM-SVDD

propose a new approache SVM-SVDD to solve data description with negative examples efficiently. We also experimentally show that, SVM-SVDD outperforms SVDD_neg on both prediction accuracy and training time.

**Acknowledgement.** This work is supported by China State Key Science and Technology Project on Marine Carbonate Reservoir Characterization (2011ZX05004-003), 973 Program (2013CB329503) and Beijing Municipal Education Commission Science, Technology Development Plan key project (KZ201210005007) and Natural Science Foundation of Shandong Province (ZR2010FM027).

# References

1. Vapnik, V.: Statistical Learning Theory. Wiley, New York (1998)
2. Freund, Y., Schapire, R.: A desicion-theoretic generalization of on-line learning and an application to boosting. In: Vitányi, P.M.B. (ed.) EuroCOLT 1995. LNCS, vol. 904, pp. 23–37. Springer, Heidelberg (1995)
3. Moya, M., Koch, M., Hostetler, L.: One-class classifier networks for target recognition applications. Technical report, SAND-93-0084C, Sandia National Labs., Albuquerque, NM, United States (1993)
4. Scholkopf, B., Williamson, R., Smola, A., Shawe-Taylor, J.: SV estimation of a distributions support. Advances in Neural Information Processing Systems 41, 42–44 (1999)
5. Tax, D., Duin, R.: Support vector data description. Machine Learning 54(1), 45–66 (2004)
6. Sanchez-Hernandez, C., Boyd, D., Foody, G.: One-class classification for mapping a specific land-cover class: Svdd classification of fenland. IEEE Transactions on Geoscience and Remote Sensing 45(4), 1061–1073 (2007)
7. Sakla, W., Chan, A., Ji, J., Sakla, A.: An svdd-based algorithm for target detection in hyperspectral imagery. IEEE Geoscience and Remote Sensing Letters 8(2), 384–388 (2011)
8. Seo, J., Ko, H.: Face detection using support vector domain description in color images. In: Proceedings of IEEE International Conference on Acoustics, Speech, and Signal Processing (ICASSP 2004), vol. 5, pp. V–729. IEEE (2004)
9. Lee, S., Park, J., Lee, S.: Low resolution face recognition based on support vector data description. Pattern Recognition 39(9), 1809–1812 (2006)
10. Luo, H., Cui, J., Wang, Y.: A svdd approach of fuzzy classification for analog circuit fault diagnosis with fwt as preprocessor. Expert Systems with Applications 38(8), 10554–10561 (2011)
11. Onoda, T., Murata, H., Yamada, S.: Non-relevance feedback document retrieval based on one class svm and svdd. In: International Joint Conference on Neural Networks, IJCNN 2006, pp. 1212–1219. IEEE (2006)
12. Veropoulos, K., Campbell, C., Cristianini, N.: Controlling the sensitivity of support vector machines. In: Proceedings of the Sixteenth International Joint Conference on Artificial Intelligence, IJCAI 1999 (1999)
13. CRLIN, http://www.csie.ntu.edu.tw/~cjlin/libsvm/
14. TAX, http://ict.ewi.tudelft.nl/~davidt/dd_tools.html
15. UCI, http://archive.ics.uci.edu/ml/

# Applying Wavelet Packet Decomposition and One-Class Support Vector Machine on Vehicle Acceleration Traces for Road Anomaly Detection

Fengyu Cong[1], Hannu Hautakangas[1], Jukka Nieminen[1], Oleksiy Mazhelis[2], Mikko Perttunen[3], Jukka Riekki[3], and Tapani Ristaniemi[1]

[1] Department of Mathematical Information Technology, University of Jyväskylä, Finland
[2] Department of Computer Science and Information Systems, University of Jyväskylä
[3] Department of Computer Science and Engineering, University of Oulu, Finland
{Fengyu.Cong,Oleksiy.Mazhelis,Tapani.Ristaniemi}@jyu.fi,
{Hannu.Hautakangas,Nieminen.Jukka}@gmail.com,
{Mikko.Perttunen,JPR}@ee.oulu.fi

**Abstract.** Road condition monitoring through real-time intelligent systems has become more and more significant due to heavy road transportation. Road conditions can be roughly divided into normal and anomaly segments. The number of former should be much larger than the latter for a useable road. Based on the nature of road condition monitoring, anomaly detection is applied, especially for pothole detection in this study, using accelerometer data of a riding car. Accelerometer data were first labeled and segmented, after which features were extracted by wavelet packet decomposition. A classification model was built using one-class support vector machine. For the classifier, the data of some normal segments were used to train the classifier and the left normal segments and all potholes were for the testing stage. The results demonstrate that all 21 potholes were detected reliably in this study. With low computing cost, the proposed approach is promising for real-time application.

**Keywords:** Anomaly detection, one-class, pothole, road, support vector machine, wavelet packet decomposition.

## 1 Introduction

In recent years, road condition monitoring has become a popular research area due to intensive and still growing traffic that puts the road surface to constant stress. Intelligent systems for detecting bad road surface conditions can assist drivers to prevent damaging vehicles and even accidents and assist road management departments in timely discovering the need of maintenance on dangerous road conditions in time. Using GPS data and the acceleration measured by the accelerometer attached on some part of the driving car have been particularly used in the pothole and other anomalies detection for road management [9], [10], [17], [19]. Fig.1 shows the image of a pothole in the road [10] and Fig.2 demonstrates the accelerometer orientation [19]. This pioneering work has many advantages, especially in reducing the cost of the road management in

C. Guo, Z.-G. Hou, and Z. Zeng (Eds.): ISNN 2013, Part I, LNCS 7951, pp. 291–299, 2013.
© Springer-Verlag Berlin Heidelberg 2013

contrast to many other systems including laser profilometer measurement [12], ground penetrating radar [14], collection and analysis of images of road segments [4], [11], [15], [18], and so on. Furthermore, with such work, real-time intelligent systems become possible.

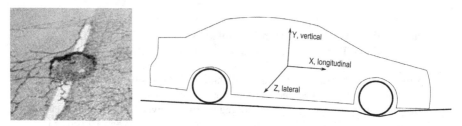

**Fig. 1.** Image of pothole [10]    **Fig. 2.** Accelerometer orientation

In [10], [17] accelerometer data were used for classifying road segments as the normal or containing anomaly by applying simple threshold values and a high-pass filter. In [19], classification was realized by machine learning methods, namely, with multi-class support vector machine (SVM) [7] on time-domain and frequency-domain features. However, these studies did not fully consider the properties of anomaly in the road condition monitoring. In this study, we propose a novel approach involving the feature extraction, feature selection and the classification with the consideration of the nature of detecting potholes in the road.

For the feature extraction, two types of information tend to be used according to the mechanism of the data formulation and the transformation of the collected data. Particularly, in the time or frequency domain, the peak amplitude of the collected data within certain duration and the power spectrum density (PSD) are often measured [10]. The peak amplitude is very sensitive and may be affected by many factors since it just contains the information within a very short duration. The PSD actually assumes the data are stationary [16], and this assumption might be correct when the road is smooth and the collected accelerations tend to be stationary. However, in case that a car passes the anomaly, for example, the pothole, the collected accelerations are transient in short duration and they are definitely not stationary any more [10]. Thus, the transformation for the non-stationary signal should be used in the feature extraction for the anomaly. From this point of view, the wavelet transformation based methods are appropriate candidates for the feature extraction [8]. Hence, the wavelet packet decomposition (WPD) is used for the feature extraction in this study [21].

The feature selection plays an important role in pattern recognition [1]. It assists to remove the redundant features and obtain the discriminative features. We try four of the generally used methods in this study. They are forward selection (FS) [1], backward selection (BS) [1], genetic algorithm (GA) [22] and principal component analysis (PCA) [1]. After comparing the performance in detecting potholes in the road, we may determine the best method for feature selection in our study.

After the feature is extracted, a machine learning based classifier is usually exploited to recognize the feature. There are mainly three ways to construct such classifiers including the supervised, the semi-supervised and the unsupervised. For an anomaly detection problem, there are often two classes of samples to recognize [2]. However, since the number of samples of the anomaly is usually much smaller than that of the normal, the supervised method is not appropriate, but the semi-supervised and the unsupervised methods are [2]. The semi-supervised approach implies that the classifier is trained with normal data, and then tested with an independent set of normal and anomaly data. This is often named as the one-class classification which is also referred as outlier detection, novelty detection or anomaly detection [2]. In this problem, the data instances that do not belong to the normal data are separated. Here, the one-class SVM [20] is applied.

In this study, accelerometer data were collected by the system used in [19]. Part of one car's normal data was used for training the classifier and the left normal data of that car and pothole data of three cars were for the testing stage. All 21 potholes were successfully detected by the proposed approach.

## 2    Method

### 2.1    Data Description

In order to make reliable analysis one must ensure high quality of the data. Further data analysis is much harder or even impossible if the data is invalid or badly corrupted. The accelerometer data in this work were collected by Perttunen et al. from the University of Oulu in Finland. The data were collected using a Nokia N95 mobile phone mounted on the car's dashboard. The N95 mobile phone has a built-in 3-axis accelerometer and a GPS receiver. The sampling frequency of the accelerometer was 38 Hz. The GPS data were not used in this study.

Three different cars were used to collect data. Each drive was also recorded using a video camera, which was attached to the head rest of the passenger's seat. The video was synchronized with accelerometer data after the drive so that it could be seen on the video when the car hits a road anomaly, for example a speed bump. This allows marking certain measurements of the accelerometer as anomalies. During the video analysis, it was noticed that it was impossible to observe when exactly the anomaly begins and ends in the level of milliseconds. It was also hard to classify the anomalies into different categories at the same time. Of course, some of the categories are easy to classify, for example speed bumps associated with potholes. But other kinds of anomaly were not easy to recognize from the video. So, the classification in this study was not targeted to recognize different potholes, but to discriminate potholes from normal road conditions, i.e., to detect potholes.

While the data were labeled, 21 segments with potholes were extracted from three drives for investigation, and 1764 normal segments from one car were used for training and testing the classifier. Normal segments of the other two roads were not used for analysis since the roads were not very smooth. The duration of each segment was three seconds and a Hamming window was used. Each segment of the pothole was arranged in the center of the segment to avoid energy leakage. Although accelerations in three directions were measured, we found the y-axis data mostly revealed the changes of accelerations when a car passed a pothole in contrast to the normal road conditions. Therefore, accelerations in this direction were chosen for analysis in this study. Fig.3 shows two segments containing the maximum and minimum powerful potholes and normal segments. The data were filtered by a 1-5 Hz band-pass filter for better visualization.

## 2.2    Feature Extraction

Feature extraction is by WPD here. The decomposition of a signal by WPD can be illustrated by successive low-pass and high-pass filters shown in Fig.4 since WPD can be carried out by an iterated application of quadrature mirror filters and followed by downsampling [21]. The variance of coefficients at each end node can be regarded as a feature of the signal [21]. For this approach, the key questions for WPD are how to determine the type of

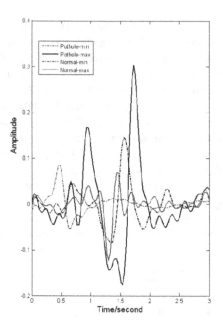

**Fig. 3.** Accelerations of potholes and normal segments

wavelets and the number of levels for decomposition and how to select coefficients at the desired end notes for formulating features [5].

Feature extraction by WPD is indeed based on the wavelet transform, which can also be regarded as a filtering process. Therefore, the frequency responses of such filter should match the spectral properties of desired signals [5], [6]. In this study, we used Daubecheis wavelet [8] since the frequency responses of this wavelet under the selected parameters for WPD match the spectral properties of accelerations of potholes. Indeed, this wavelet is very generally used for wavelet based analysis. The number of samples in each segment is 114 as the sampling frequency is 38 Hz and the duration of window is 3 seconds. For WPD, two to power of the number of levels

should approximate the number of samples [5]. Therefore, seven levels were determined in this study. And then, 128 features were exacted and distributed from 0-38 Hz (Note: sampling frequency is 38 Hz here). Furthermore, after checking the power spectrum of potholes and normal segments, we decided to select the WPD features within the frequency range 1.5-10 Hz. Subsequently, the 30 features from # 5 to # 34 among 128 features [5] were finally chosen for the further pothole detection. Then, each feature was normalized to its

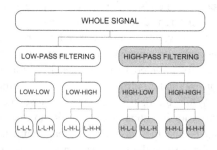

**Fig. 4.** Multi-scale decomposition for WPD [21]

standard deviation before feature selection. Toolbox of Wavelet in MATLAB (The Mathworks, Inc., Natick, MA) was used for WPD.

## 2.3     Feature Selection

Although some features were selected among all extracted features during the feature extraction process in terms of the rough evaluation of power spectrum of potholes and normal segments, there were still 30 features left for classification. Indeed, such a number is still too high for real-time application. Therefore, machine learning based feature selection can assist to reduce the computing demand for classification. Feature selection was designated to find the features best discriminating the potholes and normal segments in this study. We tried four extensively used methods here. For the completeness of the study, they are briefly introduced next. Please refer to [1] for more details.

Forward selection [1] is one of the basic feature selection methods. As a part of stepwise selection method, the idea, like in other feature selection methods, is to select a subset of features, which yields accurate enough results compared to results with all features. Forward selection begins with zero features in the model. The first feature is selected by testing each feature and then selecting the feature that yields, for example, the best classification result or the best f-value in statistical tests, such as, ANOVA (analysis of variance). This feature is the most significant feature. When a feature is selected, it is moved from unselected feature set to selected feature set. After this, the algorithm continues by comparing which feature of the remaining unselected features yields the best result with previously selected features and moves that feature from the unselected to the selected feature set. The procedure ends when there are no more features that increases the result or increases it only a bit. Another ending condition can be by a pre-defined upper number of selected features.

Another popular feature selection method is backward selection [1], which is opposite process compared to the forward selection. While forward selection begins with zero selected features, backward selection begins with all features. In every round, the algorithm tests all of the remaining features and removes the feature that decreases the results the most. This procedure continues until the result does not increase or it increases only a bit or the pre-defined number of features reaches.

The model of genetic algorithms was introduced by John Holland in 1975 [13]. Genetic algorithms are a group of computational models searching a potential solution to a specific problem using a simple chromosome-like data structure, which is inspired by evolution [1], [22]. A chromosome is a set of instructions which one algorithm will use to construct a new model or a function, such as, an optimization problem or selecting a subset of features for SVM. All features are represented as a binary vector of size $m$, where $m$ is the number of features. '1' means that a feature is part of the subset and '0' means that the feature is not part of the subset. An algorithm can be considered as a two-stage process. It begins with the current population where the best chromosomes are selected to create an intermediate population. Recombination and mutation is then applied to create the next population. This two-stage process constitutes one generation in the execution of a genetic algorithm. The algorithm begins with initial population of chromosomes. Typically initial population is chosen randomly from the original dataset. Then each chromosome is evaluated and assigned a fitness value. Chromosomes which represent better solution for the target problem are given a better fitness value than those chromosomes that provides poorer solution. Better fitness value means better reproducing chances. Reproducing may occur through crossover, mutation or reproduction operations. Please refer to [22] for more details about using an genetic algorithm for feature selection.

PCA is another frequently used method for feature selection [1]. The object of PCA is to find uncorrelated principal components that describe the dependencies between multiple variables. The principal components are ordered so that the first component explains the largest amount of variance in the data, and the second component is for the second largest variance, and so on. Generally it is expected that most of the variance in the original data set is covered by the first several principal components. PCA can be easily produced through the eigenvalue decomposition of the covariance matrix of multivariate datasets. It must be noticed that forward and backward selection and genetic algorithm do not affect data, and they are just methods to choose the best feature combinations, but PCA is for new features.

## 2.4     One-Class SVM

SVM is probably the mostly used classifier in the past decade [3]. It is a binary classifier that was invented by Cortes and Vapnik in 1995 [7]. It is based on generalized portrait algorithm and on Vapnik's research in statistical learning theory from the late 1970's. Basically SVM is intended to classify only two classes but it can be extended to support one-class and multiclass classification.

One scheme of one-class SVM is to map the input data into a high dimensional feature space and then fit all or most of the data into a hypersphere [3]. The volume of the hypersphere is minimized and all of the data samples that do not fall in the hypersphere are considered as anomalies. Another idea is the $v$-SVM that creates a decision function that separates most of the training data from the origin with a maximum margin [3]. The parameter $v$ is associated with the number of support vectors and outliers. So, one-class SVM tries to find a separating hyperplane and maximizes the distance between the two classes while two-class SVM can be solved by constructing a hypersphere that captures most of the training data and minimizes its volume or separates training data from the origin with maximum margin [3].

## 2.5     Data Processing

LIBSVM software was used [3]. We used one-class SVM with Gaussian radial basis kernel with the parameters $v$ and $\gamma$ with values 0.01 and 0.00002, correspondingly. SVM was trained with 70% of the normal data (1234 segments) and the training data was chosen randomly. Rest of the normal data (530 segments) and all 21 anomaly segments were used to test the accuracy of the constructed SVM model.

We present the result using sensitivity and specificity derived from the confusion matrix [1]. Such a matrix includes true positive (TP: 'pothole' is classified as 'pothole'), false negative (FN: 'pothole' is classified as 'normal'), false positive (FP: 'normal' is classified as 'pothole') and true negative (TN: 'normal' is classified as 'normal'). Sensitivity is equal to TP/(TP+FN), and specificity is defined as TN/(TN+FP). Their ideal values are 1 with zero FN and zero FP.

# 3    Results

All classification results that we present are the average of 1000 SVM classifications. If all 30 features were used for the classification, TP was 21, i.e., all 21 potholes were recognized as potholes, and TN was about 524 and FP was about 6, i.e., six normal segments were recognized as potholes.

For feature selection, we tested the four methods mentioned in the subsection 2.3 using different numbers of selected features. The results are shown in Fig.5. Obviously, PCA is the best for feature selection when the number of features is larger than 5. FS is the best when the number of features is larger than 2 and smaller than 6.

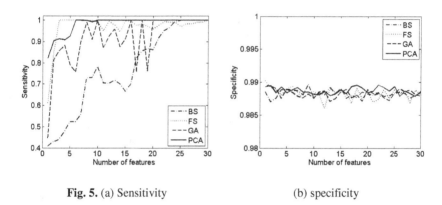

**Fig. 5.** (a) Sensitivity                         (b) specificity

# 4    Conclusion

This study has attempted to solve road surface anomaly detection problems with mobile phone's embedded accelerometer, wavelet packet decomposition for feature extraction, feature selection methods, and one-class support vector machine for classification. The achieved results are promising. As shown in Fig.5, one-class anomaly detection could be done very accurately. Our best true positive rate was 100%. These results are surprisingly good with our limited and unbalanced data sets.

In the data collection, some of the timestamps were corrupted and anomalies mislabeled due to limitation of hardware. This gives some extra challenge to data analysis. Thus, development of a proper data collection framework is essential in order to obtain reliable and stable results for real-time classification system in real production environment.

**Acknowledgments.** The work was supported by Sensor Data Fusion and Applications in Cooperative Traffic ICT SHOK project (TEKES, Finnish Funding Agency for Technology and Innovation).

# References

1. Alpaydin, E.: Introduction to machine learning, 2nd edn. MIT Press (2010)
2. Chandola, V., Banerjee, A., Kumar, A.: Anomaly detection–A Survey. ACM Computing Surveys 41 (2009)
3. Chang, C.C., Lin, C.J.: LIBSVM: A Library for Support Vector Machines. ACM Transactions on Intelligent Systems and Technology 2, 1–27 (2011)
4. Cheng, H.D., Miyojim, M.: Automatic Pavement Distress Detection System. Information Sciences 108, 219–240 (1998)
5. Cong, F., Huang, Y., Kalyakin, I., et al.: Frequency Response Based Wavelet Decomposition to Extract Children's Mismatch Negativity Elicited by Uninterrupted Sound. J. Med. Biol. Eng. 32, 205–214 (2012)
6. Cong, F., Leppanen, P.H., Astikainen, P., et al.: Dimension Reduction: Additional Benefit of an Optimal Filter for Independent Component Analysis to Extract Event-Related Potentials. J. Neurosci. Methods 201, 269–280 (2011)
7. Cortes, C., Vapnik, V.: Support-Vector Networks. Mach. Learn. 20, 273–297 (1995)
8. Daubechies, I.: Ten lectures on wavelets. Society for Industrial and Applied Mathematics (1992)
9. DuPont, E.M., Moore, C.A., Collins, E.G., et al.: Frequency Response Method for Terrain Classification in Autonomous Ground Vehicles. Auton. Robot. 24, 337–347 (2008)
10. Eriksson, J., Girod, L., Hull, B., et al.: The Pothole Patrol: Using a Mobile Sensor Network for Road Surface Monitoring. In: Proceedings of the 6th International Conference on Mobile Systems, Applications, and Services (MobiSys 2008), Breckenridge, CO, USA, pp. 29–39 (2008)
11. Georgopoulos, A., Loizos, A., Flouda, A.: Digital Image Processing as a Tool for Pavement Distress Evaluation. ISPRS Journal of Photogrammetry and Remote Sensing 50, 23–33 (1995)
12. Hesami, R., McManus, K.J.: Signal Processing Approach to Road Roughness Analysis and Measurement, pp. 1–6 (2009)
13. Holland, J.: Adaptation in natural and artificial systems. University of Michigan Press, Ann Arbor (1975)
14. Houston, D.R., Pelczarski, N.V., Esser, B., et al.: Damage Detection in Roadways with Ground Penetrating Radar 4084, 91–94 (2000)
15. Lin, J., Liu, Y.: Potholes Detection Based on SVM in the Pavement Distress Image, pp. 544–547 (2010)
16. Mitra, S.: Digital signal Processing—A computer based approach. McGraw–Hill (2005)
17. Mohan, P., Padmanabhan, V.N., Nericell, R.: Rich Monitoring of Road and Traffic Conditions using Mobile Smartphones, pp. 323–336 (2008)
18. Nejad, F.M., Zakeri, H.: A Comparison of Multi-Resolution Methods for Detection and Isolation of Pavement Distress. Expert Syst. Appl. 38, 2857–2872 (2011)

19. Perttunen, M., Mazhelis, O., Cong, F., et al.: Distributed Road Surface Condition Monitoring using Mobile Phones. In: Hsu, C.-H., Yang, L.T., Ma, J., Zhu, C. (eds.) UIC 2011. LNCS, vol. 6905, pp. 64–78. Springer, Heidelberg (2011)
20. Rajasegarar, S., Leckie, C., Bezdek, J.C., et al.: Centered Hyperspherical and Hyperellipsoidal One-Class Support Vector Machines for Anomaly Detection in Sensor Networks. IEEE Trans. Inf. Forensic Secur. 5, 518–533 (2010)
21. Schclar, A., Averbuch, A., Rabin, N., et al.: A Diffusion Framework for Detection of Moving Vehicles. Digit. Signal Prog. 20, 111–122 (2010)
22. Yang, J., Honavar, V.: Feature Subset Selection using a Genetic Algorithm. IEEE Intell. Syst. 13, 44–49 (1998)

# Aeroengine Turbine Exhaust Gas Temperature Prediction Using Process Support Vector Machines

Xu-yun Fu and Shi-sheng Zhong

School of Naval Architecture and Ocean Engineering,
Harbin Institute of Technology at Weihai,
264209 Weihai, China
{fuxy,zhongss}@hitwh.edu.cn

**Abstract.** The turbine exhaust gas temperature (EGT) is an important parameter of the aeroengine and it represents the thermal health condition of the aeroengine. By predicting the EGT, the performance deterioration of the aeroengine can be deduced in advance and its remaining time-on-wing can be estimated. Thus, the flight safety and the economy of the airlines can be guaranteed. However, the EGT is influenced by many complicated factors during the practical operation of the aeroengine. It is difficult to predict the change tendency of the EGT effectively by the traditional methods. To solve this problem, a novel EGT prediction method named process support vector machine (PSVM) is proposed. The solving process of the PSVM, the kernel functional construction and its parameter optimization are also investigated. Finally, the proposed prediction method is utilized to predict the EGT of some aeroengine, and the results are satisfying.

**Keywords:** Aeroengine, Condition monitoring, Turbine exhaust gas temperature, Process support vector machines, Time series prediction.

## 1 Introduction

The operational economy and reliability of the aeroengine are subjects of primary concern for airline companies. Both the maintenance ideology shift from mainly relying on preventation to reliability centered and the maintenance strategy shift from only time based maintenance to its combination with condition based maintenance and condition monitoring have been advanced energetically to resolve the contradiction between the aeroengine operational economy and its reliability since the 1960s. Aeroengine health monitoring and evaluation are the prerequisite and basis for the shifts of the maintenance ideology and the maintenance strategy. The possibility of inadequate maintenance and superfluous maintenance can be significantly reduced in condition of ensuring the safety of air transport by determining whether or not the aeroengine needs maintenance and how it is maintained according to its actual health condition.

The turbine exhaust gas temperature (EGT) is a significant indicator in aeroengine health condition monitoring. With the increase of the service life of the aeroengine, the EGT gradually rises. When the EGT is above the scheduled threshold which is determined by the original equipment manufacturer, the aeroengine needs to be

C. Guo, Z.-G. Hou, and Z. Zeng (Eds.): ISNN 2013, Part I, LNCS 7951, pp. 300–310, 2013.

removed timely for maintenance. Therefore, predicting the trend of the EGT has great significance to monitor the performance deterioration of the aeroengine, prognosticate the remaining life of the aeroengine, and reduce the aeroengine's failure and the maintenance costs. In practice, aeroengine health condition monitoring engineers usually also monitor the EGT margin and the EGT index. The EGT margin is the difference between the temperature at which the aeroengine must be operated to deliver the required thrust and the certification temperature. The increase in the value of EGT equals the decrease in the value of EGT margin, so monitoring EGT is indirectly monitoring EGT margin. The EGT index is not the real EGT, but the function of the real EGT and aeroengine fuel flow. This makes the EGT index reflect the performance of the aeroengine more completely, thus this paper will focus on the EGT index prediction.

The EGT index of the aeroengine is influenced by many complicated factors. It is difficult or even impossible to describe the variety of the EGT index by a determinate mathematic model. Aiming at solving this problem, the large amount of the collected EGT index data can be shrunken into a time series model. Thus, the tendency of EGT index can be predicted by some time series prediction methods. Traditional time series prediction methods are mainly based on the regression analysis [1]. The regression analysis is very mature in theory, but its accuracy is not high and its fault tolerance ability is poor. Artificial neural networks have been widely used in aeroengine condition monitoring and fault diagnosis [2] since multilayer feedforward neural networks were proved to be able to approximate any continuous function with any degree of accuracy [3,4]. As a new time series prediction method, artificial neural networks have been successful in many practical applications [5] for its good nonlinear capability, parallel distributed storage structures and high fault tolerance. Considering that time accumulation exists in time-varying systems, process neural network (PNN) was proposed in 2000 [6] and its application in time series prediction indicates that it seems to improve prediction accuracy [7]. However, both the traditional artificial neural networks and the process neural networks are all constructed around the empirical risk minimization (ERM) principle, which limits their generalization capability. Statistical learning theory shows that the ERM does not mean the minimization of the expectation risk and the overfitting is easy to happen if the ERM principle is adopted [8]. The ERM implies a premise that the number of training samples is infinite. The support vector machine is a machine learning model that follows the principle of structural risk minimization (SRM) that is rooted in VC dimension theory. Because its optimization object is SRM which ensures the minimization of both the empirical risk and the confidence interval, the support vector machine has a good generalization capability. Therefore, combined the SVM with the PNN, a method named process support vector machines (PSVM) is proposed to predict aeroengine EGT index in this paper.

The remainder of this paper is organized as follows. In section 2, the time series prediction model based on PSVM is proposed. In section 3, the solving process of the PSVM is described. In section 4 and 5, the kernel functional construction and parameter optimization of the PSVM are investigated respectively. In section 6, the proposed prediction method is utilized to predict the EGT of some aeroengine, and the prediction results are satisfying. The last section of the paper presents a conclusion.

## 2     Time Series Prediction Model Based on PSVM

Consider the training samples such as $\{(\vec{x}_i(t), y_i)\}_{i=1}^N$, where $\vec{x}_i(t)$ is the input function vector for the $i$th example and $y_i$ is the corresponding desired response. The architecture of the PSVM is depicted in Fig. 1.

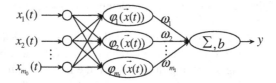

**Fig. 1.** The Topological Architecture of the PSVM

The model as shown in Fig. 1 is comprised of three layers. The first layer is the input layer, which has $m_0$ nodes. The second layer is the hidden layer, which is composed of $m_1$ nodes. The last layer is the output layer, which has only one node.

The output of the PSVM can be expressed as

$$y = f(\vec{x}(t), \vec{\omega}) = \vec{\omega}^T \vec{\varphi}(\vec{x}(t)) + b \tag{1}$$

Where $\vec{x}(t) = (x_1(t), x_2(t), \cdots, x_{m_0}(t))^T$, $\vec{\omega} = (\omega_1, \omega_2, \cdots, \omega_{m_1})^T$, $\omega_k$ is the connection weight between the output node and the $k$th node in the hidden layer, $\varphi_k(\vec{x}(t))$ is the $k$th nonlinear basis function in the hidden layer, $m_1$ is the dimensionality of the hidden space, which is determined by the number of process support vectors extracted from the training data by the solution to the constrained optimization problem, and $b$ is the bias of the output layer.

## 3     Solving Process

The first step in the time series prediction using PSVM is to define the $\varepsilon$-insensitive loss function.

$$L_\varepsilon(y, f(\vec{x}(t), \vec{\omega})) = \begin{cases} |y - f(\vec{x}(t), \vec{\omega})| - \varepsilon, & |y - f(\vec{x}(t), \vec{\omega})| > \varepsilon \\ 0, & other \end{cases} \tag{2}$$

Where $\varepsilon$ is the prescribed parameter.

On the basis of the definition of the $\varepsilon$-insensitive loss function, the time series prediction problem can be reformulated as follows by introducing two sets of nonnegative slack variables $\{\xi_i\}_{i=1}^N$, $\{\xi_i'\}_{i=1}^N$ and a regularization parameter $C$.

$$\begin{cases} \min \vec{\varphi}(\vec{\omega},\vec{\xi},\vec{\xi}') = C\sum_{i=1}^{N}(\xi_i + \xi'_i) + \frac{1}{2}\vec{\omega}^T\vec{\omega} \\ \text{s.t.} \quad y_i - \vec{\omega}^T\vec{\varphi}(\vec{x}_i(t)) - b - \varepsilon \leq \xi_i, \\ \quad\quad \vec{\omega}^T\vec{\varphi}(\vec{x}_i(t)) - y_i + b - \varepsilon \leq \xi'_i, \quad\quad i = 1, 2, \cdots, N \\ \quad\quad \xi_i \geq 0, \\ \quad\quad \xi'_i \geq 0, \end{cases} \quad (3)$$

Accordingly, the Lagrangian function can be defined as follows

$$J(\vec{\omega},\vec{\xi},\vec{\xi}',\vec{\alpha},\vec{\alpha}',\vec{\gamma},\vec{\gamma}') = C\sum_{i=1}^{N}(\xi_i + \xi'_i) + \frac{1}{2}\vec{\omega}^T\vec{\omega} - \sum_{i=1}^{N}\alpha_i\left[\vec{\omega}^T\vec{\varphi}(\vec{x}_i(t)) - y_i + b + \varepsilon + \xi_i\right] \\ - \sum_{i=1}^{N}\alpha'_i\left[y_i - \vec{\omega}^T\vec{\varphi}(\vec{x}_i(t)) - b + \varepsilon + \xi'_i\right] - \sum_{i=1}^{N}(\gamma_i\xi_i + \gamma'_i\xi'_i) \quad (4)$$

Where $\alpha_j$, $\alpha'_j$, $\gamma_j$, $\gamma'_j$ are the Lagrange multipliers.

By carrying out this optimization we have

$$\begin{cases} \vec{\omega} = \sum_{i=1}^{N}(\alpha_i - \alpha'_i)\vec{\varphi}(\vec{x}_i(t)) \\ \sum_{i=1}^{N}(\alpha_i + \alpha'_i) = 0 \\ \gamma_i = C - \alpha_i \\ \gamma'_i = C - \alpha'_i \end{cases} \quad (5)$$

Substitute (5) into (4), thus the convex functional can be got as follows

$$\begin{cases} \max Q(\vec{\alpha},\vec{\alpha}') = \sum_{i=1}^{N}y_i(\alpha_i - \alpha'_i) - \varepsilon\sum_{i=1}^{N}(\alpha_i + \alpha'_i) \\ \quad -\frac{1}{2}\sum_{i=1}^{N}\sum_{j=1}^{N}(\alpha_i - \alpha'_i)(\alpha_j - \alpha'_j)\vec{\varphi}^T(\vec{x}_i(t))\vec{\varphi}(\vec{x}_j(t)) \\ \text{s.t.} \quad \sum_{i=1}^{N}(\alpha_i - \alpha'_i) = 0 \\ \quad\quad 0 \leq \alpha_i \leq C \\ \quad\quad 0 \leq \alpha'_i \leq C \end{cases} \quad (6)$$

Equation (6) is essentially a quadratic programming problem and we can get the Lagrange multipliers through solving it. According to the Karush-Kuhn-Tucker (KKT) conditions, the product between the Lagrange multipliers and the constraints has to vanish at the point of the solution of the convex program

$$\begin{cases} \alpha_i\left(\varepsilon + \xi_i - y_i + \vec{\omega}^T\vec{\varphi}(\vec{x}_i(t)) + b\right) = 0 \\ \alpha'_i\left(\varepsilon + \xi'_i + y_i - \vec{\omega}^T\vec{\varphi}(\vec{x}_i(t)) - b\right) = 0 \\ (C - \alpha_i)\xi_i = 0 \\ (C - \alpha'_i)\xi'_i = 0 \end{cases} \quad (7)$$

From (7) we can get

$$\alpha_i * \alpha'_i = 0 \quad (8)$$

According to (7), if $\alpha_i = 0$, $\alpha'_i = 0$, and vice versa. Thus, the $\alpha_i$ and $\alpha'_i$ are bound to meet the following combinations

(1) $\alpha_i = 0$, $\alpha'_i = 0$

(2) $0 < \alpha_i < C$, $\alpha'_i = 0$

(3) $\alpha_i = 0$, $0 < \alpha'_i < C$

(4) $\alpha_i = C$, $\alpha'_i = 0$

(5) $\alpha_i = 0$, $\alpha'_i = c$

The particular data points ($\vec{x_i}(t)$, $y_i$) for which one of the (2)~(5) is satisfied are called process support vector (PSV). Especially, the data points ($\vec{x_i}(t)$, $y_i$) for which (4) or (5) is satisfied are called boundary process support vector (BPSV). The data points ($\vec{x_i}(t)$, $y_i$) for which (2) or (3) is satisfied are called normal process support vector (NPSV). Thus, the bigger $\varepsilon$ is, the fewer the number of PSV is and the lower the accuracy of the estimated function is also.

According to (7), if $0 < \alpha_i < C$ and $\alpha'_i = 0$, the bias $b$ can be expressed as follows

$$b = y_i - \left( \sum_{j=1}^{N} (\alpha_j - \alpha'_j) \vec{\varphi}(\vec{x_j}(t)) \right)^T \vec{\varphi}(\vec{x_i}(t)) - \varepsilon \tag{9}$$

If $\alpha_i = 0$ and $0 < \alpha'_i < C$, the bias $b$ can be expressed as follows

$$b = y_i - \left( \sum_{j=1}^{N} (\alpha_j - \alpha'_j) \vec{\varphi}(\vec{x_j}(t)) \right)^T \vec{\varphi}(\vec{x_i}(t)) + \varepsilon \tag{10}$$

In order to ensure the reliability of the result, the bias $b$ is calculated respectively according to all NPSVs and then its final value is their average value

$$b = \frac{1}{m_1} \sum_{i=1}^{m_1} \left( y_i - \left( \sum_{j=1}^{N} (\alpha_j - \alpha'_j) \vec{\varphi}(\vec{x_j}(t)) \right)^T \vec{\varphi}(\vec{x_i}(t)) - \varepsilon \cdot sign(\alpha_i - \alpha'_i) \right) \tag{11}$$

Substitute (5) and (11) into (1), thus, the approximating function can be expressed as follows

$$F(\vec{x}(t), \vec{\omega}) = \sum_{i=1}^{N} (\alpha_i - \alpha'_i) \vec{\varphi}^T(\vec{x}(t)) \vec{\varphi}(\vec{x_i}(t)) + b \tag{12}$$

It is obvious that the EGT index prediction problem can be solved by (12).

## 4     Kernel Functional Construction

From the (1), it is seen that the PSVM maps an input function vector $\vec{x}(t)$ into a high-dimensional feature space using mapping functional vector $\vec{\varphi}(\cdot)$ and then builds the model based on SRM. It is hard to calculate $\vec{\varphi}(\vec{x}(t))$ directly by a computer. According to the (12), the term $\vec{\varphi}^T(\vec{x}(t))\vec{\varphi}(\vec{x}_i(t))$ represents the inner product of two vectors induced in the feature space by the input function vector $\vec{x}(t)$ and the input function vector $\vec{x}_i(t)$ pertained to the $i$th example. Thus, if we can find a functional $K(\bullet)$ as follows

$$K(\vec{x}_i(t),\vec{x}_j(t)) = \vec{\varphi}^T(\vec{x}_i(t)) \cdot \vec{\varphi}(\vec{x}_j(t)) \tag{13}$$

In the high-dimensional feature space only $K(\vec{x}_i(t),\vec{x}_j(t))$ is needed to calculate and the specific form of the mapping functional vector $\vec{\varphi}(\bullet)$ is unnecessary to be known. The functional $K(\bullet)$ for which the (13) is satisfied is called the kernel functional corresponded to the kernel of the traditional SVM.

The input of PSVM, which is different from the instantaneous discrete input of the traditional SVM, is often time-varying continuous function. Thus, the functional is essential different from the kernel of the traditional SVM. The definition of the kernel functional may be formally stated as

**Definition 1:** Let $\psi$ be a subset of $L^2$ space. Then a functional $K:\psi\times\psi\rightarrow K$ is called a kernel functional of the PSVM if there exists a K -Hilbert space $H$ and a map $\vec{\varphi}:\psi\rightarrow H$ such that for all $\vec{x}(t),\vec{y}(t)\in\psi$ we have

$$K(\vec{x}(t),\vec{y}(t)) = \left\langle \vec{\varphi}\left(\vec{x}(t)\right),\vec{\varphi}\left(\vec{y}(t)\right)\right\rangle \tag{14}$$

Definition 1 can not be used to determine a specific form of a kernel functional. Mercer's theorem tell us whether or not a candidate kernel of a traditional SVM is actually an kernel in some space and therefore admissible for use in the traditional SVM. According to the function approximation theory, the kernel functional $K(\vec{x}(t),\vec{y}(t))$ can be uniformly approximated by a set of orthogonal functions in some function space. Thus, the method that determine a kernel of a traditional SVM according to Mercer's theorem can be used for reference to determine a kernel functional of a PSVM. This theorem may be formally stated as

**Theorem 1:** Let $K(\vec{x},\vec{y})$ be a continuous symmetric kernel that is defined in the closed interval $\vec{a}\leq\vec{x}\leq\vec{b}$ and likewise for $\vec{y}$. The kernel $K(\vec{x},\vec{y})$ can be expanded in the series

$$K(\vec{x},\vec{y}) = \sum_{i=1}^{\infty}\lambda_i\varphi_i(\vec{x})\varphi_i(\vec{y}) \tag{15}$$

With positive coefficients, $\lambda_i>0$ for all $i$. For this expansion to be valid and for it to converge absolutely and uniformly, it is necessary and sufficient that the condition

$$\int_{\vec{a}}^{\vec{b}}\int_{\vec{a}}^{\vec{b}} K(\vec{x},\vec{y})\psi(\vec{x})\psi(\vec{y})d\vec{x}d\vec{y}\geq 0 \tag{16}$$

Holds for all $\psi(\cdot)$ for which

$$\int_a^b \psi^2(\vec{x})d\vec{x} < \infty \tag{17}$$

The functions $\varphi_i(\vec{x})$ are called eigenfunctions of the expansion and the numbers $\lambda_i$ are called eigenvalues. The fact that all of the eigenvalues are positive means that the kernel $K(\vec{x}, \vec{y})$ is positive definite.

For $K(\vec{x}_i(t), \vec{x}_j(t))$, it is apparently

$$\sum_{i,j=1}^{N} c_i c_j K\left(\vec{x}_i(t), \vec{x}_j(t)\right) = \sum_{i,j=1}^{N} c_i c_j \left\langle \vec{\varphi}\left(\vec{x}_i(t)\right), \vec{\varphi}\left(\vec{x}_j(t)\right) \right\rangle = \left\| \sum_{i=1}^{N} c_i \vec{\varphi}\left(\vec{x}_i(t)\right) \right\| \geq 0 \tag{18}$$

Therefore, the kernel functional of the PSVM is positive definite. It implies the kernel functional also satisfy the Mercer theorem. Thus, it is not hard to determine the kernel functional of the PSVM according to the Mercer theorem.

By experience, four kernel functional may be determined as follows

(1) linear kernel functional $K\left(\vec{x}_i(t), \vec{x}_j(t)\right) = \int_0^T \vec{x}_i(t) \cdot \vec{x}_j(t)dt$

(2) polynomial kernel functional $K\left(\vec{x}_i(t), \vec{x}_j(t)\right) = \left(\int_0^T \vec{x}_i(t) \cdot \vec{x}_j(t)dt + 1\right)^p$

(3) Gauss kernel functional $K\left(\vec{x}_i(t), \vec{x}_j(t)\right) = \exp\left(-\frac{1}{2\sigma^2} \int_0^T \left\| \vec{x}_i(t) - \vec{x}_j(t) \right\|^2 dt \right)$

(4) sigmoid kernel functional $K\left(\vec{x}_i(t), \vec{x}_j(t)\right) = \tanh\left(\gamma \int_0^T \vec{x}_i(t) \cdot \vec{x}_j(t)dt + \beta \right)$

If they can not meet the needs of solving practical problems, the four kernel functionals may also be recombined until a new kernel functional is constructed to meet the needs.

## 5    Parameter Optimization

The performance of PSVM mainly refers to its generalization capability and the complexity of the learning machine. The parameters which affect the performance of PSVM mainly are the kernel functional parameters, the regularization parameter $C$ and the prescribed parameter $\varepsilon$. The interaction among these parameters affects the performance of PSVM. Therefore, the interaction must be considered to find the optimal parameters.

Particle swarm optimization (PSO), which was first described in 1995 [9], is a stochastic, population-based evolutionary computer algorithm for problem solving. As a highly efficient parallel optimization algorithm, PSO can be used to solve a large number of non-linear, nondifferentiable and multi-peak value complex optimization proposition, but the algorithm procedure is simple and the number of parameters needed to adjust is relatively small.

Let $f : R^D \to R$ be the fitness function that takes a particle's solution with several components in higher dimensional space and maps it to a single dimension metric. Let there be $n$ particles, each with associated positions $\vec{x}_i \in R^D$ and velocities $\vec{v}_i \in R^D$, $i = 1, \cdots, n$. Let $\hat{\vec{x}}_i$ be the current best position of each particle and let $\hat{\vec{g}}$ be the global best.

The complete procedure of the optimization algorithm is summarized as follows:

step1   Initialize $\vec{x}_i$ and $\vec{v}_i$ for all $i$ and $\hat{\vec{x}}_i = \vec{x}_i$, $\hat{\vec{g}} = \arg\min_{\vec{x}_i} f(\vec{x}_i), i = 1, \cdots, n$. One choice is to take $x_{ij} \in U[a_j, b_j]$ and $\vec{v}_i = 0$ for all $i$ and $j = 1, \cdots, D$, where $a_j$, $b_j$ are the limits of the search domain in each dimension, and $U$ represents the uniform distribution.

step2   Create random vectors $\vec{r}_1$, $\vec{r}_2$: $r_{1j}$ and $r_{2j}$ for all $j$, by taking $r_{1j}, r_{2j} \in U[0,1]$ for $j = 1, \cdots, m$. $\vec{v}_i = \omega \vec{v}_i + c_1 \vec{r}_1 \circ (\hat{\vec{x}}_i - \vec{x}_i) + c_2 \vec{r}_2 \circ (\hat{\vec{g}}_i - \vec{x}_i)$, $\vec{x}_i = \vec{x}_i + \vec{v}_i$. Where $\omega$ is an inertial constant and good values are usually slightly less than 1, $c_1$ and $c_2$ are constants that say how much the particle is directed towards good positions, $\circ$ operator indicates element-by-element multiplication.

step3   Update the local bests: if $f(\vec{x}_i) < f(\hat{\vec{x}}_i)$, $\hat{\vec{x}}_i = \vec{x}_i$. Update the global best: if $f(\vec{x}_i) < f(\hat{\vec{g}})$, $\hat{\vec{g}} = \vec{x}_i$.

step4   If the stop criteria has been met, stop; otherwise, go to step2.

In this paper, parameters to be optimized are the kernel functional parameters, the regularization parameters $C$ and the prescribed parameter $\varepsilon$. The mean square error (MSE) of prediction values is selected as the fitness value.

## 6    Application Test

The EGT index data used in this paper was taken from some aeroengine, and its sampling interval is about 30 flight cycles. We get a time series of the aeroengine EGT index with 75 discrete points such as $\{EI_i\}_{i=1}^{75}$, which is depicted in Fig. 2.

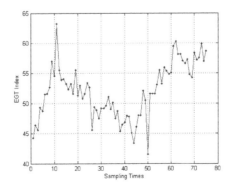

**Fig. 2.** EGT Index Time Series

Although it has reduced some stochastic noise of the raw data, the aeroengine condition monitoring software can not effectively reduce the gross error from the instrument record mistakes or the measure accidents. Data cleaning must be carried on in order to remove the gross error of the EGT index. The visualization method is adopted in this paper to clean the data of the EGT index. Fig. 2 shows that the change of the 11th and the 50th point are abnormal in EGT index time series $\{EI_i\}_{i=1}^{75}$. Therefore, these two points belong to the gross error and should be removed. Missing data after removing the gross error is completed by the following equation

$$EI_i = \frac{EI_{i-1} + EI_{i+1}}{2} \tag{19}$$

After data cleaning, the vector $(EI_{j1}, EI_{j2}, \cdots, EI_{j6})^T$ can be used to generate an input function $IF_j(t)$ by nonlinear least-squares method, where $j = 1, 2, \cdots, 69$, and $EI_{j7}$ can be used as the corresponding desired output to $IF_j(t)$. Thus, we get 69 samples such as $\{(IF_j(t), EI_{j+7})\}_{j=1}^{69}$. The samples $\{(IF_j(t), EI_{j+7})\}_{j=1}^{35}$ are utilized to train the time series prediction model based on the PSVM. The Gaussian kernel functional $K\left(\vec{x}_i(t), \vec{x}_j(t)\right) = \exp\left(-\frac{1}{2\sigma^2} \int_0^T \left\|\vec{x}_i(t) - \vec{x}_j(t)\right\|^2 dt\right)$ is selected as the kernel functional of the PSVM. The width $\sigma$ of the Gaussian kernel functional, the regularization parameter $C$ and the $\varepsilon$-insensitive parameter are confirmed by using PSO algorithm. The PSVM completed training in 37.2633 minutes. The samples $\{(IF_j(t), EI_{j+7})\}_{j=36}^{69}$ are selected to test the time series prediction model. The test results are depicted in Fig. 3 and the average relative error is 2.77%.

In order to compare the performance of the PSVM, the traditional SVM is trained by the same training samples. The Gaussian kernel $K\left(\vec{x}_i, \vec{x}_j\right) = \exp\left(-\frac{1}{2\sigma^2} \left\|\vec{x}_i - \vec{x}_j\right\|^2\right)$ is selected as the kernel of the SVM. The width $\sigma$ of the Gaussian kernel, the

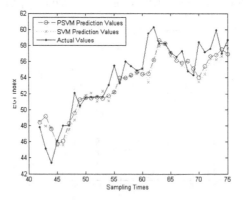

**Fig. 3.** EGT Index Prediction Results

regularization parameter $C$ and the $\varepsilon$-insensitive parameter are confirmed by using PSO algorithm. The test results are also depicted in Fig. 3 and its average relative error is 2.92%.

The results as shown in Fig. 3 indicate that the prediction accuracy of the PSVM are better than the traditional SVM. The EGT index time series predicion's average relative error using the PSVM is 2.77% and it seems to meet the actual needs. It shows that the EGT index time series prediction using the PSVM can help condition monitoring engineers determine the health status of the aeroengine accurately, and can provide some decision support for the plan of the removal of the aeroengine.

# 7    Conclusions

The aeroengine plays a significant role as the heart of an aircraft. The aeroengine health monitoring is essential in terms of the flight safety and also for reduction of the maintenance cost. The EGT is one of the most important health parameters of the aeroengine. By predicting the EGT, maintenance crew can judge the performance deterioration and find the latent gas path faults in the aeroengine in advance. To solve the problem that the traditional method is difficult to predict the EGT accurately, a prediction method based on PSVM is proposed in this paper. To validate the effectiveness of the proposed prediction method, the EGT prediction model based on PSVM is utilized to predict the EGT index of some aeroengine, and the test results indicate that the PSVM can be used as a tool to predict the health condition of the aeroengine.

**Acknowledgments.** This research was supported by the National High Technology Development Program of China (No. 2012AA040911-1), Science and technology plan of Civil Aviation Administration of China (No. MHRD201122), and Scientific Research Foundation of Harbin Institute of Technology at Weihai (No. HIT(WH) X201107).

# References

1. Chatfield, C.: The Analysis of Time Series: An Introduction, 6th edn. Chapman & Hall/CRC, BocaRaton (2003)
2. Stephen, O.T.O., Riti, S.: Advanced Engine Diagnostics using Artificial Neural Networks. Applied Soft Computing 3, 259–271 (2003)
3. Hornik, K., Stinchcombe, M., White, H.: Multilayer Feedforward Networks are Universal Approximators. Neural Networks 2, 359–366 (1989)
4. Funahashi, K.: On The Approximate Realization of Continuous Mappings by Neural Networks. Neural Networks 2, 183–192 (1989)
5. Reyes, J., Morales-Esteban, A., Martínez-Álvarez, F.: Neural networks to predict earthquakes in Chile. Applied Soft Computing 13, 1314–1328 (2013)

6. He, X.G., Liang, J.Z.: Process Neural Networks. In: Proceedings of the Conference on Intelligent Information Processing (2000)
7. Ding, G., Zhong, S.S.: Aircraft Engine Lubricating Oil Monitoring by Process Neural Network. Neural Network World 16, 15–24 (2006)
8. Vapnik, V.: The Nature of Statistical Learning Theory. Springer, New York (1995)
9. Kennedy, J., Eberhart, R.: Particle Swarm Optimization. In: International Conference on Neural Networks (1995)

# The Effect of Lateral Inhibitory Connections in Spatial Architecture Neural Network

Gang Yang, Jun-fei Qiao, Wei Li, and Wei Chai

Intelligent Systems Institute, College of Electronic Information and Control Engineering
Beijing University of Technology, Beijing, P.R. China
{hank.yang2010,isi.bjut,lw2011609}@gmail.com

**Abstract.** Based on the theories of lateral inhibition and artificial neural network (ANN), the different lateral inhibitory connections among the hidden neurons of SANN are studied. With the connect mode of activation-inhibition-activation, the SANN will obtain a higher learning accuracy and generalization ability. Furthermore, this inhibitory connection considers both the activation before and after been inhibited by surrounding neurons. The effectiveness of this inhibitory mode is demonstrated by simulation results.

**Keywords:** Lateral Inhibition, Activation, Inhibition, Spatial Architecture Neural Network.

## 1 Introduction

Lateral inhibition (LI) phenomenon is a result of mutual effect between excitability function of each neuron caused by the outside input stimulations and inhibition function caused by the surrounding neurons. It has been confirmed that LI exists in human and animal's many sensory systems such as tactile system and vision system. The physiological significance of lateral inhibition mechanism is very clear, so it has been widely used in the study of artificial neural networks (ANNs). [1-3]

The spatial architecture neural network (SANN)[4] is an ANN adopts recurrent lateral inhibition in the hidden layers. It means that, the mutual inhibitory effect of hidden neurons in SANN is caused by the output of surrounding neurons. Namely, the $j$th neuron of $l$th hidden layer may be inhibited by some other neurons of $l$th hidden layer, and the afferent inhibitory inputs from the output of excited neurons.

LI comprises two inhibitory modes: recurrent lateral inhibition (RLI) and non-recurrent lateral inhibition (N-RLI), and the difference between these two mutual inhibitions is that what caused the inhibitory effect. If it was caused by the network output, then it is the RLI; otherwise, it is the N-RLI. The neural network whose connection mode adopts the lateral inhibition mechanism is called lateral inhibition neural network (LINN). In these networks, if one neuron is excitatory after being stimulated, it will inhibit the surrounding neurons through the inhibitory connections; thereby achieve the competition between neurons.

The rest of this paper is organized as follows. In section 2, we briefly review the lateral inhibition and introduce the architecture of SANN. The detailed analysis of

C. Guo, Z.-G. Hou, and Z. Zeng (Eds.): ISNN 2013, Part I, LNCS 7951, pp. 311–318, 2013.

inhibitory influence caused by different lateral connection types are presented in Section 3. Section 4 gives the simulation results of numerical experiments, and Section 5 concludes this paper.

# 2 Review of Lateral Inhibition and SANN

## 2.1 Lateral Inhibition

Fig.1 shows the schematic diagram of inhibitory systems [5], including the recurrent inhibition and non-recurrent inhibition.

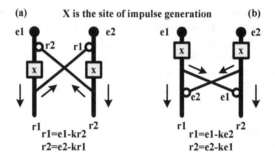

**Fig. 1.** Schematic diagram of inhibitory systems: (a) recurrent inhibition; (b) non-recurrent inhibition

Assuming that there are $n_l$ neurons in the $l$th hidden layer. The outputs of hidden neurons in SANN are determined by the external environment stimulus diminished by the lateral inhibitory influence exerted by other neurons. The output of the $j$th hidden neuron with recurrent lateral inhibition $y_{o,j}$ is given by [5-6]:

$$y_{o,j} = y_{i,j} - \sum_{r=1, r \neq j}^{n_l} \varphi_{jr}\left(y_{o,r} - \theta_{jr}\right), \quad r = 1, 2, ..., n_l \tag{1}$$

where $y_{i,j}$ is the external stimulus of $j$th neuron received, $\varphi_{jr}$ is the lateral inhibitory coefficient between $r$th neuron and $j$th neuron, $\theta_{jr}$ is the inhibiting threshold value, $r \neq j$ indicated that the self-inhibitory influence is omitted. The threshold and the inhibitory coefficient are labeled to indicate the direction of the action: $\theta_{jr}$ is the threshold value of $j$th neuron's output reach to which it begins to inhibit $r$th neuron; $\theta_{rj}$ is the reverse. In the same way, $\varphi_{jr}$ is the coefficient of the inhibitory action of $r$th neuron on $j$th neuron, $\varphi_{rj}$ is the reverse.

In the lateral inhibition network, the inhibitory influence is proportional to the inhibitory coefficients and the activation level (its response $y_{o,j}$). The inhibitory coefficients are real numbers in [0, 1], with 0 meaning no inhibition and 1 meaning full inhibition. The inhibitory coefficients from neuron $r$ to $j$ and from $j$ to $r$ usually are different. If the activity level of neuron $r$ is high, it will inhibit other neurons with large values of inhibitory coefficients, and to the contrary, the inhibited coefficient by other neurons will be small values. A neuron $r$ can inhibit other neurons such as neuron $j$ of a same layer if and only if its activation level $y_{o,r}$ is greater than or equals to the inhibiting threshold $\theta_{jr}$. The amount of inhibition between two neurons is calculated according to the activation level and the corresponding inhibitory coefficient from the inhibiting neuron to inhibited neuron. For example, neuron $j$ is the current receptor neuron, $\varphi_{jr}$ is the inhibitory coefficient from neuron $r$ to neuron $j$, $y_{o,r}$ is the activation level of neuron $r$. The inhibition amount $P(j, r)$ from neuron $r$ to neuron $j$ can be calculated according to the Eq (2).

$$P(j,r) = \begin{cases} \varphi_{jr}\left(y_{o,r} - \theta_{jr}\right) & y_{o,r} \geq \theta_{jr} \\ 0 & y_{o,r} < \theta_{jr} \end{cases} \tag{2}$$

For $j$th neuron, the total inhibitory input by the surrounding units is a linear superposition of all single inhibiting units. Then the output of inhibited neuron $j$ is given by:

$$y_{o,j} = \begin{cases} y_{i,j} - \sum_{j=1,j\neq r}^{n_i} P(j,r), & y_{o,r} \geq \theta_{jr} \\ y_{i,j}, & y_{o,r} < \theta_{jr} \end{cases} \tag{3}$$

The experimental researches of inhibition found that the distribution function of lateral inhibitory coefficient is a Gaussian distribution function [20] (such as Fig.3). The inhibitory coefficient of this paper is given by:

$$\varphi(d) = \begin{cases} \mu \exp\left[-\phi d^2\right], & d \neq 0 \\ 0, & d = 0 \end{cases} \tag{4}$$

where $\mu$ and $\phi$ are adjustable parameters.

Another important concept of lateral inhibition is the inhibitory scope. The value of inhibitory scope defines the range of an excitatory neuron inhibiting other neurons. In other words, it indicates how many surrounding neurons will be inhibited by the excitatory neuron.

From the analysis of recurrent lateral inhibition network, we can find that it is a dynamic process contains interaction, and the influence scope of each excitatory neuron is not only limited to the surrounding neurons of the same layer but will works on all hidden neurons at last.

## 2.2    Spatial Architecture Neural Network

Inspired by the biotomy and neurobiology findings and the theory of artificial neural networks, the spatial architecture neural network is proposed by Qiao et al. to obtain higher learning accuracy and generalization ability. In SANN, the first hidden layer will receive information from input neurons, and each subsequent layer neuron will get signal from all the previous layer neurons. Meanwhile, in each hidden layer the $j$th neuron will be inhibited by surrounding excited hidden neurons through the lateral inhibitory connections if and only if $\bar{o}_r \geq \theta_{jr}$. Here, $\bar{o}_r$ is the activity of $r$th neuron and will take place the representation of $y_{o,r}$ in the following context of this paper.

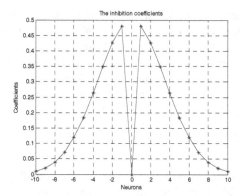

**Fig. 2.** Distribution of lateral inhibition coefficient ($\mu = 0.5$, $\phi = 0.04$). Lateral axis indicates the interaction neurons, which can be measured by the activity level or distance between two interaction neurons. And longitudinal axis denotes the corresponding lateral inhibition coefficient between two neurons. Because there is no self-inhibition in SANN, so let $\varphi_{rr} = 0$.

The connection schematic diagram of single-hidden-layer SANN is shown in Fig.3. The output of SANN is given by

$$ O = f\left(\sum_{k=1}^{m} o_k\right) = f\left(\sum_{k=1}^{m}\left[f\left(\sum_{i=1}^{n} \omega_{ki} x_i - \delta_i + \sum_{j=1}^{n_l} \omega_{kj} o_j - \delta_j\right)\right]\right) \tag{5} $$

where $f(\cdot)$ is the activation function, $\delta_j$ is the bias value and $o_j$ is the output of $j$th neuron. The output of $j$th hidden neuron is given by

$$ o_j = g\left(x_i, \omega_{ji}, \overline{o}_r, \delta\right) \tag{6} $$

which is a function of the input pattern $x_i$, hidden-input weights $\omega_{ji}$, intermediate output of surrounding excited neurons $\overline{o}_r$ and biases $\delta$.

## 3    Analysis of Different Lateral Inhibitory Connections

Based on the above analysis of recurrent lateral inhibition system and the structure of artificial neuron, the inhibitory influence exerted by surrounding excitatory neurons may works before or after being activated. So, there are three possible different connect modes of lateral inhibitory connection in SANN, which is according to the site of inhibition generation (see Fig. 2).

In the Fig.2, $u_j$ is the sum of afferent feed-forward input from the ($l$-1)th layer and spatial span input from 0 to ($l$-2)th layers; $\overline{o}_j$ is the intermediate output of $j$th neuron, actually it is the response of $u_j$; $f(\cdot)$ is the activation function, $\delta_j$ is the bias value and $o_j$ is the output of $j$th neuron. Where, $A$, $B$ and $C$ is the possible site of inhibition generation.

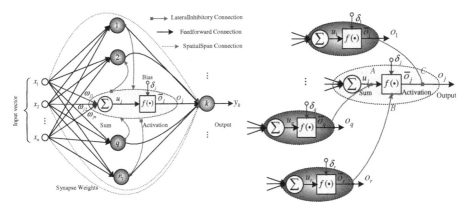

**Fig. 3.** Connection schematic diagram of single-hidden-layer SANN.

**Fig. 4.** Connection Schematic diagram of lateral inhibitory connections in $l$th hidden layer of SANN.

### 3.1    Case A: Site A, Inhibited-Activated

The first possible case of lateral inhibitory connection is that the inhibitory influence occurred on the site of $A$. The output of $j$th hidden neuron is given by:

$$o_j = f\left(u_j - \delta_j - L_j\right) = f\left(\sum_{i=1}^{n} \omega_{ji} x_i - \delta_j - \sum_{j=1, j\neq r}^{n_l} \varphi_{jr}\left(\overline{o}_r - \theta_{jr}\right)\right) \tag{7}$$

where $L_j$ indicates the total lateral inhibitory influences exerted by the surrounding hidden neurons on $j$th neuroon.

In this case, the current $j$th excited hidden neuron will be inhibited by surrounding neurons before it response to the stimulus input $u_j$.

### 3.2    Case B: Site B, Activated-Inhibited-Activated

The second possible case of lateral inhibitory connection is that the inhibitory influence occurred on the site of $B$. The output of $j$th hidden neuron is given by:

$$o_j = f\left(\overline{o}_j - L_j\right) = f\left(f\left(\sum_{i=1}^{n} \omega_{ji} x_i - \delta_j\right) - \sum_{j=1, r\neq j}^{n_l} \varphi_{jr}\left(\overline{o}_r - \theta_{jr}\right)\right) \tag{8}$$

In this case, the current $j$th excited hidden neuron will be inhibited by surrounding neurons after it responses to the stimulus input $u_j$, and then it receives the lateral inhibitory influence exerted by the surrounding neurons, and at last the $j$th neuron will be activated by $f(\cdot)$.

### 3.3    Case C: Site C, Activated-Inhibited

The third possible case of lateral inhibitory connection is that the inhibitory influence occurred on the site of $C$. The output of $j$th hidden neuron is given by:

$$o_j = \overline{o}_j - L_j = f\left(\sum_{i=1}^{n} \omega_{ji} x_i - \delta_j\right) - \sum_{j=1, r \neq j}^{n_l} \varphi_{jr}\left(\overline{o}_r - \theta_{jr}\right) \quad (9)$$

The two interaction elements of lateral inhibition in this case is similar to the above second case, namely it will receive the inhibitory influence after being activated by $f(\cdot)$. However, in this case the response after being inhibited will not be activated by $f(\cdot)$, which is the difference compared with the second case.

Without loss of generality, the intermediate output of $j$th neuron from $l$th hidden layer is given by

$$\overline{o}_j = f\left(\sum_{l=0}^{l-1}\left(\sum_{i=1}^{n_l} \omega_{ji} o_i - \delta_j\right)\right) \quad (10)$$

It can be seen that, from the representation of above three connect modes the functional relationship between the inputs and outputs are as follows respectively,

$$\begin{cases} o_j^A = F\left[f\left(\omega_{ji}, x_i, \delta_j, \varphi_{jr}, \overline{o}_r, \theta_{jr}\right)\right] \\ o_j^B = F\left[f\left(f\left(\omega_{ji}, x_i, \delta_j\right), \varphi_{jr}, \overline{o}_r, \theta_{jr}\right)\right] \\ o_j^C = F\left[f\left(\omega_{ji}, x_i, \delta_j\right), \varphi_{jr}, \overline{o}_r, \theta_{jr}\right] \end{cases} \quad (11)$$

Obviously, the second connection mode of CaseB is expressive effective than other two modes from the mapping point of view.

## 4    Simulation Examples

The three lateral inhibitory connect modes has been empirically tested on the parity-3 classification problem, which was considered as a difficult benchmark problem for neural networks. [7-9] There are $2^3$ input patterns in 3-dimensional space and the input-output of parity problem is given by

$$y = \begin{cases} 0, & \text{if } \sum_{i=1}^{N} x_i \text{ is even number} \\ 1, & \text{if } \sum_{i=1}^{N} x_i \text{ is odd number} \end{cases} \quad (12)$$

where $x_i$ is the network input vector, and $y$ is the desired output. The training data is generated from (11), while the testing data is based on 0.1 and 0.9 that satisfy the relationship of parity-3 function.

We mainly compared the learning accuracy; generalization ability and training time of SANN with these three different connect modes. All the simulations are implemented in Matlab 7.0 environment running on an ordinary PC with Intel ® Core ™ 2 Duo CPU 3.0 GHz. The sigmoid additive activation function $f(\cdot)=1/(1+e^{-x})$ has been used for activating hidden neurons, while the output neurons are linear. The max iteration number is set as 8000, the learning rate is 0.15, the momentum term is 0.7, and the desired mse is 1e-5. From the approximation point of view, 2 hidden neurons are used to solve this task in the simulation study.

**Table 1.** Training and testing MSE on the parity-3 problem and iteration numbers

| Connect Modes | Iteration Number | Training MSE | Testing MSE |
|---|---|---|---|
| CaseA1 | 1268 | 0.000018 | 0.027656 |
| CaseA2 | 2244 | 0.000019 | 0.027613 |
| CaseB1 | **443** | **0.000016** | **0.029770** |
| CaseB2 | **662** | **0.000019** | **0.027007** |
| CaseC1 | 4068 | 0.000020 | 0.469019 |
| CaseC2 | 5083 | 0.000020 | 0.239818 |

Figs. 5-7 and Table 1 are the simulation results from this experiment. As shown in Fig.5, CaseBs obtained a training mse of 0.000016 and 0.000019 with 2 hidden neurons after 443 and 662 iterations, while the biggest training mse on this problem given by CaseCs are 0.000020 after more than 4000 iterations. From Fig.5, we can also observe that the training convergent speeds of CaseBs are the best two recoreds, while the results based on CaseCs are the last records.

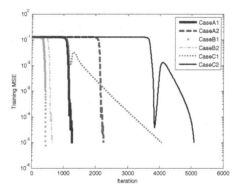

**Fig. 5.** Comparison of training mean square error on parity-3 benchmark problem based on three different lateral inhibitory connections

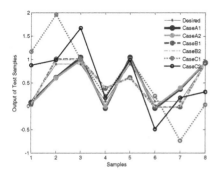

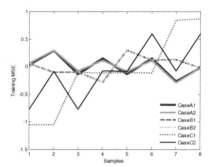

**Fig. 6.** Comparison of testing output on parity-3 benchmark problem based on three different lateral inhibitory connections

**Fig. 7.** Comparison of absolute testing error on parity-3 benchmark problem based on three different lateral inhibitory connections

Figs.6 and 7 are the outputs of three lateral inhibition modes and the error between the desired outputs. Seen from Fig.6, we can find that the output of test samples is more close to the desired output when adopt the second lateral inhibitory connection (CaseBs), while the worst results are also generated from the third connect modes (CaseCs). From the Table 1, this experiment shown that the best inhibition connect mode for SANN is the second case (CaseBs), in which the hidden neurons will be activated at last. All of the iteration number, training mse and testing mse are the lowest record in this experiment. It means that, with this inhibition, SANN will have higher accuracy and generalization ability.

# 5    Conclusions

The effectiveness of three possible lateral inhibitory connections in SANN is studied. Based on the lateral inhibition mechanism and the structure of artificial neurons, we defined the different inhibitory connections according to the site of inhibition generation. The second connect mode, whose hidden neurons will be activated after have been inhibited, will improve the learning efficient of SANN both of the learning accuracy and generalization. The introducing of Lateral Inhibition in SANN will improve the competition in the real world applications, and the efficient learning algorithm is still a research field.

**Acknowledgments.** This study is supported by the National Natural Science Foundation of China under Grant No.60873043. The author would also like to thank the anonymous reviewers for their constructive comments.

# References

1. Meir, E., von Dassow, G., Munro, E., Odell, G.M.: Robustness, Flexibility, and the Role of Lateral Inhibition in the Neurogenic Network. Current Biology 12(10), 778–786 (2002)
2. Xue, Y.B., Yang, L., Haykin, S.: Decoupled Echo State Networks with Lateral Inhibition. Neural Networks 20(3), 365–376 (2007)
3. Hu, X., Li, O.: Structure Learning of a Behavior Network for Context Dependent Adaptability. In: IAT 2006, pp. 407–410. IEEE Computer Society, Washington (2006)
4. Yang, G., Qiao, J., Bo, Y.: Research on Artificial Neural Networks with Spatial Architecture Based on Span Connection and Lateral Inhibition Mechanism. International Journal of Computational Science and Engineering 6(1-2), 86–95 (2011)
5. Ratliff, F., Hartline, H.K., Miller, W.H.: Spatial and Temporal Aspects of Retinal Inhibitory Interaction. Journal of the Optical Society of America 53(1), 110–120 (1963)
6. Hartline, H.K., Ratliff, F.: Inhibitory Interaction of Receptor Units in the Eye of Limulus. The Journal of General Physiology 40(3), 357–376 (1957)
7. Lavretsky, E.: On the Exact Solution of the Parity-N Problem Using Ordered Neural Networks. Neural Networks 13(6), 643–649 (2000)
8. Jordanov, I., Georgieva, A.: Neural Network Learning with Global Heuristic Search. IEEE Transactions on Neural Networks 18(3), 937–942 (2007)
9. Gao, H., Shiji, S., Cheng, W.: Orthogonal Least Squares Algorithm for Training Cascade Neural Networks. IEEE Transactions on Circuits and Systems I: Regular Papers 59(11), 2629–2637 (2012)

# Empirical Mode Decomposition Based LSSVM for Ship Motion Prediction

Zhou Bo[1] and Shi Aiguo[2]

[1] Dep. of academics, Dalian naval academy, Dalian, China
[2] Dept. of navigation, Dalian naval academy, Dalian, China
judyever@sina.com

**Abstract.** An empirical mode decomposition (EMD) based Lease square support vector machines (LSSVM) is proposed for ship motion prediction. For this purpose, the original ship motion series were first decomposed into several intrinsic mode functions (IMFs), then a LSSVM model was used to model each of the extracted IMFs, so that the tendencies of these IMFs could be accurately predicted. Finally, the prediction results of all IMFs are combined to formulate an output for the original ship motion series. Experiments on chaotic datasets and real ship motion data are used to test the effectiveness of the proposed algorithm.

**Keywords:** ship motion, empirical mode decomposition, Lease square support vector machines.

## 1 Introduction

Ship motion at sea is a nonlinear and non-stationary time series. The prediction of ship motion is commonly done using various forms of statistical models. Lease square support vector machines (LSSVM) are closely related to statistical models and have been proved useful in a number of regression applications.

The main motivation of this study is to propose an EMD-based LSSVM approach for ship motion prediction and compare its prediction performance with some existing forecasting techniques. The rest of this study is organized as follows. Section 2 describes the formulation process of the proposed EMD-based LSSVM learning method in detail. For illustration and verification purposes, two main series, chaos series and real ship motion data are used to test the effectiveness of the proposed methodology, and the corresponding results are reported in Section 3. Finally, some concluding remarks are drawn in Section 4.

## 2 Methodology Formulation

### 2.1 Empirical Mode Decomposition (EMD)

The empirical mode decomposition (EMD) technique, first proposed by Huang et al. (1998), is a form of adaptive time series decomposition technique using the

C. Guo, Z.-G. Hou, and Z. Zeng (Eds.): ISNN 2013, Part I, LNCS 7951, pp. 319–325, 2013.

Hilbert–Huang transform (HHT) for nonlinear and nonstationary time series data. The basic principle of EMD is to decompose a time series into a sum of oscillatory functions, namely, intrinsic mode functions (IMFs). In the EMD, the IMFs must satisfy the following two prerequisites [1]:

(1) In the whole data series, the number of extrema (sum of maxima and minima) and the number of zero crossings, must be equal, or differ at most by one, and

(2) The mean value of the envelopes defined by local maxima and minima must be zero at all points.

With these two requirements, some meaningful IMFs can be well defined. Otherwise, if one blindly applied the technique to any data series, the EMD may result in a few meaningless harmonics (Huang et al.,1999). Usually, an IMF represents a simple oscillatory mode, compared with the simple harmonic function. Using the definition, any data series $x(t)(t=1,2,......,n)$ can be decomposed according to the following sifting procedure.

(1) Identify all the local extrema, including local maxima and local minima, of $x(t)$,

(2) Connect all local extrema by a cubic spline line to generate its upper and lower envelopes, and compute the point-by-point envelope mean $m(t)$ from upper and lower envelopes.

(3)Extract the details, $imf(t) = x(t) - m(t)$ and check the properties of $imf(t)$ :

1) if $imf(t)$ meets the above two requirements, an IMF is derived and meantime replace x(t) with the residual $r_i(t) = x(t) - imf_i(t)$;

2) if c(t) is not an IMF, replace $x(t)$ with $imf_i(t)$ .

Repeat Steps (1)–(3) until the stop criterion is satisfied.

The EMD extracts the next IMF by applying the above sifting procedure to the residual term $r_1(t) = x(t) - imf_1(t)$   has at most one local extremum or becomes a monotonic function, from which no more IMFs can be

extracted. The above sifting procedure can be implemented using Matlab software. At the end of this sifting procedure, the data series x(t) can be expressed by

$$X(t) = \sum_{i=1}^{n} imf_i(t) + r_n(t) \tag{1}$$

where n is the number of IMFs, $r_n(t)$ is the final residue, which is the main trend of $x(t)$, and $imf_i(t)$ $(i=1,2,......,n)$ are the IMFs, which are nearly orthogonal to each other, and all have nearly zero means. Thus, one can achieve decomposition of the data series into n-empirical mode functions and one residue. The IMF components contained in each frequency band are different and they change with variation of time series $x(t)$, while $r_n(t)$ represents the central tendency of data series $x(t)$.

Relative to the traditional Fourier and wavelet decompositions, the EMD technique has several distinct advantages. First of all, it is relatively easy to understand and implement. Second, the fluctuations within a time series are automatically and adaptively selected from the time series, and it is robust for nonlinear and nonstationary time series decomposition. Third, it lets the data speak for themselves. EMD can adaptively decompose a time series into several independent IMF components and one residual component. The IMFs and the residual component displaying linear and nonlinear behavior depend only on the nature of the time series being studied. Finally, in wavelet decomposition, a filter base function must be determined beforehand, but it is difficult for some unknown series to determine the filter base function. Unlike wavelet decomposition, EMD is not required to determine a filter base function before decomposition. In terms of the above merits the EMD can be used as an effective decomposition tool.

## 2.2    Lease Square Support Vector Machines

LSSVM are proposed by Suykens and Vandewalle. The basic idea of mapping function is to map the data into a high dimensional feature space, and to do linear regression in this space [2].

Given a training data set of $N$ points $\{(x_i, y_i)\}_{i=1}^{N}$ with input data $x_i \in R^n$ and output data $y_i \in R$. According to the SVM theory, the input space $R^n$ is mapped into a feature space $Z$ with $\varphi(x_i)$ being the corresponding mapping function. In the feature space, we take the form (2) to estimate the unknown nonlinear function where $w$ and $b$ are the parameters to be selected.

$$y(x) = w \cdot \varphi(x) + b, \qquad w \in Z, b \in R \tag{2}$$

The optimization problem is as follows:

$$\min_{w,b,e} J(w,e) = \frac{1}{2} w^T w + \frac{1}{2} \gamma \sum_{k=1}^{N} e_k^2 \qquad \gamma > 0 \tag{3}$$

Where $\gamma$ is an regularization constant. (3) is subjected to:

$$y_k = w^T \varphi(x_k) + b + e_k, \quad k = 1 ; \cdots , N \tag{4}$$

Where $e_k$ is the error between actual output and predictive output of the kth data. The LS-SVM model of the data set can be given by:

$$y(x) = \sum_{i=1}^{N} \alpha_i \phi(x_i)^T \phi(x_j) + b \tag{5}$$

Where $\alpha_i \in R(i=1,2,......,N)$ are Lagrange multipliers. The mapping function can be paraphrased by a kernel function $K(\cdot, \cdot)$ because of the application of Mercer's theorem, which means that $K(x,x_i)(i=1,2......,N)$ are any kernel functions satisfying the Mercer condition.

$$\Omega_{ij} = \phi(x_i)^T \phi(x_j) = K(x_i,x_j), \quad \Omega \in \mathbb{R}^{N \times N} \tag{6}$$

The typical kernel functions are linear, polynomial, Radial Basis Function (RBF), MLP functions, etc. Analytical solutions of parameters $\alpha_i \in R$ and $b$ can be obtained from the equation

$$y(x) = \sum_{i=1}^{N} \alpha_i K(x,x_i) + b \tag{7}$$

Note that in the case of RBF kernels:

$$K(x,x_i) = \exp\{-\frac{\|x-x_i\|^2}{2\sigma^2}\} \tag{8}$$

There are only two additional tuning parameters , kernel width parameter $\sigma$ in (8) and   regularization parameter $\gamma$ in (3).These parameters have very important impact on the accuracy of model LS-SVM. In this study, they are automatically tuned by PSO algorithm in the training phase [3].

## 2.3    Overall Process of the EMD-Based LSSVM

Suppose there is a time series $x(t)$ t=1, 2,..., N, in which one would like to make the L-step ahead prediction, i.e. $x(t+1)$. For example, L=1 means one single-step ahead prediction and L=15 represents 15-step ahead prediction. The proposed EMD-based LSSVM forecasting method is generally composed of the following three main steps , as illustrated in Fig. 1.

(1) The original time series $x(t)$ , t=1, 2,..., N is decomposed into n IMF components, $imf_j(t)$ , j=1, 2,..., n, and one residual component $r_n(t)$ via EMD.

(2)For each extracted IMF component and the residual component, the LSSVM is used as a forecasting tool to model the decomposed components, and to make the corresponding prediction for each component.

(3)The prediction results of all extracted IMF components and the residue produced by LSSVM in the previous step are combined to generate an aggregated output, which can be seen as the final prediction result for the original time series.

To summarize, the proposed EMD-based LSSVM forecasting method is actually an"EMD–LSSVM" ensemble learning approach. That is, it is an "EMD (Decomposition)–LSSVM (Prediction)" methodology. In order to verify the effectiveness of the proposed EMD-based LSSVM ensemble methodology, two main time series, chaos time series and real ship motion data  are used for testing purpose in the next section.

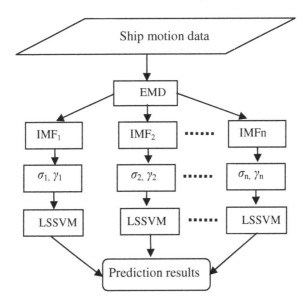

**Fig. 1.** The overall process of the EMD-based LSSVM methodology

# 3    Experimental Results

## 3.1    Real Ship Motions Prediction

In order to test the validity of EMD-LSSVM method for real ship motions prediction, we collected a series of roll and pitch time series of a real ship during her seakeeping trial using tilt sensor motion measuring system, and the sampling frequency is 5Hz. Fig.2 and Fig.3 show the prediction result sample of the real ship's roll and pitch in seakeeping trial.The means and standard deviations for the three records are shown in Table I.

## 3.2    Prediction Error Analysis

Let X be the m true values for a testing time series dataset, and X be the m predicted values obtained L-step ahead. Two error measures are used to evaluate the prediction performance. One is the root mean squared error (RMSE), which is the square root of the variance as defined in Equation 9.

$$RMSE = \sqrt{\frac{1}{L}\sum_{i=1}^{L}\left(x_{m+i} - x_{m+i}^{*}\right)^{2}} \tag{9}$$

The other error measure is the symmetric mean absolute percentage error (SMAPE), which is based on relative errors and is defined in Equation 10.

$$SMAPE = \frac{1}{L}\sum_{i=1}^{L}\frac{\left|x_{m+i} - x_{m+i}^{*}\right|}{\left(x_{m+i} + x_{m+i}^{*}\right)/2} \tag{10}$$

**Table 1.** Performance in terms of RMSE and SMAPE

| prediction horizon | RMSE interactive single-step methods | k-NN based LS-SVM | SMAPE interactive single-step methods | k-NN based LS-SVM |
|---|---|---|---|---|
| 10 | 0. 3125 | 0. 2033 | 0. 0164 | 0. 0151 |
| 40 | 0. 6957 | 0. 5486 | 0. 0346 | 0. 0193 |
| 70 | 1. 2872 | 1. 0509 | 0. 0558 | 0. 0447 |
| 100 | 2. 1120 | 1. 9198 | 0. 0832 | 0. 0676 |
| 130 | 2. 4819 | 1. 9447 | 0. 1738 | 0. 1430 |
| 160 | 2. 7642 | 2. 4520 | 0. 4193 | 0. 2446 |

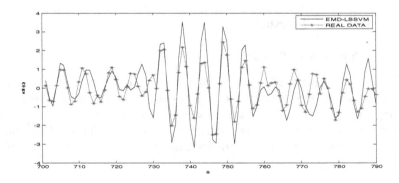

**Fig. 2.** 90 steps (18 seconds)prediction result of roll data EMD-based LSSVM algorithm (using 700 points for training)

## 4    Concluding Remarks

In this paper we proposed a forecasting method based on EMD and LSSVM algorithm. We have shown that reasonable prediction performance can be achieved in the case of nonstationary time series. The hybrid training algorithm of the EMD and LSSVM made the parameter searching process more effective.

One drawback with our prediction method is the computational cost associated with the training matrix. The results suggest that the window size does have an important effect on the quality of a LSSVM based forecaster.

**Acknowledgments.** This work is funded by the National Defense Hydrodynamic Pre-study Fund.

## References

1. Shi, A.G., Zhou, B.: Time-domain forecast based on the real ship sway data. In: Advanced Materials Research, vol. 503-504, pp. 1397–1400. Frontiers of Manufacturing Science and Measuring Technology II (2012)
2. Zhou, B., Shi, A.G.: LSSVM and hybrid particle swarm optimization for ship motion prediction. In: Proceedings of 2010 International Conference on Intelligent Control and Information Processing, pp. 183–186. IEEE Press, Dalian (2010)
3. Shi, A.G., Zhou, B.: K-nearest neighbor LS-SVM method for multi-step prediction of chaotic time series. In: Electrical & Electronics Engineering, pp. 407–409. IEEE Symposium on Digital Object (2012)

# Optimal Calculation of Tensor Learning Approaches

Kai Huang and Liqing Zhang

MOE-Microsoft Key Laboratory for Intelligent Computing and Intelligent Systems,
Department of Computer Science and Engineering, Shanghai Jiao Tong University,
Shanghai 200240, China
huangkai888888@yahoo.com.cn,
zhang-lq@cs.sjtu.edu.cn

**Abstract.** Most algorithms have been extended to the tensor space to
create algorithm versions with direct tensor inputs. However, very unfortunately basically all objective functions of algorithms in the tensor
space are non-convex. However, sub-problems constructed by fixing all
the modes but one are often convex and very easy to solve. However, this
method may lead to difficulty converging; iterative algorithms sometimes
get stuck in a local minimum and have difficulty converging to the global
solution. Here, we propose a computational framework for constrained
and unconstrained tensor methods. Using our methods, the algorithm
convergence situation can be improved to some extent and better solutions obtained. We applied our technique to Uncorrelated Multilinear
Principal Component Analysis (UMPCA), Tensor Rank one Discriminant Analysis (TR1DA) and Support Tensor Machines (STM); Experiment results show the effectiveness of our method.

**Keywords:** Tensor learning approaches, Alternating projection optimization procedure, Initial value problem.

## 1 Introduction

Many pattern recognition methods have been extended to the tensor space.
The PCA method has been extended to multilinear PCA (MPCA)[1] and
UMPCA[2]. LDA has been extended to TR1DA[3] and general tensor discriminant analysis (GTDA)[4]. ICA has been extended to an independent subspace
approach[5]. non-negative matrix factorization has been extended to a multilinear non-negative tensor factorization [6]. Support vector machine has been extended to STM[7][8], even the Minimax probability machine to a tensor Minimax
probability machine[8]. A supervised tensor learning (STL) framework has been
proposed to provide a common framework for extending vector-based methods
to tensors[8].

The objective function of methods is basically non-convex and there are a lot
of local solutions. However tensor sub-problems derived by fixing all the modes
but one are often convex and very easy to solve. So most of the algorithms use

C. Guo, Z.-G. Hou, and Z. Zeng (Eds.): ISNN 2013, Part I, LNCS 7951, pp. 326–333, 2013.

an alternating projection/optimization procedure to solve the problem. But the algorithm may have trouble converging; sometimes it may iterate between several local solutions and not converge to a global solution. So we put forward a set of calculation methods for both constrained and nonconstrained tensor problems; they improve both the calculated result and convergence behavior.

In this paper, two approximation approaches for unconstrained tensor approaches are introduced; they are alternating least squares (Section 2.1.1) and the gradient descent algorithm (Section 2.1.2). In section 2.2 our approach for constrained conditions, the NLPLSQ approach, is introduced. Section 3 describes experiments which demonstrate the good convergence of our tensor approximation approaches (Section 3.1) and the performance improvement (Section 3.2). Then a conclusion is given in Section 4.

## 2 Our Approach

The general tensor method often assigns a randomly generated value or all 1s to the initial value. The result is often very unsatisfactory. All tensor methods aim to find the optimal solution in the rank one tensor space, so the initial value can be the rank one tensor which is closest to the optimal solution of the original vector space problem.

$$\min f(a^{(1)}, \ldots, a^{(N)}) \equiv \frac{1}{2} \left\| Z - \left[\!\left[ a^{(1)}, \ldots, a^{(N)} \right]\!\right] \right\|^2 \tag{1}$$

Here the Z is a tensor produced by transforming the optimal vector apace solution using the inverse process of vectorization. For tensor algorithms, there are two cases, one with equality or inequality constraints, the other unconstrained. We propose effective methods for both cases.

### 2.1 Approaches for Unconstrained Conditions

Many tensor algorithms only optimize the target function without other constraints. The feasible domain of the algorithm is the entire range of rank one tensors. TR1DA is such an algorithm. There are two possible algorithms for such problems, alternating least squares and gradient descent. The condition here is quite similar to the canonical tensor decomposition when tensor rank is 1[9][10].

**Alternating Least Squares.** The idea of this algorithm is actually very simple. The main idea is like the idea of supervised tensor learning (STL). When calculating the value of a certain mode, fix the other modes. A problem of this form is convex and easy to solve. The equation is as follows:

$$\min_{a^{(n)}} f(a^{(1)}, \ldots, a^{(N)}) = \frac{1}{2} \left\| Z - a^{(1)} \circ \cdots \circ a^{(n)} \circ \cdots \circ a^{(N)} \right\|^2 \tag{2}$$

We can expand the equation as follows:

$$= \min_{a^{(n)}} \left\| Z_{(n)} - a^{(n)} (a^{(N)} \otimes \cdots \otimes a^{(n-1)} \otimes a^{(n+1)} \otimes \cdots \otimes a^{(1)})^T \right\|^2 \quad (3)$$

Here $\otimes$ means the kronecker product and $Z_{(n)}$ means transforming a tensor $Z$ to a matrix corresponding to mode n. The solution of this problem is as follows:

$$= Z_{(n)} \left( (a^{(N)} \otimes \cdots \otimes a^{(n-1)} \otimes a^{(n+1)} \otimes \cdots \otimes a^{(1)})^T \right)^\dagger \quad (4)$$

**Gradient Descent.** The gradient descent algorithm (GDA) can also be used. It ensures convergence to the global minimum only for a convex function on a convex set. But here the target function, although non-convex, is very close to a convex function. Therefore, the algorithm converges to the global minimum in most cases.

$$x = \left[ a^{(1)}, \ldots, a^{(n)}, \ldots, a^{(N)} \right]^T \quad (5)$$

The key point in gradient descent is to determine the gradient of the objective function for $x$,

$$\nabla f(x) = \left[ \frac{\partial f}{\partial a^{(1)}}, \ldots, \frac{\partial f}{\partial a^{(n)}}, \ldots, \frac{\partial f}{\partial a^{(N)}} \right]^T \quad (6)$$

The objective function can be written as:

$$f(x) = \underbrace{\frac{1}{2} \|Z\|^2}_{f_1(x)} - \underbrace{\left\langle Z, a^{(1)} \circ a^{(2)} \circ \cdots \circ a^{(N)} \right\rangle}_{f_2(x)} + \underbrace{\frac{1}{2} \left\| a^{(1)} \circ a^{(2)} \circ \cdots \circ a^{(N)} \right\|^2}_{f_3(x)} \quad (7)$$

The first summand does not involve the variables; therefore,

$$\frac{\partial f_1}{\partial a^{(n)}} = 0 \quad (8)$$

Here the 0 is a 0 vector of length $I_n$ while the second term can be written as

$$f_2(x) = Z \times_{m=1}^N a_r^{(n)} \quad (9)$$

$$= (Z \times_{m=1, m \neq n}^N a_r^{(m)})^T a_r^{(n)} \quad (10)$$

It is obvious with f2 written in this way:

$$\frac{\partial f_2}{\partial a^{(n)}} = (Z \times_{m=1, m \neq n}^N a_r^{(m)}) \quad (11)$$

The third summand is

$$f_3(x) = \prod_{m=1}^{N} a^{(m)^T} a^{(m)} \tag{12}$$

therefore,

$$\frac{\partial f_3}{\partial a^{(n)}} = 2( \prod_{m=1, m \neq n}^{N} a^{(m)^T} a^{(m)}) a^{(n)} \tag{13}$$

Combining these two terms yields the desired result. Another important issue is choosing an appropriate step length. If it is too small, much more iteration will be needed. If it is too big, the iteration may tend to diverge or come to an inaccurate result.

## 2.2  Approach for Constrained Condition

For a variety of tensor algorithms, there is some constraint in the calculation. That is, the feasible region is not the full rank one tensor region but has some constraints such as the left part of the equations below. The UMPCA and STM are such algorithms.

**Nonlinear Least Squares with Nonlinear Constraints.** For nonlinear least squares, there are a lot of algorithms, such as the the Gauss-Newton approach. There are also many algorithms can solve the linear least squares problem with constraints. An intuitive idea is to combine these two algorithms. First, the original problem is transformed into a linear least squares problem and then a method which solves constrained least squares problems is applied[11][12]. However there is a difficulty, For our problem, the Jacobi matrix is always singular but the Gauss-Newton method needs the Jacobi matrix to be nonsingular. Although a damped Gauss-Newton method such as the Levenberg-Marquardt algorithm can solve this problem, additional errors are introduced in the process. Fortunately, there is a very effective algorithm which can directly solve constrained nonlinear least squares problems. By introducing new variables, the original problem can be transformed into an equivalent optimization problem[13][14]. In this way, the problem that the Jacobi matrix is singular is naturally solved and does not artificially introduce error.

$$
\begin{array}{lll}
\min \frac{1}{2} \sum_{i=1}^{l} f_i(x)^2 & & \min \frac{1}{2} z^T z \\[2mm]
g_j(x) = 0 \quad j = 1, \ldots, m_e & & f_i(x) - z_i = 0 \quad i = 1, \ldots, l \\[2mm]
g_j(x) \geq 0 \quad j = m_e + 1, \ldots, m & \Longrightarrow & g_j(x) = 0 \quad j = 1, \ldots, m_e \\[2mm]
x_l \leq x \leq x_u & & g_j(x) \geq 0 \quad j = m_e + 1, \ldots, m \\[2mm]
& & x_l \leq x \leq x_u
\end{array} \tag{14}
$$

The transformed problems can be directly solved by the SQP algorithm which can achieve very good convergence.

## 3    Experiments

### 3.1    Tensor Approximation Approaches

In Figure 1, we compare the convergence property and approximation error of our tensor approximation algorithms. In Subfigure a, for different tensors, we compare the number of steps the ALS algorithm takes to converge. An increase in the number of dimensions or modes delays the convergence; it may need more iteration steps. In addition, if the number of dimensions for each mode is similar or the same, the algorithm tends to converge quickly. It always uses fewer steps and the steps of different retries tend to be the same.

In Subfigure b, the number of steps of the gradient descent algorithm for different step lengths are shown. Although this algorithm always converges to the globally optimal solution, no matter what initial value it is assigned, the step length greatly affects the step count. For different tensors, with a decrease in step length, the number of iterations grow exponentially. With more modes or a more complex tensor structure, the number of steps increases significantly.

In Subfigure c, we compare the iteration step count of NLPLSQ and SQP for UMPCA and STM, NLPLSQ converges faster than SQP and the iteration count for different retries is more stable; the variance is small.

Subfigure d shows the difference of approximation error for SQP and NLPLSQ applied to UMPCA and STM. The error is much smaller for UMPCA. NLPLSP has a lower approximation error. As for STM, the error decreases for most cases but not as much as for UMPCA. In most cases, the decease is less than 0.1.

### 3.2    Improvement for Tensor Approaches

Subfigures a,b,c of Figure 2 demonstrate the iterative convergence process of TR1DA,UMPCA and STM after using our approach for . TR1DA is for unconstrained problems and UMPCA and STM for constrained problems. Here we should pay special attention to UMPCA, the calculation of other components but the first is a constrained problem.

Subfigure a compares the iterative convergence process of TR1DA using our approach for different tensors. Using our approach, the target function value comes to a stable solution after only a few iterations, much faster than the original method. Sometimes the target function curve is close to horizontal. A tensor with more modes will make it harder to converge.

Subfigure b compares the iterative convergence process of UMPCA for the original approach, ALS and NLPLSQ. The experiment is for 2*2*2, 6*6*6 and 10*10. It is clear that a complicated tensor requires more iterations to converge. The UMPCA algorithm is special, so for the initial value setting of our approach, the target function value sometimes experiences a decrease in the first step.

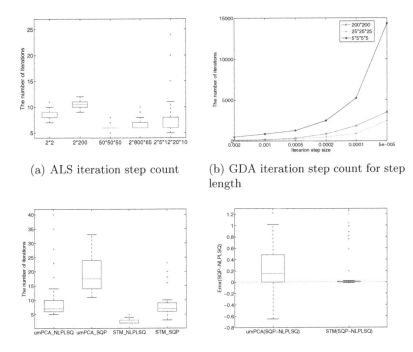

(a) ALS iteration step count

(b) GDA iteration step count for step length

(c) iteration step count for constrained condition

(d) approximation error comparison

**Fig. 1.** Tensor approximation approaches for constrained and unconstrained condition

Subfigure c compares the iterative convergence process of STM for the original approach, ALS and NLPLSQ. The experiment is on two cases 3*3*3 and 10*10. Our two approaches improve the convergence to some extent; NLPLSQ is a little bit better than ALS in most cases.

Subfigure d shows the final result improvement of TR1DA, UMPCA and STM after adopting our approach. The ALS approach improves the performance of TR1DA in most cases. UMPCA is somewhat special. ALS and NLPLSQ are better than the original approach in fewer than a third of cases. The result for ALS is better than for NLPLSQ. As for STM, ALS and NLPLSQ perform well about twice as often as the original method. NLPLSQ is better than ALS.

## 4 Conclusion

For tensor algotithsms, if we optimize a specific mode and fix all the other modes, the subproblem becomes convex. However, even so, solving methods tend to fall into local solutions and have difficulty converging. These problems are often caused by unreasonable initial value selection. The essence of tensor methods is finding the optimal solution of the original vector space within the rank one

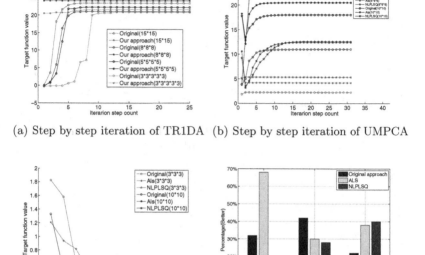

(a) Step by step iteration of TR1DA    (b) Step by step iteration of UMPCA

(c) Step by step iteration of STM    (d) Statistics of final results improvement

**Fig. 2.** The improvement given by our methods for TR1DA, UMPCA and STM

tensor limitation and this paper takes advantage of this. First, calculate the optimal solution of the optimization problem in the vector space. Then in the range of rank one tensors we calculate the tensor closest to the vector space optimal solution in the least squares sense. Of the three types of tensor methods, our approach is for TVP and TSP typed methods. Depending on whether there are constraints, we propose ALS, gradient descent algorithm (GDA) and NLPLSQ. Experiment results show that the convergence of tensor approaches is greatly improved by adopting our method; convergence becomes more stable and a better target function value is achieved.

# References

1. Lu, H., Plataniotis, K.N., Venetsanopoulos, A.N.: Multilinear principal component analysis of tensor objects for recognition. In: Proc. Int. Conf. on Pattern Recognition, pp. 776–779 (2006)
2. Lu, H., Plataniotis, K.N., Venetsanopoulos, A.N.: Uncorrelated multilinear principal component analysis through successive variance maximization. In: ICML (2008)
3. Tao, D., Li, X., Wu, X., Maybank, S.: Tensor rank one discriminant analysis-a convergent method for discriminative multilinear subspace selection. Neurocomput. 71(10-12), 1866–1882 (2008)

4. Tao, D., Li, X., Wu, X., Maybank, S.J.: General tensor discriminant analysis and gabor features for gait recognition. IEEE Trans. Pattern Anal. Mach. Intell. 29(10), 1700–1715 (2007)
5. Alex, M., Vasilescu, O., Terzopoulos, D.: Multilinear independent components analysis. IEEE Computer Society Conference on Computer Vision and Pattern Recognition 1, 547–553 (2005)
6. Shashua, A., Hazan, T.: Non-negative tensor factorization with applications to statistics and computer vision. In: Proceedings of the International Conference on Machine Learning (ICML), pp. 792–799. ICML (2005)
7. Cai, D., He, X., Wen, J.-R., Han, J., Ma, W.-Y.: Support tensor machines for text categorization. Technical report, Computer Science Department, UIUC, UIUCDCS-R-2006-2714 (April 2006)
8. Tao, D., Li, X., Hu, W., Maybank, S., Wu, X.: Supervised tensor learning. In: Proceedings of the Fifth IEEE International Conference on Data Mining, ICDM 2005, pp. 450–457. IEEE Computer Society, Washington (2005)
9. Acar, E., Kolda, T.G., Dunlavy, D.M.: An optimization approach for fitting canonical tensor decompositions. Technical report (2009)
10. Kolda, T.G., Bader, B.W.: Tensor decompositions and applications. Siam Review 51(3), 455–500 (2009)
11. Coleman, T.F., Li, Y.: A reflective newton method for minimizing a quadratic function subject to bounds on some of the variables. SIAM Journal on Optimization 6(4), 1040–1058 (1996)
12. Gill, P.E., Murray, W., Wright, M.H.: Practical Optimization, vol. 1. Academic Press (1981)
13. Schittkowski, K.: Nlplsq: A fortran implementation of an sqp-gauss-newton algorithm for least-squares optimization (2009)
14. Schittkowski, K.: Nlpqlp: A fortran implementation of a sequential quadratic programming algorithm with distributed and non-monotone line search (2010)

# Repeatable Optimization Algorithm Based Discrete PSO for Virtual Network Embedding

Ying Yuan[1], Cui-Rong Wang[2], Cong Wan[2], Cong Wang[2] and Xin Song[2]

[1] School of Information Science and Engineering, Northeastern University,
Shenyang, 11004, China
[2] School of Computer and Communication Engineering,
Northeastern University at Qinhuangdao, Qinhuangdao, 66004, China
{yuanying1121,Congw1981}@gmail.com,
wangcr@mail.neuq.edu.cn,
10000cong@163.com

**Abstract.** Aiming at reducing the link load and improving substrate network resource utilization ratio, we model the virtual network embedding (VNE) problem as an integer linear programming and present a discrete particle swarm optimization based algorithm to solve the problem. The approach allows multiple virtual nodes of the same VN can be embedded into the same physical node as long as there is enough resource capacity. It not only can cut down embedding processes of virtual link and reduce the embedding time, but also can save the physical link cost and make more virtual networks to be embedded at the same time. Simulation results demonstrate that comparing with the existing VNE algorithm, the proposed algorithm performs better for accessing more virtual networks and reducing embedding cost.

**Keywords:** network virtualization, embedding algorithm, virtual network, discrete particle swarm optimization.

## 1    Introduction

Network virtualization is the core problem of VNE. In the virtual network environment, the Internet service provider (ISP) is divided into two parts: infrastructure provider (InP) and service providers (SP) [1]. Network virtualization can provide more flexibility by separating the network provider from infrastructure provider. InPs manage the physical Infrastructure while multiple SPs will be able to create heterogeneous VNs to offer customized end-to-end services to the users by leasing shared resources from one or more InPs [2].

Virtual network embedding problem is known to be NP-hard even in the offline case [3]. Thus efficient VNE techniques that intelligently use the substrate network resources are important. Most of literatures about VNE problem formulate the VNE as an optimization problem with the embedding cost as the objective. Recently, a number of heuristic-based algorithms or customized algorithms have appeared in the relevant literature [4-8].

C. Guo, Z.-G. Hou, and Z. Zeng (Eds.): ISNN 2013, Part I, LNCS 7951, pp. 334–342, 2013.

The authors in [4] have provided a two stage algorithm for embedding the VNs. Firstly, they embedding the virtual nodes. Secondly they proceed to map the virtual links using shortest paths and multi-commodity flow (MCF) algorithms. In order to increase the acceptance ratio and the revenue, D-ViNE and R-ViNE are designed in [5]. The authors formulated the VNE problem as a mixed integer program through substrate network augmentation, and then relaxed the integer constraints to obtain a linear program. VNE-AC algorithm in [6] is a new VNE algorithm based on the ant colony meta-heuristic. They do not restrain the VNE problem by assuming unlimited substrate resources, or specific VN topologies or restricting geographic locations of the substrate core node. The authors in [7] propose VNE strategy with topology-aware node ranking. They apply the Markov Random Walk (RW) model to rank a network node based on its resource and topological attributes before mapping the virtual network components. The authors in [8] use a joint node and link mapping approach for the VI mapping problem and develop a virtual infrastructure mapping algorithm.

In this paper, we propose a VNE algorithm based on discrete particle swarm optimization (DPSO), which allows repeatable embedding over same substrate node as long as the node has enough available resources. We denote our algorithm by M-VNE-DPSO.

The rest of the paper is organized as follows. In section 2, we give the detailed description of VNE and its general model. In section 3, firstly, we give our particular model for VNE and briefly introduce DPSO, the parameters and operations of the DPSO are redefined. Then we discuss how to deal with VNE problem with DPSO. The simulation environment and results are given in Section 4. Section 5 gives the conclusion.

# 2    Network Model and Problem Description

## 2.1    Substrate Network

We model the substrate network as a weighted undirected graph and denote it by $G^S = (N^S, E^S, A_N^S, A_E^S)$, where $N^S$ is the set of substrate nodes and $E^S$ is the set of substrate link. We denote the set of loop-free substrate paths by $P^S$. The notations $A_N^S$ denote the attributes of the substrate nodes, including CPU capacity, storage, and location. The notations $A_E^S$ denote the attributes of the substrate edges, including bandwidth and delay. In this paper, each substrate node $n^s \in N^S$ is associated with the CPU capacity. Each substrate link $e^s(i, j) \in E^S$ between two substrate nodes $i$ and $j$ is associated with the bandwidth.

## 2.2    Virtual Network Request

Similar to the substrate network, a virtual network can be represented by a weighted undirected graph $G^V = (N^V, E^V, C_N^V, C_E^V)$, where $N^V$ and $E^V$ denote the set of virtual nodes and virtual link, respectively. Virtual nodes and edges are associated with constraints on resource requests, denoted by $C_N^V$ and $C_E^V$, respectively.

## 2.3    Virtual Network Embedding

When a VN request arrives, the substrate network has to determine whether to accept the request or not. If the request is accepted, the substrate network has to perform a suitable Virtual network embedding and allocate substrate resource. The VNE problem is defined by a embedding $\Gamma: G^V \mapsto (N^{S*}, P^{S*}, R_N, R_E)$ , from $G^V$ to $G^S$ , where $N^{S*} \subseteq N^S$ , $P^{S*} \subseteq P^S$ .

The process of VNE has two parts: virtual network node embedding and virtual network edge embedding.

In virtual network nodes embedding, each virtual node is hosted on substrate nodes that satisfy the requested CPU constraints:

$$\Gamma_N : N^V \mapsto N^{S*}$$
$$\Gamma_N(n^v) \in N^{S*}, \ \forall n^v \in N^V \tag{1}$$
$$Rcpu(n^v) \leq Ccpu(\Gamma_N(n^v))$$

where $Rcpu(n^v)$ is requested CPU constraints for the virtual node $n^v$ , $Ccpu(\Gamma_N(n^v))$ is the available CPU capacity of substrate node.

In virtual network edges embedding, each link embedding assigns virtual links to loop-free paths on the substrate that satisfy the required bandwidth constraints:

$$\Gamma_E : E^V \mapsto P^{S*}$$
$$\Gamma_E(e^v) \in P, \ \forall e^v \in E^V$$
$$Pbw(e^v) \leq Cbw(\Gamma_E(e^v)) \tag{2}$$
$$Cbw(\Gamma_E(e^v)) = \min_{e^s \in P^s} Cbw(e^s)$$

where $Pbw(e^v)$ is requested bandwidth constraints for the virtual edge, $Cbw(\Gamma_E(e^v))$ is the available bandwidth capacity of substrate path. The available bandwidth capacity of a substrate path $P \subseteq P^{S*}$ is qualified by the edge which has the minimum bandwidth capacity in this path.

## 2.4    System Object

The main objective of virtual network embedding is to make efficient use of the substrate network resources when mapping the virtual network into the substrate network. In this paper, we aim to decrease cost of the InP so as to support more virtual networks.

Similar to the early work in [4, 5, 6, 9], Firstly, we define the revenue and cost.
**Definition 1: Revenue**    the sum of total an virtual network request gain from InP at time $t$ as

$$R(G^V(t)) = \alpha_1 \bullet \sum_{e^v \in E^V} bw(e^v) + \alpha_2 \bullet \sum_{n^v \in N^V} cpu(n^v) \tag{3}$$

where $bw(e^v)$ is the bandwidth of the request link, $cpu(n^v)$ is the CPU capacity of the request node. $\alpha_1$ and $\alpha_2$ are weighting coefficients to balance the effect of bandwidth and CPU.

**Definition 2**: **Cost** the sum of total substrate resources allocated to that virtual network at time $t$ as

$$C(G^V(t)) = \alpha_3 \bullet \sum_{e^v \in E^V} bw(e^v) Length(P) + \alpha_4 \bullet \sum_{n^v \in N^V} cpu(n^v) \qquad (4)$$

Where $Length(P)$ is the hop count of the virtual link $e^v$ when it is assigned to a set of substrate links. $\alpha_3$ and $\alpha_4$ are weighting coefficients similarly to $\alpha_1$ and $\alpha_2$.

In a given period of time virtual network embedding algorithm should minimize the cost of substrate network and accept the largest possible number of virtual network requests. We define the long-term average cost as

$$C\!\big/\!_T = \lim_{T \to \infty} \frac{\sum_{t=0}^{T} C(G^V(t))}{T} \qquad (5)$$

# 3     M-VNE-DPSO Algorithms

## 3.1     System Model

We model the problem of embedding of a virtual network as a mathematical optimization problem using integer linear programming (ILP). We should minimize the usage of the substrate resources. Thus, the objective of our optimization problem is defined as follows:

$\varphi_{ij}^w$ is a binary variable.  $\forall w \in E^v$  $\forall i, j \in N^s$  $\varphi_{ij}^w \begin{cases} 0 & i = j \\ 1 & i \neq j \end{cases}$

Object:

$$Minimize \sum_{(i,j) \in P^s} \varphi_{ij}^w \times bw(e^w) \qquad (6)$$

Node Constraints:

$$\forall u \in N^V, \ \forall i \in N^S \quad Ccpu(i) = Cpu(i) - \sum_{n^v \to i} Cpu(n^v) \geq Rcpu(u) \qquad (7)$$

where $\sum_{n^v \to i} Cpu(n^v)$ is the total amount of CPU capacity allocated to different virtual nodes hosted on the substrate node; $Cpu(i)$ is the total amount of CPU capacity of the substrate node.

Link Constraints:

$$\forall i, j \in N^s, \ P^{ij} \in P^s, \ \forall w \in e^v \quad Cbw(P^{ij}) = \min_{e^s \in P^{ij}} Cbw(e^s) \geq Rbw(w) \qquad (8)$$

$$Cbw(e^s) = Bw(e^s) - \sum_{e^v \to e^s} Bw(e^s) \qquad (9)$$

where $Cbw(P^{ij})$ is the available bandwidth capacity of a substrate path from $i$ to $j$; $\sum_{e^v \to e^s} Bw(e^s)$ is the total amount of bandwidth capacity allocated to different virtual links hosted on the substrate link; $Bw(e^s)$ is the total amount of bandwidth capacity of the substrate link.

## 3.2    Discrete PSO for Virtual Network Embedding

Particle swarm optimization (PSO) is a population based stochastic optimization technique developed by Dr. Eberhart and Dr. Kennedy in 1995, inspired by social behavior of bird flocking or fish schooling.

During the evolutionary process, the velocity and position of particle updated as follows:

$$V^{k+1} = \omega V^k + c_1 r_1 (X_p^k - X^k) + c_2 r_2 (X_g^k - X^k) \tag{10}$$

$$X^{k+1} = X^k + V^{k+1} \tag{11}$$

Where $V^k = \left[ v_1^k, v_2^k, \cdots, v_m^k \right]$ is the velocity vector; $X^k = \left[ x_1^k, x_2^k, \cdots, x_m^k \right]$ is the position vector; $\omega$ denote the inertia weight; $r_1$ and $r_2$ denote two random variables uniformly distributed in the range of $(0, 1)$; $c_1$ and $c_2$ denote the accelerator of particle; $X_p^k$ denote the position with the best fitness found so far for the $k$th particle; $X_g^k$ denote the best global position in the swarm.

Standard PSO is not directly applicable to the optimal VNE problem, so we used variants of PSO for discrete optimization problems to solve the optimal VNE problem.

Redefine the position and velocity parameters for discrete PSO as follows:

**Definition 3:** Position $X^k = \left[ x_1^k, x_2^k, \cdots, x_m^k \right]$ a possible VNE solution, where $x_i^k$ is the number of the substrate node the $i$th virtual node embedding to. $m$ denotes the total number of nodes in virtual network $k$.

**Definition 4:** Velocity $V^k = \left[ v_1^k, v_2^k, \cdots, v_m^k \right]$ makes the current VNE solution to achieve a better solution, where $v_i^k$ is a binary variable. For each $v_i^k$, if $v_i^k = 1$, the corresponding virtual node's position in the current VNE solution should be remains; otherwise, should be adjusted by selecting another substrate node.

The operations of the particles are redefined as follows:

**Definition 5:** Addition of Position and Velocity $X^k + V^k$ a new position that corresponds to a new virtual network embedding solution. If the value of $v_i^k$ equals to 1, the value of $x_i^k$ will be kept; otherwise, the value of $x_i^k$ should be adjust by selecting another substrate node. For example, $(1, 5, 6, 3, 2) + (1, 0, 1, 0, 1)$ denotes the second and fourth virtual node embedding solutions should be adjusted.

**Definition 6:** Subtraction of Position $X^m - X^n$ a velocity vector. It indicates the differences of the two virtual network embedding solutions $X^m$ and $X^n$. If $X^m$ and $X^n$ have the same values at the same dimension, the resulted value of the corresponding dimension is 1, otherwise, the resulted value of the corresponding dimension is 0. For example, $(5, 5, 3, 2, 4) - (5, 3, 3, 2, 6) = (1, 0, 1, 1, 0)$.

**Definition 7:** Multiple of Velocity $\psi * V^m$    keep $V^m$ with probability $\psi$ in the corresponding dimension.

**Definition 8:** Addition of Multiple $\psi_1 * V^m + \psi_2 * V^n$ a new velocity that corresponds to a new virtual network embedding solution, where $\psi_1 + \psi_2 = 1$. If $V^m$ and $V^n$ have the same values at the same dimension, the resulted value of the corresponding dimension will be kept; otherwise, keep $V^m$ with probability $\psi_1$ and keep $V^n$ with probability $\psi_2$. For example, 0.3 (1, 0, 0, 1, 1) + 0.7 (1, 0, 1, 0, 1) = (1, 0, *, *, 1), where * denotes the probability of being 0 or 1. In this example, the first * is equal to 0 with probability 0.3 and equal to 1 with probability 0.7.

Because of the specificity of discrete quantity operation, we modify the particle motion equation and cancel the original inertia item. The position and velocity of particle $k$ are determined according to the following velocity and position update recurrence relations:

$$V^{k+1} = \psi_1 * (X_p^k - X^k) + \psi_2 * (X_g^k - X^k) \tag{12}$$

$$X^{k+1} = X^k + V^{k+1} \tag{13}$$

Where $\psi_1$ and $\psi_2$ are set to constant values that satisfy the inequality $\psi_1 + \psi_2 = 1$.

### 3.3     M-VNE-DPSO Algorithm Description

We embed virtual nodes using the DPSO algorithm discussed in previous sub section and embedding virtual links using the well-known FloydWarshall shortest path algorithm. Firstly, we introduce a step for substrate nodes and substrate edges initialization before virtual network embedding. The detailed steps of the initialization algorithm are shown as follows:

**Algorithm 1.** The initialization algorithm $Ini(G^{s'})$

1. For substrate network , sort the substrate nodes *NList* according to their cpu capacity $Ccpu(n^s)$ in descending order sort the substrate edges *EList* according to their bandwidth capacity $Cbw(e^s)$ in descending order;
2. For a virtual network request ,sort the virtual nodes *VNList* according to their required CPU resources $Rcpu(n^v)$ in ascending order, then $\min Rcpu(n^v) \leftarrow$ required CPU resources of the first node of *VNList* ;sort the virtual edges *VEList* according to their required bandwidth resources $Rbw(e^v)$ in ascending order, then $\min Rbw(e^v) \leftarrow$ required bandwidth resources of the first edge of *VEList* ;
3. Obtain $NList' \leftarrow NList \mid \{Ccpu(n^s) < \min Rcpu(n^v)\}$ ;
   $EList' \leftarrow EList \mid \{Chw(e^s) < \min Rbw(e^v)\}$ .

In node embedding step, the algorithm with repeatable embedding over substrate nodes is put forward, in which, multiple virtual nodes in the same virtual network can be mapped to the same substrate node if there is enough resource capacity and takes the objective function (6) as fitness function $\phi(x)$ . The detailed steps of the M-WNE-DPSO algorithm are shown as follows:

**Algorithm 2.** The M-VNE-DPSO algorithm

1. $Ini(G^{S'})$ ;
2. Initialize n=Particle Count, m=Max Iteration Count;
3. Randomly generated $X^k$ and $V^k$ ;
4. Each $X^k$ corresponds to link embedding with shortest path. Set $X_p^k$ and $X_g^k$ to these particles according to their fitness values of $\phi(x)$ . If $X^k$ is an unfeasible position, the fitness value $\phi(x)$ of this particle will be set to $+\infty$ ;
5. If $\phi(x)$ of the particle equal to $+\infty$ , re-initialize its $X^k$ and $V^k$ . Otherwise, if $\phi(X_p^k) \geq \phi(X^k)$, then set $X^k$ to be the $X_p^k$ of the particle. If ( $\phi(X_g^k) \geq \phi(X^k)$ ), then set $X^k$ to be the $X_g^k$ of the particle;
6. use formula (12) and (13) to update $X^k$ and $V^k$ ;
7. If $X_g^k$ unchanged for three times or *Iterationcount* $\geq M$ , goto step 8, otherwise goto step 4;
8. If $X_g^k = +\infty$ , there is no feasible solution. The virtual network request put into wait queue, otherwise, output the virtual network embedding solution, update the resources of substrate network.

# 4    Simulation

We implemented the M-VNE-DPSO algorithm using the CloudSim3.0.1 on a high level PC which has one Intel Core i7-3770 CPU and 20G DDR3 1600 RAM. We write a random topological generator in java to generate topologies for the underlying substrate networks and virtual networks in CloudSim. Substrate networks in our experiments have 100 nodes, each node connect to other nodes with probability 0.2, so there are about 500 links in the networks. Physical node CPU and link bandwidth capacities are 100 units. For each VN request, the number of virtual nodes was randomly determined by a uniform distribution between 4 and 10. The average VN connectivity was fixed at 50%. The CPU and bandwidth requirements of virtual nodes and links are real numbers uniformly distributed between 3 and 30 units. Each virtual network's living time uniformly distributed between 100 and 1000 time unit. We analyze the performance of the new algorithm by comparing it with the D-ViNE-SP and VEN-R-PSO [9] algorithm.

In first experiment, we simulated 1000 virtual network request for each algorithm, each test run 30 times and we have plot the average result of t accessed VN request number variation to time of the three embedding algorithms. As shown in Fig.1, M-VNE-DPSO accessed more VN requests than others at the same time and performance most smooth when the substrate network is full load, so it can finish the 1000 request faster. Another advantage of M-VNE-DPSO is that because of the repeatable nature in embedding the algorithm can easily complete all

1000 request we had submitted. And relatively in this experiment, the connectivity of virtual networks is higher than the substrate network and the number of virtual network is random, so if some virtual node's connectivity is higher than every substrate node, such "big" virtual networks won't be accessed in non-repeatable embedding, i.e. not all the VN request will be accessed in VEN-R-PSO and D-ViNE-SP.

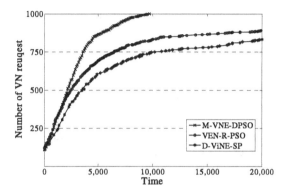

**Fig. 1.** Number or accessed virtual networks correspond to running time

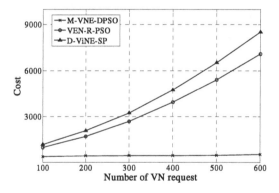

**Fig. 2.** Comparison of embedding cost for different number of VN request

The second experiment runs number 100, 200, 300, 400, 500, 600 of virtual network request respectively, every case operation 30 times and take the average cost of the three algorithms. From Fig. 2.we can see, when running 100 virtual network request, D-ViNE-SP algorithm is the highest Cost, the M-VNE-PSO algorithm less than VNE-R-PSO algorithm, along with the increase of the number of virtual network, their difference is more and more obvious. The reason is that in D-ViNE-SP, its relaxation-based approach weakens the coordination between node mapping and link mapping, which results in poor performance. M-VNE-PSO does using repeatable embedding over substrate nodes that saves the substrate link cost and makes more virtual network embedded.

## 5    Conclusions

This paper introduced a new VNE algorithm with repeatable embedding over substrate nodes base on DPSO. We increase a step initialization before virtual network embedding which helps accelerate the convergence speed. We also allow that multiple virtual nodes in the same virtual network can be mapped to the same substrate node if there is enough resource capacity. From the customer point of view, this operation can increase embedding efficiency and reduce the embedding cost; from the InP's perspective optimal resource usage strategy results in improving the resource utilization as well as accommodating more VN requests. Simulation results show that our algorithms outperform the previous approaches in terms of the accept number of VN request and the cost of substrate network.

**Acknowledgments.** This work was supported by The Central University Fundamental Research Foundation, under Grant. N110323009.

## References

1. Rahman, M.R., Aib, I., Boutaba, R.: Survivable virtual network embedding. In: Crovella, M., Feeney, L.M., Rubenstein, D., Raghavan, S.V. (eds.) NETWORKING 2010. LNCS, vol. 6091, pp. 40–52. Springer, Heidelberg (2010)
2. Nick, F., Li, X.G., Jennifer, R.: How to lease the Internet in your spare time. J. ACM SIGCOMM CCF 37(1), 61–64 (2007)
3. Mosharaf, C., Muntasir, R.R., Raouf, B.: ViNEYard: Virtual Network Embedding Algorithms With Coordinated Node and Link Mapping. J. IEEE/ACM Transactions on Networking 20(1), 206–219 (2011)
4. Minlan, Y., Yung, Y., Jennifer, R., Mung, C.: Rethinking virtual network embedding: substrate support for path splitting and migration. J. ACM SIGCOMM CCR 38(2), 17–29 (2008)
5. Mosharaf, C., Muntasir, R.R., Raouf, B.: Virtual Network Embedding with Coordinated Node and Link Embedding. In: INFOCOM 2009, pp. 783–791. IEEE Press, Rio de Janeiro (2009)
6. Ilhem, F., Nadjib, A.S., Guy, P.: VNE-AC: Virtual Network Embedding Algorithm based on Ant Colony Metaheuristic. In: IEEE ICC 2011, pp. 1–6. IEEE Press, Kyoto (2011)
7. Xiang, C., Sen, S., Zhong, B.Z., Kai, S., Fang, C.Y.: "Virtual network embedding Through Topology-Aware Node Ranking. J. ACM SIGCOMM CCR 4(2), 39–47 (2011)
8. Hong, F.Y., Vishal, A., Chun, M.Q., Hao, D., Xue, T.W.: A cost efficient design of virtual infrastructures with joint node and link mapping. Journal of Network and Systems Management 2012 20(1), 97–115 (2012)
9. Xiang, C., Zhong, B.Z., Sen, S., Kai, S., Fang, C.Y.: Virtual Network Embedding Based on Particle Swarm Optimization. J. Acta Electronica Sinica 39(10), 2240–2244 (2011)

# An Energy-Efficient Coverage Optimization Method for the Wireless Sensor Networks Based on Multi-objective Quantum-Inspired Cultural Algorithm

Yinan Guo, Dandan Liu, Meirong Chen, and Yun Liu

China University of Mining and Technology, Xuzhou, Jiangsu, China

**Abstract.** The energy-efficiency coverage of wireless sensor network is measure by the network cover rate and the node redundancy rate. To solve this multi-objective optimization problem, a multi-objective quantum-inspired cultural algorithm is proposed, which adopts the dual structure to effectively utilize the implicit knowledge extracted from the non-dominating individuals set to promote more efficient search. It has three highlights. One is the rectangle's height of each allele is calculated by non-dominated sort among individuals. The second is the crowding degree that records the density of non-dominated individuals in the topological cell measure the uniformity of the Pareto-optimal set instead of the crowding distance. The third is the update operation of quantum individuals and the selection operator are directed by the knowledge. Simulation results indicate that the layout of wireless sensor network obtained by this algorithm have larger network cover rate and less node redundancy rate.

**Keywords:** Wireless sensor network, cultural algorithm, real-coded quantum-inspired evolutionary algorithm, multi-objective optimization.

## 1 Introduction

Wireless sensor network(WSN) containing large numbers of sensor nodes monitors the targets. Its energy-efficiency coverage optimization is actually how to rationally arrange the sensor nodes. Thus it is converted to a multi-objective optimization problem(MOP). Now, many natural-inspired optimization algorithms are introduced to solve the problems, such as genetic algorithm[1], ant colony optimization[2], particle swarm algorithm[3] and quantum algorithm[4]. However, the optimization objective normally is the network coverage rate only or the weighted objective combing the network coverage rate and node redundancy rate. The conflict between objectives was not considered. So we regard maximum network coverage rate and minimum node redundancy rate as two independent objectives.

Until now, many multi-objective optimization algorithms had been presented, such as strength Pareto evolutionary algorithm II(SPEAII)[5], niched Pareto genetic algorithm (NPGA)[6], no-dominated sorting genetic algorithm II(NSGAII)[7] and multi-objective quantum evolutionary algorithms(MOQA). Meshoul [8] described the chromosome by quantum-encoded probabilistic representation. Kim[9] employed the

C. Guo, Z.-G. Hou, and Z. Zeng (Eds.): ISNN 2013, Part I, LNCS 7951, pp. 343–349, 2013.

principles of quantum computing including superposition and interference to improve the quality of the non-dominated solution set. Wei[10] decomposed MOP into many scalar optimization sub-problems. Each sub-problem simultaneously evolved based on the population composed of $q$-bit individuals. Yang[11] constructed a triploid chromosome and the chaos encoding probability amplitude. However, they do not fully utilize the implicit information during the evolution, which limits the algorithm's performances. Based on the dual structure in cultural algorithm[12], the author gave real-coded quantum-inspired cultural algorithm[13]. But it only fits for the scalar optimization problem. So a novel multi-objective real-coded quantum-inspired cultural algorithm (MORQCA) is proposed and applied to optimize WSNs' energy-efficiency coverage.

## 2     The WSN' Energy-Efficient Coverage Optimization Model

The key issues that directly influence WSN' energy-efficiency coverage optimization model lie on two aspects: (i)the sensor node's sensing model. (ii)how to evaluate the performances of WSN' energy-efficiency.

The sensor nodes' sensing model describes its monitoring ability and sensing range. In this paper, the probability sensing model is adopted. Considering the node's electric characterize and white noises, the probability for monitoring the target $o_k$ by the sensor node $s_i = \{x_i, y_i\}$, expressed by $P_{cov}(o_k, s_i)$, is exponentially decreased as the distance between them is increasing[14]. Suppose there are $N$ sensor nodes. As they all contribute to monitor the target, the ability to monitor $o_k$ decides by the joint probability of all nodes, defined as $P_{cov}(o_k) = 1 - \prod_{i \in N}(1 - P_{cov}(o_k, s_i))$

The QoS of WSN is commonly measured by the network coverage rate and the node redundancy rate. The former reflects the coverage degree of monitored area by all sensor nodes. The latter measures the uniformity degree on sensor nodes' distribution. Without loss of generality, the monitored area is evenly partitioned into discrete grid along X-Y coordination[14]. Both objectives are obtained based on the joint monitored probability of all sensor nodes:

$$f_1 = NCR = \frac{\sum_{k=1}^{a \times b} P_{cov}(o_k)}{a \times b} \tag{1}$$

$$f_2 = NRR = \frac{N\pi r^2 - \sum_{k=1}^{a \times b} P_{cov}(o_k)}{N\pi r^2} \tag{2}$$

The essential of above optimization problem is to obtain an optimal sensor nodes' arrangement, which has maximum NCR and minimum NRR. Two objectives are incompatible each other. In order to simply the computation, we covert all objectives to maximum problem.

# 3    The WSN'S Energy-Efficiency Coverage Optimization Method Based on Morqca

## 3.1    MQEA in Population Space

There are two kinds of individuals in MQEA: evolutionary individuals and quantum individuals. Each evolutionary individual denotes the sensor nodes' locations as $p^i(t)=\{s_1^i,s_2^i,\cdots,s_N^i\}=\{(x_1^i,y_1^i),(x_2^i,y_2^i)\cdots,(x_N^i,y_N^i)\}, i=1,2,\cdots n$. $n$ is population size. Each gene of $p^i(t)$ is described by a rectangle in a quantum individual $q^i(t)$, which is uniquely determined by its center and width[15] expressed by $(x_{cj}^i,x_{wj}^i,y_{cj}^i,y_{wj}^i)$. Because non-dominated rank can evaluate individuals instead of the fitness values, we present a novel method to calculate the rectangle's height of $q^i(t)$ in MOP based on non-dominated sorting method. Let $zc^i(t)$ be the number of dominating individuals.

$$x_{hj}^i(t) = \frac{zc^i(t)+1}{\sum_{k=1}^{n}(zc^k(t)+1)}$$

(3)

In MOP, we hope all optimal solutions are uniformly distributed along the Pareto front. The individuals with less non-dominated rank and less crowding degree are better and kept in next population. In this paper, the crowding degree of $p^i$ is measured by the density of certain topological cell defined in Section 3.2.

## 3.2    The Extraction and Utilization of Knowledge in Belief Space

In this paper, normative knowledge and topological knowledge are used. Normative knowledge expressed by $K_1=\langle(l_1(t),u_1(t)),(l_2(t),u_2(t))\rangle$ memorizes the extreme limits of all non-domination individuals' fitness values. $u_j$ and $l_j$ respectively the maximum and minimum values of $j$th objective for all non-dominated individuals.

Topographic knowledge $K_3=\langle C_1(t),C_2(t),\cdots;C_k(t),\cdots\rangle$ records the distribution of non-dominated individuals' fitness vector in objective space. The objective space is uniformly divided into subspace along each dimension by grid method[16]. We call the subspace cell, denoted by $C_k(t)=\langle L_k,U_k,d_k\rangle$. $L_k$ and $U_k$ are $k$th cell's bound. $d_k(t)$ is $k$th cell's density, which decided by the proportion of non-dominated individuals in $k$th cell to Pareto-optimal set. The crowding degree of $i$th individual is defined as $ar^i(t)-d_k(t),\forall x^i(t)\in C_k(t)$. $ar^i(t)<ar^l(t)$ indicates that $p^i(t)$ locates in a relatively incompact area and has larger chance to reserve in next generation.

Normative knowledge and topographic knowledge are used to influence the selection operation of evolutionary individuals and the update operation of quantum individuals so as to avoid the blind selection and obtain the uniform Pareto-optimal set close to the Pareto front. In the selection operation, the domination relationship between two individuals is expressed by $\Gamma^i(t)$ and the individuals satisfying $P_s(t)=\{p^i(t)\mid \forall i,\max_i \Gamma^i(t)\}$ are selected to the next population.

$$\Gamma^i(t) = \begin{cases} \Gamma^i(t)+1 & \left(p^i(t) \succ p'(t)\right) \cup \left(\overline{f}_k(p^i(t)) \notin [l_k, u_k]\right) \cup \left(ar^i(t) < ar'(t)\right) \\ \Gamma^i(t) & \left(p^i(t) \prec p'(t)\right) \cup \left(ar^i(t) \geq ar'(t)\right) \end{cases} \quad (5)$$

The update operation of quantum individuals' width and center are influenced by topographic knowledge so as to lead the population close to the better area. Suppose $\theta$ is the constriction factor of step size. $\delta^i(t)$ denotes the evolution degree and is defined in Section 4. If an evolutionary individual's performances become worse, the rectangle's width is enlarged so as to expand the feasible search space.

$$x_{cj}^i(t+1) = \overline{x_j^i}(t), l = \arg\min_k \left\| \overline{p^k}(t), p^i(t) \right\|, \overline{p^k}(t) \in S_p(t) \quad (6)$$

$$x_{wj}^i(t+1) = x_{wj}^i(t) \theta^{\delta^i(t)} \quad (7)$$

## 4    Analysis of the Simulation Results

Suppose the size of the monitored area is $20 \times 20$. Each sensor node is described by the probability sensing model with following parameters: $r_e = 1.5$, $\alpha_1 = \beta_1 = 1$, $\alpha_2 = 0$, $\beta_2 = 0.5$, $r = 3$. The main parameters in MORQCA and other compared algorithms are: $N = 25$, $P_m = 0.1$, $n = 40$, $n_s = 20$, $T = 1000$, $n_q = 40$, $c = 30$, $\tau = 5$. In order to quantitatively compare the algorithms' performances, Three metrics[17] including minimal spacing (SM), hyperarea (H) and purity(P) are adopted. Based on H-metric and SM-metric, the evolution degree is defined as follows. $\delta^i(t)$ is less if the convergence and uniformity of the solutions in last generation are both improved.

$$\delta^i(t) = \begin{cases} -1 & (SM(p^i(t)) < SM(p^i(t-1)) \wedge (H(p^i(t)) > H(p^i(t-1)) \\ 1 & (SM(p^i(t)) > SM(p^i(t-1)) \wedge (H(p^i(t)) < H(p^i(t-1)) \\ 0 & otherwise \end{cases} \quad (8)$$

### 4.1    Comparison of the Algorithm's Performances with Different $N$

We set the number of sensor nodes respectively are 15, 20, 25, 30 and 35. Under different number of sensor nodes, the statistical data are listed in Table.1 and shown in Fig.1 by box plot. $\mu$ and $\sigma$ are the mean and the standard deviation. Obviously, SM-metric is smallest and H-metric is largest when $N=25$, which means the pareto-optimal solutions are uniformly distributed and close to the true Pareto front. When the number of sensor nodes is more or less, the algorithm's performance becomes worse. So we choose $N=25$ in following experiments. From Fig.1(c), if the network consists of less nodes, all nodes are uniformly arranged and the node redundancy rate is less. However, the network coverage rate is not good enough. More sensor nodes can ensure the expected network coverage rate whereas they repeatedly cover the detected area. This increases the network's cost.

**Table 1.** Comparison of the performances with different number of sensor nodes

| N | | 15 | 20 | 25 | 30 | 35 |
|---|---|---|---|---|---|---|
| H | $\mu$ | 9.62E1 | 1.22E2 | 1.36E2 | 8.65E1 | 5.13E1 |
| | $\sigma$ | 6.89 | 7.13 | 6.64 | 8.18 | 6.73 |
| SM | $\mu$ | 4.02E-3 | 3.64E-3 | 3.12E-3 | 4.22E-3 | 6.98E-3 |
| | $\sigma$ | 7.37E-4 | 6.25E-4 | 7.12E-4 | 9. 05E-4 | 5.12E-4 |

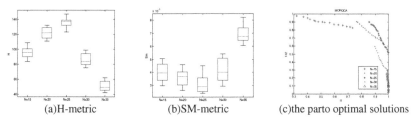

(a)H-metric          (b)SM-metric          (c)the parto optimal solutions

**Fig. 1.** Comparison of the metrics with different number of sensor nodes

## 4.2 Comparison of Different Algorithms

Under the same experimental conditions, simulation results derived from applying NSGAII, SPEAII, multi-objective cultural algorithm(MOCA), MOQA, and MORQCA are listed in Table.2. and presented in Fig.2 by box plot. The data in Table.2 show that SM-metric and H-metric of MORQCA are better than other algorithms. The reason for that is in MORQCA, the extracted knowledge is fully used to ensure the evolution toward the potential domination area and make the optimal non-dominated solutions quickly close to the better Pareto-optimal front. Besides, the knowledge-inducing selection operation operator ensures the better individuals reserved in the next generation and avoid the sightless selection. Therefore, the optimal non-dominated solutions can be uniformly distributed along the Pareto front, which makes the SM-metric less.

**Table 2.** Comparison of the metrics with different algorithms

| algorithm | | NSGAII | SPEAII | MOCA | MOQA | **MORQCA** |
|---|---|---|---|---|---|---|
| SM | $\mu$ | 3.14E-2 | 3.33E-2 | 2.25E-2 | 3.35E-2 | **6.33E-3** |
| | $\sigma$ | 3.50E-3 | 3.16E-3 | 5.81E-3 | 8.95E-3 | **8.45E-4** |
| H | $\mu$ | 7.85E1 | 8.80E1 | 1.18E2 | 1.05E2 | **1.40E2** |
| | $\sigma$ | 1.12E1 | 1.18E1 | 1.59E1 | 1.04E1 | **1.50E1** |
| P | $\mu$ | 2.73E-2 | 1.59E-2 | 5.26E-1 | 4.18E-1 | **8.61E-1** |
| | $\sigma$ | 7.75E-3 | 6.01E-3 | 6.39E-2 | 5.67E-2 | **4.67E-2** |

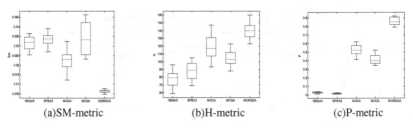

(a)SM-metric      (b)H-metric      (c)P-metric

**Fig. 2.** Comparison of the metrics with different algorithms

## 5    Conclusions

We choose maximum network cover rate and minimum node redundancy rate as the indices to measure WSN's energy-efficiency coverage and convert it into a two-objective optimization problem. Thereby a novel multi-objective quantum-inspired cultural algorithm is proposed to solve this problem. In population space, there are two highlights. One is the rectangle's height of each allele is calculated based on the number of dominated individuals. The other is the crowding degree replaces the crowding distance to measure the scatting statue of Pareto-optimal solutions. In belief space, the extracted implicit knowledge records the information about the Pareto-optimal solutions in objective space and utilized to direct the update operation for quantum individuals and the selection operator of evolutionary individuals. Experimental results show that the proposed algorithm kept the diversity of population better and obtained the better and more uniform Pareto-optimal solutions by comparing with other multi-objective optimization algorithms. That means the layout of wireless sensor network obtained by the proposed algorithm have larger network cover rate and less node redundancy rate.

**Acknowledgments.** This work was supported by National Natural Science Foundation of Jiangsu under Grant BK2010183, the Fundamental Research Funds for the Central Universities under Grant 2012LWB76, Jiangsu Overseas Research & Training Program for University Prominent Young & Middle-aged Teachers and Presidents.

## References

1. Jia, J., Chen, J., Chang, G.-R., et al.: Optimal coverage scheme based on genetic algorithm in wireless sensor networks. Control and Decision 22(11), 1289–1292 (2007)
2. Lee, J.-W., Choi, B.-S., Lee, J.-J.: Energy-efficient coverage of wireless sensor networks using ant colony optimization with three types of pheromones. IEEE Transactions on Industrial Informatics 7(3), 419–427 (2011)
3. Aziz, N.A., Mohemmed, A.W., Alias, Y.: A wireless sensor network cverageoptimization algorithm based on particle swarm optimization and voronoidiagram. In: IEEE International Conference on Networking, Sensing and Control, pp. 602–607 (2009)

4.  Hua, F., Shuang, H.: Optimal sensor node distribution based on the new quantum genetic algorithm. Chinese Journal of Sensors and Actuators 21(7), 1259–1263 (2008)
5.  Zitzler, E., Laumanns, M., Thiele, L.: SPEA2: improving the strength Pareto evolutionary algorithm,Technical Report 103, Computer Engineering andNetworks Laboratory, Swiss Federal Institute of Technology Zurich, Switzerland (2001)
6.  Horn, J., Nafpliotis, N., Goldberg, D.E.: A niched Pareto genetic algorithmfor multiobjective optimization. In: IEEE World Congress on Computational Intelligence, pp. 67–72 (1994)
7.  Deb, K.: A fast and elitist multiobjective geneticalgorithm: NSGA-II. IEEE Transaction on EvolutionaryComputation 6(2), 182–197 (2002)
8.  Meshoul, S., Mahdi, K., Batouche, M.: A quantum inspired evolutionary framework for multi-objective optimization. In: Bento, C., Cardoso, A., Dias, G. (eds.) EPIA 2005. LNCS (LNAI), vol. 3808, pp. 190–201. Springer, Heidelberg (2005)
9.  Kim, Y., Kim, J.-H., Han, K.-H.: Quantum-inspired multiobjectiveevolutionary algorithm formultiobjective 0/1 knapsack problems. In: 2006 IEEE Congress on Evolutionary Computation, pp. 9151–9156 (2006)
10. Wei, X., Fujimura, S.: Multi-objective quantum evolutionary algorithm for discrete multi-objective combinational problem. In: Proceeding of International Conference on Technologies and Applications of Artificial Intelligence, pp. 39–46 (2010)
11. Yang, X.-W., Shi, Y.: A real-coded quantum clone multi-objective evolutionary algorithm. In: Proceeding of International Conference on Consumer Electronic, Communications and Networks, pp. 4683–4687 (2011)
12. Reynolds, R.G.: An introduction to cultural algorithms. In: Proceeding of the Third Annual Conference on Evolutionary Programming, pp. 131–139 (1994)
13. Guo, Y.-N., Liu, D., Cheng, J., et al.: A novel real-coded quantum-inspired cultural algorithm. Journal of Central South University 42, 130–136 (2011)
14. Li, S.J., Xu, C.F., Pan, Y.H.: Sensor deployment optimization for detecting maneuvering targets. In: Proceedings of International Conference on Information Fusion, pp. 1629–1635 (2005)
15. Cruz, A.V.A., Vellasco, M.B.R., Pacheco, M.A.C.: Quantum-inspired evolutionary algorithm for numerical optimization. In: Proceeding of IEEE Congress on Evolutionary Computation, pp. 19–37 (2006)
16. Best, C., Che, X., Reynolds, R.G., et al.: Multi-objective cultural algorithms. In: Proceeding of IEEE Congress on Evolutionary Computation, pp. 1–9 (2010)
17. Bandyopadhyay, S., Pal, S.K., Aruna, B.: Multiobjective GAs,quantitative indices and pattern classification. IEEE Transactions on Systems, Man, and Cybernetics-Part B: Cybernetics 5(34), 2088–2099 (2004)

# Artificial Bee Colony Algorithm for Modular Neural Network

Chen Zhuo-Ming[1], Wang Yun-Xia[2], Ling Wei-Xin[2],
Xing Zhen[2], and Xiao Han-Lin-Wei[1]

[1] The Centre of Language Disorder, The First Affiliated Hospital, Jinan University,
Guangzhou, China
[2] School of Science, South China University of Technology, Guangzhou, China
1090029753@qq.com, lan_shan2007@163.com, lingweixin@21cn.com

**Abstract.** The Artificial bee colony (ABC) algorithm is simple, robust and has been used in the optimization of synaptic weights from an Artificial Neural Network (ANN). However, this is not enough to generate a robust ANN. Modular neural networks (MNNs) are especially efficient for certain classes of regression and classification problems, as compared to the conventional monolithic artificial neural networks. In this paper, we present a model of MNN based on ABC algorithm (ABC-MNN). Experiments show that, compared to the monolithic ABC-NN model, classifier designed in this model has higher training accuracy and generalization performance.

**Keywords:** Modular Neural Network, Artificial Bee Colony Algorithm, Learning Algorithm.

## 1 Introduction

ANNs are commonly used in pattern classification, function approximation, optimization, pattern matching, machine learning and associative memories. But the monolithic neural network has serious learning problems——it easily forgets initialization settings and stores the knowledge in a sparsely [1].The retrieval problem with monolithic networks can be solved by proper network design, but such scales very badly with increasing complexity.

A large amount of research in numerous problem domains is done in the past few years. In [2], the MNN which is optimized by Hierarchical Genetic Algorithm is applied for the speaker identification. In [3] MNN is used for the biometric recognition. Local experts and an integrated unit are two components which are used in the architecture. In [4], the topology and parameters of the MNN are optimized with a Hierarchical Genetic Algorithm, and it is used for human recognition. In [5], the MNNs are used for face recognition with large datasets, and its architecture is optimized by a parallel genetic algorithm. In [6] the authors propose a new approach to genetic optimization of MNNs with fuzzy response integration which is applied to human recognition. In [7], the MNN model which is optimized by PSO and trained by the OWO-HWO algorithm is applied to analog circuit fault diagnosis.

C. Guo, Z.-G. Hou, and Z. Zeng (Eds.): ISNN 2013, Part I, LNCS 7951, pp. 350–356, 2013.

In [8] the authors train an ANN by means of ABC algorithm. In [9] the authors apply this algorithm to train a feed-forward Neural Network. In [10], the authors present an ABC based synthesis methodology for ANNs, by evolving the weights, the architecture and the transfer functions of each neuron. It says that ABC algorithm is a good optimization technique for ANN. In this paper we want to verify if this algorithm performs in MNNs. As we will see, the MNNs obtained are optimal in the sense that the architecture is simple with high recognition.

The paper is organized as follows: in section 2 the basics of ABC and the ANN trained by ABC algorithm (ABC-NN) are presented. In section 3 MNNs based on ABC algorithm (ABC-MNN) are explained. In section 4 the experimental results using different classification problems are given. Finally, in section 5 the conclusions of the work are presented.

## 2    Neural Network Learning Algorithm Based on ABC(ABC-NN)

ANN is widely used in approximation and classification problems. One of the widely used ANNs is the feed-forward neural network, which is trained by means of the back-propagation (BP) algorithm [11]. This algorithm minimizes the Mean-Square Error (MSE) function given in (1).

$$e = \frac{1}{p*m}\sum_{i=1}^{p}\sum_{j=1}^{m}(d_{ij} - y_{ij})^2 \tag{1}$$

where $d \in |R^m$ and $y \in |R^m$ are respectively target and actual output of the neural network, p is the number of samples and m is the number of the output layer node. Some algorithms constantly adjust the values of the synaptic weights until the value of the error no longer decreases, but it is easy to converge to a local minimum instead of to the desired global minimum. So a powerful swarm intelligence optimization algorithm ABC is introduced to enhance the neural network training.

ABC algorithm is based on the metaphor of the bees foraging behavior which is a very simple, robust and population based stochastic optimization algorithm. ABC algorithm is proposed by Karaboga in 2005 [12] for solving numerical optimization problems. In ABC algorithm, the position of a food source represents a possible solution to the optimization problem and the nectar amount of a food source corresponds to the quality (fitness) of the associated solution. The colony of artificial bees contains three groups of bees: *Employed bees, Onlookers* and *Scout bees*. These bees have got different tasks in the colony, i. e., in the search space.

*Employed bees:* Each bee searches for new neighbor food source near their hive. After that, it compares the food source against the old one using (2). Then, it saves in their memory the best food source.

$$v_{ij} = x_{ij} + rand(0,1)(x_{ij} - x_{kj}) \tag{2}$$

where $k \in \{1,2, ... , SN\}$ and $j \in \{1,2, ... , D\}$ are randomly chosen indexes. Although $k$ is determined randomly, it has to be different from $i$. $SN$ is the number of the *Employed bees* and $D$ is the dimension of the solution.

After that, the bee evaluates the quality of each food source based on the amount of the nectar (the information) i.e. the fitness function is calculated. Providing that its nectar is higher than that of the previous one, the bee memorizes the new position and forgets the old one. Finally, it returns to the dancing area in the hive, where the *Onlooker bees* are.

*Onlooker bees:* This kind of bees watch the dancing of the employed bee so as to know where the food source can be found, if the nectar is of high quality, as well as the size of the food source. The *Onlooker bee* chooses a food source depending on the probability value associated with that food source, $p_i$ is calculated by the following expression:

$$p_i = {fit_i} \Big/ {\sum_{k=1}^{SN} fit_k} \tag{3}$$

where $fit_i$ is the fitness value of the solution $i$ which is proportional to the nectar amount of the food source in the position $i$ and $SN$ is the number of food sources which is equal to the number of Employed bees.

*Scout bees:* This kind of bees helps abandon the food source which can not be improved further through a predetermined number of cycles and produce a position randomly replacing it with the abandoned one. This operation can be defined as in (4).

$$x_{ij} = x_{min}^j + rand(0,1)(x_{max}^j - x_{min}^j) \tag{4}$$

The multi-layered NN structure is trained using ABC algorithm by minimizing the MSE function given in (1). The solution (food source) consists of the neural network's weights and bias. The fitness value of each solution is the value of the error function evaluated at this position. The pseudo-code of the ABC-NN algorithm is shown as follows:

```
Program ABC-NN(globalx)
   const   MCN=500,SN=20,goal=0.01;
   var     cycle:0..MCN;
   begin
     cycle :=0;
     Initialize the weights and bias xᵢ of the network to
       small random values by using (4), i=1,2,…,SN;
     Evaluate the MSE function's value e by (1) of the
       population xᵢ , i=1,2,…,SN;
     repeat
       cycle :=cycle+1;

       Produce new solutions Vᵢ for the employed bees by
         using (2) and evaluate them;
       Apply the greedy selection process;
       Calculate the probability values pᵢ for the
         solutions xᵢ by (3);
```

```
Produce the new solutions V_i for the onlookers from
    the solutions x_i selected depending on p_i and
    evaluate them;
Apply the greedy selection process;
Determine the abandoned solution for the scout, if
    exist, and replace it with a new randomly produced
    solution x_l by (4);
Memorize the best solution globalx achieved so far;
until  cycle = MCN or min(e) < goal
end
```

## 3    ABC-MNN Model

There exists a lot of neural network architectures in the literature that work well when the number of inputs is relatively small, but when the complexity of the problem grows or the number of inputs increases, their performance decreases very quickly. The MNN is used in such cases which work as a combination of neural networks. The idea shares conceptual links with "divide and conquer" methodology. The MNN has a hierarchical organization comprising multiple neural networks which is responsible for solving some part of the problem. The combination of estimators may be able to exceed the limitation of a single estimator.

The ABC algorithm has a strong ability to find global optimistic result. MNN is especially efficient for certain classes of regression and classification problems, as compares to the conventional monolithic artificial neural networks [13]. Combining the ABC with the MNN, a new hybrid algorithm (ABC-MNN) is proposed in this paper. The algorithm is made up of Data Division Module (DDM), *ABC-NN_i* Module and Integration Module (IM). The structure of the ABC-MNN is shown in figure 1.

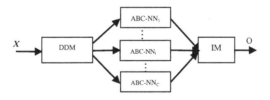

**Fig. 1.** Architecture of the ABC-MNN

Assume that the input sample is $(X, T)$, where $X = (x_1, x_2, ..., x_m)$ is input vectors, m is the number of the samples, $x \in |R^n$, $T$ is the target output and $O$ is the actual output . The dataset has $C$ classes.

Data Division Module (DDM) works to divide the dataset. For each module of ABC-NN, the dataset $X$ is divided into two categories, so it produces $C$ datasets according to the number of the categories of the dataset. Each dataset has only two categories, i.e., $i$th class and non- $i$th type. The C dataset is noted as $S_1, S_2, ... S_C$.

$S_i = \{(X, T^i) | X \text{ is the input vectors and } T^i \text{ is the new target vectors}\}$

Each dataset $S_i$ corresponds to a *ABC-NN$_i$* Module. The ABC-NN is an independent ANN which is a three-layer forward neural network like the BP neural network, and its structure is 'n-p-1', where p is the number of the hidden layer node. The input of the model *ABC-NN$_i$* is the dataset $S_i$. At this stage, ABC is used to evolve the synaptic weights of sub-neural network *ABC-NN$_i$* so as to obtain a minimum Mean Square Error (MSE) as well as a minimum classification error (CER) for the $i$th class of data. Supposing in the well trained network *ABC-NN$_i$*, using $y_{ij}$ to denote the output of the $j$th sample.

Integration Module (IM) is used to integrate the outputs of all ABC-NN modules. Using $o_j$ to denote the category of the $j$th sample and $o_j = i$ means that the $j$th sample belongs to $i$th class. Here the winner-take-all rule is used. $o_j = k$, which $y_{kj} = max(y_{1j}, y_{2j}, \dots, y_{cj})$. So the $j$th sample is identified as the $k$th class.

# 4    Experiments and Comparison

Several experiments are performed in order to evaluate the accuracy of the ABC-MNN designed by means of the proposal. The accuracy of the ABC-MNN is tested with four pattern classification problems which are taken from UCI machine learning benchmark repository [14]: Iris, Glass, Segment and Optdigits. Their characteristics are given in Table 1.

**Table 1.** Datasets characteristics

| Datasets | Observations | Features | Classes | Respective observations |
|---|---|---|---|---|
| Iris | 150 | 4 | 3 | 50,50,50 |
| Glass | 214 | 9 | 6 | 70,76,17,13,9,29 |
| Segment | 2310 | 19 | 7 | 330,330,330,330,330,330,330 |
| Optdigits | 5620 | 64 | 10 | 541,573,556,552,595,584,549,572,536,562 |

The parameters of the ABC algorithm and the network are set to the same value for all the dataset problems: Colony size (NP = 40), number of food sources NP/2, limit = 50, the maximum number of cycles is MCN = 500, the minimum of the MSE function (1) is goal=0.01 and the transfer function is sigmoid function.

20 experiments using each dataset are performed. Ten for the case of ABC-NN model and ten for the ABC-MNN model. For each experiment, each dataset is randomly divided into two sets: a training set and a testing set, this with the aim to prove the robustness and the performance of the methodology. The same parameters are used through the whole experimentation.

Once generated the ANN for each problem, we proceed to test their accuracy. Table 2 shows the best, average and worst percentage of classification for all the experiments using ABC-NN and ABC-MNN. In this Table, we can observe that the best percentage of recognition for most databases is achieved only during training phase. The accuracy slightly diminish during testing phase, but the Glass problem is more serious. However, the results obtained with the proposed methodology ABC-MNN are highly acceptable and stable. The training and testing accuracy of the ABC-MNN methodology are highly enhanced and more stable. For the worst values achieved with the ANN are also represented. Particularly, the dataset that provides the worst results is the Glass problem which is very complicate and imbalance. Nonetheless, the accuracy achieved is highly acceptable.

From these experiments, we observe that the ABC algorithm is able to find the best configuration for an ANN given a specific set of patterns that define a classification problem. The experimentation shows that the design generated by the proposal presents an acceptable percentage of classification for training and testing phase with the MNNs.

**Table 2.** Comparison of training and testing accuracy

| Dataset | | ABC-NN | | ABC-MNN | |
|---------|---------|------------|-----------|-------------|------------|
| | | Training(%) | Testing(%) | Training(%) | Testing(%) |
| Iris | best | 99.12 | 97.22 | 99.12 | 97.22 |
| | average | 97.45 | 96.66 | 98.15 | 97.22 |
| | worst | 96.49 | 94.44 | 97.36 | 97.22 |
| Glass | best | 76.54 | 75.00 | 83.95 | 76.92 |
| | average | 75.55 | 71.15 | 82.77 | 74.42 |
| | worst | 74.69 | 63.46 | 80.86 | 73.07 |
| Segment | best | 90.55 | 89.72 | 95.27 | 94.77 |
| | average | 88.69 | 87.10 | 94.98 | 93.36 |
| | worst | 86.29 | 83.97 | 94.52 | 91.81 |
| Optdigits | best | 81.46 | 78.46 | 92.93 | 89.59 |
| | average | 76.67 | 71.94 | 92.21 | 88.19 |
| | worst | 68.21 | 62.66 | 91.49 | 87.25 |

## 5    Conclusions

In this paper, a new hybrid method for combining the MNN with ABC algorithm is proposed. From the foregoing experimental researches, it is concluded that the MNNs which are evolved by ABC algorithm are characterized by satisfying approaching results and high training speed. In this work we also test the performance of the ABC algorithm. Although the ABC-MNN has exceeded the traditional algorithm in convergence speed and classification precision, it still needs to span broad activities and require consideration of multiple aspects. The classifiers of imbalanced datasets are also needed further study.

**Acknowledgement.** This work is supported by a grant from the Key Projects in the National Science & Technology Pillar Program during the Twelfth Five-year Plan Period(No. 2011BAI08B11), by grant from the Projects in the Science and Technology Program of Guangzhou(No. 2012Y2-00023).

# References

1. Auda, G., Kamel, M.: Modualr Neural Network: A Survey. Int. J. Neural Syst. 9(2), 129–151 (1999)
2. Martinez, G., Melin, P., Castillo, O.: Optimization of Modular Neural Networks using Hierarchical Genetic Algorithm Applied to Speech Recognition. In: Proceedings of International Joint Conference on Neural Networks, Canada (2005)
3. Hidalgo, D., Melin, P., Licea, G.: Optimization of modular neural networks with interval type-2 fuzzy logic integration using an evolutionary method with application to multimodal biometry. In: Melin, P., Kacprzyk, J., Pedrycz, W. (eds.) Bio-inspired Hybrid Intelligent Systems for Image Analysis and Pattern Recognition. SCI, vol. 256, pp. 111–121. Springer, Heidelberg (2009)
4. Sánchez, D., Melin, P., Castillo, O.: A New Model of Modular Neural Networks with Fuzzy Granularity for Pattern Recognition and Its Optimization with Hierarchical Genetic Algorithms. In: Batyrshin, I., Sidorov, G. (eds.) MICAI 2011, Part II. LNCS (LNAI), vol. 7095, pp. 331–342. Springer, Heidelberg (2011)
5. Fevrier, V., Patricia, M., Herman, P.: Parallel genetic algorithms for optimization of Modular Neural Networks in pattern recognition. In: IJCNN 2011, pp. 314–319 (2011)
6. Melin, P., Sánchez, D., Castillo, O.: Genetic optimization of modular neural networks with fuzzy response integration for human recognition. Information Sciences 197, 1–19 (2012)
7. Sheikhan, M., Sha'bani, A.A.: PSO-optimized modular neural network trained by OWO-HWO algorithm for fault location in analog circuits. Neural Compute Appl. (available online April 25, 2012), doi:10.1007/s00521-012-0947-9
8. Karaboga, D., Akay, B.: Artificial Bee Colony (ABC) Algorithm on Training Artificial Neural Networks. In: Proceedings of the 15th IEEE Signal Processing and Communications Applications (SIU 2007), pp. 1–4 (2007)
9. Karaboga, D., Akay, B., Ozturk, C.: Artificial bee colony (abc) optimization algorithm for training feed-forward neural networks. In: Torra, V., Narukawa, Y., Yoshida, Y. (eds.) MDAI 2007. LNCS (LNAI), vol. 4617, pp. 318–329. Springer, Heidelberg (2007)
10. Garro, B.A., Sossa, H., Vazquez, R.A.: Artificial neural network synthesis by means of artificial bee colony (abc) algorithm. In: 2011 IEEE Congress on Evolutionary Computation (CEC), pp. 331–338 (2011)
11. Akay, B., Karaboga, D.: A Modified artificial bee colony algorithm for real-parameter optimization. Information Sciences 192(1), 120–142 (2012)
12. Karaboga, D.: An idea based on honey bee swarm for numerical optimization. Computer Engineering Department, Engineering Faculty, Erciyes University. Tech. Rep. (2005)
13. Ling, W.X., Zheng, Q.L., Chen, Q.: GPCMNN: A Parallel Cooperative Modular Neural Network Architecture Based on Gradient. Chinese Journal of Computers 27(9), 1256–1263 (2004)
14. Murphy, P.M., Aha, D.W.: UCI Repository of machine learning databases, University of California, Department of Information and Computer Science, Irvine, CA, US. Tech. Rep. (1994)

# An Intelligent Optimization Algorithm for Power Control in Wireless Communication Systems

Jing Gao[*], Jinkuan Wang, BinWang, and Xin Song

School of Information Science & Engineering,NortheasternUniversity,
110004 Shenyang, China
{summergj,wjk,bwang,xsong}@126.com

**Abstract.** High instantaneous peak power of the transmitted signals is the main obstacle of orthogonal frequency division multiplexing (OFDM) systems for its application, therefore, the peak to average power ratio (PAPR) reduction has been one of the most important technologies. Among all the existing methods, partial transmit sequences (PTS) is a distortionless phase optimization technique that significantly improves PAPR performance to with a small amount of redundancy. However, the computational complexity in conventional PTS increases exponentially with the number of subblocks. In this paper, an intelligent optimization method is proposed for PTS technique to obtain good balance between computational complexity and PAPR performance. Simulation results show that the proposed method can achieve better performance compared with conventional algorithms.

**Keywords:** OFDM, Power Control, PAPR, Optimization.

## 1    Introduction

Orthogonal-frequency-division-multiplexing (OFDM) is one of the most popular modulation techniques because of high bandwidth efficiency and robustness to multi-path environments [1].

Despite the advantages, high peak-to-average power ratio (PAPR) value of the signals is a major drawback of the OFDM systems. The high PAPR value causes in-band distortion and out-of-band radiation due to unwanted saturation in the high power amplifier (HPA).Various techniques have been proposed to handle this problem in recent years, including amplitude clipping, coding, nonlinear companding transform schemes, active constellation extension, selective mapping and partial transmit sequences (PTS) [2-9].

Among these methods, PTS is a distortionless phase optimization technique that provides excellent PAPR reduction with a small amount of redundancy. Nevertheless, finding the optimum candidate requires the exhaustive search over all combinations of allowed phase factors, and the search complexity increases exponentially with the number of sub-blocks. Hence, a suboptimal PTS method is proposed in this paper, which is combined with a new evolutionary optimization algorithm known as

---

[*] Corresponding author.

C. Guo, Z.-G. Hou, and Z. Zeng (Eds.): ISNN 2013, Part I, LNCS 7951, pp. 357–366, 2013.

adaptive acterial foraging algorithm (A-BFA). The A-BFA has been successfully used to solve various kinds of optimization problems and can offer good performance in terms of solution quality and convergence speed. Therefore, the proposed method can achieve better tradeoff between PAPR performance and computational complexity.

## 2     OFDM Systems and PTS Technique

With OFDM modulation, a block of $N$ data symbols (one OFDM symbol), $\{X_n, n = 0,1,\cdots, N-1\}$ will be transmitted in parallel such that each modulates a different subcarrier from a set $\{f_n, n = 0,1,\cdots, N-1\}$. The $N$ subcarriers are orthogonal, i.e. $f_n = n\Delta f$, where $\Delta f = 1/NT$ and $T$ is the symbol period. The complex envelope of the transmitted OFDM signal can be represented as

$$x_n(t) = \frac{1}{\sqrt{N}}\sum_{n=0}^{N-1} X_n e^{j2\pi f_n t} \qquad 0 \leq t \leq NT \tag{1}$$

then the PAPR of the transmitted OFDM signal can be defined as

$$PAPR = \frac{\max_{0 \leq t \leq NT} |x_n(t)|^2}{E[|x_n(t)^2|]} = \frac{\max_{0 \leq t \leq NT} |x_n(t)|^2}{\frac{1}{NT}\int_0^{NT} |x_n(t)|^2 \, dt} \tag{2}$$

where E[•] denotes the expected value.

The complementary cumulative distribution function (CCDF) is used to measure the PAPR performance, which represents the probability of the PAPR exceeds the given threshold $PAPR_0$

$$CCDF = Pr(PAPR > PAPR_0) \tag{3}$$

### 2.1     PTS Algorithm

In a typical OFDM system with PTS technique to reduce the PAPR, the input data block $X$ is partitioned into $M$ disjoint subblocks, which are represented by the vectors $X^{(m)} = \{X_0^{(m)} X_1^{(m)} \cdots X_{N-1}^{(m)}\}$, therefore

$$X = \sum_{m=0}^{M-1} X^{(m)} \tag{4}$$

Then, the subblocks $X^{(m)}$ are transformed into M time-domain partial transmit sequences by IFFTs. These partial sequences are independently rotated by phase factors $b_m = e^{j\theta_m}$, $\theta_m \in \{\frac{2\pi k}{W}|_{k=0,1,\cdots W-1}\}$. The object is to optimally combine the $M$ subblocks to obtain the OFDM signals with the lowest PAPR

$$x = \sum_{m=0}^{M-1} b_m x^{(m)} \tag{5}$$

Assuming that there are $W$ phase angles to be allowed, thus there are $D = W^M$ alternative representations for an OFDM symbol. The block diagram of the PTS technique is shown in Fig.1.

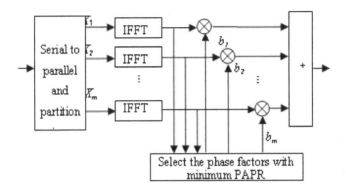

**Fig. 1.** Block diagram of the PTS technique

## 2.2    Problem Formulation

Based on the consideration above, the problem of PTS, which is trying to find the aggregate of phase factors vector $b_m$ to yield the OFDM signals with the minimum PAPR, can be considered as the combinatorial optimization problem. In other words, the objective function (6) is to minimize the PAPR of the transmitted OFDM signals. Constraint ensures the phase factors to be a finite set of values $0 \le \theta_m < 2\pi \, (0 \le m \le M - 1)$

To minimize

$$f(b) = \|x\|_\infty^2 = \left\| \sum_{m=0}^{M-1} b_m x^{(m)} \right\|_\infty^2 \tag{6}$$

Subject to    $b = \{e^{j\theta_m}\}^{M-1}$

where $\theta_m \in \{\frac{2\pi k}{W}|_{k=0,1,\cdots W-1}\}$. Since PTS with binary weighting factors $b_m \in \{\pm 1\}$, i.e. $W=2$, attains a favorable performance-redundancy tradeoff, we concentrate on this choice in the following.

Above all, the process of searching the optimal phase factors in PTS algorithm can be formulated as an optimization problem with some constrains. Therefore, an optimal combinational scheme derived from bacterial foraging, is proposed to achieve better PAPR reduction with low search numbers.

# 3    PAPR Reduction with Adaptive Bacterial Foraging Algorithm

Natural selection has a tendency to eliminate animals having poor foraging strategies and favor the ones with successful foraging strategies to propagate their genes as these are more likely to reach a successful reproduction. Poor foraging strategies are either completely eliminated or transferred into good ones after many generations are produced. This evolutionary process of foraging inspired the researchers to utilize it as an optimization tool. The E-Coli bacteria present in our intestines also practice a foraging strategy. The control system of these bacteria governing their foraging process can be subdivided into fouractions, which are chemotaxis, swarming, reproduction and elimination-dispersal.

Chemotaxis

This process is achieved by swimming and tumbling via flagella. Depending upon the rotation of flagella in each bacterium, it decides whether it should move in a predefined direction (swimming) or altogether in different directions (tumbling) in the entire lifetime.

Swarming

During the process of reaching toward the best food location, it is always desired that the bacterium which has searched the optimum path should try to produce an attraction signal to other bacteria, so that they swarm together to reach the desired location. In this process, the bacteria congregate into groups and hence move as concentric patterns of groups with high bacterial density.

Reproduction

The least healthy bacteria die and the other healthiest bacteria each split into two bacteria, which are placed in the same location. This makes the population of bacteria constant.

Elimination and dispersal

In the local environment of the bacteria, the lives of a bacteria population may change either gradually (e.g., via consumption of nutrients) or suddenly due to some other influence. All the bacteria in a local region may be killed or a group may be dispersed into a new location in the environment. They have the effect of possibly destroying the chemotaxis progress, but they also have the effect of assisting in chemotaxis, since dispersal may place bacteria near good food sources.

## 3.1    Bacterial Foraging Algorithm (BFA)

A flowchart of the BFA used in this paper is shown in Fig. 2. $J(i, j, k)$ represents the cost function where $i$ is the index of the bacterium, $j$ is the index for the chemotactic step, and $k$ is the index for the reproduction step.

The BFA can be explained as follows.There are a number of external parameters that control the behavior of the algorithm. These parameters include the number of bacteria $S$, maximum number of chemotactic loop $N_c$, maximum number of reproduction $N_{re}$, number of search space dimensions $p$, divisor of the step size $d_s$, swim length $N_s$, counter for swim length $m$, and chemotactic step sizes $C(i), i=1,2,\cdots,S$.

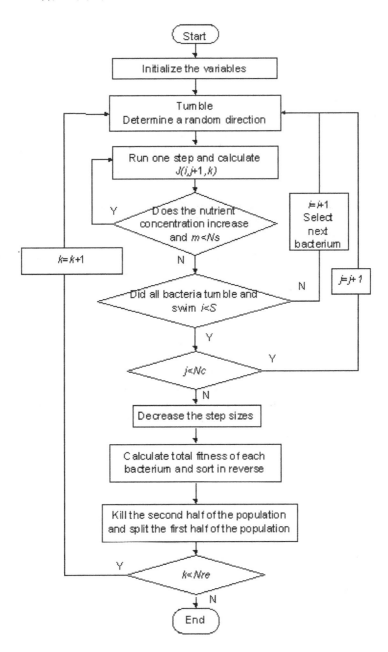

**Fig. 2.** Flowchart of BFA algorithm

The bacterium is pointed in a random direction after a tumble. To represent a tumble, a unit length random direction $\delta(j)$ is generated; this is used to define the direction of movement after a tumble.

$$\zeta^i(j+1,k)=\zeta^i(j,k)+C(i)\delta(j) \tag{7}$$

where $\zeta^i(j,k)$ represents the $ith$ bacterium at $jth$ chemotactic and $kth$ reproductive step.

Each bacterium $i=1,2,\cdots,S$, performs its swimming and tumbling tasks once in each iteration. An iteration cycle is a chemotactic process in BFA. By tumbling and swimming, bacteria try to reach the denser nutrition places which mean the lower cost function values. The bacteria which can achieve more plentiful nutrition are deemed as the healthier population members. When the chemotactic loop counter $j$ reaches the maximum iteration number $N_C$, the reproduction step is performed, and then optimization starts again. Owing to reproduction step, the least healthier bacteria die, and the other healthier bacteria split into two at the same location. The new chemotactic process continues with a new healthier bacteria population. These nested loops are performed until the reproduction loop counter $k$ reached the maximum reproduction number $N_{re}$. In this paper, the coordinates of bacteria represent the position values of phase factors. At the end of optimization process, the coordinates of bacterium with the lowest cost value (the healthiest member) are the optimal phase factors arrived by A-BFA.

## 3.2    Adaptive Bacterial Foraging Algorithm (A-BFA)

The length unit step of the basic BFA is a constant parameter which may guarantee good searching results for small optimization problems. However, when applied to complex large-scale problems with high dimensionality it shows poor performance. The run-length parameter is the key factor for controlling the search ability of the BFA. From this perspective, balancing the exploration and exploitation of the search could be achieved by adjusting the run-length unit. In this paper, we use a non-linear decreasing dynamic function to perform the swim walk instead of the constant step. This function is expressed as

$$C(i,j+1)=\left(\frac{C(i,j)-C(Nc)}{Nc+C(Nc)}\right)(Nc-j) \tag{8}$$

where $j$ is the chemotactic step and $Nc$ is the maximum number of chemotactic steps while $C(Nc)$ is a predefined parameter.

The stopping criterion of the original BFA is the maximum number of the chemotactic steps, the reproduction steps and the elimination/dispersal events. This criterion increases the computation requirements of the algorithm in some cases. In this paper, an adaptive stopping criterion is applied so that the algorithm adjusts the maximum number of iterations depending on the improvement of the cost function. The chemotaxis operation stops either when there is no improvement in the solution or when the maximum number of chemotactic steps is reached.

### 3.3     A-BFA to Reduce PAPR of OFDM Signals

As above mentioned, the steps, in which the A-BFA is presented for searching the optimal combination of phase factors in PTS algorithm. Initially, the phase factors and A-BF algorithm parameters are specified. Using randomly generated initial parameters, the efficiency of the PAPR reduction is determined by means of the A-BF algorithm. The implementation of A-BF algorithm for the PAPR reduction can be described as follows:

Step 1. Input BF parameters, $p, S, N_c, N_s, N_{re}, C(i)$, $i=1,2, \cdots, S$ must be chosen.

Step 2. Generate the positions of phase factors randomly for a population of bacteria, and specify the phase factors $b = \{e^{j\theta_m}\}^{M-1}$, $\theta_m \in \{\frac{2\pi k}{W}|_{k=0,1,\cdots W-1}\}$

Step 3. Evaluate the objective value of each bacterium in the population using Eq(6).

Step 4. Modify the positions of the phase factors for all the bacteria using tumbling/swimming process.

Step 5. Perform reproduction and elimination-dispersal operation, and modify the step adaptively with formula (8).

Step 6. If the maximum number of chemotatic, reproduction and elimination-dispersal steps are reached, then go to Step 7. Otherwise, go to Step 4.

Step 7. Output the equivalent phase factors corresponding to the overall best bacterium.

## 4     Simulation Results

In this section, we present some simulations to demonstrate the performance of the A-BFA algorithm. Assuming that $10^4$ random QAM modulated OFDM symbols were generated with $N = 256$ subcarriers, $M=8$ sub-blocks and the phase factors $b_m \in \{\pm 1\}$ $(W=2)$. In BFA-PTS algorithm, for the consideration of tradeoff between PAPR performance and computational complexity, the parameters are specified carefully. The number of bacteria is $S = 10$, maximum number of chemotactic loop is $N_C=4$, maximum number of reproduction is $N_{re} = 5$, number of search space dimensions is $p = 8$, swim length is $N_s = 4$.

Especially, the maximum number of reproduction $N_{re}$ is the most important paremeter to impose the performance of proposed algorithm, thus an appropriate reproduction number $N_{re} = 5$ is chosen to yield the best balance between PAPR reduction performance and complexity. Moreover, the performance of the proposed scheme in PAPR reduction is evaluated by the CCDF as in Fig.3 and Fig.4.

In Fig.3, the performance of PTS algorithm is analyzed, to compare PAPR reduction performance with A-BFA algorithm. When $\Pr(\mathrm{PAPR}>\mathrm{PAPR}_0)=10^{-3}$ , the $\mathrm{PAPR}_0$ of the original OFDM is 11dB, PTS and A-BFA improve it by 7.05dB and 7.6dB respectively. Obviously, for $M=8$, $W=2$, the conventional PTS technique requires $W^M =2^8 = 256$ iterations per OFDM frame, while the A-BFA technique requires $N_c \times N_s \times N_{re} = 80$ iterations per OFDM frame. It is evident that the A-BFA can achieve much lower computational complexity with relatively small PAPR performance degradation.

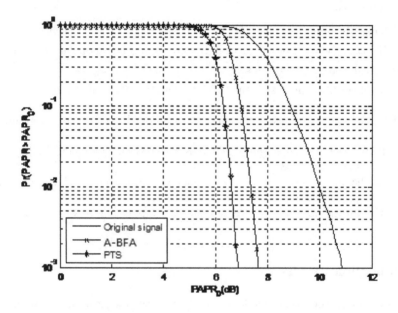

**Fig. 3.** Comparison of PAPR performance (PTS, A-BFA)

Fig.4 shows some comparisons of the PAPR reduction performance between A-BFA and IPTS. When $\Pr(\mathrm{PAPR}>\mathrm{PAPR}_0)=10^{-3}$, the $\mathrm{PAPR}_0$ of the original OFDM is 11.2dB, A-BFA is 7.55dB and IPTS is 9.2dB. It is clear that, for $M=8$, $W=2$, the IPTS technique requires $MW = 16$ iterations and the BFA-PTS technique requires $N_c \times N_s \times N_{re} = 80$ iterations per OFDM frame. Compared with the IPTS algorithm, the A-BFA algorithm can improve 1.65 dB. From the curves in Fig.4, it is shown that the A-BFA can offer better tradeoff between PAPR performance and complexity compared with IPTS.

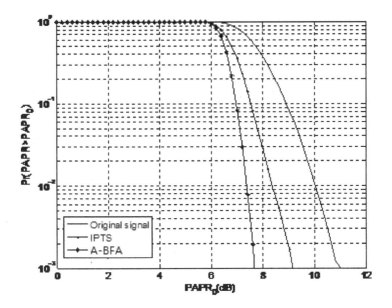

**Fig. 4.** Comparison of PAPR performance (IPTS, A-BFA)

## 5    Conclusions

The A-BFA is well known as powerful tool to reduce computational complexity and present satisfactory performance for optimization problems. Then, the A-BFA algorithm is proposed to solve the optimal phase factor search in PTS technique efficiently. Simulation results show that A-BFA scheme could achieve perfect balance between PAPR reduction performance and computational complexity compared with the conventional PTS techniques.

**Acknowledgments.** This work has been supported by the Central University Fundamental Research Funding, under Grant no.N100323006, the National Natural Science Foundation of China under Grant No. 60874108 and 61004052, and the open research fund (No.20100106) of Key Laboratory of Complex System and Intelligence Science, Institute of Automation, Chinese Academy of Sciences.

## References

1. Jiang, T., Wu, Y.: An Overview: Peak-to-Average Power Ratio Reduction Techniques for OFDM signals. IEEE Transactions on Broadcasting 54(2), 257–268 (2008)
2. Ou, J., Zeng, X., Tian, F., Wu, H.: Simplified Repeated Clipping and Filtering with Spectrum Mask for PAPR Reduction. Journal of Convergence Information Technology, AICIT 6(5), 251–259 (2011)

3. Wang, C.-L., Kum, S.-J.: Novel Conversion Matrices for Simplifying the IFFT Computation of an SLM-Based Reduction Scheme for OFDM Systems. IEEE Transactions on Communications 57(7), 1903–1907 (2009)
4. Deng, S.-K., Lin, M.-C.: Recursive Clipping and Filtering With Bounded Distortion for PAPR Reduction. IEEE Transactions on Communications 55(1), 227–230 (2007)
5. Lim, D.-W., Noh, H.-S., Jeon, H.-B., No, J.-S., Shin, D.-J.: Multi-Stage TR Scheme for PAPR Reduction in OFDM Signals. IEEE Transactions on Broadcasting 55(2), 300–304 (2009)
6. Shen, W., Sun, H., Cheng, E., Zhang, Y.: Performance Analysis of DFTSpread based OFDM Transmission System over Underwater Acoustic Channels. Journal of Convergence Information Technology, AICIT 6(7), 79–86 (2011)
7. Eonpyo, H., Youngin, P., Sangchae, L., Dongsoo, H.: Adaptive phase rotation of OFDM signals for PAPR reduction. IEEE Transactions on Consumer Electronics 57(4), 1491–1495 (2011)
8. Maryam, S., Yongjun, K., Vahid, T.: New codes from dual BCH codes with applications in Low PAPR OFDM. IEEE Transactions on Wireless Communications 10(12), 3990–3994 (2011)
9. Soobum, C., Sangkyu, P.: A new selected mapping scheme without additional IFFT operations in OFDM systems. IEEE Transactions on Consumer Electronics 57(4), 1513–1518 (2011)

# Optimized Neural Network Ensemble by Combination of Particle Swarm Optimization and Differential Evolution

Zeng-Shun Zhao[1,2,*], Xiang Feng[1], Fang Wei[1], Shi-Ku Wang[1],
Mao-Yong Cao[1], and Zeng-Guang Hou[3]

[1] College of Information and Electrical Engineering,
Shandong University of Science and Technology, Qingdao, 266590, China
[2] School of Control Science and Engineering, Shandong University, Jinan, 250061, China
[3] State Key Laboratory of Management and Control for Complex Systems,
Institute of Automation, Chinese Academy of Sciences, Beijing, 100190, P.R. China
zhaozengshun@gmail.com

**Abstract.** The Neural-Network Ensemble (NNE) is a very effective method where the outputs of separately trained neural networks are combined to perform the prediction. In this paper, we introduce the improved Neural Network Ensemble (INNE) in which each component forward neural network (FNN) is optimized by particle swarm optimization (PSO) and back-propagation (BP) algorithm. At the same time, the ensemble weights are trained by Particle Swarm Optimization and Differential Evolution cooperative algorithm(PSO-DE). We take two obviously different populations to construct our algorithm, in which one population is trained by PSO and the other is trained by DE. In addition, we incorporate the fitness value from last iteration into the velocity updating to enhance the global searching ability. Our experiments demonstrate that the improved NNE is superior to existing popular NNE.

**Keywords:** Neural Network Ensemble, Back-Propagation, Particle Swarm Optimization, Differential Evolution.

## 1 Introduction

Neural Network Ensemble (NNE) is a learning mechanism which has a collection of a finite number of neural networks trained for the same task. In Hansen and Salamon's work [1], it has been first proposed. Its main idea is that the predicting ability of a neural network system could be significantly improved by assembling a set of neural networks, for example, training many neural networks and then combining their predictions in some way [2]. But only by averaging, the combined prediction would not be effective, because in some cases, maybe some components of ensemble behave unsatisfactory. In [3], authors thought that it might be better to ensemble some components other than all of the trained neural networks; they introduced Genetic

---

* Corresponding author.

C. Guo, Z.-G. Hou, and Z. Zeng (Eds.): ISNN 2013, Part I, LNCS 7951, pp. 367–374, 2013.

algorithm based selective ensembles (GASEN), which employed genetic algorithm to evolve the weights assigned to each FNN for the best appropriate prediction. In [4], Kennedy and Eberhart put forward the binary particle swarm optimization (BiPSO) to optimize the NNE: in the BiPSO, the weight of each FNN could be zero or 1, and the ensemble problem of NNE would be transformed into selecting the best appropriate FNN set by PSO. Another version of PSO, denoted as DePSO, in which the weight of each FNN could be decimal number.

The FNN often adopts the BP algorithm to optimize the weights. However, BP may lead to a failure in finding a global optimal solution [5]. But on the other part, the gradient descending method of BP could achieve higher convergent accuracy and faster convergent speed around the global optimum.

The PSO algorithm is showed to converge rapidly during the initial stages of a global search. But around global optimum, the search process may become very slow, the improvement decreasing gradually with the searching iterating. Another shortcoming is that the particles would easily oscillate in different sinusoidal waves, converging quickly, sometimes prematurely [6] [7]. In [8], the authors proposed PSO and BP couple-algorithm to train the weights of FNN, where the hybrid algorithm could make use of both global searching ability of the PSO and local searching ability of the BP algorithm. [9] proposed a new kind of hybrid method which was based on fuzzy set theory and used PSO-BP couple algorithm to determine the weight of different FNN, then synthesized their assessment result to form the final output according to the weight.

In our paper, we propose the improved PSO-BP-NNE mode, which means that we use PSO and BP to train each component FNN, and then use the Particle Swarm Optimization and Differential Evolution cooperative algorithm(PSO-DE) to optimize the NNE. There are two stages: In the FNN training stage, firstly, we use PSO to train each component FNN, when the constrain-condition is reached, we apply the BP algorithm into training until the new termination condition reached. In the NNE training stage, we present the multi-populations cooperative optimization (PSO and DE) to train the weight of each component FNN. In addition, we introduce an improved PSO algorithm which incorporates the fitness function into the velocity updating. In our experiment, the proposed algorithm is verified superior to the general NNE which is optimized by single algorithm.

## 2    Component Neural Network Optimized by PSO and BP

The gradient descending technique proposed by Werbos [10], is widely used in optimizing and training FNN. But it has its own disadvantage which is sensitive to the initial weight vector, often leading to a different result by virtue of different weight vector. The disadvantage leads trapping in a local solution which is bias to the best solution. But it could achieve faster convergent speed around global optimum, due to the reasons above, we introduce the PSO and BP couple-algorithm to optimize the FNN. The detailed gradient descending technique is described in [10] and [11].

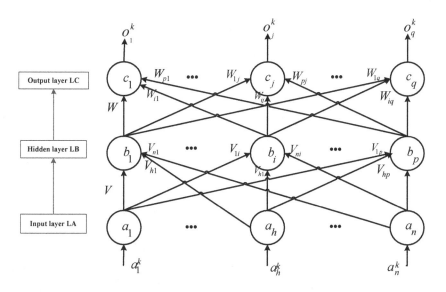

**Fig. 1.** The forward neural network architecture

The idea of BP is to make the error back-propagate to update the parameters of FNN, and the parameters include two sections: one is between the input-layer and hidden-layer, the other is between hidden-layer and output-layer. If we suppose that $X = \{x_1, x_2...x_m\}$ has $m$ input samples and $Y = \{y_1, y_2...y_n\}$ has $n$ output results. There are $p$ neurons in hidden-layer and $q$ neurons in output-layer. The thresholds of hidden neurons are $\theta^1 = \{\theta^1_1, \theta^1_2...\theta^1_p\}$ and thresholds of the output-layer are $\theta^2 = \{\theta^2_1, \theta^2_2...\theta^2_q\}$. We suppose the weights between input-layer and hidden-layer are $V = \{v_{11}, v_{12}..v_{21}..v_{np}\}$, weights between hidden-layer and output-layer are $W = \{w_{11}, w_{12}..w_{21}..w_{pq}\}$. The transition function of hidden-layer is $f(\cdot)$ and the transition function of output-layer is $g(\cdot)$.

We suppose the error between final-output and the expected output is $E$.

$$E = \frac{1}{2}\sum_{k=1}^{n}\{y_k - g[\sum_{j=1}^{p}w_{kj}f(\sum_{i=1}^{m}v_{ij}x_i - \theta^1_j) - \theta^2_k]\}^2 \tag{1}$$

We could get the updating formula of the two weights, as follows:

$$v_{ji}(t+1) = v_{ji}(t) + \Delta v_{ji} = v_{ji}(t) - \eta^1\frac{\partial E}{\partial v_{ji}} \tag{2}$$

$$w_{kj}(t+1) = w_{kj}(t) + \Delta w_{kj} = w_{kj}(t) - \eta^2\frac{\partial E}{\partial w_{kh}} \tag{3}$$

We could get the updating formula of the two thresholds, as follows:

$$\theta^1_j(t+1) = \theta^1_j(t) + \Delta\theta^1_j = \theta^1_j(t) + \eta^1\frac{\partial E}{\partial \theta^1_j} \tag{4}$$

$$\theta_k^2(t+1) = \theta_k^2(t) + \Delta\theta_k^2 = \theta_k^2(t) + \eta^2 \frac{\partial E}{\partial \theta_k^2} \tag{5}$$

Where $\eta^1, \eta^2$ is the learning rate.

But it should be noticed that the learning rate which controls the convergence to a local optimal solution is often determined by experiments or experience. If it is not ideal enough, it would easily result in oscillating of the network and could not converge rapidly.

The PSO algorithm could be described as a swarm of birds or pigeons hovering in the sky for food. We assume the pigeon swarm as a random particle swarm, and each particle stands for one bird. Every bird has its own location and flying-velocity. One swarm has $m$ particles, and the number of dimensions of every particle is $n$, denoted as $X_i = (x_{i1}, x_{i2}...x_{in})$ and $V_i = (v_{i1}, v_{i2}...v_{in})$ where $X_i$ and $V_i$ are the position and velocity of the $i-th$ particle in n dimensional space. At each step of iterations, the particles update their positions and velocities according to the two best values: One is $P_i = (p_{i1}, p_{i2}...p_{in})$, representing the previous best value of $i-th$ particle up to the current step. Another is $G_i = (g_1, g_2...g_n)$, representing the best value of all particles in the population. After obtaining the two best values, each particle updates its position and velocity according to the following equations:

$$v_{i,d}^{k+1} = w \cdot v_{i,d}^k + c_1 \cdot rand() \cdot (pbest_{i,d}^k - x_{i,d}^k) + c_2 \cdot rand() \cdot (gbest_d^k - x_{i,d}^k) \tag{6}$$

$$x_{i,d}^{k+1} = x_{i,d}^k + v_{i,d}^{k+1} \tag{7}$$

Here, rand() is a random number in the range [0, 1] generated according to a uniform probability distribution. In the general PSO, the learning actors $c_1$ and $c_2$ are positive constants, usually $c_1 = 2.8$, $c_2 = 1.3$. $w'$ is the inertial weight used to balance global and local searching. The detailed description could be referred in [12].

In the general PSO mechanism, the fitness values are used to select the best solutions. But the direct relation between the sequential iterations is usually ignored. In many cases, the values of the fitness mean the distance between the current location and the real best location. Incorporating the fitness value could enhance the global search-ability and diversity of the particle swarm.

Based on the motivation above, we introduce the improved PSO:

*if* $fitness(x_i^k) > \theta$ ( $\theta$ is the given threshold )

$\hat{v}_i^k = fitness(x_i^k) \times v_i^k$

*else* $\hat{v}_i^k = fitness(x_i^k) \times \xi \times v_i^k$ (Where $\xi$ is the acceleration constant)

*endif*

The improved updating equations list as follows:

$$v_{i,d}^{k+1} = w \cdot \hat{v}_i^k + c_1 \cdot rand() \cdot (pbest_{i,d}^k - x_{i,d}^k) + c_2 \cdot rand() \cdot (gbest_d^k - x_{i,d}^k) \tag{8}$$

$$x_{i,d}^{k+1} = x_{i,d}^k + v_{i,d}^{k+1} \tag{9}$$

The procedures for PSO–BP couple algorithm could be summarized as follows:

Step 1: Initialize the swarm of PSO: get M particles, set the initial weight $w$ and learning factor $c_1$, $c_2$, the maximal iterative generations $T_{max-pso}$ and $T_{max-BP}$.

Step 2: Evaluate the fitness of each particle of PSO, $pbest_i$ represents the previous best value of the $i-th$ particle up to the current step. $gbest$ represents the best value of all the particles in the population.

Step 3: Do step 3 until $t > T_{max-pso}$ or do step4 if the best position has not changed for several iterations, or else, return to the step 4.

Step 4: Do BP algorithm until $t > T_{max-BP}$.

Step 5: $if$    $fitness(gbest) < E$   (E is the given threshold)

      Output the prediction and MSE

    $else$  Continue to do BP for several iterations.

      Output the prediction and MSE

   $endif$

## 3    Neural Network Ensemble Optimized by DE-PSO

The authors have described the NNE in detail in[13][14]. Having obtained each refined component FNN as described in Section II, we would concentrate on how to combine the output of each component FNN.

$$\bar{f}(x) = \sum_{i=1}^{n} \bar{w}_i f_i(x) \tag{10}$$

$$\bar{w}_i = \frac{w_i}{\sum_{i=1}^{n} w_i}, 0 < w_i < 1 \tag{11}$$

where $f_i(x)$ represents the output of the $i-th$ FNN and $\bar{w}_i$ represents the importance of the $i-th$ FNN. Our idea is that, for the best appropriate prediction how to optimize $\bar{w}_i$ of each sub-network, which corresponds to the solution of the optimization problem of the particles. But in [13], the authors recommended to average the weight of each sub-network. In this paper, we introduce the multi-population cooperative algorithm which could not only avoid trapping into the local solution, but also increase the diversity of particles. Here we introduce another global-searching algorithm, differential evolution algorithm [15]. DE is also a floating-point encoded evolutionary algorithm for global optimization over continuous spaces, but it creates new candidate solutions by combining the parent individual and several other individuals of the same population. It consists of selection, crossover and mutation.

In [16] [17], authors utilized DE to optimize PSO to improve the efficiency and precision. The cooperative algorithm tends to compensate for disadvantage of the individual method and could be apt to the best solution. We also incorporate this idea, but in every iteration, the two populations optimized respectively by different algorithms would be compared and to select the best appropriate solution which determines the evolution direction.

Our architecture is as follows.

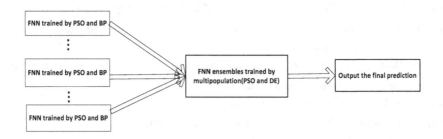

**Fig. 2.** The sketch diagram of the whole mechanism

The procedure for NNE-multi-population algorithm could be summarized as follows:

Step 1: Initialize the weight of each FNN which has been optimized by PSO and BP.

Step 2: Each particle represents a set of weights which means that each dimension represents one weight of each component FNN .The population is duplicated into 2 identical swarms.

Step 3: One swarm is optimized by PSO and the other is optimized by DE respectively.

Step 4: After each Step3, the *gbest_pso* and *gbest_DE* are calculated.

$$gbest = \max(gbest\_pso, gbest\_DE)$$

Step 5: Do the Step 3-Step 4 loop until the Max-iteration is reached.

Step 6: Output the MSE.

## 4    Experiment

To test the efficiency of the improved NNE, we perform the comprehensive experiments to compare different optimization methods. We select the input-sample set for training from $X = \{-4:0.08:4\}$ with 100 samples, we could get the expected output via the equation $y = 1.1 \times (1 - x + 2 \cdot x^2) \cdot e^{(-x^2/2)}$. We suppose the test-sample $\hat{X} = \{-3.96:0.08:3.96\}$ with 100 samples. We regard the MSE, the mean square error between the real-output and the expected-output, as the measure variable.

The performance is compared between various ensemble ways with the different component FNN and different ways to combine the output of each component FNN. In our experiment, there are three kinds of component FNN: optimized by BP, optimized by PSO, optimized by PSO and BP. The ensemble weights of NNE are optimized in five ways: simple averaging, general PSO, improved PSO, multi-population improved PSO and multi-population improved PSO and DE, which are listed in the following table.

**Table 1.** The train-MSE and test-MSE comparison between five ensemble ways with each FNN optimized by three ways

| The optimized method of NNE | Each component FNN optimized by BP | | Each component FNN optimized by PSO | | Each component FNN optimized by PSO and BP | |
|---|---|---|---|---|---|---|
| | MSE-train | MSE-test | MSE-train | MSE-test | MSE-train | MSE-test |
| Simple average | 0.4543 | 0.4154 | 0.0133 | 0.0134 | 5.1015e-007 | 4.9619e-007 |
| General PSO | 0.4324 | 0.4031 | 0.0056 | 0.0055 | 1.8851e-007 | 1.8320e-007 |
| Improved PSO | 0.3348 | 0.3095 | 0.0058 | 0.0057 | 1.4201e-007 | 1.4180e-007 |
| Multi-population Improved PSO | 0.2883 | 0.2616 | 0.0046 | 0.0043 | 7.0418e-008 | 6.7776e-008 |
| Multi-population Improve PSO and DE | 0.1997 | 0.1905 | 0.0041 | 0.0044 | 4.5633e-008 | 4.3873e-008 |

From Table I, the results which are related to individual networks optimized by different algorithm have been listed. It could see that, the individual network optimized by BP and PSO couple-algorithms does better than other algorithms. Among different NNE training algorithms, we could discover that the multi-population cooperative algorithm is superior to the other NNE trained algorithms.

## 5    Conclusion

In this paper, the superiority of individual networks optimized by different algorithm is analyzed, which reveals that in some cases the ensemble mechanism is superior to the simplex neural network. The weights of NNE also reveals the importance of individual networks, Experimental results show that multi-population cooperative algorithm is a promising ensemble approach that is superior to both averaging all and our other enumerating algorithms.

**Acknowledgments.** This work was supported in part by the National Natural Science Foundation of China (Grant Nos. 60805028, 61175076), Natural Science Foundation of Shandong Province (ZR2010FM027), China Postdoctoral Science Foundation (2012M521336), Open Research Project under Grant 20120105 from SKLMCCS, SDUST Research Fund (2010KYTD101).

# References

1. Hansen, L.K., Salamon, P.: Neural network ensembles. IEEE Trans. Pattern Analysis and Machine Intelligence 12(10), 993–1001 (1990)
2. Perrone, M.P., Cooper, L.N.: When networks disagree: Ensemble methods for hybrid neural networks. Neural Networks for Speech and Image Processing. pp. 126–142 (1993)
3. Zhou, Z.-H., Wu, J., Tang, W.: Ensembling Neural Networks: Many Could Be Better Than All. Artificial Intelligence 137(1-2), 239–263 (2002)
4. Kennedy, J., Eberhart, R.: A discrete binary version of the particle swarm optimization. In: Proceedings IEEE International Conference on Computational Cybernatics and Simulation, Piscataway, pp. 4104–4108 (1997)
5. Gori, M., Tesi, A.: On the problem of local minima in back-propagation. IEEE Trans. Pattern Anal. Mach. Intell. 14(1), 76–86 (1992)
6. Poli, R., Kennedy, J., Blackwell, T.: Particle swarm optimization. Swarm Intell. 1, 33–57 (2007), doi:10.1007/s11721-007-0002-0(2007)
7. Zhao, Z., Wang, J., Tian, Q., Cao, M.: Particle Swarm-Differential Evolution Cooperative Optimized Particle Filter. In: ICICIP, pp. 485–490 (2010)
8. Zhang, J., Zhang, J., Lok, T., Lyu, M.R.: A hybird particle swarm optimization-back-propagation algorithm for feedward neural netwrok train. Applied Mathematics and Computation 185, 1026–1037 (2007)
9. Yuan, H., Zhi, J., Liu, J.: Application of particle swarm optimization algorithm-based fuzzy BP neural network for target damage assessment. Scientific Research and Essays 6(15), 3109–3121 (2011)
10. Werbos, P.J.: Beyond regression: New tools for predictions and analysis in the behavioral science. Ph.D. Thesis, Harvard University (1974)
11. Vogl, T.P., Mangis, J.K., Rigler, A.K., Zink, W.T., Alkon, D.L.: Accelerating the convergence of the back-propagation method. Biological Cybernetics 59, 257–263 (1988)
12. Kennedy, J., Eberhart, R.: Particle swarm optimization. In: Proc. IEEE Conf. Neural Networks, Piscataway, pp. 1942–1948 (1995)
13. Optiz, D., Shavlik, J.: Actively searching for an effectively neural network ensemble. Connection Science 8(34), 337–353 (1996)
14. Valentini, G., Masulli, F.: Ensembles of Learning Machines. In: Marinaro, M., Tagliaferri, R. (eds.) WIRN VIETRI 2002. LNCS, vol. 2486, pp. 3–20. Springer, Heidelberg (2002)
15. Storn, R., Price, K.: Differential evolution - a simple and efficient heuristic for global optimization over continuous spaces. J. Global Optimization 11(4), 341–359 (1997)
16. Das, S., Abraham, A., Konar, A.: Particle Swarm Optimization and Differential Evolution Algorithms: Technical Analysis, Applications and Hybridization Perspectives. In: Liu, Y., Sun, A., Loh, H.T., Lu, W.F., Lim, E.-P. (eds.) Advances of Computational Intelligence in Industrial Systems. SCI, vol. 116, pp. 1–38. Springer, Heidelberg (2008)
17. Luitel, B., Venayagamoorthy, G.K.: Differential Evolution Particle Swarm Optimization for Digital Filter Design. In: Proc. 2008 IEEE Congress on Evolutionary Computation (CEC 2008), Crystal city, Washington, DC, USA, pp. 3954–3961 (July 2008)

# SOR Based Fuzzy K-Means Clustering Algorithm for Classification of Remotely Sensed Images

Dong-jun Xin[1,*] and Yen-Wei Chen[1,2,*]

[1] College of Computer Science and Information Technology,
Central South University of Forestry and Technology, Hunan, China
[2] College of Information Science and Engineering,
Ritsumeikan University, Japan
chen@is.ritsumei.ac.jp

**Abstract.** Fuzzy k-means clustering algorithms have successfully been applied to digital image segmentations and classifications as an improvement of the conventional k-means cluster algorithm. The limitation of the Fuzzy k-means algorithm is its large computation cost. In this paper, we propose a Successive Over-Relaxation (SOR) based fuzzy k-means algorithm in order to accelerate the convergence of the algorithm. The SOR is a variant of the Gauss–Seidel method for solving a linear system of equations, resulting in faster convergence. The proposed method has been applied to classification of remotely sensed images. Experimental results show that the proposed SOR based fuzzy k-means algorithm can improve convergence speed significantly and yields comparable similar classification results with conventional fuzzy k-means algorithm.

**Keywords:** Remotely Sensed Image, Successive Over-Relaxation, Fuzzy k-means, Classification.

## 1 Introduction

K-means algorithm is a widely used clustering technique[1], which aims to find a grouping (cluster) of unlabeled data points, whose members are more similar each other than they are to others. The centroid of each cluster is the mean of all the members in the cluster. Fuzzy k-means (FKM) was originally introduced by Bezdek[2] as an improvement of conventional k-means clustering algorithm. Unlike traditional hard clustering schemes, such as k-means [1], that assign each data point to a specific cluster, the FKM employs fuzzy partitioning such that each data point belongs to a cluster to some degree specified by a Fuzzy membership grade [3-5]. Both K-means [6] and Fuzzy K-means [7,8] algorithms have been successfully applied to classification of remotely sensed images. Classification or region segmentation of a satellite image (remotely sensed image) is an important issue for many applications, such as remote sensing (RS) and geographic information system (GIS) updating. The satellite image is a record of relative reflectance of particular wavelengths of electromagnetic

---

* Corresponding author.

C. Guo, Z.-G. Hou, and Z. Zeng (Eds.): ISNN 2013, Part I, LNCS 7951, pp. 375–382, 2013.

radiation. A particular target reflection depends on the surface feature of the target and the wavelength of the incoming radiation. Multi-spectral information has been widely used for classification of remotely sensed images [9-11]. It has been shown that the FKM is very powerful for image classification such as remotely sensed image classification than the conventional k-means clustering algorithm. The main limitation of FKM is its large computation cost. In this paper, we propose a Successive Over-Relaxation (SOR) based fuzzy k-means algorithm in order to accelerate the convergence of the algorithm. The SOR is a variant of the Gauss–Seidel method for solving a large system of linear equations, resulting in faster convergence [12,13] and it has been used for many applications such as support vector machine [14]. The proposed method has been applied to classification of remotely sensed images. Experimental results show that the proposed SOR based fuzzy k-means algorithm can improve convergence speed significantly and yields comparable similar classification results with conventional fuzzy k-means algorithm.

## 2    Fuzzy k-Means Clustering

Let $X=\{x_1,x_2,...,x_n\}$ be a set of given data and $M=\{m_1,m_2,...,m_k\}$ be a set of cluster centers. The idea of the FKM is to partition $n$ data points into $K$ clusters by minimizing the following objective function:

$$J_{FKM} = \sum_{j=1}^{k}\sum_{i=1}^{n}(\mu_j(x_i))^b d_{ij}(x_i,m_j). \tag{1}$$

where $b$ is the weighting exponent which determines the fuzziness of the clusters. Distance from $i$-th sample to $j$-th cluster center defines as:

$$d_{ij}(x_i,m_j) =\| x_i - m_j \|^2 . \tag{2}$$

$\mu_j(x_i)$ indicates the membership degree to which data point $x_i$ belongs to $j$-th cluster. It should be noted that $\mu_j(x_i)=1$ in conventional K-means algorithm.

$$\begin{cases} 0\le \mu_j(x_i)\le 1 & i=1,...,n,\ j=1,...,k \\ \sum_{j=1}^{k}\mu_j(x_i)=1 & \forall i, \end{cases} \tag{3}$$

In order to minimize the objective function, following two gradient equations are set to zero and solved for $m_j$ and $\mu_j(x_i)$, respectively.

$$\frac{\partial J_{FKM}}{\partial m_j} = 0 \quad \rightarrow \quad m_j = \frac{\sum_{j=1}^{k} \left[\mu_j(x_i)\right]^b x_i}{\sum \left[\mu_j(x_i)\right]^b}, j = 1, 2, \cdots, k \ . \tag{4}$$

$$\frac{\partial J_{FKM}}{\partial \mu_j(x_i)} = 0 \rightarrow \mu_j(x_i) = \frac{(1/d_{ij})^{1/(b-1)}}{\sum_{r}^{k} (1/d_{ij})^{1/(b-1)}} \ , i=1,2,\ldots,n \ \ j=1,2,\ldots,k. \tag{5}$$

## 3    SOR Based Fuzzy K-Means Clustering Algorithm

The SOR is devised by applying extrapolation to Gauss-Seidel(GS). This extrapolation takes the form of a weighted average between the previous iteration and the computed GS iteration successively for each component. Suppose a linear equation to be solved as

$$\begin{bmatrix} a_{1,1} & a_{1,2} & \cdots & a_{1,n} \\ a_{2,1} & a_{2,2} & \cdots & a_{2,n} \\ \vdots & \vdots & \ddots & \vdots \\ a_{n,1} & a_{n,2} & \cdots & a_{n,n} \end{bmatrix} \begin{bmatrix} x_1 \\ x_2 \\ \vdots \\ x_n \end{bmatrix} = \begin{bmatrix} b_1 \\ b_2 \\ \vdots \\ b_n \end{bmatrix} . \tag{6}$$

The solution $x_i$ can be solved by the use of SOR as

$$x_i^{(k)} = (1-w)x_i^{(k-1)} + \frac{w}{a_{ii}} \left[ -\sum_{j=1}^{i-1} a_{ij}x_j^{(k)} - \sum_{j=i+1}^{n} a_{ij}x_j^{(k-1)} + b_i \right]. \tag{7}$$

Where $w$ is the extrapolation (weighting) factor or Relaxation factor. It has be proven that $0 < w < 2$ will lead to convergence [13].
Convergence condition

$$\sum_{i=1}^{N} \left| y_i^{(k+1)} - y_i^{(k)} \right| < \varepsilon \sum_{i=1}^{N} \left| y_i^{(k+1)} \right| . \tag{8}$$

where $\varepsilon$ is an infinitesimal positive number.

We apply SOR to FKM algorithm in order to accelerate the convergence of FKM algorithm. The algorithm of our proposed SOR based FKM (SFKM) is summarized as the following steps:

Step1. Initialize $m_1^0$, $m_2^0,...,$ $m_k^0$, $\mu_j^0(x_i)$, $j=1,..., k; i=1,...,n$.

Step2. Normalize $\mu_j^0(x_i)$ by Eq.3

Step3. Update the cluster centers $m_j^{N+1}$ by the equation

$$m_j^{N+1} = w\frac{\sum_{j=1}^{k}\left[\mu_j^N(x_i)\right]^b x_i}{\sum\left[\mu_j^N(x_i)\right]^b} + (1-w)m_j^N \ . \tag{9}$$

Update $\mu_j^{N+1}(x_i)$ by the equation

$$\mu_j^{N+1}(x_i) = w\frac{(1/d_{ij})^{1/(b-1)}}{\sum_{r=1}^{k}(1/d_{ir}^N)^{1/(b-1)}} + (1-w)\mu_j^N(x_i) \ . \tag{10}$$

And

$$d_{ir}^N(x_i,m_r) = \| x_i - m_r^N \|^2 \ . \tag{11}$$

Step4. If $\| \mu^{(N+1)} - \mu^N \| < \varepsilon$, stop iterating; Otherwise, return to Step3.

## 4     Experimentation Results and Analysis

The original remotely sensed image is true-color composite image which sizes is 1024*768, which is shown in Fig.1. The number of clusters ($K$) is set to 6. In general, values of $w>1$ are used to speedup convergence of a slow-converging process, but it is not possible to compute in advance the value of w that will maximize the rate of con-vergence of SOR. We use the gradually experimental method (GEM) [13] to deter-mine the relaxation factor $w$. We experiment by setting $w$ with 1.25, 1.5, and 1.75 and discuss their performance. For each $w$, five experiments have been done with different initial parameters. Experimental environments is shown in below: Intel Core(TM)2 CPU1.66GHz; 2.5GB ram and Matlab7.0.

Typical classification results are shown in Fig.1. Fig.1(a) is the original image. Fig.1(b) is the classification result using the conventional FKM. Figs.1(c)-(e) are classification results using our proposed SFKM method with different $w$ values. Fig.1(f) shows the difference of Fig.1(b) (the conventional FKM) and Fig.1(e) (our proposed SFKM, $w=1.75$). As shown in Fig.1(f), there are no difference between Fig.1(b) (the conventional FKM) and Fig.1(e) (our proposed SFKM), i.e. our proposed SFKM can obtain the classification result of the same accuracy as the conventional FKM. The convergence of our proposed SFKM and the conventional FKM are shown in Fig.2. In Fig.2, the vertical axis is value of the objective function as shown in Eq.(1) and the horizontal axis is iteration number. Since the iteration covers a large range of numbers, the convergence is shown on a logarithmic scale. Fig.2(a) shows the convergence of the conventional FKM, while Figs.2(b)-(d) show results of our proposed SFKM with $w=1.25$, $w=1.5$ and $w=1.75$, respectively. It can be seen that our proposed SFKM is much faster than the conventional FKM. Detailed comparison of computation time is summarized in Table 1. As shown in table 1, for our proposed SFMK, each averaged iteration count for $w=1.25$, 1.5, 1.75 was 213, 124, 101 and each averaged computation time for $w=1.25$, 1.5, 1.75 was 369s, 269s, 205s, while for conventional FKM, the averaged iteration count for convergence is 343 and the averaged computation time is about 708.5s. Based on the experimental results, the better value of the extrapolation factor $w$ is 1.75. By the use of our proposed SFKM, the computation time can be reduced to about 1/3.5.

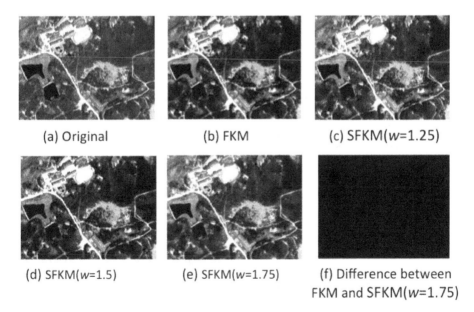

(a) Original                    (b) FKM                    (c) SFKM($w$=1.25)

(d) SFKM($w$=1.5)              (e) SFKM($w$=1.75)          (f) Difference between
                                                          FKM and SFKM($w$=1.75)

**Fig. 1.** Classification results of FKM and SFKM

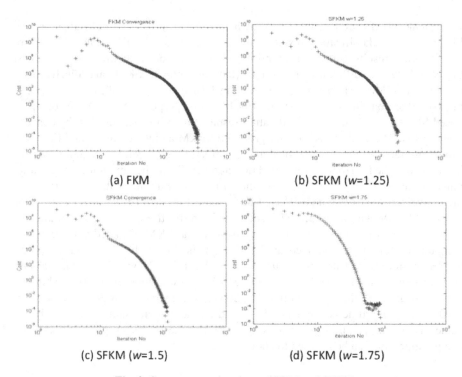

**Fig. 2.** Convergence situations of FKM and SFKM

**Table 1.** Comparison of the proposed SFKM and the conventional FKM

|   |   | *Exp* | *1* | *2* | *3* | *4* | *5* | *Avg.* |
|---|---|---|---|---|---|---|---|---|
| SFKM | w=1.25 | Iteration count | 205 | 205 | 226 | 216 | 213 | 213 |
|  |  | Runtime | 354.4s | 355.0s | 392.1s | 375.4s | 370.4s | 369.5s |
|  | w=1.5 | Iteration count | 117 | 101 | 127 | 165 | 107 | 124 |
|  |  | Runtime | 256.3s | 220.4s | 277.5s | 360.4s | 233.5s | 269.6s |
|  | w=1.75 | Iteration count | 137 | 98 | 120 | 57 | 95 | 101 |
|  |  | Runtime | 278s | 198s | 242.3s | 115.1s | 191.7s | 205s |
| FKM |  | Iteration count | 368 | 356 | 355 | 296 | 337 | 343 |
|  |  | Runtime | 762.6s | 735.1s | 734.4s | 612.2s | 698.4s | 708.5s |

# 5     Conclusions

In this paper, we proposed a Successive Over-Relaxation (SOR) based fuzzy k-means algorithm for classification of remotely sensed images. Experimental results show that the proposed SOR based fuzzy k-means algorithm can improve convergence speed significantly and yields comparable similar classification results with conventional fuzzy k-means algorithm. The computation time was shortened by ~1/3.5. The proposed method can be considered as an improvement of the conventional fuzzy k-means algorithm and it can be applied to a lot of applications.

**Acknowledgments.** This research was supported in part by 863 Project "Research on the key technology of digital forest model and visualization simulation", & distinguished professors of Hunan "Furong Scholar Program" fund, & doctor fund (07Y007).

# References

1. Hartigan, J.A., Wong, M.A.: Algorithm AS 136: A K-Means Clustering Algorithm. Journal of the Royal Statistical Society, Series C (Applied Statistics) 28, 100–108 (1979)
2. Bezdek, J.C.: Pattern Recognition with Fuzzy Objective Function Algorithm. Plenum, New York (1981)
3. Zadeh, L.A.: Fuzzy sets. Information and Control 8(3), 338–353 (1965)
4. Syed, J.K., Aziz, L.A.: Fuzzy Inventory Model without Shortages Using Signed Distance Method. Applied Mathematics & Information Sciences 1, 203–209 (2007)
5. Shi, F., Xu, J.: Emotional Cellular-Based Multi-Class Fuzzy Support Vector Machines on Product's KANSEI Extraction. Applied Mathematics & Information Sciences 6, 41–49 (2012)
6. Chitade, A.Z., et al.: Colour Based Image Segmentation Using K-Means Clustering. International Journal of Engineering Science and Technology 2, 5319–5325 (2010)
7. Chen, C.M., Chang, H.Y., et al.: A Fuzzy–Based Method for Remote Sensing Image Contrast Enhancement. In: The International Archives of the Photogrammetry, Remote Sensing and Spatial Information Sciences, vol. XXXVII, pp. 995–998 (2008)
8. Hung, C.C.A.: New Adaptive Fuzzy Clustering Algorithm for Remotely Sensed Images. In: IEEE International Geoscience and Remote Sensing Symposium, pp. 863–866 (2008)
9. Avery, T.E., Berlin, G.L.: Fundamentals of Remote Sensing and Airphoto Interpretation. Macmillan Publishing Co., New York (1992)
10. Zeng, X.Y., Chen, Y.-W., Nakao, Z.: Classification of remotely sensed images using independent component analysis and spatial consistency. Journal of Advanced Computational Intelligence and Intelligent Informatics 8, 216–222 (2004)
11. Chen, Y.-W., Han, X.-H.: Supervised Local Subspace Learning for Region Segmentation and Categorization in High-Resolution Satellite Images. In: Trémeau, A., Schettini, R., Tominaga, S. (eds.) CCIW 2009. LNCS, vol. 5646, pp. 226–233. Springer, Heidelberg (2009)
12. Saad, Y.: Iterative Methods for Sparse Linear Systems, 1st edn. PWS (1996)

13. Kahan, W.: Gauss-Seidel Methods of Solving Large Systems of Linear Equations. Ph.D. thesis, Toronto, Canada, University of Toronto (1958)
14. Mangasarian, O.L., Musicant, D.R.: Successive Overrelaxation for Support Vector Machines. IEEE Tran. on Neural Networks 10, 1032–1037 (1999)

# UMPCA Based Feature Extraction for ECG

Dong Li, Kai Huang, Hanlin Zhang, and Liqing Zhang

MOE-Microsoft Key Laboratory for Intelligent Computing and Intelligent Systems,
Department of Computer Science and Engineering, Shanghai Jiao Tong University,
Shanghai 200240, China

**Abstract.** In this paper, we propose an algorithm for 12-leads ECG signals feature extraction by Uncorrelated Multilinear Principal Component Analysis(UMPCA). However, traditional algorithms usually base on 2-leads ECG signals and do not efficiently work out for 12-leads signals. Our algorithm aims at the natural 12-leads ECG signals. We firstly do the Short Time Fourier Transformation(STFT) on the raw ECG data and obtain 3rd-order tensors in the spatial-spectral-temporal domain, then take UMPCA to find a Tensor-to-Vector Projection(TVP) for feature extraction. Finally the Support Vector Machine(SVM) classifier is applied to achieve a high accuracy with these features.

**Keywords:** ECG, Feature Extraction, Tensor, UMPCA.

## 1 Introduction

Electrocardiography(ECG) is a transthoracic interpretation of the electrical activity of human heart. ECG recognition and analysis are widely researched in recent years, with eyes to auto-diagnosis of heart disease, big data service and so forth. Many algorithms on machine learning have been proposed for ECG feature extraction, so as to achieve good classification accuracy. Zhao et al. put forth an effective method using wavelet transform and SVM classifier[1]. Alexakis et al. described an algorithm combining artificial neural networks(ANN) and Linear Discriminant Analysis(LDA) techniques for feature extraction[2]. Zhang et al. took the Principal Component Analysis(PCA) algorithm to extract feature of ECG signals[3]. Besides, An ECG feature extraction scheme using Gaussian Mixture Model(GMM) was presented by Roshan et al in [4].

The 12-leads ECG signals provide most spatial information about the heart's electrical activity in 3 approximately orthogonal directions. If the whole 12-leads information of raw ECG signals are used, more robust features can be extracted for more widely and deeply analysis. As a result, efficient representation and high classification accuracy of ECG signals can be achieved. But most researches are limitedly based on 2-leads ECG signals because of lack of 12-leads ECG database. Besides, traditional methods usually require the reshaping of multi-leads ECG data into vectors in a very high-dimensional space. This increases the computational complexity and memory demands, also discards much meaningful structural information of raw ECG data. As a result, the Tensors, which are

C. Guo, Z.-G. Hou, and Z. Zeng (Eds.): ISNN 2013, Part I, LNCS 7951, pp. 383–390, 2013.

closer to the natural structure of ECG signals, are adopted as more effective representation of ECG.

In this paper, we have proposed a tensor-based algorithm of feature extraction for the 12-leads ECG signals, which are represented by 3rd-order tensors, i.e. 12-leads signals in the spatial-spectral-temporal domain. Firstly We use STFT on the raw ecg signals to create tensors. Secondly Uncorrelated Multilinear Principal Component Analysis (UMPCA) is taken to find a Tensor-to-Vector Projection(TVP) to extract feature. Finally Support Vector Machine(SVM) with Radial Basis Function(RBF) kernel is called for classification in the feature space. We test this proposed method based on private large-scale database, and achieve high classification accuracies.

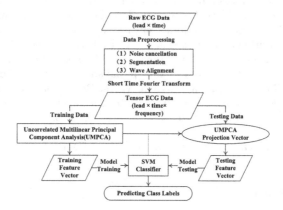

**Fig. 1.** The Framework of the Algorithm

The rest of the paper is organized as follows: Section 2 introduces the concrete procedure for feature extraction based on tensor representation using UMPCA. Section 3 demonstrates the experiment results. Section 4 is our conclusion.

## 2    UMPCA Based Algorithm

### 2.1    ECG Database

In this study, we use a large ECG database get from the local hospital, which consists of 3,000 pieces high quality 12-leads ECG records. Each piece contains 10 to 25 heartbeats, and there are 47,279 beats in total. All these records are detected from wide ranges of people: male and female, old and young, healthy and ill. We take the doctors' diagnosis as the label for the beats. Also we have 6 types of labels as follows:

### 2.2    Data Pre-process

The raw ECG signals are usually interfered with random and background noises while acquisition. In the first step of pre-process, we used wavelet transformation

**Table 1.** Database used in this study

| Label | Abbreviation | Number |
|---|---|---|
| Normal beat | N | 15991 |
| Sinus bradycardia | S | 11775 |
| Electrical axis left side | E | 6846 |
| Left ventricular hypertrophy | V | 1700 |
| Right bundle branch block beat | R | 8014 |
| Left bundle branch block beat | L | 2953 |

to cancel high-frequency noises, and used median filter to eliminate the baseline drift. The second part is R-wave alignment and beat segmentation, after that we get several pieces of ECG signals, each containing single heartbeat.

### 2.3 Short-Time Fourier Transform

The raw ECG signals only represent the features in the spatial-temporal domain. In order to transform the original signals into the spatial-spectral-temporal domain, we use Short Time Fourier Transform(STFT)[5]. We demote S as a 12-leads(lead×time) ECG signal sample, $s[l, n]$ as the discrete-time signal at time $n$ for lead $l$, then the short-time Fourier transform at time $n\triangle t$ and frequency $f$ is defined as:

$$\mathbf{STFT}\{s[l, n]\}(m, w) \equiv S(l, m, n)$$
$$= \Sigma_{m=-\infty}^{\infty} w(n - m) s(l, m) e^{-j2\pi fm} \tag{1}$$

likewise, with signal $s[n]$ and window $w[n]$. In this case, we choose to use the hann window, and transform all the raw signals to 3rd-order tensors as a result.

### 2.4 Uncorrelated Multilinear Principal Component Analysis

Uncorrelated Multilinear Principal Component Analysis(UMPCA) accepts general tensors as input for pattern classification[6]. Utilizing the Tensor-to-Vector Projection(TVP) principle, it parallelizes the successive variance maximization approach seen in the classical PCA derivation. In UMPCA, a number of Elementary Multilinear Projections(EMPs) are solved to maximize the captured variance, subject to the zero-correlation constraint.

**Multilinear Fundamentals.** We generally denote the tensor representation of ECG signals as an $N$-th order tensor $\mathbf{T} \in R^{N_1 \times N_2 \times ... \times N_n}$ with N indices $i_n, n=1,2,...,N$, and each $i_n$ addresses the $n$-mode of $\mathbf{T}$. The n-order product of $\mathbf{T}$ by a matrix $\mathbf{M}$ ,denoted by $\mathbf{T} \times \mathbf{M}$, is a tensor as:

$$(\mathbf{T} \times_n \mathbf{M})(i_1, ..., i_n - 1, j_n, i_n + 1..., i_N) = \Sigma_{i_n} \mathbf{T}(i_1, ..., i_N) \times \mathbf{M}(j_n, i_n). \tag{2}$$

The scalar product of two tensors $\mathbf{T}_1, \mathbf{T}_2 \in R^{N_1 \times N_2 \times \cdots \times N_n}$ is defined as:

$$< \mathbf{T}_1, \mathbf{T}_2 >= \Sigma_{i_1}...\Sigma_{i_N} \mathbf{T}_1(i_1,...,i_N)\mathbf{T}_2(i_1,...,i_N). \tag{3}$$

**Tensor-to-Vector Projection.** TVP is a generalized version of the projection framework firstly introduced in [7]. It consists of multiple EMPs. An EMP is a multilinear projection $\{u^{(1)^T}, u^{(2)^T}, ..., u^{(N)^T}\}$ comprised of one unit projection vector per mode. The TVP from a tensor $\mathbf{T} \in R^{N_1 \times N_2 \times \cdots \times N_n}$ to a vector $y \in R^P$ consists of P EMPs$\{u_p^{(1)^T}, u_p^{(2)^T}, ..., u_p^{(N)^T}\}$,p = 1,...,P, which can be written as:

$$y = \mathbf{T} \times_{n=1}^{1} \{u_p^{(n)^T}, n = 1, ..., N\}_{p=1}^P, \tag{4}$$

where the pth component of y is obtained from the pth EMP as:

$$y(p) = \mathbf{T} \times_1 u_p^{(1)^T} \times_2 u_p^{(2)^T} ... \times_N u_p^{(N)^T} \tag{5}$$

**Uncorrelated Multilinear Principal Component Analysis.** In our model, we have a set of M tensor object set $\{X_1, X_2, ..., X_M\}$ for training, where $X_m \in R^{I_1 \times I_2 \times \cdots \times I_N}$ and $I_n$ is the n-mode dimension of the tensor. Our objective is to determine a TVP ,which consists of P EMPs $\{u_p^{(n)^T}, n = 1, ..., N\}_{p=1}^P$, so that we can map the origin tensor space $R_1^I \otimes R_2^I \otimes ... \otimes R_N^I$ into a vector subspace $R^P$ (with $P < \Pi_{n=1}^N I_n$), as it shows in (4). We also denote $g_p$ as the pth coordinate vector, which is the representation of the training samples in the pth EMP space. Its mth component $g_p(m) = y_m(p)$.

Following the standard derivation of PCA provided in [8],the variance of the principal components is considered one at a time, starting from the first principal component that targets to capture the most variance. In the setting of TVP, we denote $S_{T_p}^y$ to measure the variance of the projected samples:

$$S_{T_p}^y = \Sigma_{m=1}^M (y_m(p) - \bar{y}_p)^2 \tag{6}$$

which is maximized in each EMP, subject to the constraint that the P coordinate vectors $\{g_p \in R^M, p = 1, 2, ...P\}$ are uncorrelated.

Generally, our objective is to determine a set of P EMPs $\{u_p^{(n)^T}, n = 1, ..., N\}_{p=1}^P$ that maximize the variance captured while producing uncorrelated features. The objective function for determining the pth EMP can be expressed as:

$$\{u_p^{(n)^T}, n = 1, ...N\} = argmax \Sigma_{m=1}^M (y_m(p) - \bar{y}_p)^2, \tag{7}$$

$$subject\ to\ u_p^{(n)^T} u_p^{(n)} = 1\ and\ \frac{g_p^T g_q}{\|g_p\| \|g_q\|} = \delta_{pq},\ p, q = 1, ..., P, \tag{8}$$

where $\delta_{pq}$ is the Kronecker delta defined as:

$$\delta_{pq} = \begin{cases} 1, & if\ p = q \\ 0, & Otherwise \end{cases} \tag{9}$$

Thus, given our tensor ECG samples $\mathbf{X}$, the corresponding UMPCA feature vector $\mathbf{y}$ is obtained as:

$$\mathbf{y} = \mathbf{X} \times_{n=1}^{1} \{u_p^{(n)^T}, n = 1, ..., N\}_{p=1}^{P}. \tag{10}$$

To solve the problem in (7), we take the successive variance maximization approach. The P EMPs $\{u_p^{(n)^T}, n = 1, ..., N\}_{p=1}^{P}$ are sequentially determined in P steps. This procedure goes as follows:

**Step 1**:Determine the 1st EMP $\{u_1^{(n)^T}, n = 1, 2, ...N\}$ by maximizing $S_{T_1}^{y}$.

**Step 2**:Determine the 2nd EMP $\{u_2^{(n)^T}, n = 1, 2, ...N\}$ by maximizing $S_{T_2}^{y}$ subject to the constraint that $g_2^T g_1 = 0$.

**Step 3**:Determine the 3rd EMP $\{u_3^{(n)^T}, n = 1, 2, ...N\}$ by maximizing $S_{T_3}^{y}$ subject to the constraint that $g_3^T g_1 = 0$ and $g_3^T g_2 = 0$.

**Step p(4,...,P)**:Determine the pth EMP $\{u_p^{(n)^T}, n = 1, 2, ...N\}$ by maximizing $S_{T_p}^{y}$ subject to the constraint that $g_p^T g_q = 0$ for q = 1,...,p-1.

Since the determination of each EMP $\{u_p^{(n)}, n = 1, ..., N\}$ is iterative in nature, and solving the projection vector in one mode requires the projection vectors in all the other modes, we use random initialization to draw each element of the n-mode projection vectors $\{u_p^{(n)}\}$ from a zero-mean uniform distribution between [-2.5,2.5], with normalization to have unit length as well. In each iteration, the update of projection vector $u_p^{n^*}$ in a given mode $n^*$ always maximizes $S_{T_p}^{y}$, so the scatter $S_{T_p}^{y}$ is non-decreasing. Therefore, UMPCA is expected to convergence over iterations. The iterative procedure terminates when

$$\frac{S_{T_{p(k)}}^{y} - S_{T_{p(k-1)}}^{y}}{S_{T_{p(k-1)}}^{y}} < \eta, \tag{11}$$

where $S_{T_{p(k)}}^{y}$ is the total scatter captured by the pth EMP obtained in the kth iteration of UMPCA and $\eta$ is a small number threshold. Alternatively, the convergence of the projection vectors should also satisfy the inequality:

$$dist(u_{p(k)}^{(n)}, u_{p(k-1)}^{(n)}) < \epsilon, \tag{12}$$

where

$$dist(u_{p(k)}^{(n)}, u_{p(k-1)}^{(n)}) = min(\|u_{p(k)}^{(n)} + u_{p(k-1)}^{(n)}\|, \|u_{p(k)}^{(n)} - u_{p(k-1)}^{(n)}\|) \tag{13}$$

and $\epsilon$ is a small number threshold which we define it as $10^{-4}$. Also we need to take the computational cost into consideration, a maximum number of iterations K(e.g. K = 30) is set to terminate iterative procedures in practice for convergence.

## 2.5   SVM-Based Classification

The high-performance Support Vector Machine(SVM)[9] is taken as our classifier. The Gaussian Radial Basis Function(RBF), is used for Multicategory

Classification. With Cross-Validation, two parameters: $\mathbf{C},\lambda(\lambda = \dfrac{1}{2\sigma^2})$, are adjusted for good classification effect. In this work, we set $\mathbf{C} = 16$ and $\lambda = 1.4142$.

## 3    Experiments and Results

To evaluate the proposed tensor-based algorithm, we test it on our lab's private database. After using STFT on each single heartbeat records and get 3rd-order tensor representation for ECG signals, we split these records into two datasets: the training set and the testing set. We use the training set for UMPCA and find a TVP, and extract features from both training set and testing set. Finally we use the training set to train the SVM Classifier Model, with which we use the testing set for classification.

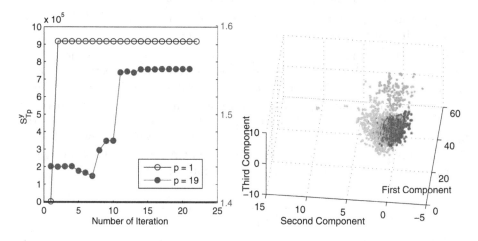

**Fig. 2.** Convergence of $S_{T_p}^y$ when training for TVP

**Fig. 3.** Data Distribution in 3D

We split the original whole records into two sets randomly: the training set consists of 24000 beats, the testing set consists of 23279 beats. In the first step of UMPCA, random initialization discussed in previous Section are tested up to 30 iterations, with the projection order fixed. Fig.2 shows the simulation results on the training sets.

Then we obtain the TVP for projection, with which we transform all the 3-order tensor data into 20-dimension feature vectors. Fig.3 shows the data distribution of three of the six classes ECG heartbeats in 3D feature space(Green for N,Brown for S,Light Blue for L), from which we know that different heartbeats are well separated. Fig.4 shows that twenty is a good number of dimension to use. Finally, we use SVM for optimal classification. Fig.5 shows result details, in which the cell in ith row and jth column means the number of heartbeat which is the ith type in fact, and predicted as the jth type in experiment. Additional, Table 2 shows the statistical results for each type of beats.

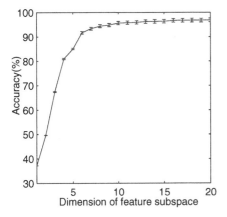

|   | N | S | E | V | R | L |
|---|---|---|---|---|---|---|
| N | 7665 | 154 | 49 | 1 | 10 | 10 |
| S | 181 | 5570 | 26 | 0 | 5 | 3 |
| E | 103 | 38 | 3178 | 3 | 14 | 8 |
| V | 8 | 10 | 0 | 800 | 13 | 1 |
| R | 14 | 9 | 10 | 0 | 4018 | 2 |
| L | 32 | 8 | 13 | 0 | 9 | 1314 |

**Fig. 4.** Accuracy in different dimensional subspace

**Fig. 5.** Test results Details

**Table 2.** Test results

| Beat Type | Training Beats | Training Accuracy | Testing Beats | Max Testing Accuracy | Testing Average | Testing Variance |
|---|---|---|---|---|---|---|
| N | 8102 | 99.56% | 7889 | 97.16% | 97.11% | 1.26e-6 |
| S | 5990 | 98.96% | 5785 | 96.28% | 96.25% | 1.21e-6 |
| E | 3502 | 99.40% | 3344 | 95.03% | 95.01% | 3.69e-6 |
| V | 869 | 99.99% | 832 | 96.15% | 96.05% | 2.51e-6 |
| R | 3961 | 99.99% | 4053 | 99.14% | 99.11% | 2.71e-7 |
| L | 1576 | 99.80% | 1376 | 95.49% | 95.43% | 2.12e-6 |

## 4 Conclusions

In this paper, we have proposed a feature extraction algorithm for 12-leads ECG signals. Computer simulations show that our approach gives the excellent performances of successful classification on 3-order tensor data in the spatial-spectral-temporal domain, accuracy is found to be 96.85%, and outperforms some other traditional vector-based methods. The good performance of the tensor-based algorithm demonstrates that it's an effective and robust data exploratory tool in 12-leads ECG signals analysis.

## References

[1] Zhao, Q.B., Zhang, L.Q.: ECG Feature Extraction and Classification Using Wavelet Transform and Support Vector Machines. In: International Conference on Neural Networks and Brain, pp. 1089–1092 (2005)

[2] Alexakis, C., Nyongesa, H.O., Saatchi, R., Harris, N.D., Davies, C., Emery, C., Ireland, R.H., Heller, S.R.: Feature Extraction and Classification of Electrocardiogram(ECG) Signals Related to Hypoglycaemia. In: Conference on Computers in Cardiology, pp. 537–540 (2003)

[3] Zhang, H., Zhang, L.Q.: ECG analysis based on PCA and Support Vector Machines. In: International Conference on Neural Networks and Brain, pp. 743–747 (2005)

[4] Martis, R.J., Chakraborty, C., Ray, A.K.: A two-stage mechanism for registration and classification of ECG using Gaussian mixture model. Pattern Recognition 42(11), 2979–2988 (2009)

[5] Allen, J.B., Rabiner, L.R.: A Unified Approach to Short-Time Fourier Analysis and Synthesis. Proceedings of IEEE 65(11), 1558–1564 (1977)

[6] Lu, H., Plataniotis, K.N., Venetsanopoulos, A.N.: Uncorrelated Multilinear Principal Component Analysis for Unsupervised Multilinear Subspace Learning. IEEE Transactions on Neural Networks 20, 1820–1836 (2009)

[7] Shashua, A., Levin, A.: Linear image coding for regression and classification using the tensor-rank principle. In: IEEE Conference on Computer Vision and Pattern Recognition, vol. 1, pp. 42–49 (2001)

[8] Jolliffe, I.T.: Principal Component Analysis, 2nd edn. Springer Serires in Statistics (2002)

[9] Bishop, C.M.: Pattern Recognition and Machine Learning. Springer-Verlag New York (2006)

# Genetic Algorithm Based Neural Network for License Plate Recognition*

Wang Xiaobin, Li Hao, Wu Lijuan, and Hong Qu

Computational Intelligence Laboratory
School of Computer Science and Engineering
University of Electronic Science and Technology of China

**Abstract.** This paper combines genetic algorithms and neural networks to recognize vehicle license plate characters. We train the neural networks using a genetic algorithm to find optimal weights and thresholds. The traditional genetic algorithm is improved by using a real number encoding method to enhance the networks weight and threshold accuracy. At the same time, we use a variety of crossover operations in parallel, which broadens the range of the species and helps the search for the global optimal solution. An adaptive mutation rate both ensures the diversity of the species and makes the algorithm convergence more rapidly to the global optimum. Experiments show that this method greatly improves learning efficiency and convergence speed.

**Keywords:** license plate recognition, genetic algorithms, neural networks, character recognition.

## 1 Introduction

Vehicle License Plate Recognition (LPR) is a significant application of digital image processing, artificial intelligence, and pattern recognition in the field of Intelligent Transportation Systems [1,2]. A complete license plate recognition system mainly consists of three parts: vehicle license plate location, character segmentation, and character recognition; of which the latter is the most critical part. Character recognition methods that are frequently used include the template matching method [3,4], the neural network method [5,6], and the support vector machine method [7]. Artificial neural networks have received a wide range of research and applications because of its good self-adaptability, self-organization, strong learning ability, association function, and fault-tolerance [8,9,10,11]. For vehicle license plate recognition, multilayer feedforward based on back-popagation is more commonly used, and some achievements have been made in this area. Error back-propagation algorithms, also known as BP algorithms, is a class of guided learning algorithms. Nevertheless, BP algorithms are based on gradient descent algorithms, its training process can easily fall into a

---

* This work was supported by National Science Foundation of China under Grant 61273308.

C. Guo, Z.-G. Hou, and Z. Zeng (Eds.): ISNN 2013, Part I, LNCS 7951, pp. 391–400, 2013.

local extreme point of the error function instead of the global optimal solution. At the same time, as the result of that setting the initial value of network is mostly dependent on the experience value, this can easily result in the training process converging too slowly or even misconvergence if the initial value is selected inappropriately. Having a global optimal search ability, genetic algorithms [12,13,14,15]can overcome the deficiencies of BP algorithms. This paper combines genetic algorithms with neural networks and makes full use of their respective advantages for vehicle license plate character recognition, with good results through testing.

## 2    Neural Network Theory Based on Genetic Algorithm

A neural network includes a learning process and an identification process. The purpose of the learning process is to find a set of optimal network weights and thresholds which minimize the error between the actual outputs and the the expected outputs. The learning process is complete when an optimal solution is found.

The genetic algorithm based neural network takes this as its learning algorithm and encodes the threshold and weights of the hidden layer and output layer of the neural network to form chromosomes inside the genetic space at first. It then uses the selection, crossover and mutation operations of genetic evolution to optimize the chromosomes. Evolution terminates when a set of weights and thresholds are found which minimize the error between actual outputs and the expected outputs. Figure 1 shows a sketch of such a network for vehicle license plate recognition.

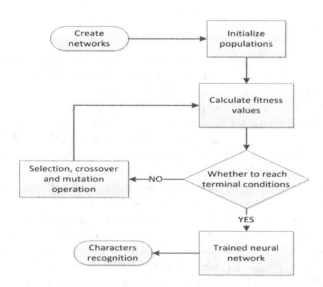

**Fig. 1.** Sketch of a neural network based on a genetic algorithm for vehicle license plate recognition

# 3    Improved Implementation of the Genetic Algorithm

Traditional genetic algorithms uses single crossover operations and variation probability. This makes genetic algorithms that use real number coding easily precocious. In order to solve this problem, this paper considers characteristics of neural networks and then improves the selection operation, crossover operation and mutation operation of neural networks.

• **Improvement of Selection Operator**

In nature, most species inheritance modes are mating between excellent individuals, but there are a few species where some good individuals mate with bad individuals, which maintains diversity. Therefore, in this paper we simulate this evolution mode, select 7/8 of parents from individuals with excellent fitness, another 1/8 from individuals with poor fitness. In order to avoid the best individual being destroyed by crossover and mutation operations, we use optimal preservation strategies; namely, the best individual of this generation is directly inherited to the next generation after each evolution.

• **Improvement of Crossover Operator**

The selection of the crossover operator, being the most important operator in the genetic algorithm and the main way to produce new individuals, has a direct impact on the performance of the algorithm. It cannot reach excellent effect if we only use one single crossover operator. In this paper, through a lot of experiments, we greatly improve the performance of the algorithm by using various crossover operators in combination. The four crossover operators: intermediate recombination, daisy chain crossover, genetic uniform crossover and linear recombination are used to cross parents at ratio $1 : 1 : 1 : 1$. Crossover probability is set to 0.7. The following discussion is mainly focused on the daisy chain crossover operator.

The daisy chain is a crossover operator which exchanges weights and thresholds of neurons to be crossed, selects two parent individuals randomly and determines a good position for them to be crossed. It then exchange neurons and finally produces two new individuals. Weights and thresholds of the hidden layer, weights and thresholds of the output layer needs to be crossed respectively. The selected position, namely, the number of neurons to be crossed, must be identical. Taking the hidden layer as an example, as show in Fig. 2: $a1$ represents the weight of the first neuron in the hidden layer of parent1, $b1$ represents the threshold of the first neuron in the hidden layer of parent1, $\beta 1$ represents weight of the first neuron in the hidden layer of parent2, $c1$ represents the threshold of the first neuron in the hidden layer of the parent2.

| Parent 1 | $\alpha 1, b1$ | $\alpha 2, b2$ | $\alpha 3, b3$ | $\alpha 4, b4$ |
|---|---|---|---|---|
| Parent 2 | $\beta 1, c1$ | $\beta 2, c2$ | $\beta 3, c3$ | $\beta 4, c4$ |
| Crossover chart | 1    0 | 1    0 | | |
| Filial generation1 | $\beta 1, c1$ | $\alpha 2, b2$ | $\beta 3, c3$ | $\alpha 4, b4$ |
| Filial generation2 | $\alpha 1, b1$ | $\beta 2, c2$ | $\alpha 3, b3$ | $\beta 4, c4$ |

**Fig. 2.** Schematic plot of daisy chain crossover

## • Improvement of Mutation Operator

The mutation operator plays a subsidiary role in genetic algorithms, it can maintain the diversity of species and avoid the genetic algorithm falling into local search. In this paper, we use a mutation operator which mutates each individual and mutates weights and thresholds respectively with self-adaptive mutation probability. This means that the mutation probability is larger at the beginning of the training phase which makes the algorithm search all local optima as much as possible. As evolution progresses, the mutation probability gradually decreases, so that the algorithm converges to the global optimum.

## 4    Implementation and Testing of the Algorithm

In order to use the combination of genetic algorithms and neural networks to recognize characters we must solve the following key issues.

1. Structure of neural networks.
2. Encoding for solution space of problem.
3. Ensure the fitness function of the genetic algorithm.

Following is a concrete analysis of these critical issues.

## • Structure of the Neural Network

In this paper, we use three-layer feedforward neural networks. According to the preceding analysis, vehicle license plate character recognition is a small classification problem. A neural network which contains a hidden layer can approximate any nonlinear function, and therefore we only need three-layer neural networks to identify characters. Because the characters on the vehicle license plate consist of Chinese characters, letters, and numbers, four neural networks are designed according to these circumstances, i.e., a Chinese character network, a letter network, a number network, as well as a letter-number network. We take each pixel in the normalized character image as an input of the neural network. The uniform size of characters is $32 \times 16$, and therefore, a total of 512 inputs. The amount of output neuron nodes is determined by the classification problem's category amount. For the Chinese character network, characters in license plates include shortened forms of names of provinces, municipalities and autonomous regions, a total of 31 different characters. It also contains 13 Chinese characters for military vehicles, as well as embassy, consulate, temporary and coach cars, all-in-all 51 Chinese characters. So the number of output nodes in the Chinese character network is 51. Numbers from 0 to 9 total of ten, so output the amount of output nodes for the number network is 10. The second character in license plate is a letter from $A$ to $Z$ (except $I$), a total of 25 letters, as a result, the number of output nodes in the letter network is 25. From the third character to the seventh character, the letter $I$ and $O$ do not appear, therefore the number of output nodes of the letter-number network is 34. The amount of hidden units is generally less than the amount of input layer nodes, combined with genetic algorithm it is generally chosen from 10 to 100. In this paper we choose 50 as the amount for the Chinese character network, 40 for the letter network, 50 for the number network, and 60 for the letter-number network. We select the unipolar sigmoid function (1) as the activation function of the hidden layer and the output layer. We also add an error correction value to each neuron of the hidden layer and output layer, also called a threshold value. Table 1 shows the amount of nodes for each layer of the neural network in this paper, and Table 2 shows the parameters settings of the neural network and genetic algorithm.

$$f(x) = \frac{1}{1 + e^{-x}} \ . \tag{1}$$

**Table 1.** The number of nodes in each layer of the neural network

|  | Chinese character network | Letter network | Number network | Letter-number network |
|---|---|---|---|---|
| Input layer | 512 | 512 | 512 | 512 |
| Hidden layer | 50 | 40 | 50 | 60 |
| Output layer | 51 | 25 | 10 | 34 |

**Table 2.** Parameters of neural network and genetic algorithm

| Type of networks | Crossover probability | Mutation probability | Target value of error | Amount of evolution generations |
|---|---|---|---|---|
| Chinese character | 0.7 | 0.1→0.0001 | 0.0001 | 1000 |
| Letter | 0.7 | 0.1→0.0001 | 0.00001 | 1200 |
| Number | 0.7 | 0.1→0.0001 | 0.00001 | 800 |
| Letter-number | 0.7 | 0.1→0.0001 | 0.00005 | 1200 |

## • Encoding of Problem Solution Space

The solution space in this paper means thresholds and weights of the network, i.e., the purpose of the network training is to find an optimal combination of weights and thresholds. Usually a binary encoding method is used, but the vehicle license plate recognition system involves many parameters and needs higher precision. However, the accuracy of binary coding is restricted by chromosome length. In this paper we adopt a real number coding so that it meets the accuracy requirement, in the meantime makes the coding significance clear.

• **Fitness Function**

To select good parents for a variety of genetic manipulations, we need to base our genetic algorithm on a good fitness function. Error is a fundamental criterion and a measure of an individual's good or bad; namely, the smaller the error is, the higher fitness the individual has. Conversely, the greater the error is, the lower fitness it has. Fitness function $F$ is related to error $E$ in this article, $F = C - E$, where $C$ is a constant and $E$ is the error, as shown in (2).

$$E = \sum (t_k - o_k)^2 . \tag{2}$$

Where $t_k$ means desired output value of the network and $o_k$ means actual output value.

## 5 Simulation Studies

Because of the limited length of this article, it is not easy to list all the weights and threshold trained by a sample network of $32 \times 16$, a total of 512 inputs. In the following, we take a sample of $7 \times 5$, a total of 35 inputs as an example to show the trained optimization curve and threshold in the hidden layer. Only the optimization curve of $32 \times 16$ is shown. Shown in Fig. 3 is a digital sample of $7 \times 5$.

```
1 1 1 1 1          1 1 1 1 1
1 0 0 0 1          1 0 0 0 0
1 0 0 0 1          1 0 0 0 0
1 0 0 0 1          1 1 1 1 1
1 0 0 0 1          0 0 0 0 1
1 0 0 0 1          0 0 0 0 1
1 1 1 1 1          1 1 1 1 1
```

**Fig. 3.** Digital sample 0 and 5 of $7 \times 5$

Algorithm parameters are as follows, amount of hidden layer neurons: 20, amount of output layer neurons: 10, population size: 400. Due to the smaller amount of outputs, we set the weights range as:$[-10,10]$ to expand the search scope of solutions. Since the subspace generated by intermediate recombinant and linear recombinant is slightly larger than the parent value, the final weights will be slightly larger than $[-10,10]$. The amount of the algorithm's genetic generations: 2000. Figure 4 is a graph of the algorithm optimizing.

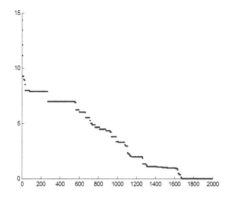

**Fig. 4.** Result diagram of $7 \times 5$ sample after digital network training

The abscissa in Fig. 4 represents the amount of evolution generations, the ordinate represents error value. As can be seen from Fig. 4, in the initial stage of evolution the curve is relatively smooth and error converges slowly. In the later stage of evolution, the error curve declines sharply. As a result of that in the initial stage of evolution the mutation probability is relatively large, the algorithm basically searches for optimal value in the entire solution space. In the later stage of evolution, mutation probability is very small, and the algorithm converges to the global optimum. After 2000 generations, error is close to 0 and algorithm converges with excellent effect. Table 3 shows the hidden layer's threshold trained by digital network.

**Table 3.** Threshold value of hidden layer after digital network training

| Number of neurons | threshold value | number of neurons | threshold value |
|---|---|---|---|
| 1 | 4.1298 | 11 | 9.6997 |
| 2 | 5.4397 | 12 | 4.2231 |
| 3 | 6.4123 | 13 | 4.7711 |
| 4 | 5.7615 | 14 | 5.3735 |
| 5 | 4.7931 | 15 | 5.4468 |
| 6 | 6.1266 | 16 | 2.1759 |
| 7 | 5.7350 | 17 | 5.1540 |
| 8 | 5.2849 | 18 | 3.9687 |
| 9 | 4.5954 | 19 | 4.2383 |
| 10 | 5.7543 | 20 | 5.6170 |

And then test the sample of $32 \times 16$, algorithm parameters are as follows: amount of hidden layer neurons: 50, amount of output layer neurons: 10, population size: 400, weights range: $[-1, 1]$. The crossover probability is 0.7. The mutation probability is at first larger as well, then it gradually decreases as

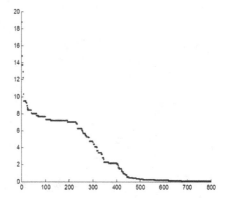

**Fig. 5.** Result diagram of 32 × 16 sample after digital network training

evolution progresses. The amount of the algorithm's genetic generations is 800. Figure 5 is a graph of the algorithm's optimization.

The abscissa in Fig. 5 represents the amount of generations of evolution; the ordinate represents error value. As can be seen from Fig. 5, after evolution of 800 generations the output error nearly reduces to 0.

We select a total of 80 vehicle license plate images with different sizes, clarity and inclination to test license plate recognition. This involves the Chinese characters representing Sichuan, Beijing, Shanxi, Zhejiang, Jiangsu, Yunnan, Henan, Guangdong, Heilongjiang and Shanxi; a total of 10 characters as well as the 26 letters from A to Z and 10 numbers from 0 to 9. This experiment is conducted along the following practice: automatically eliminate the earlier failure with no image trimming and do not include it in the follow-up experiment. Table 4 shows the recognition result of the system.

**Table 4.** Recognition results

| Treating process | Amount of correct results | Accuracy |
|---|---|---|
| Locating of license plate | 78 | 97.5% |
| Characters segmentation | 518 | 94.8% |
| Chinese character | 70 | 94.6% |
| Letter | 72 | 97.3% |
| Number | 232 | 91.6% |
| Letter-number | 354 | 89.7% |

We compare the algorithm in this paper with the BP algorithm on recognition efficiency of vehicle license plate characters. The result is shown in Table 5. It can be seen that the algorithm in this paper has much larger recognition rate than the BP algorithm for the Chinese character network and letter network. Meanwhile, the two algorithms have similar efficiency for the letter-number

**Table 5.** Comparison between the algorithm in this paper and BP algorithm

| Type of network | BP algorithm | Algorithm of this paper |
|---|---|---|
| Chinese character | 88.99% | 94.6% |
| Letter | 77% | 97.3% |
| Number | 92.3% | 91.6% |
| Letter-number | 90.4% | 89.7% |

network. This shows that the algorithm in this paper is superior to the traditional BP algorithm.

## 6 Conclusions

Through the above analysis and experimental results, the following conclusions can be drawn:

1. Using genetic algorithms in training of neural networks can achieve better results and effectively overcome the BP neural network's drawbacks of slow convergence and easily falling into local minimum values.
2. The experiments shows that using the genetic algorithm based neural network can effectively improve the accuracy of vehicle license plate recognition.

## References

1. Chen, Z.-X., Liu, C.-Y., Chang, F.-L., Wang, G.-Y.: Automatic License-Plate Location and Recognition Based on Feature Salience. Vehicular Technology 58(7), 3781–3785 (2009)
2. Ying, W., Yue, L., Jingqi, Y., Zhenyu, Z., von Deneen, K.M., Pengfei, S.: An Algorithm for License Plate Recognition Applied to Intelligent Transportation System. Intelligent Transportation Systems 12(3), 830–845 (2011)
3. Xiaoping, L., Xiaoxing, L., Shuaizong, W., Lin, Z., Yinxiang, L., Hongjian, D.: Research on the Recognition Algorithm of the License Plate Character Based on the Multi-resolution Template Matching. In: 4th International Conference on New Trends in Information Science and Service Science, pp. 341–344. IEEE Press, Gyeongju South Korea (2010)
4. Wakahara, T., Yamashita, Y.: Multi-template GAT/PAT Correlation for Character Recognition with a Limited Quantity of Data. In: 20th International Conference on Pattern Recognition, pp. 2873–2876. IEEE Press, Istanbul Turkey (2010)
5. Jianlan, F., Yuping, L., Mianzhou, C.: The research of vehicle license plate character recognition method based on artificial neural network. In: 2nd International Asia Conference on Informatics in Control, Automation and Robotics, pp. 317–320. IEEE Press, Wuhan China (2010)
6. Jianwei, G., Jinguang, S.: License Plate Recognition System Based on Orthometric Hopfield Network. In: International Conference on MultiMedia and Information Technology, MMIT 2008, pp. 594–597. IEEE Press, Three Gorges China (2008)

7. Abdullah, S.N.H.S., Omar, K., Sahran, S., Khalid, M.: License Plate Recognition Based on Support Vector Machine. In: International Conference on Electrical Engineering and Informatics, ICEEI 2009, pp. 78–82. IEEE Press, Selangor Malaysia (2009)
8. Malheiros-Silveira, G.N., Hernandez-Figueroa, H.E.: Prediction of Dispersion Relation and PBGs in 2-D PCs by Using Artificial Neural Networks. Photonics Technology Letters 24(20), 1799–1801 (2012)
9. Domenech, C., Wehr, T.: Use of Artificial Neural Networks to Retrieve TOA SW Radiative Fluxes for the EarthCARE Mission. Geoscience and Remote Sensing 49(6), 1839–1849 (2011)
10. Janakiraman, V., Bharadwaj, A., Visvanathan, V.: Voltage and Temperature Aware Statistical Leakage Analysis Framework Using Artificial Neural Networks. Computer-Aided Design of Integrated Circuits and Systems 29(7), 1056–1069 (2011)
11. Zhou, D.Q., Annakkage, U.D., Rajapakse, A.D.: Online Monitoring of Voltage Stability Margin Using an Artificial Neural Network. Power Systems 25(3), 1566–1574 (2010)
12. Zuqing, Z., Chuanqi, W., Weida, Z.: Using Genetic Algorithm to Optimize Mixed Placement of 1R/2R/3R Regenerators in Translucent Lightpaths for Energy-Efficient Design. Communications Letters 16(2), 262–264 (2012)
13. Lau, H.C.W., Chan, T.M., Tsui, W.T., Pang, W.K.: Application of Genetic Algorithms to Solve the Multidepot Vehicle Routing Problem. Automation Science and Engineering 7(2), 382–392 (2010)
14. Cabral, H.A., de Melo, M.T.: Using Genetic Algorithms for Device Modeling. Magnetics 47(5), 1322–1325 (2011)
15. Shuo, C., Arnold, D.F.: Optimization of Permanent Magnet Assemblies Using Genetic Algorithms. Magnetics 47(10), 4104–4107 (2011)

# Optimizing Fuzzy ARTMAP Ensembles Using Hierarchical Parallel Genetic Algorithms and Negative Correlation

Chu Kiong Loo[1], Wei Shiung Liew[1], and Einly Lim[2]

[1] Faculty of Computer Science and Information Technology
University of Malaya
Kuala Lumpur, Malaysia
{ckloo.um,einly_lim}@um.edu.my, liew.wei.shiung@gmail.com
[2] Faculty of Engineering
University of Malaya
Kuala Lumpur, Malaysia

**Abstract.** This study demonstrates a system and methods for optimizing a pattern classification task. A genetic algorithm method was employed to optimize a Fuzzy ARTMAP pattern classification task, followed by another genetic algorithm to assemble an ensemble of classifiers. Two parallel tracks were performed in order to assess a diversity-enhanced classifier and ensemble optimization methodology in comparison with a more straightforward method that does not rely on diverse classifiers and ensembles. Ensembles designed with diverse classifiers outperformed diversity-neutral classifiers in 62.50% of the tested cases. Using a negative correlation method to manipulate inter-classifier diversity, diverse ensembles performed better than non-diverse ensembles in 81.25% of the tested cases.

**Keywords:** genetic algorithms, pattern classification, classifier ensemble, diversity.

## 1 Introduction

The effectiveness of a pattern recognition task depends on the type of pattern classifier used, the reliability of the data set used for training the classifier to recognize specific patterns, as well as the training method itself. For a given pattern classifier, the best combination of parameter settings can be evaluated on a case-by-case basis using genetic algorithms (GA). Another common method is by creating a committee of classifiers to combine complementary information from multiple sources in order to achieve a greater degree of understanding. Genetic optimization of classifier ensembles are not a new concept in pattern recognition. Ruta and Gabrys [1] performed several different evolutionary optimization methods for classifier selection. Coupled with majority voting, their combinatory method displayed an improved accuracy and reliability when performing classification in comparison with a single-best classifier selection strategy. Similarly,

C. Guo, Z.-G. Hou, and Z. Zeng (Eds.): ISNN 2013, Part I, LNCS 7951, pp. 401–410, 2013.

a study by Zhou et al. [2] revealed that in some cases, ensembling a subset of classifiers may outperform an ensemble with all classifiers included.

In 2003, an experiment by Kuncheva and Whitaker [3] studied the relationship between several diversity measures in relation to ensemble classification accuracy. The study did not find any conclusive connections between increased ensemble diversity and ensemble classification accuracy. In addition, a 2010 study by Lofstrom et al. [4] also displayed similar results, concluding that there is little incentive for employing complex diversity algorithms in favour of simpler methods of classifier ensembling.

The proposed system used Fuzzy ARTMAP neural network architecture as a pattern learner and classifier. ARTMAP optimization was performed using genetic algorithms to generate a population of optimum ARTMAP configurations. A number of ARTMAP classifiers were then selected to form an ensemble, using a probabilistic voting strategy to combine individual decisions. The proposed methodology was tested using two methods of classifier optimization and ensemble selection. The first method used a simple GA for classifier optimization.. To test the hypothesis regarding the contribution of diversity for classifier and ensemble optimization, the second method was devised to enhance classifier and ensemble diversity. The simple GA method was replaced with a hierarchical fair-competition parallel GA (HFCPGA). A genetic-optimized negative correlation method was used in both methods for classifier selection, using a single parameter to manipulate the weight of diversity. The ensembles' pattern recognition rate was then evaluated to determine the effectiveness of the methodology.

The next section details individual parts of the proposed system. The methodology of the experiment will be covered in Section 3, and Section 4 will present the findings of the test.

## 2    Overview of System and Method

The system consists of several discrete sections integrated into a single framework for optimizing a pattern classification task. Fuzzy ARTMAP classifiers were used for learning and classifying patterns, using a given data set of training examples for supervised learning. Factors which influence the pattern recognition rate of the classifiers, such as the training sequence, feature subset, and ARTMAP vigilance, were optimized using genetic algorithms for efficient searching through the solution-space. From the resultant pool of optimum candidate FAM classifiers, a selection method chose a number of classifiers to form an ensemble that would score higher classification accuracy than any single individual. Predictive decisions from the multitude of classifiers within the ensemble were aggregated using a probabilistic voting strategy.

### 2.1    Fuzzy ARTMAP

The Fuzzy ARTMAP (FAM) neural network [5] was developed based on the principles of adaptive resonance theory [6]. The classification performance of the

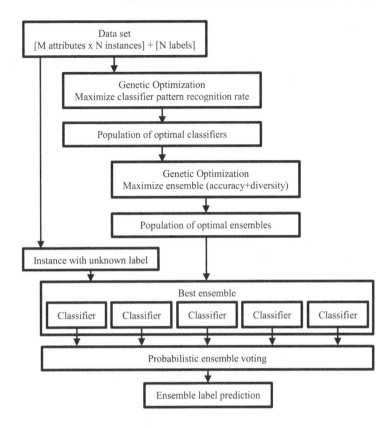

**Fig. 1.** Flowchart of the system

FAM can be influenced by several factors: the parameters used for defining the structure of the FAM, and the sequence in which data is presented to the FAM during the supervised training process. The structural parameters of the FAM that have been identified for optimization includes baseline vigilance, learning rate, and number of epochs to present a new training pattern. Discovering the best combination of structural parameters, as well as the best sequence of training data, requires an exhaustive searching method. For this purpose, a genetic algorithm was proposed to search the solution-space.

## 2.2  Hierarchical Fair-Competition Parallel Genetic Algorithms

In the context of ensembles, a prevailing understanding is that diversity between component classifiers is a key for creating an ensemble that can significantly outperform any single classifier. However, utilizing genetic algorithms to generate classifiers for an ensemble has a drawback of convergence. Pressure from the evolutionary selection process perpetuates classifiers sharing a certain combination of genetic traits, eventually causing the classifiers in the population to evolve into genetically-similar cousins.

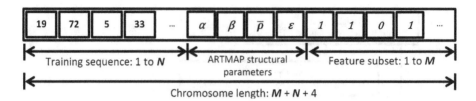

Fig. 2. Structure of a chromosome for a pattern classification task with a data set consisting of **N** vectors with **M** attributes. Subchromosomes consist of the sequence in which vectors were presented, parameters for initializing the FAM classifier, and feature selection for the data vectors.

Lee and Kim [7] proposed using HFCPGA for generating a diverse population of optimization solutions. In general, the HFCPGA method maintains multiple subpopulations of chromosomes as opposed to a single encompassing population. Chromosomes were assigned to subpopulations in a hierarchy: chromosomes that yielded classifiers with a higher pattern recognition rate were assigned to one subpopulation, while subsequent subpopulations were populated by chromosomes with decreasing fitness. Genetic operators such as crossover and reproduction were performed within each subpopulation independent of the others. Periodically, newly generated chromosomes with higher or lower than average fitness were immigrated into a higher or lower subpopulation respectively. This methodology effectively divides genetic convergence across multiple groups, while introducing new genetic material to increase diversity. A series of comparative analyses were then performed with multiple data sets, concluding with a clear demonstration of diversity measurements for HFCPGA and simple GA.

The following Table 1 compares the key differences between both genetic algorithm optimization methods.

## 2.3   Genetic-Optimized Negative Correlation

The classifier optimization step created a population of chromosomes, each representing a single configuration to create and train a FAM classifier. An ensemble was created by selecting and combining multiple classifiers. Ideally, the selected

Table 1. Comparison of differences between simple GA and HFCPGA optimization methods

| Simple GA | Hierarchical Fair-Competition Parallel GA |
|---|---|
| Single population of chromosomes. | Multiple subpopulations of chromosomes. |
| Random chromosome placement. | Chromosomes distributed among subpopulations grouped according to similar fitness. |
| Parent chromosomes selected at random from population. | Roulette-wheel selection chooses a subpopulation. Two parent chromosomes were then selected at random from the subpopulation. |

component FAMs should be diverse from each other while possessing a good predictive accuracy. Uncorrelated predictive errors from each classifier in the ensemble can then be cancelled out using a voting strategy for decision combination.

The negative correlation method [8] constructs diverse ensemble with $J$ component classifiers using an equation [7] that can be simplified into:

$$\hat{E}_J = (\frac{1}{J}) \sum_{j=1}^{J} [C_j - (\frac{\lambda}{J} \sum_{t=1}^{J} K_{jt})] \tag{1}$$

The term $C_j$ is the predictive error of the $j^{th}$ FAM classifier, while $K_{jt}$ is the inter-classifier diversity between the $j^{th}$ FAM and the $t^{th}$ FAM. The $\lambda$ term is a user-defined parameter to adjust the weight of diversity in determining the classifier selection. Setting $\lambda$ to zero eliminates the diversity requirement, while setting it to a non-zero number introduces a measure of diversity in the classifier selection process.

Given a selection of classifiers, the ensemble was initialized with the first selected classifier and the negative correlation error function was calculated, $\hat{E}_{J=1}$. As each additional classifier were added into the ensemble, the new ensemble's error function was computed, $\hat{E}_{J+1}$. If the latest classifier does not contribute toward ensemble accuracy or diversity (i.e. $\hat{E}_{J+1} \geq \hat{E}_J$), then the classifier in question was removed from the ensemble. The optimization step thus solves for FAM classifier combinations which yields the smallest error function $\hat{E}_J$ in order to achieve maximum inter-classifier diversity and minimum ensemble predictive error.

## 2.4   Probabilistic Voting

A pattern classification task usually involves a classifier assigning a single observation into one of several possible discrete categories. Misclassification occurs when the observation is sufficiently vague that the classifier assigns it into the wrong category. An ensemble of multiple diverse classifiers usually has a better classification accuracy than any single classifier, since the misclassifications were usually distributed among the residual categories. However this may not always be the case, as misclassifications can occur if a majority of the classifiers perform correlated errors on a particularly difficult observation.

The output decisions from individual FAMs in the ensemble were combined using a probabilistic voting strategy [9][10]. In general, each classifier in the ensemble is given a weightage inversely proportional to its error function $\hat{E}_J$. Consequently, instead of assigning one vote per classifier, the voting system gives greater priority to the classification decisions made by highly accurate FAMs over less accurate FAMs. If however, the best classifier makes a classification error, the voting system will still be able to choose the correct classification if support from the lesser FAMs outweighs that of the dominant FAM.

# 3  Experiment Methodology

An experiment was designed to test the pattern recognition capability of the proposed system. Data sets selected from the UCI Machine Learning Repository [11]. Details of the data set are described in Table 2. In addition, the parameters relevant to the experiment is detailed in Table 3. Both GA and HFCPGA methods were coded and executed in the Matlab programming environment, utilizing the Parallel Processing Toolbox for increased performance. Empirical testing of both methods showed no significant difference in computation time when tested under identical settings.

**Table 2.** Data sets selected for benchmarking

| Data set | Instances | Attributes | Categories |
|---|---|---|---|
| Acute Inflammations | 120 | 6 | 4 |
| Arrhythmia | 452 | 279 | 16 |
| Glass | 214 | 9 | 6 |
| Ionosphere | 351 | 34 | 2 |
| Iris | 150 | 4 | 3 |
| Planning Relax | 182 | 12 | 2 |
| Seeds | 210 | 7 | 3 |
| Wine | 178 | 13 | 3 |

The experiment methodology was performed with the following arrangements:

1. Given a training data set for a pattern recognition task, a population of chromosomes were generated.
2. The initial FAM was created using the structural parameters defined in the solution. The data set was reformatted using the defined training sequence and attribute subset. The FAM was trained and tested using leave-one-out cross-validation. The fitness of the chromosome was calculated as the percentage of tests correctly classified by the FAM.
   - For HFCPGA, the chromosomes were distributed evenly into subpopulations according to fitness. High-fitness chromosomes were assigned to higher subpopulations, while low-fitness chromosomes were shifted into lower subpopulations. This was performed only during the first generation. For subsequent generations, if there existed a chromosome in a lower subpopulation with higher fitness than a chromosome in a higher subpopulation, then these two chromosomes swap positions in their respective subpopulations. Immigration was limited to one chromosome pair per generation, and the pair with the largest difference in fitness was selected for immigration.
3. A fraction of the population with the lowest pattern classification rate were discarded. For each chromosome discarded, an offspring chromosome was created by genetic reproduction and crossover of two parent chromosomes.

**Table 3.** Experiment parameters

| Parameter name | Default | Description |
|---|---|---|
| Genetic.chromosomes | 50 | Total number of chromosomes |
| Genetic.convergence | 3 | Number of consecutive generations with no improvement before incrementing mutation rate |
| Genetic.generations | 50 | Maximum number of optimization generations |
| Genetic.maxsubset | 0.5 | Maximum fraction size of feature subset |
| Genetic.mutationrate | 0.1 | Fraction of genes per offspring to mutate |
| Genetic.maxmutation | 1.0 | Maximum mutation rate |
| Genetic.selection | 0.5 | Fraction of chromosomes to carry over to the next generation |
| Genetic.subpopulations | 5 | Number of populations to maintain for HFCPGA |
| NegCorr.maxvoters | 49 | Maximum number of classifiers selected for ensemble using genetic-optimized negative correlation selection |

- Both parent chromosomes were compared for common genetic traits, which were then passed down to the offspring chromosome.
- The remaining uncommon genes were shuffled into the offspring. Genetic mutation was then performed, using simple bit flip for the binary portion of the chromosome, swapping gene positions for the training sequence, and adding or subtracting a random fraction number for the ARTMAP parameter genes.

4. The generation counter was incremented and steps 2-3 were repeated until convergence.
5. Convergence criteria was defined thus: each generation, the mean and maximum fitness of the population of chromosomes were recorded. After three consecutive generations with no improvement in either mean or maximum fitness, the mutation parameter was increased to enhance the genetic algorithm searching process. The optimization process was terminated once the mutation rate was increased to 1.0 with no improvement in mean or maximum fitness. A maximum was also defined to terminate the genetic optimization step after the pre-determined number of generations have elapsed.
6. From the resultant population of chromosomes, a number of classifiers were selected to form an ensemble, using genetic-optimized negative correlation.
   (a) Chromosomes in the form of a binary string representing the population, with 1s and 0s to select and deselect classifiers respectively.
   (b) For each chromosome, an ensemble was created by using the negative correlation method.
   (c) Each classifier in the ensemble was fitness tested using leave-one-out cross-validation. Output of each classifier was integrated via probabilistic voting to determine ensemble output. Fitness of the chromosome was computed as the percentage of tests correctly classified by the ensemble as a single unified classifier.

(d) Steps (b) and (c) was performed once for $\lambda = 0$ and once for $\lambda = 1$ to observe the effect of introducing diversity into the classifier selection process.

## 4    Results and Discussion

For each data set, optimization was performed ten times, and the results were averaged. The experiment findings are summarized in Table 4 below.

During classifier optimization, the population's mean and maximum fitness (MeanAcc, MaxAcc) were computed after each generation of optimization. Genetic convergence was determined when the population's MeanAcc and MaxAcc does not increase in three consecutive generations, or until the maximum number of generations was achieved. Comparing GA and HFCPGA methods in this experiment, neither achieved a faster convergence consistently.

The similarity index was incorporated as a means to determine the degree of diversity of the population's classifier output. An index of 1.00 indicates that all classifiers in the population produce identical classification results. Ideally, low similarity index is desirable due to uncorrelated predictive errors which can be corrected using a voting strategy. While there was also no significant difference in similarity and mean fitness, the simple GA method displayed consistently higher maximum fitness than HFCPGA. This finding was consistent with a survey of genetic algorithm methods by Cantú-Paz [12], which established that parallel GA methods were slower to achieve convergence and usually having fitness less than that of a single population GA.

This experiment defined two static values for the $\lambda$ parameter in the negative correlation step: 0 to create ensembles based only on the merit of the classifiers' predictive accuracy, and 1 to introduce inter-classifier diversity as an equally important factor. A proposal for future experiments will be to use genetic algorithms to optimize the $\lambda$ parameter for a range of values from 0 to higher than 1. This will be especially effective for classification tasks with low recognition rates. Placing a higher priority on inter-classifier diversity rather than classifier accuracy may be able to yield better ensembles through combining complementary information rather than relying on the strengths of individual classifiers.

The experiment discovered that classifier ensembles with diversity incorporated ($\lambda = 1$) often outperformed diversity-neutral classifiers ($\lambda = 0$), in 13 out of 16 cases. Comparing GA-optimized ensembles against HFCPGA-optimized ensembles, 10 times out of 16, HFCPGA ensembles outperformed GA ensembles while in 2 cases out of 16, GA outperformed HFCPGA. A conclusion was decided that incorporating diversity into classifier and ensemble optimization was able to improve pattern recognition rates. For future experiment designs, data sets with difficult pattern classification would be more suitable in order to showcase the increase in pattern recognition rates when classifiers were incorporated into ensembles.

**Table 4.** Experiment results for classification tasks with simple GA and HFCPGA. Results taken as average over ten iterations.

| Data set | Convergence | Similarity | MeanAcc | MaxAcc | Ensemble Acc $\lambda = 0$ | $\lambda = 1$ |
|---|---|---|---|---|---|---|
| Acu.Inflam. GA | 25.00 | 0.9399 | 0.9526 | 1.000 | 1.000 | 1.000 |
| HFCPGA | 23.80 | 0.9394 | 0.9624 | 1.000 | 1.000 | 1.000 |
| Arrhythmia GA | 33.10 | 0.7229 | 0.6003 | 0.6495 | 0.6787 | 0.6981 |
| HFCPGA | 32.20 | 0.7322 | 0.6099 | 0.6570 | 0.6829 | 0.7075 |
| Glass GA | 27.80 | 0.7711 | 0.7625 | 0.7982 | 0.8813 | 0.8850 |
| HFCPGA | 28.20 | 0.7584 | 0.7525 | 0.8000 | 0.8542 | 0.8869 |
| Ionosphere GA | 35.70 | 0.9535 | 0.9322 | 0.9500 | 0.9674 | 0.9690 |
| HFCPGA | 38.20 | 0.9530 | 0.9334 | 0.9500 | 0.9677 | 0.9699 |
| Iris GA | 16.00 | 0.9892 | 0.9028 | 0.9822 | 0.9822 | 0.9874 |
| HFCPGA | 16.20 | 0.9900 | 0.9000 | 0.9815 | 0.9815 | 0.9874 |
| Plan. Relax GA | 29.00 | 0.6757 | 0.6415 | 0.7116 | 0.7473 | 0.7945 |
| HFCPGA | 27.80 | 0.6744 | 0.6416 | 0.7074 | 0.7725 | 0.7956 |
| Seeds GA | 26.30 | 0.9299 | 0.9338 | 0.9577 | 0.9772 | 0.9783 |
| HFCPGA | 26.00 | 0.9332 | 0.9353 | 0.9561 | 0.9783 | 0.9788 |
| Wine GA | 35.10 | 0.9771 | 0.9717 | 0.9901 | 0.9975 | 0.9975 |
| HFCPGA | 33.20 | 0.9757 | 0.9700 | 0.9901 | 0.9975 | 0.9988 |

## 5 Conclusion

This paper presented a methodology for creating optimized classifier ensembles. Diversity was shown to be a significant criterion during classifier and ensemble optimization. Given two identical populations optimized by simple GA and HFCPGA respectively, 62.50% of HFCPGA-optimized ensembles outperformed their GA counterparts. In addition, diversity-enforced ensembles outperformed diversity-neutral ensembles 81.25% of the time, using negative correlation method for ensemble creation.

**Acknowledgement.** This study was funded by University of Malaya Research Grant (UMRG) project RG059-11ICT, and the University of Malaya / Ministry of Higher Education High Impact Research (UM/MoHE HIR (H-22001-00-B000010)) Grant.

## References

1. Ruta, D., Gabrys, B.: Application of the Evolutionary Algorithms for Classifier Selection in Multiple Classifier Systems with Majority Voting. In: Kittler, J., Roli, F. (eds.) MCS 2001. LNCS, vol. 2096, pp. 399–408. Springer, Heidelberg (2001)
2. Zhou, Z.H., Wu, J.X., Jiang, Y., Chen, S.F.: Genetic Algorithm based Selective Neural Network Ensemble. In: 17th Int. Conf. on Artificial Intelligence, pp. 797–802. Morgan Kaufmann Publishers (2001)

3. Kuncheva, L.I., Whitaker, C.J.: Measures of Diversity in Classifier Ensembles and Their Relationship with the Ensemble Accuracy. Machine Learning 51(2), 181–207 (2003)

4. Lofstrom, T., Johansson, U., Bostrom, H.: Comparing Methods for Generating Diverse Ensembles of Artificial Neural Networks. In: 2010 IEEE Int. Conf. on Neural Networks, pp. 1–6. IEEE (2010)

5. Carpenter, G.A., Grossberg, S., Marzukon, N., Reynolds, J.H., Rosen, D.B.: Fuzzy ARTMAP: A Neural Network Architecture for Incremental Supervised Learning of Analog Multidimensional Maps. IEEE Trans. on Neural Networks 3(5), 698–713 (1992)

6. Carpenter, G.A., Grossberg, S.: Adaptive Resonance Theory. CAS/CNS Technical Report Series 008 (2010)

7. Lee, H., Kim, E., Pedrycz, W.: A New Selective Neural Network Ensemble with Negative Correlation. Applied Intelligence, 1–11 (2012)

8. Liu, Y., Yao, X., Higuchi, T.: Evolutionary Ensembles with Negative Correlation Learning. IEEE Trans. on Evolutionary Computation 4(4), 380–387 (2000)

9. Lin, X., Yacoub, S., Burns, J., Simske, S.: Performance Analysis of Pattern Classifier Combination by Plurality Voting. Pattern Recognition Letters 24(12), 1959–1969 (2003)

10. Loo, C.K., Rao, M.V.C.: Accurate and Reliable Diagnosis and Classification Using Probabilistic Ensemble Simplified Fuzzy ARTMAP. IEEE Trans. on Knowledge and Data Engineering 17(11), 1589–1593 (2005)

11. UCI Machine Learning Repository, http://archive.ics.uci.edu/ml

12. Cantú-Paz, E.: A survey of parallel genetic algorithms. Calculateurs Paralleles, Reseaux et Systems Repartis 10(2), 141–171 (1998)

# Improvement of Panchromatic IKONOS Image Classification Based on Structural Neural Network

Weibao Zou

Key Laboratory of Western China's Mineral Resources and Geological Engineering,
Ministry of Education, Chang'an University, Xi'an, China, 710054
hkzouwb@yahoo.com

**Abstract.** Remote sensing image classification plays an important role in urban studies. In this paper, a method based on structural neural network for panchromatic image classification in urban area with adaptive processing of data structures is presented. Backpropagation Through Structure (BPTS) algorithm is adopted in the neural network that enables the classification more reliable. With wavelet decomposition, an object's features in wavelet domain can be extracted. Therefore, the pixel's spectral intensity and its wavelet features are combined as feature sets that are used as attributes for the neural network. Then, an object's content can be represented by a tree structure and the nodes of the tree can be represented by the attributes. 2510 pixels for four classes, road, building, grass and water body, are selected for training a neural network. 19498 pixels are selected for testing. The four categories can be perfectly classified using the training data. The classification rate based on testing data reaches 99.91%. In order to prove the efficiency of the proposed method, experiments based on conventional method, maximum likelihood classification, are implemented as well. Experimental results show the proposed approach is much more effective and reliable.

**Keywords:** Panchromatic image classification, structural neural network, Backpropagation Through Structure, wavelet transform.

## 1 Introduction

Mapping the land cover patterns from global, regional, to local scale are critical for the scientists and the authorities to yield better monitoring of the changing world [1]. Precise monitoring of the land cover becomes indispensable for the decision makers in dealing with public policy planning and Earth resources management. Remote sensing sensors record the spectral reflectance of different land cover materials from visible to infrared wavelength, and from moderate to very high spatial resolution. Therefore, the land cover patterns can be derived from the spectral signatures using pattern recognition techniques. The demand on land cover mapping in finer scale, especially in urban area, is raised with evidence by numerous biophysical and socio-economic studies in urban heat island [2], urban sprawl pattern [3], urban environmental quality [4], urban rainfall-runoff modeling [5], urban anthropogenic heat [6], and urban air pollution [7]. Panchromatic (PAN) satellite imagery is being

C. Guo, Z.-G. Hou, and Z. Zeng (Eds.): ISNN 2013, Part I, LNCS 7951, pp. 411–420, 2013.

explored for land cover classification [8-9]. PAN satellite imagery is a single band greyscale data with high spatial resolution (1m), which recently has been used to explore its various applications such as feature extraction of roads/highways and buildings [1], and land cover mapping [2]. Despite of finer spatial resolution, PAN satellite imagery has lower image classification comparing to the multi-spectral imagery because of a single band in PAN image.

Adoption of texture analysis is the recent trend for classifying PAN remote sensing imagery [8-9] so as to compensate the lack of the spectral information. Although improved classification accuracy has been demonstrated by using statistical textures, most recently, feature-based representation of images potentially with wavelet transform offers an attractive solution to this problem. Feature-based approach involves a number of features extracted from the raw image based on wavelet transform [10-11]. The features of objects can be described by the wavelet coefficients in low-pass and high-pass bands. Both features or objects and the spatial relationship among them play important roles in characterizing the objective contents, since they convey more semantic meaning. Although wavelet transform was applied to extract features from remote sensing image, the wavelet features cannot be properly organized so that the image contents cannot be comprehensively represented. In this study, by organizing wavelet coefficients into tree structure, object contents can be represented more comprehensively at various details which are very useful for image classification.

The neural network (NN) has been increasingly applied to remote sensing image classification in the last two decades, especially multi-spectral image classification [12-15]. Generally, Maximum Likelihood Classification (MLC) is a conventional method adopted for remote sensing imagery classification. The MLC method applies probability theory to the classification task. It computes all of the class probabilities for each raster cell and assigns the cell to the class with the highest probability value. However, for PAN image, the classification rate is too low to meet requirement because of only one band data contained in the PAN image. It has been indicated that supervised classification means based on neural network is superior to the statistical approach. Neural networks for adaptive processing of data structures are of paramount importance for structural pattern recognition and classification [16]. The main motivation of the adaptive processing is that neural networks are able to classify static information or temporal sequences and to perform automatic inferring or learning [17-19]. Sperduti and Starita proposed supervised neural networks for classification of data structures [20] where the approach is based on using generalized recursive neurons and the Backpropagation Through Structure (BPTS) algorithm [16,20]. Compared with conventional backpropagation neural network, BPTS is more reliable because of its structure algorithm and its common use in numerous research [21-26]. It is recently developed and proved to be effectively for digital image classification in the field of computer graphics. In the study, BPTS algorithm is applied to the PAN IKONOS satellite image for land cover classification together with the extracted wavelet coefficients.

The following section describes the adaptive processing of data structure. Extraction of wavelet features for experimental testing is described in section 3. Then, the conventional method for remote sensing image classification is introduced in section 4. After that, the design and analysis of experimental results are reported in sections 5 and 6, respectively. Conclusions of the research are drawn in the final section.

## 2    Backpropagation through Structure Algorithm

Neural Networks for adaptive processing of data structures are of paramount importance for structural pattern recognition and classification. The main motivation of this adaptive processing is that neural networks are able to classify static information or temporal sequences and to perform automatic inferring or learning [6-7]. Sperduti and Starita have proposed supervised neural networks for classification of data structures. This approach is based on using generalized recursive neurons and the BPTS algorithm, which is more reliable because of its structure algorithm. BPTS extends the time unfolding carried out by back-propagation through time in the case of sequences. The main idea of BPTS is depicted in Fig. 1. On the left an example of an input tree is shown. On the right, the state transition network $f(\ )$ is unrolled to match the topology of the input tree. The output network $g(\ )$ is attached to the replica of $f(\ )$ associated with the root. After recursion (1) is completed, the state vector $x(r)$ at the root $r$ contains an adaptive encoding of the whole tree. The encoding results are from a supervised learning mechanism. Forward propagation (dark arrows) is bottom up and follows equation (1). Backward propagation (light arrows) is employed to compute the gradients and proceeds from the root to leaves. The gradient contributions must be summed over all the replicas of the transition network to correctly implement weight sharing. A general framework of adaptive processing of data structures was introduced by Tsoi (1998) [6] and Cho et al. (2003) [8].

The recursive network for the structural processing is as follows:

$$x(v) = f(x(ch[v]), I(v))  \tag{1}$$

where $x(v)$ is the state vector associated with node $v$; $x(ch[v])$ is a vector obtained by concatenating the components of the state vectors contained in $v$'s children. $I$ is a generic input object and can be a directed acyclic graph with a super-source. $I(v)$ denotes the label attached to vertex $v$. The transition function $f$ is implemented by a feedforward neural network, according to the following scheme:

$$a_h(v) = \omega_{h,0} + \sum_{j=1}^{N} \omega_{hj} z_j (I(v)) + \sum_{k=1}^{m} \sum_{l=1}^{n} \omega_{hkl} x_l(ch_k[v])  \tag{2}$$

$$x_h(v) = \tanh(a_h(v)), \quad h = 1, \cdots, n  \tag{3}$$

where $x_h(v)$ is the $h$-th component of the state vector at vertex $v$; $ch_k[v]$ is the $k$-th child of $v$ and $\omega$ are weights. The output predictions are:

$$o = g(x(r)) = \frac{1}{1 + e^{-\alpha}}  \tag{4}$$

where,

$$\alpha = \omega_0 + \sum_{l=1}^{n} \omega_l x_l(r)  \tag{5}$$

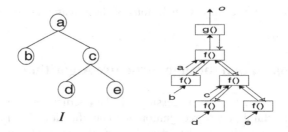

**Fig. 1.** An illustration of back propagation through structure

# 3    Description of Feature Set for Neural Network

The objective of supervised classification in remote sensing is to indentify and partition the pixels comprising the noisy image of an area according to its class, with the parameters in the model for pixel values estimated from training samples. In the paper, the spectral and frequency features are combined together as attributes for neural network. The spectral features can be obtained from the original image while the frequency features can be extracted after the image is decomposed by wavelet transform as described in the following sub-sections.

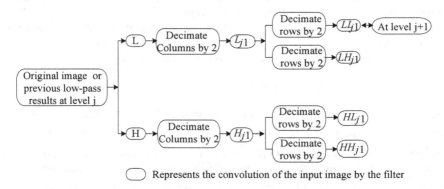

Represents the convolution of the input image by the filter

**Fig. 2.** Block chart of wavelet decomposition of an image

## 3.1    Decomposition of An Image by Wavelet Transform

With the goal of extraction of frequency features, a discrete wavelet transform can be applied to decompose a PAN image and it is summarized in Fig. 2. It is noticed the original image (or previous low-pass results at level $j$) can be decomposed into a low-pass band ($L_{j+1}$) and a high-pass band ($H_{j+1}$) through filters at level $j+1$. Through a low-pass filter and a high-pass filter, the low-pass band ($L_{j+1}$) is again decomposed into a low-pass band ($LL_{j+1}$) and a high-pass band ($LH_{j+1}$). The high-pass band ($H_{j+1}$) is decomposed into a low-pass band ($HL_{j+1}$) and a high-pass band ($HH_{j+1}$) through filters. Sub-bands $LH_{j+1}$, $HL_{j+1}$ and $HH_{j+1}$ respectively represent the characteristics of the image in the horizontal, vertical, and diagonal views. The local details of an image, such as edges of objects, are reserved in high bands. The basic energy of the image is reserved in $LL_{j+1}$. We will refer to the four sub-bands created at each level of the decomposition as $LL_{j+1}$ (Low/Low), $LH_{j+1}$ (Low/High), $HL_{j+1}$ (High/Low), and $HH_{j+1}$ (High/High). A PAN image decomposed by wavelet transform is shown in Fig. 3, which is represented by a tree structure.

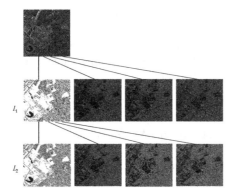

**Fig. 3.** A tree structural representation of a PAN imagery

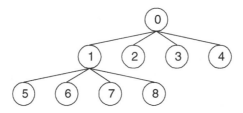

**Fig. 4.** An illustration of feature sets adopted in nodes: spatial features used in node 0; wavelet features used in nodes 1, 2, 3, 4, 5, 6, 7, and 8

## 3.2    Tree Structural Description of Features as Inputs for Neural Network

In order to conduct the image classification with BPTS NN, the image contents are represented by a tree structure as shown in Fig. 3. In the tree structure, totally, there are nine nodes illustrated in Fig. 4. Among them, the node numbered as 0 has four children nodes numbered as 1, 2, 3, and 4. The node 0 is corresponding to the original image while the four children nodes are corresponding to the four sub-bands, i.e. $LL_1$, $LH_1$, $HL_1$, and $HH_1$ sub-bands respectively. It is the similar case for node 1 that also has four children nodes, 5, 6, 7, and 8 corresponding to $LL_2$, $LH_2$, $HL_2$, and $HH_2$ sub-bands respectively.

The features obtained from spatial domain include a pixel's spectral intensity value (SIV), mean of all pixels' intensity (MEANI) in this class, standard deviation of intensity (STDI) corresponding to the mean. Another features extracted from frequency domain are a pixel's wavelet coefficient (PWC), mean of all pixels' wavelet coefficients (MEANWC) in this class, STD corresponding to the mean (STDWC). Therefore, each node contains three parameters. The node 0 includes spatial features, i.e. SIV, MEANI and STDI. The rest nodes, from node 1 to node 8, include wavelet features, i.e. PWC, MEANWC and STDWC as shown in Fig. 4.

## 4      Method of Conventional Image Classification

Generally, Maximum Likelihood Classification (MLC) is a conventional method adopted for remote sensing imagery classification. MLC applies probability theory to the classification task, in which all of the class probabilities for each raster cell are computed and the cell with the highest probability value is assigned to the class. The probability function can be expressed as:

$$p(X \mid w_i) = \frac{1}{(2\pi)^{\frac{n}{2}} |V_i|^{\frac{1}{2}}} \exp\left[ -\frac{1}{2}(X - M_i)^T V_i^{-1}(X - M_i) \right] \tag{6}$$

where $X$ is the input image vector, $M_i$ and $V_i$ are the mean vector and variance-covariance matrix for the land cover class $w_i$. After computing the posterior probability for each land cover class (i.e., $w_1$ to $w_n$ for $n$ classes), the pixel will be assigned with the specific land cover class $w_i$ which has the maximum posterior probability $P(X|w_i)$.

**Fig. 5.** A Panchromatic IKONOS image

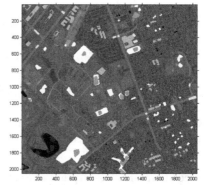

**Fig. 6.** Training and testing samples

## 5      Experimental Strategy

A PAN IKONOS satellite image with size of 2048 × 2048 pixels on Greater Toronto Area in Canada taken in 2006 is used in this study as shown in Fig. 5.

The experiment based on BPTS NN is implemented firstly. There are two parts in the experiments. One is training the neural network, and the other is testing. In order to make findings reliable, 2510 samples are selected as the training data set, among which 621 samples are for building, 297 for road, 1259 for vegetation and 333 for water bodies. The training samples for each class are displayed in the PAN image with different color shown in Fig. 6. The class of building is displayed in red, road in yellow, vegetation in green, and water bodies in blue.

Also, in Fig. 6, the testing samples for each class, which are not included in training data set, are displayed with different color, i.e. building in cyan, road in magenta, vegetation in white and water bodies in black. There are total 19498 samples selected in testing data set, among which 3568 samples are for building, 4787 for road, 8634 for vegetation and 2509 for water bodies. The numbers of pixels used for classification are listed in Table 1.

**Table 1.** Data Set for Classification

| Number of pixels in different data set | Class | | | | Total number |
|---|---|---|---|---|---|
| | Building | Road | Vegetation | Water body | |
| In training data set | 621 | 297 | 1259 | 333 | 2510 |
| In testing data set | 3568 | 4787 | 8634 | 2509 | 19498 |

# 6    Experimental Results and Analysis

## 6.1    Experimental Results

To conduct the classification by BPTS NN, the attributes of all the nodes in the tree are vectors of 3 inputs consisting of intensity, mean and STD described in Section III. In the investigation, a single hidden-layer is adopted for the neural classifier with 3 input nodes, 5 hidden nodes and 4 output nodes. The classification rate based on training data set is 100% and the rate based on testing data set is 99.91%. To compare with the conventional method, the classification rate based on MLC method for testing data set is 55.33% and the corresponding classified image is shown in Fig. 7, in which building is displayed in brown, vegetation in yellow, road in blue and water bodies in green. It is known that MLC assumes that the statistics for each class in each band are normally distributed and calculates the probability that a given pixel belongs to a specific class. Unless you select a probability threshold, all pixels are classified. Each pixel is assigned to the class that has the highest probability (i.e., the maximum likelihood). It is seen from the experimental results that the classification rates are significantly improved by the proposed method.

Since the number of hidden nodes affects the result, a series of different numbers of the hidden nodes are tested in the experiments in order to get the best result, i.e. 5, 7, 9, 11, 13 and 15. The corresponding classification rates by testing data set are listed in Table 2. 8. The variations of CPU time and mean square error (MSE) for training neural network are tabulated in Table 2, too. It is noticed that when the number of hidden nodes is 5, its classification rate on the testing set is 99.91%, which is next to the best (99.96%). Corresponding to the best rate of 99.96%, the number of hidden node is 15. It is much larger than 5, which incurs an increase in computational complexity and much more testing time. Therefore, the number 5 is a trade-off value for hidden nodes for getting a better classification rate.

**Fig. 7.** Classified image based on MLC method

**Table 2.** Classification Rates by Testing Data Set Using Different Numbers of Hidden Nodes

| Number of hidden nodes | Classification rate on testing data set by wavelet transform | Experimental data on training data set | |
|---|---|---|---|
| | | CPU time (second) | MSE |
| 5 | 99.91% | 51.20 | 0.0364 |
| 7 | 89.96% | 64.13 | 0.0279 |
| 9 | 99.48% | 49.78 | 0.0374 |
| 11 | 92.36% | 73.89 | 0.0377 |
| 13 | 99.19% | 62.69 | 0.0367 |
| 15 | 99.96% | 95.56 | 0.0319 |

## 7    Conclusion

In this paper, a method based on structural neural network is proposed for the PAN IKONOS image classification with adaptive processing of data structures. The image content is represented by a tree structural in which each node contains combined features, spatial and frequency features extracted from a PAN image.

BPTS algorithm and PAN IKONOS image decomposed by wavelet transform are first described. The spectral intensity and wavelet features are combined as the input attributes for BPTS algorithm. Based on this, 2510 pixels are used for training neural network. The classification accuracy rate up to 100% is reached. Using such a trained neural network, with another 19498 pixels, the testing is carried out and its classification rate arrives at 99.91% when the number of hidden nodes is 5. For comparison, another experiment based on MLC is conducted and the classification rate is 55.33% only. Clearly, better classification accuracy is achieved by our approach. From the experimental results and analysis, it is found a PAN satellite imagery classification using wavelet coefficients is reliable and effective.

**Acknowledgment.** This work was supported by the Special Fund for Basic Scientific Research of Central Colleges, Chang'an University (Project No. CHD2010ZD006), key project of Scientific and Technological Innovation, Chang'an University, opening fund of Key Laboratory of Western China's Mineral Resources and Geological Engineering, Ministry of Education, China and Geological Engineering, Ministry of Education and key project of the Ministry of Land & Resources, China (project No: 1212010914015), and the Discovery Grant from the Natural Sciences and Engineering Research Council of Canada (NSERC). W. Zou thanks Mr. Wai Yeung Yan who gives his contributions to the paper and Prof. Said Easa from Ryerson University who provided the panchromatic IKONOS satellite image for experimental testing.

# References

1. Cihlar, J.: Land cover mapping of large areas from satellites: Status and research priorities. Int. J. of Remote Sensing 21(6-7), 1093–1114 (2000)
2. Li, J., Song, C., Cao, L., Zhu, F., Meng, X., Wu, J.: Impacts of landscape structure on surface urban heat islands: A case study of Shanghai, China. Remote Sensing of Environment 115(12), 3249–3263 (2011)
3. Bhatta, B., Saraswati, S., Bandyopadhyay, D.: Urban sprawl measurement from remote sensing data. Applied Geography 30(4), 731–740 (2010)
4. Liang, B., Weng, Q.: Assessing urban environmental quality change of Indianapolis, United States, by the remote sensing and GIS integration. IEEE J. of Selected Topics in Applied Earth Observations and Remote Sensing 4(1), 43–55 (2011)
5. Berezowski, T., Chormański, J., Batelaan, O., Canters, F., Van de Voorde, T.: Impact of remotely sensed land-cover proportions on urban runoff prediction. International J. of Applied Earth Observation and Geoinformation 16, 54–65 (2012)
6. Zhou, Y.Y., Weng, Q., Gurney, K.R., Shuai, Y., Hu, X.: Estimation of the relationship between remotely sensed anthropogenic heat discharge and building energy use. ISPRS J. of Photogrammetry and Remote Sensing 67(1), 65–72 (2012)
7. Xian, G.: Analysis of impacts of urban land use and land cover on air quality in the Las Vegas region using remote sensing information and ground observations. Int. J. of Remote Sensing 28(24), 5427–5445 (2007)
8. Shaban, M.A., Dikshit, O.: Improvement of classification in urban areas by the use of textural features the case study of lucknow city, Uttar Pradesh. Int. J. of Remote Sensing 22(4), 565–593 (2001)
9. Zhang, Q., Wang, J., Gong, P., Shi, P.: Study of urban spatial patterns from SPOT panchromatic imagery using textural analysis. Int. J. of Remote Sensing 24(21), 4137–4160 (2003)
10. Yu, J., Ekström, M.: Multispectral image classification using wavelets: a simulation study. Pattern Recognition 36(4), 889–898 (2003)
11. Cohen, A., Ryan, R.D.: Wavelets and Multiscale Signal Processing. Chapman & Hall Press, London (1995)
12. Heermann, D., Khazenie, N.: Classification of multispectral remote sensing data using a back-propagation neural network. IEEE Trans. on Geoscience and Remote Sensing 30(1), 81–88 (1992)

13. Tzeng, Y.C., Chen, K.S.: A fuzzy neural network to SAR image classification. IEEE Trans. on Geoscience and Remote Sensing 36(1), 301–307 (1998)
14. Dai, X.L., Khorram, S.: Remotely sensed change detection based on artificial neural network. Photogrammetric Engineering & Remote Sensing 65(10), 1187–1194 (1999)
15. Yuan, H., Cynthia, F., Khorram, S.: An automated artificial neural network system for land use/land cover classification from Landsat TM imagery. Remote Sensing 1(3), 243–265 (2009)
16. Giles, C.L., Gori, M. (eds.): IIASS-EMFCSC-School 1997. LNCS (LNAI), vol. 1387. Springer, Heidelberg (1998)
17. Tsoi, A.C.: Adaptive Processing of Data Structures: An Expository Overview and Comments. Technical report, Faculty of Informatics, University of Wollongong, Australia (1998)
18. Frasconi, P., Gori, M., Sperduti, A.: A general framework for adaptive processing of data structures. IEEE Trans. on Neural Networks 9(5), 768–786 (1998)
19. Tsoi, A.C., Hangenbucnher, M.: Adaptive processing of data structures. Keynote Speech. In: Proc. of the 3rd Int. Conf. on Computational Intelligence and Multimedia Applications (ICCIMA 1999), New Delhi, India, September 23-26 (1999)
20. Sperduti, A., Starita, A.: Supervised neural networks for the classification of structures. IEEE Trans. on Neural Networks 8(3), 714–735 (1997)
21. Cho, S., Chi, Z., Wang, Z., Siu, W.: Efficient learning in adaptive processing of data structures. Neural Processing Letters 17(2), 175–190 (2003)
22. Zou, W., Lo, K.C., Chi, Z.: Structured-based neural network classification of images using wavelet coefficients. In: Wang, J., Yi, Z., Żurada, J.M., Lu, B.-L., Yin, H. (eds.) ISNN 2006. LNCS, vol. 3972, pp. 331–336. Springer, Heidelberg (2006)
23. Fu, H., Chi, Z., Feng, D., Zou, W., Lo, K.C., Zhao, X.: Pre-classification module for an all-season image retrieval system. In: Proc. of 2007 Int. Joint Conf. on Neural Networks, Orlando, Florida, USA, August 12-17 (2007)
24. Zou, W., Li, Y.: Image classification using wavelet coefficients in low-pass bands. In: Proc. of 2007 Int. Joint Conf. on Neural Networks, Orlando, Florida, USA, August 12-17 (2007)
25. Zou, W., Chi, Z., Lo, K.C.: Improvement of image classification using wavelet coefficients with structured based neural network. Int. J. of Neural Systems 18(3), 195–205 (2008)
26. Zou, W., Li, Y., Tang, A.: Effects of the number of hidden nodes used in a structured-based neural network on the reliability of image classification. Neutral Computing & Applications 18(3), 249–260 (2009)

# Local Feature Coding for Action Recognition Using RGB-D Camera

Mingyi Yuan, Huaping Liu, and Fuchun Sun

Department of Computer Science and Technology, Tsinghua University, P.R. China
State Key Laboratory of Intelligent Technology and Systems, Beijing, P.R. China

**Abstract.** In this paper, we perform activity recognition using an inexpensive RGBD sensor (Microsoft Kinect). The main contribution of this paper is that the conventional STIPs feature are extracted from not only the RGB image, but also the depth image. To the best knowledge of the authors, there is no work on extracting STIPs feature from the depth image. In addition, the extracted feature are combined under the framework of locality-constrained linear coding framework and the resulting algorithm achieves better results than state-of-the-art on public dataset.

**Keywords:** RGB-D, local feature, linear coding.

## 1 Introduction

Recognition of human actions has been an active research topic in computer vision. In the past decade, research has mainly focused on leaning and recognizing actions from video sequences captured from a single camera and rich literature can be found in a wide range of fields including computer vision, pattern recognition, machine leaning and signal processing. Recently, there are some approaches using local spatiotemporal descriptors together with bag-of-words model to represent the action, which have shown promising results. Since these approaches do not rely on some preprocessing techniques, e.g. foreground detection or body-part tracking, they are relatively robust to viewpoint, noise, changing background, and illumination. Most previous work on activity classification has focused on using 2D video. however, the use of 2D videos leads to relatively low accuracy even when there is no clutter.

On the other hand, the depth camera recently received great attentions from industrial to academic fields. Microsoft provided a popular mass-production consumer electronics device Kinect, which could provide a depth image except a color image. So some scholars begin to use Kinect for people detection and tracking[4][5].

In this work, we perform activity recognition using an inexpensive RGBD sensor (Microsoft Kinect). The main contribution of this paper is that the conventional STIPs feature are extracted from not only the RGB image, but also the depth image. To the best knowledge of the authors, there is no work on extracting STIPs feature from the depth image. In addition, the extracted feature are combined under the framework of locality-constrained linear coding framework and the resulting algorithm achieves better results than state-of-the-art on publicly dataset.

C. Guo, Z.-G. Hou, and Z. Zeng (Eds.): ISNN 2013, Part I, LNCS 7951, pp. 421–428, 2013.
© Springer-Verlag Berlin Heidelberg 2013

**Fig. 1.** STIPs features on RGB image(left) and depth image(right)

## 2    Feature Extraction

There are several schemes applied to time-consistent scene recognition issues. Some of them are statistics based approaches, such as HMM, LDM(Latent-Dynamic Discriminative Model) and so on. Others, like Space-Time Interest Points (STIPs), see the time axis as the other same dimension as the space axes and look for the features along the time axis as well. We prefer the latter schemes because the time parameter of the sample is essentially the same as the space parameters in the mathematics sense. Since we have plenty of reliable mathematics tools and feature construction schemes, the extensions of already existed feature schemes can be safely applied in such time-relevant problems. Meanwhile, those schemes can be naturally extent for more complex tasks as in our problem.

STIPs[3] is an extension of SIFT in 3-dim space and uses one of Harris3D, Cuboid or Hessian as the detector. For certain video, dense sampling is performed at regular positions and scales in space and time to get 3D patches. We perform sampling from 5 dimensions $x$, $y$, $t$, $\sigma$ and $\tau$ where $\sigma$ and $\tau$ are the spatial and temporal scale, respectively. Usually, the minimum size of a 3D patch is $18 \times 18 pixel^2$ and 10 frames. Spatial and temporal sampling are done with 50% overlap. Multi-scale patches are obtained by multiplying $\sigma$ and $\tau$ by a factor of $\sqrt{2}$ for consecutive scales. In total, there are 8 spatial and 2 temporal scales since the spatial scale is more important than the time scale. Different spatial and temporal scales are combined that each video is sampled 16 times with different $\sigma$ and $\tau$ parameters. The detector is applied in each video and locates interest points as well as the corresponding scale parameters. After that, we calculate the HOGHOF descriptors at those detected interest points and synthesize the sample features from them.

An important contribution of this paper is that the feature descriptors are extracted from both RGB image and depth image. For applying the STIPs detector and descriptor on the depth information, we scale the depth value from 16-bit unsigned integer to 8-bit unsigned integer by searching the maximum and minimum (above 0) of the depth value in the whole sample video, and transforming each depth pixel linearly as

$$d_{new} = \begin{cases} 0 & d = 0 \\ 255 \times \frac{d-v}{V-v} & d > 0 \end{cases} \tag{1}$$

where $V$ is the maximal depth of the video sample and $v$ is the minimal depth above 0 of the video sample. We save the matrices of $d_{new}$ as the gray type depth video and use it in the same way as the RGB one. For a typical video, there are about several hundred frames of RGB and depth image pairs and several thousand of STIPs descriptors detected. The STIPs descriptors are of 162 dimensional vector composed of 90-dim HOG[2] descriptor and 72-dim HOF descriptor. The HOG and HOF descriptors are computed at the detected interest point with the associated scale factors. The STIPs descriptor describes the local variation characters well in the $xy$ space as well as in the $t$ space. Fig.1 shows the features detected in one frame of both RGB image and depth image. The circles center at the interest points and the radius of the circles is proportional to the scale factor $\sigma$ of the interest point. It can be seen that the STIPs features on RGB image and depth image find different regions of the subjects because of the different pixel variation in the two types of data. In fact, the brightness of each pixel in the depth image has larger variation near the contour of the subjects, including the head, arms and legs. On the other hand, the variation of the brightness of the RGB image appears at the boundary of the texture of the subject. So the STIPs features in the RGB images disclose more detail characters of the subjects themselves while in the depth images they extract more characters from the overall contour of the subjects. In conclusion, both features are useful and equally important for classification.

## 3 Coding Approaches

A popular method for coding is the vector quantization (VQ) method, which solves the following constrained least square fitting problem:

$$\min_{\mathbf{C}} \sum_{i=1}^{M} ||\mathbf{x}_i - \mathbf{B}\mathbf{c}_i||_2^2 \text{ s.t. } ||\mathbf{c}_i||_0 = 1, ||\mathbf{c}_i||_1 = 1, \mathbf{c}_i \succeq 0, \forall i, \tag{2}$$

where $\mathbf{C} = [\mathbf{c}_1, \mathbf{c}_2, \cdots, \mathbf{c}_M]$ is the set of codes for $\mathbf{X} = [\mathbf{x}_1, \mathbf{x}_2, \cdots, \mathbf{x}_M]$. The cardinality constraint $||\mathbf{c}_i||_0 = 1$ means that there will be only one non-zero element in each code $\mathbf{c}_i$, corresponding to the quantization id of $\mathbf{x}_i$. The non-negative, $\ell_1$ constraint $||\mathbf{c}_i||_1 = 1$, $\mathbf{c}_i \succeq 0$ means that the coding weight for $\mathbf{x}_i$ is 1. In practice, the single non-zero element can be found by searching the nearest neighbor.

VQ provides effective way to treat an image as a collection of local descriptors, quantizes them into discrete "visual words", and then computes a compact histogram representation for traffic sign image classification. One disadvantage of the VQ is that it introduces significant quantization errors since only one element of the codebook is selected to represent the descriptor. To remedy this, one usually has to design nonlinear SVM as the classifier which try to compensate the quantization errors. However, using nonlinear kernels, the SVM has to pay a high training cost, including computation and storage. This means that it is difficult to scale up the algorithm to the case where $M$ is more than tens of thousands.

In our experiment, we use all the STIPs features (descriptors) detected in all the training samples as the candidate features for building the codebook. Then we apply the standard k-means approach with the nearest neighborhood scheme on the features to get a codebook with its elements being the cluster centers. For saving the computing time, and referring to the previous work[4], we set the number of the cluster to 128, 256 and 512 respectively. So the size of the codebook is a small quantity. The codebook consists of a 162-dim mean vectors or namely "words". We calculate the nearest word in the codebook to each descriptor vector and represent the vector with a new $k$-dim unit vector with the $i$-th component being 1 if the $i$-th word is nearest to the descriptor vector and the other components being 0. The new representation or namely quantified form of the descriptor vectors can be seen as a counter of the codebook words. So the sum pooling step which add all the quantified vectors together is actually the histogram of the codebook words in the video sample. Certainly, the pooling result should be normalized by the total number of the STIPs features detected in the sample for what we care about is the relative occurrence of each word in the codebook, not the absolute value. The normalized histogram cannot be classified with linear SVM scheme, but have to involve the $\chi^2$ kernel SVM.

Very recently, [6] pointed out that the sparse coding approach proposed by [7] neglected the relationship among codebook elements. Since locality is more essential than sparsity[8], [6] proposed a locality-constrained linear coding(LLC) approach. LLC incorporates locality constraint instead of the sparsity constraint, which leads to several favorable properties. Specifically, the LLC code uses the following criteria:

$$\min_{\mathbf{C}} \sum_{i=1}^{M} ||\mathbf{x}_i - \mathbf{Bc}_i||_2^2 + \lambda ||\mathbf{d}_i \odot \mathbf{c}_i||_2^2 \text{ s.t. } \mathbf{1}^T \mathbf{c}_i = 1, \forall i, \tag{3}$$

where $\odot$ denotes the element-wise multiplication, and $\mathbf{d}_i \in \mathbb{R}^K$ is the locality adaptor that gives different freedom for each basis vector proportional to its similarity to the input descriptor $\mathbf{x}_i$. Specifically,

$$\mathbf{d}_i = \exp(\frac{dist(\mathbf{x}_i, \mathbf{B})}{\sigma}) \tag{4}$$

where $dist(\mathbf{x}_i, \mathbf{B}) = [dist(\mathbf{x}_i, \mathbf{b}_1), \cdots, dist(\mathbf{x}_i, \mathbf{b}_K)]^T$, and $dist(\mathbf{x}_i, \mathbf{b}_j)$ is the Euclidean distance between $\mathbf{x}_i$ and $\mathbf{b}_j$. $\sigma$ is used for adjusting the weight decay speed for the locality adaptor. The constraint $\mathbf{1}^T \mathbf{c}_i = 1$ follows the shift-invariant requirements of the LLC code.

To solve (3), the parameters $\lambda$ and $\sigma$ should be determined, which is nontrivial task in practice. Noticing that LLC solution only has a few significant values, the authors of [6] develop an faster approximation of LLC to speedup the encoding process. Instead of solving (3), they simply use the $k$ ($k < d < K$) nearest neighbors of $\mathbf{x}_i$ as the local bases $\tilde{\mathbf{B}}_i$, and solve a much smaller linear system to get the coding vector.

The next step after the extraction of features is to refine the features of the video samples so that each sample can be represented distinctly by a single shorter length vector. According to the earlier work of Bingbing Ni in their paper[4], we apply a similar two-step scheme for the pre-processing. The first step is to merge the local features. In the referred literature, the Bag of Words (BoW) model is applied. The BoW model extracts multiple cluster centers of thousands of the features, then quantifies all the STIPs features into one of the cluster centers. The second step is to pool the features together as a single feature vector (usually with the sum pooling scheme) for that video sample.

And we perform the same process on the STIPs features of both the RGB and depth images respectively. The two result feature vectors are concatenated for classification stage. Since the BoW model generates a considerable loss during quantifying the feature into cluster center, the result feature has to be classified by nonlinear classifiers, thus increases the computational complexity in the predicting period when efficiency was more important. To overcome this demerit, we try LLC model[6] which uses a local linear coding scheme to represent the original local features by the linear combination of the $k$ nearest vectors in the codebook. The LLC features then undergo the max pooling process to generate a single feature vector for each sample and the ultimate sample features are trained and predicted with linear classifiers such as linear SVM, which has the advantage in the online part of the whole process over the BOW scheme.

## 4 Experimental Results

### 4.1 Data

We use the RGBD-HuDaAct[4] video database for testing our approach. The database is composed of 30 people playing daily activities of 13 categories including 12 named categories and 1 background category. Each sample is recorded in an indoor environment with a Microsoft Kinect sensor at a fixed position for a few seconds. The video sample consists of synchronized and calibrated RGB-D frame sequences, which contains in each frame a RGB image and a depth image, respectively from RGB and depth sensors in the same time spot. The RGB and depth images in each frame have been calibrated with a standard stereo-calibration method available in OpenCV so that the points with the same coordinate in RGB and depth images are corresponded. For each video sample, the dimension of RGB and depth images is $640 \times 480 pixel^2$. The color image is formatted as standard 24-bit RGB image, while the depth image is represented by a single channel 16-bit integer matrix, one element for each pixel.

Since the video database primarily faces the elder people's everyday living issues the categories of activities are defined in a large extent, which means each category may contain different concrete actions or different combinations of meta-actions (i.e to make a call one can stand up or sit down, with left hand or right hand) so that the recognition and classification tasks can be quite challenging. But more important, the database is the first public one involving both RGB image and depth data in one frame, which can be a justified platform to evaluate and test algorithms based on both RGB and depth data. The authors of this database proposed a baseline approach which trivially uses the depth information for classification. Our method directly fetches the features from the depth map, and merges the features from the RGB image and the depth map together. As we utilize the most part of depth information, our results can improve significantly over the baseline. Our experiments employ both BoW and LLC model on the HuDaAct database[4]. We randomly select 709 samples of 18 subjects out of total 1189 samples of 30 subjects (60% sampling) as the data set. Each subject contains videos of almost all the categories of activities the database defines, performed by certain user. We repeat the experiment for 18 times. For each round, we select one of the subjects as the testing data and the other 17 subjects as the training data. There is no duplication in the 18 splits. We record the testing results for each round and collect all the results to calculate the overall recognition rate at last. These settings are approaching to the settings by the baseline method in [4]. So we can compare our results with the baseline. The pre-processing scheme is the same as adopted by [4], which halves the dimensions of $x$, $y$ and $t$ by downsampling both in the image space and along the time axis. So the input videos are of $320 \times 240 pixel^2$ and 15 fps. We apply the original STIPs program available from [1] provided by the authors to get the HOGHOF descriptors.

## 4.2   Results

We perform the experiments on RGB videos, depth videos and both of them respectively. for each group of data, we apply both BoW and LLC model. The knn parameter of the LLC coding scheme, which decides the nonzero number of components in the LLC coded vector, is set to 10. The recognition rate is shown in Tab.1, where the results in bold font correspond to the best recognition accuracy. We can see the overall effect of experiment on depth samples are better than that on the RGB samples. And the experiment with combined features is largely improved over either of the separate ones. These facts suggest that: 1. depth sample have equal or better distinguished features for the action recognition issues; 2. combining two features can provide more comprehensive distinctions, thus overcome the limitation of the representation of one single feature.

**Table 1.** Recognition rates

| the codebook size k | 128 | 256 | 512 |
|---|---|---|---|
| BOW on RGB | 83% | 86% | 89% |
| BOW on Depth | 84% | 86% | 89% |
| BOW on RGB&Depth | **89%** | **90%** | **92%** |
| LLC on RGB | 74% | 84% | 86% |
| LLC on Depth | 78% | 86% | 88% |
| LLC on RGB&Depth | **83%** | **89%** | **91%** |

We also draw the confusion matrix of some groups of our experiment in Tab.2, Tab.3 and Tab.4, from which we can see the well classified categories and bad classified ones. The 3 tables response the results of experiments with the same parameters: the cluster number is set to 128, the coding method is nearest-neighbor and the SVM is linear. They are evaluated on different features respectively extracted from RGB, depth and both (combined feature) video samples. Almost for every category the combined feature has a better recognition rate over the result of single features. In some special cases (drink, sweep, etc.), the results have a prominent improvement. This indicates that the features from different data could be complement for each other, supporting critical distinctions from different aspects.

From the confusion matrixes we find that the *dress* and *undress* are very easily confused. This is not surprising because such sequences are indeed similar (see Fig.2 for some examples) if the temporal order is neglected. A direct remedy method is to incorporate the temporal order into the coding strategy. This remains our future work.

**Table 2.** Confusion matrix for RGB samples

| % | B | D | E | G | I | K | L | M | N | O | P | T | BG |
|---|---|---|---|---|---|---|---|---|---|---|---|---|---|
| B(bed) | **100.00** | 0.00 | 0.00 | 0.00 | 0.00 | 0.00 | 0.00 | 0.00 | 0.00 | 0.00 | 0.00 | 0.00 | 0.00 |
| D(dress) | 0.00 | **72.22** | 0.00 | 0.00 | 0.00 | 0.00 | 0.00 | 0.00 | 22.22 | 5.56 | 0.00 | 0.00 | 0.00 |
| E(exit) | 0.00 | 0.00 | **96.30** | 0.00 | 0.00 | 0.00 | 1.85 | 0.00 | 0.00 | 0.00 | 0.00 | 0.00 | 1.85 |
| G(get up) | 6.67 | 0.00 | 0.00 | **91.11** | 0.00 | 0.00 | 0.00 | 0.00 | 0.00 | 2.22 | 0.00 | 0.00 | 0.00 |
| I(sit down) | 0.00 | 0.00 | 0.00 | 0.00 | **75.93** | 9.26 | 0.00 | 0.00 | 5.56 | 0.00 | 3.70 | 1.85 | 3.70 |
| K(drink) | 0.00 | 0.00 | 0.00 | 0.00 | 0.00 | **77.78** | 0.00 | 11.11 | 0.00 | 0.00 | 11.11 | 0.00 | 0.00 |
| L(enter) | 0.00 | 0.00 | 0.00 | 0.00 | 0.00 | 0.00 | **98.15** | 0.00 | 1.85 | 0.00 | 0.00 | 0.00 | 0.00 |
| M(meal) | 0.00 | 0.00 | 0.00 | 0.00 | 0.00 | 3.70 | 0.00 | **81.48** | 0.00 | 0.00 | 14.81 | 0.00 | 0.00 |
| N(undress) | 0.00 | 18.52 | 0.00 | 0.00 | 0.00 | 0.00 | 0.00 | 0.00 | **72.22** | 5.56 | 0.00 | 1.85 | 1.85 |
| O(sweep) | 0.00 | 5.56 | 0.00 | 0.00 | 0.00 | 0.00 | 0.00 | 0.00 | 16.67 | **77.78** | 0.00 | 0.00 | 0.00 |
| P(phone) | 0.00 | 0.00 | 0.00 | 0.00 | 1.39 | 2.78 | 0.00 | 9.72 | 1.39 | 0.00 | **80.56** | 0.00 | 4.17 |
| T(stand up) | 0.00 | 0.00 | 0.00 | 1.92 | 7.69 | 0.00 | 0.00 | 0.00 | 0.00 | 0.00 | 0.00 | **90.38** | 0.00 |
| BG(background) | 0.00 | 2.22 | 0.00 | 0.00 | 0.00 | 2.22 | 0.00 | 0.00 | 6.67 | 11.11 | 6.67 | 2.22 | **68.89** |

**Table 3.** Confusion matrix for depth samples

| % | B | D | E | G | I | K | L | M | N | O | P | T | BG |
|---|---|---|---|---|---|---|---|---|---|---|---|---|---|
| B(bed) | **100.00** | 0.00 | 0.00 | 0.00 | 0.00 | 0.00 | 0.00 | 0.00 | 0.00 | 0.00 | 0.00 | 0.00 | 0.00 |
| D(dress) | 0.00 | **83.33** | 0.00 | 0.00 | 0.00 | 0.00 | 0.00 | 0.00 | 11.11 | 5.56 | 0.00 | 0.00 | 0.00 |
| E(exit) | 0.00 | 0.00 | **98.15** | 0.00 | 0.00 | 0.00 | 1.85 | 0.00 | 0.00 | 0.00 | 0.00 | 0.00 | 0.00 |
| G(get up) | 4.44 | 0.00 | 0.00 | **95.56** | 0.00 | 0.00 | 0.00 | 0.00 | 0.00 | 0.00 | 0.00 | 0.00 | 0.00 |
| I(sit down) | 0.00 | 1.85 | 0.00 | 0.00 | **79.63** | 9.26 | 0.00 | 0.00 | 1.85 | 0.00 | 1.85 | 1.85 | 3.70 |
| K(drink) | 0.00 | 0.00 | 0.00 | 0.00 | 2.78 | **80.56** | 0.00 | 9.72 | 1.39 | 0.00 | 5.56 | 0.00 | 0.00 |
| L(enter) | 0.00 | 0.00 | 5.56 | 0.00 | 0.00 | 0.00 | **94.44** | 0.00 | 0.00 | 0.00 | 0.00 | 0.00 | 0.00 |
| M(meal) | 0.00 | 0.00 | 0.00 | 0.00 | 0.00 | 11.11 | 0.00 | **83.33** | 0.00 | 0.00 | 5.56 | 0.00 | 0.00 |
| N(undress) | 0.00 | 16.67 | 0.00 | 0.00 | 0.00 | 0.00 | 0.00 | 0.00 | **74.07** | 3.70 | 0.00 | 3.70 | 1.85 |
| O(sweep) | 0.00 | 14.81 | 0.00 | 0.00 | 0.00 | 1.85 | 0.00 | 0.00 | 3.70 | **79.63** | 0.00 | 0.00 | 0.00 |
| P(phone) | 0.00 | 0.00 | 0.00 | 0.00 | 0.00 | 6.94 | 0.00 | 1.39 | 5.56 | 4.17 | **75.00** | 0.00 | 6.94 |
| T(stand up) | 0.00 | 0.00 | 0.00 | 0.00 | 0.00 | 9.62 | 0.00 | 0.00 | 1.92 | 0.00 | 3.85 | **84.62** | 0.00 |
| BG(background) | 0.00 | 8.89 | 0.00 | 0.00 | 0.00 | 2.22 | 0.00 | 2.22 | 2.22 | 4.44 | 11.11 | 0.00 | **68.89** |

**Table 4.** Confusion matrix for combined feature

| % | B | D | E | G | I | K | L | M | N | O | P | T | BG |
|---|---|---|---|---|---|---|---|---|---|---|---|---|---|
| B(bed) | **100.00** | 0.00 | 0.00 | 0.00 | 0.00 | 0.00 | 0.00 | 0.00 | 0.00 | 0.00 | 0.00 | 0.00 | 0.00 |
| D(dress) | 0.00 | **87.04** | 0.00 | 0.00 | 0.00 | 0.00 | 0.00 | 0.00 | 9.26 | 3.70 | 0.00 | 0.00 | 0.00 |
| E(exit) | 0.00 | 0.00 | **96.30** | 0.00 | 0.00 | 0.00 | 1.85 | 0.00 | 0.00 | 0.00 | 0.00 | 0.00 | 1.85 |
| G(get up) | 4.44 | 0.00 | 0.00 | **95.56** | 0.00 | 0.00 | 0.00 | 0.00 | 0.00 | 0.00 | 0.00 | 0.00 | 0.00 |
| I(sit down) | 0.00 | 0.00 | 0.00 | 0.00 | **83.33** | 5.56 | 0.00 | 0.00 | 3.70 | 0.00 | 1.85 | 1.85 | 3.70 |
| K(drink) | 0.00 | 0.00 | 0.00 | 0.00 | 0.00 | **87.50** | 0.00 | 9.72 | 0.00 | 0.00 | 2.78 | 0.00 | 0.00 |
| L(enter) | 0.00 | 0.00 | 0.00 | 0.00 | 0.00 | 0.00 | **100.00** | 0.00 | 0.00 | 0.00 | 0.00 | 0.00 | 0.00 |
| M(meal) | 0.00 | 0.00 | 0.00 | 0.00 | 0.00 | 7.41 | 0.00 | **87.04** | 0.00 | 0.00 | 5.56 | 0.00 | 0.00 |
| N(undress) | 0.00 | 12.96 | 0.00 | 0.00 | 0.00 | 0.00 | 0.00 | 0.00 | **81.48** | 3.70 | 0.00 | 0.00 | 1.85 |
| O(sweep) | 0.00 | 7.41 | 0.00 | 0.00 | 0.00 | 0.00 | 0.00 | 0.00 | 3.70 | **88.89** | 0.00 | 0.00 | 0.00 |
| P(phone) | 0.00 | 0.00 | 0.00 | 0.00 | 0.00 | 2.78 | 0.00 | 2.78 | 1.39 | 0.00 | **87.50** | 0.00 | 5.56 |
| T(stand up) | 0.00 | 0.00 | 0.00 | 0.00 | 0.00 | 7.69 | 0.00 | 0.00 | 1.92 | 0.00 | 1.92 | **88.46** | 0.00 |
| BG(background) | 0.00 | 2.22 | 0.00 | 0.00 | 0.00 | 2.22 | 0.00 | 0.00 | 4.44 | 8.89 | 8.89 | 0.00 | **73.33** |

**Fig. 2.** Snapshots of actions acted by one person. Top action: dressing. Bottom action: undressing.

## 5   Conclusion

Currently, the depth camera recently received great attentions from industrial to academic fields. In this paper, we perform activity recognition using an inexpensive RGBD sensor (Microsoft Kinect). The main contribution of this paper is that the conventional STIPs feature are extracted from not only the RGB image, but also the depth image. To the best knowledge of the authors, there is no work on extracting STIPs feature from the depth image. In addition, the extracted feature are combined under the framework of locality-constrained linear coding framework and the resulting algorithm achieves better results than state-of-the-art on publicly dataset.

**Acknowledgement.** This work is jointly supported by the National Key Project for Basic Research of China (2013CB329403), the National Natural Science Foundation of China (Grants No: 61075027, 91120011, 61210013) and the Tsinghua Selfinnovation Project (Grant No:20111081111).

## References

1. Public STIPs Binaries, http://www.di.ens.fr/~laptev/download.html
2. Dalal, N., Triggs, B.: Histograms of Oriented Gradients for Human Detection. In: IEEE Computer Society Conference on Computer Vision and Pattern Recognition, CVPR 2005, vol. 1, pp. 886–893. IEEE (2005)
3. Laptev, I.: On Space-Time Interest Points. International Journal of Computer Vision 64(2), 107–123 (2005)
4. Ni, B., Wang, G., Moulin, P.: RGBD-HuDaAct: A Color-Depth Video Database for Human Daily Activity Recognition. In: 2011 IEEE International Conference on Computer Vision Workshops (ICCV Workshops), pp. 1147–1153. IEEE (2011)
5. Sung, J., Ponce, C., Selman, B., Saxena, A.: Unstructured Human Activity Detection From RGBD Images. In: 2012 IEEE International Conference on Robotics and Automation (ICRA), pp. 842–849. IEEE (2012)
6. Wang, J., Yang, J., Yu, K., Lv, F., Huang, T., Gong, Y.: Locality-Constrained Linear Coding for Image Classification. In: 2010 IEEE Conference on Computer Vision and Pattern Recognition (CVPR), pp. 3360–3367. IEEE (2010)
7. Yang, J., Yu, K., Gong, Y., Huang, T.: Linear Spatial Pyramid Matching Using Sparse Coding for Image Classification. In: IEEE Conference on Computer Vision and Pattern Recognition, CVPR 2009, pp. 1794–1801. IEEE (2009)
8. Yu, K., Zhang, T., Gong, Y.: Nonlinear Learning Using Local Coordinate Coding. Advances in Neural Information Processing Systems 22, 2223–2231 (2009)

# Circular Projection for Pattern Recognition

Guangyi Chen[1,2], Tien Dai Bui[1], Sridhar Krishnan[2], and Shuling Dai[3]

[1] Department of Computer Science and Software Engineering, Concordia University,
Montreal, Quebec, Canada H3G 1M8
{guang_c,bui}@cse.concordia.ca
[2] Department of Electrical and Computer Engineering, Ryerson University, Toronto, Ontario,
Canada M5B 2K3
krishnan@ee.ryerson.ca
[3] State Key Lab. of Virtual Reality Technology and Systems, Beihang University,
ZipCode 100191, No 37, Xueyuan Rd., Haidian District, Beijing, P.R. China
sldai@yeah.net

**Abstract.** There are a number of methods that transform 2-D shapes into periodic 1-D signals so that faster recognition can be achieved. However, none of these methods are both noise-robust and scale invariant. In this paper, we propose a circular projection method for transforming 2-D shapes into periodic 1-D signals. We then apply a number of feature extraction methods to the 1-D signals. Our method is invariant to the translation, rotation and scaling of the 2-D shapes. Also, our method is robust to Gaussian white noise. In addition, it performs very well in terms of classification rates for a well-known shape dataset.

**Keywords:** Circular projection, Ramanujan Sums (RS), invariant features, pattern recognition, Gaussian white noise, fast Fourier transform (FFT).

## 1 Introduction

Feature extraction is an important step in pattern recognition. Among all existing feature extraction techniques, one very useful technique is to extract 1-D features from 2-D pattern images. This technique is very fast in terms of computing time when compared with those methods that extract 2-D features.

In this paper, we propose a novel method, called circular projection, which projects every radial line to one point by taking the sum of the pixel values of all pixels on this line. Our method is invariant to the translation, rotation, and scaling of the input 2-D shapes. Our experiments show that our proposed method is feasible for invariant pattern recognition in classifying a well-known shape dataset.

The organization of this paper is as follows. Section 2 proposes the circular projection of a pattern image, which is scale invariant. Section 3 briefly reviews five kinds of methods for rotation invariant feature extraction. Section 4 conducts some experiments in order to show that our proposed methods are feasible in practical pattern recognition applications. Finally, Section 5 concludes the paper.

C. Guo, Z.-G. Hou, and Z. Zeng (Eds.): ISNN 2013, Part I, LNCS 7951, pp. 429–436, 2013.

## 2    Circular Projection

Invariance and low dimension of features are of crucial significance in pattern recognition. A conventional way to obtain a smaller feature vector is to transform the 2-D pattern into a 1-D periodic signal. Existing techniques that project 2-D images to 1-D signals includes ring-projection, outer contour, line-moments, chain code, etc.

In this paper, we propose a new method, which we call it as circular projection. The method projects all pixels on the radial line that passes through the centroid with a slope of $\theta_i$ degrees. It is defined as:

$$C(\theta_i) = \sum_{j=-M}^{M} f(r_j, \theta_i)$$

where [-M, M] is the range of the sample number along the radius direction. We then normalize it as:

$$R(\theta_i) = \frac{N \times C(\theta_i)}{\sum_{i=0}^{N-1} C(\theta_i)}$$

where N is the total number of samples for the discretized variable $\theta_i$. It is easy to see that the circular projection converts the rotation of the 2-D image into circular shift in the rotation angle $\theta_i$. Also, the $R(\theta_i)$ is scale invariant due to the normalization. More importantly, the circular projection is very robust to Gaussian white noise because this kind of noise has zero mean. Fig. 1 shows two shapes and their circular projections. As we can see, the circular projection of the scaled image is the same as that of the original shape. This confirms that our proposed circular projection is scale invariant.

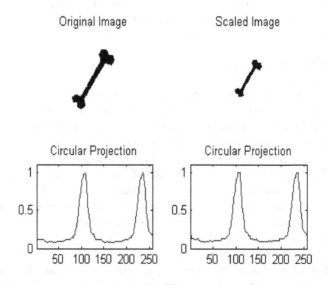

**Fig. 1.** The shape images and their circular projection

Our circular projection is both noise-robust and scale invariant. However, the line moments [4] are scale invariant but not noise robust; the ring-projection [5] is noise robust but not scale invariant; the outer-contour [7] is scale invariant but not noise robust.

# 3    Feature Extraction Methods

In this section, we study a number of feature extraction methods, which extract rotation invariant feature from the $R(\theta_i)$. This includes the TI wavelet FFT method, the FFT method, the TI Ramanujan Sums method, the orthonormal shell FFT method, and the dual tree complex wavelet (DTCWT) method.

## 3.1    TI Wavelet FFT

Translation invariant (TI) wavelet transform, also called cycle-spinning, was proposed by Coifman and Donoho [1] to suppress the Gibb's phenomenon at the discontinuous locations. The TI table is an N by D array, where D cannot be bigger than $\log_2(N)$. The $d$-th column has $N$ entries partitioned into $2^d$ boxes, each box having $N/2^d$ entries. The boxes correspond to the $2^d$ different collections of wavelet coefficients that can occur in the wavelet expansion at level $J$-$d$ under different shifts of the input vector. The fast Fourier transform (FFT) can be applied to every wavelet subband and then take the magnitude of the FFT coefficients. The resulting features are rotation invariant. The computational complexity of this method is O(N log(N)), where N is the length of the input signal..

## 3.2    Fourier Transform

The Fourier transform analyzes a signal in the time domain for its frequency content. The transform works by first transforming a function in the time domain into a function in the frequency domain. The signal can then be analysed for its frequency content because the Fourier coefficients of the transformed function represent the contribution of each sine and cosine function at each frequency.

The fast Fourier transform (FFT) can extract invariant features by taking the magnitude of the Fourier coefficients.

$$| FFT(R) | = | \sum_{i=0}^{N-1} R(\theta_i) e^{-iu\theta_i / N} |$$

where N is the dimension of the discretized variable $\theta_i$. It is easy to show that the magnitude of the FFT coefficients is invariant to the rotation angle $\theta_i$ because

$$| \sum_{i=0}^{N-1} R(\theta_i + \theta_0) e^{-iu\theta_i / N} | = | \sum_{i=0}^{N-1} R(\theta_i) e^{-iu\theta_i / N} |$$

The computational complexity of the FFT is O(N log(N)).

### 3.3    TI Ramanujan Sums (RS)

The Ramanujan Sums (RS) ([2] [3]) are the $m^{th}$ powers of $q^{th}$ primitive roots of unity, defined as

$$c_q(m) = \sum_{p=1;(p,q)=1}^{q} \exp(2i\pi \frac{p}{q} m)$$

where $(p,q) =1$ means that the greatest common divisor (GCD) is unity, i.e., $p$ and $q$ are co-primes. An alternate computation of RS can be given as

$$c_q(m) = \mu(\frac{q}{(q,m)}) \frac{\phi(q)}{\phi(\frac{q}{(q,m)})}$$

Let $q = \prod_i q_i^{\alpha_i}$ ($q_i$ prime).    Then, we have

$$\phi(q) = q \prod_i (1 - \frac{1}{q_i}) \cdot$$

The Mobius function $\mu(m)$ is defined as

$$\mu(m) = \begin{cases} 0, & \text{if } m \quad contains \ a \ square \ number \\ 1, & \text{if } m = 1 \\ (-1)^k, & \text{if } m \ is \ a \ product \ of \ k \ prime \ numbers \ . \end{cases}$$

We perform the convolution between $R(\theta_i)$ and $c_q(1:q)$, $q \in [1,Q]$, along the $r$ direction:

$$B(\theta_i, q) = conv(R(\theta_i), fliplr(c_q(1:q)))$$

where *fliplr()* is a function to flip its filter and *conv()* is the convolution operation. Let us circularly pad $q$ elements after the vector $B(\theta_i, q)$, and we obtain $\hat{B}(\theta_i, q)$. Now, convert the vector $\hat{B}(\theta_i, q)$ into a matrix with $q$ columns by the Matlab command:

$$D^q(q_i, \theta_j) = vec2mat(\hat{B}(\theta_j, q), q).$$

Next, take the sum along each column of $D^q(q_i, \theta_j)$:

$$SD(q, \theta_j) = \sum_i D^q(q_i, \theta_j) \cdot$$

Finally, we obtain the following invariant features:

$$r(q^*, \theta_j) = \max_q (| SD(q, \theta_j) | / \phi(q))$$

This method is our newly proposed method for rotation invariant pattern recognition, and it is the first method for pattern recognition by using the TI RS features.

### 3.4     Orthonormal Shell-FFT

Bui et al. developed the orthonormal shell-FFT transform [4] for invariant pattern recognition. The method applies the wavelet transform without downsampling to the $R(\theta_i)$ for a number of decomposition scales. We then apply the FFT to every wavelet scale and take the magnitude of the resulting FFT coefficients. The method is invariant to the rotation angle θ of the pattern image. We can apply a coarse-to-fine strategy to search the shape dataset and the matching is very quick because of the multiscale feature structure. The complexity of this method is O(N log(N)), where N is the length of the input signal.

### 3.5     Dual-Tree Complex Wavelet

The ordinary wavelet transform is not shift invariant due to its downsampling process. This has limited its applications in pattern recognition. The common way to overcome this limitation is to use the wavelet transform without downsampling. However, this is time-consuming.

The dual-tree complex wavelet transform (DTCWT), developed by Kingsbury [6], has the important shift invariant property, which has been used successfully in pattern recognition by Chen et al. ([7], [8], [9]). We can use this transform to extract rotation invariant features from the 1D signal $R(\theta_i)$. We take the FFT of every wavelet subband and calculate the magnitude of the FFT coefficients. In this way, we have extracted rotation invariant features. The dual-tree complex wavelets are very efficient in computational complexity $O(N)$, where $N$ is the number of samples. Since we apply the FFT to every DTCWT subband, the overall computational complexity of this method is O(N log(N)).

## 4     Experimental Results

In this paper, we conducted experiments for a well-known shape dataset [10]. The dataset is a subset of the MPEG-7 CE Shape-1 Part-B data set, which has 216 shapes in total. This dataset has 18 categories with 12 shapes in each category. The dataset is shown in Fig. 2. Each shape is matched against every other shape in the dataset. As there are 12 shapes in each category, up to 12 nearest neighbours are from the same category. We rate the performance based on the number of times the 12 nearest neighbours are in the same category. The shape classes are very distinct, but the data set shows substantial within-class variations.

Fig. 3 plots the correct recognition rates of the first 12 nearest neighbours in the same category for this dataset. We experimented with the TI wavelet-FFT features, the FFT features, the TI-RS features, the orthonormal shell-FFT features, and the DTCWT features. In general, the TI wavelet-FFT features are the best among the five feature extraction methods for this shape dataset. In addition, the TI RS method obtains relatively good classification rates, when compared with the FFT features, the orthonormal shell-FFT features and the DTCWT features.  It can be seen that our proposed methods are feasible in terms of recognition rates for classifying 2D shapes.

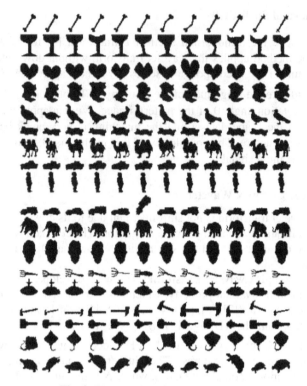

**Fig. 2.** The samples of the shape dataset

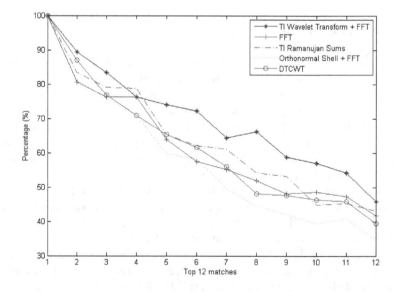

**Fig. 3.** The recognition rates of the first 12 shapes for five different methods

# 5    Conclusions and Future Work

In pattern recognition, the dimension of features is very important. A very large feature vector will require a huge amount of computation time and slow down the matching process. It is desirable to have a small feature vector while keeping the important features of the original pattern as much as possible.

In this paper, we have proposed the circular projection for invariant shape recognition. Our method is invariant to the rotation and scaling of the input shape images. Translation invariance can be achieved by moving the centroid to the image center. Experimental results have demonstrated the feasibility of the proposed method.

Future work will be done in the following ways. We would like to conduct experiments for other standard shape datasets in order to show if our proposed methods are good for practical pattern recognition applications. We would like also to develop other new algorithms for invariant pattern recognition in the fields of medical signal and image analysis.

**Acknowledgments.** This research was supported by the research grant from the Natural Science and Engineering Research Council of Canada (NSERC) and Beijing Municipal Science and Technology Plan: Z111100074811001.

# References

[1] Coifman, R.R., Donoho, D.L.: Translation invariant de-noising, Wavelets and Statistics. Springer Lecture Notes in Statistics, vol. 103, pp. 125–150. Springer, New York (1994)

[2] Ramanujan, R.: On certain trigonometric sums and their applications. Trans. Cambridge Philos. Soc. 22, 259–276 (1918)

[3] Sugavaneswaran, L., Xie, S., Umapathy, K., Krishnan, S.: Time-frequency analysis via Ramanujan sums. IEEE Signal Processing Letters 19(6), 352–355 (2012)

[4] Bui, T.D., Chen, G.Y., Feng, L.: An orthonormal-shell-Fourier descriptor for rapid matching of patterns in image database. International Journal of Pattern Recognition and Artificial Intelligence 15(8), 1213–1229 (2001)

[5] Tang, Y.Y., Li, B.F., Li, H., Lin, J.: Ring-projection-wavelet-fractal signatures: a novel approach to feature extraction. IEEE Transactions on Circuits and Systems II: Analog and Digital Signal Processing 45(8), 1130–1134 (1998)

[6] Kingsbury, N.G.: Complex wavelets for shift invariant analysis and filtering of signals. Journal of Applied and Computational Harmonic Analysis 10(3), 234–253 (2001)

[7] Chen, G.Y., Xie, W.F.: Contour-based feature extraction using dual-tree complex wavelets. International Journal of Pattern Recognition and Artificial Intelligence 21(7), 1233–1245 (2007)

[8] Chen, G.Y., Bui, T.D., Krzyzak, A.: Invariant pattern recognition using radon, dual-tree complex wavelet and Fourier transforms. Pattern Recognition 42(9), 2013–2019 (2009)

[9]  Chen, G.Y., Xie, W.F.: Pattern recognition with SVM and dual-tree complex wavelets. Image and Vision Computing 25(6), 960–966 (2007)

[10] Thakoor, N., Gao, J., Jung, S.: Hidden Markov Model-Based Weighted Likelihood Discriminant for 2-D Shape Classification. IEEE Transactions on Image Processing 16(11), 2707–2719 (2007)

# A Tensor Factorization Based Least Squares Support Tensor Machine for Classification

Xiaowei Yang, Bingqian Chen, and Jian Chen

Department of Mathematics, School of Sciences, South China University of Technology,
Guangzhou 510641, China
xwyang@scut.edu.cn

**Abstract.** In the fields of machine learning, image processing, and pattern recognition, the existing least squares support tensor machine for tensor classification involves a non-convex optimization problem and needs to be solved by the iterative technique. Obviously, it is very time-consuming and may suffer from local minima. In order to overcome these two shortcomings, in this paper, we present a tensor factorization based least squares support tensor machine (TFLS-STM) for tensor classification. In TFLS-STM, we combine the merits of least squares support vector machine (LS-SVM) and tensor rank-one decomposition. Theoretically, TFLS-STM is an extension of the linear LS-SVM to tensor patterns. When the input patterns are vectors, TFLS-STM degenerates into the standard linear LS-SVM. A set of experiments is conducted on six second-order face recognition datasets to illustrate the performance of TFLS-STM. The experimental results show that compared with the alternating projection LS-STM (APLS-STM) and LS-SVM, the training speed of TFLS-STM is faster than those of APLS-STM and LS-SVM. In term of testing accuracy, TFLS-STM is comparable with LS-SVM and is superiors to APLS-STM.

## 1 Introduction

There are two main topics of concern in the fields of machine learning, pattern recognition, computer vision and image processing: data representation and classifier design. In the past decades, numerous state-of-the-art classification algorithms have been proposed and have achieved great successes in many applications. Among these algorithms, the most prominent representative is support vector machines (SVMs) [1], which have been widely applied to text classification [2], image processing [3-6], and control [7-8]. Unfortunately, the SVM model is based on vector space and cannot directly deal with non-vector patterns. In real-world applications, image and video data are more naturally represented as second-order tensors or higher-order tensors. For example, grey level face images [9] are inherently represented as second-order tensors. Color images [10], gray-level video sequences [11], gait silhouette sequences [12-13], and hyperspectral cube [14] are commonly represented as third-order tensors. Color video sequences [15] are usually represented as fourth-order tensors. Although tensor patterns can be reshaped into vectors beforehand to meet the input requirement of SVM, several studies have indicated that this direct reshaping breaks the natural

C. Guo, Z.-G. Hou, and Z. Zeng (Eds.): ISNN 2013, Part I, LNCS 7951, pp. 437–446, 2013.
© Springer-Verlag Berlin Heidelberg 2013

structure and correlation in the original data [13], [16-17], and leads to the curse of dimensionality and small sample size (SSS) problems [18].

In the past ten years, researchers have mainly focused on data representation to address the above problems, such as tensor decomposition [19]-[21] and multilinear subspace learning [22]-[24]. Recently, several researchers [25]-[31] have suggested constructing multilinear models to extend the SVM learning framework to tensor patterns. In [25], Tao *et al.* presented a supervised tensor learning (STL) framework by applying a combination of the convex optimizations and multilinear operators, in which the weight parameter was decomposed into rank-one tensor. Based on the STL framework, Cai *et al.* [26] studied second-order tensor and presented a linear tensor least square classifier and support tensor machine (STM), Tao *et al.* [27] extended the classical linear SVM to general tensor patterns. Based on the SVM methodology within STL framework, Liu *et al.* [28] used the dimension-reduced tensors as input for video analysis. Kotsia *et al.* [29] adopted Tucker decomposition of the weight parameter instead of rank-one tensor to retain more structure information. Wolf *et al.* [30] proposed to minimize the rank of the weight parameter with the orthogonality constraints on the columns of the weight parameter instead of the classical maximum-margin criterion and Pirsiavash *et al.* [31] relaxed the orthogonality constraints to further improve the Wolf's method.

At present, the STL-based methods have two main drawbacks. On the one hand, it may suffer from local minima since the model is non-convex. On the other hand, for the non-convex optimization problems, one usually resorts to iterative techniques, which is very time-consuming. In this paper, we propose a tensor factorization based least squares support tensor machine (TFLS-STM) to overcome these two shortcomings. Firstly, we reformulate the linear least squares support vector machine (LS-SVM) model from multilinear algebra viewpoint based on alternating projection least squares support tensor machine (APLS-STM) [27] and obtain a tensor space model. The novel model works with the same principles as the linear LS-SVM but operates directly on tensor. Different from APLS-STM which derives alternating projection optimization procedure along each mode, the global optimal solution of the proposed model can be obtained by the optimization algorithms for LS-SVM. Secondly, we integrate the tensor rank-one decomposition into the model to assist the tensor inner product computation. There are two main reasons for doing this. One is that the original tensor inner product cannot exert its normal capability for capturing structural information of tensor objects because of the curse of dimensionality and SSS problems. The other is that the tensor rank-one decomposition is able to obtain more compact and meaningful representations of the tensor objects, which saves storage space and computational time. A set of experiments is conducted on six tensor classification datasets to examine the effectiveness and efficiency of TFLS-STM.

The rest of this paper is organized as follows. Section 2 covers some preliminaries including notation, basic definitions and a brief review of APLS-STM. In Section 3, TFLS-STM is proposed for tensor classification. The differences of TFLS-STM vs. LS-SVM and APLS-STM are also discussed in this section. The experimental results and analysis are presented in Section 4. Finally, Section 5 gives conclusions and future work.

## 2    Preliminaries

Before presenting our work, we first briefly introduce some notation and basic definitions used throughout this study, and then review the APLS-STM algorithm.

### 2.1    Notation and Basic Definitions

For convenience, we will follow the conventional notation and definitions in the areas of multilinear algebra, pattern recognition and signal processing [13], [32-34]. In this study, vectors are denoted by boldface lowercase letters, e.g., $\mathbf{a}$, matrices by boldface capital letters, e.g., $\mathbf{A}$, tensors by calligraphic letters, e.g., $\mathcal{A}$. Their elements are denoted by indices, which typically range from 1 to the capital letter of the index, e.g., $n = 1, \cdots, N$.

**Definition 1 (Tensor)** A tensor, also known as $N$ th$-$ order tensor, multidimensional array, $N-$ way or $N-$ mode array, is an element of the tensor product of $N$ vector spaces, which is a higher-order generalization of a vector (first-order tensor) and a matrix (second-order tensor), denoted as $\mathcal{A} \in R^{I_1 \times I_2 \times \cdots \times I_N}$, where $N$ is the order of $\mathcal{A}$, also called ways or modes. The element of $\mathcal{A}$ is denoted by $a_{i_1, i_2, \cdots, i_N}, 1 \leq i_n \leq I_n, 1 \leq n \leq N$.

**Definition 2 (Tensor product or Outer product)** The tensor product $\mathcal{X} \circ \mathcal{Y}$ of a tensor $\mathcal{X} \in R^{I_1 \times I_2 \times \cdots \times I_N}$ and another tensor $\mathcal{Y} \in R^{I_1' \times I_2' \times \cdots \times I_M'}$ is defined by

$$(\mathcal{X} \circ \mathcal{Y})_{i_1 i_2 \cdots i_N i_1' i_2' \cdots i_M'} = x_{i_1 i_2 \cdots i_N} \, y_{i_1' i_2' \cdots i_M'}. \tag{1}$$

for all values of the indices.

**Definition 3 (Inner product)** The inner product of two same-sized tensors $\mathcal{X}, \mathcal{Y} \in R^{I_1 \times I_2 \times \cdots \times I_N}$ is defined as the sum of the products of their entries, i.e.,

$$\langle \mathcal{X}, \mathcal{Y} \rangle = \sum_{i_1=1}^{I_1} \sum_{i_2=1}^{I_2} \cdots \sum_{i_N=1}^{I_N} x_{i_1 i_2 \cdots i_N} \, y_{i_1 i_2 \cdots i_N}. \tag{2}$$

**Definition 4 ($n-$ mode product)** The $n-$ mode product of a tensor $\mathcal{A} \in R^{I_1 \times I_2 \times \cdots \times I_N}$ and a matrix $\mathbf{U} \in R^{J_n \times I_n}$, denoted by $\mathcal{A} \times_n \mathbf{U}$, is a tensor in $R^{I_1 \times I_2 \times \cdots \times I_{n-1} \times J_n \times I_{n+1} \times \cdots \times I_N}$ given by

$$(\mathcal{A} \times_n \mathbf{U})_{i_1 i_2 \times \cdots i_{n-1} \times j_n \times i_{n+1} \times \cdots \times i_N} = \sum_{i_n} a_{i_1, i_2, \cdots, i_N} u_{j_n, i_n}. \tag{3}$$

for all index values.

**Remark.** Given a tensor $\mathcal{A} \in R^{I_1 \times I_2 \times \cdots \times I_N}$ and a sequences of matrices $J_n < I_n, n = 1, \cdots N$. The projection of $\mathcal{A}$ onto the tensor subspace $R^{J_1 \times J_2 \times \cdots \times J_N}$ is defined as $\mathcal{A} \times_1 \mathbf{U}^{(1)} \times_2 \mathbf{U}^{(2)} \times \cdots \times_N \mathbf{U}^{(N)}$. Given the tensor $\mathcal{A} \in R^{I_1 \times I_2 \times \cdots \times I_N}$, the matrices $F \in R^{J_n \times I_n}$ and $G \in R^{J_m \times I_m}$, one has $(\mathcal{A} \times_n \mathbf{F}) \times_m \mathbf{G} = (\mathcal{A} \times_m \mathbf{G}) \times_n \mathbf{F} = \mathcal{A} \times_n \mathbf{F} \times_m \mathbf{G}$.

**Definition 5 (Frobenius Norm)** The Frobenius norm of a tensor $\mathcal{A} \in R^{I_1 \times I_2 \times \cdots \times I_N}$ is the square root of the sum of the squares of all its elements, i.e.

$$\|\mathcal{A}\|_F = \sqrt{\langle \mathcal{A}, \mathcal{A} \rangle} = \sqrt{\sum_{i_1=1}^{I_1} \sum_{i_2=1}^{I_2} \cdots \sum_{i_N=1}^{I_N} a_{i_1 \times i_2 \times \cdots \times i_N}^2} . \tag{4}$$

**Definition 6 (Tensor rank-one decomposition)** Let $\mathcal{A} \in R^{I_1 \times I_2 \times \cdots \times I_N}$ be a tensor. If it can be written as

$$\mathcal{A} = \sum_{r=1}^{R} \mathbf{u}_r^{(1)} \circ \mathbf{u}_r^{(2)} \circ \cdots \circ \mathbf{u}_r^{(N)} = \sum_{r=1}^{R} \prod_{n=1}^{N} \circ \mathbf{u}_r^{(n)} , \tag{5}$$

where $\mathbf{u}_r^{(n)} \in R^{I_n}$, we call (5) tensor rank-one decomposition of $\mathcal{A}$ with length $R$. Particularly, if $R = 1$, it is called rank-1 tensor. If $R$ is the minimum number of rank-1 tensors that yield $\mathcal{A}$ in a linear combination, $R$ is defined as the rank of $\mathcal{A}$, denoted by $R = rank(\mathcal{A})$. Moreover, if $\mathbf{u}_i^{(n)}$ and $\mathbf{u}_j^{(n)}$ are mutually orthonormal for all $i \neq j, 1 \leq i, j \leq R, n = 1, \cdots, N$, the formula (5) is often called rank-$R$ approximation [33], [35].

## 2.2    Alternating Projection Least Squares Support Tensor Machine

Considering a training set of $l$ pairs of samples $\{\mathcal{X}_i, y_i\}_{i=1}^{l}$ for tensor binary classification problem, where $\mathcal{X}_i \in R^{I_1 \times I_2 \times \cdots \times I_N}$ are the input data and $y_i \in \{-1, +1\}$ are the corresponding class labels of $\mathcal{X}_i$, APLS-STM is composed of $N$ quadratic programming (QP) problems with equality constraints, and the $n$-th QP problem can be described in the following [27]:

$$\min_{\mathbf{w}^{(n)}, b^{(n)}, \xi^{(n)}} J(\mathbf{w}^{(n)}, b^{(n)}, \xi^{(n)}) = \frac{1}{2} \|\mathbf{w}^{(n)}\|_F^2 \prod_{1 \leq i \leq N}^{i \neq n} \|\mathbf{w}^{(i)}\|_F^2 + \frac{\gamma}{2} \sum_{j=1}^{l} (\xi_j^{(n)})^2, \tag{6}$$

s. t.

$$y_j \left( (\mathbf{w}^{(n)})^T \left( \mathcal{X}_j \prod_{1 \leq i \leq N}^{i \neq n} \times_i \mathbf{w}^{(i)} \right) + b^{(n)} \right) = 1 - \xi_j^{(n)}, j = 1, 2, \ldots, l. \tag{7}$$

Where $\mathbf{w}^{(n)}$ is the normal vector (or weight vector) of the hyperplane in the $n$-th mode space, $b^{(n)}$ is the bias, $\xi_j^{(n)}$ is the error of the $j$-th training sample corresponding to $\mathbf{w}^{(n)}$, $\gamma$ is the trade-off between the classification margin and misclassification error.

Obviously, each optimization problem defined in (6)-(7) is the standard linear LS-SVM. Mathematically, on the one hand, the optimization problem composing of these $N$ optimization models has no closed-form solution, we need to use the alternating projection algorithm to solve it; on the other hand, this optimization problem is non-convex, we cannot obtain its global optimal solution.

## 3    The Tensor Factorization Based Least Squares Support Tensor Machine

In this section, we first introduce TFLS-STM for tensor classification, and then analyze the differences of TFLS-STM vs. LS-SVM and APLS-STM.

### 3.1    TFLS-STM for Binary Classification

Let $\mathcal{W} = \mathbf{w}^{(1)} \circ \mathbf{w}^{(2)} \circ \cdots \circ \mathbf{w}^{(N)}$ and $\xi_j^2 = \max_{n=1,2,\ldots,N} \left\{ \left( \xi_j^{(n)} \right)^2 \right\}$. From the definitions of the outer product and the Frobenius norm of tensors, we know that

$$\left\| \mathcal{W} \right\|_F^2 = \prod_{n=1}^{N} \left\| \mathbf{w}^{(n)} \right\|_F^2 . \tag{8}$$

From the definitions of the $n-$mode product and the inner product of tensors, we have

$$(\mathbf{w}^{(n)})^T \left( \mathcal{X}_j \prod_{1 \le i \le N}^{i \ne n} \times_i \mathbf{w}^{(i)} \right)$$

$$= (\mathbf{w}^{(n)})^T \left( \mathcal{X}_j \times_1 \mathbf{w}^{(1)} \times_2 \mathbf{w}^{(2)} \times .. \times_{(n-1)} \mathbf{w}^{(n-1)} \times_{(n+1)} \mathbf{w}^{(n+1)} \times .. \times_N \mathbf{w}^{(N)} \right)$$

$$= \left( \mathcal{X}_j \times_1 \mathbf{w}^{(1)} \times_2 \mathbf{w}^{(2)} \times ... \times_{(n-1)} \mathbf{w}^{(n-1)} \times_n \mathbf{w}^{(n)} \times_{(n+1)} \mathbf{w}^{(n+1)} \times ... \times_N \mathbf{w}^{(N)} \right) = \left\langle \mathcal{W}, \mathcal{X}_j \right\rangle . \tag{9}$$

Based on (8) and (9), the $N$ QP problems arising in APLS-STM can be transformed into the following optimization problem in tensor space:

$$\min_{\mathcal{W}, b, \xi} J(\mathcal{W}, b, \xi) = \frac{1}{2} \left\| \mathcal{W} \right\|_F^2 + \frac{\gamma}{2} \sum_{j=1}^{l} \xi_j^2 , \tag{10}$$

s. t.

$$y_j \left( \left\langle \mathcal{W}, \mathcal{X}_j \right\rangle + b \right) = 1 - \xi_j . \tag{11}$$

It is obvious that when the input samples $\mathcal{X}_i$ are vectors, the optimization model (10)-(11) degenerates into the linear LS-SVM. Moreover, if we adopt the original input tensors to compute the tensor inner product $\langle \mathcal{X}_i, \mathcal{X}_j \rangle$, then the optimal solutions of (10)-(11) are the same as the linear LS-SVM.

Considering that the tensor rank-one decomposition can obtain more compact and meaningful representations of the tensor objects, we use tensor rank-one decomposition to assist the inner product computation. Let the rank-one decompositions of $\mathcal{X}_i$

and $\mathcal{X}_j$ be $\mathcal{X}_i = \sum_{r=1}^{R} \mathbf{x}_{ir}^{(1)} \circ \mathbf{x}_{ir}^{(2)} \circ \cdots \circ \mathbf{x}_{ir}^{(N)}$ and $\mathcal{X}_j = \sum_{r=1}^{R} \mathbf{x}_{jr}^{(1)} \circ \mathbf{x}_{jr}^{(2)} \circ \cdots \circ \mathbf{x}_{jr}^{(N)}$ respectively, then

the inner product of $\mathcal{X}_i$ and $\mathcal{X}_j$ is calculated as follow:

$$\langle \mathcal{X}_i, \mathcal{X}_j \rangle = \left\langle \sum_{r=1}^{R} \mathbf{x}_{ir}^{(1)} \circ \mathbf{x}_{ir}^{(2)} \circ \cdots \circ \mathbf{x}_{ir}^{(N)}, \sum_{r=1}^{R} \mathbf{x}_{jr}^{(1)} \circ \mathbf{x}_{jr}^{(2)} \circ \cdots \circ \mathbf{x}_{jr}^{(N)} \right\rangle$$

$$= \sum_{p=1}^{R} \sum_{q=1}^{R} \langle \mathbf{x}_{ip}^{(1)}, \mathbf{x}_{jq}^{(1)} \rangle \langle \mathbf{x}_{ip}^{(2)}, \mathbf{x}_{jq}^{(2)} \rangle \cdots \langle \mathbf{x}_{ip}^{(N)}, \mathbf{x}_{jq}^{(N)} \rangle \qquad (12)$$

Based on (12), we call the optimization model (10)-(11) TFLS-STM. It can be solved by sequential minimal optimization algorithm [36]. The class label of a testing example $\mathcal{X}$ is predicted as follow:

$$y(\mathcal{X}) = \text{sgn}(\sum_{j=1}^{l} \sum_{p=1}^{R} \sum_{q=1}^{R} \alpha_j y_j \prod_{n=1}^{N} \langle \mathbf{x}_{jp}^{(n)}, \mathbf{x}_q^{(n)} \rangle + b) \qquad (13)$$

Where $\mathbf{x}_{jp}^{(n)}$ and $\mathbf{x}_p^{(n)}$ are the elements of the rank-one decompositions of $\mathcal{X}_j$ and $\mathcal{X}$ respectively.

### 3.2    Analysis of the TFLS-STM Algorithm

In this section, we discuss the differences of TFLS-STM vs. LS-SVM and APLS-STM as follows:

(1) Naturally APLS-STM is multilinear SVM and constructs $N$ different hyperplanes in $N$ mode spaces. TFLS-STM is linear STM and constructs a hyperplane in the tensor space.

(2) The optimization problem (6)-(7) needs to be solved iteratively in APLS-STM while the optimization problem (10)-(11) of TFLS-STM only needs to be solved once.

(3) For the same training sample, the slack variables $\xi_j^{(n)}$ obtained by APLS-STM are often unequal in different mode spaces while TFLS-STM only obtains one slack value in the tensor space. In addition, for different mode spaces, the support vectors in one mode space may no longer be support vectors in another mode space in APLS-STM.

(4) For the weight parameter $\mathcal{W}$, APLS-STM only obtains its rank-one tensor while TFLS-STM obtains its more accurate presentation.

(5) Based on the previous work [37], we know that the computational complexity of LS-SVM is $O(l^3 \prod_{n=1}^{N} I_n)$, thus the computational complexity of APLS-STM is $O(l^3 NT \prod_{n=1}^{N} I_n)$ where $T$ is the loop number, and the computational complexity of TFLS-STM is $O(l^3 R^2 \sum_{n=1}^{N} I_n)$, which indicates that TFLS-STM is more efficient than LS-SVM and APLS-STM.

## 4    The Experimental Results and Analysis

In this section, six face datasets Yale32x32, Yale64x64, C7, C9, C27, and C29, which come from http://www.zjucadcg.cn/dengcai/Data/FaceData.html, are used to evaluate the performance of TFLS-STM. The detailed information about these six datasets is listed in Table 1. In our experiments, we compare LS-SVM with Gaussian kernel and APLS-STM with TFLS-STM. The optimal hyperparameters $(\sigma, \gamma)$ and the optimal rank parameter $R$ are found by the grid search and ten-fold cross validation strategy, where $\sigma \in \{2^{-4}, 2^{-3}, 2^{-2}, \cdots, 2^9\}$ , $\gamma \in \{2^0, 2^1, 2^2, \cdots, 2^9\}$ , $R \in \{3,4,5,6,7,8,9,10\}$ , the threshold parameter $\varepsilon$ in APLS-STM is set to $10^{-3}$. All the programs are written in C++ and compiled using the Microsoft Visual Studio 2008 compiler. All the experiments are conducted on a computer with Intel(R) Core(TM) i7-3770 3.4GHz processor and 16GB RAM memory running Microsoft Windows 7 SP1 V8.8. The optimal parameters, the corresponding average testing accuracy and the average training time obtained by TFLS-STM, LS-SVM and APLS-STM are reported in Table 2.

**Table 1.** The Detailed Information of Six Experimental Datasets

| Datasets | Number of Samples | Number of Classes | Size |
|----------|------------------|-------------------|------|
| Yale32x32 | 165 | 15 | 32x32 |
| Yale64x64 | 165 | 15 | 64x64 |
| C7 | 1629 | 68 | 64x64 |
| C9 | 1632 | 68 | 64x64 |
| C27 | 3329 | 68 | 64x64 |
| C29 | 1632 | 68 | 64x64 |

From Table 2, we have the following observations:
(1) In terms of testing accuracy, TFLS-STM is comparable with LS-SVM and outperforms APLS-STM. The reason is that the solutions obtained by TFLS-STM and LS-SVM are the global optimal solutions while APLS-STM perhaps gives the local solutions.

(2) In terms of training time, TFLS-STM is significantly superiors to LS-SVM and APLS-STM on all the datasets, which is identificial to the previous theoretical analysis. For example, the training speed of TFLS-STM on Yale64x64 is about 11 times and 726 times faster than LS-SVM and APLS-STM, respectively. The main reason is that TFLS-STM uses tensor rank-one decomposition to calculate the tensor inner product and save storage space.

**Table 2.** Comparison of the Results of TFLS-STM, LS-SVM and APLS-STM on Six Experimental Datasets

| Learning Machines | Datasets | $R$ | $\gamma$ | $\sigma$ | Testing Accuracy (%) | Training Time (Seconds) |
|---|---|---|---|---|---|---|
| LS-SVM | | -- | 256 | 128 | 77.00 | 0.16 |
| APLS-STM | Yale32x32 | -- | 1 | -- | 73.67 | 2.86 |
| TFLS-STM | | 4 | 512 | -- | 78.33 | 0.05 |
| LS-SVM | | -- | 512 | 64 | 84.33 | 0.59 |
| APLS-STM | Yale64x64 | -- | 8 | -- | 82.67 | 36.30 |
| TFLS-STM | | 3 | 2 | -- | 85.67 | 0.05 |
| LS-SVM | | -- | 512 | 32 | 96.32 | 182.67 |
| APLS-STM | C7 | -- | 1 | -- | 94.95 | 2138.24 |
| TFLS-STM | | 7 | 2 | -- | 96.20 | 76.52 |
| LS-SVM | | -- | 512 | 8 | 97.43 | 183.65 |
| APLS-STM | C9 | -- | 1 | -- | 96.05 | 2148.31 |
| TFLS-STM | | 4 | 1 | -- | 96.96 | 24.73 |
| LS-SVM | | -- | 512 | 16 | 96.62 | 183.21 |
| APLS-STM | C27 | -- | 2 | -- | 94.83 | 2753.91 |
| TFLS-STM | | 5 | 1 | -- | 96.10 | 41.71 |
| LS-SVM | | -- | 512 | 16 | 96.64 | 183.11 |
| APLS-STM | C29 | -- | 1 | -- | 94.78 | 2363.10 |
| TFLS-STM | | 6 | 1 | -- | 96.03 | 59.97 |

# 5    Conclusions and Future Work

In this paper, TFLS-STM has been presented for tensor classification. In TFLS-STM, the linear LS-SVM model is reformulated from multilinear algebra viewpoint and can operates directly on tensor. Furthermore, the proposed model uses more compact $R$ rank-one tensors as input data instead of the original tensors, which makes TFLS-STM have strong capabilities for capturing essential information from tensor objects and saving storage space and computational time. The experimental results show that in terms of training speed, TFLS-STM is significantly faster than LS-SVM and APLS-STM. As for the testing accuracy, TFLS-STM is comparable with LS-SVM and is superiors to APLS-STM.

In future work, we will investigate the reconstruction techniques of tensor data so that TFLS-STM can handle high-dimensional vector data more efficiently. Another interesting topic would be to design some tensor kernel so as to generalize TFLS-STM to nonlinear case. Further study on this topic will also include many applications of TFLS-STM in real-world tensor classification.

**Acknowledgements.** This work was supported by the National Natural Science Foundation of China under Grant No. 61273295.

# References

1. Vapnik, V.: The Nature of Statistical Learning Theory. Springer (1995)
2. Mitra, V., Wang, C.J., Banerjee, S.: Text Classification: A Least Square Support Vector Machine Approach. Appl. Soft Comput. 7, 908–914 (2007)
3. Luts, J., Heerschap, A., Suykens, J.A.K., Van Huffel, S.: A Combined MRI and MRSI Based Multiclass System for Brain Tumour Recognition Using LS-SVMs with Class Probabilities and Feature Selection. Artif. Intell. Med. 40, 87–102 (2007)
4. Li, J., Allinsion, N., Tao, D., Li, X.: Multitraining Support Vector Machine for Image Retrieval. IEEE T. Image Process. 15, 3597–3601 (2006)
5. Bovolo, F., Bruzzone, L., Carlin, L.: A Novel Technique for Subpixel Image Classification Based on Support Vector Machine. IEEE T. Image Process. 19, 2983–2999 (2010)
6. Adankon, M.M., Cheriet, M.: Model Selection for the LS-SVM Application to Handwriting Recognition. Pattern Recogn. 42, 3264–3270 (2009)
7. Suykens, J.A.K., Vandewalle, J., De Moor, B.: Optimal Control by Least Squares Support Vector Machines. Neural Networks 14, 23–35 (2001)
8. Pahasa, J., Ngamroo, I.: A Heuristic Training-based Least Squares Support Vector Machines for Power System Stabilization by SMES. Expert Syst. Appl. 38, 13987–13993 (2011)
9. Felsberg, M.: Low-Level Image Processing with the Structure Multivector. PhD thesis, Institute of Computer Science and Applied Mathematics Christian-Albrechts-University of Kiel, TR no. 0203 (2002)
10. Plataniotis, K., Venetsanopoulos, A.: Color Image Processing and Applications. Springer, Berlin (2000)
11. Negi, P., Labate, D.: 3-D Discrete Shearlet Transform and Video Processing. IEEE T. Image Process. 21, 2944–2954 (2012)
12. Sarkar, S., Phillips, P.J., Liu, Z.Y., Robledo, I., Grother, P., Bowyer, K.: The Human ID Gait Challenge Problem: Data Sets, Performance, and Analysis. IEEE T. Pattern Anal. 27, 162–177 (2005)
13. Lu, H., Plataniotis, K.N., Venetsanopoulos, A.N.: MPCA: Multilinear Principal Component Analysis of Tensor Objects. IEEE T. Neural Network 19, 18–39 (2008)
14. Renard, N., Bourennane, S.: Dimensionality Reduction Based on Tensor Modeling for Classification Methods. IEEE T. Geosci. and Remote 47, 1123–1131 (2009)
15. Kim, M., Jeon, J., Kwak, J., Lee, M., Ahn, C.: Moving Object Segmentation in Video Sequences by User Interaction and Automatic Object Tracking. Image Vision Comput 19, 245–260 (2001)
16. Geng, X., Smith-Miles, K., Zhou, Z.H., Wang, L.: Face Image Modeling by Multilinear Subspace Analysis with Missing Values. IEEE T. Syst., Man, Cy. B 41, 881–892 (2011)
17. Lu, H., Plataniotis, K.N., Venetsanopoulos, A.N.: Uncorrelated Multilinear Discriminant Analysis with Regularization and Aggregation for Tensor Object Recognition. IEEE T. Neural Network 20, 103–123 (2009)

18. Yan, S., Xu, D., Yang, Q., Zhang, L., Tang, X., Zhang, H.J.: Multilinear Discriminant Analysis for Face Recognition. IEEE T. Image Process. 16, 212–220 (2007)
19. Hazan, T., Polak, S., Shashua, A.: Sparse Image Coding using a 3D Non-negative Tensor Factorization. In: 10th IEEE International Conference on Computer Vision, vol. 1, pp. 50–57. IEEE Computer Society (2005)
20. Kolda, T.G., Bader, B.W.: Tensor Decompositions and Applications. SIAM Rev. 51, 455–500 (2009)
21. Bourennane, S., Fossati, C., Cailly, A.: Improvement of Classification for Hyperspectral Images Based on Tensor Modeling. IEEE Geosci. Remote S. 7, 801–805 (2010)
22. Yan, S., Xu, D., Yang, Q., Zhang, L., Tang, X., Zhang, H.J.: Discriminant Analysis with Tensor Representation. In: IEEE Conference on Computer Vision and Pattern Recognition, vol. I, pp. 526–532 (2005)
23. Cai, D., He, X.F., Han, J.W.: Subspace Learning Based on Tensor Analysis. Technical Report UIUCDCS-R-2005-2572 (2005)
24. Lu, H., Plataniotis, K.N., Venetsanopoulos, A.N.: A Survey of Multilinear Subspace Learning for Tensor Data. Pattern Recogn. 44, 1540–1551 (2011)
25. Tao, D., Li, X., Hu, W., Maybank, S.J., Wu, X.: Supervised tensor learning. In: Proceedings of the IEEE International Conference on Data Mining, Houston, Texas, USA, pp. 450–457 (2005)
26. Cai, D., He, X.F., Han, J.W.: Learning with Tensor Representation. Technical Report UIUCDCS-R-2006-2716 (2006)
27. Tao, D., Li, X., Wu, X., Hu, W., Maybank, S.J.: Supervised Tensor Learning. Knowl. Inf. Syst. 13, 1–42 (2007)
28. Liu, Y.N., Wu, F., Zhuang, Y.T., Xiao, J.: Active Post-refined Multimodality Video Semantic Concept Detection with Tensor Representation. In: Proceeding of the 16th ACM International Conference on Multimedia, pp. 91–100 (2008)
29. Kotsia, I., Patras, I.: Support Tucker Machines. In: IEEE Conference on Computer Vision and Pattern Recognition, pp. 633–640 (2011)
30. Wolf, L., Jhuang, H., Hazan, T.: Modeling Appearances with Low-Rank SVM. In: IEEE Conference on Computer Vision and Pattern Recognition, pp. 1–6 (2007)
31. Pirsiavash, H., Ramanan, D., Fowlkes, C.: Bilinear Classifiers for Visual Recognition. In: IEEE Conference on Computer Vision and Pattern Recognition, pp. 1–9 (2010)
32. Savicky, P., Vomlel, J.: Exploiting Tensor Rank-one Decomposition in Probabilistic Inference. Kybernetika 43, 747–764 (2007)
33. Wang, H., Ahuja, N.: Rank-R Approximation of Tensors Using Image-as-matrix Representation. In: IEEE Conference on Computer Vision and Pattern Recognition, vol. 2, pp. 346–353 (2005)
34. de Lathauwer, L.: Signal Processing Based on Multilinear Algebra. PhD thesis, Katholieke Univ. Leuven, Belgium (1997)
35. Zhang, T., Golub, G.H.: Rank-one Approximation to High Order Tensors. SIAM J. Matrix Anal. Appl. 23, 534–550 (2001)
36. Keerthi, S.S., Shevade, S.K.: SMO Algorithm for Least Squares SVM Formulations. Neural Comput. 15, 487–507 (2003)
37. Chu, C.T., Kim, S.K., Lin, Y.A., Yu, Y., Bradski, G., Ng, A.Y., Olukotun, K.: Map-reduce for Machine Learning on Multicore. In: Schöölkopf, B., Platt, J., Hoffman, T. (eds.) Advances in Neural Information Processing Systems, vol. 19, pp. 281–288. MIT Press, Cambridge (2007)

# A Remote Sensing Image Classification Method Based on Extreme Learning Machine Ensemble

Min Han and Ben Liu

Faculty of Electronic Information and Electrical Engineering,
Dalian University of Technology, Dalian 116023, China
minhan@dlut.edu.cn

**Abstract.** There are few training samples in the remote sensing image classification. Therefore, it is a highly challenging problem that finds a good classification method which could achieve high accuracy and strong generalization to deal with those data. In this paper, we propose a new remote sensing image classification method based on extreme learning machine (ELM) ensemble. In order to promote the diversity within the ensemble, we do feature segmentation and nonnegative matrix factorization (NMF) to the original data firstly. Then ELM is chosen as base classifier to improve the classification efficiency. The experimental results show that the proposed algorithm not only has high classification accuracy, but also handles the adverse impact of few training samples in the classification of remote sensing well both on the remote sensing image and UCI data.

**Keywords:** Remote Sensing Classification, Nonnegative Matrix Factorization-(NMF), Extreme Learning Machine (ELM), Ensemble Learning.

## 1    Introduction

Remote sensing(RS) image classification is a way to distinguish class attributes and distribution of ground objects based on the feature of material electromagnetic radiation information in the remote sensing images. It's a hot topic in the field of remote sensing. Though the remote sensing image has large number of data samples, the data types are complex and few data samples are available for training. So it is difficult to be classified well only by a single classifier [1]. Some scholars have proposed ensemble algorithms to solve the problem. Mingmin Chi proposed an ensemble learning algorithm which combined generative and discriminative models for remote sensing image classification [2]. But it worked only for hyperspectral remote sensing image which has low generalization. Xin Pan integrated the rough set with the genetic algorithm in order to reduce the number of input features to a single classifier and to avoid bias caused by feature selection [3].

There are two main factors affecting the performance of ensemble learning algorithms: the accuracy of base classifiers and the diversity among the base classifiers [4]. Therefore, how to increase the diversity within the ensemble and keep a high accuracy in the base classifiers is an urgent problem. To promote the diversity,

C. Guo, Z.-G. Hou, and Z. Zeng (Eds.): ISNN 2013, Part I, LNCS 7951, pp. 447–454, 2013.

Garcia-Pedrajas [5] gave weights to every base classifier for each training phase. However, the algorithm is sensitive to mis-indexing data which could cause over fitting. Rodriguez [6] proposed Rotation Forest (ROF) of which the diversity is improved through applying feature extraction to subsets of features for each base classifier by PCA. Decision trees are chosen as base classifiers in ROF. But they are not suitable for remote sensing image classification.

In this study, we present a remote sensing image classification method based on NMF and ELM [7] ensemble (NMF-ELM). Diversity is improved by feature segmentation and NMF [8] for each base classifier. Due to the fast capability and good generalization performance of ELM, we choose it as base classifier to promote the efficiency of remote sensing image classification. Simulation results substantiate the proposed method on both remote sensing image data sets and UCI data sets.

The rest of this paper is organized as follows. Section 2 presents the basic idea of proposed method NMF-ELM in detail. Section 3 gives two simulation results. The conclusions are given in Section 4.

## 2     ELM Based Ensemble Algorithm for RS Classification

### 2.1     Nonnegative Matrix Factorization

The remote sensing image data has a characteristic of nonnegative. When we process these data by linear representation method, decomposition results are required to be nonnegative. In this case, if we use the traditional factor analysis method (such as PCA) to process these data, it may lose the physical meaning, because its results may contain negative numbers. But the use of nonnegative matrix factorization can avoid this problem effectively.

NMF is a matrix factorization method which gives a nonnegative constraint to each element in the treated matrix. Let $F$ be a $M \times N$ matrix where each element is nonnegative. Then to decompose $F$ into $W$ and $H$:

$$F \approx WH \tag{1}$$

Donate $W$ as basic matrix in the form of a $M \times T$ matrix. Donate $H$ as coefficient matrix in the form of a $T \times N$ matrix. When $T$ is smaller than $M$, we can choose the coefficient matrix to replace the original data matrix in order to achieve dimensionality reduction. At the same time, because of the nonnegative constraint of each element in the decomposition process, they exists additive joint only. After decomposition, the matrix $W$ and $H$ can maintain the feature information of the original matrix well.

### 2.2     Extreme Learning Machine

ELM is a feedforward neural network with a simple three-layer structure: input layer, hidden layer and output layer. Let $n$ be the input layer node number, let $r$ be the hidden layer node number, and let $c$ be the output layer node number. For $N$ different samples    $(x_i, l_i)$    ,    $1 \le i \le N$    ,    where    $x_i = [x_{i1}, x_{i2}, ..., x_{in}]^T \in R^n$    ,

$l_i = [l_{i1}, l_{i2}, ..., l_{ic}]^T \in R^c$, the mathematical expression of ELM is shown in formula (2):

$$O_k = \sum_{i=1}^{r} \beta_i g(w_i \cdot x_k + b_i), k = 1, 2, ..., N \tag{2}$$

Where $O_k = [O_{k1}, O_{k2}, ..., O_{kc}]^T$ is the network output value, $w_i = [w_{i1}, w_{i2}, ..., w_{in}]$ is the weight vector connecting the $i$th hidden node and the input nodes, $\beta_i = [\beta_{i1}, \beta_{i2}, ..., \beta_{ic}]^T$ is the weight vector connecting the $i$th hidden node and the output nodes, $g(x)$ is activation function, generally set as Sigmoid function, and $b_i$ is the threshold of the $i$th hidden node.

At the beginning of training, $w_i$ and $b_i$ are generated randomly and kept unchanged. And $\beta$ is the only parameter to be trained. The mathematical expression is shown in formula (3) and (4):

$$H = \begin{bmatrix} g(w_1 \cdot x_1 + b_1) & ... & g(w_r \cdot x_1 + b_r) \\ \vdots & ... & \vdots \\ g(w_1 \cdot x_N + b_1) & ... & g(w_r \cdot x_N + b_r) \end{bmatrix}_{N \times r} \tag{3}$$

$$\beta = H^{\dagger} L \tag{4}$$

Where $H$ is called the hidden layer output matrix of the neural network, $\beta = [\beta_1, \beta_2, ..., \beta_r]^T$ is the output weight vector, $L = [l_1, l_2, ..., l_N]^T$ is the desired output vector. When $\beta$ is solved, ELM network training process is completed.

## 2.3    NMF-ELM Algorithm

NMF-ELM is an ensemble algorithm based on NMF and ELM. The main idea of NMF-ELM is to achieve good classification result through promoting diversity. The structure of NMF-ELM is show in Fig.1.And the structure of ELM base classifier is shown in Fig.2. Specifically, the diversity is improved by feature segmentation and NMF.

Let $p = [p_1, p_2, ..., p_n]^T$ be a sample point described by $n$ features. Let $P$ be the sample set containing the training objects in the form of a $n \times N$ matrix. Let $Y = [y_1, y_2, ..., y_N]$ be a vector with class labels for the data, where $y_i$ takes a value from the set of class labels $\{l_1, l_1, ..., l_c\}$, and $c$ is the number of labels. Denote by $B_1, B_2, ..., B_q$ the classifier in the ensemble and by F, the feature set, where $q$ is the number of classifiers. NMF-ELM is described as below:

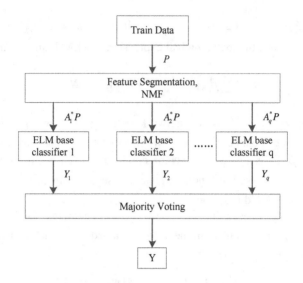

**Fig. 1.** The Structure of NMF-ELM

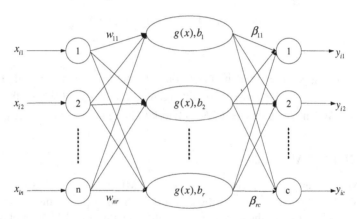

**Fig. 2.** The Structure of ELM Base Classifier

(1) Split $F$ randomly into $K$ disjointed subsets. Let $K$ be an adjustable variable. Each feature subset contains $m = n / K$ features.

(2) Denote the $j$th subset of features by $F_{i,j}$ for the training set of classifier $B_i$ ,where $1 \le i \le q, 1 \le j \le K$. After NMF, $F_{i,j} \approx W_{i,j} H_{i,j}$ .In order to improve diversity and keep the original data information as much as possible, we do not reduce the dimension of $F_{i,j}$. So $H_{i,j}$ is an $m \times N$ matrix, $W_{i,j}$ is an $m \times m$ matrix.

(3) Let $W_{i,j}^{\dagger}$ be the pseudo-inverse of $W_{i,j}$ . Organize $W_{i,j}^{\dagger}$ with coefficients in a sparse matrix $A_i$ .

$$A_i = \begin{bmatrix} W_{i,1}^\dagger & [0] & \cdots & [0] \\ [0] & W_{i,2}^\dagger & \cdots & [0] \\ \vdots & \vdots & \ddots & \vdots \\ [0] & [0] & \cdots & W_{i,k}^\dagger \end{bmatrix} \tag{5}$$

Where $[0]$ has the same dimension with $W_{i,j}^\dagger$. To calculate the training set for classifier $B_i$, we first rearrange the row of $A_i$ so that they correspond to the original features. Denote the new sparse matrix as $A_i^*$.

(4) The training set for classifier $B_i$ is $(A_i^* X, Y)$. Then the training of ELM network can be generalized by three points: (a) Randomly assign input weight $w_i$ and bias $b_i$. (b) Calculate the hidden layer output matrix $H$. (c) Calculate the output weight $\beta$.

(5) To get the final classification results, we integrated the results of all the base classifiers by majority voting.

## 3    Simulation Results

To testify the validity of the proposed algorithm NMF-ELM, Zhalong wetland remote sensing image data and 5 UCI data were used for simulation. The performance was measured by accuracy.

### 3.1    Simulation Results for Zhalong Wetland Remote Sensing Data Sets

This section would verify the performance of NMF-ELM in Zhalong wetland remote sensing image. We used the Landsat ETM+ image which contains 8 bands(band1,2,3,4,5,6,7,8). The image was 512*512 pixels. And the true color image of the remote sensing image was show in Fig.3 (a). Visual interpretation and field research showed that the Zhalong image contained five major classes: agricultural land, fire area, water, marsh, and saline-alkali soil.

We chose 10000 sample points , and took 80% sample points as the training set, the left 20% as the testing set. And we compared NMF-ELM with other algorithm on these datasets, such as Bagging algorithm [9], Adaboost algorithm [4], ROF [6], and ELM [7]. These algorithms were run on the dataset for 50 times and the mean value was considered to be the final result. The detailed results were shown in Table 1.

**Table 1.** Overall Accuracy of Zhalong Wetland （%）

|  | Bagging | Adaboost | ROF | ELM | NMF-ELM |
|---|---|---|---|---|---|
| Zhalong data | 89.5 | 76.05 | 89.51 | 84.55 | 89.65 |

As can be seen from table 1, the proposed algorithm NMF-ELM had the highest accuracy than other algorithms.

Then we used the 10000 sample points as training set to classify the whole image. The classification figure was shown in Fig.3. As can be seen from the figure, the Adaboost algorithm couldn't recognize water and mash, the ELM arose over fitting, but the NMF-ELM got the best result.

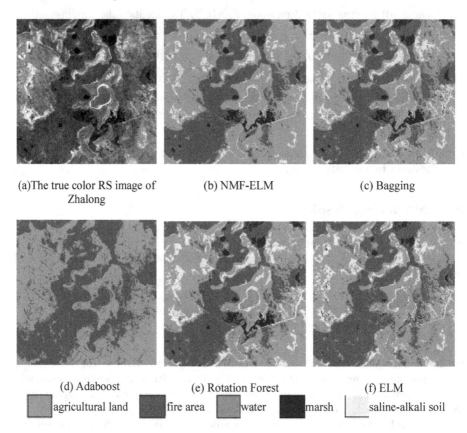

(a)The true color RS image of    (b) NMF-ELM                (c) Bagging
         Zhalong

(d) Adaboost          (e) Rotation Forest              (f) ELM

☐ agricultural land    ■ fire area  ☐ water  ■ marsh  ☐ saline-alkali soil

**Fig. 3.** Different Classification Results by Different Classfiers

## 3.2    Simulation Results for UCI Data Sets

To testify the generalization of the proposed algorithm NMF-ELM, 5 UCI data sets were used for simulation. The features of 5 UCI data sets were shown in Table 2. We took NMF-ELM with other algorithms on the UCI data set, such as Bagging algorithm [9], Adaboost algorithm [4], and ROF algorithm [6]. These algorithms were run on the data for 50 times and the mean value was taken to be the final result. The result was shown in Fig.4-8.

As can be seen from fig.4-8, no matter how many base classifiers were used in the ensemble, the proposed algorithm NMF-ELM got the highest accuracy on 5 UCI data sets. It is illustrated that the NMF-ELM was with strong generalization and stability.

**Table 2.** The Feature of 5 UCI Data

|  | Instances | Attributes | Labels | Attributes Types |
|---|---|---|---|---|
| Balance scale | 625 | 4 | 3 | Categorical |
| Diabetes | 768 | 8 | 2 | Categorical, Integer |
| Heart_S | 270 | 13 | 2 | Integer |
| Tic-tac-toe | 958 | 9 | 2 | Categorical |
| Zoo | 101 | 17 | 7 | Categorical, Integer |

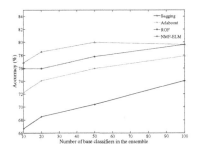 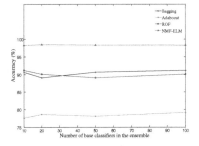

**Fig. 4.** Performance Curves of Methods on Balance-scale

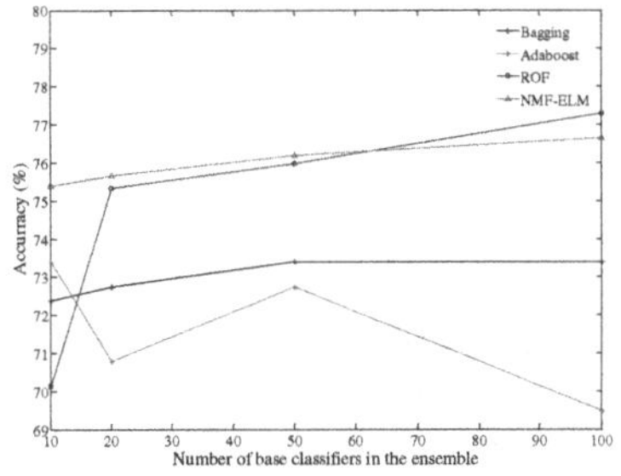

**Fig. 5.** Performance Curves of Methods on Diabetes

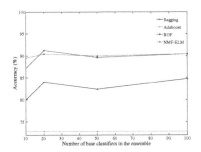

**Fig. 6.** Performance Curves of Methods on Heart-statlog

**Fig. 7.** Performance Curves of Methods on Tic-tac-toe

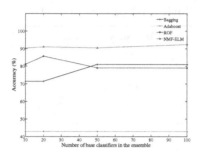

**Fig. 8.** Performance Curves of Methods on Zoo

# 4    Conclusions

To solve the problem of few training samples available in the remote sensing image classification, we proposed an ELM based ensemble algorithm (NMF-ELM). The ensemble diversity is promoted by feature segmentation and NMF. And the classification efficiency is improved by choosing ELM as base classifier. Simulation results show that the proposed algorithm can settle the problem well. With this algorithm, higher classification accuracy can be achieved and stronger generalization can be obtained.

# References

1. Jin, T., Minqiang, L., Fuzan, C., et al.: Coevolutionary learning of neural network ensemble for complex classification tasks. Pattern Recognition 45(4), 1373–1385 (2012)
2. Chi, M., Kun, Q., Benediktsson: Ensemble Classification Algorithm for Hyperspectral Remote Sensing Data. IEEE Geoscience and Remote Sensing Letters 6(4), 762–766 (2009)
3. Pan, X., Zhang, S.: Ensemble remote sensing classifier based on rough set theory and genetic algorithm. In: Eighteenth International Conference on Geoinformatics, pp. 1–5 (2010)
4. Jin, T., Minqiang, L., Fuzan, C., et al.: Coevolutionary learning of neural network ensemble for complex classification tasks. Pattern Recognition 45(4), 1373–1385 (2012)
5. Garcia-Pedrajas, N.: Supervised projection approach for boosting classifiers. Pattern Recognition 42(9), 1742–1760 (2009)
6. Rodriguez, J.J., Kuncheva, L.I., Alonso, C.J.: Rotation Forest: A New Classifier Ensemble Method. IEEE Transactions on Pattern Analysis and Machine Intelligence 28(10), 1619–1630 (2006)
7. Huang, G.-B., Zhu, Q.-Y., Siew, C.-K.: Extreme learning machine: Theory and applications. Neurocomputing 70(1-3), 489–501 (2006)
8. Lee, D.D., Seung, H.S.: Learning the parts of objects by nonnegative matrix factorization. Nature 401, 788–791 (1999)
9. Bühlmann, P.: Handbook of Computational Statistics, pp. 985–1022. Springer, Berlin (2012)

# Model Identification of an Unmanned Helicopter Using ELSSVM

Xinjiu Mei[*] and Yi Feng

School of Control Science and Engineering, Dalian University of Technology, Dalian,
Liaoning, 116024, China
mxj1223@yahoo.cn, fengyi@dlut.edu.cn

**Abstract.** The dynamic model of unmanned helicopter is a coupled nonlinear
system. With respect to the identification problem for this model, extended least
squares support vector machine (ELSSVM) is proposed. ELSSVM extends the
solution space of structure parameters to improve the convergence performance.
Base width of kernel function and regularization parameter of ELSSVM are
minimized by differential evolution (DE). As compared to the traditional
identification method for helicopter dynamic model, the proposed method omits
the linear process and the trained model is closer to the helicopter dynamic
model. The data-driven based experiments show that the proposed method takes
a short training time and has a high identification accuracy.

**Keywords:** Extended least squares support vector machine, coupled nonlinear
system, differential evolution, model identification, unmanned helicopter.

## 1 Introduction

A flight control system with autonomous flight ability is essential for an unmanned
helicopter. If the dynamic model can accurately reflect characteristics throughout the
whole flight envelope, the flight control system will present an ideal performance.
The traditional modeling approach for an unmanned helicopter flight dynamic system
needs a large number of wind tunnel experiments. All those works take a lot of time
and resources [1]. In order to brief the modeling process and reduce the cost, system
identification becomes an ideal substitute. The flight dynamic model of unmanned
helicopter is a highly coupled nonlinear system. In the traditional identification modeling method of helicopter, "small perturbation" assumption is used to linearize the
nonlinear dynamic model. Then linear system identification method is introduced to
identify those unknown model parameters [2]. Unfortunately, the linear process results in a loss of model accuracy.

In the field of nonlinear system identification, neural network (NN) has been widely used. When NN is used to identify the dynamic model of helicopter, we do not
need to know about coupled relationships of system statuses detailly, the linearization

---

[*] Corresponding author.

C. Guo, Z.-G. Hou, and Z. Zeng (Eds.): ISNN 2013, Part I, LNCS 7951, pp. 455–464, 2013.

process can also be omitted. However, NN has the problem of local minimum point and "over learming" phenomenon. In addition, the structure of NN is generally determined by experience. In order to get a higher prediction accuracy, NN needs a large number of training samples [3]. Support vector machine (SVM) can simply the modeling process as well. This theory is based on VC dimension theory and structural risk minimization principle. It can overcome the problem of local minimum point and "over learning" phenomenon [4]. Further, the least squares support vector machine (LSSVM) simplifies the computation complexity of SVM. It improves the training speed and needs fewer training samples [5]. While, the regularization parameter (RP) and the kernel based width (KBW) of LSSVM is directly given before training process. It needs a profound knowledge of the trained model. There is no theoretical to guarantee the prediction accuracy.

In this paper, the extended least squares support vector machine (ELSSVM) is proposed. In essence, the identification of unmanned helicopter dynamic model can be regarded as a function regression problem. ELSSVM is used to establish an model to approximate the model. In ELSSVM, two parameters are added into the solution space of structure parameters. In consideration of computation time and solving precision, differential evolution (DE) is introduced to compute those parameters. DE was proposed by Storn and Price. It is a heuristic optimization algorithm [6]. It has a high computation accuracy and a fast optimizing speed. And its global convergence guarantees the convergence performance of ELSSVM. The rest of this paper is organized as follows. Section 2 presents the mechanism model of a single-rotor helicopter with tail rotor in hovering state. In section 3, we present in detail the theory of ELSSVM. In addition, the training procedure and convergence performance are presented and analyzed. In section 4, when the helicopter hovers stably in the air, flight status and control information are collected. They are sent back to ground station through wireless transmission equipment. After preprocessing the collected data, ELSSVM is used to train and test the model. Finally, a conclusion is given in section 5.

## 2 Dynamic Model Analysis of Unmanned Helicopter

In this paper, the identification object is a single-rotor helicopter with tail rotor. It weights 300kg. We will identify its model in hovering state.

Helicopter can be regarded as an ideal rigid body. In the orthogonal body axes system, fixed at the center of gravity of the whole aircraft, helicopter owns six Degrees Of Freedom (DOF). Statuses in inertial coordinate system include three translation velocity components, three rotational velocity components and three Euler angles [7].

Based on theoretical mechanics and moment of momentum theorem [8], it is easy to establish motion equations (1):

Where, $p$, $q$, $r$ represent angular velocity of roll, pitch and yaw. $u$, $v$, $w$ represent longitudinal speed, later speed and vertical speed. $\theta$, $\phi$, $\psi$ represent pitch angle, roll angle and heading angle. $I_x$, $I_y$, $I_z$ represent inertia moment. $I_{xz}$ represents the product of inertia. $M_x, M_y, M_z$ represent the sum of turn torque. $M$ represents the quality of helicopter. $F_x, F_y, F_z$ represent total force in three body axes.

$$\begin{cases} I_x \dot{p} = (I_y - I_z)qr + I_{xz}(\dot{r} + pq) + M_x \\ I_y \dot{q} = (I_z - I_x)pr + I_{xz}(r^2 - p^2) + M_y \\ I_z \dot{r} = (I_x - I_y)pq + I_{xz}(\dot{p} - qr) + M_z \\ \dot{u} = (rv - qw) + F_x/M \\ \dot{v} = (pw - ru) + F_y/M \\ \dot{w} = (qu - pv) + F_z/M \\ \dot{\psi} = q \sin\phi \sec\theta + r \cos\phi \sec\theta \\ \dot{\theta} = q \cos\phi - r \sin\phi \\ \dot{\phi} = p + q \sin\phi \tan\theta + r \cos\phi \tan\theta \end{cases} \tag{1}$$

In equations (1), force and moment are relevant to not only control variables but also flight statuses. They can be expressed as follows (take $F_x$ for example):

$$F_x = f(u, v, w, p, q, r, \phi, \theta, \delta_e, \delta_a, \delta_c, \delta_r) \tag{2}$$

There are four more control variables in the above equation. $\delta_e$ is longitudinal control input, $\delta_a$ is lateral control input, $\delta_c$ is vertical control input, $\delta_r$ is yawing control input. As $\psi$ has almost no effect on the other statuses, the dynamic model can be simplified as a system with 8 statuses and 4 control variables.

## 3　ELSSVM Method

### 3.1　LSSVM Algorithm

Define training samples as $\{x_k, y_k\}(k = 1,...,l)$, where $x_k \subset R^n$, $y_k \subset R$. At the k-th sample point, $x_k$ represent system statuses and control variables, $y_k$ represents output. The input space can be mapped to a higher dimensional feature space by nonlinear mapping function $\varphi(\cdot)$. The purpose of LSSVM is to construct a function as follow:

$$y(x) = w^T \varphi(x) + b \tag{3}$$

Where, weight vector $w \subset R^n$ and offset $b \subset R$. The objective function of LSSVM can be described as follow:

$$\min_{w,b,\xi_i} \quad J(w,\xi) = \frac{1}{2}\|w\|^2 + \frac{1}{2}\gamma\sum_{i=1}^{l}\xi_i^2$$

$$s.t. \quad \begin{cases} y_i = w^T\varphi(x_i)+b+\xi_i \\ i=1,2,\cdots,l \end{cases} \tag{4}$$

In the above equation, $\gamma$ is error penalty coefficient. $\xi_i$ is error. Its dual problem of Lagrange polynomial is equivalent to the following equation:

$$L(w,b,\xi,\alpha) = J(w,\xi) - \sum_{i=1}^{l}\alpha_i\left\{w^T\varphi(x_i)+b+\xi_i - y_i\right\} \tag{5}$$

In the above dual Lagrange polynomial, $\alpha_i$ is Lagrange multiplier. After derivation calculus to polynomial (5), we get equations (6):

$$\begin{bmatrix} 0 & \vec{1}^T \\ \vec{1} & K+\gamma^{-1}I \end{bmatrix}\begin{bmatrix} b \\ \alpha \end{bmatrix} = \begin{bmatrix} 0 \\ Y \end{bmatrix} \tag{6}$$

Assume $A = K+\gamma^{-1}I$. After one step derivation, we get equations (7).

$$\begin{cases} b = \dfrac{\vec{1}^T A^{-1}Y}{\vec{1}^T A^{-1}\vec{1}} \\ \alpha = A^{-1}(Y-b\vec{1}) \end{cases} \tag{7}$$

Assume $K_{kj} = \varphi(x_k)^T \cdot \varphi(x_j)$, then equation (8) can be established as follow:

$$f(x) = \sum_{i=1}^{l}\alpha_i K(x_i,x)+b \tag{8}$$

The kernel function in equation (8) is $K(x_i,x) = \exp\{-\|x_i - x\|_2^2/\sigma^2\}$. The above kernel function must satisfy Mercer condition [9].

## 3.2    Extended Solution Space of ELSSVM

LSSVM has simplified the modeling process of helicopter dynamic model. It also makes a signification improvement on SVM. But it still has a bad convergence performance. In ELSSVM, RP and KBW are added into the solution space of structure par-ameters to improve the convergence performance.

Different $\gamma$ and $\sigma$ will establish different models. In order to evaluate models, it is necessary to define an uniform performance index.

Define $X = \{x_k, y_k\}(k = 1,...,l)$ as the training samples, equation (8) represents prediction response of the i-th sample. Where $x_k \subset R^n$, $y_k \subset R$. Define $e = y_i - f(x_i)$. The energy function is established as follow:

$$J = \frac{1}{2}e^T \cdot e \tag{9}$$

Equation (9) primely reflects the error between prediction response and actual output. By minimizing equation (9) using DE, a convergent model will be established.

The individual of DE is defined as $(\gamma, \sigma)$. Define population size as $NP$. The dimension of each individual is $D$. The maximum number of generations is $M$. The i-th individual in the g-th generation is $x_{i,g}$. After mutation operation, we can get $v_{i,g}$:

$$v_{i,g} = x_{p1,g} + F \cdot (x_{p2,g} - x_{p3,g}) \tag{10}$$

$p1, p2, p3 \in (1,\cdots,NP)$, $p1 \neq p2 \neq p3 \neq i$. Mutagenic factor $F \in (0,1)$ controls the magnitude of differential vector. $x_{p1,g}$ is the based vector.

Crossover operation increases the diversity of population. Hybridize intermediate individual with target individual, we get the candidate individual.

$$u_{ji,g} = \begin{cases} v_{ji,g}, rand_j(0,1) \leq CR \ \ or \ \ j = rand_j(1,D) \\ x_{ji,g}, otherwise \end{cases} \tag{11}$$

$i = 1,\cdots,NP$, $j = 1,\cdots,D$, $rand_j(1,D)$. Cross-factor $CR \in [0,1]$ is the probability that individual component replaces target individual.

Select the fitness value of candidate individual. Then decide whether to replace the current target individual with the candidate individual.

$$v_{i,g+1} = \begin{cases} u_{i,g+1}, J(u_{i,g+1}) \leq J(x_{i,g}) \\ x_{i,g} \ \ , otherwise \end{cases} \tag{12}$$

$\varepsilon$ is a sufficiently small positive constant. When $J < \varepsilon$, $v_{i,g+1}$ is the optimal solution and iteration process ends. Otherwise, repeat the above process.

As the description above, DE does not need the gradient information of individual objective function. So the individual objective function does not need to be continuous and drivable [10]. After the mutation and crossover operations, new individuals take one-to-one competition with their parents. The smaller individual will become new parents in the next generation. Compared with genetic algorithm, DE omits the time-consuming selection operation and takes less computation time.

### 3.3    Training Procedure of ELSSVM

Here is the training procedure of ELSSVM:

Step 1: Initialize the population;

Step 2: Train the model of each individual in this generation using LSSVM;

Step 3: Use the trained model to predict the response of training samples. Prediction information and actual output make up the energy function $J$ ;

Step 4: If $J < \varepsilon$ , the training process ends. Else, continue;

Step 5: When iteration reachs the $M$-th generation, the individual with minimum $J$ in this generation is the final answer. Else, continue;

Step 6: Go on the mutation, crossover and elimination operations. Use the achieved RP and KBW to update the population in next generation;

Step 7: Return to the step 2.

Each iteration procedure is regarded as a map. The map is a random contraction operator with an unique random fixed point. Based on the random functional theory, the whole iteration process is asymptotically convergent [11]. All in all, DE algorithm in each iteration operation guarantees the convergence performance of ELSSVM.

## 4    Experiment and Analysis

### 4.1    Data Preprocessing

In order to identify the flight dynamic model of this unmanned helicopter, training samples must sufficiently reflect characters of the helicopter. According to this demand, control variables and flight statuses are sampled at the sampling period of 0.1s. The information will be sent back to the ground station via wireless devices. In the data collecting process, there are a lot of interferences. Those interferences will affect the identification accuracy. So it is necessary to preprocess the experimental data.

Firstly. Eliminate effects of high frequency noise. After spectrum analysis of the collected data, design a low pass filter to filter high frequency noise. Cutoff frequency is designed at 5HZ.

Secondly. Due to the bandwidth limitation of transfer data, the sampling frequency is too low. In order to reflect system characteristics sufficiently, spline interpolation method is used to smooth the samples.

Thirdly. Three translational velocities are collected in North-East-Down coordinate system. Translational velocities in dynamic equations are in body coordinate system. So coordinate transformation must be implemented before model identification.

## 4.2    Model Identification

Take 100 samples as training set and another 100 as test set. Define initial population size $NP = 20$, crossover probability $CR = 0.6$ and crossover factor $F = 0.5$. Maximum number of iteration procedure is 100. The radial bases function (RBF) works as the kernel function. Take equation (9) as error criterion.

ELSSVM, LSSVM and back propagation neural network (BP) are used to train the model. In order to eliminate accidental factors, perform the identification experiment 10 times using each method respectively. Then calculate the mean value of preformance indicators. ELSSVM and LSSVM have the same initial structure parameters. The only difference is that ELSSVM has an extended solution space. In the model trained by BP, there are 12 nodes in the input layer, 25 nodes in the hidden layer and 1 node in the output layer. The node function here is S-function. The experiment results are listed in the Table 1.

**Table 1.** Parameter table of experiment results

| System status | Method | Training time | Training error | Test error |
|---|---|---|---|---|
| Longitudinal velocity | BP | 13.13s | $1.82 \times 10^{-4}$ | $3.30 \times 10^{-3}$ |
| | LSSVM | 0.13s | $1.47 \times 10^{-8}$ | $2.37 \times 10^{-4}$ |
| | ELSSVM | 1.05s | $1.27 \times 10^{-9}$ | $6.04 \times 10^{-5}$ |
| Lateral velocity | BP | 12.34s | $9.99 \times 10^{-5}$ | $6.50 \times 10^{-3}$ |
| | LSSVM | 0.13s | $7.29 \times 10^{-8}$ | $1.40 \times 10^{-3}$ |
| | ELSSVM | 1.00s | $8.62 \times 10^{-10}$ | $6.56 \times 10^{-5}$ |
| Vertical velocity | BP | 9.72s | $9.93 \times 10^{-5}$ | $9.94 \times 10^{-4}$ |
| | LSSVM | 0.11s | $3.41 \times 10^{-8}$ | $4.17 \times 10^{-4}$ |
| | ELSSVM | 1.10s | $1.59 \times 10^{-9}$ | $1.57 \times 10^{-5}$ |
| Roll rate | BP | 10.44s | $5.80 \times 10^{-3}$ | $2.81 \times 10^{-2}$ |
| | LSSVM | 0.11s | $1.17 \times 10^{-6}$ | $4.39 \times 10^{-2}$ |
| | ELSSVM | 1.08s | $5.94 \times 10^{-8}$ | $1.50 \times 10^{-3}$ |
| Pitch rate | BP | 11.67s | $1.80 \times 10^{-3}$ | $2.39 \times 10^{-2}$ |
| | LSSVM | 0.11s | $4.88 \times 10^{-7}$ | $2.70 \times 10^{-2}$ |
| | ELSSVM | 1.18s | $2.63 \times 10^{-8}$ | $8.20 \times 10^{-4}$ |
| Yaw rate | BP | 5.27s | $2.00 \times 10^{-3}$ | $7.14 \times 10^{-2}$ |
| | LSSVM | 0.11s | $2.44 \times 10^{-6}$ | $1.61 \times 10^{-1}$ |
| | ELSSVM | 1.09s | $2.51 \times 10^{-7}$ | $6.40 \times 10^{-3}$ |
| Roll | BP | 11.95s | $9.77 \times 10^{-5}$ | $2.90 \times 10^{-3}$ |
| | LSSVM | 0.11s | $1.54 \times 10^{-7}$ | $1.57 \times 10^{-2}$ |
| | ELSSVM | 1.07s | $1.40 \times 10^{-8}$ | $9.72 \times 10^{-4}$ |
| Pitch | BP | 12.33s | $9.89 \times 10^{-5}$ | $7.10 \times 10^{-3}$ |
| | LSSVM | 0.13s | $1.03 \times 10^{-7}$ | $8.50 \times 10^{-3}$ |
| | ELSSVM | 1.09s | $1.10 \times 10^{-8}$ | $1.46 \times 10^{-4}$ |

It can be concluded from Table 1, ELSSVM take a shorter training time than BP and LSSVM. In the training process of BP, the back propagation has to be performed iteratively until error criterion is met. Whereas, the structure parameters of LSSVM and ELSSVM can be calculated just by one step. ELSSVM has an extended solution space which takes extra time to optimize RP and KBW. Hence, compared to LSSVM, ELSSVM takes a litter longer training time. In extended solution space of ELSSVM, DE is used to seek RP and KBW. Thus, ELSSVM has higher training accuracy and smaller prediction error.

After training the model respectively by BP ,LSSVM and ELSSVM, test set is used to inspect the prediction performance. Here are linear velocity curves of the three trained model.

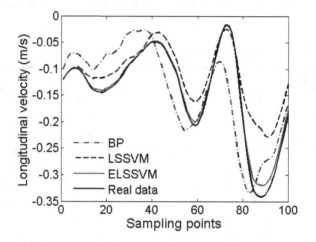

**Fig. 1.** Curves of longitudinal velocity

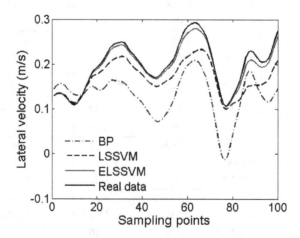

**Fig. 2.** Curves of lateral velocity

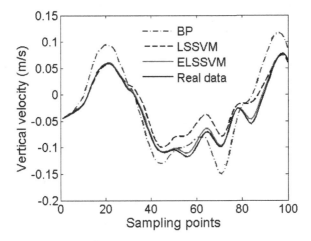

**Fig. 3.** Curves of vertical velocity

Figs.1–3 present linear velocities of three trained models. The models are impelled by test set. The curves impelled by test set can reflect prediction performance directly. As DE is introduced to seek the optimal RP and KBW, the prediction model trained by ELSSVM is closer to the objective model. From the figures, we can see that the linear velocity curve of ELSSVM is closer to the real data curve. It can be concluded that the model trained by ELSSVM has a better prediction performance and convergence performance than the other two.

## 5    Conclusion

Unmanned helicopter is a multi-DOF machine. It is a highly coupled nonlinear system. It is hard to achieve a high precision model using the traditional identification method. In this paper, ELSSVM is proposed to train the dynamic model. As compared to the traditional identification method, ELSSVM simplifies the identification process and extends the solution space of structure parameters. Experiment results show that the proposed method has higher training accuracy and prediction accuracy. It also solves the bad convergence performance problem of LSSVM. In essence, establishing the prediction model by ELSSVM is to solve differential equations using implicit solution method. Whereas, the training process of LSSVM is to solve differential equations by explicit solution method. Implicit solution has a better convergence performance than explicit solution in solving differential equation. In one sense, the works in this paper provide a new strategy to solve the identification problem of highly coupled nonlinear model with multi-DOF.

## References

1.  Padfield, G.D.: Helicopter Flight Dynamics, vol. 2, p. 680. Blackwell, Oxford (2007)

2. Kim, S.K., Tilbury, D.M.: Mathematical modeling and experimental identification of a model helicopter. In: AIAA Modeling and Simulation Technologies Conference and Exhibit, Boston, August 10-12 (1998)
3. Zhou, J.G., Qin, G.: Application of BP Neural Network Forecast Models Based on Principal Component Analysis in Railways Freight Forecas. In: International Conference on Computer Science and Service System, pp. 2201–2204 (2012)
4. Veillard, A., Racoceanu, D., Bressan, S.: pRBF Kernels: A Framework for the Incorporation of Task-Specific Properties into Support Vector Methods. In: International Conference on Machine Learning and Applications, pp. 156–161 (2012)
5. Huang, G., Song, S.J., Wu, C.: Robust Support Vector Regression for Uncertain Input and Output Data. IEEE Trans. Neural Network and Learning System 23(11), 1690–1700 (2011)
6. Price, K., Storn, R.M., Lampinen, J.A.: Differential Evolution: A Practical Approach to Global Optimization. Springer, Heidelberg (2005)
7. Klein, V., Moreli, E.A.: Aircraft System Identification Theory and Practice. AIAA Education Series. American Institute of Aeronautics and Astronautics, New York (2006)
8. Chen, M., Ge, S.S., Ren, B.: Robust attitude control of helicopter with actuator dynamics using neural networks. Control Theory and Application 4, 2837–2854 (2010)
9. Vapnik, V.: The Nature of Statiscal Learning Theory. Springer, New York (1999)
10. Feoktistov, V.: Differential Evolution In Search of Solutions. Springer, Heidelberg (2006)
11. Sun, C.F., Zhou, H.Y., Chen, L.Q.: Improved differential evolution algorithms. In: IEEE International Conference on Computer Science and Automation Engineering, vol. 3, pp. 142–145 (2012)

# A Feature Point Clustering Algorithm Based on GG-RNN

Zhiheng Zhou[1,2], Dongkai Shen[1], Lei Kang[1], and Jie Wang[3]

[1] College of Electronics & Information Engineering, South China University of Technology,
Guangzhou, 510641, China
[2] Desay Corporation, Huizhou, 516006, China
[3] Institute of Acoustics and Lighting Technology,
Guangzhou University, Guangzhou, 510006, China
zhouzh@scut.edu.cn, shendkwhy@gmail.com

**Abstract.** In the field of object recognition in computer vision, feature point clustering algorithm has become an important part of the object recognition. After getting the object feature points, we make the feature points in clustering in the use of GG-RNN clustering algorithm, to achieve multi-part of the object clustering or the multi-object clustering. And the GG-RNN clustering algorithm we propose innovatively, is merged with the grayscale and gradient information based on Euclidean distance in the similarity calculation. Compared with the distance description of basic RNN algorithm, the similarity calculation of high-dimensional description of GG-RNN will improve the accuracy of the clustering in different conditions.

**Keywords:** clustering, the reciprocal nearest neighbors, similarity, grayscale, gradient.

## 1 Introduction

Object recognition has occupied an important position in the field of computer vision recently. Object recognition is divided into overall object identification and comprehensive object identification of each part. After the extract of image feature point, using the clustering algorithm to cluster the feature point to obtain a multi-part or multi-object and recognize the object by synthesizing the above clustering result. There are many clustering algorithms such as the K-means clustering [1] and the RNN clustering [2] which is the basic algorithm of our GG-RNN method. In this field, the latest research is advanced by the ISM team which proposed the implicit shape model. They used the feature point clustering to locate the different parts of object for the recognition in the next step [3] .RNN clustering algorithm has the advantage that you don't need to preset the number of clusters to get the clustering of feature points. With this advantage, we can find the target even though the number of target is unknown. However, the RNN clustering method effects mainly depend on the similarity metrics. Single linkage, Complete linkage, Average linkage [4] form the three common linkage metrics according to the different inter-cluster distance metric. But the above three linkage metrics are just based on the distance information and lacked the other

C. Guo, Z.-G. Hou, and Z. Zeng (Eds.): ISNN 2013, Part I, LNCS 7951, pp. 465–470, 2013.

information in the image. This disadvantage leads the clustering can't bind other image information to achieve a good clustering performance. To overcome these shortcomings, this paper combines the image grayscale and gradient information with the basic distance information, proposing the new similarity measure method named GG-RNN, which is short for RNN algorithm based on Grayscale and Gradient. With the GG-RNN method, we make the object clustering in complicated condition successfully which provide a powerful support for the late object recognition.

## 2    RNN Clustering Theory

The RNN clustering algorithm was introduced in [2].The reciprocal nearest neighbors [5] means that vector $x_i$ is the nearest neighbor of vector $x_j$, and vector $x_j$ is the nearest neighbor of vector $x_i$, i.e. vector $x_i$ and vector $x_j$ are RNN. Nearest neighbor means that vector $x_j$ which taken from the entire set of vectors has the nearest distance to vector $x_i$ in the whole space except the vector $x_i$ itself, therefore vector $x_j$ is the nearest neighbor of vector $x_i$, i.e. vector $x_j$ is the NN of vector $x_i$ .There are the basic steps of RNN algorithm as follows:

Step1: A given data set V, and an empty NN linked list L, and R stored the data other than those in the L, C stored the clusters. Randomly taken $x_1$ from R and put it into the linked list L; then find $x_2$ which is the NN of $x_1$ ,and put it into the linked list L; find $x_3$ which is the NN of $x_2$ next time, and put into the linked list L; according to the above step , finally we can find the NN of $x_{l-1}$ is $x_l$ ,and put it into the linked list L; and at the same time we find the NN of $x_l$ is $x_{l-1}$ ,therefore $x_l$ and $x_{l-1}$ are RNN. Stop the iteration at this time. So the final the NN chain can be represented as $\{x_1, x_2 = NN(x_1),..., x_{l-1} = NN(x_l), x_l = NN(x_{l-1})\}$ ,and it has length $l$ .Noticed that the distance between two neighbor vectors in NN chain is getting smaller and smaller [3], same as the increasingly high degree of similarity. Therefore the distance between $x_{l-1}$ and $x_l$ which are in the end of this chain is the Minimum distance, and the similarity between them is also the highest similarity, i.e. R=V\L after this iteration.

Step2: Calculate the similarity of $x_{l-1}$ and $x_l$ in the link L, i.e. MinSim and compare with a preset similarity threshold, i.e. Thres. If MinSim is less than Thres ,means that the clustering result of NN chain is not ideal, we will classify each data of the chain into a cluster (a class) ,and remove them from set V. If MinSim is greater than Thres, means that the clustering result of NN chain is meet the requirements, $x_{l-1}$ and $x_l$ will be agglomerated in a new cluster $x_s$ , removed from L simultaneously, and put into R. Then continue look for the NN of $x_{l-2}$ in the new R, and repeat the previous procedure until find a similarity meet the requirement of NN-chain or link L is cleared.

Step3: Randomly take new vector from new R, and repeat the above step 1 and step 2 until the set V is empty. Finish clustering and end.

## 3    Similarity Calculation and GG-RNN

In the above process, the similarity threshold Thres is the key point determining clustering results because it constrained the clustering degree. In the traditional calculation, similarity is represented by the Euclidean distance between two clusters, and there are three distance calculation methods as follows, Single linkage, Complete linkage, Average linkage [4].

Single linkage means that the cluster distance equals to the minimum distance between two feature points from all feature points, is defined as

$$D(X,Y) = \min_{x_i \in X \, y_j \in Y} \left( d\left(x_i, y_j\right) \right) \tag{1}$$

Complete linkage means that the cluster distance equals to the maximum distance between two feature points bagging from all feature points, is defined as

$$D(X,Y) = \max_{x_i \in X \, y_j \in Y} \left( d(x_i, y_j) \right) \tag{2}$$

Average linkage means that the cluster distance equals to the average distance between two feature points from all feature points, is defined as

$$D(X,Y) = \frac{1}{XY} \sum_{x_i \in X} \sum_{y_j \in Y} d\left(x_i, y_j\right) \tag{3}$$

In the above method, the object clustering of images won't achieve a good performance with the single constraint factor, Euclidean distance. Taking this reason into consider, two factors add into the new similarity calculation: grayscale values and gradient magnitude, which are described the grayscale, edge feature and texture feature.

Here, we list the similarity calculation of the traditional method and the improved method as follows:

The traditional similarity calculation is defined as:

$$D=d \tag{4}$$

The improved similarity calculation is defined as:

$$D = a \times d + b \times I + c \times grad \tag{5}$$

Make width represent the width of image, height represent the height of image, $I(x_i, y_i)$ represent the grayscale values of image, $grad(x_i, y_i)$ represent the gradient magnitude of image.

The square of normalized distance value is defined as:

$$d = [(x_i - x_j) / width]^2 + [(y_{i-}y_j) / height]^2 \tag{6}$$

The square of normalized grayscale difference value is defined as:

$$I = [I(x_i, y_i) - I(x_j, y_j)]^2 / \{\max[I(x, y)] - \min[I(x, y)]\}^2 \tag{7}$$

The square of normalized gradient magnitude difference is defined as:

$$grad = [\mathrm{grad}(x_i, y_i) - \mathrm{grad}(x_j, y_j)]^2 / \{\max[grad(x, y)] - \min[grad(x, y)]\}^2 \tag{8}$$

The weights of variable $d, I, grad$ are defined as a, b, c, and with change of the weights get a better clustering performance. The algorithm with above three variables is called RNN algorithm based on Grayscale and Gradient, i.e. GG-RNN

## 4    Experimental Comparison of Clustering Algorithms

In this section we present an experimental comparison between the basic RNN and the GG-RNN clustering algorithm. We use the harris corner as feature point of images, and cluster the harris corner with the basic RNN and GG-RNN algorithm. We locate each cluster with a rectangle box. According many times emulation, we show parts of clustering results as follows. In each row of below figures, the first image is the original image, the second image is the result of harris corner detecting, the third image is the result of basic RNN clustering based on distance, the fourth image is the result of improved RNN clustering.

Fig.1 represents the comparison between the basic RNN based on distance and the RNN based on grayscale and distance. With the grayscale information, the clustering performance at the bottom of this image has been improved that the corner on the jacket has been clustered as same cluster with the grayscale factor. The third image clustering with a weight, i.e. a=1, b=0,c=0. The fourth image clustering with a weight, i.e. a=1, b=0.2, c=0

**Fig. 1.** Basic RNN based on distance and the improved RNN based on grayscale and distance

Fig.2 represents the comparison between the basic RNN based on distance and the RNN based on gradient and distance. With the gradient information, the clustering performance at the middle top of this image has been improved that the corner of the black vehicle has been clustered as the same cluster with the gradient factor .The third image clustering with a weight, i.e. a=1,b=0,c=0. The fourth image clustering with a weight, i.e. a=1, b=0, c=0.2.

**Fig. 2.** Basic RNN based on distance and the improved RNN based on gradient and distance

Fig.3 represents the comparison between the basic RNN based on distance and the improved RNN based on grayscale, gradient and distance which we call GG-RNN. With the gradient information, the fourth image shows the clustering performance getting improved obviously. The third image clustering with a weight, i.e. a=1, b=0, c=0. The fourth image clustering with a weight, i.e. a=1, b=0.4, c=1.

**Fig. 3.** Basic RNN based on distance and GG-RNN

## 5    Summary

The paper discuss about the RNN clustering algorithm which clusters the image feature point of multi-object and multi-part. Take the disadvantage that Euclidean distance is the only information for clustering into consideration, we propose an improved RNN algorithm combined with grayscale and gradient information for feature point clustering. We obtain the conclusion that GG-RNN is effective after many times emulations. Compared with the basic RNN, GG-RNN gets a good and exact performance in the image feature point clustering. And the exact feature point clustering also provide a support for the late object recognition which is worth to mention in the later research.

**Acknowledgements.** This work was supported by National Natural Science Foundation of China (gra-nt 61003170), China Postdoctoral Science Foundation (grant 2012M511561), Funda-mental Research Funds for the Central Universities SCUT (grant    2012ZZ0033),Guan-gdong    Province    Science    Foundation(grant S2011010001936), Guangzhou Pearl River New Star Special of Science and Technology (grant2011J2200015).

# References

1. MacQueen, J.: Some methods for classification and analysis of multivariate observations. In: Proceedings of the 5th Berkeley Symposium on Mathematical Statistics and Probability, pp. 281–297 (1967)
2. De Rham, C.: La classification hiérarchique ascendante selon la méthode des voisins réciproques. Les Cahiers de l'Analyse des Données 5(2), 135–144 (1980)
3. Leibe, B., Mikolajczyk, K., Schiele, B.: Efficient clustering and matching for object class recognition. In: BMVC (2006)
4. Huang, S., Li, X.: Based on Improving the Effect of RNN Clustering Algorithm Research. Journal of Taiyuan Normal University 11(2), 72–75 (2012)
5. López-Sastre, R.J., Oñoro-Rubio, D., Gil-Jiménez, P., Maldonado-Bascón, S.: Fast reciprocal nearest neighbors clustering. Signal Process. 92(1), 270–275 (2012)

# Local Fisher Discriminant Analysis with Locally Linear Embedding Affinity Matrix

Yue Zhao and Jinwen Ma*

Department of Information Science, School of Mathematical Sciences
And LMAM, Peking University, Beijing, 100871, China
jwma@math.pku.edu.cn

**Abstract.** Fisher Discriminant Analysis (FDA) is a popular method for dimensionality reduction. Local Fisher Discriminant Analysis (LFDA) is an improvement of FDA, which can preserve the local structures of the feature space in multi-class cases. However, the affinity matrix in LFDA cannot reflect the actual interrelationship among all the neighbors for each sample point. In this paper, we propose a new LFDA approach with the affinity matrix being solved by the locally linear embedding (LLE) method to preserve the particular local structures of the specific feature space. Moreover, for nonlinear cases, we extend this new LFDA method to the kernelized version by using the kernel trick. It is demonstrated by the experiments on five real-world datasets that our proposed LFDA methods with LLE affinity matrix are applicable and effective.

**Keywords:** Dimensionality reduction, Fisher discriminant analysis (FDA), Kernel method, Local affinity matrix.

## 1 Introduction

In data analysis and information processing, dimensionality reduction approach is often used for simple but effective data representation. Actually, it tries to find a linear projection such that the data in a higher dimensional space can be effectively reexpressed in a lower dimensional space [1],[2]. Actually, Fisher Discriminant Analysis (FDA) [3] and Principal Component Analysis [4] are popular dimensionality reduction methods. Here we focus on the FDA approach to dimensionality reduction. In principle, FDA tries to make the between-class scatter be maximized while the within-class scatter be minimized in the lower dimensional space. However, FDA has two major drawbacks: (1). If the samples in one class come from several separate clusters, it cannot behave properly and leads to an unreasonable result [3]. (2). It cannot work well for a nonlinear classification problem [5].

In order to get rid of the first drawback, Masashi [6] proposed the Local Fisher Discriminant Analysis (LFDA), which makes the close samples still be close in the dimensionality reduction feature space via the Locality Preserving

---

* Corresponding author.

C. Guo, Z.-G. Hou, and Z. Zeng (Eds.): ISNN 2013, Part I, LNCS 7951, pp. 471–478, 2013.

Projection (LPP) mechanism [7]. But the used affinity matrix cannot reflect the actual interrelationship among all the neighbors at each local sample point. So, its performance is still limited.

As for the second drawback, the Kernel Fisher Discriminant Analysis (KFDA) was used [5],[8]. Its main idea is to map the input space into a higher dimensional feature space in which the corresponding sample classes are linearly separated so that we can conduct the FDA method in this projected feature space.

In this paper, we propose a new LFDA approach to data dimensionality reduction for a classification problem. The within-class scatter matrix as well as the between-class scatter matrix are redefined through a special weighted matrix as the affinity matrix is solved by the locally linear embedding (LLE) [9] method to preserve the particular local structures of the specific feature space. Moreover, for nonlinear classification cases, we extend this new LFDA method to the kernelized version by using the kernel trick. It is demonstrated by the experiments on five real-world datasets that our proposed LFDA methods with LLE affinity matrix are applicable and effective.

The rest of this paper is organized as follows. In Section 2, we give a brief review of FDA and LFDA. In Section 3, we present the new LFDA approach. Its kernelized version is presented in Section 4. Their experimental results and comparisons are given in Section 5. We give a brief conclusion in Section 6.

## 2    Review of FDA and LFDA

For clarity, we give some mathematical notations used throughout the paper. $x_i \in \mathbb{R}^d$ and $z_i \in \mathbb{R}^r (1 \leq r < d)$ are the $i$-th input data and its corresponding low dimensional projection or embedding $(i = 1, 2, \cdots, n)$, where $n$ is the number of samples, $d$ is the dimensionality of the input data and $r$ is the reduced dimensionality. $y_i \in \{1, 2, \cdots, c\}$ are the associated class labels, and $c$ is the number of classes. $n_l$ is the number of samples in class $l$, thus $\sum_{l=1}^{c} n_l = n$. $X$ is defined as the matrix of collection of all samples, i.e., $X = (x_1, x_2, ..., x_n)$.

### 2.1    FDA

We begin with the review of Fisher Discriminant Analysis (FDA) [3]. The within-class scatter matrix $S_w$ and the between-class scatter matrix $S_b$ are defined as follows.

$$S_w = \sum_{l=1}^{c} \sum_{i:y_i=l} (x_i - \mu_l)(x_i - \mu_l)^T; \tag{1}$$

$$S_b = \sum_{l=1}^{c} n_l (\mu_l - \mu)(\mu_l - \mu)^T, \tag{2}$$

where $\mu_l$ is the mean of the samples in class $l$ and $\mu$ is the mean of all samples.

The main idea of FDA is just to find out a group of $r$ projection vectors such that in the embedding feature space, the within-class scatter along each projection vector is minimized while the between-class scatter along each projection vector is maximized. Mathematically, such a projection vector $V_{FDA}$ can be solved by

$$V_{FDA} = \text{argmax} \frac{V^T S_b V}{V^T S_w V}, \tag{3}$$

where $V$ is an arbitrary projection vector.

Let $\{\varphi_k\}_{k=1}^d$ be the generalized eigenvectors associated with the generalized eigenvalues $\lambda_1 \geq \lambda_2 \geq ... \geq \lambda_d$ of the generalized eigenvalue problem: $S_b \varphi = \lambda S_w \varphi$. In this way, the $k$-th solution $V_{FDA}$ of Eq.(3) is just $\varphi_k$. Thus, the required transformation matrix is $T_{FDA} = (\varphi_1, \varphi_2, ..., \varphi_r)$. That is, for $x_i$, its embedded vector $z_i$ is given by $z_i = T_{FDA}^T x_i$.

## 2.2   LFDA

We further review the Local Fisher Discriminant Analysis (LFDA) [6]. In order to preserve the local structure of the data, we can define the local within-class scatter matrix $\tilde{S}_w$ and the local between-class scatter matrix $\tilde{S}_b$ as follows.

$$\tilde{S}_w = \frac{1}{2} \sum_{i,j=1}^n P_{ij}^w (x_i - x_j)(x_i - x_j)^T; \tag{4}$$

$$\tilde{S}_b = \frac{1}{2} \sum_{i,j=1}^n P_{ij}^b (x_i - x_j)(x_i - x_j)^T, \tag{5}$$

where

$$P_{ij}^w = \begin{cases} \frac{A_{ij}}{n_l}, & \text{if } y_i = y_j = l, \\ 0, & \text{if } y_i \neq y_j, \end{cases} \quad P_{ij}^b = \begin{cases} (\frac{1}{n} - \frac{1}{n_l})A_{ij}, & \text{if } y_i = y_j = l, \\ \frac{1}{n}, & \text{if } y_i \neq y_j, \end{cases}.$$

$A$ is the affinity matrix where its element $A_{ij}$ denotes the affinity degree between $x_i$ and $x_j$. In principle, $A_{ij}$ tends to be 0 as $x_i$ and $x_j$ tend to be far away. Generally, $A$ can be computed by $A_{ij} = \exp\left(-\frac{\|x_i - x_j\|^2}{\sigma_i \sigma_j}\right)$, where $\sigma_i = \|x_i - x_i^k\|$, $x_i^k$ is the $k$-th neighbor of $x_i$.

Just as FDA, the projection vector $V_{LFDA}$ of LFDA can be defined by $V_{LFDA} = \text{argmax} \frac{V^T \tilde{S}_b V}{V^T \tilde{S}_w V}$ and the transformation matrix can be solved in the same way.

Since the affinity matrix only uses the information of the $k$-th neighbor and does not consider the interrelationship among all the neighbors, it is sensitive to the outliers. Besides, LFDA does not take the affinity of samples in different classes into consideration.

## 3   Proposed LFDA with LLE Affinity Matrix

We begin to reformulate the FDA in a matrix manner.

$$S_w = \sum_{l=1}^c \sum_{i:y_i=l} (x_i - \mu_l)(x_i - \mu_l)^T = X(D - W)X^T$$

where $D = (D_{ij})_{d \times d}$ with $D_{ij} = 1$, and $W = (W_{ij})_{d \times d}$ is a weight matrix given by

$$W_{ij} = \begin{cases} \frac{1}{n_l}, & \text{if } y_i = y_j = l; \\ 0, & \text{if } y_i \neq y_j. \end{cases}$$

Similarly, we then have $S_b = X(W - B)X^T$, where $B$ is the $d \times d$ matrix with its all elements being $1/n$.

It should be noted that $W$ is symmetrical and the sum of each row of $W$ is 1, which can be considered as the affinity or weight matrix of LFDA. In this sense, the within-class scatter matrix and between-class scatter matrix of our new LFDA approach are defined as follows.

$$\overline{S}_w = X(D - \widetilde{W})X^T; \overline{S}_b = X(\widetilde{W} - B)X^T,$$

where $\widetilde{W}$ serves as the new affinity or weight matrix. Moreover, it can be solved as the reconstruct matrix in the locally linear embedding (LLE) method [9],[10]. In this way, $\widetilde{W}$ can preserve the local structure of the input data in a more precise manner. For clarity, we introduce the reconstruction matrix of LLE as follows.

Suppose that each input data point and its neighbors lie on or near a locally linear patch of the manifold. We can calculate the reconstruction matrix $W$ as $\widetilde{W} = \arg\min \sum_i^n |x_i - \sum_{j=1}^k \widetilde{W}_{ij} x_{i_j}|^2$, subject to $\sum_j \widetilde{W}_{ij} = 1$ and $\widetilde{W}_{ij} = 0$ if $x_j$ is not the $k$-th or less nearest neighbor of $x_i$, where $x_{i_j}$ is the $j$-th neighbor of $x_i$.

For each input data point $x_i$, let $C^i$ be the local covariance matrix with $C_{jl}^i = (x_i - x_{i_j})^T (x_i - x_{i_l})$. Using the Lagrange multiplier method, we can get the weight vector of $x_i$, $\widetilde{W}_i$ by $\widetilde{W}_i = \frac{(C^i)^{-1}e}{e^T(C^i)^{-T}e}$, where $e$ is the vector whose elements are all 1. Then we can get $\widetilde{W}$ by combining those $\widetilde{W}_i$ together.

Actually, the reconstruction matrix $\widetilde{W}$ represents the local correlation of the data. The small absolute value of $\widetilde{W}_{ij}$ shows that $x_j$ contributes little to the reconstruction of $x_i$, which certainly means little affinity between $x_i$ and $x_j$. So, it is reasonable to use the reconstruction matrix as the affinity matrix in our new LFDA approach. Since $\widetilde{W}$ is a sparse matrix based on the distance information, we obtain its elements with neglecting the label constraints. $\widetilde{W}$ takes into consideration the interrelationship of all the input data points in a neighborhood. Thus, it may be more precise than the affinity matrix used in the conventional LFDA.

In the same way, the projection vector $\overline{V}$ of the proposed LFDA with LLE reconstruction matrix information can be defined as

$$\overline{V} = \arg\max \frac{V^T \overline{S}_b V}{V^T \overline{S}_w V}. \tag{6}$$

Therefore, the transformation matrix should be $\overline{T} = (\varphi_1, \varphi_2, ..., \varphi_r)$, where $\{\varphi_k\}_{k=1}^d$ are the generalized eigenvectors associated with the eigenvalues $\lambda_1 \geq \lambda_2 \geq ... \geq \lambda_d$ of the generalized eigenvalue problem: $\overline{S}_b \varphi = \lambda \overline{S}_w \varphi$. In this way, for each $x_i$, the corresponding projection vector $z_i$ is given by $z_i = \overline{T}^T x_i$.

# 4   Kernelized Version of Proposed LFDA Approach

We further present the kernelized version of our proposed LFDA with LLE information. Let $\phi : z \in \mathbb{R}^d \to \phi(z) \in \mathbb{F}$ be a nonlinear mapping from the input space to a higher dimensional feature space $\mathbb{F}$. The main idea of kernel methods is to make a nonlinear classification problem be linear in a higher dimensional feature space so that the classification problem can be effectively and efficiently solved. Generally, the kernelization of a conventional method makes use of a kernel function $\kappa(.)$ which serves as the inner product in the higher dimensional space, i.e., $\kappa(x, y) = < \phi(x), \phi(y) >$.

The generalized eigenvalue problem of our proposed LFDA approach with LLE information can be expressed by

$$X(\widetilde{W} - B)X^T \varphi = \lambda X(D - \widetilde{W})X^T \varphi, \tag{7}$$

where $\lambda$ is the eigenvalue and $\varphi$ is the corresponding eigenvector.

Then, in the higher dimensional space, Eq.(7) can be generalized as

$$S_b^\phi \widetilde{\varphi} = \widetilde{\lambda} S_w^\phi \widetilde{\varphi} \tag{8}$$

where $S_b^\phi$ and $S_w^\phi$ are the scatter matrix in the kernel space, $S_b^\phi = \Phi(\widetilde{W} - B)\Phi^T$, $S_w^\phi = \Phi(D - \widetilde{W})\Phi^T$, where $\Phi$ is the projection matrix of the input data in the higher dimensional space, that is, $\Phi = [\Phi_1, \Phi_2, \cdots, \Phi_n]$, where $\Phi_i = \phi(x_i)$. According to the reproducing kernel theory [11], $\widetilde{\varphi}$ is the linear combination of the $\phi(x_i)$, that is $\widetilde{\varphi} = \Phi\alpha$.

Let $K$ be the kernel matrix, where $K_{ij} = < \phi(x_i), \phi(x_j) >$. Then we multiply Eq.(8) by $\Phi^T$ and have

$$K(\widetilde{W} - B)K\alpha = \lambda K(D - \widetilde{W})K\alpha. \tag{9}$$

Let $\alpha_k$ for $k = 1, 2, \cdots, n$ be the generalized eigenvectors associated with the generalized eigenvalues $\lambda_1 \geq \lambda_2 \geq ... \geq \lambda_n$ of Eq.(9). Then, the embedded or projected vector of $x_i$ in the dimensionality reduction feature space is given by

$$z_i = (\alpha_1, \alpha_2, ..., \alpha_r)^T \begin{pmatrix} K(x_1, x_i) \\ K(x_2, x_i) \\ \vdots \\ K(x_n, x_i) \end{pmatrix}.$$

# 5   Experimental Results

In this section, we test the proposed LFDA approach with LLE affinity matrix as well as its kernelized version on the classifications of five challenging real-world datasets, being compared with some typical dimensionality reduction methods.

We use the Iris, wine, Wisconsin Diagnostic Breast Cancer (WDBC) and Wisconsin Breast Cancer (WBC) datasets selected from UCI Machine Learning

Repository [12]. Actually, there are 16 missing values in WDBC and we just set them as zero. We also use the USPS handwritten digit (USPS-HD) dataset. For simplicity, we just pick up 200 images randomly for each digit from the dataset. Some basic numbers are listed in Table 1. For clarity, the numbers of training and test sample points are denoted as $n_{training}$ and $n_{test}$.

**Table 1.** The basic numbers of five real-world datasets in the experiments

| Dataset | $c$ | $d$ | $r$ | $n$ | $n_{training}$ | $n_{test}$ |
|---------|-----|-----|-----|-----|----------------|------------|
| Iris | 3 | 4 | 2 | 150 | 120 | 30 |
| wine | 3 | 13 | 6 | 178 | 142 | 36 |
| WBC | 2 | 9 | 2 | 699 | 420 | 279 |
| WDBC | 2 | 30 | 8 | 569 | 350 | 219 |
| USPS-HD | 10 | 256 | 40 | 2000 | 1400 | 600 |

We firstly implement the proposed LFDA approach with LLE affinity matrix (referred to as LFDA-LLE) and its kernelized version (referred to as KLFDA-LLE) for dimensionality reduction. For comparison, we also implement the FDA, LFDA, PCA and LLE approaches. We then implement SVM [13],[14] for supervised classification. As for KLFDA-LLE, the kernel function is selected as the Gaussian kernel function where $\sigma$ is given by experience. In order to test the performance and stability of these dimensionality reduction approaches, we implement the above procedure on a couple of randomly selected training and test sets with the fixed numbers $n_{training}$ and $n_{test}$ for 30 times. The average Classification Accuracy Rates (CARs) of the SVM algorithms with these six approaches are listed in Table 2. Moreover, the sketches of the average CARs of the SVM algorithms with those approaches along the reduced dimensionality are shown in Figure 1.

**Table 2.** The average classification accuracy rates (CARs) of the SVM algorithms with six dimensionality reduction approaches

| Dataset | PCA | LLE | FDA | LFDA | LFDA-LLE | KLFDA-LLE |
|---------|-----|-----|-----|------|----------|-----------|
| Iris | 0.9417 | 0.9433 | 0.9500 | 0.9733 | 0.9733 | 0.9667 |
| wine | 0.9537 | 0.7722 | 0.9722 | 0.9269 | 0.9896 | 0.9639 |
| WBC | 0.9484 | 0.9222 | 0.9614 | 0.9530 | 0.9670 | 0.9753 |
| WDBC | 0.9479 | 0.9161 | 0.9227 | 0.9259 | 0.9727 | 0.9748 |
| USPS-HD | 0.8817 | 0.7600 | 0.8733 | 0.9223 | 0.9210 | 0.9250 |

It can be seen from Table 2 that the proposed LFDA approach with LLE affinity matrix is better than LFDA on wine data, WBC data and WDBC data. And it performs equally as LFDA on Iris data but it is sightly weaker than LFDA on USPS-HD data. In the very high dimensionality cases such as WDBC data

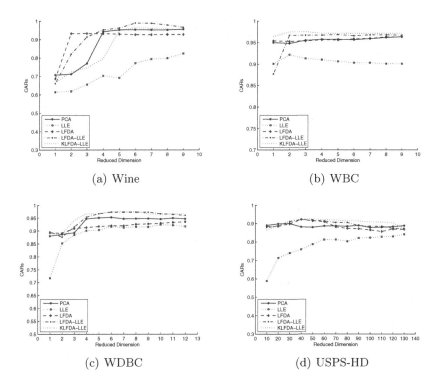

**Fig. 1.** The sketches of the average CARs of the SVM algorithm with five dimensionality reduction approaches along the reduced dimensionality

and USPS-HD data, the kernelized version becomes better than both LFDA and LFDA-LLE. On the other hand, it can be seen from Figure 1 that when the reduced dimensionality is very low, our proposed approach is not as good as LFDA, but can quickly increase to the maximum value and keeps stable as the reduced dimensionality increases. On the other hand, LFDA and FDA cannot increase so high, but reduce obviously on USPS-HD data. It can be also seen that PCA and LLE usually behave poorly. The possible reason may be that both these two approaches are unsupervised dimensionality reduction methods, while the others are supervised methods.

## 6    Conclusions

We have investigated the data dimensionality reduction problem and proposed a new local Fisher discriminant analysis with locally linear embedding affinity matrix in order to preserve the local structures of the input feature space. Actually, the within-class scatter matrix $S_w$ and the between-class scatter matrix $S_b$ are redefined and the affinity or weight matrix is solved through the Locally Linear Embedding (LLE) method. Moreover, the proposed LFDA approach is extended to the kernelized version for nonlinear classification. It is demonstrated by the experiments on five real-world datasets that our proposed dimensionality reduction approaches are more effective and stable in comparison with PCA, LLE, FDA, and LFDA.

**Acknowledgments.** This work was supported by the Natural Science Foundation of China for Grant 61171138 and BGP Inc., China national petroleum corporation.

# References

1. Hinton, G.E., Salakhutdinov, R.R.: Reducing the Dimensionality of Data with Neural Networks. Science 313(5786), 504–507 (2006)
2. Yan, S., Xu, D., Zhang, B., Zhang, H.J., Yang, Q., Lin, S.: Graph Embedding and Extenaions: A General Framework for Dimensionality Reduction. IEEE Transactions on Pattern Analysis and Machine Intelligence 29, 40–51 (2007)
3. Fukunnaga, K.: Introduction to Statistical Pattern Recognition, second editon. Academic Press, Boston (1990)
4. Hotelling, H.: Analysis of a Complex of Statistical Variable into Principal Components. Journal Educational Psychology 24(7), 498–520 (1933)
5. Müller, K., Mika, S., Rätsch, G., Tsuda, K., Schölkopf, B.: An Introduction to Kernel-Based Learning Algorithms. IEEE Transactions on Neural Networks 12(2), 181–201 (2001)
6. Sugiyama, M.: Dimensionality Reduction of Multimodal Labeled Data by Local Fisher Discriminant Analysis. Journal of Machine Learning Research 8, 1027–1061 (2007)
7. He, X., Niyogi, P.: Locality preserving projections. In: Thrun, S., Saul, L., Schölkopf, B. (eds.) Advances in Neural Information Processing Systems 16. MIT Press, Cambridge (2004)
8. Baudat, G., Anouar, F.: Generalized Discriminant Analysis Using a Kernel Approach. Neural Computation 12, 2385–2404 (2000)
9. Roweis, S.T., Saul, L.K.: Nonlinear Dimensionality Reduction by Locally Linear Embedding. Science 290, 2323–2326 (2000)
10. Tenenbaum, J.B., Silva, V., Langford, J.C.: A Global Geometric Framework for Nonlinear Dimensionality Reduction. Science 290, 2319–2323 (2000)
11. Taylor, J.S., Cristianini, N.: Kernel Methods for Pattern Analysis. Cambridge University Press, London (2004)
12. UCI Machine Learning Repository, http://mlearn.ics.uci.edu/databases
13. Suykens, J.A.K., Vandewalle, J.: Least Squares Support Vector Machine Classifiers. Neural Processing Letters 9(3), 293–300 (1999)
14. LibSVM, http://www.csie.ntu.edu.tw/~cjlin/libsvm

# A Facial Expression Recognition Method by Fusing Multiple Sparse Representation Based Classifiers

Yan Ouyang and Nong Sang

Institute for Pattern Recognition and Artificial Intelligence/ Huazhong University of Science and Technology, 430087 Wuhan, China
oyy_01@163.com, nsang@hust.edu.cn

**Abstract.** We develop a new method to recognize facial expressions. Sparse representation based classification (SRC) is used as the classifier in this method, because of its robustness to occlusion. Histograms of Oriented Gradient (HOG) descriptors and Local Binary Patterns are used to extract features. Since the results of HOG+SRC and LBP+SRC are complimentary, we use a classifier combination strategy to fuse these two results. Experiments on Cohn-Kanade database show that the proposed method gives better performance than existing methods such as Eigen+SRC, LBP+SRC and so on. Furthermore, the proposed method is robust to assigned occlusion.

**Keywords:** *Terms*—facial expression recognition, HOG descriptors, Local binary pattern, Classifiers combination, Sparse representation based classification (SRC), assigned occlusion.

## 1    Introduction

The facial expressions recognition (FER), though an interesting application, has been tackled by several researches. In psychology, human facial expression is categorized into six classes of expression (happiness, sadness, disgust, surprise, anger and fear) [1]. There're many state of the art published work [2],[3],[4],[5],[6].In practical application, facial expression of human may be occluded by hands, shoulders and so on. So, until now on, the recognition of facial expressions is still a challenge work.

Inspired by the human vision nervous system can easily recognize the facial expressions with partial occlusions. The latest work to consider about occlusion problems in the application of facial expression recognition by simulating the biological visual nervous system are [7] and [8]. They represent two ideas to solve the occlusion problems. One is simulating the process of biological visual cortex to extract local features such as [7]. So the final classifier can robust to occlusion problem by the help of local features. The other is simulating biological visual perception system to build occlusion robust classifiers such as Sparse Representation based Classification (SRC)[8],[9],[10],[11].In [7] , face images are divided into many local patches and Gabor features are extracted in each patches, the final judgment is made by using NN classifier. There are two aspects worth improving in this method. First, Haar features

C. Guo, Z.-G. Hou, and Z. Zeng (Eds.): ISNN 2013, Part I, LNCS 7951, pp. 479–488, 2013.

are proved to give better performance than Gabor features when recognizing facial expressions [2],[12]. The other is that SRC method is proved to show better robustness than NN when solving occlusion problem [7],[13].In present , SRC is wildly used in the recognition of facial expressions[8],[9],[10],[11]. These work show that SRC has the ability to deal with occlusion problems.

As mentioned above, the suitable strategy to solve occlusion problem is the combination of local features and occlusion robust classifiers. The feature selection is directly affecting the success. There're two factors must be considered in the choice of local features. One is the local features should contain expression information as much as possible. The other is the local features should be illumination invariance. In psychology, facial expression is the combination of action units [2].The shape information and texture information can describe the movement of face components. So, in practical, the HOG descriptors and LBP may be the best choice. In this paper, we use HOG descriptors and LBP conjunction with SRC separately to recognize facial expressions. So, there're two judgment vectors. The result in these judgment vectors is complementary. For example, LBP+SRC gives better performance than HOG+SRC when recognize disgust facial expression. Since, this paper uses a method [17] to fuse these two judgment vectors. The strategy of proposed method is shown in Fig.1. The experiments on Cohn-Kanade database [15] show that the proposed method outperforms existing methods for FER and robust to occlusion. In our experiment, we test the proposed method on face image with eye occlusion and mouth occlusion.

The rest of the paper is organized as follows. Section 2 introduces the process of HOG descriptor extraction. Section 3 introduces the process of LBP feature extraction. Section 4 shows the proposed method. Section 5 shows the experiment results. Section 6 concludes the paper.

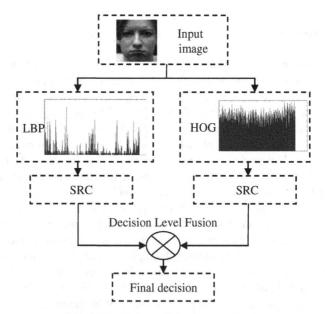

**Fig. 1.** The diagram of proposed method

## 2    Sparse Representation Based Classification

As mentioned in [13], the Algorithm of Sparse Representation-based Classification is shown as below:

Step1: Input a matrix of training samples $A = [A_1, A_2, ..., A_k]$ for k classes, a test sample $y$

Step 2: Normalize the columns of A to have unit $l^2$-norm.

Step 3: Solve the $l^1$-minimization problem:

$$\hat{x}_1 = \arg\min_x \|x\|_1 \ subject\ to\ Ax = y \tag{1}$$

Step 4: Compute the residuals

$$r_i(y) = \|y - A\delta_i(\hat{x}_1)\|_2 \ \ for\ \ i = 1, ..., k \tag{2}$$

where $\delta_i(x) = [0, ..., x_1^i, x_2^i, \cdots, x_{n_i}^i, ..., 0,]$, a coefficient vector is whose entries are zero except those associated with the ith class.

Step 5: Output:

$$identity(y) = \arg\min_i r_i(y) \tag{3}$$

## 3    HOG Descriptors Extraction Strategy

As mentioned in [14], the general process of HOG descriptors can be described as Fig.2.It can be seen that HOG descriptors extract shape information of face images. There're many parameters need to be considered: the size of spatial cells, orientation bins and the size of spatial blocks. As mentioned in [2] and [7], two neighboring cells of face image are interrelated. And the relationships will strongly influence the success of facial expression recognition. So, we choose six spatial cell segmentations: (1) face images are divided into 961 cells, cell size is 4×4 and neighboring cells are 50% overlapping; (2) face images are divided into 256 cells ,cell size is 4×4 and neighboring cells are no overlapping; (3) face images are divided into 225 cells, cell size is 8×8 and neighboring cells are 50% overlapping; (4) face images are divided into 64 cells, cell size is 8×8 and neighboring cells are no overlapping; (5) face images are divided into 49 cells, cell size is 16×16 and neighboring cells are 50% overlapping; (6) face images are divided into 16 cells, cell size is 16×16 and neighboring cells are no overlapping. Another important parameter is orientation bin numbers. We choose six bin numbers N=2, 6,8,9,12,18 with space range [0,360]. The rest parameters are described as below: (1) The orientation gradients are calculated by using 3×3 Sobel mask without Gaussian smoothing; (2) Each spatial block contains four cells; (3) L2-norm method is used to implement block normalization.

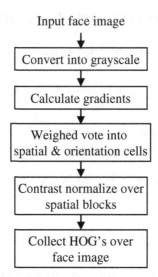

**Fig. 2.** The process of HOG descriptors extraction

## 4    LBP Feature Extracting Strategy

As mentioned in [3], [9], [16], the process of LBP feature extraction can be described as Fig 3.There are two key parameters: (1) LBP operator; (2) The division of local patches. We choose four LBP operators as shown in Fig 4. The divisions of local patches are introduced as below: (1) the size of patch is 4×4 and neighboring patches are 0%, 25% and 50% overlapping separately. (2) The size of patch is 8×8 and neighboring patches are 0%, 25% and 50% overlapping separately. So we totally get six divisions.

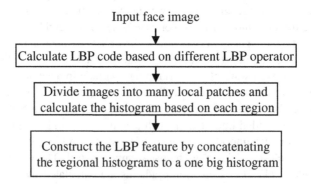

**Fig. 3.** The general process of extracting LBP features

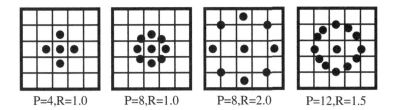

P=4,R=1.0          P=8,R=1.0          P=8,R=2.0          P=12,R=1.5

**Fig. 4.** LBP operators

## 5     Proposed Method

The proposed method uses HOG descriptors, LBP as features and SRC as classifier. First use HOG+SRC and LBP+SRC to judge the test samples of facial expressions separately. Then use these two results vectors as the posteriori probability of test samples to make the final decision more accurate [17]. The theory of classifier combination in [17] can be briefly described as below.

According to the Bayesian theory, given measurements $x_i$, i = 1, ... ,R of different classifier method,  the pattern , Z, should be assigned  to class $w_j$ provided the a posteriori probability of that interpretation is maximum, i.e..

$$assign \quad Z \rightarrow w_j \quad if \tag{4}$$

$$P\left(w_j | x_1,...., x_R\right) = \max_k P\left(w_k | x_1,...x_R\right) \tag{5}$$

Where R represents the number of classifiers and k represents the class numbers. There're six combination rules in [17] based on this basic theory. The applicable condition of these rules is that the given measurements of classifiers are complementary.

So, in briefly, the process of the proposed method can be introduced as below.

Step 1: Input a dictionary D which constructed by training samples
Step 2: Input a test sample y and apply feature transformation $P_1$ (HOG descriptors transformation), $P_2$ (LBP feature transformation) to both y and D
Step 3: Calculate the  $x$  by solving $l_1$-minimization problem:

$$\min\|x\|_1 \quad s.t. \quad P_j Dx = P_j y \quad (j =1, 2) \tag{6}$$

Step 4: Judge the class where test sample y belongs to:

$$identify(y) = \arg\min_i \left\|P_j y - D\delta_i\left(P_j x\right)\right\|_2 \tag{7}$$

Step 5: Go to step 2 until all the test samples are classified.
Step 6: Calculate two results vector $Q_1$ and $Q_2$ ($Q_1$ is the result judging by HOG+SRC, $Q_2$ is the result judging by using LBP+SRC).
Step 7: Use the combination rules in [17] to fuse the two results and generate final decision of each test sample.

# 6     Experiment

Experiments are conducted on the Cohn-Kanade facial expression database [15]. We select 339 image sequences from 94 subjects. From each sequence, we select the last three frames to organize our dataset and normalize the entire dataset to 64×64 pixels by using the same method as [2]. There're many person-independent validation method, such as 10-fold [7] and so on. In this paper we use the same person-independent validation method as [2]. We randomly select 66% subjects to organize the dictionary and the rest subjects are used as test samples.

## 6.1     HOG Descriptors Parameter Selection

As mentioned above, we need to find some suitable parameters when extracting HOG descriptors. Fig 5 shows the results of different parameter selection. It can be seen that when face images are divided into 961 local patches and bin number set to 9, the experiment result on our dataset is the best. The best accurate rate is 94.1%.

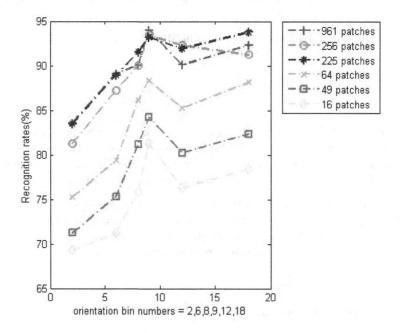

**Fig. 5.** Recognition rates based on different parameters

## 6.2     LBP Parameter Selection

As shown in Fig 6, it can be seen that the best performance is 92.398%.The size of patch is 4×4 and neighboring patches are 25% overlapping. LBP (P=8, R=2.0) operator is chosen.

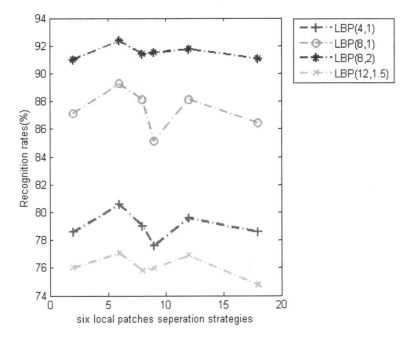

**Fig. 6.** Recognition rates based on different parameters

## 6.3    Experiment on Clean Face Image

In this part, we test the proposed method on our dataset which contain only clean face image. Table1 and Table 2 show the confusion matrix of HOG+SRC and LBP+SRC. It can be seen that LBP+SRC gives better performance than HOG+SRC when recognizing disgust expression. So, we use the combination strategy in [17] to combine these two result vectors. Though [17] introduces six combination rules, table 3 shows the fusion results based on the six combination rules.

**Table 1.** The confusion matrix(%) of HOG+SRC

|  | Anger | Disgust | Fear | Happi-Ness | Sadness | surprise |
|---|---|---|---|---|---|---|
| Anger | 84.62 | 0 | 0 | 0 | 15.38 | 0 |
| Disgust | 0 | 93.33 | 0 | 0 | 6.67 | 0 |
| Fear | 0 | 13.33 | 86.67 | 0 | 0 | 0 |
| Happi-ness | 0 | 0 | 0 | 100 | 0 | 0 |
| Sadness | 0 | 0 | 0 | 0 | 100 | 0 |
| Surprise | 0 | 0 | 0 | 0 | 0 | 100 |

**Table 2.** The confusion matrix(%) of LBP+SRC

|  | Anger | Disgust | Fear | Happi-ness | Sadness | surprise |
|---|---|---|---|---|---|---|
| Anger | 84.62 | 0 | 0 | 0 | 15.38 | 0 |
| Disgust | 4.44 | 95.56 | 0 | 0 | 0 | 0 |
| Fear | 5 | 0 | 80.00 | 11.67 | 3.33 | 0 |
| Happi-Ness | 0 | 0 | 1.23 | 98.77 | 0 | 0 |
| Sadness | 3.03 | 0 | 0 | 0 | 96.67 | 0 |
| Surprise | 0 | 0 | 1.23 | 0 | 0 | 98.77 |

**Table 3.** The fusion results(%) by using different combination rules

|  | Sum | Product | Max | Min | Median | Majority rote |
|---|---|---|---|---|---|---|
| Rate | 95.64 | 95.33 | 94.42 | 94.79 | 95.23 | 93.96 |

We compared our method to the existing FER method based on SRC. For example, RAW+SRC [8], Gabor+SRC[8], Eigen+SRC [8]. Table 4 shows the comparison results.

**Table 4.** The perfomance of the method based on SRC on our dataset

| Methods | Accuracy (%) |
|---|---|
| RAW+SRC | 85.70 |
| Gabor+SRC | 81.34 |
| Eigen+SRC | 80.32 |
| Ours | 95.64 |

We also compared the proposed method to other automatic FER methods. The results are shown in Table 5. Because of different dataset and validation method, the results in table 5 are not directly comparable. But through the comparison, it can be seen that our method is efficient.

**Table 5.** Comparison with different method based on the CK database

|  | Subjects | Classes | Measures | Rate(%) |
|---|---|---|---|---|
| [2] | 96 | 6 | — | 92.3 |
| [3] | 96 | 6 | 10-fold | 92.1 |
| [4] | 97 | 6 | 5-fold | 90.9 |
| [7] | 94 | 7 | 10-fold | 91.51 |
| ours | 94 | 6 | — | 95.64 |

## 6.4     Experiment on Face Image with Assigned Occlusion

The advantage of SRC is the robustness to occlusion. So, in this part, we test our method on the face image with two kinds of occlusions. One is eye region occlusion; the other is mouth region occlusion as shown in Figure 7. We compared our method to the existing methods based on SRC. From table 6, it can be seen that our method is outperform than the other method based on SRC.

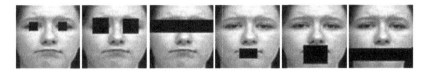

**Fig. 7.** Examples of face image with assigned occlusion. (1) Eye occluded with small occlusion. (2) Eye occluded with medium occlusion. (3) Eye occluded with large occlusion. (4) Mouth occluded with small occlusion. (5) Mouth occluded with medium occlusion. (6) Mouth occluded with large occlusion.

**Table 6.** Recognition rates(%) on the occlusion face image

|  | Eye small | Eye Medium | Eye Large | Mouth Small | Mouth Medium | Mouth Large |
|---|---|---|---|---|---|---|
| Ours | 93.34 | 89.23 | 85.12 | 85.45 | 79.65 | 72.58 |
| RAW+ SRC | 83.54 | 75.45 | 60.12 | 75.35 | 69.25 | 51.56 |
| Gabor+ SRC | 76.32 | 68.23 | 60.23 | 72.12 | 63.12 | 52.12 |
| Eigen+ SRC | 71.23 | 62.23 | 51.12 | 70.23 | 64.12 | 50.23 |

## 7    Conclusion

A Novel approach for facial expression recognition by fusion different classifier based on Sparse Representation is proposed. The experiment on Cohn-Kanade database shows that the proposed algorithm gives better performance than the existing state of art work. But the time cost of proposed method is much higher than the existing methods based on SRC.

## References

1. Izard, C.E.: The face of emotion. AppletonCentury-Crofts, New York (1971)
2. Yang, P., Liu, Q.S., Metaxas, D.N.: Exploring facial expressions with compo-sitional features. In: IEEE International Conference on Computer Vision and Pattern Recognition, pp. 2638–2644 (2010)
3. Shan, C., Gong, S., McOwan, P.W.: Robust facial expression recognition using local binary patterns. In: IEEE International Conference on Image Processing, pp. 370–373 (2005)
4. Yeasin, M., Builot, R., Sharma, R.: From facial expression to level of interest: a spatiotemporal approach. In: Conference on Computer Vision and Pattern Recognition, pp. 922–927 (2004)

5. Pantic, M., Rothkrantz, L.J.M.: Automatic analysis of facial expressions: The state of the art. IEEE Trans. on PAMI. Intell. 22, 1424–1445 (2000)
6. Ou, J., Bai, X.B., Pei, Y., Ma, L., Liu, W.: Automatic facial expression recognition using Gabor filter and expression analysis. In: International Conference on Computer Modeling and Simulation, pp. 215–218 (2010)
7. Wenfei, G., Xiang, C., Venkatesh, Y.V., Huang, D., Lin, H.: Facial expression recognition using radial encoding of local Gabor features and classifier synthesis. Pattern Recognition 45(1), 80–91 (2012)
8. Cotter, S.F.: Sparse representation for accurate classification of corrupted and oc-cluded facial expressions. In: Proc. ICASSP, pp. 838–841 (2010)
9. Huang, M.W., Wang, Z.W., Ying, Z.L.: A new method for facial expression recognition based on Sparse representation plus LBP. In: International Congress on Image and Signal Processing (CISP), pp. 1750–1754 (2010)
10. Huang, M.W., Ming, Z.L.: The performance study of facial expression recognition via sparse representation. In: International Conference on Machine Learning and Cybernetics, ICMLC, pp. 824–827. IEEE Press, Jiangmen (2010)
11. Cotter, S.F.: Recognition of occluded facial expressions using a fusion of localized sparse representation classifiers. In: Digital Signal Processing Workshop and IEEE Signal Processing Education Workshop (DSP/SPE), pp. 4–7 (2011)
12. Whitehill, J., Omlin, C.W.: Haar features for facs au recognition. In: 7th International Conference on Automatic Face and Gesture Recognition, pp. 97–101. IEEE Press, Bell-ville (2006)
13. Wright, J., Yang, A.Y., Ganesh, A., Sastry, S.S., Ma, Y.: Robust face recognition via sparse representation. IEEE Trans. on PAMI 31(2), 210–227 (2009)
14. Dalal, N., Triggs, B.: Histograms of oriented gradients for human detection. In: IEEE Conf. on Computer Vision and Pattern Recognition, pp. 886–893 (2005)
15. Kanade, T., Cohn, J.F., Tian, Y.: Comprehensive database for facial expression analysis. In: International Conference on Automatic Face and Gesture Recognition, pp. 46–53 (2000)
16. Ojala, T., Pietikaninen, M., Maenpaa, T.: Multiresolution Gray-Scale and Rotation Invariant Texture Classification with Local Binary Patterns. IEEE Trans. on PAMI (24), 971–987 (2002)
17. Josef, K., Mohamad, H., Robert, P.W.D., Jiri, M.: On combining classifiers. IEEE Trans. on PAMI (20), 226–239 (1998)

# Image Data Classification Using Fuzzy c-Means Algorithm with Different Distance Measures

Dong-Chul Park

Dept. of Electronics Engineering, Myong Ji University, Korea
parkd@mju.ac.kr

**Abstract.** Fuzzy c-Means algorithms(FCMs) with different distance measures are applied to an image classification problem in this paper. The distance measures discussed in this paper are the Euclidean distance measure and divergence distance measure. Different distance measures yield different types of Fuzzy c-Means algorithms. Experiments and results on a set of satellite image data demonstrate that the classification model employing the divergence distance measure can archive improvements in terms of classification accuracy over the models using the FCM and SOM algorithms which utilize the Euclidean distance measure.

## 1  Introduction

Recent increase in the use of large image database requires an automatic classification tool which yields efficient content-based classification of image data. In developing an automatic image classification tool, a clustering algorithm can be utilized as a part of the classification tool. Several clustering algorithms have been adopted for this purpose. Among them, k-Means algorithm [1,2] and Self Organizing Map (SOM)[3] have been widely applied to the problem of clustering data in practice for their simplicity. The k-Means algorithm and adaptive k-Means algorithm can be considered as the basis of SOM. In order to improve the performance of k-Means algorithm and SOM, Fuzzy C-Means (FCM) clustering algorithm is later introduced[4,5]. Note that these algorithms are all designed with Euclidean distance measure. In order to deal with nonlinear classification problems, kernel methods are introduced[6,7]. The kernel methods project the input data into a feature space with a higher dimension and nonlinear problems in input data space are solved linearly in feature space. The kernel methods, consequently, can improve the classification accuracy. Unlike SOM, an object can be assigned to all available classes with different certainty grades in FCM. By doing so, the FCM algorithm is more robust when compared with SOM or k-Means algorithm.

Divergence distance measure has been also combined with the FCM algorithm for utilizing statistical information from image data[8]-[10]. By combining the ideas of FCM and divergence distance measure for the satellite image classification problem, a higher image classification accuracy can be expected.

The remainder of this paper is organized as follows: Section 2 summarizes the conventional FCM with Euclidean distance measure. The FCM with divergence

C. Guo, Z.-G. Hou, and Z. Zeng (Eds.): ISNN 2013, Part I, LNCS 7951, pp. 489–496, 2013.
© Springer-Verlag Berlin Heidelberg 2013

distance measure is summarized in Section 3. Experiments and results on a set of satellite image data are reported in Section 4. Finally, conclusions are given in Section 5.

## 2   Fuzzy c-Means(FCM)with Euclidean Distance Measure

A clustering algorithm's purpose is to group of similar data and separate dissimilar ones. FCM has been widely used as a clustering algorithm. Bezdek first defines a family of objective functions $J_m, 1 < m < \infty$ and generalized the *fuzzt ISODATA* by establishing a convergence theorem for that family of objective functions [4]. For FCM, the objective function is defined as :

$$J_m(U, v) = \sum_{k=1}^{n} \sum_{i=1}^{c} (\mu_{ki})^m (d_i(x_k))^2 \qquad (1)$$

where the following notations are used:

- $\bullet$ $d_i(x_k)$ : distance from the input data $x_k$ to $v_i$, the center of the cluster $i$
- $\bullet$ $\mu_{ki}$ : membership value of the data $x_k$ to the cluster $i$
- $\bullet$ $m$ : weighting exponent, $m \in 1, \cdots, \infty$
- $\bullet$ $n$ and $c$ : number of input data and clusters.

Note that the distance measure used in FCM is the Euclidean distance.

The condition for minimizing the objective function by Bezdek are summarized with the following equations:

$$\mu_{ki} = \frac{1}{\sum_{j=1}^{c} \left( \frac{d_i(x_k)}{d_j(x_k)} \right)^{\frac{2}{m-1}}} \qquad (2)$$

$$v_i = \frac{\sum_{k=1}^{n} (\mu_{ki})^m x_k}{\sum_{k=1}^{n} (\mu_{ki})^m} \qquad (3)$$

The FCM finds the optimal values of prototypes iteratively by applying the above two equations alternately.

## 3   FCM with Divergence Distance Measure

In clustering algorithms, it is very important to choose a proper distance measure since the chosen distance measure affects the performance of an algorithm[5]. When a data vector is given as a Gaussian Probability Density Functions(GPDF), the Divergence distance measure can be a promising choice[5,11]. The Divergence distance measure is also called as *Kullback-Leibler Divergence*. Given two

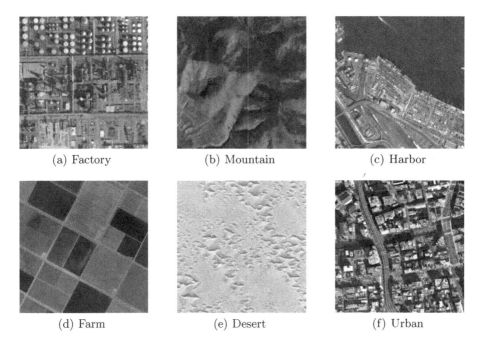

(a) Factory          (b) Mountain          (c) Harbor

(d) Farm          (e) Desert          (f) Urban

**Fig. 1.** Examples of different category data

GPDFs, $x = (x_i^\mu, x_i^{\sigma^2})$ and $v = (v_i^\mu, v_i^{\sigma^2})$, $i = 1, \cdots, d$ , the Divergence distance measure is defined as:

$$D(x, v) = \sum_{i=1}^{d} \left( \frac{x_i^{\sigma^2} + (x_i^\mu - v_i^\mu)^2}{v_i^{\sigma^2}} + \frac{v_i^{\sigma^2} + (x_i^\mu - v_i^\mu)^2}{x_i^{\sigma^2}} - 2 \right)$$

$$= \sum_{i=1}^{d} \left( \frac{(x_i^{\sigma^2} - v_i^{\sigma^2})^2}{x_i^{\sigma^2} v_i^{\sigma^2}} + \frac{(x_i^\mu - v_i^\mu)^2}{x_i^{\sigma^2}} + \frac{(x_i^\mu - v_i^\mu)^2}{v_i^{\sigma^2}} \right) \qquad (4)$$

where the following notations are used:

- $\bullet x_i^\mu$ : $\mu$ value of the $i^{th}$ component of $x$
- $\bullet x_i^{\sigma^2}$ : $\sigma^2$ value of the $i^{th}$ component of $x$
- $\bullet v_i^\mu$ : $\mu$ value of the $i^{th}$ component of $v$
- $\bullet v_i^{\sigma^2}$ : $\sigma^2$ value of the $i^{th}$ component of $v$

By combining the divergence measure and FCM, the FCM with divergence measure (D-FCM) is introduced[10]. D-FCM calculates and updates their parameters at each application of each data vector while FCM calculates the center parameters of the clusters after applying all the data vectors. The advantage of this iterative application and updating the center parameters was reported. The updating strategy for the mean and variance of center values follows the procedure of [8]. When a data vector $x$ is applied, the membership for the data vector to the cluster $i$ is calculated by the following:

$$\mu_i(x) = \frac{1}{\sum_{j=1}^{c} (\frac{D(x,v_i)}{D(x,v_j)})^2} \tag{5}$$

After finding the proper membership grade from an input data vector $x$ to each cluster $i$, the D-FCM updates the mean and variance of each prototype as follows:

$$v_i^\mu(n+1) = v_i^\mu(n) - \eta\mu_i^2(x)(v_i^\mu(n) - x^\mu) \tag{6}$$

$$v_i^{\sigma^2}(n+1) = \frac{\sum_{k=1}^{N_i}(x_{k,i}^{\sigma^2}(n) + (x_{k,i}^\mu(n) - v_i^\mu(n))^2)}{N_i} \tag{7}$$

where the following notations are used:

- • $v_i^\mu(n)$ or $v_i^{\sigma^2}(n)$ : the mean or variance of the cluster $i$ at the time of iteration $n$
- • $x_{k,i}^\mu(n)$ or $x_{k,i}^{\sigma^2}(n)$ : the mean or variance of the $k^{th}$ data in the cluster $i$ at the time of iteration $n$
- • $\eta$ and $N_i$ : the learning gain and the number of data in the cluster $i$

## 4    Experiments and Results

In order to evaluate the performances of FCMs with Euclidean distance measure and Divergence distance measure, a set of satellite image data is collected. The data set consists of six different image categories representing different areas as shown in Fig. 1: factory area, mountain area, harbor area, farming area, desert area, and urban area. Each category contains 100 images. From each class of data set, 80 images were randomly chosen for training classifiers while the remaining images were used for testing classifiers. Localized image representation method is adopted for extracting texture information from image data. The contents of a image data is presented by a collection of localized features which is computed at different points of interest in image space. The localized interest of points in image space is determined by the used of sliding windows or blocks. By combining these extracted localized features, the feature for the entire image is obtained.

For the texture information of an image, Gabor filters [13] and wavelet filters [14] are frequently adopted for extracting the frequency domain information. In case of the applications that requires a real-time operation, these filters are not good choice because of their high computational complexity. Tools with less computational complexity are required for this case and the Discrete Cosine Transform (DCT) is one of them for feature extracting methods suitable for

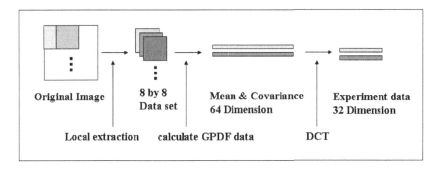

**Fig. 2.** Feature extraction procedure [9]

our purpose. The DCT extracts the frequency information from an image data. Extraction of the localized frequency information form a image is summarized in Fig. 2. The feature extraction procedure works basically in sliding window fashion and calculates the DCT coefficients from local points of interest. Its window size is $8 \times 8$ and yields 64 DCT coefficients. Out of 64 DCT coefficients, 32 lower half frequency coefficients are used for our experiments for each block of image data. By combining the extracted frequency information from image blocks, a feature vector is obtained for each data. The distribution of its feature vectors formed from each image data is then used for finding the prototype of each image category by using a clustering algorithm. Each prototype represents a category with its mean vector and covariance matrix. For evaluating a classifier, the following Bayesian classifier is utilized.

$$Genre(x) = \arg \max_{i} P(x|v_i) \tag{8}$$

$$P(x|v_i) = \sum_{i=1}^{M} c_i \aleph(x, \mu_i) \tag{9}$$

$$\aleph(x, \mu_i) = \frac{1}{\sqrt{(2\pi)^d}} e^{-0.5(x-\mu_i)^T (x-\mu_i)} \tag{10}$$

where the following notations are used:

- $\bullet$ $M$: the number of code vectors
- $\bullet$ $c_i$: the weight of the code vectors
- $\bullet$ $d$ : the number of dimensions of the feature vectors (d=32)
- $\bullet$ $\mu_i$ : the mean of the $i - th$ group of the class's data

In order to evaluate different distance measure used for FCM, experiments on a set of satellite image data are performed with conventional FCM, FCM with Divergence distance measure, and SOM. Fig. 3 summarizes their performance in terms of classification accuracies for the three classifiers. Note that the conventional FCM and SOM use the Euclidean distance measure and utilize only the

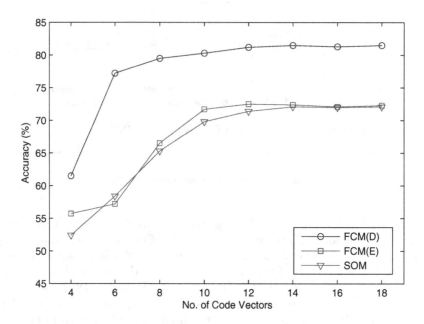

**Fig. 3.** Overall classification accuracies using different algorithms

mean vector value from GPDF data for each image data. Meanwhile, FCM with Divergence distance measure utilizes the covariance matrix information from GPDF in addition to the mean vector value when calculating the distance between two GPDFs. Experiments are performed by varying the number of code vectors from 4 to 18. The average classification accuracies obtained from SOM, FCM with Euclidean distance measure (FCM(E)), and FCM with Divergence distance measure (FCM(D)) over different numbers of code vectors are 66.7%, 67.5%, and 78.0%, respectively. Fig.3 summaries the classification accuracies among different classifiers with different numbers of code vectors. As can be seen from Fig.3, the classification accuracy for a classifier increases as the number of code vectors grows and saturates when the number of code vectors reaches at 10. Table 1 shows the classification accuracy for each category for different classifiers when each classifier uses 14 code vectors. As can be seen from Fig. 3 and Table 1, the classifier that uses Divergence distance measure outperforms the classifiers that use the Euclidean distance measure. It is somewhat obvious because the Divergence distance measure utilizes the covariance information in GPDF while the Euclidean measure distance does not. However, it is worthwhile to mention that how the covariance information can be utilized in FCM with the Divergence distance measure. In order to improve the classification accuracy, the future study should include other feature extraction methods than the DCT.

Table 1. Classification accuracy (%) of different algorithms using 12 code vectors

|         | Factory | Mountain | Harbor | Farm | Desert | Urban | Overall |
|---------|---------|----------|--------|------|--------|-------|---------|
| FCM(D)  | 75.8    | 74.2     | 88.5   | 74.0 | 84.3   | 93.3  | **81.5%** |
| FCM(E)  | 71.4    | 64.5     | 76.1   | 67.4 | 71.3   | 83.7  | **72.4%** |
| SOM     | 73.3    | 62.5     | 75.4   | 65.5 | 71.1   | 84.7  | **72.1%** |

# 5 Conclusion

In this paper, a comparative study on the use of different distance measures for satellite image classification method has been performed. The distance measures discussed in this paper are the Euclidean distance measure and divergence distance measure. Different distance measures requires updating formulas for different types of Fuzzy c-Means algorithms. Experiments and results on a set of satellite image data demonstrate that the classification model employing the divergence distance measure can archive improvements in terms of classification accuracy over the models using the FCM and SOM algorithms which utilize the Euclidean distance measure. It is somewhat obvious because the Divergence distance measure utilizes the covariance information in GPDF while the Euclidean measure distance does not. Other distance measures than the Divergence distance measure will be included in future work.

**Acknowledgment.** This work was supported by National Research Foundation of Korea Grant funded by the Korean Government (2010-0009655) and by the IT R&D program of The MKE/KEIT (10040191, The development of Automotive Synchronous Ethernet combined IVN/OVN and Safety control system for 1Gbps class).

# References

1. Hartigan, J.: Clustering Algorithms. Wiley, New York (1975)
2. Darken, C., Moody, J.: Fast adaptive k-means clustering: Some empirical resultes. In: Proc. of Int. Joint Conf. Neural Networks, vol. 2, pp. 238–242 (1990)
3. Kohonen, T.: The Self-Organizing Map. Proc. IEEE 78, 1464–1480 (1990)
4. Bezdek, J.: A convergence theorem for the fuzzy isodata clustering algorithms. IEEE Trans. on Pattern Analysis and Machine Intelligence PAMI-2(1), 1–8 (1980)
5. Bezdek, J.: Pattern Recognition with Fuzzy Objective Function Algorithms. Plenum, New York (1981)
6. Jonh, S., Nello, C.: Kernel methods for pattern analysis. Cambridge University Press (2004)
7. Muller, K.: An introduction to kernel-based learning algorithms. IEEE Trans. on Neural Networks 12, 181–201 (2001)
8. Park, D.-C., Nguyen, D.-H., Beack, S.-H., Park, S.: Classification of Audio Signals Using Gradient-Based Fuzzy c-Means Algorithm with Divergence Measure. In: Ho, Y.-S., Kim, H.-J. (eds.) PCM 2005. LNCS, vol. 3767, pp. 698–708. Springer, Heidelberg (2005)

9. Park, D., Woo, D.: Image classification using Gradient-Based Fuzzy c-Means with Divergence Measure. In: Proc. Int. Joint Conference on Neural Networks, pp. 2520–2524 (2008)

10. Park, D.-C.: Satellite image classification using a divergence-based fuzzy c-means algorithm. In: Elmoataz, A., Mammass, D., Lezoray, O., Nouboud, F., Aboutajdine, D. (eds.) ICISP 2012. LNCS, vol. 7340, pp. 555–561. Springer, Heidelberg (2012)

11. Fukunaga, K.: Introduction to Statistical Pattern Recognition. Academic Press Inc. (1990)

12. Park, D.-C., Kwon, O.-H., Centroid, C.J.: neural network with a divergence measure for gpdf data clustering. IEEE Trans. on Neural Networks 19(6), 948–957 (2008)

13. Daugman, J.G.: Complete Discrete 2D Gabor Transform by Neural Networks for Image Analysis and Compression. IEEE Trans. on Acoust., Speech, and Signal Processing 36, 1169–1179 (1988)

14. Pun, C.M., Lee, M.C.: Extraction of Shift Invariant Wavelet Features for Classification of Images with Different Sizes. IEEE Trans. on Pattern Analysis and Machine Intelligence 26(9), 1228–1233 (2004)

15. Huang, Y.L., Chang, R.F.: Texture Features for DCT-Coded Image Retrieval and Classification. In: Proc. IEEE Int. Conf. on Acoustics, Speech, and Signal Processing, vol. 6, pp. 3013–3016 (1999)

# Global Matching to Enhance the Strength of Local Intensity Order Pattern Feature Descriptor

Hassan Dawood, Hussain Dawood, and Ping Guo[*]

Image Processing and Pattern Recognition Laboratory
Beijing Normal University, Beijing 100875, China
{hasandawod,hussaindawood2002}@yahoo.com, pguo@ieee.org

**Abstract.** Local intensity order pattern feature descriptor is proposed to extract the feature of image recently. However, it did not provide the global information of an image. In this paper, a simple, efficient and robust feature descriptor is presented, which is realized by adding the global information to local intensity features. A descriptor, which utilizes local intensity order pattern and/or global matching, is proposed to gather the global information with local intensity order. Experimental results shows that the proposed hybrid approach outperform over the state-of-the art feature extraction method like scale-invariant feature transform, local intensity order pattern and DAISY for standard oxford dataset.

**Keywords:** Local Intensity Order Pattern, Global matching, Feature extraction, Image classification, Interest point based feature.

## 1 Introduction

Feature extraction method plays an important role in the field of image processing, such as in object recognition [1], remote sensing [2], medical imaging [3], image retrieval [4], wide base line matching [5], panoramic image stitching [6] and so on. Many methods have been used to detect the interest regions, among them, commonly used for affine covariant region detection includes IBR (Intensity-Based Region) [5], Harris-affine [7], Hessian-affine [8], MSER (Maximally Stable Extremal Region) [9] and EBR (Edge- Based Region) [5] methods. Currently, most popular feature extraction method is SIFT (Scale Invariant Feature Transform), which is proposed by Lowe [1] based on histogram. Krystian *et al.* [11] proposed an extension of SIFT descriptor named as GLOH (Gradient Location-Orientation Histogram). SURF (Speeded-Up Robust Features) is presented by Bay *et al.* [12], in which they got the speed gain due to the usage of integral images, which drastically reduce the number of operations for simple box convolutions, independent of the chosen scale. Lazebnik *et al.* proposed a rotation invariant descriptor called the RIFT (Rotation Invariant Feature Transform) [19]. Tola *et al.* proposed a method called DAISY [13], which creates a histogram of

---

[*] Corresponding author.

C. Guo, Z.-G. Hou, and Z. Zeng (Eds.): ISNN 2013, Part I, LNCS 7951, pp. 497–504, 2013.
© Springer-Verlag Berlin Heidelberg 2013

gradient orientations and locations, the increase in speed comes from replacing weighted sums used by the earlier descriptors by sums of convolutions, which can be computed very quickly and from using a circularly symmetrical weighting kernel.

Bin Fan [14] proposed a descriptor by combining the intensity order with gradient distribution in multiple support regions. Heikkila *et al.* [15] proposed a texture feature called center-symmetric local binary pattern (CS-LBP) by combining the strengths of SIFT and local binary pattern (LBP) [16] texture operator. Weber local descriptor [17] consists of two components: differential excitation and orientation. Dalal and Triggs proposed a histogram of oriented gradients (HOG) [18] feature descriptor also. Many variants of LBP have been recently proposed and have achieved considerable success in various tasks. Ahonen *et al.* exploited the LBP for face recognition [20]. Rodriguez and Marcel proposed adapted LBP histograms for face authentication [21]. Local intensity order pattern (LIOP) [24] is proposed to encode the local ordinal information of each pixel and the overall ordinal information is used to divide the local patch into sub-regions, which are used for accumulating LIOPs. However, by using such features for recognition has a low recognition rate.

In this paper, we propose an efficient global matching scheme where the LIOP is used for feature extraction. Global matching is obtained by computing the minimum distance in all directions. Also, by adopting the global matching technique effectively, we can reduce the requirement of using the large training sets.

The rest of paper is organized as follows: Section 2 briefly reviews the LIOP; Section 3 presents the proposed global matching method with dissimilarity matrix, Section 4 presents the detailed experiments and results. Finally in Section 5 the conclusion is given.

## 2     LIOP

In this section, we will briefly review LIOP [24] descriptor. In LIOP, the overall intensity is used to divide the local patch into sub-regions named as ordinal bins and the intensity relationship of neighboring sample points is calculated. LIOP descriptor is constructed by concatenating all bins.

Assuming $p^N = \{(p_1, p_2, ..., p_N): p_i \in R\}$ is the set of $N$-dimensional vectors and set of all possible permutations of integer $\{1, 2,...,N\}$, and the mapping $\gamma$ : from $N$-dimensional vector to $N$-dimensional permutation vector is defined as $P^N \rightarrow \prod^N$, where $P \in P^N$ and $\pi \in \prod^N$. The mapping can be defined as

$$\gamma(P) = \pi, P \in P^N, \pi \in \prod^N \qquad (1)$$

Where $\pi = (i_1, i_2, i_3, ..., i_N)$ and $P_{i_1} \leq P_{i_2} \leq ... \leq P_{i_N}$. The mapping $\gamma$ divides $P^N$ into $N!$ partitions. For a given permutation $\pi \in \prod^N$, the corresponding partition of $P^N$ is defined as

$$S(\pi) = \{P : \gamma(P) = \pi, P \in P\}. \qquad (2)$$

From Eq. 2 the equivalence equation can be written as:

$$P, P' \in S(\pi)$$
$$\pi = (i_1, i_2, ..., i_N) \Leftrightarrow P_{i_1} \leq P_{i_2} \leq ... \leq P_{i_N}, P'_{i_1} \leq P'_{i_2} \leq ... \leq P'_{i_N}. \tag{3}$$

By using the index table, the feature mapping function $\phi$ is defined to map a permutation $\pi$ to an $N!$-dimensional feature vector $V_{N!}^i$ whose elements are zero except for the $i$-th element which is 1. Where $\phi$ is defined as

$$\phi(\pi) = V_{N!}^{Ind(\pi)}, \pi \in \Pi^N. \tag{4}$$

From the above definitions, Assuming $P(x)$ is a $N$-dimensional vector which consist the intensity of $N$ neighboring sample points of $x$ in the local patch, the LIOP of the point $x$ is defined as:

$$LIOP(x) = \phi(\gamma(P(x)))$$

$$= V_{N!}^{Ind(\gamma(P(x)))} \tag{5}$$

$$= \left( 0,...,0 \underset{(Ind(\gamma(P(x))))}{1} ,0,...0 \right),$$

where $P(x) = (I(x_1), I(x_2), ..., I(x_N)) \in P^N$ and $I(x_i)$ denotes the intensity of $i$-th neighboring sample point $x_i$. The mathematical definition of LIOP is

$$\text{LIOP descriptor} = (des_1, des_2, ..., des_B),$$

$$des_i = \sum_{x \in bin_i} LIOP(x). \tag{6}$$

Where $B$ represents the ordinal bins. The dimension of the descriptor is $N! \times B$. Due to the Gaussian noise effect on intensity a weighting function is proposed to improve the robustness of the LIOP descriptor, which is defined as:

$$w(x) = \sum_{i,j} sgn\left(\left|I(x_i) - I(x_j)\right| - T_{lp}\right) + 1, \tag{7}$$

Where sgn() is the sign function and $T_{lp}$ is a present threshold. So the LIOP descriptor with weighting is written as:

$$\text{LIOP descriptor} = (des_1, des_2, ..., des_B) \tag{8}$$

$$des_i = \sum_{x \in bin_i} w(x) LIOP(x),$$

## 3    Global Matching with Dissimilarity Metric

The rotation invariant methods have been proposed in literature [28,29], where LIOP extracts the features from the local region. However, it is not able to correctly recognize those images in the same class. For example, the LIOP histogram of two images from the same class with scale variation is calculated, because each image exhibits different information, the LIOP histogram of these two images is different and they will possible be misclassified into different class. This is due to the loss of global information if we only take the locally rotation invariant feature.

We can compute the difference between train image and test image by nonparametric static test. By using different type, we can check the dissimilarity between two histograms like chi-square statics, likelihood ratio and histogram intersection [16]. The *TR* is used to stand for the training images and *TE* express the testing images in the datasets.

$$D(TR,TE) = \sum_{n=1}^{N} \frac{\left(TR_n - TE_n\right)^2}{TR_n + TE_n} \quad . \tag{9}$$

Where $N$ is the number of bins and $TR_n, TE_n$ represents the training and testing image in corresponding $n$-th bin. Global matching can be implemented by exhaustive search, where by comparing the histogram of sample image and model image we can get the minimum distance between two images [25].

## 4    Experimental Results

The Oxford dataset [26] is used to evaluate the proposed descriptor. The Oxford dataset contains images with different geometric and photometric transformations of structured and textured scenes. It contains six different transformations like scale change, image blur, viewpoint change, illumination change, image rotation and JPEG compression, which are used in this work. Fig.1 shows some samples from this Oxford dataset.

**Fig. 1.** Typical images used for the evaluation from Oxford dataset. Where Graf (Zoom, Viewpoint change, Rotation), Wall (Viewpoint), Ubc (JPEG compression), Trees (Image blur, Textured scene).

For interest region detection, we have used affine covariant region known as Harris-affine (haraff) [7]. We compared proposed descriptor with some well know state-of-art feature descriptor, such as SIFT [1], DAISY [13] and LIOP [24].The performance of our descriptor is evaluated on the basis as in [10], in which it depends upon the number of correct matches and number of false matches. The results are presented with recall and precision curves, while the number of correct matches and ground truth correspondences is determined by the overlap error [8].

The performance of different methods in terms of precision and recall is evaluated. Also the average precision and recall for each class has been calculated to confirm that if our proposed global matching method performs well or not for variations of images in every class.

Affine covariant region adopted in our experiments is Harris-affine [7]. *K*-means clustering algorithm is used to get the histograms from LIOP features. The experimental results are shown in Fig. 2 and 3 for the case of haraff. We can observe from Fig. 2 that for only "leuven" class our proposed global matching does not perform well in term of precision, and the recall is good in most of the cases except "Gref" in Fig. 3.

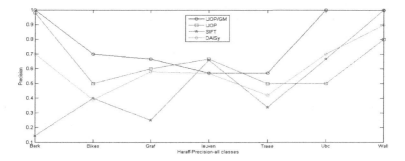

**Fig. 2.** Precision of Oxford dataset by using haraff

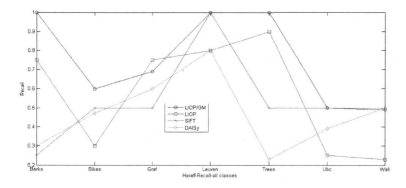

**Fig. 3.** Recall by using Haraff affine covariant region on Oxford dataset

The recall and precision are defined as:

$$recall = \frac{w_c}{w_H},$$

$$precision = \frac{w_C}{w_{Auto}}.$$

Where $w_H$ is the manually classified images in the test data set, and automatically classified images are $w_{Auto}$. The $w_c$ express the correct classified images. From these experiments, it can be seen that our proposed global matching method performs better on histograms.

## 5    Conclusion

By using hybrid approach, a novel descriptor is proposed to exploit the local and global information of image in this paper. The LIOP histograms are used to measure the dissimilarity between images. Global matching is obtained by computing the minimum distance in all directions. The experimental results on different image transformations demonstrated that the proposed global rotation invariant matching scheme outperforms the state-of-the-art methods in terms of recall and precision.

**Acknowledgments.** The research work described in this paper is fully supported by the grants from the National Natural Science Foundation of China (Project No. 90820010, 60911130513). Prof. Ping Guo is the author to whom all correspondence should be addressed.

## References

1. Lowe, D.G.: Distinctive image features from scale-invariant keypoints. International Journal of Computer Vision 60(2), 91–110 (2004)
2. Cohen, F.S., Fan, Z., Attali, S.: Automated inspection of textile fabrics using textural models. IEEE Trans. on Pattern Analysis and Machine Intelligence 13(8), 803–808 (1991)
3. Ji, Q., Engel, J., Craine, E.: Texture analysis for classification of cervix lesions. IEEE Trans. on Medical Imaging 19(11), 1144–1149 (2000)
4. Nister, D., Stewenius, H.: Scalable recognition with a vocabulary tree. In: Proc. Computer Vision and Pattern Recognition, vol. 2, pp. 2161–2168 (2006)
5. Tuytelaars, T., VanGool, L.: Matching widely separated views based on affine invariant regions. International Journal of Computer Vision 59(1), 61–85 (2004)
6. Brown, M., Lowe, D.G.: Automatic panoramic image stitching using invariant features. International Journal of Computer Vision 74(1), 59–73 (2007)
7. Mikolajczyk, K., Schmid, C.: Scale & affine invariant interest point detectors. International Journal of Computer Vision 60(1), 63–86 (2004)

8. Mikolajczyk, K., Tuytelaars, T., Schmid, C., Zisserman, A., Matas, J., Schaffalitzky, F., Kadir, T., Gool, L.V.: A comparison of affine region detectors. International Journal of Computer Vision 65, 43–72 (2005)

9. Matas, J., Chum, O., Martin, U., Pajdla, T.: Robust wide baseline stereo from maximally stable extremal regions. In: Proc. British Machine Vision Conference, vol. 1, pp. 384–393 (2002)

10. Yang, D., Guo, P.: Improvement of Image Modeling with Affinity Propagation Algorithm for Semantic Image Annotation. In: Leung, C.S., Lee, M., Chan, J.H. (eds.) ICONIP 2009, Part I. LNCS, vol. 5863, pp. 778–787. Springer, Heidelberg (2009)

11. Mikolajczyk, K., Schmid, C.: A performance evaluation of local descriptors. IEEE Trans. on Pattern Analysis and Machine Intelligence 27(10), 1615–1630 (2005)

12. Bay, H., Tuytelaars, T., Van Gool, L.: SURF: Speeded Up Robust Features. In: Leonardis, A., Bischof, H., Pinz, A. (eds.) ECCV 2006, Part I. LNCS, vol. 3951, pp. 404–417. Springer, Heidelberg (2006)

13. Tola, E., Lepetit, V., Fua, P.: Daisy: An efficient dense descriptor applied to wide-baseline stereo. IEEE Trans. on Pattern Analysis and Machine Intelligence 32(5), 815–830 (2010)

14. Fan, B., Wu, F.C., Hu, Z.Y.: Aggregating gradient distributions into intensity orders: A novel local image descriptor. In: Proc. Computer Vision and Pattern Recognition, pp. 2377–2384 (2011)

15. Heikkil, M., Pietikinen, M., Schmid, C.: Description of interest regions with local binary patterns. Pattern Recognition 42(3), 425–436 (2009)

16. Ojala, T., Pietikainen, M., Maenpaa, T.: Multiresolution gray-scale and rotation invariant texture classification with local binary patterns. IEEE Trans. on Pattern Analysis and Machine Intelligence. 24(7), 971–987 (2002)

17. Chen, J., Shiguang, S., He, C., Zhao, G.Y., Pietikainen, M., Chen, X.L., Gao, W.: WLD: A Robust Local Image Descriptor. IEEE Trans. on Pattern Analysis and Machine Intelligence. 32(9), 1705–1730 (2010)

18. Dalal, N., Triggs, B.: Histograms of Oriented Gradients for Human Detection. In: Proc. Computer Vision and Pattern Recognition, vol. 1, pp. 886–893 (2005)

19. Lazebnik, S., Schmid, C., Ponce, J.: A Sparse Texture Representation Using Local Affine Regions. IEEE Trans. on Pattern Analysis and Machine Intelligence 27(8), 1265–1278 (2005)

20. Ahonen, T., Hadid, A., Pietikainen, M.: Face Description with Local Binary Patterns: Application to Face Recognition. IEEE Trans. on Pattern Analysis and Machine Intelligence 28(12), 2037–2041 (2006)

21. Rodriguez, Y., Marcel, S.: Face Authentication Using Adapted Local Binary Pattern Histograms. In: Leonardis, A., Bischof, H., Pinz, A. (eds.) ECCV 2006. LNCS, vol. 3954, pp. 321–332. Springer, Heidelberg (2006)

22. Kyrki, V., Kamarainen, J.K.: Simple Gabor feature space for invariant object recognition. Pattern Recognition Letter 25(3), 311–318 (2004)

23. Kingsbury, N.G.: Rotation-invariant local feature matching with complex wavelets. In: Proc. European Signal Processing Conference, pp. 4–8 (2006)

24. Wang, Z.H., Fan, B., Wu, F.C.: Local Intensity Order Pattern for Feature Description. In: Proc. International Conference on Computer Vision, pp. 603–610 (2011)

25. Guo, Z.H., Zhang, L., Zhang, D.: Rotation invariant texture classification using LBP variance (LBPV) with global matching. Pattern Recognition 43(3), 706–719 (2010)

26. http://www.robots.ox.ac.uk/vgg/research/affine/

27. http://vision.ia.ac.cn/Students/wzh/publication/liop/index.html

28. Goswami, B., Chan, C.H., Kittler, J., Christmas, B.: Local ordinal contrast pattern histograms for spatiotemporal, lip-based speaker authentication. In: Proc. In Biometrics: Theory Applications and Systems, pp. 1–6 (2010)
29. Gupta, R., Patil, H., Mittal, A.: Robust order-based methods for feature description. In: Proc. Computer Vision and Pattern Recognition, pp. 334–341 (2010)

# An Approach of Power Quality Disturbances Recognition Based on EEMD and Probabilistic Neural Network

Ling Zhu, Zhigang Liu, Qiaoge Zhang, and Qiaolin Hu

School of Electrical Engineering, Southwest Jiaotong University, Chengdu 610031, China
zhuling198910@126.com

**Abstract.** Based on intrinsic mode functions (IMFs), standard energy difference of each IMF obtained by EEMD and probabilistic neural network (PNN), a new method is proposed to the recognition of power quality transient disturbances. In this method, ensemble empirical mode decomposition (EEMD) is used to decompose the non-stationary power quality disturbances into a number of IMFs. Then the standard energy differences of each IMF are used as feature vectors. At last, power quality disturbances are identified and classified with PNN. The experimental results show that the proposed method can effectively realize feature extraction and classification of single and mixed power quality disturbances.

**Keywords:** power quality disturbances, ensemble empirical mode decomposition (EEMD), intrinsic mode function (IMF), probabilistic neural network (PNN).

## 1    Introduction

With the wide applications of electronic devices in power system, the quality of electric power has become an important issue for electric utilities. Economic losses to sensitive power users caused by transient power quality disturbances are the common problems in power system. Therefore, the classification and recognition of transient power quality disturbances is very important to improve power quality. In recent years, there have been several methods used for feature extraction of transient power quality disturbances, such as Fourier transform[1], wavelet transform[2], Hilbert-Huang transform[3], Wigner-Ville distribution[4], S-transform[5], etc. Besides, kinds of classifiers also have been designed to identify transient disturbance, such as: expert system[6], support vector machines[7], artificial neural networks [8] ,and so on.

HHT including empirical mode decomposition (EMD) and Hilbert transform, which is a self-adaptive method for analyzing nonlinear and non-stationary signals, has been gradually applied to detecting power quality disturbances[9-10]. Recently empirical mode decomposition (EMD) has proven to be quite available in a extensive range of applications for extracting features from signals generated in noisy nonlinear and non-stationary processes. As useful as EMD proved to be, it still leaves some difficulties unresolved, such as mode mixing. To overcome the frequency separation problem, Zhaohua Wu and N.E. Huang proposed a new noise-assisted data analysis

C. Guo, Z.-G. Hou, and Z. Zeng (Eds.): ISNN 2013, Part I, LNCS 7951, pp. 505–513, 2013.
© Springer-Verlag Berlin Heidelberg 2013

(NADA) method named Ensemble EMD (EEMD), which collate the portion of each IMF by adding noise to deal with mode mixing[11]. Therefore, EEMD has been more and more widely application to extract characteristics of different disturbances[12].

Artificial Neural Network (ANN) is a kind of artificial intelligence, which has strong power in self-study, self-accommodate and nonlinear transition[8]. It has been used in pattern recognition, optimal control, statistical computation, power system and many other fields. Probabilistic neural networks (PNN) is a kind of artificial neural network, which is simple in learning rule and rapid in training. Besides, PNN has very strong anti-interference, strong capability of nonlinear approximation with high reconstructing accuracy and better classifying performance[13-14]. In the paper, EEMD is used to decompose the power quality signals with disturbances into IMFs (Intrinsic Mode Functions) to extract feature vectors and PNN is used to classify and recognize the types of the disturbances.

## 2    EEMD and PNN

### 2.1    EMD and EEMD

EMD was proposed by Huang in 1998 as an adaptive time-frequency data analysis method in nonlinear and non-stationary signals[9-10]. EMD is used to decompose the complex signals into IMFs, which include the intrinsic features of the signals. Each IMF is defined by two principal characteristics: the number of its extreme must be equal to, or differ by at most one and its mean is zero.

The steps comprising the EMD method are as follows [9].

1) Identify local maxima and minima of the distorted signal $X(t)$ and perform interpolation method between the maxima and the minima to obtain the envelopes, then compute the mean of the envelopes $m_1$ to calculate the difference $h_1$.

$$h_1 = X(t) - m_1 \tag{1}$$

If $h_1$ is an IMF, it means that the first IMF is obtained, which consists of the highest frequency components present in the original signal.

2) If $h_1$ is not an IMF, then repeat steps 1 on $h_1$ instead of $X(t)$, until the new $h_1$ satisfies the conditions of an IMF recorded as $c_1$.

3) Compute the residue, $r_1$.

$$r_1 = X(t) - c_1 \tag{2}$$

4) If the residue is larger than the threshold of error tolerance, repeat steps 1-3 to obtain the second IMF and residue. Repeat the steps 1-4 until the residual signal is a monotone function.

$$r_n = r_{n-1} - c_n \tag{3}$$

If orthogonal IMFs are obtained in this iterative manner, the original signal may be reconstructed as:

$$X(t) = \sum_{i=1}^{n} c_i + r_n \tag{4}$$

The final residue exhibits any general trends followed by the original signal, and each IMF represent different frequency components of the original signal.

Though EMD is proved to be useful in many conditions, it still leaves some difficulties unresolved, such as mode-mixing. Zhaohua Wu and N. E. Huang put forward EEMD, a new noise-assisted data analysis approach to solve the scale separation problem of EMD. EEMD is developed below [11-12]:

1) Add a white noise series to the target data;
2) Decompose the data with added white noise into IMFs;
3) Repeat step 1 and 2 again, but with different white noise series each time;
4) Obtain the means of corresponding IMFs as the final result.

## 2.2    PNN (Probabilistic Neural Network)

PNN is a classification network proposed by Dr. DF Specht in 1989, which is based on Bayesian decision theory and probability function estimation theory[13-14]. Because it has shorter training time and higher classification accuracy than traditional neural networks, PNN is an effective training artificial neural network for many classification problems. The PNN neural network consists of three layers namely input, hidden and output layers as shown in **Fig.1.**

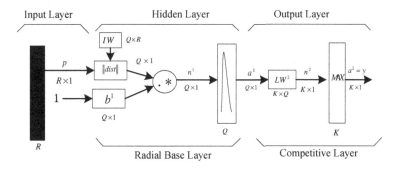

**Fig. 1.** The architecture of PNN model

Where $R$ represents the number of the input vector, $Q$ represents the number of the target samples which equals to the number of the neurons in radial base layer, $K$ represents the categories of the input vector which is equal to the neurons in competitive layer. The input layer as the first layer is formed by the source nodes, the second layer is hidden layer including the radial basis function neurons and the third layer is output layer including the competition neurons. The number of radial basis neurons equals the number of training samples, while the number of input layer nodes depends on the dimension of the input vector.

The PNN is similar to the RBF in the hidden layer and the PNN generally takes form of the Gauss Function as the radials function [14]:

$$a(n) = radbas(n) = e^{-n^2} \tag{5}$$

The distance $\|dist\|$ between the weight vector $IW$ and the input vector $P$ of every neuron in the hidden layer connected with the input layer is multiplied by a threshold $b = [-\log(0.5)]^{0.5}/spread$ to form the self input.

The relationship between the input layer and the hidden layer is

$$a_j^1 = f = radbas(\|W - p\|b_j^1) \quad j = 1, 2, ..., Q \tag{6}$$

Where $a_j^1$ is the output of the $j$ th hidden layer neuron, $p = [p_1, p_2, ..., p_R]$ is the input vector, $b_j^1 (j = 1, 2, ..., Q)$ is the $j$ th PNN width, $a^1 = [a_1^1, a_2^1, a_3^1, ..., a_Q^1]$ is the output of the hidden layer. The relationship between the hidden layer and the output layer is

$$a_m^2 = \sum_{j=1}^{Q} w_{jm}^2 a_j^1, m = 1, 2, ..., K \tag{7}$$

where $w_{jm}^2$ is the output layer weight $j$ th hidden layer neuron output acts on $m$ th competition neuron. $a^2 = [a_1^2, a_2^2, a_3^2, ..., a_K^2]$ is the output of the output layer and $y = MAX(a^2)$ is the output vector, which set the max element as 1 and other elements as 0.

## 3    Classification Based on EEMD and PNN

In this paper, EEMD is used to decompose the disturbance signal into layers to get the IMFs and then obtain the standard energy difference represented as $F1 \sim F9$ of the IMFs as follows.

$$F_i = \sqrt{\frac{1}{N_i} \sum_{k=1}^{N_i} |d_i(k)|^2} - \sqrt{\frac{1}{N_i} \sum_{k=1}^{N_i} |d_{(pure)i}(k)|^2} \quad i = 1, 2, 3...N \tag{8}$$

where $d_1 \sim d_9$ represent the IMF components of the disturbance signal and $d_{(pure)1} \sim d_{(pure)9}$ represent the IMF components of the normal signal.

Transient power quality problems can be divided into six categories such as voltage swell, voltage sag, voltage interruption, transient pulse, transient oscillation and harmonics. The paper's goal is to classify the six kinds of transient disturbances using EEMD and PNN. There are three steps to finish the classification shown in **Fig.2**.

Step 1: EEMD is used to decompose the disturbance and obtain the IMF components. Some examples used EEMD is shown from **Fig.3** to **Fig.5**.

It can be seen from **Fig.3** to **Fig.5** that the IMF component will have some essential differences when the different disturbance is different. Therefore, we select IMF component coefficient to obtain the characteristics of the different signals.

Step 2: The IMF component coefficient energy differences are calculated according to Eq.(8) and selected as the characteristic quantities.

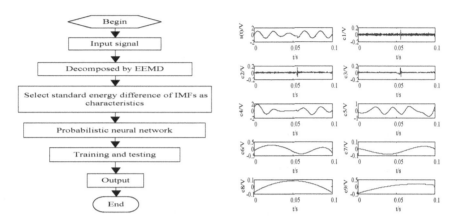

**Fig. 2.** The flow chart of proposed algorithm          **Fig. 3.** EEMD of voltage sag

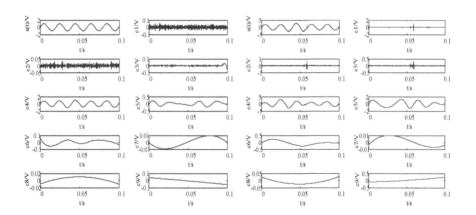

**Fig. 4.** EEMD of voltage signal     **Fig. 5.** EEMD of transient pulse+ voltage swell

**Fig.6** shows the difference between the standard energy differences of the six kinds of signals. The IMF component coefficient energy differences are quite different. That is the reason why we select them as the characteristic quantities.

Step 3: Classification with PNN. Input the characteristic quantities into the PNN. After training and testing, output the results and calculate the accuracy.

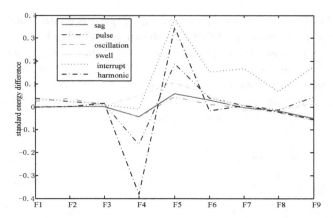

**Fig. 6.** The curve of the characteristic extraction

# 4    Simulation Analysis

## 4.1    Extract the Characteristic Quantities

There are a variety of power quality disturbances and power quality disturbances are complex and hard to obtain. We use the MATLAB to simulate the disturbance signals according to standard of the IEEE[15] .The functions of each disturbance are shown in table I, Twhere $\alpha$ is the amplitude of the disturbance; $\beta$ is fundamental frequency multiples; $\gamma$ is for pulse amplitude; $t_1$ is the starting moment ; $t_2$ is the end moment; $c$ is attenuation coefficient; $u(t)$ is unit step function; $\omega_0$ is fundamental angular frequency; $T$ is cycle.

**Table 1.** The characteristics and main types of power quality

| Disturbance type | Mathematical model | Model's parameters |
|---|---|---|
| Voltage swell | $u(t) = \left(1 + \alpha\left(u(t_2) - u(t_1)\right)\right)\sin(\omega_0 t)$ | $\alpha = 0.1 \sim 0.9$ ; $T < t_2 - t_1 < 9T$ |
| Voltage sag | $u(t) = \left(1 - \alpha\left(u(t_2) - u(t_1)\right)\right)\sin(\omega_0 t)$ | $\alpha = 0.1 \sim 0.9$ ; $T < t_2 - t_1 < 9T$ |
| Voltage interruption | $u(t) = \left(1 - \alpha\left(u(t_2) - u(t_1)\right)\right)\sin(\omega_0 t)$ | $\alpha = 0.9 \sim 1$ ; $T < t_2 - t_1 < 9T$ |
| Harmonics | $u(t) = \sin(\omega_0 t) + \alpha_3 \sin(3\omega_0 t) + \alpha_5 \sin(5\omega_0 t) + \alpha_7 \sin(7\omega_0 t)$ | $0.05 < \alpha_3 < 0.15 \; 0.02 < \alpha_5 < 0.1$ $0.02 < \alpha_7 < 0.1$ |
| Transient oscillation | $u(t) = \sin(w_0 t) + \alpha \cdot e^{-c(t-t_1)} \cdot \sin(\beta w_0 t) \cdot (u(t_2) - u(t_1))$ | $\alpha = 0.1 \sim 0.8$ ; $c = 2.5 \sim 5$ ; $0.5T < t_2 - t_1 < 3T$ ; $\beta = 5 \sim 10$ |
| Transient pulse | $u(t) = \sin(\omega_0 t) + \gamma\left[\varepsilon(t_2) - \varepsilon(t_1)\right]$ | $1 < \gamma < 3; 1ms < t_2 - t_1 < 3ms$ |
| Voltage signal | $u(t) = \alpha\sin(\omega_0 t)$ | $\alpha = 1$ ; $\omega_0 = 50Hz$ |

## 4.2    Classification and Recognition

(1)The classification of a single disturbance
The groups of transient data can be produced in MATLAB according to the mathematical models shown in **Table.1**, 400 samples of each disturbance (200 samples for training, others for testing) under different SNR from 30dB to 50dB are simulated to test the accuracy of the proposed method. The simulation results are shown in Table 2.

**Table 2.** The classification result of a single disturbance in different noises

| Disturbance type | SNR/dB | number of right samples | number of wrong samples | Accuracy/% |
|---|---|---|---|---|
| | 30 | 194 | 6 | 97 |
| Voltage swell | 40 | 196 | 4 | 98 |
| | 50 | 197 | 3 | 98.5 |
| | 30 | 192 | 8 | 96 |
| Voltage sag | 40 | 194 | 6 | 97 |
| | 50 | 195 | 5 | 97.5 |
| | 30 | 190 | 10 | 95 |
| Voltage interruption | 40 | 193 | 7 | 96.5 |
| | 50 | 195 | 5 | 97.5 |
| | 30 | 193 | 7 | 96.5 |
| Harmonics | 40 | 194 | 6 | 97 |
| | 50 | 196 | 4 | 98 |
| | 30 | 194 | 6 | 97 |
| Transient oscillation | 40 | 197 | 3 | 98.5 |
| | 50 | 198 | 2 | 99 |
| | 30 | 193 | 7 | 96.5 |
| Transient pulse | 40 | 196 | 4 | 98 |
| | 50 | 197 | 3 | 98.5 |

As Table 2 shows, no matter the low SNR(as 50dB) or the high SNR(as 30dB), the accuracy of the method proposed in this paper is high(the lowest is 95%). Simulation result shows that the method is very satisfied for the application under different noise environments.

(2)The classification of mixed disturbance
In order to analyze whether the method is suitable to identify the compound disturbance, six categories of mixed disturbances are selected to experiment and the results are shown in Table 3.

As is shown in Table 3, the algorithm also has high classification accuracy when it is applied to the mixed disturbances. In addition, due to the small amplitude and noise affecting, several cases can not be completely accurately classified.

**Table 3.** The classification result of mixed disturbances

| Disturbance types | Number of right samples | Number of wrong samples | Accuracy /% |
|---|---|---|---|
| swell + pulse | 195 | 5 | 97.5 |
| swell + oscillation | 193 | 7 | 96.5 |
| sag + pulse | 194 | 6 | 97 |
| sag + oscillation | 196 | 4 | 98 |
| harmonics + pulse | 193 | 7 | 96.5 |
| harmonics + oscillation | 190 | 10 | 95 |

# 5    Conclusions

This paper proposed a new method for power quality disturbances recognition based on EEMD and PNN. The features of power quality disturbances are extracted by EEMD, six types of disturbances and their compound disturbances are classified precisely by the proposed method. Simulation results show that the method is effective especially under high noise environments. Otherwise, the principle of the method is very simple and the further work will be focused on improving the classification efficiency.

**Acknowledgments.** This project was supported by National Natural Science Foundation of China (U1134205, 51007074) and Fundamental Research Funds for Central Universities (SWJTU11CX141) in China.

# References

1. Wang, G.-B., Xiang, D.-Y., Ma, W.-M.: Improved Algorithm for Noninteger Harmonics Analysis Based on FFT Algorithm and Neural Network. Proceedings of the CSEE 28(4), 102–108 (2008)
2. Lin, S., He, Z., Luo, G.: A wavelet energy moment based classification and recognition method of transient signals in power transmission lines. Power System Technology 32(20), 30–34 (2008)
3. Li, T., Zhao, Y., Li, N., et al.: A new method for power quality detection based on HHT. Proceedings of the CSEE 25(17), 52–56 (2005)
4. Bulus-Rossinia, L.A., Costanzo-Casoa, P.A., Duchowicza, R., et al.: Optical pulse compression using the temporal Radon–Wigner transform. Optics Communications 283, 2529–2535 (2010)
5. Yi, J.-L., Peng, J.-C., Tan, H.-S.: A summary of S-transform applied to power quality disturbances analysis. Power System Protection and Control 39(3), 141–147 (2011)
6. Styvaktakis, E., Boolen, M.J., Gu, I.Y.H.: Expert system for classification and analysis of power system events. IEEE Transactions on Power Delivery 17(2), 423–428 (2002)
7. Song, X.-F., Chen, J.-C.: Classification method of dynamic power quality disturbances based on SVM. Electric Power Automation Equipment 26(4), 39–42 (2006)
8. Yao, J.-G., Guo, Z.-F., Chen, J.-P.: A New Approach to Recognize Power Quality Disturbances Based on Wavelet Transform and BP Neural Network. Power System Technology 36(5), 139–144 (2012)

9. Huang, N.E., Shen, Z., Long, S.R., et al.: The empirical mode decomposition and the Hilbert spectrum for nonlinear and non-stationary time series analysis. Proceedings of the Royal Society of London Series A 45(4), 903–995 (1998)
10. Huang, N.E., Shen, Z., et al.: The empirical code decomposition and the Hilbert spectrum for non-linear and non-stationary time series analysis. Royal Society 454, 915–995 (1998)
11. Wu, Z., Huang, N.E.: Ensemble Empirical Mode Decomposition: A Noise Assisted Data Analysis Method. Center for Ocean-Land-Atmosphere Studies (2009)
12. Zhang, Y., Liu, Z.: Application of EEMD in power quality disturbance detection. Electric Power Automation Equipment 31(12), 86–91 (2011)
13. Kou, J., Xiong, S., Wan, S., Liu, H.: The Incremental Probabilistic Neural Network. In: Sixth International Conference on Natural Computation, pp. 1330–1333 (2010)
14. Wahab, N.I.A., Mohamed, A.: Transient Stability Assessment of a Large Actual Power System using Probabilistic Neural Network with Enhanced Feature Selection and Extraction. In: International Conference on Electrical Engineering and Informatics, pp. 519–524 (2009)
15. Chowdhury, H.B.: Power quality. IEEE Potentials 20(2), 5–11 (2001)

# Chinese Text Classification Based on Neural Network

Hu Li[*], Peng Zou, and WeiHong Han

School of Computer, National University of Defense Technology, Changsha 410073,
Hunan Province, P.R. China
lihu@nudt.edu.cn

**Abstract.** Text classification is widely used nowadays. In this paper, we proposed a combination feature reduction method to reduce feature space dimension based on inductive analysis of existing researches. Neural network was then trained and used to classify new documents. Existing researches mainly focus on the classification of the English text, but we focused on classification of Chinese text instead in this paper. Experimental results showed that the proposed feature reduction method performed well, and the neural network needed less terms to achieve the same accuracy compared with other classifiers.

**Keywords:** Neural network, Text classification, Chinese text.

## 1 Introductions

With the rapid development of information technology, electronic documents on the Internet increase rapidly, and the amount of network data is growing exponentially. So, how to deal with these massive data has become an important issue. One of the most effective ways is to divide these data into different categories. Text classification is widely used, such as classification of web pages, spam filtering and etc.

There are many methods for text classification, but most of them are based on the Vector Space Model (VSM) [1] proposed by G.Salton in the 1960s. In VSM, each document is expressed as a high-dimensional vector, and elements in the vector are the so-called feature items or terms. The similarity between two documents is calculated by the cosine angle between these two vectors. VSM simplifies the classification process into the operation between vectors, so that the complexity of the problem is greatly reduced, thereby obtaining a wide range of applications. However, as the amount of data increases, the number of terms growths rapidly, and the computing time increase as well. In this case, it is particularly important to find a reasonable strategy to select useful terms. Thereby, people propose a lot of feature reduction or selection methods, including the Document Frequency (DF), the Mutual Information (MI), Information Gain (IG), CHI-squared distribution (CHI), Principal Component Analysis (PCA) and the like. These methods examine the contribution of terms for text classification accuracy from different angle, and select different terms based on various strategies. Another important work for text classification is to assign each selected term a

---

[*] Corresponding author.

C. Guo, Z.-G. Hou, and Z. Zeng (Eds.): ISNN 2013, Part I, LNCS 7951, pp. 514–522, 2013.

reasonable weight. A good method should give those terms high weight whose class discrimination ability is strong, while give those terms low weight whose class discrimination ability is weak.

Since documents have been transformed into vectors, training and classification operation will be carried. Commonly used methods include Naive Bayes classifier, decision tree, support vector machine, k-nearest neighbor, neural network and etc. These methods have their own advantages and are suitable for different occasions. Neural network can fully approximate arbitrarily complex nonlinear relationship, so it is suitable for text classification problem which has complex classification plane. But neural network is of high time complexity, therefore, it is very necessary to reduce the feature space dimension prior to use neural network.

This paper is structured as follows. Section 2 gives a review of researches against text classification based on neural network. We discussed key steps within text classification based on neural network in section 3. Experiments and results analysis are showed in section 4. We concluded the paper and showed acknowledgement in the end.

## 2    Related Works

There are plenty of researches focuses on the text classification. Yang [2] analyzed and compared several feature selection method, namely DF, IG, MI, CHI and TS (Term Strength). Experiment results on English text showed that IG and CHI performed better than other methods. Savio [3] compared DF, CF-DF, TFIDF and PCA on a subset of the Reuters-22173 using neural network as the classifier. Results showed that PCA performed best. Hansen [4] used neural network ensemble technology to enhance the performance of the neural network, and cross-validation method was used to optimize the parameters and the structure of the neural network. Waleed [5] compared the performance of neural network and SVM, and found that the neural network performed fairly with SVM, but required less training documents. Saha [6] used neural network to classify English text in case that number of categories was unknown. Cheng [7] used singular value decomposition to reduce the dimension of feature space. At the same time, multi-layer perceptron and back propagation neural network were used to train and classify new documents. Vassilios [8] proposed a novel neural network, namely sigma Fuzzy Lattice Neural network and compared it with other classification algorithms, including K Nearest Neighbor and Naive Bayes Classifiers. Nerijus [9] used the decision tree to initialize the neural network, and test results on the Reuters-21578 corpus showed that the method was effective. Taeho [10] proposed an improved neural network model which directly connected the input layer and output layer together, and update weights of the neural network only in the case of misclassification.

The premise of the neural network to classify the text is that the text be represented as an acceptable input form of the neural network. Therefore, the conversion work is very important and lots of methods have been discussed in details [2-3]. In order to assign reasonable weight to each selected terms, Term Frequency Inverse Document Frequency (TFIDF) is commonly used. The TFIDF method determines a term's weight by investigating the distribution of terms within and between categories. It was recognized as the benchmark weight calculation method.

Above all, most of researches are on the English text, rarely using Chinese corpus. In addition, most of existing researches using only one feature reduction or selection method while combination methods are lack of research. Therefore, we focus on using a combination feature reduction strategy to reduce the feature space dimension and using neural network to classify Chinese text in this paper. Seen from existing researches, CHI and PCA performed best in feature selection or reduction, but PCA is very time-consuming. So, we use CHI to select the initial term set, and then use PCA for further dimension reduction. We use revised TFIDF method to calculate the weights of terms in this paper for simplicity.

# 3     Text Classification Based on Neural Network

In order to more clearly illustrate the procedure of text classification using neural network, we first describe the vector space model, and then we focus on two feature reduction methods namely CHI and PCA. Next, we show the term weight calculation method used in the paper. Finally, we describe the structure of neural network to be used.

## 3.1     Vector Space Model

Vector space model (VSM) expresses the document as a vector of terms having different weights, thus, simplifies the classification problem into the computation within the vector space, and greatly reduced the complexity of the problem. VSM expresses the document $d_i$: $d_i = (t_{i1}, t_{i2}, ..., t_{im})$, where $t_{ij}$ is the $j^{th}$ term of document $d_i$, and $m$ is the number of terms. Different terms within the document play different role, so they should be assigned different weight. Then document $d_i$ can be further expressed as $d_i = (w_{i1}, w_{i2}, ..., w_{im})$, where $w_{ij}$ is the weight of the $j^{th}$ term of document $d_i$.

From the definition of VSM, we can see that the number of terms and the weight of each term is the core factors of the text classification problem, which respectively, corresponding to the feature selection or reduction methods and the term weight calculation method. Next, we would analyze two feature selection or reduction methods and one term weight computation method.

## 3.2     CHI Method

CHI (CHI-squared distribution) method assumes that terms and categories are not independent, but rather follow chi-squared distribution with $k$ degrees of freedom. The higher CHI of a term respected to a category, the stronger correlation between them. CHI of term $t_i$ respected to category $c_j$ can be calculated as follows:

$$\chi^2(t_i, c_j) = \frac{N \times (AD - CB)^2}{(A+C) \times (B+D) \times (A+B) \times (C+D)} \tag{1}$$

Where $N$ is the number of documents within the dataset, $A$ is number of documents belonging to the class $c_j$ and containing the term $t_i$. $B$ is number of documents not belonging to the class $c_j$ but containing the term $t_i$. $C$ is number of documents belonging to the class $c_j$ but not containing the term $t_i$, $D$ is number of documents not belonging to the class $c_j$ and not containing the term $t_i$. For multi-class classification problem, firstly, CHI of term $t_i$ respected to each category is computed, and then CHI of term $t_i$ respected to the entire dataset is computed as follows:

$$\chi^2_{max}(t_i) = \max_{i=1}^{n} \chi^2(t_i, c_j) \tag{2}$$

Where n is the number of categories. Terms of CHI lower than the threshold will be filtered out, only those terms of CHI higher than the threshold be selected as the finally term set.

### 3.3    PCA Method

PCA (Principal component analysis) is a commonly used linear dimensionality reduction method which uses some linear projection to map high-dimensional data into low-dimensional space. PCA retains most of the original data features using less low-dimensional data by minimizing the variance on the projected data. PCA linear transformation concentrate different attributes into a few composite indicator (principal component), which is a linear combination of attributes. Different principal components are mutually orthogonal and thus redundant attributes are removed, achieving the purpose of dimensionality reduction.

### 3.4    Weight Computation Based on TFICF

Conventional methods generally use TFIDF to compute the weight of a term. The TFIDF method assumes that the importance of a term is proportional to the times it appeared in the document, but inversely proportional to the times it appeared in different documents. Therefore, those terms that appeared more than once in a document should be given a higher weight, while those appeared in many documents should be given a lower weight. Usually, we need to compute the term weight relative to the entire dataset, not only within a category. Therefore, in this paper, we draw on the concept TFICF (Term Frequency Inverse Class Frequency) proposed in [11], and simplified it in our experiment.

For term $t_i$ appeared in class $c_j$, its TFICF value can be computed in two parts as follows:

$$tf(t_{ij}) = \frac{n_{ij}}{\sum_k n_{kj}} \tag{3}$$

Where $n_{ij}$ is the times term $t_i$ occurred in category $c_j$, the denominator is the total times of all terms occurred in category $c_j$.

$$icf(t_i) = \log \frac{|C|}{|\{j \mid t_i \in c_j\}| + 1}$$  (4)

Where $|C|$ is the total number of categories in the dataset, $|\{j \mid t_i \in c_j\}|$ is the number of categories contains term $t_i$. To avoid division by zero in case term $t_i$ does not appear in any category, the denominator was pulsed by 1.

Finally, we can calculate the weight of term $t_i$ respected to category $c_j$ as follows:

$$tficf(t_{ij}) = tf(t_{ij}) \cdot icf(t_i)$$  (5)

### 3.5    Back Propagation Network

Through the aforementioned feature selection or reduction methods and weight calculation method, documents to be classified could be represented as relatively low-dimension vectors. Next, we use a three-layer feed forward back propagation neural network (BP) for text classification. The network is composited by three layers, including an input layer, a hidden layer and an output layer. Each layer contains a number of neurons. The number of neurons in the input layer is usually the dimension of the input vector and the number of neurons of the output layer is the category size. Hidden layer neurons have a greater impact on learning and generalization ability of the neural network, but there still lack of unified method for determining the number of hidden layer neurons. In this paper, we use "rule of thumb" called the Baum-Haussler rule [12], which states as follows.

$$N_{hidden} \le (N_{train} \cdot E_{tolerance}) / (N_{inputs} + N_{outputs})$$  (6)

Where $N_{hidden}$ is the number of hidden neurons, $N_{train}$ is the number of training patterns, $E_{tolerance}$ is the error we desire of the network, $N_{inputs}$ and $N_{outputs}$ are the number of input and output neurons respectively. In this paper, we set the number of hidden neurons as 40 according to our experiments.

Text classification based on neural network consists of two phases: network training and classification. During the network training phase, each vector is entered to the input layer, and then it passes the hidden layer; the output of the output layer is the result. The difference between the result and expected result is then back propagated to modify the connection weights, so that the difference get diminished or even disappeared. During the classification phase, test document is entered and the outputs of each output layer neurons are compared. The neural of the maximum value is the corresponding category of the test document.

The above procedure can be formalized like this. Suppose the number of categories is |C|, where $C$ is the dataset and $d$ is a document belongs to category $k$. During the training phase, the corresponding expected output should be a |C|-dimensional vector $(a_1, a_2, \cdots, a_{|C|})$, where $a_i = 0(i \neq k), a_k = 1$. If the expected output is not consistent with the actual output, then adjust the connection weights. During the classification stage, the test document $t$ is entered and the corresponding output is also a I $C$ I-dimensional output vector $(a_1, a_2, \cdots, a_{|C|})$. Suppose $\max(a_1, a_2, \cdots, a_{|C|}) = a_r$, then test document $t$ would be assigned to category $r$.

# 4 Experiments

## 4.1 Experiment Environment and Prepare Work

All experiments were done on a PC with Intel(R) Core(TM) i5-1320 CPU @2.5GHz, 10G Memory, 500G Hard Disk Capacities. We used Fudan University's Chinese Text Classification Corpus[1] in our experiment. The corpus contains 20 categories and is of 101MB size. The corpus contains 19637 documents, in which 9804 as training documents and 9833 as test documents. Train set and test set do not overlap each other.

We did Chinese word segmentation by Chinese lexical analysis system ICTCLAS[2] (Java version) which was developed by the Institute of Computing, Chinese Academy of Sciences. In order to validate the proposed method, open source tool Weka[3], which was developed by the University of Waikato, was used. We did programming based on Weka's API. All programs were written in Java under Eclipse Juno Service Release 1.The statistic information of the dataset after preprocess is show in table 1.

**Table 1.** Statistics of the dataset

| Total document# | Train document# | Test document# | Category# | Term# |
|---|---|---|---|---|
| 19637 | 9804 | 9833 | 20 | 55126 |

## 4.2 Evaluation Method

Commonly used evaluation methods for text classification include precision, recall and F1 value [2-3]. F1 combines precision and recall together and is a comprehensive assessment method. Therefore, we also use F1 as one of the evaluation criteria in this paper. In addition, we also take the running time of the program into consideration. Because even if the classification accuracy is very high, too long run time will limit the scope it can be used. Run time was measured from reading the dataset to printing the results. Because the Chinese text segmentation was done for preparation, and simply run only once, we ignored the time for segmentation.

---

[1] The corpus is available at http://www.nlpir.org/download
[2] http:// www.ictclas.org
[3] http://weka.wikispaces.com

### 4.3    Results and Analysis

At first, we tested the effectiveness of the combination feature reduction method. Feature space dimension was reduced to 500 by CHI at first, and then, neural network was trained and used to classify test documents. Since the original feature space dimension in our experiments is very high, PCA could not finish calculation within the limited time, therefore, it was not used separately for feature reduction. Then the comparison between CHI and CHI + PCA was conducted, and the dimension varies from 50 to 500 with an incensement of 50, the number of epochs to train the neural network is set as 200, and the result is show in figure 1.

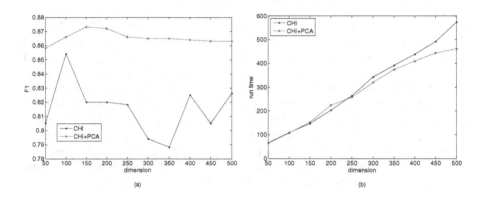

**Fig. 1.** F1 (a) and run time (b) of CHI and CHI + PCA

As can be seen from figure 1, the combination method performed better. CHI + PCA performed best when the dimension is 150, which lead to a 99% reduction relative to the original feature space dimension. Both methods' run time increased with the dimension. Since CHI + PCA choose more discriminated terms and reduced noises, it performed better than CHI. The best result CHI can achieve is lower than CHI + PCA can, therefore, one cannot improve the performance of CHI by simply increasing the dimension.

Next, we compared neural network (NN) with other classifiers, including Naïve Bayes (NB) and Support Vector Machine (SVM), the result is shown in figure 2.

Figure 2 shows that NN performed better than SVM when the dimension is lower than 450 and the result is comparable to SVM, which is considered to be one of the best classifiers. When the dimension is greater than 450, F1 value of NB decreased with the dimension, therefore, the proposed combination feature reduction method is not suitable to NB. In addition, with the increase of the dimension, time consumed by NB and SVM grown slowly, but the time consumed by NN growth rapidly. Therefore, for these time-sensitive applications, e.g. online classification, SVM is a more reasonable choice.

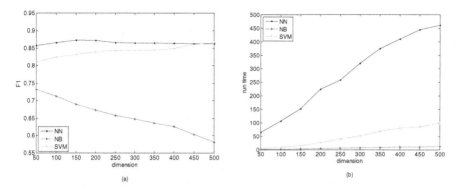

**Fig. 2.** F1 (a) and run time (b) of different classifiers

In addition, the training times of the neural network has greater impact on classification accuracy. In this experiment, we fixed the dimension on 100 and changed the training times from 1 to 500 with an increment of 50, the result is show in figure 3.

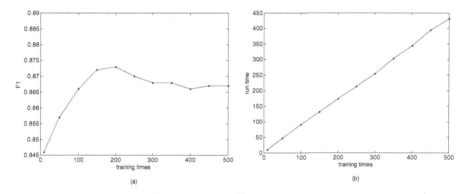

**Fig. 3.** F1 (a) and run time (b) against training times

We can see from figure 3 that with the increase of training times, F1 increased at first and reached peek at 200, then decreased when the train times greater than 300. With too little training times, the neural network cannot reach a reasonable connection weight between neurons, therefore, get poor result. On the contrary, training too many times will make neural network over-fitting and thus performed bad on test data. Therefore, it is crucial to find a reasonable training times when use neural network on text classification. In addition, we can see that the run time increased proportionally to the training times, which means time consumed in each training cycle is equally.

## 5    Conclusions

There have been a lot of researches on text classification based on neural network. Based on the analysis of existing researches, we proposed a combination feature reduction method. Terms selected by the proposed method were then used as the input of the neural network. The proposed method was verified on the actual Chinese text corpus and the results showed that the method can effectively reduce the feature space dimension. Besides, neural network needed less terms to achieve the same accuracy compared with other classifiers.

**Acknowledgements.** The authors' work was sponsored by the National High Technology Research and Development Program (863) of China(2010AA012505, 2011AA010702, 2012AA01A401 and 2012AA01A402), the Nature Science Foundation of China (60933005, 91124002), Support Science and Technology Project of China (2012BAH38B04, 2012BAH38B06) and Information Safety Plan (242) of China (2011A010). The authors would like to thank to the Chinese lexical analysis system provided by Institute of Computing, Chinese Academy of Sciences; Corpus provided by Fudan University; and the open free source provided by Weka.

# References

1. Furnkranz, J.: Round Robin classification. Mach. Learn. Res. 2, 721–747 (2002)
2. Yang, Y., Pedersen, J.P.: A Comparative Study on Feature reduction in Text Categorization. In: Proceedings of 14th International Conference on Machine Learning, pp. 412–420. Morgan Kaufmann, San Francisco (1997)
3. Lam, S.L.Y., Lee, D.L.: Feature Reduction for Neural Network Based Text Categorization. In: 6th International Conference on Database Systems for Advanced Applications, pp. 195–202. IEEE Computer Society, Washington, DC (1999)
4. Hansen, L.K., Salamon, P.: Neural Network Ensembles. IEEE Trans. Pattern Anal. Mach. Intell. 12(10), 993–1001 (1990)
5. Waleed, Z., Sang, M.L., Silvana, T.: Text classification: neural networks vs support vector machines. Industrial Management & Data Systems 109(5), 708–717 (2009)
6. Diganta, S.: Web Text Classification Using a Neural Network. In: 2011 Second International Conference on Emerging Applications of Information Technology (EAIT), pp. 57–60. IEEE Press, Kolkata (2011)
7. Cheng, H.L.: Neural Network for Text Classification Based on Singular Value Decomposition. In: 7th IEEE International Conference on Computer and Information Technology, pp. 47–52 (2007)
8. Kaburlasos, V.G., Fragkou, P., Kehagias, A.: Text Classification Using the σ-FLNMAP Neural Network. In: International Joint Conference on Neural Networks, vol. 2, pp. 1362–1367 (2001)
9. Nerijus, R., Ignas, S., Vida, M.: Text Categorization Using Neural Networks Initialized with Decision Trees. Informatica 15(4), 551–564 (2004)
10. Taeho, J.: NTC (Neural Text Categorizer): Neural Network for Text Categorization. International Journal of Information Studies 2(2), 83–96 (2010)
11. Keim, D.A., Oelke, D., Rohrdantz, C.: Analyzing document collections via context-aware term extraction. In: Horacek, H., Métais, E., Muñoz, R., Wolska, M. (eds.) NLDB 2009. LNCS, vol. 5723, pp. 154–168. Springer, Heidelberg (2010)
12. Baum, E.B., Haussler, D.: What size net gives valid generalization. Neural Computation 1(1), 151–160 (1989)

# The Angular Integral of the Radon Transform (aniRT) as a Feature Vector in Categorization of Visual Objects

Andrew P. Papliński

Monash University, Australia
Andrew.Paplinski@monash.edu

**Abstract.** The recently introduced angular integral of the Radon transform (aniRT) seems to be a good candidate as a feature vector used in categorization of visual objects in a rotation invariant fashion. We investigate application of aniRT in situations when the number of objects is significant, for example, Chinese characters. Typically, the aniRT feature vector spans the diagonal of the visual object. We show that a subset of the full aniRT vector delivers a good categorization results in a timely manner.

**Keywords:** Radon transform, categorization of visual objects, Chinese characters, Self-Organizing Maps, incremental learning.

## 1 Introduction

The paper is continuation of [20,22,21] where we considered categorization of visual objects based on the angular integral of the Radon transform [25] (aniRT). The principal problem in such categorization is the selection of the features that can represent a visual object in a way invariant to rotation, scaling, translation (RST), changes in illumination and the viewpoint. In this context, the main attraction of the Radon transform is its ability to deliver rotational invariance of the visual object. We will discuss this aspect in the section that follows. In this paper, we consider a simple case of one visual object in the image. This can be considered as a special case of more general invariant visual recognition based on a variety of local descriptors (e.g. [14,5,19]). Three groups of methods, namely, the Scale Invariant Feature Transform [16] (SIFT), PCA-SIFT [12] and the "Speeded Up Robust Features" (SURF) [2], seem to dominate the field. A comparison between the three methods is given in [11]. In [7] a rotation-invariant kernels are discussed in application to shape analysis. An interesting method (Ciratefi) that delivers the RST-invariance based on the template matching is presented in [13,1].

The prime application of the Radon transform has been in computer tomography. Relatively recently, Radon transform has been applied in a variety of image processing problems. Typically, Radon transform is used in conjunction with other transforms, wavelet and Fourier included. Magli et al. [17] and Warrick and

C. Guo, Z.-G. Hou, and Z. Zeng (Eds.): ISNN 2013, Part I, LNCS 7951, pp. 523–531, 2013.

Delaney [26] seem to initiate the use of Radon transform in combination with wavelet transform. More recently, a similar combination of transforms has been used in rotation invariant texture analysis [10,29], and in shape representation [28]. Other approach to rotation invariant texture analysis uses Radon transform in combination with Fourier transform [27]. Chen and Kégl [3] consider feature extraction using combination of three transforms: Radon, wavelet and Fourier. In [15], texture classification is performed by using a feature descriptor based on Radon transform and an affine invariant transform. Miciak [18] describes a character recognition system based on Radon transform and Principal Component Analysis. Hejazi et al. [8] present discrete Radon transform in rotation invariant image analysis. Close to our considerations are object identification problems discussed by Hjouj and Kammler in [9].

To our knowledge, the application of the angular integral of the Radon transform (aniRT) in the rotational invariance categorization problem has originally been presented in [20,22,21]. In those papers we used the aniRT feature vector in conjunction with the Self-Organizing Maps to categorize a relatively small number, say 30, of visual objects. We used the full aniRT feature vector that has the number of components equal to the diagonal of the visual objects.

In this paper we aim at testing the aniRT feature vector when the number of visual objects is significant, say 20000. We use a black-an-white rendering of Chinese Characters as our test visual objects. Using all the components, say 100, of the aniRT vector would be very time-demanding in the visual categorization process.

## 2    An Angular Integral of Radon Transform (aniRT)

We consider first a real continuous function of two real variables $x, y$ (or one complex variable $z$) $f(z)$, like an image, where $z = x + jy$. The Radon transform, $Rf(\theta, s)$, (see [25]) of such a function is an integral of $f(z)$ over straight lines $z(t) = e^{j\theta}(s + jt)$ (see Fig. 1).

$$Rf(\theta, s) = \int_{-\infty}^{+\infty} f(e^{j\theta}(s + jt))dt \qquad (1)$$

where $\theta$ is the slope of the line (or ray), $s$ is the distance of the line from the origin, and $t \in (-\infty, +\infty)$ is the line parameter, such that for $t = 0$ the line goes through the point $z(0) = se^{j\theta}$ as indicated in Fig. 1. Descriptively, we say that each 2-D point of the Radon transform, $Rf(\theta, s)$, is calculated as a summation (an integral) of the values of the function $f(z)$, e.g. pixels of an image, along the ray $z(t) = e^{j\theta}(s + jt)$, which is a straight line located at the distance $s$ from the origin (the image centre) and the slope $\theta$.

In the next step we consider the formation from the Radon transform, $Rf(\theta, s)$, a feature (or signature) function, say, $h(s)$, that can be used to categorize the function $f(x, y)$ in general, and images in particular, in a rotation-invariant

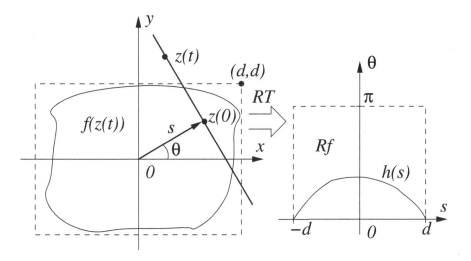

**Fig. 1.** The Radon Transform of a continuous function

fashion. This can be easily achieved by integrating the Radon transform wrt to the angular variable $\theta$:

$$h(s) = \int_0^\pi Rf(\theta, s)d\theta \tag{2}$$

With reference to the right-hand side of Fig. 1 we note that the integral is performed with respect to the angular variable. Such a feature (signature) function, $h(s)$, retains some characteristics of the original function $f(x, y)$, but the angular dependency is removed, hence, providing the rotational invariance. In a discrete case, $h(s)$, becomes a vector that spans the diagonal of the image, $s \in [-d, +d]$. We will be referring to this vector as the angular integral of the Radon Transform (aniRT).

Before we consider an effective way of calculating aniRT, $h(s)$, we recall a fundamental property of the Radon transform which says that the Radon transform at the point $(\theta, s)$ of a function rotated by the angle $\alpha$ is equal to the Radon transform of the original, un-rotated function calculated at the point $(\theta + \alpha, s)$:

$$Rf(\theta, s; \alpha) = \int_{-\infty}^{+\infty} f(e^{j(\theta+\alpha)}(s + jt))dt = Rf(\theta + \alpha, s) \tag{3}$$

In practical terms it means that instead of rotating the rays $z(t)$ we can rotate the image $f(z)$ in order to calculate the Radon transform at the point $(s, \theta)$. Typically, rotation of an image for the purpose of visualization on a rectangular grid implies the need for interpolation. However, in order to calculate the Radon transform, or specifically the aniRT vector, $h(s)$, we just rotate the grid, that is, the coordinates of the pixels, without modifying the pixels values.

Let us consider some details of the computational procedure. Firstly, the rotation angles are quantized and the quantization step, that is, the smallest rotation angle $\delta$, should be at the order of (see Fig. 2)

$$\delta = \arctan \frac{2}{\rho} \tag{4}$$

where $\rho = \max(r, c)$ and $r$ and $c$ are the number of rows and columns in the image, respectively. The rotation angles, $\theta$, for which the Radon transform is calculated are:

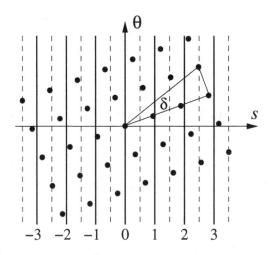

**Fig. 2.** Calculation of the aniRT vector

$$\theta_k = k \cdot \delta , \text{ for } k = 1, 2, \ldots, \left\lfloor \frac{\pi}{\delta} \right\rfloor \tag{5}$$

Now, consider coordinates of all pixels, $z_i = x_i + jy_i$, for $i = 1, 2, \ldots, N$ where $N$ is to total number of image pixels. We can pre-calculate all rotated image coordinates, that is,

$$\zeta_{ik} = z_i e^{j\theta_k} \tag{6}$$

With reference to Fig. 2 we notice that the Radon transform integral of eqn. (1) becomes a sum of pixel values $f(\zeta_{ik})$ along the vertical direction. Such a sum gives one point of the Radon transform $Rf(\theta, s)$ where $s = \text{round}(\text{real}(\zeta))$, is the rounded value of the real part of the rotated pixels coordinates. It is now obvious that to calculate the aniRT vector, $h(s)$, we sum pixels values, $f(\zeta_{ik})$, for all angles as in eqn. (2).

In Fig. 3 we present the aniRT vectors, $h(s)$, calculated for 10 randomly selected Chinese characters. The characters are rendered in the Microsoft JhengHei font of the size 28. The number of pixels is $70 \times 70$, hence the size (the number of components) of the aniRT vectors is 98. We observe two basic properties of the aniRT vectors:

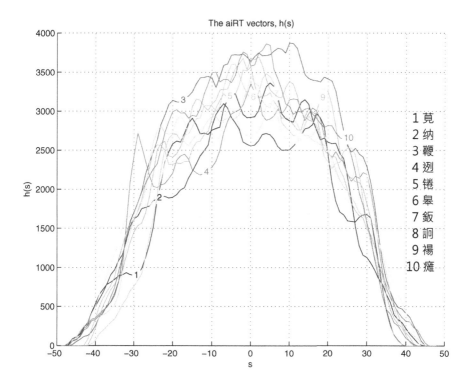

**Fig. 3.** The aniRT vectors, $h(s)$, for 10 randomly selected Chinese characters

- they are significantly different for different characters
- the main differentiation occurs in the central part of the vectors.

The obvious question now is how many components we need to differentiate a given set of images. Before we investigate this issue, in Fig. 4 we present the aniRT vectors calculated for 20,000 Chinese characters. We create 10 random permutations of the Chinese characters and investigate how many components of the aniRT vectors are needed to have a unique representation of the characters, The results are presented in Table 1. A bit unexpectedly, we observe from the last row of the Table 1, the number of components to make the selection of 20,000 characters uniquely represented is just 3. The result seems to be perfectly correct if we consider the fact that the components of the aniRT vectors are coded by

**Table 1.** The minimum number of components of the aniRT vectors to differentiate specified number of characters

| cmp | number of characters | | | | | | | | | |
|---|---|---|---|---|---|---|---|---|---|---|
| 1 | 2 | 2 | 2 | 2 | 2 | 2 | 2 | 2 | 2 | 2 |
| 2 | 6 | 58 | 8 | 72 | 107 | 100 | 14 | 85 | 67 | 107 |
| 3 | 1102 | 168 | 669 | 2137 | 836 | 892 | 951 | 1061 | 2106 | 825 |

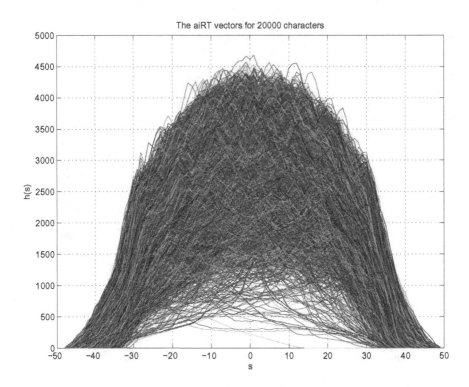

**Fig. 4.** The aniRT vectors, $h(s)$, for 20,000 Chinese characters

four 13-bit numbers (the maximum value is greater than $4096 = 2^{12}$ — see Fig. 4), giving the total number of bits being 15, just enough to code 20,000 characters.

## 3   Example of Categorization of Chinese Characters with the Incremental Self-Organizing Map (iSOM)

Unlike our previous usage of aniRT vectors in categorization of visual objects ([20,22,21]) where the all components of the vector were utilized, here we demonstrate the result of categorization based on small number of components as discussed in the previous section. For this purpose we will create a self-organizing map for a random selection of 200 characters out of the total number of 20,000 characters as in Fig. 5. We will use the version of the incremental SOM as presented in [21]. Similar, but non-incremental SOMs for categorization are also presented in [23,6,4,24]. The main characteristic features distinguishing our SOMs from the most commonly used are as follows:

– Instead being located on a rectangular grid, our neuronal units are we randomly distributed inside a unit circuit. The neuronal units are represented by the yellow dots in Fig. 5.

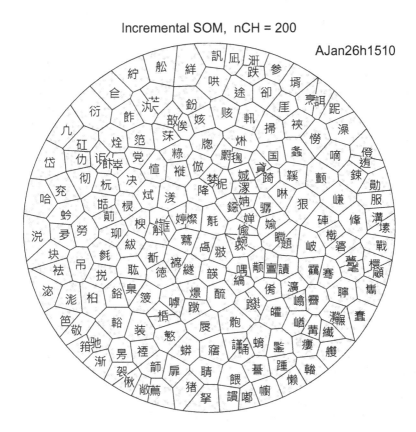

**Fig. 5.** iSOM for 200 Chinese characters

- A fix number (stochastically), 16 in the example, of neuronal units per character is maintained. This is to simulate a redundance observed in our brain to represents mental objects.
- All vectors are normalised to be unity length and located on the respective hyperspheres.
- We use the "dot-product" version of the Kohonen learning law.
- Although not explicitly shown here, the learning is done in an incremental fashion, adding one character at a time as in [21].

With reference to Fig. 5 it needs to be said that the topological grouping of Chinese characters is not an "intelligent" one, e.g. characters with the same radicals being grouped together. Instead the grouping is based on the similarity of the angular integral of the Radon Transform, or rather the four central components of the aniRT. In general, the objective of the example is to demonstrate that such a small number of components is enough to spread apart 200 characters in the SOM. As discussed in [22] the aniRT vectors ensure the rotational invariance of the representation of visual objects.

# 4   Concluding Remarks

We have investigated some properties of the recently introduced feature vector for images, aniRT, which is calculated as an angular integral of the Radon Transform of images. We have demonstrated a novel method of calculating the aniRT vector based on rotation of the image grid. Such a method gives a significant time advantage when an aniRT vector is calculated for a large number of images as in our example with Chinese characters. Subsequently, we have shown that to differentiate 20,000 characters only four components of the aniRT vectors are required. Finally, we have demonstrated the formation of a SOM consisting of 200 characters using 4 components of the aniRT feature vector.

# References

1. Araújo, S.A., Kim, H.Y.: Color-Ciratefi: A color-based RST-invariant template matching algorithm. In: Proc. 17th Int. Conf. Systems, Signals and Image Proc., pp. 1–4. Rio de Janeiro (2010)
2. Bay, H., Tuytelaars, T., Van Gool, L.: SURF: Speeded up robust features. In: Leonardis, A., Bischof, H., Pinz, A. (eds.) ECCV 2006, Part I. LNCS, vol. 3951, pp. 404–417. Springer, Heidelberg (2006)
3. Chen, G., Kégl, B.: Feature extraction using Radon, wavelet and Fourier transform. In: Proc. IEEE Int. Conf. Syst. Man and Cybernetics, pp. 1020–1025 (2007)
4. Chou, S., Papliński, A.P., Gustafsson, L.: Speaker-dependent bimodal integration of Chinese phonemes and letters using multimodal self-organizing networks. In: Proc. Int. Joint Conf. Neural Networks, Orlando, Florida, pp. 248–253 (August 2007)
5. Fehr, J.: Local rotation invariant patch descriptors for 3D vector fields. In: Proc. Int. Conf. Pat. Rec., Istanbul, pp. 1381–1384 (2010)
6. Gustafsson, L., Papliński, A.P.: Bimodal integration of phonemes and letters: an application of multimodal self-organizing networks. In: Proc. Int. Joint Conf. Neural Networks, Vancouver, Canada, pp. 704–710 (July 2006)
7. Hamsici, O.C., Martinez, A.M.: Rotation invariant kernels and their application to shape analysis. IEEE Tran. PAMI 31(11), 1985–1999 (2009)
8. Hejazi, M., Shevlyakov, G., Ho, Y.S.: Modified discrete Radon transforms and their application to rotation-invariant image analysis. In: Proc. IEEE Workshop Mult. Sig. Proc., pp. 429–434 (2006)
9. Hjouj, F., Kammler, D.W.: Identification of reflected, scaled, translated, and rotated objects from their Radon projections. IEEE Trans. Img. Proc. 17(3), 301–310 (2008)
10. Jafari-Khouzani, K., Soltanian-Zadeh, H.: Rotation-invariant multiresolution texture analysis using Radon and wavelet transforms. IEEE Trans. Img. Proc. 14(6), 783–795 (2005)
11. Juan, L., Gwun, O.: A comparison of SIFT, PCA-SIFT and SURF. Int. J. Img. Proc. 3(4), 143–152 (2009)
12. Ke, Y., Sukthankar, R.: PCA-SIFT: A more distinctive representation for local image descriptors. In: Proc. Conf. Comp. Vis. Pat. Rec., Washington, pp. 511–517 (2004)

13. Kim, H.Y., de Araújo, S.A.: Grayscale template-matching invariant to rotation, scale, translation, brightness and contrast. In: Mery, D., Rueda, L. (eds.) PSIVT 2007. LNCS, vol. 4872, pp. 100–113. Springer, Heidelberg (2007)
14. Kingsbury, N.: Rotation-invariant local feature matching with complex wavelets. In: Proc. 14th Europ. Sig. Proc. Conf., Florence, Italy, pp. 1–5 (2006)
15. Liu, G., Lin, Z., Yu, Y.: Radon representation-based feature descriptor for texture classification. IEEE Trans. Img. Proc. 18(5), 921–928 (2009)
16. Lowe, D.G.: Distinctive image features from scale-invariant keypoints. Int. J. Comp. Vision 60(2), 91–110 (2004)
17. Magli, E., Presti, L.L., Olmo, G.: A pattern detection and compression algorithm based on the joint wavelet and Radon transform. In: Proc. IEEE 13th Int. Conf. Dig. Sig. Proc., pp. 559–562 (1997)
18. Miciak, M.: Character recognition using Radon transformation and principal component analysis in postal applications. In: Proc. Int. Multiconf. Comp. Sci. Info. Tech., pp. 495–500 (2008)
19. Nelson, J.D.B., Kingsbury, N.G.: Enhanced shift and scale tolerance for rotation invariant polar matching with dual-tree wavelets. IEEE Trans. Img. Proc. 20(3), 814–821 (2011)
20. Papliński, A.P.: Rotation invariant categorization of visual objects using Radon transform and self-organizing modules. In: Wong, K.W., Mendis, B.S.U., Bouzerdoum, A. (eds.) ICONIP 2010, Part II. LNCS, vol. 6444, pp. 360–366. Springer, Heidelberg (2010)
21. Papliński, A.P.: Incremental self-organizing map (iSOM) in categorization of visual objects. In: Huang, T., Zeng, Z., Li, C., Leung, C.S. (eds.) ICONIP 2012, Part II. LNCS, vol. 7664, pp. 125–132. Springer, Heidelberg (2012)
22. Papliński, A.P.: Rotation invariant categorization of colour images using Radon transform. In: Proc. WCCI–IJCNN, pp. 1408–1413. IEEE (2012)
23. Papliński, A.P., Gustafsson, L.: Multimodal feedforward self-organizing maps. In: Hao, Y., Liu, J., Wang, Y.-P., Cheung, Y.-M., Yin, H., Jiao, L., Ma, J., Jiao, Y.-C. (eds.) CIS 2005. LNCS (LNAI), vol. 3801, pp. 81–88. Springer, Heidelberg (2005)
24. Papliński, A.P., Gustafsson, L., Mount, W.M.: A recurrent multimodal network for binding written words and sensory-based semantics into concepts. In: Lu, B.-L., Zhang, L., Kwok, J. (eds.) ICONIP 2011, Part I. LNCS, vol. 7062, pp. 413–422. Springer, Heidelberg (2011)
25. Radon, J.: Radon transform, http://en.wikipedia.org/wiki/Radon_transform
26. Warrick, A., Delaney, P.A.: Detection of linear features using a localized Radon transform with a wavelet filter. In: Proc. ICASSP, pp. 2769–2772 (1997)
27. Xiao, S.S., Wu, Y.X.: Rotation-invariant texture analysis using Radon and Fourier transforms. J. Phys.: Conf. Ser. 48, 1459–1464 (2007)
28. Yao, W., Pun, C.M.: Invariant shape representation by Radon and wavelet transforms for complex inner shapes. In: Proc. IEEE Int. Conf. Inform. Autom., pp. 1144–1149 (2009)
29. Yu, G., Cao, W., Li, Z.: Rotation and scale invariant for texture analysis based on Radon transform and wavelet transform. In: Proc. 3rd ICPCA, pp. 704–708 (2008)

# A Fast Algorithm for Clustering with MapReduce[*]

Yuqing Miao, Jinxing Zhang, Hao Feng, Liangpei Qiu, and Yimin Wen

School of Computer science and Engineering, Guilin University of Electronic Technology,
Guilin 541004, China
myq@ustc.edu.cn, jinxing_zhang@sina.cn,
{fengh,ymwen}@guet.edu.cn, qlp_1018@126.com

**Abstract.** MapReduce is a popular model in which the dataflow takes the form of
a directed acyclic graph of operators. But it lacks built-in support for iterative
programs, which arise naturally in many clustering applications. Based on micro-cluster and equivalence relation, we design a clustering algorithm which can
be easily parallelized in MapReduce and done in quite a few MapReduce rounds.
Experiments show that our algorithm not only runs fast and obtains good accuracy but also scales well and possesses high speedup.

**Keywords:** Micro-Cluster, Equivalence relation, Clustering, MapReduce, Data
mining.

## 1 Introduction

Cloud computing is a new computing model which is attracting more and more attention from the research and application domain of data mining. Hadoop is an
open-source cloud computing platform which provides a distributed file system
(HDFS) [1] and implements a computational paradigm named MapReduce [2]. MapReduce is designed for processing parallelizable problems across huge datasets using
a large number of computers. A MapReduce procedure mainly consists of three stages:
map, shuffle and reduce. The map phrase preliminarily processes the input data and
produces intermediate results which are stored in local disk. In the shuffle phrase, the
intermediate results are copied and transferred to one or more machines over the network. The reduce phrase copes with the copied data and output the final results. It is
easy to notice that MapReduce does not support iterative algorithms effectively, as
each iteration might cause large amount of intermediate data and large bandwidth
consumption.

Based on the ideas of k-means algorithm[3] and micro-cluster structure[4], combined with our equivalence relation, we proposed a new clustering algorithm called
BigKClustering. First, BigKClustering divides the dataset into many groups, just like
k-means does in its first two steps of the first iteration, and constructs one micro-cluster

---

[*] The work described in this paper is supported by Foundation of Guangxi Key Laboratory of
Trustworthy Software, China (kx201116) and Educational Commission of Guangxi Province,
China(201204LX122).

C. Guo, Z.-G. Hou, and Z. Zeng (Eds.): ISNN 2013, Part I, LNCS 7951, pp. 532–538, 2013.

corresponding to each group. Then, all the micro-clusters that are closed enough will be connected and put into the same group by the equivalence relation. Finally, the center of each group will be calculated and that will be the center of a real cluster in the dataset. We conduct comprehensive experiments to evaluate our algorithm. Our key observations are: 1) BigKClustering can be implemented in MapReduce naturally and done among three MapReduce rounds; 2) BigKClustering runs very fast and obtains high clustering quality; 3) BigKClustering has pretty good scaleup and speedup, and its time cost is quite stable.

## 2    Related Work

Micro-cluster is a technique of data summarization for clustering data streams and large dataset. It is an extension of the Clustering Feature (CF) [5].Given N d-dimensional data points in a cluster: $\{X_i\}$ where $i = 1, 2, ..., N$, CF is defined as a vector: $CF = (N, LS, SS)$, where N is the number of data points in the cluster, LS the linear sum of the N data points, SS the square sum of the N data points. Based on this vector, the mean and variance of all the data in a cluster can be calculated. Therefore, we can use a micro-cluster to represent a series of data and its main distribution features. For more information about the applications of micro-cluster, please refer to [6].

Up till now, Clustering based on MapReduce model has been studied in many literatures. Papadimitriou [7] proposed a distributed co-clustering framework on Hadoop called DisCo, providing practical methods for distributed data pre-processing and co-clustering. Ene et al. [8] focused on k-center and k-median and developed fast clustering algorithms with constant factor approximation guarantees. They use sampling strategy to decrease the data size and ran a time consuming clustering algorithm such as k-means on the resulting dataset. Their algorithms could run in a constant number of MapReduce rounds but the performance did not always outperform their counterparts. Bahman et al. [9] proposed a parallel algorithm to generate initial centers for k-means, thus dramatically improved the performance of k-means, both in terms of quality and convergence properties.

## 3    The Algorithm

In this section, we show how our algorithm BigKClustering works. First, we give the definitions of micro-cluster and equivalence relation. Next, we introduce BigKClustering in detail. Finally, we discuss a MapReduce realization of BigKClustering.

### 3.1    Definitions

**Micro-cluster:** Given a group of d-dimensional points $X=(x_1,...,x_n)$, A micro-cluster is defined as a $(3d + 2)$-dimensional vector $(n_i, CF1_i, CF2_i, Center_i, max_i)$ , wherein i can be regarded as the micro-cluster id; $n_i, CF1_i$ and $CF2_i$ each correspond to N,LS and SS in a CF vector; $Center_i$ is the point we choose arbitrarily from X as one of the centers to set part the total dataset, like an initial center for k-means, don't confuse it with the average of X. $max_i$ is the longest distance between all the points in X to $Center_i$.

**Equivalence Relation:** Given a set of micro-clusters MC, to any two micro-clusters $S_i, S_j \in MC$, if their distance is not greater than a given threshold d(also called the connection distance), or there is a bunch of micro-clusters $(S_i, S_{t1}, S_{t2}, \ldots, S_j)$ among which the distance between any two adjacent micro-clusters is not greater than d, then we say $S_i, S_j$ have an equivalence relation. In our work, the distance of two micro-clusters is $dis(S_i, S_j) = dis(Center_i, Center_j) - max_i - max_j$ where Euclidean distance are used between two points. If the distance is a negative value, then it is set to be zero. If the distance is zero and the distance of $Center_i$ to $Center_j$ is less than $max_i$ or $max_j$, we also think that $S_i, S_j$ have an equivalence relation. All the micro-clusters that have an equivalence relation will be connected and form a group of micro-clusters.

## 3.2    The BigKClustering Algorithm

BigKClustering (Algorithm 1) can be seen as the sequential version of our algorithm. The MapReduce implementation will be discussed in the next section.

Typically, a clustering algorithm uses iterative strategy to obtain better result. During the iterations, the original data points are regrouped again and again and the compactness of some data points may be easily destroyed, thus reducing the clustering quality. Our algorithm intends to group together those most compact points in the dataset and then constructs series of micro-clusters. Every micro-cluster is used as a single point. In this way, those compact data points will not be parted during the clustering process and always belong to one cluster, which most existing clustering algorithms fail to do.

After the construction of micro-clusters, the equivalence relation is employed to connect the micro-clusters and then groups of micro-clusters are formed. The center of the group will be treated as the center of a real cluster. As to the connection distance, we first initialize it with the mean value (m) of all the max values contained in micro-clusters and then adjust it until we get the right number of clusters.

---

**Algorithm 1.** BigKClustering (ds, BigK, k)

---

Input: dataset (ds); number of micro-clusters (BigK); number of real clusters (k).
Output: clusters.

1: Arbitrarily choose BigK centers from ds.
2: Assign each point in ds to its closest center. This divides ds into BigK parts $p_1 \cdots p_{BigK}$.
3: Construct BigK micro-clusters $s_1, s_2, \ldots, s_{BigK}$, each corresponds to one data part.
4: Calculate the initial connection distance m.
5: Groups of micro-clusters $\leftarrow$ joinToGroups $(< s_1, s_2, \ldots, s_{BigK} >, k, m)$
6: Calculate the centers of all the groups and take them as the centers of real clusters.
7: Assign each point in ds to its closest center and give it a cluster label.

---

joinToGroups is a function to generate expected number of groups of micro-clusters, its pseudo codes are described as follows:

---

---

**Algorithm 2.** JoinToGroups  $(< s_1, s_2, \ldots, s_{BigK} >, k, d)$

---

Input: list of micro-clusters ($<s_1, s_2, \ldots, s_{BigK} >$);   number of clusters (k); connection distance (d).

Output: groups of micro-clusters.

   1) For i=1 to BigK do
          j=i-1; flag=true;
    1.1) if(j != 0)
        For k=1 to j do
       1.1.1) if(dis($s_i,s_k$)==0)
             if(dis(Center$_i$,Center$_j$)$\leq$max$_i$ || dis(Center$_i$,Center$_j$)$\leq$max$_j$)
               flag=false; break;
       1.1.2) else if (dis($s_i,s_k$)$\leq$d)
             flag=false; break;
    1.2) if(flag==true)
        give $s_i$ a new group id;
    1.3) else
        add $s_i$ to the same group as $s_k$;
   2) If the number of groups is k, then turn to step 4). Otherwise, turn to step 3).
   3) Adjust d, turn to step 1).
   4) return all the groups.

---

### 3.3   MapReduce Implementation

BigKClustering can be easily paralleled in MapReduce. Step 1 through step 3 can be done in one MapReduce Job: each mapper working on an input split outputs a series of intermediate key/value pairs among which the value is a data point and the key is the point closest center; the reducers take charge of constructing micro-clusterings based on their input records, one record one micro-cluster. All micro-clusters are put together in a file on HDFS at last. Step 4 through step 6 can be done with another Job which consists of only one mapper and one reducer: the mapper read all the micro-clusters from HDFS and calculates the initial connection distance; the reducer is responsible for connecting the micro-clusters based on the equivalence relation and calculating the centers for the real clusters. Step 7 is quite similar to step 2, assigns each data point to its nearest cluster center and gives it a label.

As can be seen from the above, our parallel algorithm effectively avoids the iterative operations shared by many traditional clustering algorithms, thus substantially reduces the intermediate values and network traffics, which are caused by MapReduce framework. Moreover, the running time of our parallel algorithm mostly costs in the construction of micro-clusters which can be completely paralleled. So, our algorithm can run quite fast.

# 4    Experiments

In this section, we present the experimental setup for evaluating BigKClustering. All the algorithms are run on a Hadoop cluster of 3 nodes, each with two Intel-Core 3.1GHz processors and 4GB of memory.

## 4.1    Datasets

To evaluate the accuracy and running time of our algorithm, we generate a set of synthetic datasets with eight real clusters. Each dataset consists of 10,000 vectors and the number of data dimensions of each dataset corresponds to 2、4、8 respectively. All the vectors follow normal distribution and are standardized in [0, 1]. To evaluate the scaleup and speedup of our algorithm, we generate three 10-dimensional datasets, the number of their records correspond to 1*108 (≈1GB) 、2*108、3*108, respectively.

## 4.2    Results and Analysis

We adopt clustering cost to evaluate the accuracy of our algorithm, as k-meansII [9] did in their experiments. We compare our algorithm against k-meansII and the parallel implementation of k-means. All experimental results are averaged over 10 runs. The results of the algorithms are displayed in Fig.1 through Fig.4.

Fig.1 to Fig.2 shows that BigKClustering not only runs faster but also achieves lower cost than its counterparts. Fig.3 and Fig.4 show that BigKClustering scales very well and possesses high speedup. In Fig.2, we notice that the running time of our algorithm is more than the compared algorithms when the number of data dimensions is 2. The reason is that the running time of our algorithm is mainly attributed by the time cost of constructing micro-clusters which is determined by the number of micro-clusters, while the running time of other two algorithms are mainly attributed by the time cost of iterations which are greatly affected by the number of data dimensions. This also well explains that the running time of our algorithm is quite stable, while the running time of the other two algorithms changes dramatically as the number of data dimensions increase.

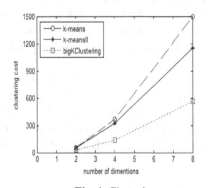
**Fig. 1.** Clustering cost

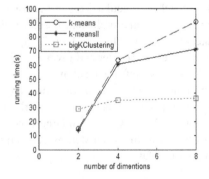
**Fig. 2.** Time cost

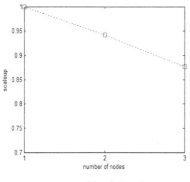

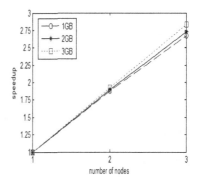

**Fig. 3.** Scaleup                                **Fig. 4.** Speedup

### 4.3    Parameters Discussion

Theoretically, the larger the number BigK of micro-clusters is, the higher the quality of clustering might be. Experiments show that our algorithm can obtain high accuracy with relatively short running time when BigK is set to be around 60.As to the connection distance d, we initialize it with m and then adjust it by using iterative strategy. For example, if BigK is proved to be smaller than expected (i.e. the number of groups of micro-clusters generated is greater than k) after the first iteration of connecting micro-clusters, it will be set to be 2*m; if BigK is then proved to be greater than expected after the following iteration, it will be set to be 1.5*m. The same process will continue until k groups of micro-clusters are generated. Experiments demonstrate that reaching k groups of micro-clusters always just needs a few rounds of iteration.

## 5    Conclusions

With huge dataset emerged, most of the existed algorithms or old models can not meet practical needs any more. In this paper we proposed BigKClustering, a new efficient clustering algorithm which bases on micro-cluster and equivalent relation. The algorithm is simple and can be easily paralleled in MapReduce with only three MapReduce rounds. Experimental results show that BigKClustering not only runs fast and achieves high clustering quality but also scales well and possesses high speedup.

## References

1. Apache Hadoop, http://hadoop.apache.org
2. Dean, J., Ghemawat, S.: MapReduce: simplified data processing on large clusters. In: Proceedings of Operating Systems Design and Implementation, San Francisco, CA, pp. 137–150 (2004)
3. Lloyd, S.: Least squares quantization in PCM. IEEE Transactions on Information Theory 28(2), 129–136 (1982)

4. Aggarwal, C., Han, J., Wang, J., Yu, P.: A Framework for Clustering Evolving Data Streams. In: Proceedings of the 29th International Conference on Very Large Data Bases, Berlin, German, pp. 81–92 (2003)
5. Zhang, T., Ramakrishnan, R., Livny, M.: BIRCH: an efficient data clustering method for very large databases. In: Proceedings of the 1996 ACM SIGMOD International Conference on Management of Data, Montreal, Quebec, Canada, pp. 103–114 (1996)
6. Amini, A., Wah, T.: Density micro-clustering algorithms on data streams: A review. In: International Conference on Data Mining and Applications (ICDMA), Hong Kong, China, pp. 410–414 (2011)
7. Papadimitriou, S., Sun, J.: DisCo: Distributed Co-clustering with Map-Reduce: A Case Study Towards Petabyte-Scale End-to-End Mining. In: Proceedings of the 8th IEEE International Conference on Data Mining, Pisa, Italy, pp. 512–521 (2008)
8. Ene, A., Im, S., Moseley, B.: Fast clustering using MapReduce. In: Proceedings of the 17th ACM SIGKDD International Conference on Knowledge Discovery and Data Mining, California, USA, pp. 681–689 (2011)
9. Bahmani, B., Moseley, B., Vattani, A., Kumar, R., Vassilvitskii, S.: Scalable k-means++. Proceedings of the VLDB Endowment 5(7), 622–633 (2012)

# Gaussian Message Propagation in *d*-order Neighborhood for Gaussian Graphical Model

Yarui Chen[1], Congcong Xiong[1], and Hailin Xie[2]

[1] School of Computer Science and Information Engineering,
Tianjin University of Science and Technology, Tianjin 300222, P.R. China
[2] School of Information Technical Science
Nankai University

**Abstract.** Gaussian graphical models are important undirected graphical models with multivariate Gaussian distribution. A key probabilistic inference problem for the model is to compute the marginals. Exact inference algorithms have cubic computational complexity, which is intolerable for large-scale models. Most of approximate inference algorithms have a form of message iterations, and their computational complexity is closely related to the convergence and convergence rate, which causes the uncertain computational efficiency. In this paper, we design a fixed parameter linear time approximate algorithm — the Gaussian message propagation in *d*-order neighborhood. First, we define the *d*-order neighborhood concept to describe the propagation scope of exact Gaussian messages. Then we design the algorithm of Gaussian message propagation in *d*-order neighborhood, which propagates Gaussian messages in variable's *d*-order neighborhood exactly, and in the $(d + 1)$th-order neighborhood partly to preserve the spread of the Gaussian messages, and computes the approximate marginals in linear time $O(n \cdot d^2)$ with the fixed parameter *d*. Finally, we present verification experiments and comparison experiments, and analyze the experiment results.

**Keywords:** Gaussian graphical model, Probabilistic inference, Message propagation, *d*-order neighborhood.

## 1 Introduction

Gaussian graphical models are basic undirected graphical models with multivariate Gaussian distribution and conditional independence assumptions [1,2], and have wide application in image analysis, natural language processing, time-series analysis etc [3,4]. The key problem of probabilistic inference for the Gaussian graphical model is to compute the marginals of variables [5]. For tree-like models, exact inference algorithms, such as Gaussian elimination, belief propagation or junction tree algorithms, can present the marginals in linear time [6]. For general graphical models, these exact inference algorithms have cubic $O(n^3)$ computational complexity [6]. For large-scale models with more complex graphs, arising in oceanography, 3D-tomography, and seismology, the cubic computational complexity becomes computationally prohibitive [7]. Then various approximate

C. Guo, Z.-G. Hou, and Z. Zeng (Eds.): ISNN 2013, Part I, LNCS 7951, pp. 539–546, 2013.
© Springer-Verlag Berlin Heidelberg 2013

inference algorithms have been developed, such as loopy belief propagation [8], mean field method [9,10].

The loopy belief propagation algorithm propagates the belief messages in the models with cycles directly, and provides successful approximation in some applications [8]. But the algorithm may converges to local optimum, may even fail to converge for general models, and its computational complexity is closely related to the convergence and convergence rate [11]. The mean field algorithm propagates the variational messages to approximate the marginals, and research shows that if the variational message converge, the algorithm can compute the correct mean parameter [10]. This series of message iteration algorithms have the computational complexity $O(m \cdot n \cdot N)$, but the iteration number $N$ is closed related to the convergence of the algorithms. So these algorithms have uncertain computational efficiency.

In this paper, we design a fixed parameter linear time $O(n \cdot d^2)$ approximate inference algorithm— Gaussian message propagation in $d$-order neighborhood (GaussianMP-$d$). First, we define the $d$-order neighborhood concept to describe the propagation scope of Gaussian message, and show the message propagation process in the $d$-order neighborhood of univariable (GaussianVariableMP-$d$), which propagates the Gaussian messages in variable's $d$-order neighborhood exactly, and in the $(d+1)$th-order neighborhood partly to preserve the computational complexity increasing. Then we design the GaussianMP-$d$ algorithm based on the GaussianVariableMP-$d$ unit, which executes the GaussianVariableMP-$d$ for variables in the elimination order $I$ and the reverse order $I'$ respectively to compute all the Gaussian messages, and calculates the approximate marginals with these Gaussian messages. Finally, we present verification experiments and comparison experiments to demonstrate the efficiency and flexibility of the GaussianMP-$d$ algorithm.

## 2    Gaussian Elimination Process

The Gaussian graphical model is a undirected graphical model based on graph $G = (V, E)$, where the vertex set $V$ denotes the Gaussian random variable set $x = \{x_1, \cdots, x_n\}$, and the edge set $E$ reflects the conditional independences among variables. The probability distribution of Gaussian graphical model is

$$p(x) = \exp\left\{\langle h, x \rangle + \frac{1}{2}\langle J, xx^\mathrm{T} \rangle - A(h, J)\right\},$$

$$A(h, J) = \log \int_\Xi \exp\left\{\langle h, x \rangle + \frac{1}{2}\langle J, xx^\mathrm{T} \rangle\right\}\mathrm{d}x.$$

Where the $h = [h_1, \cdots, h_n]^\mathrm{T}$, $J = [J_{ij}]_{n \times n}$ are the model parameters, $\langle, \rangle$ denotes dot product operation, $A(h, J)$ is the log partition function, $\Xi = \{(h, J) \in \mathbb{R}^n \times \mathbb{R}^{n \times n} | J \prec 0, J = J^\mathrm{T}\}$ is the constraint set of parameter $(h, J)$.

An important inference problem for the Gaussian graphical model is to compute the marginals $p(x_i)$. Gaussian elimination algorithm is an exact inference method with variable elimination/marginalization. The distribution of

$x_U = x \setminus x_s$ can be computed by eliminating $x_s$, and the corresponding model parameters $J_U^*, h_U^*$ are

$$J_U^* = J_{U,U} - J_{U,s} J_{s,s}^{-1} J_{s,U},$$
$$h_U^* = h_U - J_{U,s} J_{s,s}^{-1} h_s.$$

During the elimination process of the single variable $x_s$, the parameters of the neighbors $\{t \mid t \in N(s)\}$ have only been changed. The parameter update formulas of the neighbor variable $\{t \mid t \in N(s)\}$ are

$$J_{tt} \leftarrow J_{tt} + \left( -\frac{J_{st}^2}{J_{ss}} \right), \qquad h_t \leftarrow h_t + \left( -\frac{J_{st}}{J_{ss}} h_s \right).$$

The update formula of the edges $\{(t, u) \mid t, u \in N(s)\}$ is

$$J_{tu} \leftarrow J_{tu} + \left( -\frac{J_{st} J_{su}}{J_{ss}} \right).$$

If there is no edge between node $t$ and $u$, an edge $(t, u)$ would be added with parameter $J_{tu} \leftarrow \left( -\frac{J_{st} J_{su}}{J_{ss}} \right)$. Obviously, the neighborhoods $\{t \mid t \in N(s)\}$ form a complete graphs with $m = |N(s)|$ nodes, and the computational complexity of elimination of the single variable $x_s$ is $O(m^2)$. The scale of the complete graph becomes larger with the elimination of variables, and the computational complexity of elimination of single variable is trend to $O(n^2)$. Then the computational complexity of the Gaussian elimination algorithm is $O(n^3)$.

# 3 Gaussian Message Propagation in $d$-order Neighborhood

In this section, we define the concept of $d$-order Gaussian elimination neighborhood, and design the algorithm of Gaussian message propagation in $d$-order neighborhood (GaussianMP-$d$ Algorithm).

## 3.1 $d$-order Gaussian Elimination Neighborhood

**Definition 1 (dth-order Neighborhood).** *For the Gaussian graphical model $G$, let $N_d(i)$ denote the dth-order Gaussian elimination neighborhood (abbreviated to dth-order neighborhood) of node $i$, which is defined recursively as following:*

1. *The 1-st order neighborhood $N_1(i)$ is the set of neighbors of the node $i$, that is $N_1(i) = \{j \mid (i, j) \in E\}$.*
2. *Let $s, t \in N_1(i)$. If the edge $(s, t)$ is added during Gaussian elimination of variable $x_i$, we label*

$$s \in N_2(t), \quad t \in N_2(s).$$

3. Let $u \in N_a(i)$, $v \in N_b(i)$. If the edge $(u, v)$ is added during Gaussian elimination of variable $i$, we label

$$u \in N_{a+b}(v), \quad v \in N_{a+b}(u).$$

**Definition 2 (d-order Neighborhood).** *For the Gaussian graphical model $G$, the $d$-order neighborhood is the union set of $N_d(i), d = 1, \cdots, d$, that is*

$$\cup N_d(i) = N_1(i) \cup N_2(i) \cup \cdots \cup N_d(i).$$

The Gaussian message propagation in the $d$-order neighborhood of the variable $x_i$ (GaussianVariableMP-$d$) is to propagate the Gaussian messages in the $d$-order neighborhood exactly, and in the $(d+1)$th-order neighborhood partly through avoiding adding new edges to decrease the computational complexity. Let $I$ denote the node set in elimination order, $I_{\mathrm{elim}}$ the set of nodes eliminated in elimination order, $I_{\mathrm{left}}$ the set of nodes left in elimination order, the GaussianVariableMP-$d$ algorithm contains the following four steps, and he formal description is shown in Algorithm 1.

1. Update the parameters $\{\hat{J}_{ii}, \hat{h}_i\}$ of the variable $x_i$ with the Gaussian messages propagated from the variables in $I_{\mathrm{elim}}$, that is

$$\hat{J}_{ii} = J_{ii} + \sum_{j \in (\cup N_{d+1}(i)) \cap I_{\mathrm{elim}}} \triangle J_{j \to i},$$

$$\hat{h}_i = h_i + \sum_{j \in (\cup N_{d+1}(i)) \cap I_{\mathrm{elim}}} \triangle h_{j \to i}, \tag{1}$$

where $\triangle J_{j \to i}, \triangle h_{j \to i}$ are the Gaussian messages from node $j$ to $i$.
Then update the parameters $\{\hat{J}_{ik}(k) \mid k \in (\cup N_{d+1}(i)) \cap I_{\mathrm{left}}\}$ corresponding to the edges $\{(i, k) \mid k \in (\cup N_{d+1}(i)) \cap I_{\mathrm{left}}\}$, that is

$$\hat{J}_{ik} = J_{ik} + \sum_{j \in (\cup N_{d+1}(i)) \cap I_{\mathrm{elim}}} \triangle J_{j \to ik}, \tag{2}$$

where $\triangle h_{j \to ik}$ denote the message from the $j$ to the edge $(i, k)$.
2. Compute the Gaussian messages $\{\triangle J_{i \to s}, \triangle h_{i \to s} \mid s \in (\cup N_d(i)) \cap I_{\mathrm{left}}\}$ in the $d$-order neighborhood exactly, that is

$$\triangle J_{i \to s} = -\hat{J}_{is} \hat{J}_{ii}^{-1} \hat{J}_{is}, \quad \triangle h_{i \to s} = -\hat{J}_{is} \hat{J}_{ii}^{-1} \hat{h}_i. \tag{3}$$

For $\forall \ s, t \in (\cup N_d(i)) \cap I_{\mathrm{left}}$, if the edge $(s, t) \notin E$, we add the edge $(s, t)$. Then compute the message from $i$ to the edge $(s, t)$, that is

$$\triangle J_{i \to st} = -\hat{J}_{is} \hat{J}_{ii}^{-1} \hat{J}_{it}. \tag{4}$$

3. Compute the messages $\{\triangle J_{iu}, \triangle h_{iu} \mid u \in N_{n+1}(i) \cap I_{\mathrm{left}}\}$ in the $(d+1)$th-order neighborhood partly, that is

$$\triangle J_{i \to u} = -\hat{J}_{iu} \hat{J}_{ii}^{-1} \hat{J}_{iu}, \quad \triangle h_{i \to u} = -\hat{J}_{iu} \hat{J}_{ii}^{-1} \hat{h}_i. \tag{5}$$

Let $v \in (\cup N_d(i)) \cap I_{\text{left}}$, if there is an edge $(u,v) \in E$, we compute the message $\triangle J_{i \rightarrow uv}$ from node $i$ to the edge $(u,v)$, that is

$$\triangle J_{i \rightarrow uv} = -\hat{J}_{iu} \hat{J}_{ii}^{-1} \hat{J}_{iv}. \tag{6}$$

Obviously, we didn't add new edge in this step, which decreases the computational complexity.

4. Add the node $i$ to $I_{\text{elim}}$, and delete the node $i$ from $I_{\text{left}}$.

---

**Algorithm 1.** GaussianVariableMP-$d$ of variable $x_i$

---

**Data**: Gaussian graphical model $G$, variable $x_i$, $I$, $I_{\text{elim}}$, $I_{\text{left}}$
**Result**: $\{\triangle J_{i \rightarrow j}, \triangle h_{i \rightarrow j}, \triangle J_{i \rightarrow jk}\}$
**begin**
    Label the neighborhoods of variable $x_i$ from $N_1(i)$ to $N_{d+1}(i)$;
    Update the parameters $\hat{J}_{ii}, \hat{h}_i$ with (1);
    **for** $k \in (\cup N_{d+1}(i)) \cap I_{\text{left}}$ **do**
        | Update the parameters $\hat{J}_{ik}$ with (2);
    **end**
    **for** $s \in (\cup N_d(i)) \cap I_{\text{left}}$ **do**
        | Update the messages $\triangle J_{i \rightarrow s}, \triangle h_{i \rightarrow s}$ with (3);
    **end**
    **for** $\forall s,t \in (\cup N_d(i)) \cap I_{\text{left}}$ **do**
        **if** $(s,t) \notin E$ **then**
            | add $(s,t) \in E$;
        **end**
        Update the messages $\triangle J_{i \rightarrow st}$ with (4);
    **end**
    **for** $u \in N_{n+1}(i) \cap I_{\text{left}}$ **do**
        Update the messages $\triangle J_{i \rightarrow u}, \triangle h_{i \rightarrow u}$ with (5);
        **for** $v \in (\cup N_d(i)) \cap I_{\text{left}}$ **do**
            **if** $(u,v) \in E$ **then**
                | Update the messages $\triangle J_{i \rightarrow uv}$ with (6);
            **end**
        **end**
    **end**
    Add the node $i$ to $I_{\text{elim}}$,   delete the node $i$ from $I_{\text{left}}$.
**end**

---

## 3.2   GaussianMP-$d$ Algorithm

The GaussianMP-$d$ Algorithm is to execute the GaussianVariableMP-$d$ for variables in the elimination order $I$ and its inverse elimination order $I'$ respectively, and compute the approximate marginal distributions with the Gaussian messages in the $d$-order neighborhood. Here, we select a cutset $P = \{x_{A_1}, x_{A_2}, \cdots\}$ of the Gaussian graphical model $G = (V, E)$, which is also the cutset of the elimination order $I$. Then, we can compute the approximate marginal distribution of $\boldsymbol{x}_A$, that is

$$\tilde{p}(\boldsymbol{x}_A) \propto \exp\left\{\langle \tilde{h}_A, \boldsymbol{x}_A \rangle + \frac{1}{2}\langle \tilde{J}_A, \boldsymbol{x}_A \boldsymbol{x}_A^{\mathrm{T}} \rangle \right\}. \tag{7}$$

Where the model parameters $\tilde{h}_A = [\tilde{h}_a]^{\mathrm{T}}, \tilde{J}_A = [\tilde{J}_{a,b}]_{|A|\times|A|}$ can be computed with the these messages, that is

$$\tilde{h}_a = h_a + \sum_{j \in \cup N_{d+1}(i), j \notin A} \triangle h_{j \to a},$$

$$\tilde{J}_{aa} = J_{aa} + \sum_{j \in \cup N_{d+1}(i), j \notin A} \triangle J_{j \to a}, \tag{8}$$

$$\tilde{J}_{ab} = J_{ab} + \sum_{\substack{k \in \cup N_{d+1}(a), \\ k \in \cup N_{d+1}(b), k \notin A}} \triangle J_{k \to ab}.$$

Generally speaking, the probability inference in subset $\boldsymbol{x}_A$ is trackable. The approximate marginal distributions of variables $x_a \in \boldsymbol{x}_A$ can be computed with the Gaussian elimination algorithm exactly. The formal description of GaussianMP-$d$ algorithm is shown in Algorithm 2. The computational complexity of the GaussianMP-$d$ algorithm is fixed parameter linear time $O(n \cdot d^2)$.

---

**Algorithm 2.** GaussianMP-$d$ Algorithm

---

**Data**: Gaussian graphical model $G$
**Result**: $\{\tilde{p}(x_i) \mid x_i \in \boldsymbol{x}\}$
**begin**
    Select a elimination ordering $I$;
    **for** $x_i \in I$ **do**
      | Run GaussianVariableMP-$d$ algorithm;
    **end**
    **for** $x_i \in I'$ **do**
      | Run GaussianVariableMP-$d$ algorithm;
    **end**
    Select a cutset $P$;
    **for** *variable subset* $\boldsymbol{x}_A \in P$ **do**
      Compute approximate marginal distribution $\tilde{p}(\boldsymbol{x}_A)$ with (7),(8);
      Run exact belief propagation algorithm in $\boldsymbol{x}_A$;
      Output $\{\tilde{p}(x_{a_1}) \mid x_{a_1} \in \boldsymbol{x}_A\}$;
    **end**
    Output $\{\tilde{p}(x_i) \mid x_i \in \boldsymbol{x}\}$.
**end**

---

## 4   Experiments

In this section, we compare the approximate marginal distributions with some numerical experiments. We experiment with a $8 \times 8$ two dimension lattice Gaussian graphical model. Specially, we generate the attractive Gaussian model $G_1$

with model parameters $J_{ij} \in (0, 20)$, $h_i \in (0, 30)$), and the repulsive Gaussian model $G_2$ with model parameters $J_{ij} \in (-20, 0)$, $h_i \in (-30, 0)$), which all satisfy the parameter conditions $\boldsymbol{J} = \boldsymbol{J}^\mathrm{T}, \boldsymbol{J} \succ 0$.

For the model $G_1$, we select the elimination order from left to right, and from bottom up, and execute the GaussianMP-$d(d = 1, 2, 3)$ algorithms respectively. The experiment results are shown in Figure 1, which show that the approximate marginal distributions become tighter as the neighborhood order $d$ increasing, and the GaussianMP-$d$ algorithms present the low bounds of the parameters compared with the exact values.

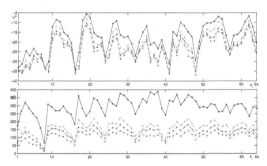

**Fig. 1.** The comparisons of approximate marginals of the GaussianMP-$d(d = 1, 2, 3)$ algorithms for $G_1$. The solid lines denote exact model parameters of marginals, the dashed lines, dashdot lines and dotted lines denote the model parameters based on the GaussianMP-$d$(d=1,2,3) algorithms respectively.

For the Gaussian graphical model $G_1, G_2$, we first execute the GaussianMP-$d(d = 2, 3)$ algorithms respectively, then run the mean field method with full factorial free distribution. The experiment results are shown in Figure 2, which shows that: (1) The GaussianMP-$d(d = 2, 3)$ algorithms have higher computational accuracy on the parameters $\{J_{ii} \mid i \in V\}$ than the mean field method has. (2) The mean field algorithm provides some better values for parameters $\{h_i\}$, also some worse values. Conversely, the GaussianMP-$d$ $(d = 2, 3)$ algorithms provide more stable approximate values for parameters $\{h_i \mid i \in V\}$.

# 5    Conclusions

For the Gaussian graphical models, we have defined the $d$-order neighborhood concept, and designed the GaussianMP-$d$ algorithm with fixed parameter linear $O(n \cdot d^2)$ computational complexity. The $d$-order neighborhood concept describes the propagation scope of Gaussian messages, which reveals that the Gaussian messages become less accurate as the neighborhood order $d$ increases. Based on this, the GaussianMP-$d$ algorithm makes full use of the messages in the $d$-order neighborhood, and obtains the linear running time at the cost of the accuracy of the left neighborhoods. The order parameter $d$ also provides a trade-off criterion for the computational complexity and the approximate accuracy.

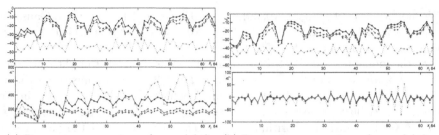

(a) Parameter comparisons for model $G_1$. (b) Parameter comparisons for model $G_2$.

**Fig. 2.** Accuracy comparisons of the approximate marginal distributions for model $G_1$ and $G_2$. The solid lines denote the exact values of parameters, the dashed lines and the dashdot lines denote the approximate values based on the GaussianMP-$d(d = 2, 3)$, the dotted lines the approximate values based on the mean field algorithm.

**Acknowledgments.** This work is supported in part by Tianjin Science and Technology Development of High School (Grant No. 20110806) and Scientific Research Foundation of Tianjin University of Science and Technology (Grant No. 20110404).

# References

1. Cozman, F.G.: Graphical models for imprecise probabilities. International Journal of Approximate Reasoning 39(2-3), 167–184 (2005)
2. Rue, H., Held, L.: Gaussian Markov random fields: Theory and applications. Chapman & Hall, London (2005)
3. de Camposa, L.M., Romero, A.E.: Bayesian network models for hierarchical text classification from a thesaurus. International Journal of Approximate Reasoning 50(7), 932–944 (2009)
4. Malioutov, D.: Approximate inference in Gaussian graphical models. PhD thesis, Massachusetts Institute of Technology (2008)
5. Bishop, C.M.: Pattern recognition and machine learning, ch. 10, pp. 461–522. Springer (2006)
6. Jordan, M.I., Weiss, Y.: Graphical models: probabilistic inference. In: Arbib, M.A. (ed.) The Handbook of Brain Theory and Neural Networks, pp. 490–496. The MIT Press (2002)
7. Wainwright, M.J., Jordan, M.I.: Graphical models, exponential families, and variational inference. Foundations and Trends in Machine Learning 1(1-2), 1–305 (2008)
8. Yedidia, J.S., Freeman, W.T., Weiss, Y.: Understanding belief propagation and its generalization. Technical Report TR2001-22, Mitsubishi Electric Research Labs (January 2002)
9. Winn, J.: Variational message passing and its applications. PhD thesis, University of Cambridge (2003)
10. Winn, J., Bishop, C.M.: Variational message passing. Journal of Machine Learning Research 6, 661–694 (2005)
11. Johnson, J.K.: Convex relaxation methods for graphical models: lagrangian and maximum entropy approaches. PhD thesis, Massachusetts Institute of Technology (2008)

# Fast Image Classification Algorithms Based on Random Weights Networks

Feilong Cao*, Jianwei Zhao, and Bo Liu

Department of Information and Mathematics Sciences, China Jiliang University,
Hangzhou 310018, Zhejiang Province, P.R. China
feilongcao@gmail.com

**Abstract.** Up to now, rich and varied information, such as networks, multimedia information, especially images and visual information, has become an important part of information retrieval, in which video and image information has been an important basis. In recent years, an effective learning algorithm for standard feed-forward neural networks (FNNs), which can be used classifier and called random weights networks (RWN), has been extensively studied. This paper addresses the image classification algorithms using the algorithm. A new algorithm of image classification based on the RWN and principle component analysis (PCA) is proposed. The proposed algorithm includes significant improvements in classification rate, and the extensive experiments are performed using challenging databases. Compared with some traditional approaches, the new method has superior performances on both classification rate and running time.

**Keywords:** Image classification, PCA, Random weights networks.

## 1 Introduction

Image information systems are becoming increasingly important with the advancements in broadband networks, high-powered workstations etc. Large collections of images are becoming available to the public, from photo collection to web pages, or even video databases[23]. Since visual media requires large amounts of memory and computing power for processing and storage, there is a need to efficiently index and retrieve visual information from image database. In recent years, image classification has become an interesting research field in applications (see [9], [15]). How to index efficiently large number of color image, classification plays an important and challenging role(see [20], [5]).

In general, as shown in Fig. 1, the fist step of image classification is to define an effective representation of the image, which includes sufficient information for the image for future classification. The second step of image classification is to classify a new image with the chosen representation(see [22], [1], [18]). Our main focus of this research is to find suitable classifier for the classification.

* The research was supported by the National Natural Science Foundation of China (Nos. 61272023, 61101240) and the Zhejiang Provincial Natural Science Foundation of China (No. Y6110117).

C. Guo, Z.-G. Hou, and Z. Zeng (Eds.): ISNN 2013, Part I, LNCS 7951, pp. 547–557, 2013.

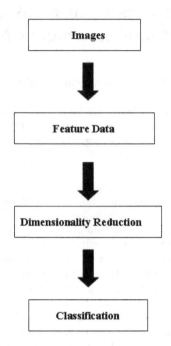

**Fig. 1.** Main procedures in a image classification system

In the investigation of image classification, there have been various classifiers, both linear and nonlinear, such as feed-forward neural network (FNN), support vector machine (SVM), polynomial classifier, and fuzzy rule-based system. In the these classifiers, FNN seems one of the most popular techniques. Although the FNN is being intensively studied for many years, most of them may be classified as different variations of the perceptron to recognize image characters.

SVM was introduced by Vapnik [21] as popular method for classification (see [19]). The working mechanism of the SVM is to learn a separating hyperplane to maximize the margin and produce a good generalization capability. Currently, SVM has been successfully applied in many areas such as face detection, handwritten digit recognition, etc(see [17], [12], [7]).

The learning algorithm for FNN, called random weights networks (RWN), which can be used in regression and classification problems. It is well-known that traditional FNN approaches, such as BP algorithms, usually face difficulties in manually tuning control parameters, but RWN can avoid such issues and reaches good solutions analytically, and its learning speed of RWN is faster than other traditional methods. The reason for which is that RWN randomly chooses the input weights and bias of the FNN instead of tuning.

In this article, we will propose an algorithm for image classification based on principle component analysis (PCA) and RWN. Our work combines the two algorithms to eliminate the inherent shortcomings of some previous methods. We will perform extensive experiments to demonstrate superiority of our proposed technique over existing methods.

We organize the study in the following manner. In section 2, we will briefly introduce the PCA and RWN. In Section 3, we will propose our image classification algorithm, and use small scale image database and big stand database in the experiment. Conclusions of this article will be presented in final section.

## 2    Review for PCA and RWN

### 2.1    PCA Algorithm

Given a $t$-dimensional vector representation of each image, the PCA (see [19], [24]) can be used to find a subspace whose basis vectors correspond to the maximum-variance directions in the original space.

Let $W$ represent the linear transformation that maps the original $t$-dimensional space onto a $f$-dimensional feature subspace where normally $f \ll t$. The new feature vectors

$$y_i \in W^T x_i, \ i = 1, \ldots, N. \tag{1}$$

The columns of $W$ are the eigenvalues $e_i$ obtained by solving the eigenstructure decomposition

$$\lambda e_i = Q e_i, \tag{2}$$

where

$$Q = X X^T \tag{3}$$

is the covariance matrix and $\lambda_i$ the eigenvalue associated with the eigenvector $e_i$. Before obtaining the eigenvectors of $Q$: 1) the vectors are normalized such that $\|x_i\| = 1$ to make the system invariant to the intensity of illumination source, and 2) the average of all images is subtracted from all normalized vectors to ensure that the eigenvector with the highest represents the dimension in the eigenspace in which variance of vectors is maximum in a correlation sense.

### 2.2    Random Weights Networks

The standard FNN with single hidden layer, $\tilde{N}$ hidden nodes, and activation function $g(x)$ are mathematically modeled as

$$\sum_{i=1}^{\tilde{N}} \beta_i g_i(x) = \sum_{i=1}^{\tilde{N}} \beta_i g(w_i \cdot x + b_i), \tag{4}$$

where $w_i = [w_{i1}, w_{i2}, \ldots, w_{in}]^T$ is the weight vector connecting the $i$th hidden node and the input nodes, $\beta_i = [\beta_{i1}, \beta_{i2}, \ldots, \beta_{im}]^T$ is the weight vector connecting the $i$th hidden node and output nodes, and $b_i$ is the threshold of the $i$th hidden node. And $w_i \cdot x_j$ denotes the inner product of $w_i$ and $x_j$.

Since FNN can approximate complex nonlinear mappings directly from the input samples (see [4], [6], [10], [2], [3]) and provide models for a large class of natural and artificial phenomena that are difficult to handle using classical parametric techniques, they have been extensively used in many fields. The traditional FNN with $\tilde{N}$ hidden nodes and activation function $g(x)$ can interpolate these $N$ samples with zero error, which means that for $N$ distinct samples $(x_i, t_i)$, where $x_i = [x_{i1}, x_{i2}, \ldots, x_{in}]^T \in R^n$ and $t_i = [t_{i1}, t_{i2}, \ldots, t_{im}]^T \in R^m$, there holds $\sum_{j=1}^{\tilde{N}} \|o_j - t_j\| = 0$, i. e., there exist $\beta_i$, $w_i$ and $b_i$ such that

$$\sum_{i=1}^{\tilde{N}} \beta_i g(w_i \cdot x_j + b_i) = t_j, j = 1, \ldots, N. \tag{5}$$

The above $N$ equations can be written compactly as

$$H\beta = T \tag{6}$$

where

$$
\begin{aligned}
&H(w_i, \ldots, w_{\tilde{N}}, b_1, \ldots, b_{\tilde{N}}, x_1, \cdots, x_N) \\
&= \begin{bmatrix}
g(w_1 \cdot x_1 + b_1) & \cdots & g(w_{\tilde{N}} \cdot x_1 + b_{\tilde{N}}) \\
\vdots & \ddots & \vdots \\
g(w_1 \cdot x_N + b_1) & \cdots & g(w_{\tilde{N}} \cdot x_N + b_{\tilde{N}})
\end{bmatrix}_{N \times \tilde{N}}
\end{aligned}
\tag{7}
$$

$$
\beta = \begin{bmatrix} \beta_1^T \\ \vdots \\ \beta_{\tilde{N}}^T \end{bmatrix}_{\tilde{N} \times m} \quad \text{and} \quad T = \begin{bmatrix} t_1^T \\ \vdots \\ t_N^T \end{bmatrix}_{N \times m}. \tag{8}
$$

$H$ is called the hidden layer output matrix of the neural network; the $i$th column of $H$ is the $i$th hidden node output with respect to inputs $x_1, x_2, \ldots, x_N$.

Traditionally, in order to train an FNN, one may wish to find specific $\hat{w}_i, \hat{b}_i, \hat{\beta}(i = 1, \ldots, \tilde{N})$ such that

$$\|H(\hat{w}_1, \ldots, \hat{w}_{\tilde{N}}, \hat{b}_1, \ldots, \hat{b}_{\tilde{N}})\beta - T\| = \min_{w_i, b_i, \beta} \|H(w_1, \ldots, w_{\tilde{N}}, b_1, \ldots, b_{\tilde{N}})\beta - T\| \tag{9}$$

which is equivalent to minimizing the cost function

$$E = \sum_{j=1}^{N} \left( \sum_{i=1}^{\tilde{N}} \beta_i g(w_i \cdot x_j + b_i) - t_j \right)^2. \tag{10}$$

In the conventional FNN theory, all the parameters of network (e.g., the hidden layer parameters $(w_i, b_i)$ and the output weights $\beta_i$) are required freely adjustable. According to the theory, hidden layer parameters $(w_i, b_i)$ and input weights need to be tuned properly for given trained samples.

A famous algorithm to tune and find these parameters is so-called back-propagation (BP) algorithm, which uses gradient descent method to adjust input and output weights and biases values of hidden layer nodes by solving the least square problem for the training samples.

Although the BP algorithm is popular for solving practical problems, its success depends upon the quality of the training data. Specially, the performance of the BP algorithm in function approximation becomes unsatisfactory when gross errors are present in the training data. It is clear that gradient descent-based BP algorithm is generally very slow due to improper learning steps or may easily converge to local minima. And many iterative learning steps may be required by such learning algorithms in order to obtain better learning performance. If the input weights and biases are chosen randomly, i.e., the the input weights and biases are considered as random variables which obey the uniform distribution on $(0, 1)$, then it is possible to improve the learning speed compared with the traditional BP algorithms, and the least square model of linear equation with hidden layer output matrix can be used to estimate the output weights of FNN by calculating the Moore-Penrose generalized inverse. This idea can be found in some prior articles [16], [11], [14], and [13], and was named as so-called Extreme Learning Machine (ELM) in [8]. In fact, this idea is an algorithm on random weights network (RWN), which can be summarized as follows.

**Algorithm of Random Weights Network (RWN):**

Given a training set $\aleph = \{(X_i, t_i)| \in R^n, t_i \in R^m, i = 1, 2, \ldots, N\}$ and activation function $g(x)$, hidden node number $\tilde{N}$.

**Step 1:** Randomly assign input weight $w_i$ and bias $b_i$ $(i = 1, 2, \ldots, \tilde{N})$;

**Step 2:** Calculate the hidden layer output matrix $H$.

**Step 3:** Calculate the output weight $\beta$ by $\beta = H^\dagger T$, where $T = (t_1, t_2, \ldots, t_n)^T$.

We have the following important properties:

(1) Minimum training error. The special solution $\hat{\beta} = H^\dagger T$ is one of the least-squares solution of a general linear system $H\beta = T$, meaning that the smallest training error can be reached by this special solution:

$$\|H\hat{\beta} - T\| = \|HH^\dagger T - T\| = \min_\beta \|H\beta - T\|, \tag{11}$$

where $H^\dagger$ is the Moore-Penrose (MP) generalized inverse of $H$. Although almost all learning algorithms wish to reach the minimum training error, most of them cannot reach it because local minimum or infinite training iteration is usually not allowed in applications.

(2) Smallest norm of weights. Furthermore, the special solution $\hat{\beta} = H^\dagger T$ has the smallest norm among all the least squares solution of $H\beta = T$:

$$\|\hat{\beta}\| = \|H^\dagger T\| \le \|\beta\|, \qquad \forall \beta \in \left\{\beta : \|H\beta - T\| \le \|Hz - T\|, \forall z \in R^{\tilde{N} \times N}\right\}. \tag{12}$$

(3) The minimum norm least-squares solution of $H\beta = T$ is unique, which is $\hat{\beta} = H^\dagger T$.

Unlike traditional BP algorithms, RWN algorithm has concise architecture, no need to tune input weights and biases. Particularly, its speed is usually faster than that of BP algorithm. In order to calculate the MP generalized inverse of $H$, there have been several methods, such as orthogonal projection, orthogonalization method, iterative method, and Singular Value Decomposition (SVD) etc..

## 3    The Proposed Classification Method

The proposed method is based on PCA, and dimensionally reduced coefficients are used to train and test an RWN classifier (see Fig. 2). Images from each database are converted into gray level image before use. Each database is randomly divided into training and testing sets so that 40% of images of each subject are used as prototypes and remaining images are used during testing phase. We use grayscale histogram as the feature of a picture, and each vector size of $1 \times U(U = 256)$ represents an image. Then PCA is used to generate low dimensional features $1 \times U' \ll 1 \times U$, for each sample, dimensionally reduced feature sets are randomly selected for training of an RWN, whereas, remaining features of the same database are used to test. Note that we do not assume any a priori knowledge of the scene, background, object location, and illumination conditions.

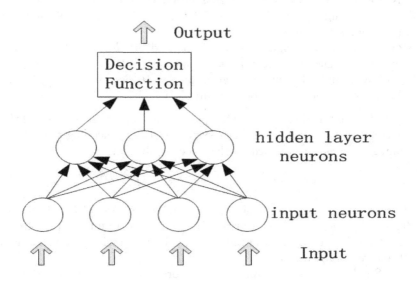

**Fig. 2.** Architecture of an RWN classifier

Our classification algorithm is summarized as follows:

**Our Fast Algorithm:** Given a image datebase;

**Step 1:** Images from each database are converted into gray level image and, we use grayscale histogram as the feature of a picture;

**Step 2:** PCA is used to generate low dimensional features;
**Step 3:** Dimensionally reduced feature sets are randomly selected for training of an RWN;
**Step 4:** Using the RWN classifier to classify the remaining images features.

### 3.1 Experiment on Image Database

To evaluate the performance of proposed method for image classification, we perform a number of experiments using our special database. This database includes 4 classes of images as shown in Fig. 3. And each class has 10 images. In experiment, each image are converted into gray image. These images are downloaded from http://www.msri.org/people/ members/eranb/.

**Fig. 3.** Samples in our image database

The experiment is implemented in MATLAB R2009 version on a PC with 2.71G MHz CPU and 1.75 GB SDRAM. The parameters of the SVM are set as follows: The Gaussian Radial Basis Function (RBF) kernel $\exp(-\|x-x_j\|^2/2\sigma^2)$ is used as the kernel of the SVM classifier and the active function of RWN. And we take $\sigma = 5$, $C = 100$ in (6), and $\tilde{N} = 1600$. We carried out 20 experiments for RWN, and used the average values as the final results. The experimental results are shown in Table 2.

Table 1. Comparison of RWN and SVM in image database

|  | Training time | Testing time | Testing accuracy |
|---|---|---|---|
| RWN | 0.8133 | 0.6234 | 0.8229 |
| SVM | 1.8922 | 2.8862 | 0.8067 |

## 3.2 Experiment on Standard Large Scale Image Database

Maybe the particular example in previous section is not enough to show the efficiency of our algorithm. In this section we will use large scale database to test the proposed algorithm. The UCI database were created by Vision Group, University of Massachusetts. It can be downloaded from: http://www.archive.ics.uci.edu/ ml/ datasets/Image +Segmentation). The instances were drawn randomly from a database of 7 outdoor images. The images were segmented to create a classification for every pixel. Each instance is a $3 \times 3$ region. The first data is 210×19, and the second data is 2100×19. The parameters of the SVM are set as follows: The Gaussian RBF kernel was used as the kernel of the SVM classifier and the active function of RWN. And the parameters were taken as $C = 1200$, $\sigma = 5$, and $\tilde{N} = 2000$.

Table 2. Comparison of RWN and SVM

|  | Training time | Testing time | Testing accuracy |
|---|---|---|---|
| RWN | 1.15 | 0.83 | 0.8366 |
| SVM | 2.2969 | 3.6875 | 0.8167 |

Before experiment, we drop those data which can not represent the features clearly. That is, we drop the 4th, 5th, 6th columns. Then we classify these features with SVM and RWN, respectively. Here, we carried out the experiments 20 times, and the experimental results are shown in Fig. 4, 5 and 6, respectively. Table 2 is the case of average of the results. From those results, we found that though RWN is not stable as SVM, it is more effective than SVM, and the accuracy and time are both better than those of SVM.

## 4    Conclusion

This paper proposed a new algorithm for image classification. We first took the grayscale histogram as the features of the images, and then we used PCA to

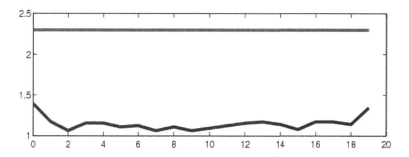

**Fig. 4.** Comparison of training time

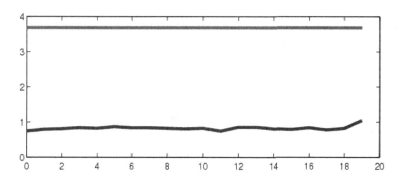

**Fig. 5.** Comparison of testing time

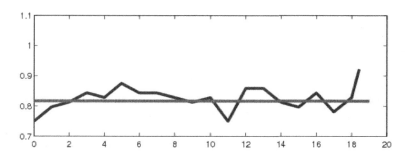

**Fig. 6.** Comparison of testing accuracy

reduce the dimension of the feature vectors. Finally, we input the low dimensional feature data to the RWN classifier for learning an optimal model. The remaining image was tested by the learned RWN classifier. Experimental results showed that the proposed algorithm is more superior than popular SVM in both the accuracy and the learning speed.

# References

1. Chapelle, O., Haffner, P., Vapnik, V.N.: Support vector machines for histogram-based image classification. IEEE Trans. Neural Networks 10, 1055–1064 (1999)
2. Chen, T.P., Chen, H.: Approximation capability to functions of several variables, nonlinear functionals, and operators by radial basis function neural networks. IEEE Trans. Neural Networks 6, 904–910 (1995)
3. Chen, T.P., Chen, H.: Universal approximation to nonlinear operators by neural networks with arbitrary activation functions and its application to dynamical systems. IEEE Trans. Neural Networks 6, 911–917 (1995)
4. Cybenko, G.: Approximation by superposition of sigmoidal function. Mathematics of Control, Signals and Systems 2, 303–314 (1989)
5. Dettig, R.L., Landgrebe, D.A.: Classification of multispectral image data by extraction and classification of homogeneous objects. IEEE Trans. Geoscience Electronics 14, 19–26 (1976)
6. Funahashi, K.I.: On the approximate realization of continuous mappings by neural networks. Neural Networks 2, 183–192 (1989)
7. Guo, G., Li, S.Z., Chan, K.: Face recognition by support vector machines. In: Proceedings of the Foruth IEEE International Conference on Automatic Face and Gesture Recognition, pp. 196–201. IEEE Press, Grenoble (2000)
8. Huang, G.-B., Zhu, Q.Y., Siew, C.K.: Extreme learning machine: Theory and applications. Neurocomputing 70, 489–501 (2006)
9. Huang, K., Aviyente, S.: Wavelet feature selection for image classification. IEEE Trans. Image Processing 17, 1709–1719 (2008)
10. Hornik, K.: Approximation capabilities of multilayer feedforward networks. Neural Networks 4, 251–257 (1991)
11. Igelnik, B., Pao, Y.H.: Stochastic choice of basis functions in adaptive function approximation and the functional-link net. IEEE Trans. Neural Networks 6, 1320–1329 (1995)
12. Li, Z., Tang, X.: Using support vector machine to enhance the performance of bayesian face recognition. IEEE Trans. Information Forensics and Security 2, 174–180 (2007)
13. McLoone, S., Irwin, G.: Improving neural network training solutions using regularisation. Neurocomputing 37, 71–90 (2001)
14. McLoone, S., Brown, M.D., Irwin, G., et al.: A hybrid linear nlinear training algorithm for feedforward neural networks. IEEE Trans. Neural Networks 9, 669–684 (1998)
15. Mohammed, A.A., Minhas, R., Jonathan Wu, Q.M., Sid-Ahmed, M.A.: Human face recognition based on multidimensional PCA and extreme learning machine. Pattern Recognition 44, 2588–2597 (2011)
16. Pao, Y.H., Park, G.H., Sobajic, D.J.: Learning and generalization characteristics of the random vector functional-link net. Neurocomputing 6, 163–180 (1994)
17. Phillips, P.J.: Support vector machines applied to face recognition. In: Proceedings of Advances in Neural Information Processing Systems $\Pi$, Denver, Colordo, USA, pp. 113–141 (2001)
18. Taqi, J.S.M., Karim, F.: Finding suspicious masses of breast cancer in mammography images using particle swarm alogrithm and its classification using fuzzy methods. In: The International Conference on Computer Communication and Informatics, pp. 1–5 (2012)

19. Turk, M., Pentland, A.: Eigenfaces for recognition. Journal of Cognitive Neuroscience 2, 71–86 (1991)
20. Tzeng, Y.C., Fan, K.T., Chen, K.S.: A parallel differential box-counting algorithm applied to hyperspectral image classification. IEEE Trans. Geoscience and Remote Sensing Letters 5, 272–276 (2012)
21. Vapnik, V.N.: The Nature of Statistical Learning Theory. Springer Verlag Press, New York (1995)
22. Zhang, B., Wang, Y., Wang, W.: Batch mode active learning for multi-label image classification with informative label correlation mining. In: IEEE Workshop on Applications of Computer Vision, pp. 401–407 (2012)
23. Zhao, D.Q., Zou, W.W., Sun, G.M.: A fast image classification algorithm using support vector machine. In: The 2nd Internatinal Conference on Computer Technology and Development, pp. 385–388 (2010)
24. Zhao, W., Chellappa, R., Phillips, P.J., Rosenfeld, A.: Face recognition: A literature survey. ACM Computing Surveys (2003)

# A Convolutional Neural Network for Pedestrian Gender Recognition

Choon-Boon Ng, Yong-Haur Tay, and Bok-Min Goi

Universiti Tunku Abdul Rahman, Kuala Lumpur, Malaysia
{ngcb,tayyh,goibm}@utar.edu.my

**Abstract.** We propose a discriminatively-trained convolutional neural network for gender classification of pedestrians. Convolutional neural networks are hierarchical, multilayered neural networks which integrate feature extraction and classification in a single framework. Using a relatively straightforward architecture and minimal preprocessing of the images, we achieved 80.4% accuracy on a dataset containing full body images of pedestrians in both front and rear views. The performance is comparable to the state-of-the-art obtained by previous methods without relying on using hand-engineered feature extractors.

**Keywords:** Gender recognition, convolutional neural network.

## 1 Introduction

Classifying the gender of a person has received increased attention in computer vision research in recent years. There are a number of possible applications, such as in human-computer interaction, surveillance, and demographic collection. While there has been quite a number of works on recognizing gender from facial information alone, less work has been done on using cues from the whole body. In certain situations, using the face may not be possible for privacy reasons, or due to insufficient resolution. Another simpler reason would be that, from the back view of a person, the face is not visible. Most facial gender recognition systems rely on constrained environments, for example frontal or near-frontal view of the head. Thus, we believe there are merits to using the whole body. In particular, this paper focuses on pedestrian gender recognition using computer vision.

Let us consider how we might be able to identify the gender of a pedestrian based on the whole human body rather than relying on the face alone. Due to differences between the male and female anatomy, the body shape can act a strong cue. However, clothing may cause occlusion, such as loose-fitting clothes that make the body shape less obvious. There are clothes for different gender, but similar types are also worn by both, such as long pants or T-shirts. Hairstyle acts a strong cue in the majority of cases, but hair length can be a source of confusion. Despite all this, humans in most situations have the ability to distinguish gender accurately.

The first investigation into gender recognition based on the human body was presented by Cao et al. [1]. Their parts-based method used Histogram of Oriented

C. Guo, Z.-G. Hou, and Z. Zeng (Eds.): ISNN 2013, Part I, LNCS 7951, pp. 558–564, 2013.

Gradients (HOG) features to represent small patches of the human body image. These patches are overlapping partitions and used as weak features for a boosting type classifier. Their method gave better classification results that using only raw images with an Adaboost or random forest classifier. Collins et al. [2] proposed descriptors using dense HOG features computed from a custom edge map. This was combined with color features captured from a histogram computed based on the hue and saturation values of the pixels. Guo et al. [3] used biologically-inspired features derived from Gabor filters followed by manifold learning, with linear SVM as classifier. Best results were obtained by first classifying the view (front, back, or mixed) and followed by a gender classifier for each view. Bourdev et al. [4] used random patches called *poselets*, represented with HOG features, color histogram and skin features Their method relied on using a heavily annotated training dataset and context information.

In this paper, we present a discriminatively-trained convolutional neural network (CNN), inspired by the work of LeCun et al.[5] for gender classification of pedestrians in full body images. CNN is a hierarchical neural architecture that integrates feature extraction and classification in a single framework. It is able to automatically learn the features from the training data, instead of relying on the use of hand-crafted features. CNNs have been successfully applied to various pattern recognition tasks such as handwriting recognition [5], face recognition [6], face detection [7], traffic sign classification [8] and action recognition [9].

Our proposed CNN has a straightforward architecture, without incorporating recently proposed architectural elements such as local contrast normalization [10], rectifying nonlinearities [10] and multistage features [11]. Despite its simplicity, we were able to achieve competitive performance comparable to the state-of-the-art for pedestrian gender recognition.

The remainder of this paper is organized as follows. In section 2, the architecture of the convolutional neural network is introduced and explained. In section 3, the dataset used and the details of our proposed CNN is described. The experiment results are presented and compared with other methods in section 4. Finally, in section 5, conclusions are drawn and future work proposed.

## 2     Convolutional Neural Network (CNN)

Convolutional neural networks are a class of biologically-inspired, multi-layered neural networks. It models the human visual cortex using several layers of non-linearities and pooling. Other classes of such models, inspired by the hierarchical nature of the mammalian primary visual cortex, include the Neocognitron [12] and HMAX model [13]. CNNs are able to automatically learn feature detectors which are adapted to the given task and, to a certain degree, invariant of scale, translation and deformation. The architecture of CNN attempts to realize these invariances using three main ideas, namely local receptive fields, shared weights and spatial subsampling [5].

Traditional CNN typically includes two different kinds of layers inspired by the simple and complex cells in the visual cortex. The convolution layer contains feature maps obtained from convolution of filters with the previous layer's feature maps. The subsampling layer is produced from downsampling each of the feature map.

Figure 1 shows the architecture of our proposed convolutional neural network, which is comprised of 7 layers. The first convolution layer C1 consists of a number of feature maps obtained by convolution of the input with a set of filters, and which are then passed through a squashing activation function. Each unit of a feature map shares the same set of weights for the filter. The connection of each unit to the units located in a small neighbourhood in the previous layer implements the idea of local receptive fields to extract features. The shared weights enable features to be detected regardless of their location in an image. Weight sharing also reduces greatly the number of trainable parameters to achieve better generalization ability.

Let $W_{i,j}$ be the filter of size $n \times m$ which connects the $i$-th feature map from the previous layer $I_i$ to the $j$-th feature map $C_j$ and $b_j$ the corresponding trainable bias. The feature map is obtained as following:

$$C_j = \sigma(\sum_{i \in S} W_{i,j} \otimes I_i + b_j)$$

where $\otimes$ denotes the convolution operation and $S$ denotes the set of all or selected feature maps from the previous layer. The squashing activation function $\sigma$ which introduces non-linearities is normally either a sigmoid $\sigma(x) = 1/(1 + e^{-x})$ or hyperbolic tangent function $\sigma(x) = \tanh(x)$. If the size of a feature map is $h \times w$, then convolution with filter of size $n \times m$ will produce an output of size $(h - n + 1) \times (w - m + 1)$, disregarding border effects.

The layer S2 is obtained by downsampling each feature map in layer C1. In contrast to [5] which uses an averaging operation, we use the so-called max pooling operation [13][14], where the feature map is partitioned into non-overlapping $p \times p$ sub-regions and the maximum value is output from each sub-region. Thus each feature map is downsampled by a factor of $p$, e.g. the size is halved if $p = 3$. The spatial downsampling operation introduces invariance to small translations. Incidentally, it also reduces the computational complexity for the next convolution layer as the feature map's size is reduced.

In a similar manner, the convolution layer C3 is obtained followed by max-pooling layer S4. Note that each feature map in layer C3 can be either connected to all the feature maps from layer S2 or a subset of it. Layer S4 is fully connected to the units of layer F5, which is a layer of neuron units similar to the hidden layer of a neural network. The output layer contains logistic regression units for classification.

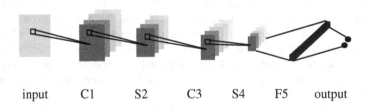

input     C1     S2     C3     S4     F5     output

**Fig. 1.** Architecture of our proposed CNN

# 3     Experiment

## 3.1     Dataset

We evaluate the gender classification ability of the CNN on the MIT Pedestrian data-set [15]. There are a total of 924 colour images of male and female pedestrians, in frontal and rear view. The size of the images is 64x128 pixels, with the person's body aligned to the center and the distance between the shoulders and feet approximately 80 pixels. Cao et al. [1] provided gender labels for 888 images (the remaining were indistinguishable) consisting of 600 males and 288 females. The breakdown accord-ing to the pose is 420 frontal views and 468 rear views. Figure 2 shows some exam-ples of the images from the dataset.

As preprocessing, the images were cropped to 54x108 by removing the border pix-els equally before resizing down to 40x80. Generally better results were obtained compared to without cropping. This could be due to the border pixels containing only background clutter, hence providing miscues. Furthermore, we assume pedestrian detectors would provide a tighter bounding box than compared to the images from this dataset. The images were then converted to grayscale and scaled down to values in the range [0,1] before being used as input to train the CNN.

## 3.2     The Proposed CNN

The detail of the architecture used in our experiments is as follows. Layer C1 contains 10 features maps and uses 5x5 filters, hence when the input image is 40x80, the size of each feature map is 36x76. After downsampling using 2x2 max pooling, each fea-ture map in layer S2 is 18x38. Layer C3 contains 20 features maps with the size 14x34 produced from 5x5 filters. Each feature map in this layer is connected to all the feature maps in layer S2. Layer S4 contains 7x17 feature maps obtained from 2x2 max pooling. All units of the feature maps are connected to each of the 25 neuron units in layer F5. Finally, the output layer has two units for binary classification. We use hyperbolic tangent activations in both the convolution and hidden layers. The total number of free parameters that are learnt by training is 64,857.

**Fig. 2.** Examples of pedestrian images from the MIT pedestrian dataset [15]

### 3.3    Training and Evaluation

Our CNN was implemented using Python with Theano library [16] and trained using mini-batch stochastic gradient descent with learning rate decay. The weights were randomly initialized from a uniform distribution in the range $[-\sqrt{(6/f)}, \sqrt{(6/f)}]$, where $f$ equals the total number of input and output connections, following the suggestion in [17].

During each iteration, a batch of images from the training set was presented to the CNN and the weights were updated by backpropagation. Validation was performed using the validation set after each epoch. The minimum validation error was taken as the best result. We used five-fold cross validation and the mean of the validation results was taken to determine the overall measure of accuracy.

## 4      Results and Analysis

Table 1 shows the results in comparison with other works on gender recognition from human body, evaluated on the same dataset. Our trained CNN achieved an average accuracy of 80.4 % on the validation set which is comparable to the best result by Guo et al. [3]. It should be noted that their method employs a view classifier, followed by a gender classifier for each different view (frontal, rear and mixed). Without using a view classifier, the accuracy is 79.2 %.

**Table 1.** Comparison of gender classification accuracy on the MIT pedestrian dataset

| Method | Accuracy (%) |
|---|---|
| Cao et al [1] | 75.0 |
| Collins et al. [2] | 76.0 (frontal view only) |
| Guo et al. [3] | 80.6 |
| *Our method* | *80.4* |

In contrast, our method goes not require a separate view classifier. The CNN integrates feature extraction and classification in a single framework, where the features are learnt. Furthermore, our CNN uses a relatively small number of feature maps compared to recent works using CNN for recognition tasks [8][18], thus requires less computational intensity.

Interestingly, the method of Guo et al. [3] also used a biologically inspired architecture, with features derived using Gabor filters and max pooling operation. This corresponds to layer C1 and S2 of the CNN. The difference is that Gabor filters can be considered as hard-wired, engineered features, while the CNN learns the optimal features.

Figure 3 shows some examples of classification errors made by the CNN. The first three images on the left are misclassified males and the rest are misclassified females.

(a) M → F                                      (b) F → M

**Fig. 3.** Examples of misclassified images (a) Male misclassified as female (b) Female misclassified as male

## 5    Conclusion and Future Work

In this paper, we have presented a convolutional neural network for gender classification of pedestrians in full body images. We achieved 80.4% accuracy on the MIT pedestrian dataset, an accomplishment matching or even better than those using handcrafted features. In our present approach, we use only fully supervised training with a simple, basic CNN architecture. For future work, we plan to implement improvements to our CNN architecture and explore the use of unsupervised pretraining to improve its classification performance.

**Acknowledgements.** The authors wish to thank the following for their support: UTAR Staff Scholarship Scheme, UTAR Research Fund, and EV-Dynamic Sdn. Bhd.

## References

1. Cao, L., Dikmen, M., Fu, Y., Huang, T.: Gender recognition from body. In: Proceeding of the 16th ACM International Conference on Multimedia (2008), pp. 725–728 (2008)
2. Collins, M., Zhang, J., Miller, P.: Full body image feature representations for gender profiling. In: 2009 IEEE 12th International Conference on Computer Vision Workshops, ICCV Workshops, pp. 1235–1242 (2009)
3. Guo, G., Mu, G., Fu, Y.: Gender from body: A biologically-inspired approach with manifold learning. In: Zha, H., Taniguchi, R.-i., Maybank, S. (eds.) ACCV 2009, Part III. LNCS, vol. 5996, pp. 236–245. Springer, Heidelberg (2010)
4. Bourdev, L., Maji, S., Malik, J.: Describing People: A Poselet-Based Approach to Attribute Classification. In: 2011 IEEE International Conference on Computer Vision (ICCV), pp. 1543–1550 (2011)
5. LeCun, Y., Bottou, L., Bengio, Y., Haffner, P.: Gradient-based learning applied to document recognition. Proceedings of the IEEE 86(11), 2278–2324 (1998)
6. Lawrence, S., Giles, C.L., Tsoi, A.C., Back, A.D.: Face recognition: a convolutional neural-network approach. IEEE Transactions on Neural Networks / A Publication of the IEEE Neural Networks Council 8(1), 98–113 (1997)
7. Osadchy, M., Cun, Y., Miller, M.: Synergistic face detection and pose estimation with energy-based models. The Journal of Machine Learning Research 8, 1197–1215 (2007)

8. Ciresan, D., Meier, U., Masci, J., Schmidhuber, J.: A committee of neural networks for traffic sign classification. In: The 2011 International Joint Conference on Neural Networks (IJCNN), vol. 1(1), pp. 1918–1921 (2011)

9. Ji, S., Xu, W., Yang, M., Yu, K.: 3D Convolutional Neural Networks for Human Action Recognition. IEEE Transactions on Pattern Analysis and Machine Intelligence 35(1), 221–231 (2013)

10. Jarrett, K., Kavukcuoglu, K., LeCun, Y.: What is the best multi-stage architecture for object recognition? In: 2009 IEEE 12th International Conference on Computer Vision, pp. 2146–2153 (2009)

11. Sermanet, P., LeCun, Y.: Traffic sign recognition with multi-scale convolutional networks. In: The 2011 International Joint Conference on Neural Networks (IJCNN), pp. 2809–2813 (2011)

12. Fukushima, K.: Neocognitron: A hierarchical neural network capable of visual pattern recognition. Neural Networks 1(2), 119–130 (1988)

13. Riesenhuber, M., Poggio, T.: Hierarchical models of object recognition in cortex. Nature Neuoscience 2(11), 1019–1025 (1999)

14. Ranzato, M., Huang, F.J., Boureau, Y.-L., LeCun, Y.: Unsupervised Learning of Invariant Feature Hierarchies with Applications to Object Recognition. In: 2007 IEEE Conference on Computer Vision and Pattern Recognition, pp. 1–8 (2007)

15. Oren, M., Papageorgiou, C., Sinha, P., Osuna, E., Poggio, T.: Pedestrian detection using wavelet templates. In: Proceedings of IEEE Computer Society Conference on Computer Vision and Pattern Recognition, pp. 193–199 (1997)

16. Bergstra, J., et al.: Theano: A CPU and GPU Math Compiler in Python. In: Proceedings of the Python for Scientific Computing Conference, SciPy (2010)

17. Glorot, X., Bengio, Y.: Understanding the difficulty of training deep feedforward neural networks. In: Proceedings of the International Conference on Artificial Intelligence and Statistics (AISTATS 2010), vol. 9, pp. 249–256 (2010)

18. LeCun, Y., Kavukcuoglu, K., Farabet, C.: Convolutional networks and applications in vision. In: Proceedings of 2010 IEEE International Symposium on Circuits and Systems (ISCAS), pp. 253–256 (2010)

# The Novel Seeding-Based Semi-supervised Fuzzy Clustering Algorithm Inspired by Diffusion Processes

Lei Gu[1,2]

[1] Guangxi Key Laboratory of Wireless Wideband Communication & Signal Processing,
Guilin, China, 541004
[2] School of Computer Science and Technology,
Nanjing University of Posts and Telecommunications, Nanjing, China, 210023
gulei@njupt.edu.cn

**Abstract.** Semi-supervised clustering can take advantage of some labeled data called seeds to bring a great benefit to the clustering of unlabeled data. This paper uses the seeding-based semi-supervised idea for a fuzzy clustering method inspired by diffusion processes, which has been presented recently. To investigate the effectiveness of our approach, experiments are done on three UCI real data sets. Experimental results show that the proposed algorithm can improve the clustering performance significantly compared to other semi-supervised clustering approaches.

**Keywords:** Semi-supervised clustering, Seeding, Fuzzy clustering, Diffusion Processes, Neighborhood Graph.

## 1 Introduction

With the rapid developments of computer science, pattern recognition has played and important role in our life. Data clustering is one of the popular pattern recognition techniques. The aim of data clustering methods is to divide data into several homogeneous groups called clusters, within each of which the similarity or dissimilarity between data is larger or less than data belonging to different groups[1]. It has been used in a wide variety of fields, ranging from machine learning, biometric recognition, image matching and image retrieval, to electrical engineering, mechanical engineering, remote sensing and genetics.

Unsupervised clustering partitions all unlabeled data into a certain number of groups on the basis of one chosen similarity or dissimilarity measure[2,3]. Different measure of the similarity or dissimilarity can lead to various clustering methods such as k-means[4], fuzzy c-means[5], mountain clustering, subtractive clustering[6] and neural gas[7]. In these traditional clustering algorithms, k-means(KM), which can be easily implemented, is the best-known squared error- based clustering algorithm. Recently, a novel fuzzy clustering approach inspired by diffusion processes(DifFUZZY) was presented in [8]. Its main idea is that the concepts from diffusion processes in graphs is applied to the fuzzy clustering method[8]. Experiments on some data sets show that the DifFUZZY in [8] is valid and can have encouraging performance.

C. Guo, Z.-G. Hou, and Z. Zeng (Eds.): ISNN 2013, Part I, LNCS 7951, pp. 565–573, 2013.

Semi-supervised clustering can also divide a collection of unlabeled data into several groups. However, a small amount of labeled data is allowed to be applied to aiding and biasing the clustering of unlabeled data in semi-supervised clustering unlike the unsupervised clustering, and so a significant increase in clustering performance can be obtained by the semi-supervised clustering[9]. The popular semi-supervised clustering methods are composed of two categories called the similarity-based and search-based approaches respectively[10]. In similarity-based methods, an existing clustering algorithm employs a specific similarity measure trained by labeled data. In search-based methods, the clustering algorithms modify the objective function under the aid of labeled data such that better clusters are found[10]. A number of semi-supervised clustering approaches published until now belongs to the search-based methods. For example, [11] presented a semi-supervised clustering with pairwise constraints and [10] gave an active semi-supervised fuzzy clustering. It is noticeable that semi- supervised clustering by seeding was proposed in [9]. [9] introduced a clustering method viewed as the semi-supervised variants of k-means and also called Seed-KMeans(S-KM). The S-KM can apply some labeled data called seeds to the initialization of the k-means clustering.

In this paper, the seeding-based semi-supervised clustering technique is introduced into the DifFUZZY clustering algorithm. The proposed method(S-DifFUZZY) uses some labeled data for affecting the structure of the $\sigma$-neighborhood graph and the corresponding matrix. This way can lead to the increase of the clustering accuracies proven by final experimental results on three UCI real data sets.

The remainder of this paper is organized as follows. Section 2 reports the DifFUZZY clustering. In Section 3, the proposed S-DifFUZZY clustering method is formulated. Experimental results are shown in Section 4, and Section 5 gives our conclusions.

## 2     The DifFUZZY Clustring Method

The DifFUZZY clustering method is composed of main three operations. Assume that $X = \{x_1, x_2, \cdots, x_n\}$ is a set of $n$ unlabeled data in the $d$-dimensional space $R^d$ and $X$ can be divided into $k$ clusters. Firstly, some core clusters should be found in the first operation of the DifFUZZY and this operation is outlined as follows[8]:

Step1. Let $\sigma = t'$.

Step2. Construct the $\sigma$-neighborhood graph. In this graph, each data point from $X$ is used as one node and two data point $x_i$ and $x_j$ can be connected by an edge when they satisfy the following condition where $\| \bullet \|$ is the Euclidean norm:

$$\frac{\|x_i - x_j\|}{\max\limits_{p=1,2,\cdots,n; q=1,2,\cdots,n} \left( \|x_p - x_q\| \right)} < \sigma \tag{1}$$

Step3. Let $E(\sigma)$ be equal to the number components of the $\sigma$-neighborhood graph which should contain at least $S$ vertices.

Step4. Let $\sigma = \sigma + t''$. If $\sigma > 1$, then goto Step5; otherwise goto Setp2.

Step5. Let $\sigma^* = \min\left( \arg\max_{\sigma \in (0,1]} \left( E(\sigma) \right) \right)$.

Step6. $E(\sigma^*)$ components of the $\sigma^*$-neighborhood graph which should contain at least $S$ vertices is regarded as the core clusters.

Step7. Label all data points in each core cluster. All data points without core clusters will be assigned to core clusters at the third operation of the DifFUZZY.

In [8], let $t' = 0.001$ and $t'' = 0.05$. Moreover, $S$ is the mandatory parameter of the DifFUZZY and Adjusting it to make $E(\sigma^*) = k$.

Secondly, three matrices should be calculated in the second operation of the DifFUZZY. The second operation is given as follows[8]:

Step1. Let a set $L = \{l_1, l_2, \cdots, l_h\}$, $t = 1$ and $\beta_t = l_t$.

Step2. Compute each element $m_{i,j}(\beta_t)$ of the matrix $M(\beta_t)$ according to the following Eq.(2) ($i = 1, 2, \cdots, n$, $j = 1, 2, \cdots, n$):

$$m_{i,j}(\beta_t) = \begin{cases} 1 & \text{if } x_i \text{ and } x_j \text{ belong to} \\ & \text{one same core cluster} \\ \exp\left( -\dfrac{\|x_i - x_j\|^2}{\beta_t} \right) & \text{otherwise} \end{cases} \tag{2}$$

Step3. Set $H(\beta_t) = \sum_{i=1}^{n} \sum_{j=1}^{n} m_{i,j}(\beta_t)$.

Step4. Let $t = t + 1$. If $t > h$, then let $t = 1$ goto Step5; otherwise let $\beta_t = l_t$ and goto Step2.

Step5. If $H(\beta_t) > \min_{i=1,2,\cdots,h} (H(\beta_i)) + \gamma_1 \left( \max_{i=1,2,\cdots,h} (H(\beta_i)) - \min_{i=1,2,\cdots,h} (H(\beta_i)) \right)$, then let $\beta^* = \beta_t$ and goto Step6; otherwise let $t = t + 1$ and goto Step5.

Step6. Based on $\beta^*$, calculate each elements $m_{i,j}(\beta^*)$ of the matrix $M(\beta^*)$ by the above Eq.(2).

Step7. Let the matrix $Z$ be a diagonal matrix and its diagonal elements can compute according to the following equation:

$$z_{i,i} = \sum_{j=1}^{n} m_{i,j}(\beta^*) \qquad i = 1, 2, \cdots, n \tag{3}$$

Step8. Calculate the matrix $Q$ by the following equation:

$$Q = I + (M - Z)\frac{\gamma_2}{\max\limits_{i=1,2,\cdots,n}(z_{i,i})} \tag{4}$$

where $I \in R^{n \times n}$ is the identity matrix.

Note that the default values of two parameters $\gamma_1$ and $\gamma_3$ are 0.3 and 0.1 respectively in [8]. Furthermore, the $L = \{l_1, l_2, \cdots, l_h\}$ can be generated by a Matlab's function called logspace in [8].

Finally, all data points without the core clusters are labeled in the third operation of the DifFUZZY. This operation is given as follows[8]:

Step1. Set $\theta$ be the second largest eigenvalue of $Q$ and compute the integer parameter $\mu$ according to the following equaltion:

$$\mu = \left\lfloor \frac{\gamma_3}{|\log \theta|} \right\rfloor \tag{5}$$

where $\gamma_3 = 1$.

Step.2 Let $t = 1$. Assume that $c_1, c_2, \cdots, c_k$ are $k$ core clusters and $\hat{x}_1, \hat{x}_2, \cdots, \hat{x}_g$ are data points without $k$ core clusters. Note that core clusters is also the clusters of the DifFUZZY.

Step3. Let $s = 1$.

Step4. Find a data point $x^*$ from $c_s$, which distance from $\hat{x}_t$ is minimum.

Step5. If $x^*$ is the $r$th element of $X$ and $\hat{x}_t$ is the $v$th element of $X$, then after substitute the $r$th row for the $v$th row and substitute the $r$th column for the $v$th column in the above matrix $M(\beta^*)$, the new matrix $\hat{M}(\beta^*)$ is gained.

Step6. According to $\hat{M}(\beta^*)$, calculate the matrices $\hat{Z}$ and $\hat{Q}$ respectively by the Eq.(3) and Eq.(4).

Step7. Compute the diffusion distance between $\hat{x}_t$ and the cluster $c_s$ by the following equation:

$$G(\hat{x}_t, c_s) = \left\| Q^\mu q - \hat{Q}^\mu q \right\| \tag{6}$$

where $q$ is a $n$-dimensional vector, the $v$th element of $q$ is equal to 1 and other elements are equal to 0.

Step8. Let $s = s + 1$. If $s > k$, then goto Step9; otherwise goto Step4.

Step9. Based on the following Eq.(7), the membership value of $\hat{x}_t$ in each cluster $c_s$.

$$f\left(\hat{x}_t, c_s\right) = \frac{G\left(\hat{x}_t, c_s\right)^{-1}}{\sum_{\rho=1}^{k} G\left(\hat{x}_t, c_\rho\right)^{-1}}, \qquad s = 1, 2, \cdots, k \qquad (7)$$

Step10. Label $\hat{x}_t$ according to the maximum membership value.

Step11. Let $t = t + 1$. If $t > g$, then goto Step12; other goto Step3.

Step12. End the DifFUZZY.

## 3    The Proposed S-DifFUZZY Algorithm

First, the generation method of seeds is described. Given the number of clusters $k$ and a nonempty set $X = \{x_1, x_2, \cdots, x_n\}$ of $n$ unlabeled data in the $d$-dimensional space $R^d$, the clustering algorithm can partition $X$ into $k$ clusters. Let $W$ also called the seed set and generated randomly be the subset of $X$ and for each $x_\alpha$ ($x_\alpha \in W$), its label be given by means of supervision. This way of producing seeds is similar to the S-KM clustering approach[9].

Secondly, the proposed S-DifFUZZY clustering algorithm is outlined as follows:

Step1.    Assume    that    $W$    contains    $B$    labeled    data    points    and $W = \{x_\alpha | \alpha = 1, 2, \cdots, B\}$.

Step2. Do three operations of the DifFUZZY clustering algorithm in turn. But there are two differences between the DifFUZZY algorithm and the novel S-DifFUZZY approach. Firstly, the method of constructing the σ−neighborhood graph is changed and the Eq.(1) is not used. If two data point $x_i$ and $x_j$ can be connected by an edge, then the value of the function $\Phi\left(x_i, x_j\right)$ is equal to 1; otherwise the value of $\Phi\left(x_i, x_j\right)$ is equal to 0. $\Phi\left(x_i, x_j\right)$ can be computed by the Eq.(8). Secondly, the method of computing each element $m_{i,j}\left(\beta_t\right)$ of the matrix $M\left(\beta_t\right)$ is changed into the following two steps:

i)      Apply the Eq.(2) to calculate each element $m_{i,j}\left(\beta_t\right)$ of the matrix $M\left(\beta_t\right)$. Let $\phi$ is equal to the minimum value of all $m_{i,j}\left(\beta_t\right)$.

ii)     Use the E.q(9) for adjusting the value of some elements in $M\left(\beta_t\right)$.

Step3. End the S-DifFUZZY.

Because the proposed S-DifFUZZY utilizes some labeled data point called seeds to change the structure of the σ−neighborhood graph and the value of some elements of the important matrix $M(\beta_t)$ , this novel algorithm can gain the better performance than the unsupervised DifFUZZY clustering.

$$\Phi(x_i,x_j)=\begin{cases} 0 & x_i,x_j \in W \text{ and the label of } x_i \text{ is not same with } x_j \\ \\ 1 & x_i,x_j \in W \text{ and the label of } x_i \text{ is same with } x_j \\ \\ 1 & \dfrac{\|x_i-x_j\|}{\max\limits_{p=1,2,\cdots,n;\, q=1,2,\cdots,n}\|x_p-x_q\|}<\sigma, \quad x_i,x_j \notin W \\ \\ 0 & \dfrac{\|x_i-x_j\|}{\max\limits_{p=1,2,\cdots,n;\, q=1,2,\cdots,n}\|x_p-x_q\|}\geq\sigma, \quad x_i,x_j \notin W \end{cases} \tag{8}$$

$$m_{i,j}(\beta_t)=\begin{cases} \phi & x_i,x_j \in W \text{ and the label of } x_i \text{ is not same with } x_j \\ \\ 1 & x_i,x_j \in W \text{ and the label of } x_i \text{ is same with } x_j \end{cases} \tag{9}$$

## 4    Experimental Results

To demonstrate the effectiveness of the above S-DifFUZZY algorithm, we compared it with the unsupervised KM, DifFUZZY method and the semi-supervised S-KM approach, on three UCI real data sets[12] referred to as the Tae, Sonar and Haberman respect tively. The Tae data set contains 151 cases with 5-dimensional feature from three classes. The Sonar data set is a 60-dimensional data set with 208 instances of two classes. The Haberman data set collects 306 3-dimensional cases belonging to two classes. All experiments were done by Matlab on WindowsXP operating system.

For the semi-supervised S-KM and S-DifFUZZY, on each data set, we randomly generated $Z\%$ ($Z=0,10,20,\cdots,100$ ) of the data set as seeds. Since true tables are known, the clustering accuracies $V\%$ on unlabeled data, which is the remaining $(100-Z)\%$ of the data set, could be quantitatively assessed. Therefore, the clustering accuracies $Y\%$ of the whole data set consisting of unlabeled data and labeled seeds could be calculated by $V\%\cdot(100-Z)\%+Z\%$ . If $Z=0$ , the semi-supervised clustering methods became the unsupervised algorithms, and If $Z=100$, all data of the whole data set was considered as labeled seeds and $Y=100$ . On each data set the semi-supervised S-DifFUZZY and S-KM were run 20 times for different $Z$ ($Z=10,$

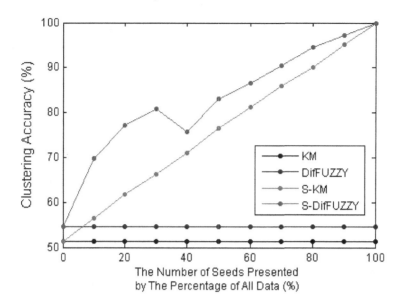

**Fig. 1.** Comparison of clustering accuracies on the Haberman data set

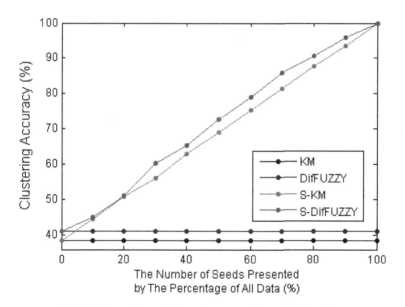

**Fig. 2.** Comparison of clustering accuracies on the Tae data set

20,···,90) and we report in Fig.1, Fig.2 and Fig.3 the average accuracies $Y\%$ of the whole data set obtained. Furthermore, we should make the number of components of the σ-neighborhood graph is equal to the number clusters by selecting the minimum

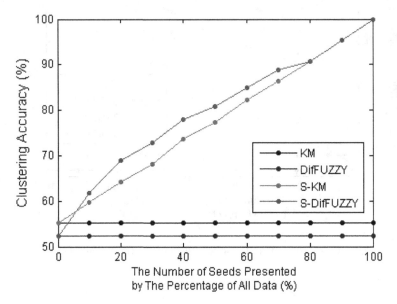

**Fig. 3.** Comparison of clustering accuracies on the Sonar data set

value of the parameter $S$ from $[2,+\infty)$ for the DifFUZZY and S-DifFUZZY clustering algorithms. The experimental results of the KM shown in all figures are averaged on 20 independent runs for each data set.

As shown in Fig.1, Fig.2 and Fig.3, firstly, the unsupervised KM and Dif-FUZZY method is not affected by the seeds because it never employs the seeds. Secondly, because the seeds are applied the S-DifFUZZY, there are the drastic distinctions between the clustering accuracies of the DifFUZZY and S-DifFUZZY algorithms. Finally, although the clustering accuracies of the S-KM are improved with the increase of the seeds, we can from the Fig.1, Fig.2 and Fig.3 that the proposed S-DifFUZZY can achieve the better performance than the S-KM when an same amount of labeled seeds is used.

## 5    Conclusions

In this paper, we propose the novel seeding-based semi-supervised fuzzy clustering algorithm inspired by diffusion processes. This semi-supervised method is also called the S-DifFUZZY. Compared with the unsupervised DifFUZZY, main idea of the S-DifFUZZY is applied to have an important effect on the corresponding matrix and the structure of the σ-neighborhood graph. Experiments are carried out on three UCI real data sets. In comparison to the S-KM and DifFUZZY, our proposed method has been demonstrated their superiority.

**Acknowledgements.** This work is supported by Foundation of Guangxi Key Laboratory of Wireless Wideband Communication & Signal Processing, Guilin University of Electronic Technology, Guangxi, China (No.21107), and this work is also supported by the Scientific Research Foundation of Nanjing University of Posts and Telecommunications (No.NY210078).

# References

1. Fillippone, M., Camastra, F., Masulli, F., Rovetta, S.: A survey of kernel and spectral methods for clustering. Pattern Recognition 41(1), 176–190 (2008)
2. Jain, A.K., Murty, M.N., Flyn, P.J.: Data clustering: a review. ACM Computing Surveys 32(3), 256–323 (1999)
3. Xu, R., Wunsch, D.: Survey of clustering algorithms. IEEE Transactions on Neural Networks 16(3), 645–678 (2005)
4. Tou, J.T., Gonzalez, R.C.: Pattern recognition principles. Addison-Wesley, London (1974)
5. Bezdek, J.C.: Pattern recognition with fuzzy objective function algorithms. Plenum Press, New York (1981)
6. Kim, D.W., Lee, K.Y., Lee, D., Lee, K.H.: A kernel-based subtractive clustering method. Pattern Recognition Letters 26(7), 879–891 (2005)
7. Martinetz, T.M., Berkovich, S.G., Schulten, K.J.: Neural-gas network for vector quantization and its application to time-series prediction. IEEE Transactions on Neural Networks 4(4), 558–569 (1993)
8. Ormella, C., Anastasios, M., Sandhya, S., Don, K., Sijia, L., Philip, K.M., Radek, E.: DifFUZZY: a fuzzy clustering algorithm for complex datasets. International Journal of Computational Intelligence in Bioinformatics and Systems Biology 1(4), 402–417 (2010)
9. Basu, S., Banerjee, A., Mooney, R.J.: Semi-supervised clustering by seeding. In: Proceedings of the Nineteenth International Conference on Machine Learning, pp. 27–34 (2002)
10. Grira, N., Crucianu, M., Boujemaa, N.: Active semi-supervised fuzzy clustering. Pattern Recognition 41(5), 1834–1844 (2008)
11. Basu, S., Banjeree, A., Mooney, R.J.: Active semi-supervised for pairwise constrained clustering. In: Proceedings of the 2004 SIAM International Conference on Data Mining, pp. 333–344 (2004)
12. UCI Machine Learning Repository,
    http://www.ics.uci.edu/~mlearn/MLSummary.html

# Fault Detection for Nonlinear Discrete-Time Systems via Deterministic Learning

Junmin Hu, Cong Wang*, and Xunde Dong

College of Automation Science and Engineering
and Center for Control and Optimization
South China University of Technology, Guangzhou, 510641
{hu.junmin,dong.xunde}@mail.scut.edu.cn,
wangcong@scut.edu.cn

**Abstract.** This paper presents a fault detection scheme for nonlinear discrete-time systems based on the recently proposed deterministic learning (DL) theory. The scheme consists of two phases: the learning phase and the detecting phase. In the learning phase, the discrete-time system dynamics underlying normal and fault modes are locally accurately approximated through deterministic learning. The obtained knowledge of system dynamics is stored in constant RBF networks. In the detecting phase, a bank of estimators are constructed using the constant RBF networks to represent the learned normal and fault modes. By comparing the set of estimators with the monitored system, a set of residuals are generated, and the average $L_1$ norms of the residuals are used to compare the differences between the dynamics of the monitored system and the dynamics of the learning normal and fault modes. The occurrence of a fault can be rapidly detected in a discrete-time setting.

**Keywords:** Deterministic learning, fault detection, nonlinear discrete-time systems, neural networks.

## 1 Introduction

The design and analysis of fault detection and isolation (FDI) for nonlinear systems are important issues in modern engineering systems. Over the past decades, much progress has been achieved for FDI of nonlinear continuous-time systems with structured and unstructured modeling uncertainties [1]. However, only limited results has been obtained for fault diagnosis of nonlinear discrete-time systems. In [2], an online approximation based fault detection and diagnosis scheme for multiple state or output faults was proposed for a class of nonlinear MIMO discrete-time systems. The faults considered could be incipient or abrupt, and are modeled using input and output signals of the system. In [3], a distributed fault detection and isolation approach based on adaptive approximation was proposed for nonlinear uncertain large-scale discrete-time dynamical systems. Local

---

* Corresponding author.

C. Guo, Z.-G. Hou, and Z. Zeng (Eds.): ISNN 2013, Part I, LNCS 7951, pp. 574–581, 2013.

and global FDI capabilities are provided due to the utilization of specialized fault isolation estimators and a global fault diagnoser.

In adaptive approximation based fault detection and isolation of general nonlinear systems, however, convergence of the employed neural network (NN) weights to their optimal values and accurate NN approximation of nonlinear fault functions was less investigated. Recently, a deterministic learning (DL) theory was proposed for NN approximation of nonlinear dynamical systems exhibiting periodic or recurrent trajectories [4]. By using localized radial basis function (RBF) NNs, it is shown that almost any periodic or recurrent trajectory can lead to the satisfaction of a partial PE condition, which in turn yields accurate NN approximation of the system dynamics in a local region along the periodic or recurrent trajectory. Based on deterministic learning (DL) theory, [5] proposed a new method for rapid detection of oscillation faults generated from nonlinear continuous-time systems.

In this paper, we present a fault detection scheme for nonlinear discrete-time systems via DL. A class of faults generated from nonlinear discrete-time systems with unstructured modeling uncertainty will be considered. As it is usually impossible to decouple fault functions from the unstructured modeling uncertainty, these two terms are combined together as a whole as the general fault function. The scheme consists of a learning phase and a detecting phase. Firstly, in the learning phase, we will show that the general fault function can be locally-accurately approximated by using an extension of DL algorithm for discrete-time systems. Specifically, the system dynamics underlying normal and fault modes are locally accurately approximated. Secondly in the detecting phase, by utilizing the learned knowledge obtained through DL, a bank of estimators are constructed using the constant RBF networks to represent the trained normal and fault modes. By comparing the set of estimators with the monitored system, a set of residuals are generated, the average $L_1$ norms of the residuals can be used for measuring the differences between the dynamics of the monitored system and the trained normal and fault modes. The occurrence of a fault can be detected according to the smallest residual principle. Detectability analysis is also carried out to show the feasibility of the fault detection scheme.

## 2 Problem Formulations

Consider the following class of uncertain nonlinear discrete-time system:

$$x(k+1) = f(x(k), u(k)) + \eta(x(k), u(k)) + \beta(k - k_0)\phi^s(x(k), u(k)) \quad (1)$$

where $k$ is the discrete-time instant, $x(k) = [x_1(k), \cdots, x_n(k)]^T \in R^n$ is the state vector of the system, $u(k) = [u_1(k), \cdots, u_m(k)]^T \in R^m$ is the control input vector, $f(x(k), u(k)) = [f_1(x(k), u(k)), \cdots, f_n(x(k), u(k))]^T$ is unknown smooth nonlinear vector field representing the dynamics of the normal model, $\eta(x(k), u(k))$ stands for the uncertainties including external disturbances, modeling errors and possibly discretization errors. The fault $\beta(k - k_0)\phi^s(x(k), u(k))$ is the deviation in system dynamics due to fault $s(s = 1, \cdots, M)$, and $\beta(k - k_0)$

represents the fault time profile, with $k_0$ being the unknown fault occurrence time. When $k \leq k_0$, $\beta(k - k_0) = 0$, and when $k \geq k_0$, $\beta(k - k_0) = 1$, $i = 1, \cdots, n$.

The system state is assumed to be observable, and the system input is usually designed as a function of the system states. The state sequence $(x(k))_{k=0}^{\infty}$ of (1) with initial condition $x(0)$ is defined as the system trajectory. The trajectory in normal mode is denoted as $\varphi^0(x(0))$ or $\varphi^0$ for conciseness of presentation, and the trajectory in fault mode $s$ is denoted as $\varphi^s(x(0))$ or $\varphi^s$ for conciseness of presentation. Assume that the system trajectories for both normal and fault modes are recurrent trajectories [6], the class of recurrent trajectory comprise the most important types (though not all types) of trajectories generated from nonlinear discrete dynamical systems (see [6] for a rigorous definition of recurrent trajectory). The system states and controls are assumed to be bounded for both normal and fault modes and $\eta(x(k), u(k))$ is bounded in a compact region by some known function $\bar{\eta}_x(x(k), u(k))$, i.e., $\|\eta(x(k), u(k))\| \leq \bar{\eta}(x(k), u(k))$, $\forall(x(k), u(k)) \in \Omega \subset R^n \times R^m$, $\forall k \geq 0$.

In system (1), since the uncertainty $\eta(x(k), u(k))$ in the state equation and fault $\beta(k - k_0)\phi^s(x(k), u(k))$ $(s = 0, 1, \cdots, M)$ cannot be decoupled from each other, we consider the two terms together as an undivided term, and define the term as the general fault function as in [5]

$$\psi^s(x(k), u(k)) = \eta(x(k), u(k)) + \beta(k - k_0)\phi^s(x(k), u(k)) \qquad (2)$$

where $s = 0, 1, \cdots, M$, $\phi^s(x(k), u(k))$ represents the $s$th fault belongs to the set of fault functions. For simplicity of presentation, the normal mode is represented by fault mode $s = 0$, with $\phi^0(x(k), u(k)) = 0$.

## 3   Learning and Representation of Faults

In this section, we investigate learning and representation of faults generated from nonlinear discrete-time system (1) by using an extension of DL algorithm.

For both normal and fault modes, combined with (2), (1) is described by

$$x(k + 1) = f(x(k), u(k)) + \psi^s(x(k), u(k)) \qquad (3)$$

Construct the following discrete-time dynamical RBF network:

$$\hat{x}(k + 1) = f(x(k), u(k)) + A(\hat{x}(k) - x(k)) + \hat{W}^{sT}(k + 1)S(x(k)) \qquad (4)$$

where $A = diag\{a_1, a_2, \cdots, a_n\}$ is a diagonal matrix, with $0 < |a_i| < 1$ being design constant, $\hat{x}(k) = [\hat{x}_1(k), \hat{x}_2(k), \cdots, \hat{x}_n(k)]^T$ is the state vector of (4), $x(k) = [x_1(k), \cdots, x_n(k)]^T$ is the state vector of system (1), The Gaussian RBF network $\hat{W}^{sT}S(x(k)) = [\hat{W}_1^{sT}S(x(k)), \cdots, \hat{W}_n^{sT}S(x(k))]$, $s = 0, 1, 2, \cdots, M$ are used to approximate the general fault function (2). The NN weights $\hat{W}(k + 1)$ are updated similarly as in [7]:

$$\hat{W}_i(k + 1) = \hat{W}_i(k) - \frac{\alpha P(e_i(k) - a_i e_i(k - 1))S(x(k - 1))}{1 + \lambda_{max}(P)S^T(x(k - 1))S(x(k - 1))} \qquad (5)$$

where $e_i(k) = \hat{x}_i(k) - x_i(k)$ is the state error, $0 < \alpha < 2$ is the learning gain for design, $P = P^T > 0$, $\lambda_{max}(P)$ denotes the largest eigenvalue of the matrix $P$. From (3) and (4), we have $e_i(k+1) = a_i e_i(k) + \tilde{W}_i^T(k+1)S(x(k)) - \epsilon$, where $\epsilon = \psi^s(x(k), u(k)) - W^{*T}S(x(k))$ is the NN approximation error, $W^*$ is the ideal weight.

The following theorem presents the learning of the general fault function.

**Theorem 1.** *Consider the close-loop system consisting of the nonlinear discrete-time system (1), the dynamical RBF network (4) and the NN update law (5). For both normal and fault modes of (1), we have that all the signals in the closed-loop remain bounded, and the general fault function $\psi^s(x(k), u(k))$ of system (1) is locally-accurately approximated along the trajectory $\varphi^s(x(0))$ by $\hat{W}^T S(\varphi^s)$ as well as by $\bar{W}^T S(\varphi^s)$, where*

$$\bar{W} = \frac{1}{k_b - k_a + 1} \sum_{k=k_a}^{k_b} \hat{W}(k) \tag{6}$$

*with $(k_a, k_b)$ representing a time segment after the transient process.*

Based on the convergence result of $\hat{W}$, we can obtain a constant vector of neural weights $\bar{W}$ according to (6), such that $\psi^s(x(k), u(k)) = \hat{W}^T S(\varphi^s) + \epsilon_1 = \bar{W}^T S(\varphi^s) + \epsilon_2$, where $\epsilon_1$ and $\epsilon_2$ are the approximation errors using $\hat{W}^T S(\varphi^s)$ and $\bar{W}^T S(\varphi^s)$, respectively. It is clear that after the transient process, $||\epsilon_1| - |\epsilon_2||$ is small.

## 4   Detection of Faults

### 4.1   Residual Generation and Decision Scheme

In the detecting phase, the monitored system is described by

$$x(k+1) = f(x(k), u(k)) + \eta(x(k), u(k)) + \beta(k - k_0)\phi^{s'}(x(k), u(k)) \tag{7}$$

where $\phi^{s'}(x(k), u(k))$ represents the deviation of system dynamics due to an unknown fault.

For the monitored system (7), the general fault function is described by

$$\psi'(x(k), u(k)) = \eta(x(k), u(k)) + \beta(k - k_0)\phi^{s'}(x(k), u(k)) \tag{8}$$

Consider the monitored system (7), by utilizing the learned knowledge obtained in the learning phase, a dynamical model is constructed as follows:

$$\bar{x}^h(k+1) = f(x(k), u(k)) + B(\bar{x}^h(k) - x(k)) + \bar{W}^{h^T}S(x(k)) \tag{9}$$

where $h = 0, 1, \ldots, M$, $\bar{x}^h(k) = [\bar{x}_1^h(k), \ldots, \bar{x}_n^h(k)]^T \in R^n$ is the state of the dynamical model, $x(k) = [x_1(k), \cdots, x_n(k)]^T \in R^n$ is the state of monitored system (7), $B = diag\{b_1, \ldots, b_n\}$ is a diagonal matrix which is kept the same for all normal and fault models, with $0 < b_i < 1$.

By combining the monitored system (7) and the dynamical model (9), the following residual system is obtained:

$$\tilde{x}_i^h(k+1) = b_i \tilde{x}_i^h(k) + (\bar{W}_i^{h^T} S_i(x(k)) - \psi_i'(x(k), u(k))), \ i = 1, \ldots, n \quad (10)$$

where $\tilde{x}_i^h(k) \triangleq \hat{x}_i^h(k) - x_i(k)$ is the state estimation error (residual), and $\bar{W}_i^{h^T} S_i(x(k)) - \psi_i'(x(k), u(k))$ is the difference of system dynamics between the monitored system and the $h$th estimator.

The average $L_1$ norm $\|\tilde{x}_i^h(k)\|_1 = \frac{1}{T} \sum_{j=k-T}^{k-1} |\tilde{x}_i^h(j)|$ is introduced for the decision of a fault, where $k \geq T$, $T \in Z_+$ is the preset period constant of the monitored system.

*Fault detection decision scheme:* Compare $\|\tilde{x}_i^s\|_1$ ($i = 1, \ldots, n$), for all $s \in \{0, 1, \ldots, M\}$ with each other. If, there exists a mode $s^*$, ($s^* \in \{0, 1, \ldots, M\}$) such that for all $i = 1, \ldots, n$, $\|\tilde{x}_i^{s^*}\|_1 < \|\tilde{x}_i^s\|_1$, ($s \in \{0, 1, \ldots, M\} \setminus \{s^*\}$), then the occurrence of a fault is deduced and the monitored system is recognized as similar to the $s^*$th mode.

## 4.2 Fault Detectability Conditions

The detectability analysis of fault detection for nonlinear discrete-time systems is stated in the following theorem.

**Theorem 2.** *(Fault detectability) Consider the residual system (10). For all* $s \in \{0, 1, \ldots, M\}$, $i \in \{1, \ldots, n\}$ *and* $k \geq T$, *if the following conditions hold:*

*1) for all* $s \in \{0, 1, \ldots, M\} \setminus s^*$, *there exists at least one interval* $I = [T_a, T_b - 1] \subseteq [k - T, k - 1]$ *such that*

$$|\psi_i^s(s(k_\tau), u(k_\tau)) - \psi_i'(s(k_\tau), u(k_\tau))| \geq \mu_i + \xi_i^* \quad \forall k_\tau \in I \quad (11)$$

*where* $\mu_i > (\epsilon_i^* + \xi_i^*)$, $l := T_b - T_a > \underline{l} := \frac{\mu_i + 2(\epsilon_i^* + \xi_i^*)}{2\mu_i + (\epsilon_i^* + \xi_i^*)} T$

*2)* $b_i$ *satisfies*

$$0 < b_i < \left( \frac{\mu_i - (\epsilon_i^* + \xi_i^*)}{5\mu_i + (\epsilon_i^* + \xi_i^*)} \right)^{\frac{l}{l - \underline{l} - 1}} \quad (12)$$

*where* $\mu_i > (\epsilon_i^* + \xi_i^*)$, $l := T_b - T_a > \underline{l} := \frac{\mu_i + 2(\epsilon_i^* + \xi_i^*)}{2\mu_i + (\epsilon_i^* + \xi_i^*)} T$.

*Then, the fault will be detected. i.e.,* $\|\tilde{x}_i^{s^*}\|_1 < \|\tilde{x}_i^s\|_1$, *holds for all* $k \geq T$.

*Proof.* Firstly, we prove that $\|\tilde{x}_i^{s^*}(k)\|_1 < \frac{\epsilon_i^* + \xi_i^*}{1 - b_i}$ for all $k \geq T$.

If the monitored system (7) is similar to the $s^*$th mode, according to the definition of similarity for discrete-time systems [8]. we have

$$\max_{x(k) \in \Omega_{\varphi^{s^*}}} \left| \psi_i'(x(k), u(k)) - \psi_i^{s^*}(x(k), u(k)) \right| < \epsilon_i^*, \ i = 1, \ldots, n \quad (13)$$

and

$$\max_{x(k) \in \Omega_{\varphi^{s^*}}} \left| \psi_i'(x(k), u(k)) - \bar{W}_i^{s^* T} S_i((x(k), u(k)) \right| < \epsilon_i^* + \xi_i^*, \ i = 1, \ldots, n \quad (14)$$

where $\epsilon_i^* > 0$ is the similarity measure between the two dynamical systems. $\xi_i^*$ and $\epsilon_i^*$ are respectively given by [8] and (13).

The error system with respect to the $s^*$th model satisfies $\tilde{x}_i^{s^*}(k+1) = b_i \tilde{x}_i^{s^*}(k) + [\bar{W}_i^{s^*T} S_i(x(k)) - \psi_i'(x(k), u(k)))]$. The solution of the above difference equation is

$$\tilde{x}_i^{s^*}(k) = b_i^k \tilde{x}_i^{s^*}(0) + \sum_{j=0}^{k-1} b_i^{k-1-j} \left[ \bar{W}_i^{s^*T} S_i(x(j)) - \psi_i'(x(j), u(j))) \right] \qquad (15)$$

By combining (14) and $0 < b_i < 1$ , we have $|\tilde{x}_i^{s^*}(k)| < b_i^k |\tilde{x}_i^{s^*}(0)| + (\epsilon_i^* + \xi_i^*) \sum_{j=0}^{k-1} b_i^{k-1-j} < b_i^k |\tilde{x}_i^{s^*}(0)| + \frac{\epsilon_i^* + \xi_i^*}{1-b_i}$. Since $x_i(0)$ is available, let $\tilde{x}_i^{s^*}(0) = 0$, $\tilde{x}_i^{s^*}(0) = 0$. Then, we have $|\tilde{x}_i^{s^*}(k)| < \frac{\epsilon_i^* + \xi_i^*}{1-b_i}$.

Thereby, we have for all $k \geq T$,

$$\|\tilde{x}_i^{s^*}(k)\|_1 = \frac{1}{T} \sum_{j=k-T}^{k-1} |\tilde{x}_i^{s^*}(j)| < \frac{1}{T} \sum_{j=k-T}^{k-1} \frac{\epsilon_i^* + \xi_i^*}{1-b_i} = \frac{\epsilon_i^* + \xi_i^*}{1-b_i} \qquad (16)$$

Secondly, we turn to prove that $\|\tilde{x}_i^s(k)\|_1 > \frac{\epsilon_i^* + \xi_i^*}{1-b_i} > \|\tilde{x}_i^{s^*}\|_1$ for all $k \geq T$.
For $k_\tau \in I$,

$$|\tilde{x}_i^s(k_\tau)| = \left| b_i^{k_\tau - T_a} \tilde{x}_i^s(T_a) + \sum_{j=T_a}^{k_\tau - 1} b_i^{k_\tau - 1 - j} \left[ \bar{W}_i^{s^T} S_i(x(j)) - \psi_i'(x(j), u(j))) \right] \right| \qquad (17)$$

From the local region is described by [8] and (12), we have

$$\left| \bar{W}_i^{s^T} S_i(x(k_\tau)) - \psi_i'(x(k), u(k))) \right|$$
$$\geq |\psi_i^s(x(k_\tau), u(k_\tau)) - \psi_i'(x(k_\tau), u(k_\tau))| - \left| \bar{W}_i^{s^T} S_i(x(k_\tau)) - \psi_i^s(x(k_\tau), u(k_\tau)) \right|$$
$$\geq \mu_i + \xi_i^* - \xi_i^* = \mu_i$$
$$\qquad (18)$$

Let $I'$ be defined as $I' = \left\{ k_\tau \in I : |\tilde{x}_i^s(k_\tau)| < \frac{2\mu_i + (\epsilon_i^* + \xi_i^*)}{3(1-b_i)} \right\}$. Then, it can be proven that there at most exists one time interval $I' = [T_a', T_b' - 1]$ in $I$ and

$$l' = T_b' - T_a' \leq \left( \frac{\ln\left( \frac{\mu_i - (\epsilon_i^* + \xi_i^*)}{[5\mu_i + (\epsilon_i^* + \xi_i^*)]} \right)}{\ln b_i} \right) + 1 = \left( \frac{\ln\left( \frac{b_i[\mu_i - (\epsilon_i^* + \xi_i^*)]}{[5\mu_i + (\epsilon_i^* + \xi_i^*)]} \right)}{\ln b_i} \right) \text{ and let } l' \text{ denote}$$

the length of the time interval $I'$.

The magnitude of $|\tilde{x}_i^s(k_\tau)|$ in the time interval $I$ can be discussed in the following cases:

1) If $|\tilde{x}_i^s(T_a)| \geq \frac{2\mu_i + (\epsilon_i^* + \xi_i^*)}{3(1 - b_i)}$, and $\bar{W}_i^{s^T} S_i(x(k)) - \psi_i'(x(k), u(k))$ has the same sign with $\tilde{x}_i^s(T_a)$, then from (17) and (18), we have

$$
\begin{aligned}
|\tilde{x}_i^s(k_\tau)| &= \left| b_i^{k_\tau - T_a} \tilde{x}_i^s(T_a) + \sum_{j=T_a}^{k_\tau - 1} b_i^{k_\tau - 1 - j} \left[ \bar{W}_i^{s^T} S_i(x(j)) - \psi_i'(x(j), u(j)) \right] \right| \\
&\geq \left| b_i^{k_\tau - T_a} \tilde{x}_i^s(T_a) + \mu_i \sum_{j=T_a}^{k_\tau - 1} b_i^{k_\tau - 1 - j} \right| \geq \frac{2\mu_i + (\epsilon_i^* + \xi_i^*)}{3(1 - b_i)}
\end{aligned}
\tag{19}
$$

Thus, $|\tilde{x}_i^s(k_\tau)| \geq \frac{2\mu_i + (\epsilon_i^* + \xi_i^*)}{3(1 - b_i)}$ holds for all $k_\tau \in I$. Therefore, $I' = \emptyset$ and $l' = 0$.

2) If $|\tilde{x}_i^s(T_a)| < \frac{2\mu_i + (\epsilon_i^* + \xi_i^*)}{3(1 - b_i)}$, and according to (17) and (18), we have for all $k_\tau \in I$

$$
\begin{aligned}
|\tilde{x}_i^s(k_\tau)| &= \left| b_i^{k_\tau - T_a} \tilde{x}_i^s(T_a) + \sum_{j=T_a}^{k_\tau - 1} b_i^{k_\tau - 1 - j} \left[ \bar{W}_i^{s^T} S_i(x(j)) - \psi_i'(x(k), u(k)) \right] \right| \\
&\geq \left| \mu_i \frac{1 - b_i^{k_\tau - T_a}}{1 - b_i} \right| - \left| b_i^{k_\tau - T_a} \tilde{x}_i^s(T_a) \right| \\
&\geq \frac{\mu_i}{1 - b_i} - \frac{5\mu_i + (\epsilon_i^* + \xi_i^*)}{3(1 - b_i)} b_i^{k_\tau - T_a}
\end{aligned}
\tag{20}
$$

Assume that there exists a time interval $I' = [T_a', T_b' - 1] \subseteq I$ with $T_a' = T_a$, then consider (20), we have

$$
\frac{2\mu_i + (\epsilon_i^* + \xi_i^*)}{3(1 - b_i)} \geq |\tilde{x}_i^s(T_b' - 1)| \geq \frac{\mu_i}{1 - b_i} - \frac{5\mu_i + (\epsilon_i^* + \xi_i^*)}{3(1 - b_i)} b_i^{T_b' - 1 - T_a'}
\tag{21}
$$

Solving the above inequality for $T_b'$ yields, we have $l' = T_b' - T_a' \leq \frac{\ln\left( \frac{\mu_i - (\epsilon_i^* + \xi_i^*)}{5\mu_i + (\epsilon_i^* + \xi_i^*)} \right)}{\ln b_i} +$ 1. From (12), which implies that $l > \underline{l} + \frac{\ln\left( \frac{b_i[\mu_i - (\epsilon_i^* + \xi_i^*)]}{5\mu_i + (\epsilon_i^* + \xi_i^*)} \right)}{\ln b_i} \geq \underline{l} + l'$. Thus we have $|\tilde{x}_i^s(T_b - 1)| \geq \frac{2\mu_i + (\epsilon_i^* + \xi_i^*)}{3(1 - b_i)}$ and $|\tilde{x}_i^s(T_b')| \geq \frac{2\mu_i + (\epsilon_i^* + \xi_i^*)}{3(1 - b_i)}$, it can be proved that $\bar{W}_i^{s^T} S_i(x(k)) - \psi_i'(x(k), u(k))$ has the same sign with $|\tilde{x}_i^s(T_b')|$. By using the analysis result of case 1), we have $|\tilde{x}_i^s(k_\tau)| \geq \frac{2\mu_i + (\epsilon_i^* + \xi_i^*)}{3(1 - b_i)}$ holds for all $k_\tau \in [T_b', T_b - 1]$. Therefore, $I' = [T_a', T_b' - 1]$ and $l' \leq \frac{\ln\left( \frac{b_i[\mu_i - (\epsilon_i^* + \xi_i^*)]}{5\mu_i + (\epsilon_i^* + \xi_i^*)} \right)}{\ln b_i}$.

3) In the case that $|\tilde{x}_i^s(T_a)| \geq \frac{2\mu_i + (\epsilon_i^* + \xi_i^*)}{-3b_i}$, and $\bar{W}_i^{s^T} S_i(X(k)) - \psi_i'(x(k), u(k))$ has a different sign with $|\tilde{x}_i^s(T_a)|$, if there exists a time $T_a' \in I$ such that $|\tilde{x}_i^s(T_a')| < \frac{2\mu_i + (\epsilon_i^* + \xi_i^*)}{-3b_i}$, then using analystic result of case 2), we have $I' = [T_a', T_b' - 1]$ and $l' \leq \frac{\ln\left( \frac{b_i[\mu_i - (\epsilon_i^* + \xi_i^*)]}{5\mu_i + (\epsilon_i^* + \xi_i^*)} \right)}{\ln b_i}$. If such time $T_a'$ does not exist, then $I' = \emptyset$ and $l' = 0$.

From the above discussion, we can summarize that there at most exists one time interval $I' = [T'_a, T'_b - 1]$ and $l' \leq \frac{\ln\left(\frac{b_i[\mu_i - (\epsilon_i^* + \xi_i^*)]}{5\mu_i + (\epsilon_i^* + \xi_i^*)}\right)}{\ln b_i}$.

Thus, in light of (16), we have $\|\tilde{x}_i^s(k)\|_1 = \frac{1}{T}\sum_{j=k-T}^{k-1} |\tilde{x}_i^s(j)| \geq \frac{1}{T} l \frac{2\mu_i + (\epsilon_i^* + \xi_i^*)}{-3b_i}$, for all $k \geq T$, with (11), we have $\|\tilde{x}_i^s(k)\|_1 \geq \frac{\mu_i + 2(\epsilon_i^* + \xi_i^*)}{3(1-b_i)} \geq \frac{\epsilon_i^* + \xi_i^*}{3(1-b_i)} > \|\tilde{x}_i^{s^*}(k)\|_1$. This ends the proof.    □

*Remark 1.* The represented mathematically rigorous proof of fault detectability conditions for discrete-time case is more succinct than [5].

## 5    Conclusions

In this paper, a approach has been proposed for fault detection of nonlinear discrete-time systems based on the recently proposed deterministic learning (DL) theory. The discrete-time system dynamics underlying normal and fault modes are locally accurately approximated through DL. The obtained knowledge of system dynamics is stored in constant RBF networks. Then, a bank of estimators are constructed using the constant RBF networks to represent the learning normal and fault modes. A set of residuals are generated by comparing the set of estimators. The occurrence of a fault can be rapidly detected according to the smallest residual principle and detectability condition is also given.

**Acknowledgments.** This work was supported by the National Science Fund for Distinguished Young Scholars (Grant No. 61225014), and by the National Natural Science Foundation of China (Grant No. 60934001).

## References

1. Persis, C.D., Isidori, A.: A geometric approach to nonlinear fault detection and isolation. IEEE Transactions on Automatic Control 46(6), 853–865 (2001)
2. Thumati, B.T., Jagannathan, S.: A Model-Based Fault-Detection and Prediction Scheme for Nonlinear Multivariable Discrete-Time Systems With Asymptotic Stability Guarantees. IEEE Transactions on Neural Networks 21(3), 404–423 (2010)
3. Riccardo, M.G., Ferrari, G., Parisini, T., Polycarpou, M.M.: Distributed fault detection and isolation of large-scale discrete-time nonlinear systems: an adaptive approximation approach. IEEE Transactions on Automatic Control 57(2), 275–290 (2012)
4. Wang, C., Hill, D.J.: Deterministic learning theory for identification, recognition and control. CRC Press (2009)
5. Wang, C., Chen, T.R.: Rapid detection of small oscillation faults via deterministic learning. IEEE Transactions on Neural Networks 22(8), 1284–1296 (2011)
6. Shilnikov, L.P., et al.: Methods of Qualitative Theory in Nonlinear Dynamics. World Scientific, Singapore (2001)
7. Yuan, C.Z., Wang, C.: Deterministic learning of Sampled-Data Nonlinear Systems. Science China (2012) (in press)
8. Wang, C., Chen, T.R., Liu, T.F.: Deterministic learning and data-based modeling and control. Acta Automatica Sinica 35, 693–706 (2009) (Chinese)

# Loose Particle Classification Using a New Wavelet Fisher Discriminant Method

Long Zhang[1], Kang Li[1], Shujuan Wang[2], Guofu Zhai[2], and Shaoyuan Li[3,*]

[1] School of Electronics, Electrical Engineering and Computer Science
Queen's University Belfast, Belfast BT9 5AH, UK
[2] School of Electrical Engineering and Automation
Harbin Institute of Technology, 150001, China
[3] Department of Automation, Shanghai Jiaotong University
Shanghai 200240, China

**Abstract.** Loose particles left inside aerospace components or equipment can cause catastrophic failure in aerospace industry. It is vital to identify the material type of these loose particles and eliminate them. This is a classification problem, and autoregressive (AR) model and Learning Vector Quantization (LVQ) networks have been used to classify loose particles inside components. More recently, the test objects have been changed from components to aerospace equipments. To improve classification accuracy, more data samples often have to be dealt with. The difficulty is that these data samples contain redundant information, and the aforementioned two conventional methods are unable to process redundant information, thus the classification accuracy is deteriorated. In this paper, the wavelet Fisher discriminant is investigated for loose particle classifications. First, the fisher model is formulated as a least squares problem with linear-in-the-parameters structure. Then, the previously proposed two-stage subset selection method is used to build a sparse wavelet Fisher model in order to reduce redundant information. Experimental results show the wavelet Fisher classification method can perform better than AR model and LVQ networks.

**Keywords:** Loose particle classification, wavelet Fisher discriminant, subset selection.

## 1 Introduction

The reliability of all electronic components and equipment used for space applications is vital for the success of the missions where they perform complex monitoring, navigation and control functions. Hence, all the components and

---

* This work was supported by the RCUK under grant EP/G042594/1 and EP/G059489/1, and partially supported by the Chinese Scholarship Council, National Natural Science Foundation of China under grant 51077022 and 61271347, and Project of Science and Technology Commission of Shanghai Municipality under Grant 11ZR1413100.

C. Guo, Z.-G. Hou, and Z. Zeng (Eds.): ISNN 2013, Part I, LNCS 7951, pp. 582–593, 2013.

equipment must be tested and screened strictly according to the standards and specifications for space applications. One critical factor is loose particles in form of foreign materials being left or trapped within components or equipment during pre-seal handling and assembly process. These loose particles, such as tin granular, aluminum shot, and wire skin, can be dislodged and freed during launch or in zero gravity and may cause system failures. For example, they can be dislodged to bare bonding wires, short-circuiting the parts in usage. A number of reports have been filed that loose particles caused some serious faults and even catastrophic failures in space projects, leading to huge economic losses and social impacts. One of the well-known examples is the failure of the Delta Launch Vehicle Program in 1965 caused by a loose bit of wire within an electronic component [1]. This has raised the importance of screening devices for loose particles to ensure that all components and equipment for space applications must be particle free.

Following MIL-STD-883 standard [2, 3], Particle Impact Noise Detection (PIND) test was designed as a non-destructive test in order to determine the presence of the loose particles left inside components like relays, transistors and micro-switches [4, 5]. This is achieved by sensing the energy released when loose particles strike the interior, the wall or elements within the components. For details, a shaker induces a series of mechanical shocks and vibrations to free or dislodge particles clinging to the interior of the components while an acoustic emission transducer mounted on the shaker is used to detect the sound energy of the loose particle impact. The captured acoustic signals are fed into a bandpass filter to filter out the shaker frequency and background noises. The resulting signals are then sent either to an oscilloscope or an audio circuit with speakers or headphones. Operators determine the presence of the loose particle either by watching the waveforms or by listening to the audio signals. The whole structure of the PIND system is illustrated in Figure 1.

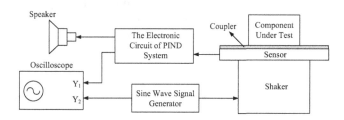

**Fig. 1.** Typical configuration of PIND test system

The PIND test system is a semi-automatic apparatus as the detection accuracy depends on both experience of the operator in identifying loose particle signals. Human factors like the operator's fatigue, negligence and lack of experience may easily lead to wrong conclusions [6]. Though the PIND test could detect the existence of large particles effectively, a drawback is its low accuracy for the detection of tiny particles. To address this problem, Fourier transform was employed for reducing noises like background noise, electrical noise and mechanical noise in the frequency domain [7]. However, the Fourier transform is only localized in frequency domain and could not capture the high frequency loose particle signals within a shorter time span. To capture the missed signals with shorter time span, a wavelet decomposition algorithm was used for loose particle detection to distinguish the loose particle signals with other noises in the wavelet domain. This method was proved useful for detecting loose particles left inside aerospace electrical relays [8].

Another key issue is that the type information about the loose particle material, such as tin, aluminium, and wire is of great importance to manufacturers for eliminating sources of loose particles [8,9]. Once the cause and source of loose particles are known, proper actions should be implemented to eliminate particles in the components. It is however difficult and complex for the identification of material type due to the nonlinear relationships between different types of materials and their signals with respect to different shapes and weights. To simplify the material identification, a linear method based on autoregressive (AR) model was proposed to extract features in order to classify the signals with different materials [9]. The results were not satisfactory due to the fact that nonlinear characteristics existing in loose particle signals were ignored. Further investigation using Learning Vector Quantization (LVQ) neural networks was used to improve the classification accuracy for loose particles left inside the aerospace power supply [6].

More recently, the detection objects have been extended to aerospace equipment, rather than components like electrical relays and semiconductors. The conventional PIND test is incapable of shaking the aerospace equipment which are large in volume and heavy in weight. Hence, a large shaker with higher drive power is used to free particles. To adopt to this change, four acoustic sensors were used for capturing the signal from loose particle vibration instead of one sensor for conventional tests. This results in redundant information from four sensors. The requirements to reduce redundant information as well as to retain useful information have to be dealt with. However, the previously mentioned two conventional methods do not consider reducing the redundant information, leading to a low classification accuracy. Further experimental results has confirmed that the conventional methods like AR model and LVQ neural networks could not meet the new challenges.

To improve the classification accuracy, this paper investigates the kernel Fisher discriminant for classification of loose particles left inside aerospace equipment. It has been shown that kernel Fisher discriminant is very competitive with many other classification methods like kernel principal component analysis, radial basis function networks and support vector machines [10,11]. The key issue for Fisher discriminant is to select a kernel like polynomial kernel, Gaussian kernel and wavelet kernel. This paper uses the wavelet kernel for loose particle classification, considering that the loose particle signal is a series of oscillation attenuation pulses which can be captured by wavelet. The kernel Fisher discriminant is a generalization of linear Fisher discriminant analysis. More specifically, the loose particle signals are first mapped into wavelet feature space using a wavelet kernel mapping and then linear discriminant analysis is subsequently performed in this kernel space.

A critical issue in the kernel Fisher discriminant is that its training time increases significantly as the number of training data increases. This may prevent its applications to large data sets. To address this problem, the kernel Fisher discriminant can be formulated as a linear-in-the-parameter model where all the training parameters can be listed and form a candidate term pool. Constructing thus a model is a linear regression problem with the aim to select a representative subset from all the candidates. Typical methods for such a problem include orthogonal least squares (OLS) methods and the fast recursive algorithm (FRA) [12] based forward and backward subset selection algorithms. In forward selection, significant model terms are added one by one until some criterion is met while in backward selection, insignificant model terms are deleted from the whole term pool one by one [13, 14]. However, these two methods are fast and greedy methods, they may not construct a parsimonious or compact model. To improve model compactness, the two-stage algorithm integrating forward selection and backward refinement using FRA method has been proposed [15]. In the first stage, an initial model is built by a novel fast forward selection approach [12]. Then, in the second stage, each previously selected term is reviewed and compared. If the significance of previously selected model term is less than any remaining term, it is replaced by the most significant one in the candidate term pool. This process stops when no term needs to be replaced. Further, a new two-stage algorithm using OLS which is computationally even more efficient than the previously proposed two-stage algorithm has been proposed recently [16].

The main purpose of this paper is to improve classification accuracy for loose particles left inside aerospace equipment. Wavelet Fisher discriminant is employed to solve this nonlinear classification problem. Further, the previously proposed two-stage algorithm is used to build a parsimonious Fisher model. It has been shown that the wavelet Fisher discriminant is able to produce better classification results compared with AR model and LVQ networks.

## 2    Loose Particle Impact Signals

The loose particle detection system is as shown in Figure 2. The whole system consists of a vibration generator, the aerospace equipment to be tested, and a data acquisition subsystem in which four acoustic sensors are used to sample the impact signals and then convert them into electronic signals. The bandwidth of the data acquisition subsystem is from $100kHz$ to $200kHz$ with the gain of $60dB$. Following Shannon sampling law, the sampling rate of $500kHz$ was employed.

The test objective is a rectangular aerospace equipment. Its volume is $150mm \times 120mm \times 100mm$ and the wall thickness is $2mm$. To carry out experiments, three different material particles including wire skin, aluminum shot and tin granular were prepared. Their weights range from $0.5mg$ to $6mg$, and their shapes are close to sphere.

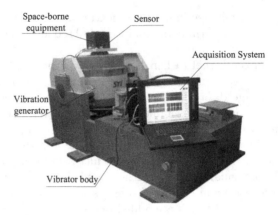

**Fig. 2.** The PIND test system

The sampling rate and sampling time are $500kHz$ and $5s$, respectively. There-fore, the total number of data for each sampling period is $2.5MB$. Typical data in a sampling period is shown in Figure 3 and more details are given in Fig-ure 4. The main characteristics of loose particle signals is that it consists of a series of oscillation attenuation pulses. In previous work, it has been shown that the wire particles have lower frequency distribution around $40Hz$ while the aluminium and tin particles have higher frequency around $80Hz$ and $100Hz$, respectively [6]. However, they distributions are overlapped, which gives rise wrong classification conclusions. Mapping these features into high dimensions can improve classification accuracy.

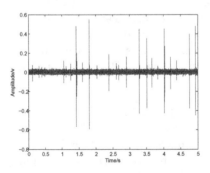

**Fig. 3.** The loose particle signals

## 3    Wavelet Fisher Discriminant

The Fisher discriminant is a typical binary classifier. Given data samples $\{\mathbf{x}_1, ..., \mathbf{x}_{n_1}, \mathbf{x}_{n_1+1}, ..., \mathbf{x}_n, \mathbf{x}_i \in \Re^N\}$ with $n_1$ samples for class $\mathbf{c}_1$ and $n_2$ samples for class $\mathbf{c}_2$, the total number of data samples is $n = n_1 + n_2$, and $N$ is the dimension of each data sample. The nonlinear Fisher discriminant first maps its input

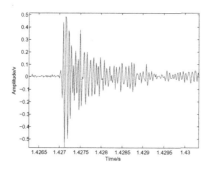

**Fig. 4.** Detailed information of the loose particle signals

samples into some high dimensional space using a nonlinear function column vector $\phi$ and then a linear discriminant analysis is carried out on the mapped feature space [11]. To implement this, two matrices have to be calculated, which are given by

$$\mathbf{S}_B^{\phi} = (\mathbf{m}_1^{\phi} - \mathbf{m}_2^{\phi})(\mathbf{m}_1^{\phi} - \mathbf{m}_2^{\phi})^T \tag{1}$$

and

$$\mathbf{S}_W^{\phi} = \sum_{i=1,2} \sum_{\mathbf{x} \in \mathbf{c}_i} (\phi(\mathbf{x}) - \mathbf{m}_i^{\phi})(\phi(\mathbf{x}) - \mathbf{m}_i^{\phi})^T \tag{2}$$

where

$$\mathbf{m}_i^{\phi} = \frac{1}{n_i} \sum_{\mathbf{x} \in \mathbf{c}_i} \phi(\mathbf{x}). \tag{3}$$

$\mathbf{S}_B^{\phi}$ and $\mathbf{S}_W^{\phi}$ are known as between-class scatter matrix and within-class matrix, respectively. $\mathbf{m}_i$ is the mean for each class in the nonlinear space. The nonlinear Fisher discriminant aims to maximize

$$J(\mathbf{\Omega}) = \frac{\mathbf{\Omega}^T \mathbf{S}_B^{\phi} \mathbf{\Omega}}{\mathbf{\Omega}^T \mathbf{S}_W^{\phi} \mathbf{\Omega}} \tag{4}$$

by finding a proper vector $\mathbf{\Omega} \in \Re^N$. This can be achieved by solving the following linear equation

$$\mathbf{\Omega} = \left(\mathbf{S}_W^{\phi}\right)^{-1} (\mathbf{m}_1^{\phi} - \mathbf{m}_2^{\phi}) \tag{5}$$

However, if the dimension of $\phi(\mathbf{x})$ is very high, (5) may be ill-conditioned or training can be computationally very expensive. In this paper, the number of data in each sampling period is $2.5MB$, hence, it is impractical to calculate this equation. To overcome this limitation, the nonlinear Fisher discriminant can be formulated as a least square problem using kernel method [11]. The kernel operation is defined as

$$k(\mathbf{x}_i, \mathbf{x}_j) = \phi(\mathbf{x}_i)^T \phi(\mathbf{x}_j) \tag{6}$$

For wavelet kernel, it is given by [17]

$$k(\mathbf{x}_i, \mathbf{x}_j) = \prod_{l=1}^{N} h\left(\frac{x_{il} - x_{jl}}{a}\right) \tag{7}$$

where $\mathbf{x}_i$ and $\mathbf{x}_j$ are two input vectors with the length of $N$. The $h$ is a wavelet function with dilation parameter $a$. In this paper, Mexican Hat wavelet function is used

$$h(\mathbf{x}) = (d - \|\mathbf{x}\|^2)e^{-\frac{\|\mathbf{x}\|^2}{2}} \tag{8}$$

where $d$ is the dimension of the input $\mathbf{x}$. Here, $d = 2$ as the input has two dimensions for two classes.

Wavelet Fisher discriminant is a kernelized version of linear discriminant analysis where it maps the input data into wavelet feature space then a linear discriminant analysis is performed in the mapped feature space. It can be also formulated as a least squares problem, which is shown as [11, 18]

$$\begin{bmatrix} 1 & k(\mathbf{x}_1, \mathbf{x}_1) & \ldots & k(\mathbf{x}_1, \mathbf{x}_n) \\ 1 & k(\mathbf{x}_2, \mathbf{x}_1) & \ldots & k(\mathbf{x}_2, \mathbf{x}_n) \\ \vdots & \vdots & \vdots & \vdots \\ 1 & k(\mathbf{x}_n, \mathbf{x}_1) & \ldots & k(\mathbf{x}_n, \mathbf{x}_n) \end{bmatrix} \begin{bmatrix} t_0 \\ \alpha_1 \\ \vdots \\ \alpha_n \end{bmatrix}$$
$$+ \begin{bmatrix} e_0 \\ e_1 \\ \vdots \\ e_n \end{bmatrix} = \begin{bmatrix} y_1 \\ \vdots \\ y_1 \\ y_2 \\ \vdots \\ y_2 \end{bmatrix} \tag{9}$$

The output $y_1$ and $y_2$ represent two classes. Re-write the equation (9) in matrix form

$$\mathbf{P}\mathbf{\Theta} + \mathbf{e} = \mathbf{y} \tag{10}$$

where $\mathbf{P}$ can be obtained explicitly using the kernel operation. The $\mathbf{\Theta} = \{t_0, \alpha_1, ..., \alpha_n\}$ is unknown vector with classification threshold $t_0$ and discriminant parameters $\{\alpha_1, ..., \alpha_n\}$. Its least squares solution is

$$\hat{\mathbf{\Theta}} = (\mathbf{P}^T\mathbf{P})^{-1}\mathbf{P}^T\mathbf{y} \tag{11}$$

However, as the column terms in $\mathbf{P}$ are usually redundant and correlated, they may result in ill-conditioned equations or a non-sparse classifier. A useful solution is to select a subset $\mathbf{P}_s$ of $\mathbf{P}$ instead of using the whole matrix $\mathbf{P}$, which can improve model parsimony and classification accuracy. For details, the subset is given by

$$\mathbf{P}_s = \begin{bmatrix} 1 & k(\mathbf{x}_1, \mathbf{x}_{i_1}) & \ldots & k(\mathbf{x}_1, \mathbf{x}_{i_s}) \\ 1 & k(\mathbf{x}_2, \mathbf{x}_{i_1}) & \ldots & k(\mathbf{x}_2, \mathbf{x}_{i_s}) \\ \vdots & \vdots & \vdots & \vdots \\ 1 & k(\mathbf{x}_n, \mathbf{x}_{i_1}) & \ldots & k(\mathbf{x}_n, \mathbf{x}_{i_s}) \end{bmatrix}. \tag{12}$$

Forward selection is a widely used algorithm for selecting a typical subset of $\mathbf{P}$. However, it is not optimal. To improve its compactness, the previously proposed two-stage method combining both the forward selection and backward refinement is employed.

Anther problem is that the wavelet Fisher discriminant is a binary classifier and it can not directly deal with multi-class classification problems. For loose particle classification in this paper, three classes of materials, namely the tin, aluminium and wire need to be distinguished. To achieve this, the binary tree approach is used in this paper. For details, this paper first performs classification of two coarse types, namely the metallic material (tin and aluminium) and non-metallic material (wire), and then it classifies tin and aluminium classes. In other words, two binary classifiers are used.

## 4   Two-Stage Orthogonal Least Squares

In this paper, the previously proposed two-stage orthogonal least squares (OLS) method is used to build a sparse wavelet Fisher discriminant model. The first stage is equivalent to the forward selection where the model terms are included into an initial model one by one, leading to a sub-model $\mathbf{P}_s = \{\mathbf{p}_{i_1}, ..., \mathbf{p}_{i_s}\}$. The second stage refines the initial model by replacing those insignificant terms compared with the terms left in the term pool. Both first and second stages perform a series of orthogonal decomposition of $\mathbf{P}$, which is given by

$$\mathbf{P} = \mathbf{WA} \tag{13}$$

where the $\mathbf{W}$ is an orthogonal matrix and $\mathbf{A}$ is an upper triangular matrix.

After this decomposition, the Fisher model can be expressed as

$$\mathbf{y} = (\mathbf{PA}^{-1})(\mathbf{A\Theta}) + \mathbf{e} = \mathbf{Wg} + \mathbf{e} \tag{14}$$

where $\mathbf{g} = [g_1, g_2, \ldots, g_M]^T$ is the orthogonal weight vector. So the original model weight vector $\mathbf{g}$ can be calculated by [19]

$$\mathbf{g} = (\mathbf{W}^T\mathbf{W})^{-1}\mathbf{W}^T\mathbf{y}. \tag{15}$$

The error is

$$\mathbf{e} = \mathbf{y} - \mathbf{Wg} = \mathbf{y} - \mathbf{W}(\mathbf{W}^T\mathbf{W})^{-1}\mathbf{W}^T\mathbf{y}. \tag{16}$$

The model parameters $\mathbf{\Theta}$ can be computed using backward substitution

$$\left.\begin{aligned} \theta_M &= g_M \\ \theta_i &= g_i - \sum_{k=i+1}^{M} \alpha_{ik}\theta_k, i = M-1, \ldots, 1 \end{aligned}\right\}. \tag{17}$$

The orthogonal decomposition can be carried out by Gram-Schmidt (GS) method. The GS procedure computes one column of $\mathbf{A}$ at a time and factorizes $\mathbf{P}$ as follows:

$$\left.\begin{aligned} \mathbf{w}_1 &= \mathbf{p}_1 \\ \alpha_{ik} &= \frac{< \mathbf{w}_i, \mathbf{p}_k >}{< \mathbf{w}_i, \mathbf{w}_i >}, \ 1 \le i < k \\ \mathbf{w}_k &= \mathbf{p}_k - \sum_{i=1}^{k-1} \alpha_{ik}\mathbf{w}_i \end{aligned}\right\} k = 2, \ldots, M \tag{18}$$

and
$$g_i = \frac{< \mathbf{w}_i, \mathbf{y} >}{< \mathbf{w}_i, \mathbf{w}_i >}, i = 1, \ldots, M. \tag{19}$$

In additional to the orthogonal decomposition, a criterion is used for model term selection. In this paper, the error reduction ratio (ERR) is used for measuring the term contribution. The ERR due to $\mathbf{w}_i$ is defined as [19]

$$\begin{aligned} [err]_i &= g_i^2 \frac{< \mathbf{w}_i, \mathbf{w}_i >}{< \mathbf{y}, \mathbf{y} >} \\ &= g_i \frac{< \mathbf{w}_i, \mathbf{y} >}{< \mathbf{y}, \mathbf{y} >}. \end{aligned} \tag{20}$$

Given the ERR criterion, the two-stage OLS is shown as follows:

(1)First stage - forward selection:

The forward selection computes each term contribution and then selects the most significant one at each step. More specifically,

At the 1st step, for $1 \leq i \leq M$, compute

$$\left. \begin{aligned} \mathbf{w}_1^{(i)} &= \mathbf{p}_i \\ g_1^{(i)} &= \frac{< \mathbf{w}_1^{(i)}, \mathbf{y} >}{< \mathbf{w}_1^{(i)}, \mathbf{w}_1^{(i)} >} \\ [err]_1^{(i)} &= g_1^{(i)} \frac{< \mathbf{w}_1^{(i)}, \mathbf{y} >}{< \mathbf{y}, \mathbf{y} >} \end{aligned} \right\}. \tag{21}$$

Find the largest ERR

$$[err]_1^{(i_1)} = max\{[err]_1^{(i)}, \ 1 \leq i \leq M\} \tag{22}$$

and select the model term associated with the number $i_1$ as the first important term

$$\mathbf{w}_1 = \mathbf{w}_1^{(i_1)} = \mathbf{p}_{i_1} \tag{23}$$

At the $k$th step, for $1 \leq i \leq M, \ i \neq i_1, \ldots, i \neq i_{k-1}$, calculate

$$\left. \begin{aligned} \alpha_{jk}^{(i)} &= \frac{< \mathbf{w}_j, \mathbf{p}_i >}{< \mathbf{w}_j, \mathbf{w}_j >}, 1 \leq j < k \\ \mathbf{w}_k^{(i)} &= \mathbf{p}_i - \sum_{j=1}^{k-1} \alpha_{jk}^{(i)} \mathbf{w}_j \\ g_k^{(i)} &= \frac{< \mathbf{w}_k^{(i)}, \mathbf{y} >}{< \mathbf{w}_k^{(i)}, \mathbf{w}_k^{(i)} >} \\ [err]_k^{(i)} &= g_k^{(i)} \frac{< \mathbf{w}_k^{(i)}, \mathbf{y} >}{< \mathbf{y}, \mathbf{y} >} \end{aligned} \right\}. \tag{24}$$

Find the largest ERR

$$[err]_k^{(i_k)} = max\{[err]_k^{(i)}, \ i \neq i_1, \ldots, i_{k-1}\} \tag{25}$$

and select the term associated with the number $i_k$ as the $k$the term

$$\mathbf{w}_k = \mathbf{w}_k^{(i_k)} = \mathbf{p}_{i_k} - \sum_{j=1}^{k-1} \alpha_{jk}^{i_k} \mathbf{w}_j. \tag{26}$$

The procedure is terminated at the $M_s$th step when

$$1 - \sum_{j=1}^{M_s} [err]_j < \rho \tag{27}$$

where $0 < \rho < 1$ is a preset tolerance. The fist stage builds an initial model with regressors $\{\mathbf{p}_{r_1}, ..., \mathbf{p}_{r_i}, ..., \mathbf{p}_{r_{M_s}}\}$, where $r_i$ is the index in the original candidate pool. For computational convenience for the second stage, the indexes of terms in the initial model are renamed as $\{\mathbf{p}_1, ..., \mathbf{p}_i, ..., \mathbf{p}_{M_s}\}$ and others remaining in the candidate pool are renames as $\{\mathbf{p}_{M_s+1}, ..., \mathbf{p}_i, ..., \mathbf{p}_M\}$.

(2)Second stage - backward model refinement:

The second stage refines the initial model by reselecting the terms. As the contributions of the latter selected terms are less than whose of the former ones in terms of ERR, the refinement begins from backward direction. Swap the $\mathbf{w}_n, n = M_s - 1, ..., 1$ to the last column. To be specific,

$$\left. \begin{array}{l} \mathbf{w}'_n = \mathbf{p}_{n+1} - \displaystyle\sum_{i=1}^{n-1} \alpha_{i(n+1)} \mathbf{w}_i \\[3mm] \mathbf{w}'_j = \mathbf{p}_{j+1} - \displaystyle\sum_{i=1}^{n-1} \alpha_{i(j+1)} \mathbf{w}_i - \displaystyle\sum_{i=n}^{j-1} \frac{<\mathbf{w}'_i, \mathbf{p}_{j+1}>}{<\mathbf{w}'_i, \mathbf{w}'_i>} \\[3mm] \cdot \mathbf{w}'_i, \ j = n+1, ..., M_s - 1 \\[3mm] \mathbf{w}'_{M_s} = \mathbf{p}_n - \displaystyle\sum_{i=1}^{n-1} \alpha_{in} \mathbf{w}_i - \displaystyle\sum_{i=n}^{M_s-1} \frac{<\mathbf{w}'_i, \mathbf{p}_n>}{<\mathbf{w}'_i, \mathbf{w}'_i>} \mathbf{w}'_i \end{array} \right\}. \tag{28}$$

Re-select the last column, for $M_s \le i \le M$, calculate

$$\left. \begin{array}{l} \mathbf{w}_{M_s}^{(i)}{}' = \mathbf{p}_i - \displaystyle\sum_{j=1}^{n-1} \alpha_{ji} \mathbf{w}_j - \displaystyle\sum_{j=n}^{M_s-1} \frac{<\mathbf{w}'_j, \mathbf{p}_i>}{<\mathbf{w}'_j, \mathbf{w}'_j>} \mathbf{w}'_j \\[4mm] g_{M_s}^{(i)}{}' = \dfrac{<\mathbf{w}_{M_s}^{(i)}{}', \mathbf{y}>}{<\mathbf{w}_{M_s}^{(i)}{}', \mathbf{w}_{M_s}^{(i)}{}'>} \\[4mm] [err]_{M_s}^{(i)}{}' = g_{M_s}^{(i)}{}' \dfrac{<\mathbf{w}_{M_s}^{(i)}{}', \mathbf{y}>}{<\mathbf{y}, \mathbf{y}>} \end{array} \right\}. \tag{29}$$

Find

$$[err]_{M_s}^{(i_{M_s})}{}' = max\{[err]_{M_s}^{(i)}{}', \ M_s \le i \le M\}. \tag{30}$$

and

$$\mathbf{w}'_{M_s} = \mathbf{w}^{(i_{M_s})'}_{M_s} \tag{31}$$

The backward procedure refines all the terms selected in the first stage. If some of them are replaced, the new model can be refined again using the same procedure and this stops until no insignificant term can be further replaced.

## 5    Classification Results

In this study, the collected data was split into training and testing data for classifying three kinds of loose particles. For each type of loose particles 250 samples were used, with first 150 for training and the remaining for testing. The weights of materials range from $0.5mg$ to $6mg$. In the training process, two Fisher models were built to classify three materials. The first model is to classify the wire from others and the second one is to classify the aluminium from tin. Both the dilation parameter $a = 1$ in the wavelet and the training tolerance $\alpha = 0.02$ were determined by trial-and-error. In order to illustrate the effectiveness of wavelet Fisher discriminant, the AR classifier and LVQ neural networks were also built for classifications. The results are shown in Table 1 and they illustrate that the wavelet Fisher discriminant can give higher classification accuracy than the conventional AR model and LVQ neural networks.

**Table 1.** Comparison of classification results

| Loose particle | Fisher | AR | LVQ |
|----------------|--------|-----|-----|
| Wire | 90% | 75% | 85% |
| Aluminum | 86% | 62% | 81% |
| Tin | 85% | 60% | 80% |

## 6    Conclusion

This paper has proposed to use the wavelet Fisher discriminant for loose particle classification. First, the wavelet Fisher discriminant has been formulated as a linear-in-the-parameters model which can be regarded as a least squares problem. Then, the two-stage subset selection method has been employed to build a sparse wavelet Fisher model, which improves the model compactness and performance. Finally, the classification results have shown that the wavelet Fisher discriminant has outperformed the conventional AR model and LVQ neural networks.

## References

1. Baily, C., Corporation, D.: Particle impact noise detection as a non-destructive test for determining integrity of electronic components. Acoustic Emission Applications in the Electronic Industry (1986)

2. Zhang, H., Wang, S., Zhai, G.: Research on particle impact noise detection standard. IEEE Transactions on Aerospace and Electronic Systems 44(2), 808–814 (2008)
3. MIL-STD-883G. Department of Defense Test Method Standard Microcircuits (2006)
4. Slyhous, S.: Particle impact noise detection (pind) combines vibration, shock, and acoustics for determining reliability of electronic components. Spectral Dynamics, 1–13 (1990)
5. Adolphsen, J., Krgdis, W., Timmins, A.: A survey of particle contamination in electric devices. Nato Science Series Sub Series III Computer and Systems Sciences (1976)
6. Wang, S., Chen, R., Zhang, L., Wang, S.: Detection and material identification of loose particles inside the aerospace power supply via stochastic resonance and lvq network. Transactions of the Institute of Measurement and Control 34(8), 947–955 (2012)
7. Dong, J., Hong, W., Zhao, X.: A research on the automatic test system for the remainder particle signals in spacecraft relays. Journal of Electronic Measurment and Instrument 12(4), 41–46 (1998)
8. Wang, S., Gao, H., Zhai, G.: Research on auto-detection for remainder particles of aerospace relay based on wavelet analysis. Chinese Journal of Aeronautics 20(1), 74–80 (2007)
9. Wang, S., Gao, H., Zhai, G.: Research on feature extraction of remnant particles of aerospace relays. Chinese Journal of Aeronautics 20(6), 253–259 (2007)
10. Mika, S., Ratsch, G., Weston, J., Scholkopf, B., Mullers, K.: Fisher discriminant analysis with kernels. In: Proceedings of the 1999 IEEE Signal Processing Society Workshop on Neural Networks for Signal Processing IX, pp. 41–48 (1999)
11. Billings, S., Lee, K.: Nonlinear fisher discriminant analysis using a minimum squared error cost function and the orthogonal least squares algorithm. Neural Networks 15(2), 263–270 (2002)
12. Li, K., Peng, J., Irwin, G.W.: A fast nonlinear model identification method. IEEE Transcations on Automatic Control 8(50), 1211–1216 (2005)
13. Gomm, J., Yu, D.: Selecting radial basis function network centers with recursive orthogonal least squares training. IEEE Transactions on Neural Networks 11(2), 306–314 (2000)
14. Miller, A.: Subset Selection in Regression, 2nd edn. CRC Press (2002)
15. Li, K., Peng, J., Bai, E.: A two-stage algorithm for identification of non-linear dynamic systems. Automatica 42(7), 1189–1197 (2006)
16. Zhang, L., Li, K., Bai, E.W., Wang, S.J.: A novel two-stage classical gram-schmidt algorithm for wavelet network construction. In: 16th IFAC Symposium on System Identification, vol. 16(1), pp. 644–649 (2012)
17. Zhang, L., Zhou, W., Jiao, L.: Wavelet support vector machine. IEEE Transactions on Systems, Man, and Cybernetics, Part B: Cybernetics 34(1), 34–39 (2004)
18. Deng, J., Li, K., Irwin, G., Harrison, R.: A fast automatic construction algorithm for kernel fisher discriminant classifiers. In: 2010 49th IEEE Conference on Decision and Control (CDC), pp. 2825–2830 (2010)
19. Chen, S., Billing, S., Luo, W.: Orthogonal least squares methods and their application to non-linear system identification. International Journal of Control 50(5), 1873–1896 (1989)

# $L_1$ Graph Based on Sparse Coding for Feature Selection

Jin Xu[1], Guang Yang[2], Hong Man[2], and Haibo He[3]

[1] Operations and Information Management Department
University of Pennsylvania, PA, USA
jinxu@wharton.upenn.edu
[2] Department of Electrical and Computer Engineering,
Stevens Institute of Technology, NJ, USA
hman@stevens.edu
[3] Department of Electrical, Computer, and Biomedical Engineering
University of Rhode Island, RI, USA
he@ele.uri.edu

**Abstract.** In machine learning and pattern recognition, feature selection has been a very active topic in the literature. Unsupervised feature selection is challenging due to the lack of label which would supply the categorical information. How to define an appropriate metric is the key for feature selection. In this paper, we propose a "filter" method for unsupervised feature selection, which is based on the geometry properties of $\ell_1$ graph. $\ell_1$ graph is constructed through sparse coding. The graph establishes the relations of feature subspaces and the quality of features is evaluated by features' local preserving ability. We compare our method with classic unsupervised feature selection methods (Laplacian score and Pearson correlation) and supervised method (Fisher score) on benchmark data sets. The classification results based on support vector machine, k-nearest neighbors and multi-layer feed-forward networks demonstrate the efficiency and effectiveness of our method.

**Keywords:** Sparse coding, Unsupervised learning, Neural network, Feature selection.

## 1 Introduction

Feature selection is an important technique in dealing with high-dimensional data. It is reported that 1.8 zettabytes (or 1.8 trillion gigabytes) of data has been created in the year 2011. How to manage the huge data to avoid the "curse of dimensionality" is essential in real applications. Feature selection, as a powerful tool of dimension reduction, has been successfully applied in pattern recognition [1] and computer version [2]. Functionally, feature selection can be divided into three groups: filter model, wrapper model and embedded model. Filter is the most popular model in recent research, as it has low computational cost and is robust in theoretic analysis. Depends on the class labels, feature selection can be implemented in supervised fashion or unsupervised fashion. Most existing filter models are in supervised fashion. In real world applications, the class labels are always scarce [3]. It is meaningful to design a filter feature selection method in unsupervised fashion.

C. Guo, Z.-G. Hou, and Z. Zeng (Eds.): ISNN 2013, Part I, LNCS 7951, pp. 594–601, 2013.

Sparse coding (representation) has been studied widely in recent literature. It reconstructs a signal (data) through a linear combination of a minimum set of atom vectors from a dictionary. In detail, a signal (data) $\mathbf{y} \in \mathbb{R}^m$ can be sparse represented through $\mathbf{y} = \mathbf{D}\mathbf{x}$, the dictionary is $\mathbf{D} \in \mathbb{R}^{m \times d}$. The correspondent coefficient $\mathbf{x} \in \mathbb{R}^d$ is sparse (the dominating elements are zeros). Spare coding has many effective applications in real-world data, such as image denoising [4], information imputation [5], dictionary learning [6] and blind source separation [7].

In this paper, we utilize the relations, established by sparse coding between the data (signals) and dictionary atoms, to build $\ell_1$ graphs. The graph has the properties in local preserving ability. We evaluate these properties to rank features and establish a new unsupervised filter model for feature selection.

The main contributions of the paper are summarized as follows:

- A graph is established through $\ell_1$-Norm Regularization. The linear relations between the signal and the dictionary atoms are shown on the graph.
- The features' local preserving ability is evaluated through spectral graph theory [8]. The unsupervised filter model is established based on the ability to perform feature selection.
- The proposed method is applied to UCI [9] benchmark data sets (binary-category and multiple-category). A 2-D visualization case study is carried out and compared with classic filter feature selection methods (Fisher score [10], Laplacian score [11] and Pearson correlation [10]). Intensive experiments of feature based classifications are conducted to demonstrate the efficiency and effectiveness of our method.

## 2  Feature Score Based on $\ell_1$ Graph

A data set $\mathbf{Y} = \{\mathbf{y}_1, \mathbf{y}_2, \cdots, \mathbf{y}_i, \cdots, \mathbf{y}_n\} \in \mathbb{R}^{n \times m}$ is assumed. Our proposed method utilizes the property of *self-characterization* in the data sets. In detail, a data (signal) can be represented by other data from the same data set through

$$\mathbf{y}_i = \mathbf{Y}\mathbf{x}_i, \quad x_{ii} = 0 \tag{1}$$

where $\mathbf{x}_i \triangleq [x_{i1} \ x_{i2} \cdots x_{in}]$ and constraint $x_{ii} = 0$ avoids the trivial solution of characterizing a data as a linear combination of itself. This formula is naturally transferred to sparse coding when we want to choose as less as possible data to represent $\mathbf{y}_i$. Therefore we assume the dictionary of sparse coding is the whole data set. Then the constrain of $\ell_0$ norm is:

$$min\|\mathbf{x}_i\|_0, \quad s.t. \quad \mathbf{y}_i = \mathbf{Y}\mathbf{x}_i \tag{2}$$

It is known that the solving of above equation is NP hard. With the relative work of Restricted Isometry Property (RIP) [12], the $\ell_0$ norm can be transferred to $\ell_1$ form and solved with $\ell_1$-regularized least squares method [13]:

$$\widehat{\mathbf{x}_i} = arg\min\{\|\mathbf{y}_i - \mathbf{Y}\mathbf{x}_i\|_2^2 + \lambda\|\mathbf{x}_i\|_1\} \tag{3}$$

We summarize the sparse coding method for all data with the matrix form through:

$$min\|\mathbf{X}\|_1, \quad s.t. \quad \mathbf{Y} = \mathbf{Y}\mathbf{X}, \quad diag(\mathbf{X}) = 0 \tag{4}$$

Inspired by the work of sparse subspace clustering [14], the similarity matrix of $\ell_1$ graph can be defined as:

$$\mathbf{W} = |\mathbf{X}| + |\mathbf{X}^T|, \quad diag(\mathbf{W}) = 0 \tag{5}$$

it means a node (signal) $i$ is connected with node $j$ by an edge with the weight $|x_{ij}| + |x_{ji}|$. Based on the graph established above, our proposed feature score $S$ based on the spectral graph theory [8] is computed as:

1. First, $\ell_1$ graph $\mathbf{G}$ ($\mathbf{G}_{ij} = \mathbf{W}_{ij}$) is built based on similarity matrix (5) with nodes (signals) ($\mathbf{Y} = \{\mathbf{y}_1, \mathbf{y}_2, \cdots, \mathbf{y}_i, \cdots, \mathbf{y}_n\}$).

2. For each feature $F_z$, the feature sets are $F_z = \mathbf{f_z}$, then $S_z$ can be computed as

$$S_z = \frac{\tilde{\mathbf{f}}_z^T L \tilde{\mathbf{f}}_z}{\tilde{\mathbf{f}}_z^T Q \tilde{\mathbf{f}}_z} \tag{6}$$

where $Q = diag(G1), 1 = [1, \cdots, 1]^T, L = Q - G$, and $\tilde{\mathbf{f}}_z$ is a classic normalization through:

$$\tilde{\mathbf{f}}_z = \mathbf{f}_z - \frac{\mathbf{f}_z^T Q 1}{1^T Q 1} 1 \tag{7}$$

The step 2 is based on the local property of each feature, $G_{ij}$ evaluates the similarity between the $i$-th and $j$-th data (nodes). In detail, when two nodes have heavily weighted edge, the good feature should have close value between these two nodes. The heuristic criterion [11] for selecting features is to minimize the function:

$$S_z = \frac{\sum_{ij}(f_{zi} - f_{zj})^2 G_{ij}}{Var(\mathbf{f}_z)} \tag{8}$$

where $Var(\mathbf{f}_z)$ is the variance for $z$-th feature, and $f_{zi}, f_{zj}$ are $z$-th feature value for node $i, j$. Based on some simple calculation, we could obtain:

$$\sum_{ij}(f_{zi} - f_{zj})^2 G_{ij} = 2\sum_{ij} f_{zi}^2 G_{ij} - 2\sum_{ij} f_{zi} G_{ij} f_{zj}$$
$$= 2\mathbf{f}_z^T Q \mathbf{f}_z - 2\mathbf{f}_z^T G \mathbf{f}_z = 2\mathbf{f}_z^T L \mathbf{f}_z \tag{9}$$

By the spectral graph theory, the $Var(\mathbf{f}_z)$ is calculated as:

$$Var(\mathbf{f}_z) = \sum_i \tilde{f}_{zi}^2 Q_{ii} = \tilde{\mathbf{f}}_z^T Q \tilde{\mathbf{f}}_z \tag{10}$$

Also, it is easy to get

$$\tilde{\mathbf{f}}_z^T L \tilde{\mathbf{f}}_z = \mathbf{f}_z^T L \mathbf{f}_z \tag{11}$$

Based on equations (8)(9)(10)(11), the selection criteria (6) is interpreted. When all the features are assigned the score, the feature selection is carried out based on the score ranking.

# 3   Experimental Evaluation

In this section, the empirical experiments are conducted on ten data sets from UCI Repository [9] to demonstrate the effectiveness of our method. There are six binary data sets and four multiple categorical data sets. The detail information for the data sets are listed in Table 1

**Table 1.** UCI Data sets

| Name | Features | Training size | Testing size | Class |
|---|---|---|---|---|
| car | 6 | 864 | 864 | 2 |
| pima | 8 | 384 | 384 | 2 |
| tic-tac-toe | 9 | 479 | 479 | 2 |
| yeast | 7 | 742 | 742 | 2 |
| hill valley | 100 | 303 | 303 | 2 |
| vehicle silhouettes | 18 | 473 | 473 | 2 |
| wine | 13 | 89 | 89 | 3 |
| image segmentation | 19 | 1165 | 1165 | 7 |
| wine quality | 11 | 2449 | 2449 | 6 |
| libras | 90 | 180 | 180 | 15 |

In the experiment, each data set is randomly separated into two equal parts. One part is used in training and the rest part is used in testing. We used the training data to build the model for feature selection. Five filter feature selection models are utilized for comparison : our proposed unsupervised Filter model via $L_1$ graph, Pearson correlation (supervised and unsupervised fashion), Fisher score (supervised filter model) and Laplacian score (unsupervised filter model). We use FL1, PCS, PCU, Fisher and Lap as abbreviations to denote these 5 methods in the experiments.

## 3.1   Case Study of 2-D Visualization

A simple case study for data wine is shown. Totally, there are 13 features for data wine, such as "Alcohol", "Magnesium" and "Proline". We use five filter methods based on training data (with size 89) and apply on the testing data. Each method chooses two features for 2-D visualization on testing data. The results are shown in Fig.1. Two features are selected by 5 different methods and plotted in each subfigure. It can be observed that the feature "Flavanoids" and feature "Color intensity" selected by FL1 method are crucial for discrimination.

## 3.2   Feature Based Classification

When selected features are more than two, we used the feature based classification to compare the feature selection methods. The experiment is conducted five times and mean outputs are obtained. The target selected features size is from one to around 80%

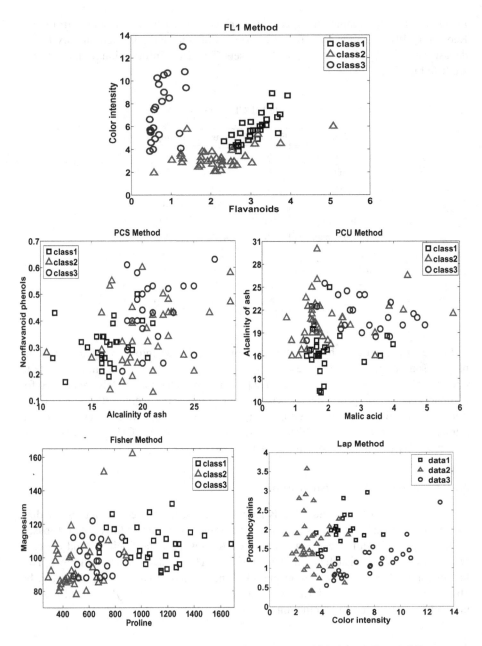

**Fig. 1.** Data Wine plotted in 2-D with selected features. 5 methods have selected different two features.

**Table 2.** Mean accuracy in low dimension (in %)

| Data set | LibSVM | | | | | 5-NN | | | | | NeuralNet | | | | |
|---|---|---|---|---|---|---|---|---|---|---|---|---|---|---|---|
| | FL1 | PCS | PCU | Fisher | Lap | FL1 | PCS | PCU | Fisher | Lap | FL1 | PCS | PCU | Fisher | Lap |
| car | **73.7** | 69.1 | 73.0 | 70.3 | 70.8 | **70.7** | 65.1 | 70.7 | 69.6 | 69.6 | **74.9** | 68.9 | 73.6 | 70.4 | 71.4 |
| pima | **71.6** | 64.9 | 65.3 | 67.8 | 67.3 | **68.6** | 61.4 | 64.0 | 64.6 | 65.8 | **71.3** | 64.6 | 66.8 | 67.5 | 67.3 |
| tic-tac-toe | **63.4** | 62.4 | 61.0 | 59.7 | 59.7 | 62.9 | 61.0 | 66.6 | **68.9** | 67.5 | 65.4 | 63.9 | 68.0 | **71.6** | 69.3 |
| yeast | 72.7 | 71.1 | 72.1 | **72.8** | 71.0 | **70.7** | 66.7 | 69.6 | 67.4 | 67.8 | 73.4 | 71.2 | **73.8** | 73.2 | 71.8 |
| hill valley | **55.9** | 50.8 | 47.9 | 50.9 | 54.8 | **52.3** | 50.8 | 50.6 | 49.4 | 52.1 | **60.5** | 54.6 | 50.3 | 53.9 | 58.8 |
| vehicle | 73.6 | 74.8 | 74.5 | 74.4 | **75.0** | 77.2 | 78.2 | 77.8 | **78.6** | 77.6 | 78.8 | 78.7 | 78.4 | 78.3 | **79.4** |
| wine | 82.6 | 69.8 | 66.6 | 75.1 | 75.4 | **83.6** | 69.3 | 68.7 | 79.4 | 75.5 | **86.7** | 72.8 | 71.0 | 79.4 | 77.7 |
| image seg | 91.1 | 91.3 | 90.8 | 90.1 | **91.5** | 92.8 | **94.8** | 92.0 | 91.2 | 93.7 | 93.4 | **95.4** | 92.4 | 91.6 | 94.1 |
| wine quality | 46.7 | **55.7** | 51.3 | 48.9 | 45.7 | 48.8 | **52.9** | 49.3 | 49.7 | 47.4 | 51.7 | **55.8** | 51.3 | 51.2 | 48.9 |
| libras | **54.0** | 40.6 | 35.8 | 35.9 | 48.4 | **57.5** | 43.2 | 43.4 | 44.7 | 52.5 | **74.4** | 56.6 | 60.0 | 61.1 | 68.2 |
| Mean Wins | **6** | 1 | 0 | 1 | 2 | **6** | 2 | 0 | 2 | 0 | **5** | 2 | 1 | 1 | 1 |

of whole feature size to give comprehensive comparison. In order to show the classification performances, we use three classic classifiers: k nearest neighbor ($k = 5$ in the experiment), LibSVM and multi-layer feed-forward networks. For briefness, we only plot two data sets (one binary-category data set and one multi-category data set) results. We abbreviate the classifiers as LibSVM, 5-NN and NeuralNet in the figures.

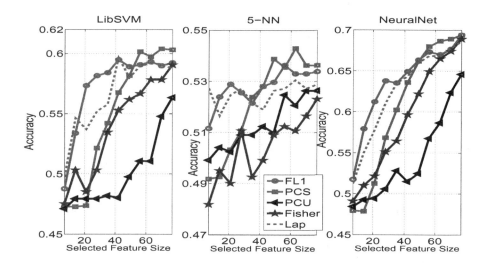

**Fig. 2.** Comparison of feature based classification accuracies for data hill valley

Fig.2 shows the comparison results for data "hill valley". When the selected features size is smaller than 40, FL1 results rank first among all the competitors with all three classifiers. And FL1 outputs rank second when the selected features size is larger than 40. In the case of multi-category data "libras" in Fig.3, the performances of FL1 rank first in most cases. It is important to note that the PCS and Fisher are supervised feature selection methods. And the performance of FL1 is competitive to PCS and Fisher in the most feature sizes.

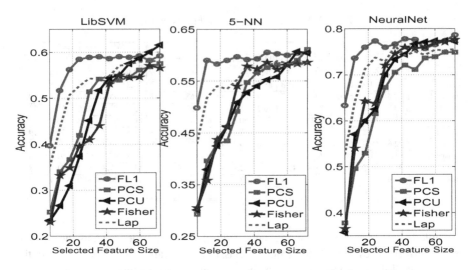

**Fig. 3.** Comparison of feature based classification accuracies for data libras

In order to give comprehensive comparison of different feature selection methods on multiple data sets, the mean accuracy in low dimension (from feature size one to around 40% of whole feature sizes) are calculated based on each data set and each classifier. Table 2 shows the detail mean outputs and the comparison results. The highest accuracy for each classifier is highlighted. It can be observed that FL1 can win 6, 6 and 5 times of 10 data sets with LibSVM, 5-NN and NeuralNet respectively.

## 4   Conclusion

We present a new filter feature selection method based on sparse coding in unsupervised fashion. Our approach aims to use $\ell_1$ graph to evaluate the local property for individual feature. Experimental comparisons with related filter methods have demonstrated that our method is effective in terms of visualization and classification. Future research work will focus on increasing dimension of the data set, statistical analysis among different filter models and improving the theoretical framework of $\ell_1$ graph for local structure.

**Acknowledgement.** This work was supported in part by the National Science Foundation (NSF) under Grant ECCS 1053717, Army Research Office (ARO) under Grant W911NF-12-1-0378, Defense Advanced Research Projects Agency (DARPA) under Grant FA8650-11-1-7148 and Grant FA8650-11-1-7152. This work is also partially supported by the Rhode Island NASA EPSCoR Research Infrastructure and Development (EPSCoR-RID) via NASA EPSCoR grant number NNX11AR21A (through a sub-award from Brown University).

## References

1. Mitra, P., Murthy, C.A., Pal, S.K.: Unsupervised feature selection using feature similarity. IEEE Transactions on Pattern Analysis and Machine Intelligence 24, 301–312 (2002)

2. Mutch, J., Lowe, D.G.: Multiclass object recognition with sparse, localized features. In: IEEE Computer Society Conference on Computer Vision and Pattern Recognition, pp. 11–18 (2006)
3. Xu, J., He, H., Man, H.: DCPE Co-Training for Classification. Neurocomputing 86, 75–85 (2012)
4. Elad, M., Aharon, M.: Image denoising via sparse and redundant representations over learned dictionaries. IEEE Transactions on Image Processing 15(12), 3736–3745 (2006)
5. Xu, J., Yin, Y., Man, H., He, H.: Feature selection based on sparse imputation. In: The 2012 International Joint Conference on Neural Networks (IJCNN), pp. 1–7 (2012)
6. Xu, J., Man, H.: Dictionary Learning Based on Laplacian Score in Sparse Coding. In: Perner, P. (ed.) MLDM 2011. LNCS, vol. 6871, pp. 253–264. Springer, Heidelberg (2011)
7. Li, Y., Amari, S., Cichocki, A., Ho, D.W.C., Xie, S.: Underdetermined blind source separation based on sparse representation. IEEE Transactions on Signal Processing 54(2), 423–437 (2006)
8. Chung, F.R.K.: Spectral Graph Theory. Regional Conference Series in Mathematics, vol. 92 (1997)
9. Frank, A., Asuncion, A.: UCI Machine Learning Repository. University of California, School of Information and Computer Science, Irvine (2010),
   http://archive.ics.uci.edu/ml
10. Guyon, I., Elisseeff, A.: An Introduction to Variable and Feature Selection. Journal of Machine Learning Research 3, 1157–1182 (2003)
11. He, X., Cai, D., Niyogi, P.: Laplacian score for feature selection. In: Proc. Advances in the Neural Information Processing Systems 18, Vancouver, Canada (2005)
12. Candes, E., Tao, T.: Near optimal signal recovery from random projections and universal encoding strategies. IEEE Trans. Inform. Theory 52, 5406–5425 (2006)
13. Kim, S., Koh, K., Lustig, M., Boyd, S., Gorinevsky, D.: An interior-point method for largescale l1-regularized least squares. IEEE Journal of Selected Topics in Signal Processing 1(4), 606–617 (2007)
14. Elhamifar, E., Vidal, R.: Sparse Subspace Clustering. In: IEEE International Conference on Computer Vision and Pattern Recognition (2009)

# An Image Segmentation Method for Maize Disease Based on IGA-PCNN

Wen Changji[1,2] and Yu Helong[1,*]

[1] College of Computer Science and Technology, Jilin Agricultural University,
Changchun, China
chagou2006@163.com
[2] Key Laboratory of Symbolic Computing and Knowledge Engineering of Ministry
of Education, College of Computer Science and Technology,
Jilin University, Changchun, China

**Abstract.** The image segmentation of plant diseases is one of the critical technical aspects of digital image processing technology for Disease Recognition. This paper proposes an improved pulse coupled neural network based on an improved genetic algorithm. An objective evaluation function is defined based on linear weighted function with maximum Shannon entropy and minimum cross-entropy. Through adaptive adjustment of crossover probability and mutation probability, we optimized the parameters of pulse coupled neural network based on the improve genetic algorithm. The improved network is used to segment the color images of Maize melanoma powder disease in RGB color subspaces. Then combined with the results by color image merger strategy, we can get the terminal results of target area. The experimental results show that this method could segment the disease regions better and set complexity parameters simplier.

**Keywords:** Maize melanoma powder disease, Genetic Algorithm, Pulse Coupled Neural Network, Image segmentation.

## 1    Introduction

Plant disease is a main reasons that affect the growth of the corn and other major crops. In order to improve the quality and the quantity of the crops ,there has grown up an urgent need for disease diagnosis and effective treatment.Normaly, the experts have to go to diagnose the diseases in the field ,which wasting time and effort . Meanwhile, due to the the effects of the subjective factors and the External environment condition, the diagnosis would easily lead to the wrong answer by human beings.The digital image processing technology has fast,accurate and objectivity merit so it has become a possible way to take the place of human vision to diagnose the deseaes. Based on the digital image processing technology ,crop disease diagonosis include: disease image preprocessing, image segmentation, feature extraction, pattern recognition and other

---

[*] Corresponding author.

C. Guo, Z.-G. Hou, and Z. Zeng (Eds.): ISNN 2013, Part I, LNCS 7951, pp. 602–611, 2013.

major steps. Effective segmentation affects the accuracy of the disease image feature extraction and pattern recognition.So the image segmentation is the key of distinguishing crop diseases effectively . General image segmentation algorithm including threshold segmentation algorithm, spatial clustering algorithm, region growing algorithm, mathematical morphology method and a variety of computational intelligence method, etc. But because of background information compleity of crop disease images , disorder disease spots , color uneven distribution, disease spots fuzzy boundary ,vein noise and interference from enviromental light, so there is no good robust and simple applicable universal method[1-3]. At the same time, the conventional disease recognition method is based on gray image segmentation processing, a large number of pathological image color texture information lossed, then it will affect the accuracy of the disease spot image feature extraction and pattern recognition .

Genetic Algorithm (Genetic Algorithm, GA) was put forward by the university of Michigan's J.H olland professor in 1975 for the first time, through the modeling of biological evolution process optimization.As a bionic Algorithm for searching, it is widely used in combinatorial optimization, machine learning, signal processing, adaptive control , artificial life and many other fields [2]. Pulse coupled neural network (PCNN) was found by Eckhorn and colleagues Which were in the research of imitation cat, monkey and small mammals visual cortical imaging mechanism, and furtherly simplified and amended by Johnson after the amendment proposed an artificial network model [3]. This model has been widely used in image denoising, image smoothing, image edge detection, image segmentation, image enhancement and image fusion, etc [4]. This paper, using the optimized characteristics of genetic algorithm global search and the PCNN visual bionic characteristics, a kind of improved genetic algorithm combined with PCNN had been proposed to control the optimized network parameters, and the proposed network model was applied to the image segmentation of complicated background corn diseases.

Structure of this paper is as follows: the first part mainly introduces research status of crop disease recognition and disease image segmentation. Meanwhile, we put forward the main idea in this paper; The second part gave the detailed introduction——the algotithm realization of PCNN based on an improved genetic algorithm.At the same time,this part gave the segmentation results merging strategy in the RGB color subspaces using the proposed algorithm in this paper; Third and forth, the paper discussed the simulation results and conclusion.

## 2    The Disease Image Segmentation Based on the Improving Genetic Pulse Coupled Neural Network Model

### 2.1    The Basic Principle of PCNN

Pulse couple neural network is also known as the third generation of neural network, this model is a kind of feedback type network which is proposed on the basis of the mechanism of the imitation biological visual cortex. The network includes receptive fields, modem domains and pulse excitation domains.The mechanism of mathematics expression of classic PCNN neural network is as follows[5]:

$$F_{ij}(n) = S_{ij}$$

$$L_{ij}(n) = \sum w_{ijkl} Y_{kl}(n-1) \tag{1}$$

$$U_{ij}(n) = F_{ij}(n)(1 + \beta L_{ij}(n))$$

$$Y_{ij}(n) = \begin{cases} 1 & U_{ij}(n) > \theta_{ij}(n) \\ 0 & else \end{cases}$$

$$\theta_{ij}(n) = exp(-a_\theta)\theta_{ij}(n-1) + V_\theta Y_{ij}(n-1)$$

In the Equ.1, $S$ is input excitation of some neuron of PCNN, $F$ is the input corresponding to the neuron, $L$ is neurons link inputs, $U$ is neuron activities of the internal network of PCNN, $Y$ is the output of PCNN, $\theta$ is a dynamic threshold value of pulse activity, $w$ is a link weight maxtrix of internal network of PCNN.

Due to the characteristics of connection of PCNN, the neurons with similar input excitation also will be activated and output pulse in the next iteration process around the neuron which is triggered and output pulse. That is, the neurons are fired which correspond to pixels with similar gray value. So PCNN has been widely used in the field of digital image processing. But in the process of using PCNN for specific image processing, the network parameters still need artificial interactive control or to be determined through experience with a large number of experiments for a long time . For the foregoing reasons ,there are some limitations in the network model in practice. Currently, the typical work for adjusting PCNN parameters mainly cover the Ref. [5, 6 7]. However, there are some limitations that the coupling coefficient and the threshold attenuation amplitude of PCNN still need to adjustment in the Ref.[5,6].In the Ref.[7], they use classical genetic algorithm to optimize connection coefficient $\beta$ , amplitude coefficient $V_\theta$ and incentive pulse attenuation coefficient $a_\theta$ . The Ref.[7] has both reference meaning and real value to improve the network performance.But, because of some limitations in the classical genetic algorithm , there is a certain gap between the actual effect and the expected results in practice.

According to the Ref.[7] which use genetic algorithm to optimize the parameters of PCNN (GA - PCNN) ,we proposed an algorithm by which the target fitness function , crossover probability and mutation probability have been improved. By this method ,the performance of PCNN has acquired enormous clout and the algorithm in this paper has been applied to the color image segmentation .

## 2.2    An Improved Genetic Algorithm

Genetic algorithm is a kind of global convergence algorithm which simulates some processes of biological evolution,and it was widely used as a powerful optimization tools in the 10 years. However, because crossover probability and mutation probability of classical genetic algorithm have been fixed in the evolutionary process, it tend to prone into local optimal prematurity. So In this paper ,we used the algorithm to adaptively adjust crossover probability and mutation probability in the Ref.[8]. The ideological basis for the algorithm is that the group fitness function mean is promoted with the evolution of population, in the meanwhile, the variance of individuals are prone to become more and more smaller under the guidance of the consistency principle, that is, the mean and variance of individual fitness functions become more and more smaller. Thus, we adopt the degressive mode on the setting crossover probability and mutation probability in the course of evolution to ensure the individual difference at the beginning of the evolution and the stable at the end of it.. Specific adjustment principle is as follows.

$$P_{ci} = P_{c0} \times \frac{S_i / S_0}{C_i / C_0} \tag{2}$$

$$P_{mi} = P_{m0} \times \lambda \times \left( 1 - \left( \frac{S_i / S_0}{C_i / C_0} \right) \right) \tag{3}$$

In the Equ.2,3, $P_{c0}$ and $P_{m0}$ are respectively the initial value of crossover probability and mutation probability. $S_i$ and $C_i$ are respectively mean and standard deviation of fitness function of the ith generation, meanwhile $S_0$ and $C_0$ are respectively mean and standard deviation of fitness function of the first generation, $\lambda \in (0,1)$.

## 2.3    An Improved Genetic Pulse Coupled Neural Network (IGA-PCNN)

In this paper,we defined objective evaluation function based on linear weighted function with maximum shannon entropy and minimum cross-entropy.Through adaptive adjustment of crossover probability and mutation probability,we optimized the parameters of pulse coupled neural network based on the improve genetic algorithm — connection coefficient $\beta$, amplitude coefficient $V_\theta$ and incentive pulse attenuation coefficient $a_\theta$. The algorithm is as follows.

- Population Initialization and Coding Description

Individual coding uses binary encoding principle and specific coding is as follows.The length of individual chromosome is 55 bits , the individual chromosome is composed of connection coefficient coding, amplitude coefficient coding and incentive pulse attenuation coefficient coding respectively. The chromosome forms the initial individuals in the group.

**Table 1.** Gene cluster definition

| Gene definition | minimum | maximum | Coding description |
|---|---|---|---|
| Connection coefficient $\beta$ | 0.0001 | 400 | 19 |
| Amplitude coefficient $V_\theta$ | 0.0001 | 400 | 19 |
| Pulse attenuation coefficient $a_\theta$ | 0.0001 | 100 | 17 |

- Fitness Function Definition

In this paper,we defined a fitness function based on linear weighted function with maximum shannon entropy and minimum cross-entropy.The definition of the fitness function is as follows.

$$\text{Fitval} = \rho H_1(p) - (1 - \rho)H_2(p)$$
$$H_1(p) = p_1 \log_2 p_1 \quad p_0 \log_2 p_0$$
$$H_1(p) = p_1 \log_2 \frac{p_0}{p_1} + p_0 \log_2 \frac{p_1}{p_0}$$

(4)

In the Equ.4, $H_1(p)$ and $H_2(p)$ are used to express maximum shannon entropy and minimum cross-entropy of the part of fitness function. $p_1$ and $p_0$ are used to show the probability which the outputs value of PCNN is 1 and 0 in turn. $\rho$ is a weighted coefficient of fitness function, as well as $\rho \in [0,1]$ .On special occasions,the fitness function is defined as maximum shannon entropy,if $\rho$ takes 1.Otherwise we defined minimum cross-entropy as the fitness function.

- Genetic Algorithm Operation

Genetic algorithm make the individualities of the initial population and the population evolving toward optimal direction of the fitness function by implementing selection, crossover and mutation operation. In this paper,we choice each individuality by the roulette way and store the optimum individuality of every generation. The algorithm drive population to evolve to optimal direction and avoid population ripening ,by using Adaptive crossover probability and mutation probability and operation of two-point crossover and single-point mutation.

- Algorithm Terminal Conditions

For the terminal conditions of the population in evolutionary process,we set the max evolutionary generation or set a threshold. The algorithm terminated when the different value of the optimum fitness value is less than the given threshold. We can ensure that population evolution is terminated, the individuality is optimal when the algorithm meets the above condition arbitrarily.

## 2.4     The Strategy of Color Disease Image Segmentation

Aiming at the color image segmentation, the general strategy that the target image is segmented into results correspongding to a certain color space, and then the results are combined by some combination strategie. Then we can obtain the final result of color image segmentation.RGB color space is more commonly expression of color space. Because of higher correlation of the components-R\G\B, the result of color image segmentation corresponding to the R\G\B can preserve a lot of detail information of the original image and increase the fault tolerance in merging process.In R\G\B subspace,we use IGA - PCNN proposed in this paper to segment color image into components corresponding to subspace, in order to get the best segmentation result of the R/G/B components.The selective hige probability merger strategy proposed by Y Tan, D Zhou [9] is used to   merge the result corresponding to the R\G\B subspace,and the final result is the interest target of color disease image segmentation.

# 3     Experiment and Analysis

## 3.1     Experiment

In the corn plots of JiLin Agricultural University, We use the camera type of panasonic SZ1 to get a group of photos by automatic exposure photograph in natural lighting conditions.We save photos with JPEG format and cut photos to size of 640*480 with Photoshop7.0. Meanwhile we use Matlab2011b as our development platform in ThinkPad core i5 with 4G memory.Then our method has been used to segment Maize disease images compare to the result with GA-PCNN and Minimum cross entropy thresholding method(MCET)[10].

Our contrast experimen need some necessary setting.Firstly,The parameters settings of GA-PCNN proposed in the Ref.[8]. are as follows.

- Population size $N$ :30
- Number of genetion $G$ :50
- Fitness deviation of four generations is less than 0.001
- Crossover probability $P_c$ :0.7
- Mutation probability $P_m$ :0.01

Secondly, We only list difference parameter Settings.The parameters settings of our algorithm are as follows.

- Population size $N$ :30
- Initial crossover probability $P_{c0}$ : 0.1
- Initial mutation probability $P_{m0}$ : 0.01
- Number of genetion $G$ :100
- Weighted coefficient of fitness function  $\rho$ :0.5

Thirdly,we choosed two common images of maize melanoma powder disease which have different external morphological characteristics.The experiment results are as follows.

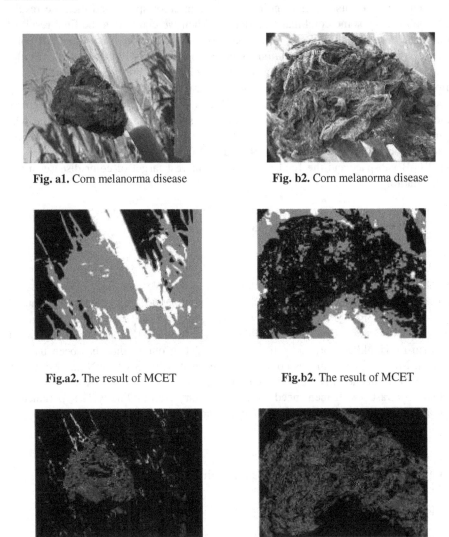

Fig. a1. Corn melanorma disease          Fig. b2. Corn melanorma disease

Fig.a2. The result of MCET          Fig.b2. The result of MCET

Fig. a3. The result of GA-PCNN in Ref.[8]     Fig. a4. The result of GA-PCNN in Ref.[8]

Fig. 1. The experience results of Corn melanorma disease

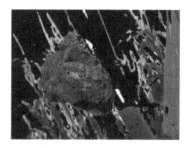 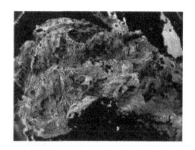

**Fig. a4.** The result of IGA-PCNN          **Fig. a4.** The result of IGA-PCNN

**Fig. a.** The first experience          **Fig. b.** The second experience

**Fig. 1.** *(Continued)*

## 3.2   Analysis

Through the contradistinctive test, we synthetic evaluation this algorithm from subjective visual evaluations and the algorithm settings.First of all, from the experiment results we can see, by the GA-PCNN algorithm and the IGA-PCNN algorithm segment the RGB color space subgraph, then merge down the segmentation extracted from the subgraph RGB color space. The target domain features,including disease profile features, disease visual color discriminant features are obviously superior to the traditional method which segmentation processing on the minimum cross entropy thresholding in gray image.In addition, compared the corn melanoma powder segmentation effects from GA-PCNN algorithm and IGA-PCNN algorithm proposed in this paper, except for the external morphological characteristics in details, IGA-PCNN algorithm got better effect,the color texture feature information also got a more satisfactory results.This paper proposes a model modified genetic pulse coupled neural network which made a less artificial regulation and higher degree of automation.

From the perspective of quantitative analysis, the method presented   in reference [13] is adopted to compare the experement results of  cross-entropy threshold,GA-PCNN and IGA-PCNN.The computing equations of comparative results are as follows in reference[13].

$$\text{Match:} \qquad ((N_1 - |N_2 - N_1|)/N_2) \times 100\% \qquad (5)$$

$$\text{Error:} \qquad ((N_2 - N_1)/(m \times n)) \times 100\% \qquad (6)$$

$$\text{Accuracy:} \qquad \text{Accuracy=Match-Error} \qquad (7)$$

In the above equations, $N_1$ is used to express pixels values of object area by using the three methods respectively, $N_2$ is used to express pixel values of object area by manual segmentation. $m \times n$ is used to express the image size. In this paper, we did the experiments ten times by each algorithm and got the mean of the results. The comparative experiment results showed that the algorithm of IGA-PCNN proposed in this paper performance better than others .

**Table 2.** Comparative experiment results

| Groups | The method of cross-entropy threshold | | | The method of GA-PCNN | | | The method of IGA-PCNN | | |
|---|---|---|---|---|---|---|---|---|---|
| | Match | Error | Accuracy | Match | Error | Accuracy | Match | Error | Accuracy |
| a1 | 93.47 | 11.86 | 81.61 | 95.24 | 3.13 | 93.11 | 96.76 | 1.06 | 95.70 |
| b1 | 93.7 | 12.47 | 81.29 | 95.10 | 3.26 | 91.84 | 95.89 | 1.57 | 94.32 |

## 4    Conclusion

In this paper,we proposed an improved genetic pulse coupled neural network . The parameters of PCNN have been optimized by the use of an improved genetic algorithm which is proposed in this paper.The advantages of the improved algorithm are as follows.Firstly,it reduces the degree of artificial participation,meanwhile improves the automation degree of algorithm.Secondly,Using this algorithm, the target area appearances of maize desease images are more meticulous in contrast to other methods and the color texture feature has been retained better in the target area of maize desease images.Therefore, the algorithm proposed in this paper is very significant for feature extraction and disease recognition. To sum up, the proposed algorithm has better application value and practical significance.

**Acknowledgements.** This research was conducted with the support of Jilin province science and technology development plan item(NCS 201105068),Project development projects in Key Laboratory of Ministry of Education(NCS 93K172012K11), Scientific research fund of Jilin Agricultural Unversity.

## References

1. Li, G., Ma, Z., Huang, C., Chi, Y., Wang, H.: Segmentation of color images of grape diseases using K-means clustering algorithm. Transactions of the Chinese Society of Agricultural Engineering 26(12), 32–37 (2010)
2. Kurugollu, F., Sankur, B., Harmanci, A.E.: Color image segmentation using histogram multithresholding and fusing. Image and Vison Computing 19, 915–928 (2001)
3. Papamarkos, N., Strouthopoulos, C., Andreadis, I.: Multithresholding of color and gray-level images through a neural network technique. Image and Vison Computing 18, 213–222 (2000)
4. Rechenberg, I.: Cybemetic solution path of an experimental problem. Roy Airer Estable, lib trans 1222 Hants, UK; Farnborough (1985)

5. Johnson, J.L., Padgett, M.L.: PCNN Models and Application. IEEE Trans. on Neural Networks 10(3), 480–498 (1999)
6. Ma, Y., Dai, R., Li, L.: Image segmentation of embryonic plant cell using pulse-coupled neural network. Chinese Science Bulletin 47(2), 167–172 (2002)
7. Ma, Y., Dai, R., Li, L.: Automated image segmentation using pulse coupled neural networks and imgae's entropy. Journal of China Institute of Communications (1), 46–51 (2002)
8. Gu, X., Guo, S., Yu, D.: A new approach for image segmentation based on unit-linking PCNN. In: Proceeding of the 2002 International Conference on Machine Learning and Cybernetics, vol. 1, pp. 175–178 (2002)
9. Ma, Y., Qi, C.: Study of Automated PCNN System Based on Genetic Algorithm. Journal of System Simulation (3), 722–725 (2006)
10. Wen, C., Wang, S., Yu, H., Su, H.: Improved genetic resonance matching network learning algorithm. Journal of Huazhong University of Science and Technology (Natural Science Edition) 39(11), 47–50 (2011)
11. Tan, Y., Zhou, D., Zhao, D., Nie, R.: Color image segmentation and edge detection using Unit-Lingking PCNN and image entropy. Computer Engineering and Applications 41(12), 174–178 (2009)
12. Li, C.H., Lee, C.K.: Minimum cross-entropy thresholding. Pattern Recognition 26, 617–625 (1993)
13. Camargo, A., Smith, J.S.: An image processing based algorithm to automatically identify plant disease visual symptoms. Biosystems Engineering 102(1), 9–12 (2009)

# A Vector Quantization Approach for Image Segmentation Based on SOM Neural Network

Ailing De and Chengan Guo[*]

School of Information and Communication Engineering,
Dalian University of Technology, Dalian, Liaoning 116023, China
deailing@mail.dlut.edu.cn, cguo@dlut.edu.cn

**Abstract.** In the existing segmentation algorithms, most of them take single pixel as processing unit and segment an image mainly based on the gray value information of the image pixels. However, the spatially structural information between pixels provides even more important information of the image. In order to effectively exploit both the gray value and the spatial information of pixels, this paper proposes an image segmentation method based on Vector Quantization (VQ) technique. In the method, the image to be segmented is divided into small sub-blocks with each sub-block constituting a feature vector. Further, the vectors are classified through vector quantization. In addition, the self-organizing map (SOM) neural network is proposed for realizing the VQ algorithm adaptively. Simulation experiments and comparison studies have been conducted with applications to medical image processing in the paper, and the results validate the effectiveness of the proposed method.

**Keywords:** image segmentation, spatial structure, SOM neural network, vector quantization.

## 1 Introduction

Image segmentation can be viewed as a process of decomposing an image into several meaningful parts for further analysis [1]. It is one of the most fundamental areas in image processing and computer vision, and has been a hot topic of general interest in the field. There are about seven existing categories of segmentation methods, including statistical probability of pixel characteristic based methods [2], region related segmentation methods, clustering methods, graph theory based methods, level set methods [3], hybrid methods [4], and fuzzy theory based methods. Most of the existing methods focus more on the gray value information than on the geometrical structure information of the pixels [5, 6]. Nevertheless, from the perspective of human vision, human vision perception of image content is more based on percipience of local regions and their spatial relationships than on pixel gray value information. Therefore, it is desirable to design a segmentation method that can effectively utilize both spatial structural information of local regions and gray value information of pixels.

---

[*] Corresponding Author.

C. Guo, Z.-G. Hou, and Z. Zeng (Eds.): ISNN 2013, Part I, LNCS 7951, pp. 612–619, 2013.

In this paper we conduct research on the segmentation issue from the perspective that takes local regions as processing unites and utilizes both gray value and spatial structure information in local regions. First, the image to be segmented is divided into a series of sub-blocks with each sub-block being viewed as a vector that contains certain structural pattern. Then, an edge detection algorithm is performed to divide the sub-blocks into two patterns: edge sub-block patterns and non-edge sub-block patterns. Further, the vector quantization (VQ) approach is proposed to cluster the vectors according to the structural patterns of the non-edge sub-block pattern vectors. Finally, the pixels of the edge sub-blocks are further processed and classified into the best matched category of neighboring non-edge pattern clusters based on the VQ results.

VQ technology originally stems from researches on encoding discrete data vectors for the purpose of data compression and communication. In this paper, VQ is introduced into image segmentation to cluster sub-block patterns for segmenting images. In addition, a SOM network model is proposed to adaptively implement the codebook design for the VQ method. Many experiments have been conducted to test the proposed method in the paper, including practical applications to the segmentation of medical MRI images. Comparison studies with some state-of-the-art segmentation methods, e.g., the improved fuzzy C-mean method (FCM), are also conducted in the paper. Validity of the proposed method is confirmed by the experiment results.

The sequel of the paper is organized as follows: Section 2 describes the proposed segmentation approach, including the classification scheme of edge and non-edge pattern vectors, the vector quantization algorithm for image segmentation, and the SOM model for designing the VQ codebook. Experiments with applications to medical image segmentation and results analyses are given in Section 3. Section 4 draws conclusions and points out the possible direction of the paper.

## 2    Image Segmentation by SOM-Based Vector Quantization

The overall scheme of the new image segmentation method consists of the following 5 computational modules:

(1) Divide the image to be segmented into small sub-blocks of $n \times n$ pixels, each sub-block constituting a vector with $n^2$ elements and each element corresponding to a pixel of the sub-block.

(2) Classify the sub-block vectors into two patterns, called the edge pattern and non-edge pattern, by using an edge detection algorithm. The edge patterns are the sub-blocks that contain edge pixels of sub-regions in the image, and the non-edge patterns are the sub-blocks without edge pixels (i.e., the relatively smoothing sub-regions of the image).

(3) Cluster the vectors (sub-blocks) of non-edge patterns into $K_c$ classes by using vector quantization (VQ) method.

(4) Implement the VQ algorithm adaptively by using an SOM network.

(5) Process the edge pattern sub-blocks based on the VQ results in the following steps:
   (i) Compare the gray value of each pixel in the edge pattern sub-blocks with the mean values of the neighboring non-edge pattern sub-blocks;
   (ii) Set the pixel value of the edge pattern sub-blocks with the closest mean value of the neighboring non-edge pattern sub-blocks.

The main ideas and algorithms for these modules will be presented in the sequel parts of the section.

## 2.1    Vector Construction and Classification of Edge and Non-edge Patterns

As the first step of the vector-quantization-based image segmentation approach, the image to be segmented is divided into nonoverlapping sub-blocks of $n \times n$ pixels. Suppose that $[f(i, j)]_{n \times n}$ $(i = i_{k+1}, ..., i_{k+n}; j = j_{k+1}, ..., j_{k+n})$ is a sub-block of the $M \times N$ image $[f(i, j)]_{M \times N}$, the vector $X(k)$ for the sub-block is constructed in the following form:

$$X(k) = [f(i_{k+1}, j_{k+1}), ..., f(i_{k+1}, j_{k+n}), ..., f(i_{k+n}, j_{k+1}), ..., f(i_{k+n}, j_{k+n})]^T \quad (1)$$

In this way, the image $[f(i, j)]_{M \times N}$ can be expressed with the vector set

$$X = \{ X(k); k = 1, ..., K_X \} \quad (2)$$

Where $K_X = MN/n^2$.

The sub-block vectors are then classified into two patterns, the edge pattern and non-edge pattern, by using the following edge detection algorithm based on the wavelet modulus maximum edge detection [9,10].

The two wavelets used in the paper are the partial derivatives of the two-dimensional Gaussian function $\theta(x, y) = \dfrac{1}{2\pi\sigma^2} e^{-(x^2 + y^2)/2\sigma^2}$:

$$\phi^1(x, y) = \frac{d}{dx} \theta(x, y) = \frac{-x}{2\pi\sigma^4} e^{-(x^2 + y^2)/2\sigma^2}, \quad \phi^2(x, y) = \frac{d}{dy} \theta(x, y) = \frac{-y}{2\pi\sigma^4} e^{-(x^2 + y^2)/2\sigma^2} \quad (3)$$

By setting the wavelets scale factor to $a = 2^j$, we have that

$$\phi^1_{2^j}(x, y) = \frac{1}{2^{2j}} \phi^1 \left[ \frac{x}{2^j}, \frac{y}{2^j} \right], \quad \phi^2_{2^j}(x, y) = \frac{1}{2^{2j}} \phi^2 \left[ \frac{x}{2^j}, \frac{y}{2^j} \right] \quad (4)$$

Then the horizontal and vertical edge information for the pixel $f(x, y)$ of an image can be calculated by the following convolution operations:

$$W^1_{2^j} f(x, y) = f(x, y) \otimes \phi^1_{2^j}(x, y), \quad W^2_{2^j} f(x, y) = f(x, y) \otimes \phi^2_{2^j}(x, y). \quad (5)$$

The modulus and angle parameters, $M_{2^j} f(x, y)$ and $A_{2^j} f(x, y)$, are then computed based on the wavelets convolution results:

$$M_{2^j} f(x, y) = \sqrt{\left| W^1_{2^j} f(x, y) \right|^2 + \left| W^2_{2^j} f(x, y) \right|^2}, \quad A_{2^j} f(x, y) = \arctan \left[ \frac{M^2_{2^j} f(x, y)}{M^1_{2^j} f(x, y)} \right] \quad (6)$$

The main steps of the proposed classification algorithm for classifying the edge and non-edge pattern vectors based on the wavelet features are as follows:

(i) Compute the threshold parameter $T_M$ based on the mean and variance values, $m_M$ and $\sigma_M^2$, of $M_{2j}f(x,y)$ for the image to be segmented:

$$T_M = m_M + \alpha \sigma_M \qquad (7)$$

Where $\alpha$ ($\alpha \geq 0$) is the adjustment parameter to be determined in experiments.

(ii) Extract the maximum direction, $A_{max}(x,y)$, of the angle values $A_{2j}f(x,y)$ in its 8-neighborhood area for each pixel.

(iii) For each sub-block, compute the mean value, $\bar{M}(x,y)$, of the 3 modulus $M_{2j}f(x,y)$ along the vertical direction of the maximum angle $A_{max}(x,y)$.

(iv) Classify the vectors of the sub-blocks by compare the mean value $\bar{M}(x,y)$ with the threshold $T_M$:

If $\bar{M}(x,y) > T_M$, the vector for the sub-block $[f(i,j)]_{n \times n}$ is classified to the edge pattern;

Otherwise, the vector is classified to the non-edge pattern.

## 2.2    Image Segmentation by Vector Quantization

After the sub-block vectors have been classified to edge pattern and non-edge pattern, we then perform segmentation on the non-edge pattern vectors using vector quantization (VQ) technique.

The vector quantization [11] is the mapping process from a $k$-dimensional space $R^k$ to a finite subset of $k$-dimensional vectors $C$, i.e., $Q: X \in R^k \rightarrow C \subset R^k$. Where $C = \{C(k); k = 1, ..., K_C\}$ is called the codebook, and $C(k)$ is called the code word. For our vector-quantization-based segmentation approach, there are two key problems to be solved: one is how to design a good code book $C$ for the specific image to be segmented, and the other is how to segment an image using a given code book. The code book design algorithm will be presented in the next subsection. We describe the segmentation algorithm here in the following.

Suppose now that the vector set $X = \{X(k); k = 1, ..., K_X\}$ has been constructed for the image $[f(i,j)]_{M \times N}$ to be segmented by using the method given in Section 2.1 and the codebook $C = \{C(k); k = 1, ..., K_C\}$ has also been obtained. The classification of edge pattern and non-edge pattern for the vectors has also been made. Our VQ-based segmentation algorithm is as follows:

(i) Compute the distances between the vector $X(k)$ and all the code words $C(k)$ in the codebook $C$:

$$d_j(k) = \| X(k) - C(j) \|, \quad j = 1, ..., K_C \qquad (8)$$

Where $X(k)$ is the non-edge pattern vector of $X = \{X(k); k = 1, ..., K_X\}$.

(ii) Find the minimal distance $d_q$ in all the $d_j$'s:

$$d_q(k) = \min_j \{d_j(j)\} \qquad (9)$$

(iii) Quantize the vector $X(k)$ to the code word $C(q)$. This means that the sub-block $[f(i, j)]_{n \times n}$ represented by $X(k)$ in the image $[f(i, j)]_{M \times N}$ has been segmented to the sub-block $C(q)$.

(iv) Goto step (i) until all the non-edge pattern vectors have been processed.

It can be seen that by performing the above algorithm, the non-edge pattern sub-blocks of the image have been segmented to $K_C$ classes with each class represented by $C(k)$, $k = 1, ..., K_C$ .

### 2.3    Codebook Design for Vector Quantization Based on SOM Network

The key problem of VQ is the design of its codebook. The classical LBG algorithm [11] has the drawback of high computational complexity. The self-organizing feature mapping (SOM) network has been successfully applied to solving unsupervised learning and clustering analysis problems [12,13]. In this paper we will apply the SOM network to the codebook design for our VQ-based image segmentation method. The architecture of the SOM network used in the paper is shown in Fig.1.

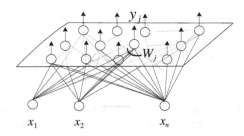

**Fig. 1.** Architecture of the SOM network used for VQ

In Fig.1, the weight vector between input $[x_1, x_2, \cdots, x_m]^T$ and output neuron $y_j$ is denoted by $W_j = [w_{j1}, w_{j2}, \cdots, w_{jm}]^T$. In the method, the input vector $[x_1, x_2, \cdots, x_m]^T$ is just the non-edge pattern vector $X(k)$ of the image $X = \{X(k); k=1, ..., K_X\}$ to be segmented, the weight vector $W_j = [w_{j1}, w_{j2}, \cdots, w_{jm}]^T$ to be designed by training process is the codeword $C(j)$ of $C = \{C(j); j=1, ..., K_C\}$.

Based on the above idea, the algorithm for training weights (the VQ codebook) is as follows:

(i) Initialization:

$W_j(0) \leftarrow$ randomly selected vectors from the training set, i.e., the non-edge pattern vectors of $X = \{ X(k); k = 1, ..., K_X \}$; $\quad k \leftarrow 0$.

(ii) Input $X(k)$, and update $W_j$ in the following method:

(ii-1) Find the weight vector $W_q$ that has the minimal distance with $X(k)$ :

$$d_q(k) = \min_j \left\{ \| W_j(k) - X(k) \| \right\} = \| W_q(k) - X(k) \| \tag{10}$$

(ii-2) Define $N_q(t_k)$ the neighborhood of $W_q$, and update the weights in $N_q(t_k)$, without updating the weights not in $N_q(t_k)$ :

$$W_j(k+1) = W_j(k) + \alpha(k)[X(k) - W_j(k)], \quad j \in N_q(k) \tag{11}$$

Where $\alpha(t_k)$ is the learning rate with $0 < \alpha(t_k) < 1$.

(iii) $k \leftarrow k+1$, and goto (ii), until the network converged.

Having the above training process finished, the trained weight vectors $W_j = [w_{j1}, w_{j2}, \cdots; w_{jm}]^T$ ($j = 1, ..., K_C$) will be taken as the codebook $C = \{ C(j); j = 1, ..., K_C \}$, and will be used for segmenting the image as discussed in Section 2.2.

## 2.4    Further Processing of the Edge Pattern Vectors Based on VQ Results

After the non-edge pattern vectors have been segmented by using the SOM-based VQ method given above, we further process the edge pattern sub-block vectors based on the VQ results in the following way:

(i) Compute the difference values between the pixel of the edge pattern sub-block and the mean value of its neighboring non-edge sub-block vectors that have been segmented by the VQ method.

(ii) Find the minimum absolute difference value, and classify the pixel to the class with the minimum difference value by replacing the pixel value with the mean value of the best matched neighboring vector.

# 3    Experimental Results

To verify the effectiveness of the SOM-based VQ segmentation method, many experiments have been conducted with applications to segmentations of human brain MRI images in the paper. All the anatomical components of the MRI brain images, including the skull, cerebrospinal fluid (csf), gray matter, white matter, and background, are segmented in the experiments. Accordingly, the improved fuzzy C-mean method (FCM), also known as FLICM proposed recently by Stelios and Vassilios [16] is also implemented for comparison study in the experiments.

Fig.2 shows a group of the experiment results on an MRI image of human brain segmented by using the new segmentation method. In the figure, Fig.2(a) is the original MRI image, in which 6 classes of anatomical components exist, including the skull, cerebrospinal fluid (csf), gray matter, white matter, background and a brain tumor in the centre part of the image. We expect to segment out all the 6 components from the image. Fig.2(b) - Fig.2(g) show the experiment results by using the new method.

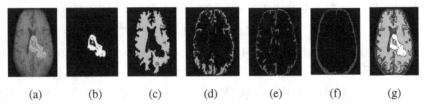

(a)        (b)        (c)        (d)        (e)        (f)        (g)

**Fig. 2.** Experiment result using the proposed method. (a) the original MRI human brain image with a brain tumor in the centre part; (b) the segmented tumor; (c) segmented white matter; (d) segmented gray matter; (e) segmented cerebrospinal fluid; (f) segmented skull; (g) the combination result of all the segmented parts

We can see from Fig.2 that the outcomes of the proposed segmentation method have fulfilled our expectation.

Meanwhile, the FLICM method [16] is also conducted on the same MRI image in the experiment. The result is shown in Fig.3.

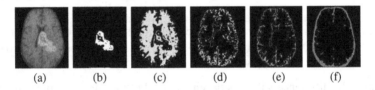

(a)        (b)        (c)        (d)        (e)        (f)

**Fig. 3.** Experiment result by FLICM method. (a) the original MRI human brain image with a brain tumor in the centre part; (b) the segmented tumor; (c) segmented white matter; (d) segmented gray matter; (e) segmented cerebrospinal fluid; (f) segmented skull

By comparing the experimental results of Fig.2 and Fig.3, we can see that the new method works well and outperforms the FLICM in segmenting the 6 parts in the human brain MRI image, each part being segmented more accurate by the new method than by FLICM.

Many other experiments have been done and similar results were obtained in the work, which will be omitted here due to the limitation of the paper pages.

## 4    Conclusions

In this paper, we proposed a new method for segmenting images by using the SOM-based vector quantization technique. The new segmentation approach can effectively utilize both the gray value and the spatial information of image pixels. Validation of the new method is confirmed by the experimental results conducted in the paper with successful applications in medical images segmentation.

It is noticed that the computational complexity of the proposed method is quite high for the SOM training process. Thus, more efficient algorithms, such as parallel algorithms, for training the SOM network need to be developed to speedup the computational time. This is the problem for further study of the paper.

# References

1. Zhang, Y.J.: Image Segmentation, pp. 6–58. Science Press, Beijing (2001)
2. Gao, X.L., Wang, Z.L., Liu, J.W.: Algorithm for Image Segmentation using Statistical Models based on Intensity Features. Acta Optica Sinica 31(1), 1–6 (2011)
3. Zhao, J., Shao, F.Q., Zhang, X.D.: Vector-valued Images Segmentation based on Improved Variational GAC Model. Control and Decision 26(6), 909–915 (2011)
4. Wu, Y., Xiao, P., Wang, C.M.: Segmentation Algorithm for SAR Images based on the Persistence and Clustering in the Contourlet Domain. Acta Optica Sinica 30(7), 1977–1983 (2010)
5. Veksler, O.: Image Segmentation by Nested Cuts. In: Proc. of IEEE Conference on Computer Vision and Pattern Recognition, pp. 339–344. IEEE Press (2000)
6. Wang, S., Lu, H.H., Yang, F.: Superpixel Tracking. In: Proc. of IEEE International Conference on Computer Vision, pp. 1323–1330. IEEE Press (2011)
7. Luo, S.Q., Zhou, G.H.: Medical Image Processing and Analysis, pp. 67–68. Science Press, Beijing (2003)
8. Zhang, Q., Lu, Z.T., Chen, C.: Spinal MRI Segmentation based on Local Neighborhood Information and Gaussian Weighted Chi-square Distance. Chinese Journal of Biomedical Engineering 30, 358–362 (2011)
9. Luo, Z.Z., Shen, H.X.: Hermite Interpolation-based Wavelet Transform Modulus Maxima Reconstruction Algorithm's Application to EMG De-noising. Journal of Electronics & Information Technology 31(4), 857–860 (2009)
10. Liu, B., Huang, L.J.: Multi-scale Fusion of Well Logging Data Based on Wavelet Modulus Maximum. Journal of China Coal Society 35(4), 645–649 (2010)
11. Linde, Y., Buzo, A., Gray, R.: An Algorithm for Vector Quantizer Design. IEEE Trans. on Communications 18(1), 84–95 (1980)
12. Costa, J.A.F., de Andrade Netto, M.L.: Automatic Data Classification by a Hierarchy of Self-organizing Maps. In: Proc. IEEE International Conference on System, Man and Cybernetics, vol. 5, pp. 419–424 (1999)
13. Kohonen, T.: The self-organizing maps. Proceedings of the IEEE 78(9), 1464–1480 (1990)
14. Gustafson, D., Kessel, W.: Fuzzy Clustering with a Fuzzy Covariance matrix. In: Decision and Control including the 17th Symposium on Adaptive Processes, vol. 17(1), pp. 761–766 (1978)
15. Babuska, R., Verbruggen, H.: Constructing Fuzzy Models by Product Space Clustering. In: Fuzzy Model Identification, pp. 53–90 (1997)
16. Stelios, K., Vassilios, C.A.: Robust Fuzzy Local Information C-means Clustering Algorithm. IEEE Transactions on Image Processing 19(5), 1328–1337 (2010)

# A Fast Approximate Sparse Coding Networks and Application to Image Denoising

Jianyong Cui, Jinqing Qi, and Dan Li

School of Information and Communication Engineering,
Dalian University of Technology, Dalian, China
JianyongCui@mail.dlut.edu.cn,
jinqing@dlut.edu.cn

**Abstract.** Sparse modeling has proven to be an effective and powerful tool that leads to state of the art algorithms in image denoising, inpainting, super-resolution reconstruction, etc. Although various sparse modeling algorithms have been proposed, a major problem of these algorithms is computationally expensive which prohibits them from real-time applications. In this paper, we propose a simple and efficient approach to learn fast approximate sparse coding networks as well as show its application to image denoising. Our experiments demonstrate that the pre-learned network is over 200 times faster than sparse optimization algorithm, and yet obtain approving result in image denoising.

**Keywords:** Image denoising, Sparse coding, Neural networks, Online gradient method.

## 1    Introduction

Removing the noise of images is a conventional issue, which has aroused much interest for years. Noise is inevitable introduced during the acquisition stage, thus denoising is usually the first step to enhance the performance of degraded images. Different algorithms are depended on noise models applied. In most cases, the noise is assumed to be an additive random noise. Additive Gaussian white noise is the common type. The model of noisy image can be depicted below:

$$y = x + v \tag{1}$$

Where y is the noisy image, and x is the ground truth. v is the noise, which is usually the Gaussian noise, as discussed above. Image denoising aims to remove the noise from y [2]. And the denoised image is as close as possible to the ground truth. We long for better quality of images with less cost, especially the wicked situations. How to obtain the less noisy images from these contaminated data has been a significant subject for researchers.

Despite the great development of digital image techniques, image denoising is still a challenge problem. The algorithm based on dictionary learning and sparse representation is an outstanding one [1],[3]. In sparse coding, the test examples are represented as a sparse linear combination of the trained dictionary which is learned by training

C. Guo, Z.-G. Hou, and Z. Zeng (Eds.): ISNN 2013, Part I, LNCS 7951, pp. 620–626, 2013.

samples. Using a small number of atoms parsimoniously chosen out of an over-complete dictionary, sparse coding techniques effectively represent a natural image. By training an appropriate dictionary as well as corresponding sparse coefficients with K-SVD, samples of data can be represented or reconstructed by dictionary [4]. Because of the high quality of dictionary and sparse coefficients, noise can be removed during this progress. Even though K-SVD algorithm performs well, the slow speed of predicting the sparse coefficients through learned dictionary limits its application.

Our motivation is to propose a fast approximate sparse encoder in which not sparse coding performance but also sparse coefficients inference speed are taken into account. Although Gaussian noise is assumed in application to image denoising, the proposed approach is also valid to other types of noise.

The paper is organized in the following parts: in section 2 we look back on the concerned benchmark work: K-SVD and the exact algorithm is presented in section 3. Then section 4 gives our experimental results. At last, we will conclude in section 5.

## 2    K-SVD Algorithm

The denoising algorithm based on dictionary learning is one of the most important branches. By training on the data, a dictionary can be obtained. With an over-complete dictionary made up of standard signal atoms, data to be processed is depicted by sparse linear combinations of these atoms belonging to dictionary [1]. Data is preprocessed for the convenience of algorithm, and usually, images are cut into a large amount of patches.

In order to achieve sparse representation of signals, Elad proposed an algorithm of dictionary learning: K-SVD. Under strict sparse constrains, they find an adapted dictionary to represent each member of the training set. K-SVD is an iterative algorithm. It alternates between computing the sparse codes of samples based on current dictionary, and updating the dictionary atoms according to modified coefficients. The modification of dictionary columns is accompanying with the update of coefficients. Thus, they converge at the same time [5]. Using patches from the contaminated image or ground truth, they train the dictionary by solving a L-0 norm problem.

$$\hat{\alpha} = \arg\min_{\alpha} \left\| D\alpha - x \right\|_2^2 + \mu \left\| \alpha \right\|_0 \tag{2}$$

$$\hat{\alpha} = \arg\min_{\alpha} \left\| D\alpha - x \right\|_2^2 + \mu \left\| \alpha \right\|_1 \tag{3}$$

The original problem is a L-0 norm one, but in this circumstance, it equals to L-1 norm formula (3), $\alpha$ is the sparse coefficient and D is the dictionary, $\mu$ is the parameter of sparsity.

Because of trained dictionary's strong representative ability, noise can be restrained during training progress. The noise level changed with the update of an output image. Usually, before finding the expected output image, several more iterations of dictionary update and representation computation are carried out. They proposed a simple method for the denoising problem. The method is based on local operations and uses sparse decomposition of every patch on the redundant dictionary.

# 3    Fast Approximate Sparse Encoder

## 3.1    Neural Networks

Image denoising using neural network is not a new area for research. Many customized networks have been used to solve the denoising problem, for instance, convolutional neural networks: CNNs [6]. Only a small set of patches are used for training, but robust to the strong noise level. Harold C. Burger et al propose a multi-layer perceptron architecture for image denoising [7], each layer has sufficient units to achieve a better performance. In addition to Gaussian noise, the multi-layer perceptron architecture is still effective for other types of noise, such as stripe noise, speckle noise.

Back propagation neural network is the most widely used neural network. It applies a smooth shrinkage function, and one hidden layer or more can be used. The neighboring layers are connected by the weights. In addition, BP network is a forward network, i.e. the signals are transferred layer by layer. But error is propagated backward, according to error of ideal output and current output, during training. In the propagation, the weights are updated layer by layer from the input layer to the output. Our proposed algorithm also uses this structure.

The network is completed by seeking a minimum of a composite function from the input to output space, known as the network function [8]. Given a training set $\{(x1, t1), (x2, t2)\ldots (xm, tm)\}$. When the input data xi from the training examples, is offered to the network, the network produces an output oi. The output is different from the target ti. By using a learning algorithm, the output and the target are identical. That is, we seek a minimum of error function of network, defined as

$$E = \frac{1}{2}\sum_{i=1}^{m}\left\|o_i - t_i\right\|^2 \qquad (4)$$

The back propagation is a method of finding a local minimum of error function. After lots of iterations, a network adapted to the train data is produced.

## 3.2    Fast Approximated Predictor

The fast approximated predictor is based on the classical BP network. Our proposed algorithm abandons the dictionary learned by K-SVD, which is strongly representative and redundant. Instead, we learn a single layer BP network to seek relationship between original signals and the sparse coefficients. To be precise, the network is a set of weights between layers. These weights play a similar role to the dictionary in the process of K-SVD algorithm. K-SVD learns a dictionary coupled with sparse coefficients. The dictionary can be considered as a backward direction transform. But noise removal is a problem that images not only need to be reconstructed, but to be represented. While in the stage of representation, sparse coefficients are acquired by means of OMP algorithm [9]. Unlike the K-SVD dictionary learning, our architecture is directly used to predict the approximate sparse coefficients or to restore the original signals. So our proposed is a bidirectional predictor.

In the training section, traditional solution is using batch learning. In this way, the gradient contribution accumulated for all data points in the training set before updating the weights. The other approach is online learning, where the weights are updated immediately after seeing every data point. Especially when the training set is redundant, it is usually much faster. The noise in the gradient can help to escape from local minima. Online gradient method is a valid tool for this kind of problem, so we apply it to our algorithm to iterate to find its optimal solution.

Figure 1 is the flowchart of the proposed methods. As illustrated in figure 1, X and $\alpha$ are the training set and coefficients. With the help of K-SVD algorithm, a dictionary and the sparse coefficients are learned in the first stage. And by the means of dictionary and coefficients, a direct and back network is learned, meanwhile. Besides, the problem we solve is a large-scale data processing. We take place of the dictionary by the network weight: W, to find the structural hidden spaces and the mapping. Even the coefficients predicted by the network are not sparse, the speed is much faster.

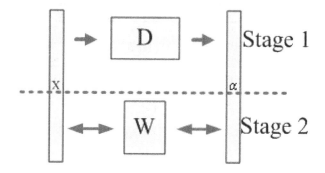

**Fig. 1.** Flowchart of our algorithm. X represents the training set. D and $\alpha$ are learned by K-SVD algorithm at the stage 1. Then with the help of training set and corresponding coefficients, a bidirectional network: W is trained between the two spaces at the stage 2. When going through the trained network, the noise is removed from noisy image.

## 4    Experimental Results

The algorithms including ours, dressed above, are patches-wise level. Two sizes of images are used. Images which have the resolution of 256×256, are cut into the blocks of 6×6 and, the number of overlap pixels is 4. Thus, sum of samples for training is 15876. The ones of 512×512 are divided into 8×8 blocks, and the number of overlap pixels is 4. Sum of the total samples is 16129. While in the test stage, samples from images of 256×256 are still the size of 6×6. The overlap pixels are 5. The size of samples from images of 512×512 remains 8×8, but the overlap pixels are 6.

The size of dictionary is set to 100, and the sparsity is 3. The network selects a single hidden layer one. BP network parameters are very sensitive to the training data, so we employ some common tricks. The first one is data normalization. Each sample subtracts the mean value, and then is normalized by its 2-norm. The other is the problem of hidden units. The number is set to 36, and learning rate is set to 0.03. Initial weights are random selected around zero.

Learning a dictionary is sometimes accomplished through learning on a noise-free dataset. We implement the experiments by training the ground truth of images, and test various levels of noise as well as different types of noise. Experiments are completed on a PC with a Pentium Dual-Core 3.2G CPU. Considering speed is an important aspect, table 1 gives the performance of the two algorithms. The principal time costs a lot, during the process which is used to predict coefficients. Denoising by K-SVD spends much time encoding a sparse coefficient. But time of this step drops a lot in our proposed algorithm, owing to the quality of BP network. Our method is about 200 times faster than the K-SVD at least. This performance may be acceptable for the application in our daily life and technique field.

**Table 1.** Elapse time of comparisons for predicting the coefficients. The noise variance sigma=35, and the type of noise is Gaussian white noise.

| images | K-SVD(s) | Our method(s) | Speed-up Ratio |
|---|---|---|---|
| Lena | 124.00 | 0.38 | 326.32 |
| Barbara | 107.18 | 0.37 | 289.67 |
| Peppers | 99.65 | 0.43 | 231.74 |
| Goldhill | 100.62 | 0.31 | 324.58 |

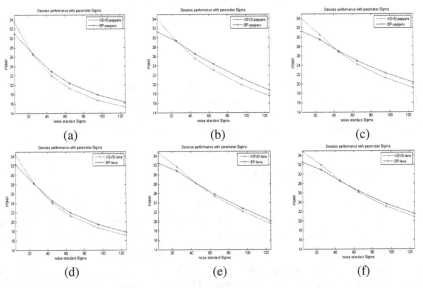

**Fig. 2.** PSNR values of K-SVD and our method with the increase of noise level. The top row is the performance for image "Peppers", and the bottom is for "Lena". Each column show the results tested by Gaussian noise, Salt and peppers noise and Speckle noise, respectively.

Our predictor is more obvious with the increase of noise level, compared to K-SVD. Figure 2 shows the experiments results. As seen from the curves of two algorithms in each figure, both methods perform well in low level of noise. But along with the increase of noise, PSNR slows down fast below Sigma=40. And performance tends to flat curves, when the noise continues to rise. For our algorithm, quality is inferior to the K-SVD in low noise level. As noise level updates, the proposed surpasses the other slightly. Even though the superiority is small, it has a tendency to keep the advantage.   And it is nearly same situation for all the types of noise, as the columns of figure depicting. A few examples are shown in Figure 3.

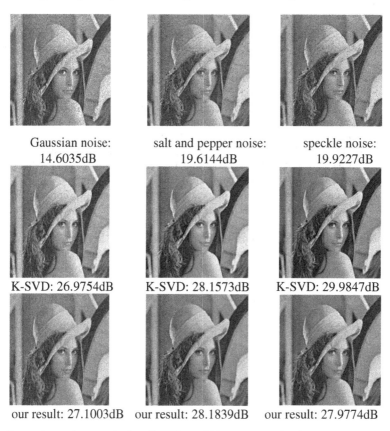

| Gaussian noise: 14.6035dB | salt and pepper noise: 19.6144dB | speckle noise: 19.9227dB |
| K-SVD: 26.9754dB | K-SVD: 28.1573dB | K-SVD: 29.9847dB |
| our result: 27.1003dB | our result: 28.1839dB | our result: 27.9774dB |

**Fig. 3.** Comparison of our method to K-SVD on different kinds of noise. The top row show the images contaminated by Gaussian noise, salt and pepper noise and speckle noise. The performance of K-SVD is described in the middle. And our results are slightly better than K-SVD under this noise level.

## 5    Conclusions

Our motivation is to train a fast approximate sparse encoder in which not sparse coding performance but also sparse coefficients inference speed are taken into account. In this paper, we propose a simple and efficient approach to learn fast approximate sparse coding networks as well as show its application to image denoising. Our experiments demonstrate that the pre-learned network is over 200 times faster than sparse optimization algorithm, and yet obtain approving result in image denoising. By training a neural network, the learned weights can achieve the performance of K-SVD algorithm. Although Gaussian noise is assumed in application to image denoising, the proposed approach is also valid to other types of noise.

# References

1. Aharon, M., Elad, M., Bruckstein, A.: K-SVD: An Algorithm for Designing Overcomplete Dictionaries for Sparse Representation. IEEE Transactions on Signal Processing 54(11), 4311–4322 (2006)
2. Motwani, M.C., Gadiya, M.C., Motwani, R.C., Harris, F.C.: Survey of Image Denoising Techniques. In: Proceedings of GSP 2004, Santa Clara, CA (September 2004)
3. Sadeghipour, Z., Babaie-Zadeh, M., Jutten, C.: An adaptive thresholding approach for image denoising using redundant representations. In: IEEE International Workshop on Machine Learning for Signal Processing, MLSP 2009, September 1-4, pp. 1–6 (2009)
4. Aharon, M., Elad, M., Bruckstein, A.: K-SVD: design of dictionaries for sparse representation. In: Proceedings of the Workshop on Signal Processing with Adaptative Sparse Structured Representations (SPARS 2005), Rennes, France, pp. 9–12 (2005)
5. Elad, M., Aharon, M.: Image Denoising Via Sparse and Redundant Representations Over Learned Dictionaries. IEEE Transactions on Image Processing 15(12), 3736–3745 (2006)
6. LeCun, Y., Bottou, L., Bengio, Y., Haffner, P.: Gradient-based learning applied to document recognition. Proceedings of the IEEE 86(11), 2278–2324 (1998)
7. Burger, H.C., Schuler, C.J., Harmeling, S.: Image denoising: Can plain neural networks compete with BM3D? In: 2012 IEEE Conference on Computer Vision and Pattern Recognition (CVPR), June 16-21, pp. 2392–2399 (2012)
8. Rojas, R.: Neural Networks: A Systematic Introduction. Springer, Berlin (1996)
9. Tropp, J.A.: Greed is good: Algorithmic results for sparse approximation. IEEE Trans. Inform. Theory 50(10), 2231–2242 (2004)

# Residual Image Compensations for Enhancement of High-Frequency Components in Face Hallucination

Yen-Wei Chen[1, 2], So Sasatani[2], and Xianhua Han[2]

[1] College of Computer Science and Information Technology,
Central South Univ. of Forestry and Technology, Hunan, China
[2] College of Information Science and Eng., Ritsumeikan University, Shiga, Japan

**Abstract.** Recently a growing interest has been seen in single-frame super-resolution techniques, which are known as example-based or learning based super-resolution techniques. Face Hallucination is one of such techniques, which is focused on resolution enhancement of facial images. Though face hallucination is a powerful and useful technique, some detailed high-frequency components cannot be recovered. In this paper, we propose a high-frequency compensation framework based on residual images for face hallucination method in order to improve the reconstruction performance. The basic idea of proposed framework is to reconstruct or estimate a residual image, which can be used to compensate the high-frequency components of the reconstructed high-resolution image. Three approaches based on our proposed framework are proposed. Experimental results show that the high-resolution images obtained using our proposed approaches can improve the quality of those obtained by conventional face hallucination method.

**Keywords:** Super Resolution, Facial Image, Face Hallucination, Residual Image, High-Frequency Components.

## 1    Introduction

There is a high demand for high-resolution (HR) images such as video surveillance, remote sensing, medical imaging and so on because high resolution images can reveal more information than low resolution images. However, it is hard to improve the image resolution by replacing sensors because of the high cost, hardware physical limits. Super resolution image reconstruction (SR) is one promising technique to solve the problem. SR can be broadly classified into two families of methods: (1) The classical multi-frame super-resolution [1], and (2) the single-frame super-resolution, which is also known as example-based or learning-based super-resolution [2-4]. In the classical multi-image SR, the HR image is reconstructed by combining subpixel-aligned multi-images (LR images). In the learning-based SR, the HR image is reconstructed by learning correspondence between low and high resolution image patches from a database.

Face Hallucination is one of learning-based SR techniques, which is focused on resolution enhancement of facial images [5-7]. Though face hallucination is a power-

C. Guo, Z.-G. Hou, and Z. Zeng (Eds.): ISNN 2013, Part I, LNCS 7951, pp. 627–634, 2013.
© Springer-Verlag Berlin Heidelberg 2013

ful and useful technique, some detailed high-frequency components cannot be recovered. In this paper, we propose a high-frequency compensation framework based on residual images for face hallucination method in order to improve the reconstruction performance. The basic idea of proposed framework, which is presented in our previous paper [8], is to reconstruct or estimate a residual image that can be used to compensate the high-frequency components of the reconstructed high-resolution image. This paper is an extension of our previous paper [8] and three approaches based on our proposed framework are studied.

The paper is organized as follows. In Section 2, we describe the conventional face hallucination method. Our proposed residual image compensation methods are presented in Section 3. Section 4 presents experimental results and quantitative evaluation. Section 5 summarizes our conclusions.

## 2    Face Hallucination

The face hallucination method is one of learning-based SR methods, which is proposed for resolution enhancement of facial images [5-7]. In this section, we briefly introduce the basic concept of face hallucination [8], which is shown in Fig.1.

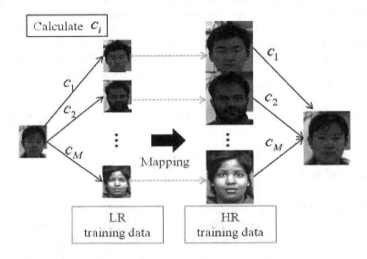

**Fig. 1.** The basic concept and schematic diagram of face hallucination

The basic idea of face hallucination is that a face image can be reconstructed from other face images by linear combination because all facial images have a similar structure. In face hallucination, an input LR image can be represented as a linear sum of the LR training images along with some learned coefficients. Due to the correlation between the LR and HR images in the training dataset, the output HR image can also be calculated by finding the linear sum of the corresponding HR images using the same coefficients.

We represent a two-dimensional face image using a column vector of all pixel values, and $\mathbf{X}_l$ represent the input LR face image. HR training images are denoted by $[\mathbf{H}_1, \mathbf{H}_2, \cdots, \mathbf{H}_M]$, and the corresponding LR training images are $[\mathbf{L}_1, \mathbf{L}_2, \cdots, \mathbf{L}_M]$, where $M$ is the number of training image pairs. First, we interpolate the LR training images and the input (test) LR image to the resolution space of the HR training images, denoted by $[\tilde{\mathbf{L}}_1, \tilde{\mathbf{L}}_2, \cdots, \tilde{\mathbf{L}}_M]$ and $\tilde{\mathbf{X}}_l$, respectively. $\tilde{\mathbf{X}}_l$ may be represented by a linear sum of interpolated training LR images using Eq. 1:

$$\tilde{\mathbf{X}}_l = c_1 \tilde{\mathbf{L}}_1 + c_2 \tilde{\mathbf{L}}_2 + \cdots + c_M \tilde{\mathbf{L}}_M \tag{1}$$

where $\mathbf{C} = [c_1, c_2, \cdots, c_M]$ are the weight coefficients, satisfying the following constraint:

$$c_1 + c_2 + \cdots + c_M = 1 \tag{2}$$

The optimal weights can be calculated by minimizing the error in reconstructing the input LR image $\tilde{\mathbf{X}}_l$ from training LR images. This error is defined in Eq. 3. After substitution of the constraints in Eq. 2 into Eq. 3, the weight vector may be obtained as Eq. 4.

$$\varepsilon(\mathbf{C}) = \left| \mathbf{X}_h - \tilde{\mathbf{X}} \right|^2 = \left| \mathbf{X}_h - \sum_{i=1}^{M} c_i \tilde{\mathbf{L}}_i \right|^2 \tag{3}$$

$$\mathbf{C} = \frac{\mathbf{S}^{-1} \mathbf{1}}{\mathbf{1}^T \mathbf{S}^{-1} \mathbf{1}}$$
$$\left( \mathbf{S} = (\tilde{\mathbf{X}}_l \mathbf{1}^T - \tilde{\mathbf{L}})^T (\tilde{\mathbf{X}}_l \mathbf{1}^T - \tilde{\mathbf{L}}) \right) \tag{4}$$

After obtaining the coefficients for reconstructing the input LR image with LR training images as given in Eq. 1, we replace $\tilde{\mathbf{L}}$ with $\mathbf{H}$ using the same coefficients $\mathbf{C}$. Subsequently, the HR image $\mathbf{X}_h$ can be obtained using the equation:

$$\mathbf{X}_h = c_1 \mathbf{H}_1 + c_2 \mathbf{H}_2 + \cdots c_M \mathbf{H}_M \tag{5}$$

## 3    Three Approaches for High Frequency Compensations

Though face hallucination is a powerful and useful technique, some detailed high-frequency components cannot be recovered. In this paper, we propose a high-frequency compensation framework based on residual images for face hallucination method in order to improve the reconstruction performance. The basic idea of proposed framework is to reconstruct or estimate a residual image, which can be used to compensate the high-frequency components of the reconstructed high-resolution image as shown in Fig. 2.

The HR and LR training image pairs $[\mathbf{H}_1, \mathbf{H}_2, \cdots, \mathbf{H}_M]$ and $[\mathbf{L}_1, \mathbf{L}_2, \cdots, \mathbf{L}_M]$ already exist in conventional face hallucination. With conventional face hallucination, for each LR training image $\mathbf{L}$, the other $M-1$ training image pairs are used to obtain the approximated HR image $\hat{\mathbf{H}}$. The HR training residual image is the difference between the original HR image and the reconstructed HR image $\hat{\mathbf{H}}$, while the LR residual image is the difference between the original LR image and the downsampled version of the reconstructed HR image. With the two training pair datasets, three approaches are proposed for high frequency compensation, which are shown in Figs.3-5.

HR Image          HR Residual          Final Image
                    Image

**Fig. 2.** Framework for recovering high-frequency components in face hallucination

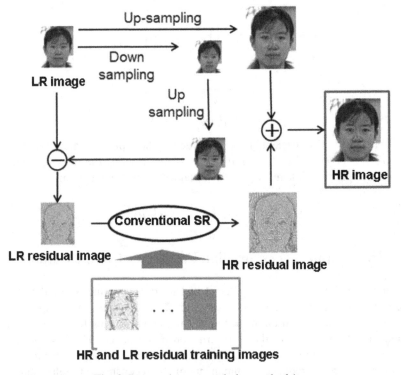

**Fig. 3.** Proposed super-resolution method 1

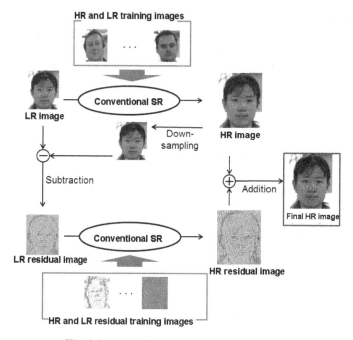

**Fig. 4.** Proposed super-resolution method 2 [8]

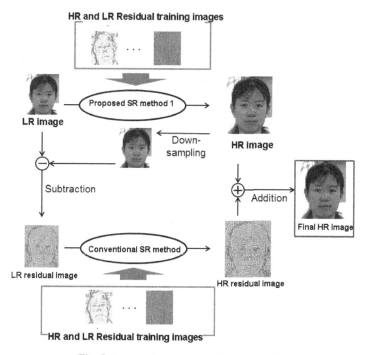

**Fig. 5.** Proposed super-resolution method 3

**Proposed Method 1:** We first use the conventional interpolation method to obtain a HR image and calculate the LR residual image between the input LR image and the downsampled reconstructed HR image. Then we reconstruct the HR residual image from the LR residual image using training residual image pairs. Finally we merge the HR residual and the interpolated HR images.

**Proposed Method 2 [8]:** We first use the conventional face hallucination method to obtain a HR image and calculate the LR residual image between the input LR image and the downsampled reconstructed HR image. Then we reconstruct the HR residual image from the LR residual image using training residual image pairs. Finally we merge the HR residual and the reconstructed HR images.

**Proposed Method 3:** We first use our proposed SR method 1 to obtain a HR image and calculate the LR residual image between the input LR image and the downsampled reconstructed HR image. Then we reconstruct the HR residual image from the LR residual image using training residual image pairs. Finally we merge the HR residual and the reconstructed HR images.

## 4    Experimental Results

In order to validate the effectiveness of our proposed methods, we apply our proposed three methods to two face databases. The first one is our developed MaVIC database (Multi-angle View, Illumination and Cosmetic Facial Image Database), which contains 99 aligned images of different persons and the size of each image is 320 × 400 [9]. The second one is C&P database provided by Cohn Kanade [10] and Pie [11], which contains 165 imperfectly aligned frontal face images, and each image size is 264×320. Each image in database is used as a HR image. We first generate the corresponding LR image by downsampling the original image (the HR image), which is a quarter of the size of the original HR image. Then we have a pair of HR and LR images for training.   The leave-one-out method is used in our experiments. In each database, we select one LR image randomly as a test image and its HR image is used as a groundtruth image for quantitative evaluation. Other image pairs are used for training. Our proposed three methods are used for HR reconstruction of the LR test image. In order to make a comparison, the conventional face hallucination method and the Bi-cubic interpolation method are also used for reconstructions. For each method, a total of 15 experiments with a different test image are performed. The Peak Signal-to-Noise Ratio (PSNR) [dB] is used as a quantitative measure for evaluation of the HR reconstruction performance. For C&P's imperfectly aligned facial datasets, our proposed patch face hallucination method [8] is used with a patch size of 3×3 and a 1×1 patch that overlaps with adjacent patches.

Typical experimental results are shown in Fig.6 and Fig.7. Figure 6 is for MaVIC database, while fig.7 is for C&P database. The averaged PSNR over 15 experiments for each method is summarized in Table 1. It can be seen that the reconstructed high-resolution images obtained using our proposed approaches are much better than those obtained by conventional face hallucination method and Bi-Cubic interpolation method and the proposed method 3 shows the best performance among three proposed methods.

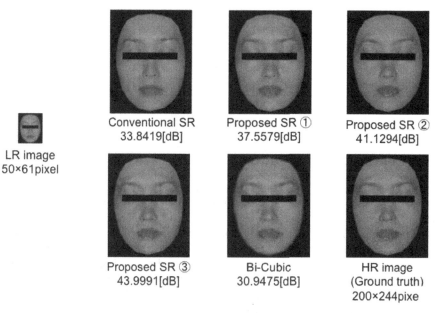

LR image
50×61pixel

Conventional SR
33.8419[dB]

Proposed SR ①
37.5579[dB]

Proposed SR ②
41.1294[dB]

Proposed SR ③
43.9991[dB]

Bi-Cubic
30.9475[dB]

HR image
(Ground truth)
200×244pixe

**Fig. 6.** Typical reconstructed images for the MaVIC database

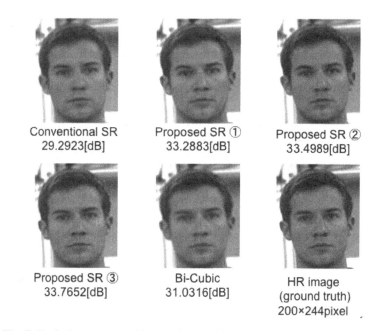

LR image
50×61pixel

Conventional SR
29.2923[dB]

Proposed SR ①
33.2883[dB]

Proposed SR ②
33.4989[dB]

Proposed SR ③
33.7652[dB]

Bi-Cubic
31.0316[dB]

HR image
(ground truth)
200×244pixel

**Fig. 7.** Typical reconstructed images for the C&P database

**Table 1.** Comparison of the averaged PSNR

|                  | MaVIC | C&P  |
| ---------------- | ----- | ---- |
| Conventional SR  | 33.5  | 31.2 |
| Proposed SR 1    | 38.6  | 32.9 |
| Proposed SR 2    | 41.3  | 33.4 |
| Proposed SR 3    | 43.7  | 33.8 |
| Bi-Cubic         | 31.2  | 30.5 |

# 5     Conclusions

We proposed a residual image compensation framework to improve the reconstruction quality for face hallucination. The basic idea of our proposed framework was to reconstruct or estimate a residual image, which can be used to compensate the high-frequency components of the reconstructed high-resolution image. Three approaches based on our proposed framework were proposed. The effectiveness of our proposed methods has been demonstrated on two face database: MaVIC and C&P. The reconstructed high-resolution images obtained using our proposed approaches are much better than those obtained by conventional face hallucination method and Bi-Cubic interpolation method and the proposed method 3 shows the best performance among three proposed methods.

# References

1. Park, S.C., Park, M.K., Kang, M.G.: Super Resolution Image Reconstruction: A Technical Overview. IEEE Signal Processing Magazine 20, 21–36 (2003)
2. Freeman, W.T., Jones, T.R., Pasztor, E.: Example-based super-resolution. IEEE Computer Graphics and Applications 22(2), 56–65 (2002)
3. Hang, H., Yeung, D.Y., Xiong, Y.: Super-resolution through neighbor embedding. In: CVPR, vol. 1, pp. 275–282 (2004)
4. Qiao, J., Liu, J., Chen, Y.M.: Joint Blind Super-Resolution and Shadow Removing. IEICE Trans. Information & Systems E90-D(12), 2060–2069 (2007)
5. Ma, X., Zhang, J., Qi, C.: An Example-Based Two-Step Face Hallucination Method through Coefficient Learning. In: Kamel, M., Campilho, A. (eds.) ICIAR 2009. LNCS, vol. 5627, pp. 471–480. Springer, Heidelberg (2009)
6. Wang, X., Tang, X.: Face Hallucination and Recognition. In: Kittler, J., Nixon, M.S. (eds.) AVBPA 2003. LNCS, vol. 2688, pp. 486–494. Springer, Heidelberg (2003)
7. Wang, X., Tang, X.: Hallucinating Face by Eigentransformation. In: ICIP 2003 (2003)
8. Sasatani, S., Han, X.-H., Igarashi, T., Ohashi, M., Iwamoto, Y., Chen, Y.-W.: High Frequency Compensated Face Hallucination. In: Proc. of IEEE International Conference on Image Processing, pp. 1561–1564 (2011)
9. Chen, Y.-W., Fukui, T., Qiao, X., Igarashi, T., Nakao, K., Kashimoto, A.: Multi-angle View, Illumination and Cosmetic Facial Image Database (MaVIC) and Quantitative Analysis of Facial Appearance. In: da Vitoria Lobo, N., Kasparis, T., Roli, F., Kwok, J.T., Georgiopoulos, M., Anagnostopoulos, G.C., Loog, M. (eds.) SSPR&SPR 2008. LNCS, vol. 5342, pp. 411–420. Springer, Heidelberg (2008)
10. Kanade, T., Cohn, J.F., Tian, Y.: Comprehensive database for facial expression analysis. In: Proc. IEEE Face and Gesture 2000, pp. 46–53 (2000)
11. Sim, T., Baker, S., Bsat, M.: The CMU pose, illumination, and expression (PIE) database. In: Proc. IEEE Face and Gesture 2002, pp. 53–58 (2002)

# Recognition Approach of Human Motion with Micro-accelerometer Based on PCA-BP Neural Network Algorithm

Yuxiang Zhang, Huacheng Li, Shiyi Chen, and Liuyi Ma

Xi'an Research Inst of Hi-Tech, Xi'an 710025, China
yuxiangz@tom.com, {huachengL135,15109111712}@163.com

**Abstract.** A human motion recognition method based on micro-acceleration sensor technology is put forward in this paper. Acceleration information acquire system is designed, which is including a tri-axial accelerometer, a micro-processor, a wireless transmission module and power supply program. The signal preprocessing and methods of feature extraction is analyzed. What's more, the experiment of human hand motion recognition based on BP neural network is carried out, results show that method proposed have recognition rate of 90%, compare the characteristics without processing and through principal component analysis (PCA) respectively after the identification experiment, the results show that the latter improve recognition effect and speed up convergence rate.

**Keywords:** motion recognition, acceleration sensor, feature extraction, PCA-BP neural network.

## 1  Introduction

In recent years, with the rapid development of micro-electromechanical systems, which have advantages of small size, low power consumption, low prices. The identification methods based on statistical pattern recognition is widely used in human motion tracking research[1-4]. Shiguang Yi used hidden Markov model for human motion recognition, motions such as walking, climbing stairs, identify activities is studied, action recognition rate between 90-100% [5]. Tong Zhang developed a wearable sensor and took advantage of the first-order support vector machine algorithm to monitor elder fall [6]. Neural network based on visual or wearable sensor in physical activity recognition have a wide range of applications[7][8].

In this essay, two aspects use neural network algorithm based on the sensor technology to recognize human motion are carried out. Firstly, micro-acceleration sensor module is designed and preprocessing of motion acceleration methods is studied. Secondly, BP neural network model as a human action recognition algorithm based on principal component analysis is proposed.

The structure of this article is organized as follows: The second section describes the hardware of sensor module design, the third part is preprocessing and feature

C. Guo, Z.-G. Hou, and Z. Zeng (Eds.): ISNN 2013, Part I, LNCS 7951, pp. 635–641, 2013.
© Springer-Verlag Berlin Heidelberg 2013

extraction methods of acceleration signal, the fourth section describes the PCA-BP neural network model and structure, the fifth part introduce human motion recognition experiments and results. The sixth part is the summary of full essay.

## 2    Design of Sensor Module

According the acceleration amplitude of human body in daily activities is at the range of ± 2g, the frequency of the movement is around 0 ~ 10Hz. This paper we selected a three-axis acceleration sensor LIS3LV02DQ, which is produced by ST Company, it contains sensing unit and IC interface, and the signal received form the sensing unit can be transmitted through I2C/SPI interface to an external device. It is very small (7 × 7 × 1.8mm), which can be embedded into hardware development board, having 640Hz frequency bandwidth and can be collected the 3D acceleration values range from ± 2g or ± 6g [9].

Data processing devices play the role of real-time data acquisition and storage of human motion acceleration. Enhanced 32-bit microcontroller STM32F103RBT6 produced by ST corporation is selected. The devices are designed for the requirements of high-performance, low-cost, low-power embedded system [10]. Human motion acceleration signal transmit in wireless access, the wireless data transmission module nRF24L01 module and NetUSB-24L11 module is selected to meet the synchronization of data transmission and reception of human action. Figure 1 is the object picture of wireless module.

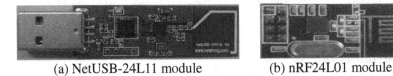

(a) NetUSB-24L11 module          (b) nRF24L01 module

**Fig. 1.** Wireless data transmission module

In order to ensure the entire acquisition system is lightweight and small size, a portable battery-powered system is used for power supply, and a separately powered plan is taken into account for this system. The switching power L6920 produced by ST corporation is chosen for the microcontroller and wireless data transmission module power supply. The output voltage range form 3V ~ 4.2V by lithium battery converted with DC-DC way, by this way can obtain stable 3.3V voltage for the power supply for microcontroller and wireless data transmission module, thus the power consumption of the system is reduced; so as to ensure pure power for acceleration sensor, a low dropout linear regulator power supply LM1117 is selected, which can ensure the accuracy and stability of the sensor supply. Figure 2 and Figure 3 is the electro circuit of wireless module respectively.

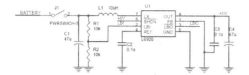

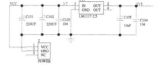

**Fig. 2.** Electro circuit of L6920          **Fig. 3.** Electro circuit of LM1117

Below figure 4 and 5 is the sensor module front and rear physical map:

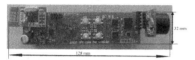

**Fig. 4.** Sensor module front map          **Fig. 5.** Sensor module rear map

# 3 Data Pre-processing and Feature Extraction

## 3.1 Data Preprocessing:

In order to obtain a clean motion acceleration signal, the pretreatment is necessary. Pretreatment including: denoising resampling motion signal, characteristic extraction.

**Denoising:** A five order Moving Average Filter (MAF) is selected for smoothing and denoising original acceleration signal of human motion, even though it is simple, but it is optimal for suppress random noise and keep the steep edge.

$$y_s(i) = \frac{1}{2N+1}\left(y(i+N) + y(i+N-1) + \ldots + y(i-N)\right) \tag{1}$$

Where $y_i$ is the original value of the signal, N=5 in this passage, $y_s(i)$ is result of the i point signal after filtering. Figure 6 shows motion acceleration signal filtering result after the five order moving average filter, the above part is the raw acceleration signal, the below part is the moving average filter result, the noise in acceleration has been removed and the details and peak information also have been retained.

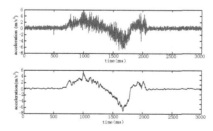

**Fig. 6.** Filter result of acceleration signal

**Motion Signals Extraction:** Since the acceleration sensor LIS3LV02DQ having little drifts in the static condition, the action signal and non-action signal are distinguished well after differential operation. Set acceleration Root-Mean-Square (RMS) value to $0.25m/s^2$ as threshold to judge whether the differential signal have action signal in time window. Sliding the time window, if the RMS value of signal after differential in time window is greater than $0.25m/s^2$, that point is considered as motion start point, while moving the time window, judge acceleration RMS value is less than $0.25m/s^2$ in next time sliding window as the end point of this motion.

**Resampling:** The sensor acquisition system sampling frequency is 1KHz, different people do the same action, those smartness, the signal have short length, other slowness, the signal length is long, thus before the feature extraction and action recognition, it is necessary to ensure the motion signal have equivalent length. Pretreatment of resampling can achieve this purpose.

## 3.2    Feature Extraction

Studies show that after the original data feature extraction, the recognition rate can improve as much as 40%, so accurate select and extract feature information from acceleration is particular significance to recognition. By investigating the method used in characteristics extraction in pattern recognition system [11]. This essay extracted the following four time domain characteristics (mean, variance, signal amplitude area, RMS value, and axis correlation coefficient) for recognition algorithm.

## 4    PCA-BP Neural Network Recognition Algorithm

The structure of BP neural network is shown in Figure 7, it is generally composed by Three-Layer structure, the input layer, hidden layer and output layer, each layer containing one or more input nodes, hidden layer may be multilayer structure, two adjacent layer neurons is connected by adjustable weights.

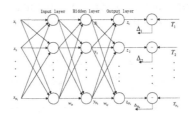

**Fig. 7.** Structure of BP neural network

The input information passed the input layer through the hidden layer, and ultimately transmit to output layer, if the output of the output layer is not equal to the expectation output value, the error back propagation process begin. Network weights are adjusted in error feedback process, the weight constantly revised so that the network output is closed to expectation output. BP neural network is widely applied because of using error back propagation algorithm. However, when this network meet complex problem, huge network structure will cause slow convergence speed, poor generalization ability. The principal component analysis is proposed to deal with feature data, aim at reducing dimension, removing correlation between character vector of the feature data and simplifying network.

If we define the acceleration characteristics extracted is $p$ dimension vector, $X = (x_1, x_2, \ldots x_p)$ 'is consists of $p \times p$ dimension random vector, the purpose of principal component analysis is converted $p$ dimension vectors to $k$ ($k \leq p$)dimension linear combination vector, while new $k$ dimension vector is mutually independent and can keep main information of $p$ dimension before converted, if we get $A = (A_1, A_2, \ldots A_k)$ as the result of PCA , A is given by:

$$\begin{cases} A_1 = b_{11}x_1 + b_{21}x_2 + \ldots + b_{p1}x_p \\ A_2 = b_{12}x_1 + b_{22}x_2 + \ldots + b_{p2}x_p \\ \ldots \\ A_k = b_{1k}x_1 + b_{2k}x_2 + \ldots + b_{pk}x_p \end{cases} \tag{2}$$

After matrix transformations (2), the selection of principal components is based on the value of cumulative variance contribution rate and variance contribution rate, which reflect new variable explain ability to all original variables, the larger its value, the more important the variable is. Given $\lambda_1, \lambda_2 \ldots \lambda_p$ are characteristic values of covariance matrix $D(X)$, the contribution rate of the $j$ main component can be explained as follows:

$$\lambda_j \Big/ \sum_{i=1}^{p} \lambda_i \tag{3}$$

The cumulative contribution rate of the former m principal components given as:

$$\sum_{i=1}^{m} \lambda_j \Big/ \sum_{i=1}^{p} \lambda_i \tag{4}$$

The selection of main component lie on the cumulative contribution ratio, the larger it is, the smaller data information loss, generally the standard of value $m$ is make the cumulative contribution rate reach 85% or more.

## 5    Experiments and Results

According to the algorithm proposed, the recognition based on motion acceleration is carried out, six motion(waving hand forward and backyard, arms waving around, fetching hand forward, rotation wrist clockwise, rotation counterclockwise, still) are recognition action. Sensor is put at the hand to collect motion data. The recognition research is carried out in Matlab7.1 software operation environment. Six motion data put into BP neural network after preprocessing, the structure of BP neural network is

13-20-1, each motion date collected 60 times, 30 times as training samples, another 30 times as test samples. Group learning way is adopted. Figure 8 is the training error curve gained by traingdx algorithm, from Figure 8 we can see that the network ran to step 4387, it reached the initialize accuracy requirements. Save the trained network, and simulate the test sample, network output 1 x 180 data values is obtained, which is shown in Figure 9.

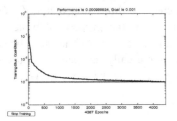

**Fig. 8.** Error curve of traingdx algorithm

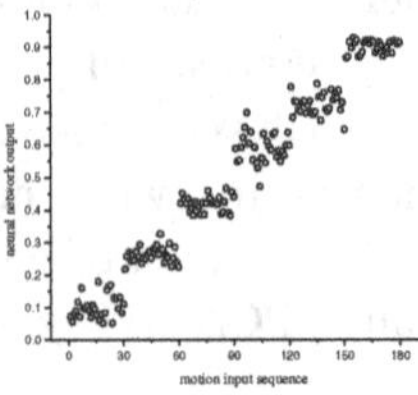

**Fig. 9.** Recognition result of test samples

In figure 9, each step can be regarded as an action type of motion, it can be clearly found that the input samples of each operation have a good classification and recognition. It is also appear some samples recognition wrong. However, the network ensures a good recognition effect for human motion.

In order to test whether the principal components obtained contain full 13 dimensions characteristics and have equivalent recognition effect. The structure of the PCA-BP neural network 4-6-1 is established, the transfer function is a hyperbolic tangent function, adaptive change learning rate and momentum components algorithm is chosen as training function and mean square error is 0.001. The error performance chart is shown as Figure 10. At 1285 step iteration network achieve required accuracy. Compared with figure 8, the principal component analysis process reduces network complexity and speeds network convergence rate.

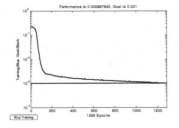
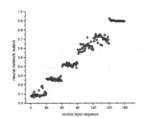

**Fig. 10.** Error curve of traingdx algorithm with PCA

**Fig. 11.** Recognition result of test samples with PCA

Similarly the test samples is simulated after the network learning well, test samples outputs can be draw as figure 11. Compared with figure 9, we can know that the four principal components through principal component analysis not only include all the information of the 13 features, but also obtain more focused and accurate recognition results, which are verified the proposed method correctness and validity.

## 6    Summary and Forwards

Based on the characteristics of micro-sensor technology, human motion acceleration acquisition module is designed in this passage, which is consists of a tri-axis accelerometer, a microprocessor, a wireless transmission module and power supply module. The acceleration signal is preprocessed and the mean, variance, amplitude of signal area, RMS value of acceleration signals are extracted. BP neural network algorithm have been selected as identification algorithm, two groups sample untreated and principal component analysis is compared for recognition experiment, the final results show that the action recognition rate as much as 90%, which is verify accuracy and reliability of proposed method.

**Acknowledgements.** This work was financially supported by the National Natural Science Foundation of China (51275517).

## References

[1] Ashik Eftakhar, S.M., Tan, J.K.: Hyongseop Kim Multiple Persons. In: Action Recognition by Fast Human Detection SICE Annual Conference, pp. 1639–1644 (2011)

[2] Hägg, J., Akan, B., Çürüklü, B.: Gesture Recognition Using Evolution Strategy Neural Network, pp. 245–248 (2008)

[3] Zhu, C., Sheng, W.: Wearable Sensor-Based Hand Gesture and Daily Activity Recognition for Robot-Assisted Living. IEEE Transactions on systems, Man and Cybernetics-Part A Systems and Humans 41(3), 569–573 (2011)

[4] (EB/OL), http://www.wii.com

[5] Zhang, T., Wang, J., Xu, L., Liu, P.: Fall Detection by Wearable Sensor and One-class SVM Algorithm. In: Huang, D.-S., Li, K., Irwin, G.W. (eds.) ICIC 2006. LNCIS, vol. 345, pp. 858–863. Springer, Heidelberg (2006)

[6] Shi, G., Zou, Y., Jin, Y.: Towards HMM based Human Motion Recognition using MEMS Inertial Sensors. In: Proceedings of the 2008 IEEE International Conference on Robotics and Biomimetics, pp. 1762–1766 (2008)

[7] Randell, C., Muller, H.: Context awareness by analysing accelerometer data. In: Proc. ISWC, Atlanta, GA, USA, pp. 175–176 (2000)

[8] Yang, J.Y., Wang, J.S., Chen, Y.P.: Using acceleration measurements for activity recognition:an effective learning algorithm for constructing neural classifier. Pattern Recognition Letters 29, 2213 (2008)

[9] Datasheet, http://www.st.com/STM32F103x6

[10] Datasheet, http://www.st.com/LIS3LV02DQ

[11] Lester, J., Choudhury, T., Borriello, G.: A practical approach to recognizing physical activities. In: Fishkin, K.P., Schiele, B., Nixon, P., Quigley, A. (eds.) PERVASIVE 2006. LNCS, vol. 3968, pp. 1–16. Springer, Heidelberg (2006)

# A Stable Dual Purpose Adaptive Algorithm for Subspace Tracking on Noncompact Stiefel Manifold[*]

Lijun Liu, Yi Xu, and Qiang Liu

School of Science, Dalian Nationalities University, Dalian 116600, P.R. China

**Abstract.** Starting from an extended Rayleigh quotient defined on the noncompact Stiefel manifold, in this paper, we present a novel dual purpose subspace flows for subspace tracking. The proposed algorithm can switch from principal subspace to minor subspace tracking with a simple sign change of its stepsize parameter. More interestingly, the proposed dual purpose gradient system behaves the same invariant property as that of the well-known Chen-Amari-Lin system. The stability of the discrete version of the proposed subspace flow is guaranteed by an additional added stabilizing term. No tunable parameter is required for the proposed algorithm as opposed to the modified Oja algorithm. The strengths of the proposed algorithm is demonstrated using a $defacto$ benchmark example.

## 1 Introduction

Subspace tracking plays a crucial role in a variety of adaptive subspace-based methods [7]. The problem of subspace tracking can be formulated as follows. Let $x(t)$ be a sequence of $n \times 1$ random data vectors with correlation matrix $C = E[x(t)x^T(t)]$ where $E[\cdot]$ denotes the expectation operator. We aim at tracking the $k$-dimensional $(k < n)$ principal subspace (PS) or minor subspace (MS) in an online manner, which incrementally updated an $n \times k$ matrix $W$ in a sample-by-sample manner.

In the last decades, a large number of algorithms have been proposed for PS or MS tracking separately ([12,1,13]). Recently there is considerable interest in constructing the dual purpose subspace tracking algorithms, in which the PS or MS subspace are adaptively tracked in the sense of simply switching the sign of the corresponding stepsize learning parameters [11,3,5,10,4].

The first dual purpose gradient flow was the modified version of Oja's flow by Chen [3]

$$\frac{dW(t)}{dt} = \pm \left\{ CW(t) - W(t)W^T(t)CW(t) \right\} + \beta W(t) \left\{ I - M(t)^T M(t) \right\}, \quad (1)$$

where $\beta > 0$ can be viewed as a penalty parameter. When $\beta = 0$, the modified is happened to be the well-known Oja's PS flow [12]. The main problem of (1) is that the penalty parameter $\beta$ should be selected as greater than the largest eigenvalue of the correlation matrix $C$, which definitely requires the prior information on the eigenstructure,

[*] Supported by National Natural Science Foundation of China ( No. 61002039 ) and The Fundamental Research Funds for the Central Universities ( No. DC10040121, DC12010216).

C. Guo, Z.-G. Hou, and Z. Zeng (Eds.): ISNN 2013, Part I, LNCS 7951, pp. 642–649, 2013.

thus make it not applicable in cases the whole signal process is unobtainable. There exists the same problem for the dual system proposed by Manton et al. [11]

$$\frac{dW(t)}{dt} = \pm CW(t)N + \beta \left\{ N - W^T(t)W(t) \right\}, \tag{2}$$

for dual purpose principal and minor components extraction, where $N \in R^{n \times k}$ is a positive diagonal matrix.

In view of Lie group theory, Chen, Amari and Lin [5] proposed the following dual purpose subspace flow by means of the natural gradient defined on the orthogonal matrix group $SO(n)$,

$$\frac{dW(t)}{dt} = \pm \left\{ CW(t)W^T(t)W(t) + W(t)W^T CW(t) \right\}. \tag{3}$$

Recently, Kong et al. [10] successfully proposed the following dual gradient flow

$$\frac{dW(t)}{dt} = \pm \left[ CW(t) - W(t) \left\{ W^T(t)W(t) \right\}^{-1} W^T(t)CW(t) \right] \times \left\{ W^T(t)W(t) \right\}^{-1} + W(t) \left[ I - \left\{ W^T(t)W(t) \right\} \right]. \tag{4}$$

Notice that its feasible domain is assumed to be $\Omega = \{W | 0 < W^T CW < \infty, W^T W \neq 0\}$, which also requires information on the data correlation matrix. When $C$ is assumed to be symmetric positive definite, in fact, the condition that $W^T CW > 0$ is equivalent to that $W$ is of full column rank. When it comes to the indefinite matrix, these two conditions are not equivalent.

In this paper, we present a novel dual purpose subspace tracking algorithm based on the optimization of an extended Rayleigh quotient defined on the noncompact $n \times k$ matrix manifold. It can handle both the principal and minor subspace tracking with a simple sign change of the stepsize parameter. The penalized version of the proposed algorithm does not require any prior information on the eigenstructure of the correlation matrix to adjust its stepsize parameter. Orthogonality constraint is finally guaranteed without explicit step by step re-orthogonalization of the weight matrix $W(t)$ during the learning process.

## 2    Derivation of the Novel Dual Gradient Flow for Subspace Tracking

Notice that for $w \in R^n$, it is well-known that Rayleigh quotient $r(w) = \frac{w^T Cw}{w^T w}$ is maximized (resp. minimized) when and only when $w$ is equal to the principal (resp. minor) eigenvector of the symmetric matrix $C$. In order to computing the principal and minor subspace of $C$, in this paper, we consider the matrix form extended Rayleigh quotient (ERQ) [9,2]

$$J_{ERQ}(W) = \text{tr} \left( W^T RW (W^T W)^{-1} \right) \tag{5}$$

defined on the noncompact Stiefel manifold

$$\mathrm{ST}(n, k) = \{W \in R^{n \times k} | \mathrm{rank}(W) = k\} \tag{6}$$

of all full rank $n \times k$ matrices.

Throughout this paper, $C$ is assumed to have the following decomposition

$$C = V \,\mathrm{diag}(\lambda_1, \lambda_2, \cdots, \lambda_n) V^T \tag{7}$$

where $\lambda_1 \geq \lambda_2 \geq \cdots \geq \lambda_n$ are the eigenvalues of $C$ and $V = [v_1 \, v_2 \, \cdots \, v_n]$ is the orthonormal eigenvector matrix of $C$. Let $V_i$ and $L_i$ be the matrix consisting of the first $i$ and last $i$ columns of $V$ respectively. In other words, $V_i$ is composed of the $i$ eigenvectors of $C$ corresponding to the $i$ largest eigenvalues $\lambda_1, \lambda_2, \cdots, \lambda_i$, and $L_i$ is composed of the $i$ eigenvectors of $C$ corresponding to the $i$ smallest eigenvalues $\lambda_{n-i+1}, \lambda_{n-i+2}, \cdots, \lambda_n$.

For the proposed extended Rayleigh quotient, we have the following desired results.

**Theorem 1.** *$W$ is a critical point of $J_{ERQ}(W)$ on $\mathrm{ST}(n, k)$ if and only if $W = \widetilde{V}_k M$, where $\widetilde{V}_r \in R^{N \times k}$ consists of the $k$ eigenvectors of $C$ and $M$ is any nonsingular $k$-by-$k$ matrix.*

*Proof.* The gradient of $J_{ERQ}$ with respect to $W$ is easily obtained as

$$\frac{1}{2}\nabla J(W) = \left[AW - W(W^T W)^{-1} W^T AW\right](W^T W)^{-1} \tag{8}$$

If $W = \widetilde{V}_k M$, where $M$ is any nonsingular $k$-by-$k$ matrix and $\widetilde{V}_k$ contains any $k$ distinct orthonormal eigenvectors of $C$, it is straightforward to show $\nabla J(W) = 0$.

Conversely, $\nabla J(W) = 0$ implies that $AW = W(W^T W)^{-1} W^T AW$, which means that $AW \in \mathrm{Span}(W)$, or equivalently $\mathrm{Span}(AW) \subset \mathrm{Span}(W)$. So we reach the conclusion that $W$ is an invariant subspace of $A$. But any $k$−dimensional invariant subspace of $C$ is spanned by $k$ eigenvectors $v_{i_1}, \cdots, v_{i_k}$. Denote $\widetilde{V}_k = [v_{i_1} \cdots v_{i_k}]$, hence $W$ is a critical point only it is of the form $W = \widetilde{V}_k M$, where $M$ is any nonsingular $k$-by-$k$ matrix. The proof is completed.

The following result directly follows from Theorem 1.

**Theorem 2.** *On manifold $\mathrm{ST}(n, k)$, the extended Rayleigh quotient $J_{ERQ}(W)$ reaches its global maximum when and only when $W = V_k M$ for arbitrary nonsingular $k$-by-$k$ matrix $M$. $J_{ERQ}(W)$ reaches its global minimum when and only when $W = L_k M$ for arbitrary nonsingular $k$-by-$k$ matrix $M$. All the other critical points are saddle points of $J_{ERQ}(W)$. Moreover the minimum and maximum value of $J_{ERQ}(W)$ are*

$$\min_{W \in \mathrm{ST}(n,k)} J_{ERQ}(W) = \lambda_{n-k+1} + \lambda_{n-k+2} + \cdots + \lambda_n, \tag{9}$$

$$\max_{W \in \mathrm{ST}(n,k)} J_{ERQ}(W) = \lambda_1 + \lambda_2 + \cdots + \lambda_k. \tag{10}$$

*Remark 1.* The above theorem shows that the extended Rayleigh quotient $J_{ERQ}$ has a global maximum (respectively, minimum) at which the columns of $W$ span

$k-$dimensional principal (respectively, minor) eigen-subspace of $C$ and no other local extremum. This implies that one can search the global extremum point of $J_{ERQ}$ by iterative methods.

Therefore minimizing $J_{ERQ}(W)$ on $\mathrm{ST}(n, k)$ will automatically result in a solution $W$ in the $k$-dimensional minor subspace of $C$. Maximizing $J_{ERQ}(W)$ on $\mathrm{ST}(n, k)$ will definitely result a $W$ in the $k$-dimensional principal subspace. In this sense, the proposed extended Rayleigh quotient is an ideal cost candidate for our dual purpose subspace flow.

Further define the Riemannian metric on $\mathrm{ST}(n, k)$ as

$$< \Omega_1, \Omega_2 >:= \mathrm{tr}\ (\Omega_1(W^T W)^{-1}\Omega_2^T), \forall \Omega_1, \Omega_2 \in T_W\ \mathrm{ST}(n, k), \qquad (11)$$

where $T_W\ \mathrm{ST}(n, k)$ denotes the tangent space at point $W \in \mathrm{ST}(n, k)$.

The following results list the Riemannian gradient formula on $\mathrm{ST}(n, k)$.

**Theorem 3.** *Consider the cost function $J_{ERQ}$ given in (5) defined on the noncompact Stiefel manifold $\mathrm{ST}(n, k)$ given by (6). With respect to the induced Riemannian metric in (11), the gradient can be calculated as*

$$\mathrm{grad}\ J = CW - W(W^T W)^{-1}W^T CW \qquad (12)$$

*Proof.* In fact, the directional derivative of (15) in the direction $\Omega \in R^{n \times k}$ can be computed as

$$\frac{1}{2}DJ_W(\Omega) = \mathrm{tr}\ (\Omega^T CW(W^T W)^{-1}) + \frac{1}{2}\mathrm{tr}\ (W^T CW(W^T W)^{-1}(\Omega^T W + W^T \Omega)(W^T W)^{-1})$$

$$= \mathrm{tr}\ (\Omega^T CW(W^T W)^{-1}) + \mathrm{tr}\ (\Omega^T W(W^T W)^{-1}(W^T CW)(W^T W)^{-1})$$

$$= \mathrm{tr}\ [\Omega^T\ (CW(W^T W)^{-1} - W(W^T W)^{-1}(W^T CW)(W^T W)^{-1})] \qquad (13)$$

As such, the gradient of $J_{ERQ}(W)$ with respect to Riemannian metric (11) is readily obtained as

$$\mathrm{grad}\ J = CW - W(W^T W)^{-1}W^T CW, \qquad (14)$$

which completes the proof.

Therefore, the novel gradient flow of $J_{ERQ}(W)$ for PS and MS tracking is derived as

$$\frac{dW(t)}{dt} = \pm\left\{CW(t) - W(t)\left(W^T(t)W(t)\right)^{-1}W^T(t)CW(t)\right\}, \qquad (15)$$

where "+" is used for PS tracking, and "−" is for MS tracking.

Direct computation shows that

$$\frac{d}{dt}\left\{W^T(t)W(t)\right\} = 0 \qquad (16)$$

along the gradient flow given in (15), which is summarized as the following proposition.

**Theorem 4.** $W^T(t)W(t)$ *is invariant under the dynamical evolvements in* (15).

The above theorem shows that $W^T(t)W(t) = W^T(0)W(0)$ for any $t \geq 0$. Also a direct consequence of this theorem is that solution $W(t)$ for initial value problem of (15) exists for $t \in [0, \infty)$. Thus if we choose $W(0) = \begin{bmatrix} D^{1/2} \\ 0 \end{bmatrix} \in R^{n \times k}$ of a diagonal matrix $D = \text{diag}\{d_1, d_2, \cdots, d_k\}$ with $d_i > 0$, then the proposed dual flow (15) have an invariant submanifold

$$S_D = \{W | W^T W = D\}. \tag{17}$$

Therefore, starting from any initial orthognormal matrix on the Stiefel manifold $St(n, k) = \{W \in R^{n \times k} | W^T W = I_k\}$, the weight flow $W(t)$ of the gradient system (15) evolves on $St(n, k)$ for all $t \geq 0$.

In order to maintain the stability of the proposed algorithm [8], similar to the approach in [4], in this paper we present the following stable flow for dual subspace tracking.

$$\frac{dW(t)}{dt} = \pm \left[ CW(t) - W(t) \left[ W^T(t)W(t) \right]^{-1} W^T(t)CW(t) \right] + W(t) \left[ D - W^T(t)W(t) \right], \tag{18}$$

where " $+$ " is for the PS tracking and " $-$ " is for MS tracking. Obviously, the equilibrium as well as the invariant property of the original dynamical system (15) are unchanged by this additional term. The extra added term $W(t)[D - W^T(t)W(t)]$ helps the gradient flow $W(t)$ to conform to the invariant submanifold constraint $W \in S_D$. If we set $D = I_k$, $S_D$ is just the orthonormal constraint of the extracted principal and minor subspace. The effectiveness of the stabilizing term is further confirmed in the section of computer simulation.

## 3    Computer Simulation

In this section, we present some simulation results to show the effectiveness of the proposed algorithm. A de facto benchmark example is used to evaluate the performance of our algorithm. Assume the signal vector $x(t)$ is generated by the following noisy model,

$$x(t) = s(t) + n(t), \tag{19}$$

where $s(t)$ is a sequence of independent jointly Gaussian random vectors with correlation matrix [1]

$$C = \begin{bmatrix} 0.9 & 0.4 & 0.7 & 0.3 \\ 0.4 & 0.3 & 0.5 & 0.4 \\ 0.7 & 0.5 & 1.0 & 0.6 \\ 0.3 & 0.4 & 0.6 & 0.9 \end{bmatrix} \tag{20}$$

and $n(t)$ is white Gaussian noise with $SNR = 0.01$dB. The purpose is to adaptively track the $k = 2$ dimensional principal and minor subspace from $\{x(t)\}$ by using the proposed algorithm (15). Initial weight matrix $W(0)$ is set to the first $k$ columns of the $n \times n$ identity matrix. For the stepsize $\mu > 0$, we follow the search-then-converge style [6] as $\mu = \sigma/(1 + t/T)$, where $T = \text{Iter}_{\text{num}}/2$ is one half of the iteration number $\text{Iter}_{\text{num}}$ and $\sigma = 0.01$.

The performances of the four considered algorithms are observed during the learning phase through the following two parameters:

– A measure of the orthogonality of $W$, defined as:

$$\gamma(t) \overset{\text{def}}{=} \|W^T(t)W(t) - I\|_{\text{F}}, \tag{21}$$

where $\| \cdot \|_{\text{F}}$ denotes the Frobenius norm.
– A measure of how much $W$'s columns are deviated from the principal or minor subspace, defined as:

$$\rho(t) \overset{\text{def}}{=} \|(I - VV^T)W\|_{\text{F}} \tag{22}$$

with $V \in R^{4 \times 2}$ being a orthonormal base of the principal subspace for PSA( respectively, minor subspace for MSA) of $x(t)$.

**Results and Discussion**

We first consider the task of principal subspace tracking. The results are shown in Fig. 1. It can be seen from Fig. 1(a), both the original algorithm and the penalized version successfully extract the desired principal subspace after a sufficiently large iterations. Meanwhile, as for the orthogonality of the weight matrix $W$, the penalized version outperforms the non-penalized one, as is shown from Fig. 1(b).

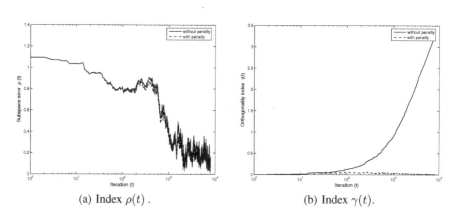

(a) Index $\rho(t)$.          (b) Index $\gamma(t)$.

**Fig. 1.** Performance of PS tracking by the of the proposed algorithm

The minor subspace tracking ability of the proposed algorithm is illustrated in Fig. 2. Again we see that the effectiveness of the novel dual purpose algorithm for MS tracking. Also the orthogonality of the weight matrix $W$ is guaranteed with the additional constraint term in (18), which is verified in Fig. 2(b).

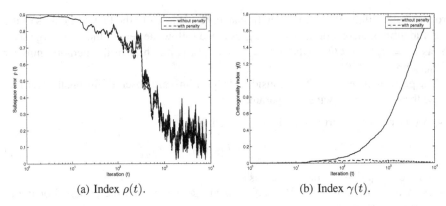

(a) Index $\rho(t)$.                    (b) Index $\gamma(t)$.

**Fig. 2.** Performance of MS tracking of the proposed algorithm

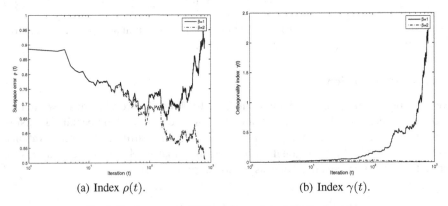

(a) Index $\rho(t)$.                    (b) Index $\gamma(t)$.

**Fig. 3.** Performance of MS tracking of the MOJA algorithm

By the modified Oja algorithm (MOJA), by setting $\beta = 1$, Fig. 3(a) and Fig. 3(b) clearly show that both $\rho(t)$ and $\gamma(t)$ diverges for MS task. To see the reason why MOJA failed in this occasion for MS tracking, in fact, theoretically speaking $\beta > 0$ should be larger than the largest largest eigenvalue of $C$ in order to extract the minor subspace. In our case, the noisy Gaussian data has a correlation matrix with the largest eigenvalue $\lambda_{max} \approx 2.3096$. Thus for $\beta = 1$ in our case, the MOJA failed to extract the minor space. Alternatively, if we set $\beta = 2$, the MOJA algorithm can successfully extract the minor subspace, as is shown in Fig. 3. It is interesting that the proposed ERQ based algorithm is stable for any $\beta > 0$, which shows its superiority over the MOJA algorithm.

## 4   Conclusion

A novel dual purpose adaptive algorithm for principal and minor subspace tracking is proposed in this paper. Different from most existing works in this area, we derived the desired gradient flow on noncompact Riemannian manifold $\mathrm{ST}(n, k)$ from an extended Rayleigh quotient. It guarantees the orthonomality of the tracked subspace and achieves high numerical stability with an additional stabilizing term. Moreover, different from the modified OJA algorithm, the proposed algorithm does not need a prior information on the eigenvalue of the correlation matrix.

# References

1. Abed-Meraim, K., Attallah, S., Chkeif, A., Hua, Y.: Orthogonal Oja Algorithm. IEEE Signal Processing Letters 7(5), 116–119 (2000)
2. Absil, P.A., Mahony, R., Sepulchre, R., Dooren, P.V.: A Grassmann-Rayleigh Quotient Iteration for Computing Invariant Subspaces. SIAM Review 44(1), 57–73 (2002)
3. Chen, T.: Modified Oja's Algorithms for Principal Subspace and Minor Subspace Extraction. Neural Processing Letters (1), 105–110 (1997)
4. Chen, T., Amari, S.I.: Unified stabilization approach to principal and minor components extraction algorithms. Neural Networks 14, 1377–1387 (2001)
5. Chen, T., Amari, S.I., Lin, Q.: A unified algorithm for principal and minor components extraction. Neural Networks 11, 385–390 (1998)
6. Darken, C., Chang, J., Moody, J.: Learning rate schedules for faster stochastic gradient search. In: Neural Networks for Signal Processing, pp. 3–12 (1992)
7. Delmas, J.: Subspace tracking for signal processing. In: Adali, T., Haykin, S. (eds.) Adaptive Signal Processing: Next Generation Solution, pp. 211–270. Wiley (2010)
8. Douglas, S.C., Kung, S., Amari, S.I.: A self-stabilized minor subspace rule. IEEE Signal Processing Letters 5(12), 328–330 (1998)
9. Helmke, U., Moore, J.B.: Optimization and dynamical systems. Springer, Heidelberg (1993)
10. Kong, X., Hu, C., Han, C.: A Dual Purpose Principal and Minor Subspace Gradient Flow. IEEE Transactions on Signal Processing 60(1), 197–210 (2012)
11. Manton, J.H., Helmke, U., Mareels, I.M.: A dual purpose principal and minor component flow. Systems & Control Letters 54(8), 759–769 (2005)
12. Oja, E.: Principal components, minor components, and linear neural networks. Neural Networks 5(6), 927–935 (1992)
13. Xu, L.: Least mean square error reconstruction principle for self-organizing neural-nets. Neural Networks 6(5), 627–648 (1993)

# Invariant Object Recognition Using Radon and Fourier Transforms

Guangyi Chen[1], Tien Dai Bui[1], Adam Krzyzak[1], and Yongjia Zhao[2]

[1] Department of Computer Science and Software Engineering, Concordia University,
1455 de Maisonneuve West, Montreal, Quebec, Canada H3G 1M8
{guang_c,bui,krzyzak}@cse.concordia.ca
[2] State Key Lab. of Virtual Reality Technology and Systems, Beihang University, ZipCode
100191, No 37, Xueyuan Rd., Haidian District, Beijing, P.R. China
activezyj@126.com

**Abstract.** In this paper, an invariant algorithm for object recognition is proposed by using the Radon and Fourier transforms. It has been shown that this algorithm is invariant to the translation and rotation of pattern images. The scaling invariance can be achieved by the standard normalization techniques. Our algorithm works even when the center of the pattern object is not aligned well. This advantage is because the Fourier spectra are invariant to spatial shift in the radial direction whereas existing methods assume the centroids are aligned exactly. Experimental results show that the proposed method is better than the Zernike's moments, the dual-tree complex wavelet (DTCWT) moments, and the auto-correlation wavelet moments for one aircraft database and one shape database.

**Keywords:** Radon transform, Fourier transform, Zernike's moments, object recognition, pattern recognition, Gaussian white noise.

## 1    Introduction

Moment invariants [1] have been a very active research topic since Hu [2] proposed the first moments for 2D pattern recognition in 1962. Khotanzad and Hong [3] studied the magnitudes of Zernike's moments in rotational invariant recognition of characters and shapes. They concluded that Zernike's moments are better than Hu's moments. Chen and Xie [4] developed the auto-correlation wavelet moments and the dual-tree complex wavelet (DTCWT) moments for recognizing 2D pattern images.

In this paper, we propose a new algorithm for object recognition by using a combination of the Radon transform and the Fourier transform. The new algorithm is invariant to translation and rotation of the object to be recognized. Scaling invariance can be achieved by standard normalization techniques. Experimental results show that the proposed algorithm is better than the Zernike's moments, the auto-correlation wavelet moments, and the DTCWT moments for recognizing 2D shapes. This indicates that the proposed algorithm in this paper is a practical approach for invariant object recognition.

C. Guo, Z.-G. Hou, and Z. Zeng (Eds.): ISNN 2013, Part I, LNCS 7951, pp. 650–656, 2013.

The organization of this paper is as follows. Section 2 proposes an algorithm for invariant pattern recognition. Section 3 conducts some experiments to verify whether this algorithm is better or not. Finally, section 4 concludes the paper.

## 2    Proposed Method

The Radon transform of an image $f(x,y)$ is defined as

$$R(t,\theta) = \sum_{x,y} f(x,y)\delta(t - x\cos\theta - y\sin\theta) \cdot$$

The Radon transform has the shift invariant property, i.e., a translation of $f(x,y)$ by a vector $(x_0,y_0)$ results in a shift of its translation in the variable $t$ by a distance $d = x_0\cos\theta + y_0\sin\theta$. In addition, a rotation of the image by an angle $\theta_0$ implies a shift $\theta_0$ of the transform in the variable $\theta$. We can take the 1D FFT along the $t$ direction of $R(t,\theta)$ and then take the magnitude of these coefficients. We can then take the 1D FFT along the $\theta$ direction of the resulting image. In this way, we have extracted translation and rotation invariant features from the pattern object. Note that scaling invariance can be achieved by standard normalization techniques in [5].

We summarize our proposed method in the following algorithm:

(a) Find the centroid of the input pattern image $f(x,y)$, and move the centroid to the center of the image containing the pattern.
(b) Scale the image by means of a standard normalization technique so that the scale variance can be eliminated.
(c) Take the Radon transform $R(t,\theta)$ of the normalized pattern image obtained from (b).
(d) Take the 1D FFT along the $t$ direction of $R(t,\theta)$ and take the magnitude of these coefficients.
(e) Take the 1D FFT along the $\theta$ direction of the resulting image from (d) and calculate the magnitude of the coefficients.
(f) Classify the input pattern into one of the existing classes by using the extracted invariant features.

The main contribution of this paper is that we have proposed a new invariant algorithm for object recognition by using a combination of the Radon transform and the Fourier transform. The new algorithm is invariant to translation, rotation, and scaling of the input pattern image. More importantly, it is very robust to Gaussian white noise. Our algorithm does not need to align the centroids of the patterns because the Fourier spectra are invariant to spatial shifts. Experimental results conducted in the next section show that the proposed algorithm is better than the Zernike's moments, the auto-correlation wavelet moments, and the DTCWT moments for recognizing 2D shapes under different noise levels.

Note that we can replace the 1D FFT in the above algorithm by sparse FFT ([6], [7]), which is faster than the standard FFT when the input signal is sparse in the frequency domain. Even though our proposed method is very simple, it outperforms the Zernike's moments, the DTCWT moments and the autocorrelation wavelet moments for the two shape databases. This means that our proposed method in this paper can be used in many real-life applications.

Wang et al. [8] proposed a scaling and rotation invariant method for pattern recognition by using the Radon and the Fourier-Mellin transforms. The method requires the centroid of the image to be fixed, which is not true in real-life applications. Our proposed algorithm in this paper can overcome this drawback by taking the 1D FFT along the rows and columns of the pattern image, respectively. In addition, our algorithm should be faster than [6] because our method uses 1D FFT whereas [6] uses the Fourier-Mellin transform.

## 3    Experimental Results

We have tested a database of 20 aircrafts as show in Fig. 1. For each aircraft, we test ten rotation angles 30°, 60°, 90°, 120°, 150°, 180°, 210°, 240°, and 270°. Fig. 2 and 3 show the images after each processing step of this algorithm for an undistorted image and an distorted image, respectively. We tested the performance of our proposed method on noisy data. We generate the noisy images with different orientations by adding Gaussian white noise to the noise-free images. The signal-to-noise ratio (SNR) is defined as

$$SNR = \frac{\sqrt{\sum_{i,j}(f_{i,j} - avg(f))^2}}{\sqrt{\sum_{i,j}(n_{i,j} - avg(n))^2}}$$

where f is the noise-free image, n is the added white noise, and avg(f) is the average value of the image f . We selected to use the Zernikes's moments for up to the $12^{th}$ order. The correct recognition rates for the Zernike's moments, the autocorrelation wavelet moments, the dual-tree complex wavelet moments, and the proposed method are shown in Tables 1, 2, 3 and 4, respectively. From these four tables, it can be seen that our proposed method is better than the Zernike's moments, the dual-tree complex wavelet moments, the auto-correlation wavelet moments.

We also tested the second database (216 shapes), which has 18 categories with 12 shapes in each category [9]. The database is shown in Fig. 5. Each shape is matched against every other shape in the database. As there are 12 shapes in each category, up to 11 nearest neighbours are from the same category. We rate the performance based on the number of times the 11 nearest neighbours are in the same category. The results are shown in Table 5 by using the Zernike's moments, the autocorrelation wavelet moments, the DTCWT moments, and the proposed method. It can be seen that the proposed method outperforms other methods compared in this paper.

**Table 1.** The recognition rates (%) of the Zernike's moments for the aircraft database with different rotation angles and SNR's

| SNR | 30° | 60° | 90° | 120° | 150° | 180° | 210° | 240° | 270° |
|---|---|---|---|---|---|---|---|---|---|
| 0.5 | 90 | 85 | 90 | 90 | 95 | 95 | 95 | 85 | 100 |
| 1.0 | 100 | 100 | 100 | 100 | 100 | 100 | 100 | 100 | 100 |
| 2.0 | 100 | 100 | 100 | 100 | 100 | 100 | 100 | 100 | 100 |
| 4.0 | 100 | 100 | 100 | 100 | 100 | 100 | 100 | 100 | 100 |

**Table 2.** The recognition rates (%) of the auto-correlation wavelet moments for the aircraft database with different rotation angles and SNR's

| SNR | 30° | 60° | 90° | 120° | 150° | 180° | 210° | 240° | 270° |
|---|---|---|---|---|---|---|---|---|---|
| 0.5 | 85 | 85 | 80 | 80 | 85 | 85 | 85 | 90 | 80 |
| 1.0 | 100 | 100 | 100 | 100 | 100 | 100 | 100 | 100 | 100 |
| 2.0 | 100 | 100 | 100 | 100 | 100 | 100 | 100 | 100 | 100 |
| 4.0 | 100 | 100 | 100 | 100 | 100 | 100 | 100 | 100 | 100 |

**Table 3.** The recognition rates (%) of the dual-tree complex wavelet moments for the aircraft database with different rotation angles and SNR's

| SNR | 30° | 60° | 90° | 120° | 150° | 180° | 210° | 240° | 270° |
|---|---|---|---|---|---|---|---|---|---|
| 0.5 | 95 | 90 | 100 | 100 | 90 | 100 | 100 | 95 | 100 |
| 1.0 | 100 | 100 | 100 | 100 | 100 | 100 | 100 | 100 | 100 |
| 2.0 | 100 | 100 | 100 | 100 | 100 | 100 | 100 | 100 | 100 |
| 4.0 | 100 | 100 | 100 | 100 | 100 | 100 | 100 | 100 | 100 |

**Table 4.** The recognition rates (%) of the proposed method for the aircraft database with different rotation angles and SNR's

| SNR | 30° | 60° | 90° | 120° | 150° | 180° | 210° | 240° | 270° |
|---|---|---|---|---|---|---|---|---|---|
| 0.5 | **100** | **100** | **100** | **100** | **100** | **100** | **100** | **100** | **100** |
| 1.0 | 100 | 100 | 100 | 100 | 100 | 100 | 100 | 100 | 100 |
| 2.0 | 100 | 100 | 100 | 100 | 100 | 100 | 100 | 100 | 100 |
| 4.0 | 100 | 100 | 100 | 100 | 100 | 100 | 100 | 100 | 100 |

**Table 5.** The top 11 matches in percentage (%) of the second database by using Zernike's moments, the autocorrelation wavelet moments, the DTCWT moments, and the proposed method. It can be seen that the proposed method is better than every method compared in this paper.

| Method | Top 11 matches (%) | | | | | | | | | | |
|---|---|---|---|---|---|---|---|---|---|---|---|
| Zernike | 100 | 93.52 | 91.67 | 86.11 | 81.02 | 79.17 | 72.69 | 71.76 | 66.67 | 65.28 | 56.48 |
| Auto-Corr | 100 | 92.13 | 86.11 | 84.24 | 78.70 | 73.61 | 72.22 | 61.59 | 62.04 | 56.02 | 48.61 |
| DTCWT | 100 | 85.19 | 81.02 | 71.76 | 70.37 | 60.65 | 60.19 | 54.63 | 45.83 | 47.22 | 34.72 |
| **Proposed** | **100** | **95.83** | **93.52** | **91.20** | **84.72** | **84.26** | **77.31** | **75.46** | **73.61** | **73.61** | **60.65** |

**Fig. 1.** The aircraft database used in the experiments

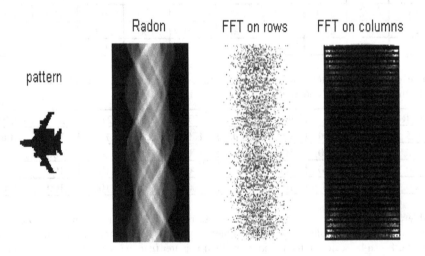

**Fig. 2.** The images of the proposed method for processing an aircraft

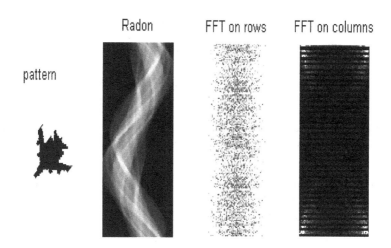

**Fig. 3.** The images of the proposed method for processing a distorted aircraft

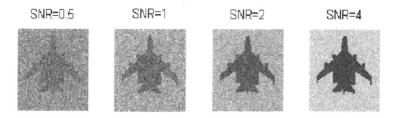

**Fig. 4.** The noisy images

**Fig. 5.** A database of shapes selected from the MPEG test database

## 4    Conclusions

In this paper, we have presented a new invariant algorithm for object recognition, based on the combination of the Radon function and the Fourier transform. The proposed algorithm has the ability to classify many types of image objects. In addition, it does not require to align the centroid of the pattern images because the Fourier spectra are invariant to spatial shifts. Experimental results have shown that the proposed algorithm is better than the Zernike's moments, the auto-correlation wavelet moments, and the DTCWT moments for an aircraft database and a shape database.

**Acknowledgments.** This research was supported by the research grant from the Natural Science and Engineering Research Council of Canada (NSERC) and Beijing Municipal Science and Technology Plan: Z111100074811001.

## References

1. Prokop, R.J., Reeves, A.P.: A survey of moments-based techniques for unoccluded object representation and recognition. CVGIP: Graphical Models Image Processing 54(5), 438–460 (1992)
2. Hu, M.K.: Visual pattern recognition by moment invariants. IRE Transactions on Information Theory 8, 179–187 (1962)
3. Khotanzad, A., Hong, Y.H.: Invariant image recognition by Zernike moments. IEEE Transactions on Pattern Analysis and Machine Intelligence 12(5), 489–497 (1990)
4. Chen, G.Y., Xie, W.F.: Wavelet-based moment invariants for pattern recognition. Optical Engineering 50(7), 077205 (2011)
5. Chen, G.Y., Bhattacharya, P.: Invariant pattern recognition using ridgelet packets and the Fourier transform. International Journal of Wavelets, Multiresolution and Information Processing 7(2), 215–228 (2009)
6. Hassanieh, H., Indyk, P., Katabi, D., Price, E.: Simple and Practical Algorithm for Sparse Fourier Transform. In: SODA (January 2012)
7. Hassanieh, H., Indyk, P., Katabi, D., Price, E.: Nearly Optimal Sparse Fourier Transform. In: STOC (May 2012)
8. Wang, X., Xiao, B., Ma, J.F., Bi, X.L.: Scaling and rotation invariant analysis approach to object recognition based on Radon and Fourier-Mellin transforms. Pattern Recognition 40, 3503–3508 (2007)
9. Sebastian, T.B., Klein, P.N., Kimia, B.B.: Recognition of Shapes by Editing Shock Graphs. In: International Conference on Computer Vision, ICCV (2001)

# An Effective Method for Signal Extraction from Residual Image, with Application to Denoising Algorithms

Min-Xiong Zhou, Xu Yan, Hai-Bin Xie, Hui Zheng, and Guang Yang[*]

Shanghai Key Laboratory of Magnetic Resonance, Physics Department,
East China Normal University, Shanghai, China
gyang@phy.ccnu.edu.cn

**Abstract.** To minimize image blurring and detail loss caused by denoising, we propose a novel method to exploit residual image. Firstly, we apply Non-local Means (NLM) filter to original image to get the denoised image and store the weights used for averaging. Secondly, we filter the residual image with the stored weights. Then a Gaussian filter is applied to the denoised residual image before we add the results to image denoised by NLM to recover the lost image details. Different from previous methods, our method uses the structure information in the original image and can be used to extract lost image details from residual images with very low SNR. An analysis on the mechanism of the signal extraction method is given. Quantitative evaluation showed that the proposed algorithm effectively improved accuracy of NLM filter. In addition, the residual of the final results contained fewer observable structures, demonstrating the effectiveness of the proposed method to recover lost details.

**Keywords:** residual, signal extraction, Non-Local Means(NLM), denoise, image enhancement.

## 1 Introduction

Image denoising is a popular topic in image processing field. One way to evaluate the performance of a denoising algorithm is to use the residual [1-4], which is defined as the difference between original image $u$ and the denoised image $D(u)$ :

$$R = u - D(u) \qquad (1)$$

The purpose of denoise algorithms is to remove noise while preserving image signal, thus an ideal residual image should be dominated by random noise with no visible image structures. However, in practice, residual images also contain structured signals, indicating detail losses in the denoising process. Therefore, we can use residual images to evaluate the denoising performance simply by inspecting the visibility of remaining signal in them. The residual has also been used to estimate noise level[5], control iterations in iterative algorithm[6], and optimize algorithm parameters[7].

One interesting application of the residual is to extract remaining signal from it and add the extracted signal back to the denoised image to compensate the loss of image

---

* Corresponding author.

C. Guo, Z.-G. Hou, and Z. Zeng (Eds.): ISNN 2013, Part I, LNCS 7951, pp. 657–663, 2013.

details. Jeng et.al first denoised the noisy image with Gaussian algorithm, and extracted signal in residual by applying Kalman filter to the residual image [8]. Dominique et.al adopted a similar strategy but used TV (Total Variation) and Wiener filters respectively for denoising and residual filtering[2]. The numerical and visual inspection showed that these compensation methods can improve the original denoising methods.

However, when these methods are applied to residual images from state-of-the-art denoising algorithms, such as the popular Non-Local Means (NLM) filter [4], the results are hardly satisfying. This is because the residual images of NLM filter contain fewer visible image structures (Fig. 1) [3, 4, 9] and it is quite challenging to denoise residual images with such a low SNR (Signal-to-Noise Ratio).

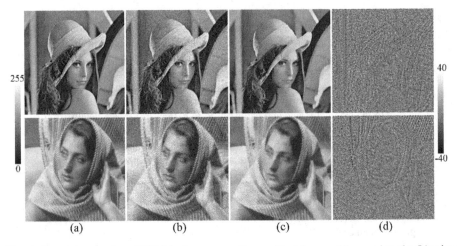

        (a)              (b)              (c)              (d)

**Fig. 1.** Denoising results of NLM and corresponding residual image. (a)ground truth; (b)noisy image (original image) with noise level 7%; (c) denoise image; (d)Residual image.

To more effectively extract signal from such residual images, we propose a new approach, which adopts the weights of NLM algorithm used in denoising the original image to average the residual image. Results from both quantitative and qualitative evaluations showed that our method can extract signals from low SNR residual images more effectively and can significantly improve denoising performance of NLM. Furthermore, since the weights used to denoise residual image have already been calculated during the NLM filtering of the noisy image, filtering of the residual image is compultionally efficient.

## 2    Method

### 2.1    Non-Local Means (NLM)

NLM is a spatial domain filter, which replaces each pixel $P(i)$ in the image with a weighted average of every pixel $P(j)$ in its "search region" $(\Omega)$:

$$NLM(P(i)) = Z_0 \sum_{\forall j \in \Omega} \omega(i, j) P(j) \tag{2}$$

where $Z_0$ is the normalization coefficient, defined as:

$$Z_0 = 1 / \sum_{\forall j \in \Omega} \omega(i, j) \tag{3}$$

The average weight $\omega(i,j)$ assigned to $P(j)$ is based on the similarity between the neighborhoods of pixel i and j, named Nef(i) and Nef(j) respectively:

$$\omega(i, j) = \exp(-K_\rho \| Ne_f(i) - Ne_f(j) \|^2 / h^2), (i \neq j) \tag{4}$$

where $h$ is a parameter controlling the degree of smoothing, and is normally set proportionally to standard deviation of noise. $K_\rho$ is a Gaussian kernel of standard deviation $\rho$. The search region ($\Omega$) can cover the whole image (thus non-local) but a limited radius $t$ is commonly adopted with regard to computational efficiency[3]. The distance between the center pixel and itself is simply set to the minimum distance found in the neighborhood.

Similar to other weighted-averaging filter, larger weights are assigned to pixels with higher similarity. What makes NLM different is that it uses the similarity between the neighborhoods of the pixels instead of the distance between pixels themselves. Thus it can make use of the redundant information in texture pattern in the image for robust denoising.

## 2.2 Information Extraction Method

To extract remaining structured information from the residual of NLM, we propose to use NLM weights calculated from original image to filter the residual image. We do this out of the following considerations: 1) SNR of the original image is much higher than that of the residual which means structured information can be easily extracted; 2) the structured information in original image is correlated to the remaining signal but no to the noise in the residual image; 3) NLM weights have been calculated beforehand, thus using them to denoise the residual will   incur minimum computational cost.

Firstly, we filter the residual $R$ using NLM weights $\omega$ of original image:

$$\hat{R}(i) = Z_0 \sum_{\forall j \in \Omega_i} \omega(i, j) R(j) \tag{5}$$

Where $\hat{R}$ is the filtered residual. Then we apply Gaussian filter to the filtered residual image for further noise suppression before adding the result back to the denoised image $D(u)$

$$\hat{D}(u) = D(u) + Gauss_\sigma(\hat{R}) \tag{6}$$

where $\hat{D}$ is the result of final denoised image, and $Gauss_\sigma$ is a Gaussian filter with STD $\sigma$.

## 2.3 Analysis of Filtered Residual

To discuss the performance of different methods on residual filtering, we quantified the respective influence of the filtering to the signal and noise in the residual. Suppose $G$ is the signal (ground truth) in the original image and $N$ is the noise. We filter the original image$(G+N)$ with the NLM weights $W_1$ and acquire residual image $R$:

$$R = (G + N) - W_1(G + N) \tag{7}$$

Because of the linearity of the weighted-averaging, $R$ is equivalent to:

$$R = G - W_1(G) + N - W_1(N) \tag{8}$$

where The first two terms represent remaining signal in residual after original NLM denoising. After filtering the residual with another set of averaging weights $W_2$, we get:

$$W_2(R) = W_2(G + N) - W_2(W_1(G + N))$$
$$= W_2(G) - W_2(W_1(G)) + W_2(N) - W_2(W_1(N)) \tag{9}$$

where the first two terms are originated from image signal, representing the signal component in filtered residual, while the latter two terms are originated from random noise and named noise component.

Three filters, namely NLM filter, Gaussian filter and proposed filter, are compared in regard to their ability to extract information from residual image, and the results are shown in Fig. 2 (see section 4.1 for parameters used). It can be seen clearly that our method acquired the clearest image pattern in the signal component with no apparent artifacts in the noise component.

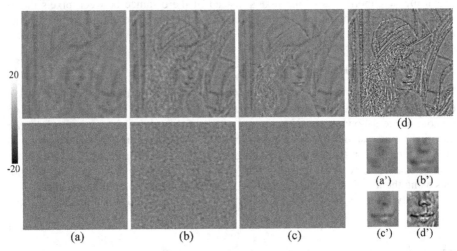

**Fig. 2.** Comparison of signal (upper) and noise (lower) components in residual image after filtering of the residual image. From left to right are results from (a) NLM filter, (b) Gaussian filter, and (c) proposed filter and (d) the actual signal in the residual image. (a' - d') are magnified regions from (a - d, upper) respectively. Noisy image with 7% noise is used.

## 3    Experiment

We used nature images Lena and Barbara (256×256) in our experiments. The dynamic range of these images are 0-255. We added 5 different levels of Gaussian noise with standard deviation of 3%, 5%, 7%, 9%, 12% respectively of the original images. Images used to demonstrate the visual results were those with 7% noise.

## 3.1    Signal Extraction

In this section, we compared the performance of signal extraction ability of proposed method with other filters. Firstly, we applied NLM to the noisy image, and acquired denoised image and the residual (Fig. 1). Then we filtered the residual by NLM filter, Gaussian filter, AD (anisotropic diffusion) filter, and the proposed method.

The parameters of NLM filter were set as recommended in [3], namely t = 5, f = 2, h = 1.2σ. The parameters of Gaussian filter (standard deviation of Gaussian kernel = 2) and AD filter (delta = 1/8, kappa = 0.25σ and iteration number = 60) were set as optimized value via exhaustive search. For Gaussian filter in our method, we used a 3×3 kernel with standard deviation of 2.

Experimental results showed that our method outperformed other filters by extracting more information from the residual images (Fig. 3).

## 3.2    Compensation

The extracted signal was added back to the denoised image to compensate detail loss caused by denoising.

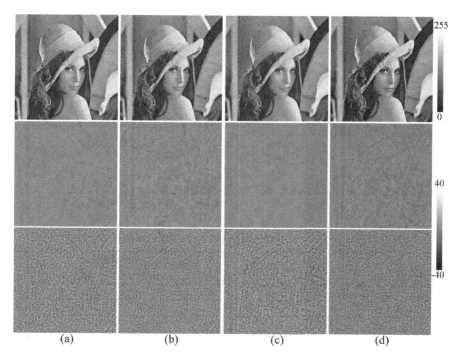

**Fig. 3.** Comparison of different filters with respect to their information extraction ability in the context of whole denoising process. Shown from upper to lower are the final denoised image with filtered residual added back (upper), filtered residual images (middle), and the final residual images showing the difference between the ground truth and the final denoised image (lower) respectively. From left to right, (a) – (d) are results from NLM filter, Gaussian filter, AD filter and our filter respectively.

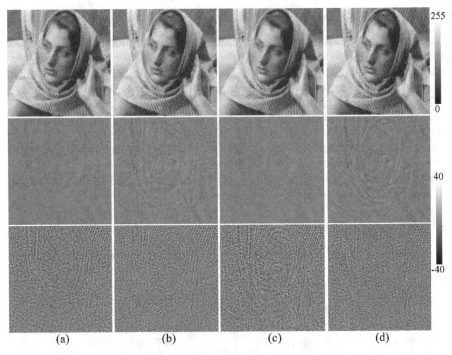

**Fig.3.** (*continued*)

The performance of these different filters was compared quantitatively using mean squared error (MSE), which is defined as:

$$MSE = \frac{1}{n}\sum_{i=1}^{n}(I(i) - Q(i))^2 \tag{10}$$

where $I$ and $Q$ denote ground truth and the final denoised image.

**Table 1.** MSE of NLM denoised images with and without signal compensation

|         | Noise Level | Original | Compensation Method | | | |
|---------|-------------|----------|------|----------|------|------|
|         |             |          | NLM | Gaussian | AD | Our |
| Lena    | 3%  | 21.25  | 21.31  | 20.40  | 20.47  | **19.80** |
|         | 5%  | 26.89  | 23.40  | 23.66  | 25.85  | **21.99** |
|         | 7%  | 62.21  | 72.41  | 60.49  | 62.25  | **59.49** |
|         | 9%  | 87.55  | 105.41 | 84.72  | 86.83  | **83.78** |
|         | 12% | 129.68 | 165.62 | 127.26 | 130.70 | **126.16** |
| Barbara | 3%  | 23.48  | 23.84  | 22.63  | 23.25  | **22.30** |
|         | 5%  | 29.11  | 25.93  | 26.01  | 28.36  | **23.59** |
|         | 7%  | 75.36  | 83.91  | 73.07  | 75.16  | **70.39** |
|         | 9%  | 110.68 | 126.05 | 107.82 | 110.55 | **104.21** |
|         | 12% | 171.15 | 200.53 | 166.81 | 172.17 | **162.51** |

We calculated final residual images for visual comparison(Fig. 3), which showed the difference between noisy image and the final denoised image after compensation. The final residual image from our method contained less structured information compared to those from other filters. Quantitative results based on MSE (Table 1) also proved that our method achieved best denoising results at all noise levels.

## 4    Conclusions

In this paper we propose a new way to filter the residual images of denoising method. It can extract lost image details even from residual images of very low SNR. The proposed filter is especially useful when combined with NLM filter because NLM filter normally produces residual images with less structured information and there is no need to calculate the weights separately. However, the proposed approach can also be used to recover lost image details to enhance other state-of-the-art denoising methods.

## References

1. Shan, Q., Jia, J., Agarwala, A.: High-quality motion deblurring from a single image. ACM Transactions on Graphics 27(3) (2008)
2. Brunet, D., Vrscay, E.R., Wang, Z.: The Use of Residuals in Image Denoising. In: Kamel, M., Campilho, A. (eds.) ICIAR 2009. LNCS, vol. 5627, pp. 1–12. Springer, Heidelberg (2009)
3. Manjon, J.V., et al.: MRI denoising using non-local means. Med. Image Anal. 12(4), 514–523 (2008)
4. Buades: A review of image denoising algorithms, with a new one. Multiscale Model. Simul. (2005)
5. Chuang, K.S., Huang, H.K.: Assessment of noise in a digital image using the join-count statistic and the Moran test. Phys. Med. Biol. 37(2), 357–369 (1992)
6. Osher, S., Burger, M., Goldfarb, D., Xu, J.J., Yin, W.T.: An iterative regularization method for total variation-based image restoration. Multiscale Modeling & Simulation 4(2), 460–489 (2005)
7. Rajwade, A., Rangarajan, A., Banerjee, A.: Automated Filter Parameter Selection using Measures of Noiseness. In: Canadian Conference Computer and Robot Vision, pp. 86–93 (2010)
8. Jeng, F.-C., Woods, J.W.: Inhomogeneous gaussian image models for estimation and restoration. IEEE Trans. Acoust., Speech, Signal Proc. 36(8), 1305–1312 (1988)
9. Coupé, P., Yger, P., Barillot, C.: Fast non local means denoising for 3D MR images. In: Larsen, R., Nielsen, M., Sporring, J. (eds.) MICCAI 2006, Part II. LNCS, vol. 4191, pp. 33–40. Springer, Heidelberg (2006)

# Visual Attention Computational Model Using Gabor Decomposition and 2D Entropy

Qi Lv, Bin Wang, and Liming Zhang

Department of Electronic Engineering, Fudan University
No. 220, Handan Road, Shanghai, 200433, China
{11210720024,wangbin,lmzhang}@fudan.edu.cn

**Abstract.** Visual attention is an important mechanism as it can be applied to many branches of computer vision and image processing such as segmentation, compression, detection, tracking and so on. Based on both capabilities and defects of existing models, the paper proposes a computational saliency-oriented model from the perspective of frequency domain. A saliency map can be generated by two main steps: firstly Gabor wavelet decomposition of the input image at certain levels is used to produce the feature components, and then these components are selected and fused in the sense of 2D entropy. The proposed algorithm outperforms most of state-of-the-art algorithms at human fixation prediction for both psychological patterns and natural images including salient objects with arbitrary sizes. Beyond that, biological plausibility of Gabor filter makes our approach more reliable and adaptive to various stimuli.

**Keywords:** Visual Attention, Extended Classical Receptive Field (ECRF), Gabor Decomposition, 2D Entropy.

## 1 Introduction

The highly evolved human vision system enables us to rapidly attend to the locations which are conspicuous within an image. It is attention mechanism that facilitates us to locate salient regions in a scene. Tsotsos [1] defines attention as the process by which the brain controls and tunes information processing. Therefore, the most general method to model attention is to simulate the functions of human brain. By devising such artificial visual attention model, visual tasks can be easily accomplished by the machines.

Borji and Itti [2] divide existing computational models into two kinds: filter ones and neural ones. Considering filter models, bottom-up model which is stimuli-based and top-down model which is task-driven are two main branches. Among different kinds of models, saliency map [3], a topological map containing global conspicuity information, is frequently assumed and utilized as it directly demonstrates the attended regions spatially. Most of state-of-the-art models are both bottom-up schemed and saliency map based, so is ours.

Itti *et al.* [4] proposes a bottom-up model which calculates saliency by linear filtering and center-surround difference upon low-level feature maps including

C. Guo, Z.-G. Hou, and Z. Zeng (Eds.): ISNN 2013, Part I, LNCS 7951, pp. 664–673, 2013.
© Springer-Verlag Berlin Heidelberg 2013

intensity, color opponents and orientation, then combines different feature maps across different scales to obtain saliency map. This algorithm is of high computational cost and fails to deal with large scale salient regions. Another model is implemented in frequency domain [5] in which Quaternion combines intensity, two color opponents (BY and RG) and motion feature to achieve Phase spectrum Fourier Transform (PQFT). This model suffers the drawback that only edges are highlighted due to not taking amplitude information into account. In [6], the author introduces a Frequency-Tuned Saliency (FTS) algorithm which retains most of the frequency components of images in order to realize better segmentation. It calculates saliency by simple subtraction between Gaussian filtered image and its global mean. This model, compared with the previous two, outstands when large scale salient regions are dominant but fails when orientation features (e.g., psychological patterns) lead to saliency or when salient objects relatively small. This is because operations involved in FTS are isotropic and retaining most frequency parts may only be suitable for large salient objects.

We find out that Itti's model fails to locate large salient objects as it removes most low frequency parts. And FTS is merely effective for large objects but incapable to locate small ones as it retains most of the frequency. Inspired by these two ideas, we introduce Extended Classical Receptive Field (ECRF) [7,8] to solve both problems above.

The rest of paper is organized as follows. In section 2, we demonstrate that saliency can be calculated by extracting information locating in certain bands of frequency domain. Gabor wavelet decomposition is then introduced to achieve feature extraction. Next, in section 3, the 2D entropy metric is utilized to select and fuse feature maps. Then we formulate the algorithm in section 4. Experimental results and comparisons are demonstrated in section 5 with analysis on various models. Finally, conclusions are made on our work as well as future work.

## 2   From ECRF Model to Gabor Decomposition

Rodieck et al. [8] proposes the classical Difference of Gaussian (DoG) model to depict the center-surround response structure of receptive field of retinal ganglion cells. It is known as the Classical Receptive Field (CRF) which can be simulated by the difference between two Gaussian functions. This DoG structure is adopted in many bottom-up models [4,6]. Physiologists have found out, however, that centre-surround CRF can be modulated by a larger region, i.e. the photoreceptor cells outside the CRF of ganglion cells [9]. These areas are regarded as non-Classical Receptive Field (nCRF). Combining both CRF and nCRF, Ghosh et al. [10] model the receptive field as a combination of three zero-mean Gaussians at three different scales:

$$\text{ECRF}(\sigma_1, \sigma_2, \sigma_3) = A_1 \frac{1}{\sqrt{2\pi}\sigma_1} e^{-\frac{x^2}{2\sigma_1^2}} - A_2 \frac{1}{\sqrt{2\pi}\sigma_2} e^{-\frac{x^2}{2\sigma_2^2}} + A_3 \frac{1}{\sqrt{2\pi}\sigma_3} e^{-\frac{x^2}{2\sigma_3^2}}, \quad (1)$$

where ECRF represents the response function, $\sigma_1$, $\sigma_2$ and $\sigma_3$ represent the scales of the center, the antagonistic surround and the extended disinhibitory surround respectively, and $A_1$, $A_2$ and $A_3$ represent the corresponding amplitudes.

We think that ECRF may solve both problems of Itti's model and FTS model because it can choose frequency bands including low frequency as well as high band-pass frequency part of input image by modulation of its parameters. That is why human beings can easily pay attention to salient objects regardless of their sizes. This idea, however, leaves another problem in real implementation: how to automatically modify these parameters in Eq.(1) to suit various stimuli with different size objects in natural image or psychological patterns as human beings do. An idea in this paper is adopting wavelet transform in different orientations and multiple scales to decompose input image to different frequency bands, and then selecting one or two corresponding bands according to certain criterion. This idea can partly simulate the ability of ECRF and also achieves the same effect of frequency band selection. Considering different categories of wavelet functions, we select Gabor function to carry out decomposition as it is related to the processes of primary visual cortex.

Conventionally, wavelet decomposition of images with Mallat algorithm [11] always mixes the two diagonal direction sub-maps into one map. This property may hinder saliency detection in many cases. Therefore, we choose 2D Gabor filter rather than Fast Wavelet Decomposition algorithm to accomplish "wavelet decomposition" in order to obtain more information on orientations. Four high-frequency filters with orientations $\{0°, 45°, 90°, 135°\}$ and one low-frequency filter, which is virtually a Gaussian filter, together amount to five 2D Gabor filters. These five Gabor filters can almost cover the whole frequency domain. They are shown in Fig.1.

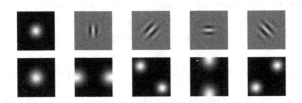

**Fig. 1.** Five Gabor 2D filters (top row) with their corresponding amplitude spectrum (bottom row). From left to right: low-frequency part, high-frequency part with $0°$, $45°$, $90°$, $135°$.

## 3   Fusion and Selection via 2D Entropy

Input image is decomposed at several different levels or scales, in other words, these five Gabor filters convolve the image at one level and then the output of low-frequency Gabor filter is subsampled to yield the input of next level. After decomposition, dozens of feature maps are obtained. These maps contain valuable local saliency information, so a strategy is required to fuse and select them to generate a significant saliency map. This fusion and selection process simulates the process of parameters adjustment of ECRF model. After combination and elimination of different bands, the effect of ECRF is achieved.

2D entropy [12,13] is a useful metric to measure the clutter degree of an image. It is originally used to determine the optimal threshold for segmentation. To calculate 2D entropy, a 2D gray-level histogram taking spatial relations into account is formed in advance by comparing the original image $f(x,y)$ and the averaging filtered version $g(x,y) = m * f(x,y)$, where $m$ is a 2D mean filter which can be 3×3 pixels. The 2D histogram is an $L \times L$ square matrix, where $L$ represents number of gray levels. A pixel located at $(x,y)$ refers to $i$th gray level in $f(x,y)$ and $j$th gray level in $g(x,y)$ respectively. After scanning all pixels in image, $r_{ij}$ denotes the number of pixels which are at $i$th gray level of $f(x,y)$ and at $j$th level of $g(x,y)$. The element of 2D histogram $p_{ij}$ is calculated as follows:

$$p_{ij} = \frac{r_{ij}}{MN},  \qquad (2)$$

where $MN$ represents total pixel number of input image.

Then the 2D entropy can be calculated based on the generated 2D histogram. According to the definition of 2D entropy, the 2D histogram mainly takes edge change into account since uniform regions scarcely alter their grey level after averaging filtering. If a map is topologically compact, which means less edge information, then the averaged map may still contains relatively less edge information. On the other hand, when a scene is cluttered, averaging filtering may lead to excessive edge information which accordingly generates relatively greater value of 2D entropy. So, the smaller the 2D entropy value is, the more significance the corresponding map represents. One example is shown in Fig.2.

 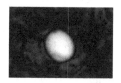 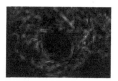

**Fig. 2.** Illustration of 2D entropy. From left to right: original image, one feature map with 2D entropy 1.6932, another map with value of 3.5245.

## 4   Details of Algorithm

According to these processing methods, we combine them as a whole model to generate the saliency map. The steps of saliency map calculation are as follows.

Four broadly-tuned color channels [4] are constructed according to the input RGB image with $r$, $g$ and $b$ referring to its red, green and blue components respectively:

$$R = r - (g + b)/2,  \qquad (3)$$

$$G = g - (r + b)/2,  \qquad (4)$$

$$B = b - (r + g)/2,  \qquad (5)$$

$$Y = (r+g)/2 - |r-g|/2 - b, \tag{6}$$

As color information is processed in primary visual cortex in an antagonistic way, two color pairs, red-green and blue-yellow, are obtained here:

$$RG = R - G, \tag{7}$$

$$BY = B - Y, \tag{8}$$

Besides, the intensity channel is averaged over the original components:

$$I = (r+g+b)/3, \tag{9}$$

These three channels I, RG and BY are regarded as input signals for the rest operations. As mentioned before, Gabor decomposition is utilized to substitute band-pass filter in each channel. The 2D Gabor function is:

$$g(x, y; \lambda, \theta, \sigma) = \exp(-\frac{x'^2 + y'^2}{2\sigma^2}) \cos(2\pi \frac{x'}{\lambda}), \tag{10}$$

where $x' = x \cos\theta + y \sin\theta$, $y' = -x \sin\theta + y \cos\theta$. And four orientations are selected as $\theta = \{0°, 45°, 90°, 135°\}$. The scale is $\sigma = 7/5$ . The low-pass 2D Gabor filter sets $\lambda$ a large number like $\lambda_{low} = 2.5^{10}$, and 4 other high-pass ones set $\lambda_{high} = 2.5$. The sizes of these filters are 15×15 pixels (Fig.1).

In our experiments, we decompose the input image repeatedly until the height of decomposed image is less than 64 pixels. Besides, feature maps extracted at the first level, which means the original size, are discarded as they contain less significant information even most can be considered as noise. That is, we first filter the input image with the low-pass Gabor filter and then subsample it. The half-sized image is considered as noise-reduced and it is then processed with the five 2D Gabor filters. Therefore, in each channel for each level except the first one, we obtain four orientation maps and one low frequency map.

The next operation is to select and fuse feature maps. First of all, 2D entropy metric for each feature map is calculated. In order to calculate the entropy, the 2D histogram should be computed in advance as described in section 3. Before calculating the 2D histogram, we use Gaussian filter with $\sigma = 0.02 \times width$ to smooth the feature maps. Then, the 2D entropy is calculated as:

$$2D\_entropy = -\sum_{i=1}^{L}\sum_{j=1}^{L} p_{ij} \log p_{ij}, \tag{11}$$

where $p_{ij}$ is calculated according to Eq.(2).

Moreover, like [14,15,16], we also take *retinal eccentricity* or *center-bias* effect into account which means that one image is weighted more when it responds more intensely and actively in center than in surround. A parameter $cb_i$ is defined for each map as:

$$cb_i = \sum_{m=1}^{M}\sum_{n=1}^{N} K(m,n) \cdot Norm(fm_i(m,n)), \tag{12}$$

where $K$ is a Gaussian kernel with the same size of feature map and the scale parameter is $\sigma_x = N/6$ and $\sigma_y = M/6$. $Norm()$ is normalization operation and $fm_i$ refers to different feature map.

This parameter is used together with 2D entropy to acquire the Modified 2D entropy (M-2D entropy):

$$me_i = \frac{2D\_entropy_i}{cb_i}, \tag{13}$$

This value is an important metric to fuse feature maps and select the optimal levels/scales. Then, we can calculate the M-2D entropy value for each feature map of each channel. For these feature maps, we fuse those at the same level to obtain a more comprehensive map, which means that four orientation maps at same level (same resolution) are integrated with each weight equaling the reciprocal of its M-2D entropy value. The low frequency map is left intact. For instance, if the intensity channel of an input image is decomposed at 4 levels, then 12 (4×3) orientation feature maps and 3 (1×3) low frequency feature maps are obtained except the first level. By fusing orientation feature maps, we will get 6 (3 + 3) comprehensive maps for the intensity channel.

From the perspective of frequency domain, the comprehensive maps contain different frequency bands of the input image from the very low frequency to nearly half of the highest frequency. But each frequency band is extracted with different weights on various orientations. As we mentioned before, the other half frequency band is discarded as noise components.

So what we have to do next is to pick up optimal frequency bands which contain most significant saliency information to construct the saliency map. This is to simulate the process of ECRF. The criterion of selection is to calculate the M-2D entropy of each comprehensive map as well. We first select one comprehensive map with the lowest entropy value $e_{min}$. Then we find out another map with the second lowest entropy value and check whether the value is less than $1.1 \times e_{min}$. If it is not, we only get one optimal band. However if it is the other case, we get 2 optimal bands if these two bands are not overlapped and one represents low frequency while the other represents high frequency. In this case, we combine these two maps linearly.

At last, 3 local saliency maps are obtained for I, RG and BY channels. The weight for each map is also calculated as the reciprocal of its M-2D entropy. These 3 maps are interpolated to reach the original resolution in order to make a pixel-wise weighted summation.

## 5   Experimental Results and Discussions

To make a comprehensive evaluation on our model, the testing databases include both artificial patterns/images and natural images. Natural images contain not only small sized salient objects but also large ones. Comparisons are performed between several models including NVT [4], PQFT [5], FTS [6] and HFT [14]. Among all of these models, PQFT resizes the input to resolution of 64×64 and HFT resizes to 128×128 as optimal defaults while others do not carry out resolution adjustment as well as our approach. However, resize of input image may probably lead to irreversible information loss.

## 5.1 Quantitative and Subjective Evaluation

For artificial patterns or psychological images, we just list results in several cases to make a subjective evaluation on each model.

For natural data, we compare the output of each model with the so-called ground truth in a quantitative way, more than just listing dozens of results. The ground truth data are based on human visual behavior and mainly include two types: fixation maps and labeled area maps. A fixation map [2] is record of human fixation within one image by eye tracking apparatus. Data of this kind are binary maps, with logical 1 (fixation points) dotted over the whole image. Ground truth maps of the other kind are also binary maps, but with consistent areas indicating logical 1 which are labeled by a number of subjects. For both types of ground truth, we choose AUC (Area under ROC Curve) metric to measure the performance of each model's capability to predict human fixation.

According to [17], the processes of *center-bias* and Gaussian smooth on saliency maps have great impact on the results of AUC metric. Here, the *center bias* operation is to take the Hadamard product of the saliency map and a Gaussian mask as the result. Besides, by adopting Gaussian filtering to the saliency map, better visual effects and more consistency of salient region can be obtained. Consequently, for an impartial and unbiased comparison, we process the saliency maps generated by each model with such operations if they are not implemented by the model itself.

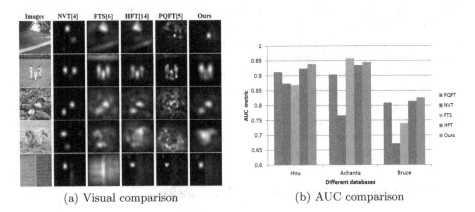

(a) Visual comparison                    (b) AUC comparison

**Fig. 3.** Visual comparison and AUC comparison for 5 models

## 5.2 Saliency Detection for Natural Images

For natural images testing, all models are running on databases including Bruce's fixation data [18], Hou's labeled data [19] and Achanta's labeled data [6]. They contain 120, 58 and 1000 natural images respectively. All saliency results are first Gaussian smoothed with parameter setting as $\sigma = 0.04 \times width$. After smoothing, the saliency map is multiplied by a Gaussian kernel with parameter $\sigma_x = width/4$ and $\sigma_y = height/4$ to achieve the effect of *center-bias*. Some results are shown in Fig.3(a) and a quantitative comparison is plotted in Fig.3(b).

Fig.3(a) illustrates that our model is proved to be effective on both small salient regions and large ones while PQFT and NVT only highlight small objects or edges. FTS fails when orientation information is required or when salient objects are relatively small. The results of HFT are also not as satisfying as ours (first and fourth row in Fig.3(a)).

Additionally, our model outperforms the others in terms of AUC metric as shown in Fig.3(b). We would like to point out that the HFT model is defective due to its resize of input image and fixed weight of I, RG and BY channel. For another, the NVT model suffers the problem that center-surround operations exclude much of the low frequency part, which actually contributes a lot when the salient region is relatively large. Considering PQFT, it totally discards the amplitude spectrum and only phase information is utilized for saliency map construction, which leads to only edges being popped out. For FTS, as it is only effective on its own database (most images with large salient areas) but fails others, it indicates that orientation is an important feature and retaining most of frequency components is neither necessary nor sufficient to calculate saliency.

## 5.3 Saliency Detection for Artificial Images

Different types of psychological patterns, construct another test bench. These images are important criterion to measure the performance of attention models.

In Fig.4, we can see that our model can deal with most cases of psychological patterns. It turns out that our model shows good potential on these patterns as well as HFT, but NVT and PQFT are less effective and FTS have the poorest performance. Nonetheless, considering cases like conjunction (last column), it may require more high-level features or top-down knowledge which are not contained in our model.

# 6    Conclusion and Future Work

We proposed a saliency model from the perspective of frequency band decomposition though implemented in spatial domain. Two main steps are: Gabor decomposition of input image at different levels generating dozens of feature maps and 2D entropy which is chosen as metric to select optimal scales (or frequency bands) and integrate feature maps. Our approach turns out to be superior compared with others on various kinds of stimuli, including artificial

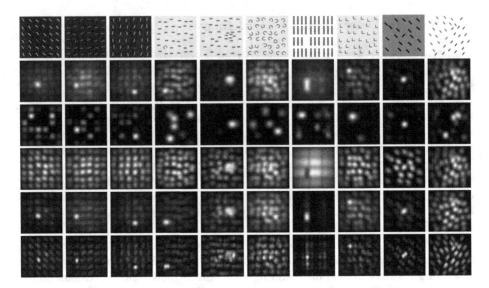

**Fig. 4.** Saliency maps of psychological images. First row: patterns, second row: our method, third row: NVT[4], fourth row: FTS[6], fifth row: HFT[14], last row: PQFT[5].

images and natural ones with large or small salient areas. Besides, although this is a bottom-up model, top-down manners or prior knowledge can be more easily included. As images are divided into different channels, scales and orientations, diverse weights can be assigned to feature maps when specific tasks are involved. However, our model still suffers a couple of drawbacks. Firstly, this algorithm requires a bit more computational cost compared with PQFT or FTS. For another, 2D entropy as a measure to fuse and select feature maps lacks of biological support.

The future work is to discover biological basis of 2D entropy, or to replace it with other more biologically plausible methods. So far, whether behavior like 2D entropy calculation exists in human visual system remains unknown. What's more, top-down mechanism turns out to be more important as we have interest in airports or buildings detection in remote sensing images with attention models.

**Acknowledgments.** This work is supported in part by the National Natural Science Foundation of China (Grant No. 61071134) and the Research Fund for Doctoral Program of Higher Education of China (Grant No. 20110071110018). We would like to thank Judd for her help on model evaluation. We also thank Bruce, Achanta and Hou for sharing their databases.

# References

1. Tsotsos, K.: A Computational Perspective on Visual Attention. MIT Press (2011)
2. Borji, A., Itti, L.: State-of-the-art in Visual Attention Modeling. IEEE Trans. on Pattern Analysis and Machine Intelligence 35(1), 185–207 (2012)

3. Koch, C., Ullman, S.: Shifts in Selective Visual Attention: Towards the Underlying Neural Circuitry. Human Neurobiol. 4, 219–227 (1985)
4. Itti, L., Koch, C., Niebur, E.: A Model of Saliency-based Visual Attention for Rapid Scene Analysis. IEEE Trans. on Pattern Analysis and Machine Intelligence 20(11), 1254–1259 (1998)
5. Guo, C., Ma, Q., Zhang, L.: Spatio-Temporal Saliency Detection Using Phase Spectrum of Quaternion Fourier Transform. In: Proc. CVPR (2008)
6. Achanta, R., Hemami, S., Estrada, F., Susstrunk, S.: Frequency-Tuned Salient Region Detection. In: Proc. CVPR (2009)
7. Li, C., Zhou, Y., Pei, X., Qiu, F., Tang, C., Xu, X.: Extensive Disinhibitory Region Beyond the Classical Receptive Field of Cat Retinal Ganglion Cells. Vision Research 32, 219–228 (1992)
8. Rodieck, R., Stone, J.: Analysis of Receptive Fields of Cat Retinal Ganglion Cells. Journal of Neurophysiology 28, 833–849 (1965)
9. Ikeda, H., Wright, M.: Functional Organization of the Periphery Effect in Retinal Ganglion Cells. Vision Research 12, 1857–1879 (1972)
10. Ghosh, K., Sarkar, S., Bhaumik, K.: A Possible Explanation of the Low-level Brightness-contrast Illusions in the Light of An Extended Classical Receptive Field Model of Retinal Ganglion Cells. Neuroscience Reserach Supplements 150, S141–S150 (1989)
11. Mallat, S.: A Theory for Multiresolution Signal Decomposition: The Wavelet Representation. IEEE Trans. on Pattern Analysis and Machine Intelligence 2(7), 674–693 (1989)
12. Abutaleb, A.: Automatic Thresholding of Gray-level Pictures Using Two-dimensional Entropy. Computer Vision, Graphics, and Image Processing 47(1), 22–32 (1989)
13. Yang, C., Chung, P., Chang, C.: Hierarchical Fast Two-dimensional Entropic Thresholding Algorithm Using A Histogram Pyramid. Optical Engineering 35(11), 3227–3241 (1996)
14. Li, J., Levine, M., et al.: Visual Saliency Based on Scale-Space Analysis in the Frequency Domain. IEEE Trans. on Pattern Analysis and Machine Intelligence 6(1), 1–16 (2012)
15. Zhang, L., Tong, M., Marks, T., Shan, H., Cottrell, G.: SUN: A Bayesian Framework for Saliency Using Natural Statistics. Journal of Vision 8(7) (2008)
16. Harel, J., Koch, C., Perona, P.: Graph-based Visual Saliency. In: Advances in Neural Information Processing Systems (2007)
17. Judd, T., Ehinger, K., Durand, F., Torralba, A.: Learning to Predict Where Humans Look. In: Proc. ICCV (2009)
18. Bruce, N., Tsotsos, J.: Saliency Based on Information Maximization. In: Advances in Neural Information Processing Systems (2006)
19. Hou, X., Zhang, L.: Saliency Detection: A Spectral Residual Approach. In: Proc. CVPR (2007)

# Human Detection Algorithm Based on Bispectrum Analysis for IR-UWB Radar

Miao Liu[*], Sheng Li[*], Hao Lv[*], Ying Tian, Guohua Lu, Yang Zhang, Zhao Li, Wenzhe Li, Xijing Jing, and Jianqi Wang[**]

School of Biomedical Engineering, the Fourth Military Medical University, Xi'an 710032, China
Sheng@mail.xjtu.edu.cn

**Abstract.** Impulse-radio Ultra-wide band (IR-UWB) radar plays an important role in searching and detecting human target in particular situations, such as counterterrorism, post-disaster search and rescue and so on. It mainly takes the advantages of its good penetrability through obstacles and high range resolution. It detects human target mainly by detecting the respiratory signal. As the higher order spectrum is immune to the Gaussian noise, a new algorithm based on the bispectrum analysis for human detection behind the wall is proposed. The results of the through-wall experiments show the algorithm has a better performance than the conventional PSD-based algorithm.

**Keywords:** human detection, UWB, bispectrum.

## 1    Introduction

As a rapidly developing technology, IR-UWB radar plays an important role in contactless human being detection in many areas, such as counterterrorism, post-disaster search and rescue and so on. It takes the advantages of the IR-UWB's high range resolution and its good penetrability through nonmetallic materials in the lower frequencies [1-3].

Considering the outstanding applications in military, civilian and medicine, many research groups are studying human detection with the UWB radar. Scientists in this area have been trying to detect the human beings more efficiently and accurately by developing the radar prototypes and raising different respiratory-motion detection (RMD) algorithms [1-9]. The human target is mainly detected by detecting and identifying the respiratory response. The IR-UWB pulses travel though the wall and reach to the human body and then are reflected backwards by the human chest and then the received echoes contain the respiratory signals.

The traditional life detection algorithms are generally based on the energy of respiratory response [5-9], those algorithms mainly focus on the power spectral density (PSD), which is the Fourier transform of the autocorrelation function [10]. However,

---

[*] The first three authors contributed equally to this work and should be regarded as co-first authors.

[**] Jianqi Wang is the corresponding author.

C. Guo, Z.-G. Hou, and Z. Zeng (Eds.): ISNN 2013, Part I, LNCS 7951, pp. 674–681, 2013.

this type of algorithm is only suitable for high signal to noise ratio (SNR) situation without the noise in the same frequency range as respiration. Typically, the displacement of chest caused by respiration is between 0.1 mm and several millimeters, depending on different person, and that caused by heartbeat is even usually below 0.1 mm [11]. So the magnitude of the respiratory response itself is weak. Also, the reflected signal carried by the electromagnetic waves is always contaminated by strong noise and intensive attenuation will appear as the radar signal travels through complex constructions [12]. The above two reasons make the SNR of the UWB echoes always low. As disturbance to the useful signal, noise always spreads in a large frequency range and mixes with the useful IR-UWB respiratory signal. Therefore, the algorithms based on the PSD may not that suitable to be applied to detect human objects behind the wall.

For Gaussian process, the spectrum of order higher than two is identically zero. Then a nonzero higher order spectrum denotes a non-Gaussian process. So the higher order spectrum analysis (HOSA) is usually applied to study of the non-Gaussian signal contaminated by Gaussian noise, without considering the SNR or whether the noise is in the same range as the wanted signal [13]. As non-Gaussian signal, the quasi-periodic respiratory signal is suitable to be analyzed by the higher order spectrum method to analyze the IR-UWB echoes and finally be used to detect the human being automatically and accurately. In this paper, we use the bispectrum, which is the third order spectrum. The preferable performance of the algorithm is proved by experiments and compared with the conventional PSD-based algorithm.

The rest of the paper is organized as follows: In Section 2, the methods in this paper are depicted. The results and discussion of the experiments are presented in Section 3 and conclusion is made in Section 4.

## 2    Methods

### 2.1    Measurement System

A set of IR-UWB radar system with center frequency of 400 MHz and bandwidth of 100 MHz, which ensures good penetrability and high range resolution, is used. It complies with the definition of UWB according to the federal communication commission (FCC). The pulse repetition frequency (PRF) is 250 KHz and the transmit power is 5 mW. In addition, the radar is controlled by a laptop and the data stream is transported through Wi-Fi.

The IR-UWB radar echoes are received by the receiver antenna, sampled by the AD converter and then are stored in the laptop in the form of data matrix. Each received waveform along the range consists of 4096 points and then the sampling rate of the radar system is approximately 61 Hz, which satisfies the Nyquist sampling theorem for the frequency of the human's respiration is 2-3.5 Hz. For the stored matrix, the time-axis related to range along each received waveform is termed fast-time, which is in the order of nanoseconds, while the time-axis along the measurement interval is termed slow-time, which is in the order of seconds [14].

## 2.2    Experiments

The block diagram of the experiment using the IR-UWB radar to detect human object is shown in Fig. 1. A human object with normal breath stands still behind a 24 cm brick wall and stay away from the radar system 3 m and 4.5 m separately.

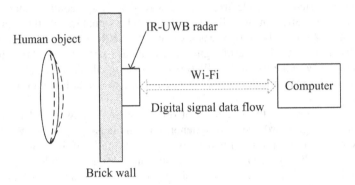

**Fig. 1.** The block diagram of the experiment using IR-UWB to detect human object

The stored data is analyzed by bispectrum analysis algorithm and PSD-based algorithm, a conventional algorithm which is used to compare the performance with the proposed algorithm.

The PSD-based algorithm is depicted by the flowchart in Fig. 2. In order to improve the SNR, accumulation in range, digital components (DC) and linear trends (LT) removal and low-pass filter in slow time dimension are employed before calculating the PSD. Considering the respiration signal is, in most cases, no more than 0.5 Hz, the PSD in the frequency range below the 0.5 Hz is accumulated, and then the accumulated value represents the amplitude of the point [15].

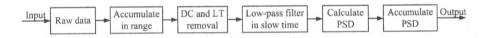

**Fig. 2.** The flowchart of PSD-based algorithm

## 2.3    RMD Algorithm Based on Bispectrum Analysis

Bispectrum analysis is a kind of HOSA with the order is three and it is the Fourier transform of the third-order cumulant. A concise introduction to the bispectrum analysis will be given as following [13]:

Consider a real discrete-time sequence x(n), which mean is zero. The third-order cumulant estimation of x(n) is defined:

$$C_{3,x}(\tau_1, \tau_2) = E(x(n)x(n+\tau_1)x(n+\tau_2)) . \tag{1}$$

The bispectrum of x(n) is

$$B(\omega_1,\omega_2) = \frac{1}{2\pi} \sum_{\tau_1=-\infty}^{\infty} \sum_{\tau_2=-\infty}^{\infty} C_{3,x}(\tau_1,\tau_2) e^{-(j\omega_1\tau_1+j\omega_2\tau_2)} \ . \tag{2}$$

That is to say, the bispectrum of x(n) is the Fourier transform of its third-order cumulant. Besides, the bispectrum of x(n) can be also given by

$$B(\omega_1,\omega_2) = H(\omega_1)H(\omega_2)H^*(\omega_1+\omega_2) \ , \tag{3}$$

where $H(\omega)$ denotes the Fourier transform of x(n).

Specific to the algorithm presented in this paper, the bispectrum estimation is the main part and the flowchart of our algorithm is shown in Fig. 3.

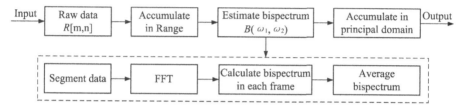

**Fig. 3.** The flowchart of the proposed signal processing

In order to reduce the computation complexity, range accumulation is employed, which can also, in some degree, increases the SNR of received signal. This step can be written as

$$R_1[l,n] = \frac{1}{Q} \sum_{m=(l-1/2)Q}^{(L+1/2)Q-1} R[m,n] \ , \tag{4}$$

where x=1, 2,..., X, Q is the window size in the range dimension, and L=⌊M/Q⌋ is the compression result in the same dimension, while ⌊a⌋ denotes the largest integer less than a [15].

After the accumulation in range dimension, a matrix R1 (l, n) will be obtained, where the l=1,2,..., L, which denotes the range information and the n=1,2,..., N , which denotes the time information. Then the bispectrum of each row signal rl (n) of the matrix is estimated. We estimate the bispectrum as following steps;

Segment data into K frames and each frame has a length of M samples with 50% of overlap.

Apply FFT algorithm to each frame and we get X(k) (λ), where λ=0, 1, ... , M/2 and k=1, 2, ... , K.

Calculate bispectrum of each frame according to Equation 3. We get bk(ω1,ω2), where k=1,2, ..., K.

Lastly, the bispectrum of each point signal is the average of the bispectrums of K frames:

$$B(\omega_1,\omega_2) = \frac{1}{K} \sum_{k=1}^{K} b_k(\omega_1,\omega_2) \ . \tag{5}$$

Bispectrum has extraordinary symmetry property. In order to reduce the computation complexity and maintain the entire information of the bispectrum at the same time;

analysis will focus on the Triangular OAB in Fig. 4, which is called the principal domain of the bispectrum plane [13]. Accumulation is implemented for the Gaussian noise will be zero in the domain, while the quasi-periodic respiratory signal will not. Then the point where the largest accumulated bispectrum appears can be regarded as the position where the human target locates.

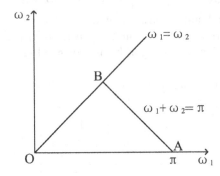

**Fig. 4.** The principal domain of the bispectrum plane

## 3     Results and Discussion

The results are shown in Fig. 5. Fig. 5(a) shows the results of the bispectrum analysis, while Fig. 5(b) shows that of the conventional PSD-based algorithm (the human object stands away from the radar 3 m) Fig. 5(c) shows the results of the bispectrum analysis, while Fig. 5(d) shows that of the conventional PSD-based algorithm (the human object stands away from the radar 4.5 m).

It can be seen from the results that the bispectrum estimation algorithm has better performance than the PSD-based algorithm. The value of the point where the human object stands is significantly higher than the other points. We can see from the figures, the respiratory response causes a relatively large value zone in both the results of the two algorithms, which can be called the affected area, and the rest of the zone can be called the non-affected area. As the two algorithms both detect the point where the largest value exists and locate the human object, values in non-affected area should be suppressed as much as possible. A parameter K is also defined to evaluate the performance. The value of K is the quotient of the highest value in the affected area over the highest value in the non-affected area. The peaks in the non-affected area will be great disturbances to the automatic detection, so the larger the K is, the better the performance is. For the bispectrum analysis algorithm, the value of the K can reach to 10, while for the PSD-based algorithm, the value can only get 2 or 3.

In addition, the affected area in the results of the bispectrum analysis is narrower than that of the PSD-based algorithm, which can be clearly seen especially when the human object is 4.5 m away from the radar. The affected area' width influences the accuracy of the detection, the narrower the affected area is, the more accurate the detection is.

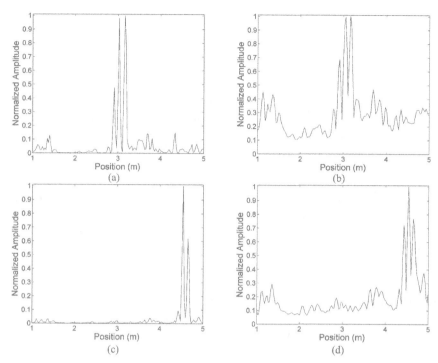

**Fig. 5.** (a) The normalized amplitude of accumulated bispectrum along the range dimension with the human object stands 3 m away from the radar (b) The normalized amplitude of accumulated PSD along the range position with the human object stands 3 m away from the radar (c) The normalized amplitude of accumulated bispectrum along the range dimension with the human object stands 4.5 m away from the radar (d) The normalized amplitude of accumulated PSD along the range position with the human object stands 4.5 m away from the radar.

The reason that causes the differences between the two algorithms is mainly that there is noise in the same frequency where the respiratory response belongs. The PSD-based algorithm cannot eliminate this kind of noise in the essence of its theory. The density of the noise is treated as a component of the useful signal in the PSD-based algorithm. Narrowing the accumulated frequency range can, in some degree, prevent the noise density be added. This kind of method needs the exact frequency range of the respiratory response, and either larger or smaller frequency range will bring departure. However, the prior knowledge of the respiratory frequency can hardly obtain for the parameter varies from person to person. And when the system is applied in search and rescue or other emergent situations, it is impossible to get the respiration frequency beforehand. Regarding the bispectrum analysis algorithm, it is immune to the Gaussian noise essentially, with no need to care about if the frequency range of the noise and that of the respiratory response are overlapped.

In conclusion, the results show the bispectrum analysis is superior over the conventional PSD-based algorithm in detecting human object as its immune characteristic to the Gaussian noise.

## 4    Conclusion

In this paper, a low power IR-UWB is employed to detect human object behind the wall and a new algorithm based on bispectrum analysis is proposed. The performance of the algorithm is compared with that of the conventional PSD-based algorithm. Experiments are carried out to test the algorithm and the results show the novel algorithm has much better performance than the contrast algorithm, which shows promising applications of the algorithm to be used in complex situations.

In the future, experiments of complicated scenarios should be carried out to testify the algorithm. In addition, as discussed above, the affected area of the results of the novel algorithm is narrower than that of the PSD-based algorithm. This kind of feature can be good for multi-human objects detection. Further researches will be carried out later.

**Acknowledgments.** This work was supported by the National Natural Science Foundation of China (Grant No. 61271102) and National Science &Technology Pillar Program (Grant No. 2012BAI20B02).

## References

1. Zetik, R., Crabbe, S., Krajnak, J., Peyerl, P., Sachs, J., Thomä, R.: Detection and localization of persons behind obstacles using M-sequence through-the-wall radar. In: Proceedings of SPIE, pp. 145–156 (2006)
2. Sachs, J., Aftanas, M., Crabbe, S., Drutarovsky, M., Klukas, R., Kocur, D., Nguyen, T., Peyerl, P., Rovnakova, J., Zaikov, E.: Detection and tracking of moving or trapped people hidden by obstacles using ultra-wideband pseudo-noise radar. In: European EuRAD 2008, pp. 408–411 (2008)
3. Ossberger, G., Buchegger, T., Schimback, E., Stelzer, A., Weigel, R.: Non-invasive respiratory movement detection and monitoring of hidden humans using ultra wideband pulse radar. In: 2004 International Workshop on Ultra Wideband Systems, Joint with Conference on Ultrawideband Systems and Technologies, Joint UWBST & IWUWBS, pp. 395–399 (2004)
4. Levitas, B., Matuzas, J.: UWB Radar for Human Being Detection Behind the Wall. In: International Radar Symposium, IRS 2006, pp. 1–3 (2006)
5. Zaikov, E., Sachs, J., Aftanas, M., Rovnakova, J.: Detection of trapped people by UWB radar. In: German Microwave Conference (GeMIC), pp. 1–4 (2008)
6. Yarovoy, A., Ligthart, L., Matuzas, J., Levitas, B.: UWB radar for human being detection. IEEE Aerospace and Electronic Systems Magazine 21, 10–14 (2006)
7. Venkatesh, S., Anderson, C.R., Rivera, N.V., Buehrer, R.M.: Implementation and analysis of respiration-rate estimation using impulse-based UWB. In: IEEE Military Communications Conference, MILCOM 2005, pp. 3314–3320 (2005)

8. Nezirovic, A., Yarovoy, A.G., Ligthart, L.P.: Signal processing for improved detection of trapped victims using UWB radar. IEEE Transactions on Geoscience and Remote Sensing 48, 2005–2014 (2010)
9. Li, Y., Jing, X., Lv, H., Wang, J.: Analysis of characteristics of two close stationary human targets detected by impulse radio UWB radar. Progress in Electromagnetics Research 126, 429–447 (2012)
10. Stoica, P., Moses, R.L.: Introduction to spectral analysis. Prentice Hall, Upper Saddle River (1997)
11. Singh, M., Ramachandran, G.: Reconstruction of sequential cardiac in-plane displacement patterns on the chest wall by laser speckle interferometry. IEEE Transactions on Biomedical Engineering 38, 483–489 (1991)
12. Chernyak, V.: Detection problem for searching survivors in rubble with UWB radars. In: European Radar Conference, EuRAD 2008, pp. 44–47 (2008)
13. Nikias, C.L., Raghuveer, M.R.: Bispectrum estimation: A digital signal processing framework. Proceedings of the IEEE 75, 869–891 (1987)
14. Lazaro, A., Girbau, D., Villarino, R.: Analysis of vital signs monitoring using an IR-UWB radar. Progress in Electromagnetics Research 100, 265–284 (2010)
15. Li, W., Jing, X., Li, Z., Wang, J.: A new algorithm for through wall human respiration monioring using GPR. In: 2012 14th International Conference on Ground Penetrating Radar (GPR), pp. 947–952 (2012)

# Author Index